Lecture Notes in Electrical Engineering

Volume 533

Lecture Notes in Electrical Engineering (LNEE) is a book series which reports the latest research and developments in Electrical Engineering, namely:

- Communication, Networks, and Information Theory
- Computer Engineering
- Signal, Image, Speech and Information Processing
- Circuits and Systems
- Bioengineering
- Engineering

The audience for the books in LNEE consists of advanced level students, researchers, and industry professionals working at the forefront of their fields. Much like Springer's other Lecture Notes series, LNEE will be distributed through Springer's print and electronic publishing channels.

More information about this series at http://www.springer.com/series/7818

Subhash Chandra Bera

Microwave Active Devices and Circuits for Communication

Springer

Subhash Chandra Bera
Space Applications Centre
Indian Space Research Organization (ISRO)
Ahmedabad, Gujarat, India

ISSN 1876-1100 ISSN 1876-1119 (electronic)
Lecture Notes in Electrical Engineering
ISBN 978-981-13-3003-2 ISBN 978-981-13-3004-9 (eBook)
https://doi.org/10.1007/978-981-13-3004-9

Library of Congress Control Number: 2018958497

This Springer imprint is published by the registered company Springer Nature Singapore Pte Ltd.
The registered company address is: 152 Beach Road, #21-01/04 Gateway East, Singapore 189721, Singapore

To Him
who gives me inspiration and patience

Foreword

It has been a pleasure for me to see this book titled ‘Microwave Active Devices and Circuits for Communication’ written by Subhash Chandra Bera based on his long experience on the design and development of on-board communication circuits and systems at Space Applications Centre, ISRO, Ahmedabad.

This book provides extensive coverage in the field of microwave engineering for graduate students, practical circuit designers and researchers. This book begins with basics of device physics and ends with the design of microwave communication systems through detailed design, analysis and realization of different circuits and systems. Apart from classical topics in microwave active devices such as p-i-n diode, Schottky diode, step recovery diode, BJT, HBT, MESFET, HFET and various microwave circuits such as switch, phase shifter, attenuator, amplifier, multiplier, mixer, the coverage extends to some modern topics such as Class-F power amplifier, direct frequency modulator, linearizer, equalizer. The written text of each topic is supplemented with suitable diagrams, and each chapter has various types of solved problems for clear understanding.

It is hoped that this book will be very useful for graduate students, circuit designers and researchers.

Ahmedabad, India
May 2018

Tapan Misra
Distinguished Scientist
Director, Space Applications
Centre (ISRO)

Preface

This book is about active devices and circuits for microwave communications appropriate for undergraduate and postgraduate students, practical circuit designers and researchers in the field of electronics and communication engineering. It presents the design and analysis of various linear and nonlinear circuits for microwave communication systems after discussing the working principle and behaviour of microwave diodes and transistors. This book begins with the basics of device physics and ends with the design of microwave communication systems through detailed design, analysis and realization of different circuits and systems. Though this book is written focusing on microwave communications, much of the material of this book is entirely generic and will be useful for other microwave applications.

This book starts with describing the diodes which are omnipresent in all solid-state circuits across all the frequency ranges starting from DC to terahertz even extending to optical frequency range as an isolated diode or as a part of other devices. Proper understanding about diodes leads to ease of understanding about other devices such as different types of bipolar and unipolar transistors in their homo-junction and hetero-junction structures which are discussed subsequently. Schottky, p-i-n, step recovery and tunnel diodes are popularly used in linear and nonlinear microwave communication circuits such as variable attenuator, power limiter, phase shifter, linearizer, frequency multiplier, mixer for their high-frequency performance and simplicity in operation. With the advancement of material technologies, there are various modern transistors such as hetero-junction bipolar and field effect transistors based on GaAs, InP and GaN technologies. These devices offer very high-frequency performance with low noise figure and high-power-handling capabilities with different device structures and circuit configurations. DC biasing circuits for the diodes and transistors are very relevant which determine the RF performance including temperature behaviour of the circuits. This book also addresses the temperature behaviour and compensation mechanism of the microwave circuits for practical applications.

This book has evolved from the author's design and development experience on microwave circuits and systems for Indian national satellite communication systems at Space Applications Center, Indian Space Research Organization (ISRO), India. This book is believed to be a useful one for students and microwave circuit designers. Any suggestions for improvement of the quality of this book would be highly appreciated.

Ahmedabad, India Subhash Chandra Bera

Contents

About the Author

Subhash Chandra Bera received his B.Sc. degree (with honours) in physics from Presidency College, Calcutta, and B.Tech. and M.Tech. degrees in radio physics and electronics from the Institute of Radio Physics and Electronics, University of Calcutta. He also received a Ph.D. degree in microwave and antenna engineering from Gujarat University, India. Since 1994, he has been with the Space Applications Centre, Indian Space Research Organization (ISRO), Ahmedabad, India, where he has been involved in the design and development of microwave circuits and systems for various INSAT and GSAT series of Communication Payload projects as well as GAGAN/IRNSS Navigation Payload projects. He has developed several state-of-the-art microwave subsystems that are operational in various Indian national satellites. His research interests include microwave active circuits in general and solid-state power amplifiers, channel amplifiers, linearizers, limiters, attenuator and equalizers in particular for spacecraft use. Currently, he is Division Head of Satcom and Navigation Systems Engineering Division, Space Applications Centre (ISRO), Ahmedabad, and Associate Project Director of GSAT-24 and GSAT-31 Communication Payloads. He has authored more than 50 research publications in international journals and presented numerous papers at national and international conferences and symposiums. He has also delivered invited talks in the field of microwave active devices and circuits at various workshops and symposiums. He has also been granted four patents.

Chapter 1
Introduction

Microwave frequency region is generally considered from 1–30 GHz in electromagnetic spectrum which corresponds to a free-space wavelength (λ_o) of 30–1 cm, respectively. Frequencies within the range of 30–300 GHz ($\lambda_o = 10-1$ mm) are considered as mmwave frequency band and submillimetre wave frequency band is considered beyond 300 GHz (λ_o <1 mm). The microwave and higher frequency circuits and systems show characteristics which are not seen in the lower frequency range. The differences in characteristics are due to the smaller wavelength of the microwave and higher frequency range which becomes comparable to the size of the circuit elements, i.e. size of components (resistors, capacitors, inductors, diodes and transistors) and even the transmission line elements. Due to the smaller wavelength compared to the circuit elements, the phase of the signal varies along the dimension of the elements, consequently, the circuit design and analysis techniques, power generation, amplification as well as measurement techniques are different from those at lower frequencies. Thus, materials, structures of the active and passive components, topology, and realization of circuits at microwave frequency are different from those at lower frequencies. Microwave frequencies are used in communication, navigation, remote sensing, radar applications, industrial measurements, and even heating and drying of food products. This book will focus on active devices and circuits for microwave communications. However, much of the materials covered by the book are entirely generic and will be useful for other applications.

1.1 Microwave Communications

Microwave frequencies have distinct advantages over lower frequencies due to its higher percentage bandwidth, higher directive coverage and higher penetration capability through the atmosphere. It is easy to design a circuit with 10% bandwidth, for example, to get 1 GHz bandwidth around 10 GHz centre frequency. Whereas, it is very difficult to design a circuit of 200% bandwidth to achieve the same bandwidth

S. C. Bera, *Microwave Active Devices and Circuits for Communication*, Lecture Notes in Electrical Engineering 533, https://doi.org/10.1007/978-981-13-3004-9_1

around 500 MHz centre frequency. Another advantage for microwave communication is the ability to use smaller aperture antenna with high directivity because antenna beam-width is inversely proportional with the ratio of antenna aperture size to the wavelength of the signal to be transmitted/received. Due to the availability of higher percentage bandwidth and highly directive coverage, it is possible for very high data rate communication of the order of multi-Gbps. With the maturity of microwave technology, it is used as the major trunk channel for long distance communication. Microwave communication is easier over difficult terrain and removes the requirement of land acquisition as required for cable systems. Infrastructure needs for short-term communication systems are possible without the high cost of installing permanent network cabling that will only be required for a limited time. These are the reasons for using microwave communication in spite of the difficulty in implementation of microwave circuits and limitation of communications within the line-of-sight range. As the lower frequency bands become congested, the communication systems need to shift towards higher and higher frequency bands to meet the capacity and data rate targets. Higher microwave and millimetre wave frequency bands are less congested and can potentially provide multi-gigahertz spectrum [1, 2]. This is the reason that satellite communication systems already using the higher microwave and mmwave frequency bands for high throughput applications. It is also envisaged that the next-generation (5G) mobile communication system needs to use higher microwave and mmwave frequency bands to meet the requirement of spectrum [3].

Block diagram of a wireless communication system such as mobile communication is shown in Fig. 1.1. It consists of a transmitter, a receiver, and a channel. RF section of the transceiver mainly consists of low-noise amplifier (LNA), power amplifier (PA), up- and downconverters, IF amplifiers, phase shifters, modulators, switches, and filters. Signal quantization, coding and decoding are performed in the digital domain. Practically, at each end of the link, a transmitter and a receiver, i.e. integrated transceiver is used. In a transmitter, the signal modulates a carrier using a particular modulation scheme and transmitted through a transmit antenna, whereas the receiver recovers the information from the received signal through the receiving antenna. In case of mobile communication system, at least one of the transceivers is mobile. It may be a handheld unit used by a pedestrian or at on-board a vehicle that can move at high speed.

The transmitter consists of a modulator, an upconverter, filters and PA. Baseband signals are modulated by the modulator then filtered, amplified and upconverted by the upconverter to a higher frequency carrier. The upconverted signal further amplified, filtered and finally amplified by a PA and transmitted by the transmitting antenna. The receiver consists of a low-noise amplifier, downconverter, bandpass filters and demodulator. The receiver shall perform the signal amplification and filtering functionality to meet the system sensitivity, channel selectivity, spurious response and power control requirements to meet the required signal-to-noise ratio.

Unlike lower frequencies, microwave frequencies are not reflected and practically not absorbed by the ionosphere on account of their small wavelength, and thus they penetrate the atmosphere without significant attenuation except at certain absorp-

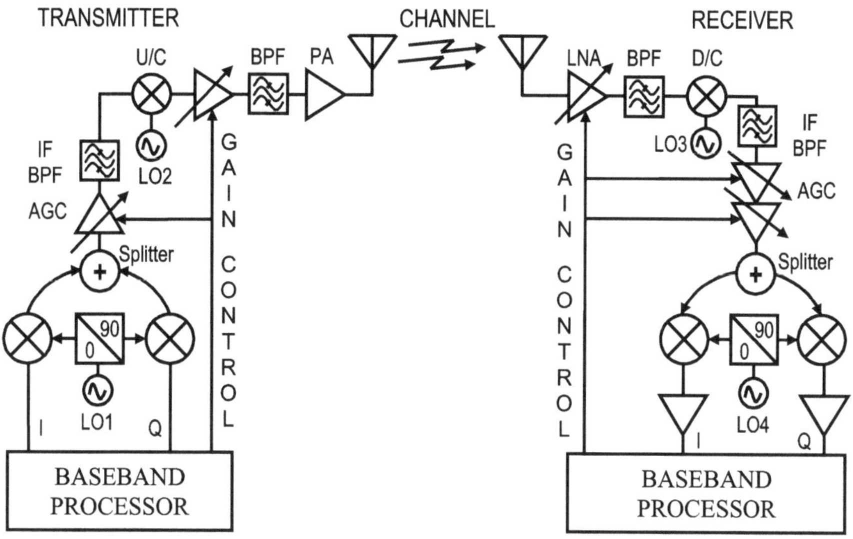

Fig. 1.1 Block diagram of a mobile communication system

tion bands of atmospheric gases & subject to clear weather. This leads to the use of microwave frequencies for satellite communication. Presently, microwave frequencies are widely used in satellite communication for telephony, broadcasting, television systems and various data communication systems. Satellite microwave communication is also very useful for disaster-prone regions due to the possibility of quick and easy establishment of ground infrastructure considering the prior existence of space segment for satellite communication. Block diagram of a satellite microwave communication system is shown in Fig. 1.2. The system consists of ground and space segments. Both the segments consist of transmitter, receiver and antenna systems like any other communication systems.

A ground transmitter section starts with the processing of information to be sent which is then suitably modulated. The filtered modulated signal is then upconverted (frequency translated) to microwave frequency by using a mixer and followed by several stages of amplification. Finally, the microwave signal is amplified by a high power amplifier (HPA) and transmitted by a highly directive antenna towards satellite. Generally, the high power amplifier operates in power back-off condition to minimize the addition of noise due to nonlinearities of the power amplifier. Sometimes, a linearizer is used before the HPA to improve the linearity of the transmitter. In general, the signal is modulated by a low-frequency carrier and then upconverted to microwave frequency as shown in the block diagram. Thus, the upconverter and amplifier section operate in microwave frequency band. However, the direct modulation scheme also can be used to modulate the microwave carrier frequency directly by the baseband signal. In this case modulator, upconverter and amplifier sections operate in the microwave frequency band.

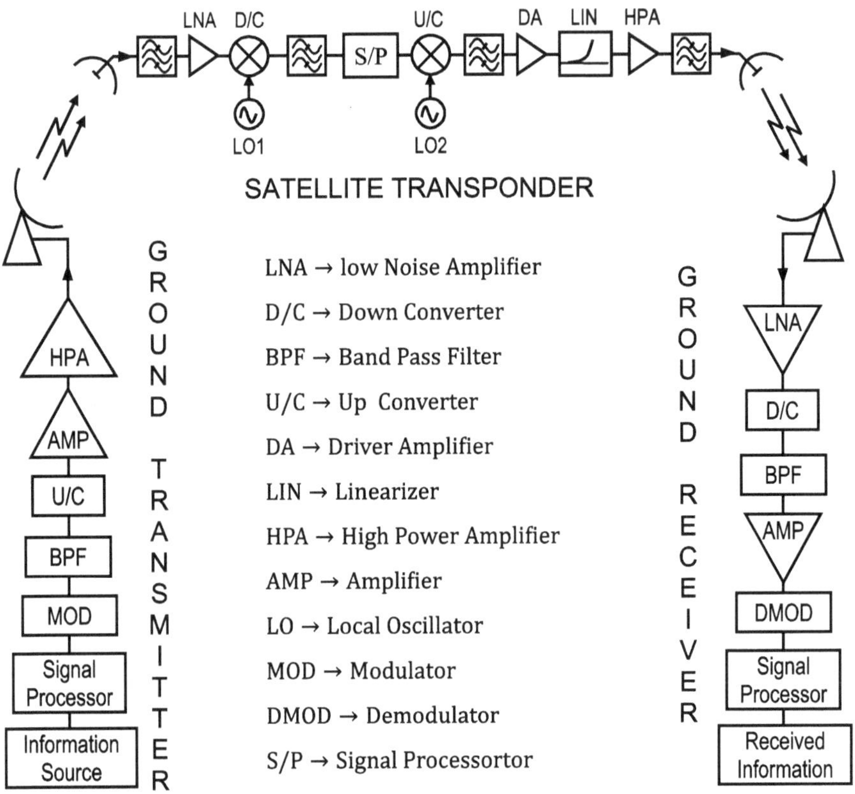

Fig. 1.2 Block diagram of satellite microwave communication system

The space segment is a repeater in space with a sizable amount of gain and frequency translation, sometimes with signal processing also. It consists of receiving antenna followed by a preselect filter to select only the desired frequency band transmitted by the ground transmitter and rejecting undesired frequency bands. Since the received signal is very feeble (of the order of pW), the signal is first amplified by a low-noise amplifier (LNA) and then downconverted using a mixer circuit. The output of the downconverter is filtered and processed by a signal processing (S/P) unit. The signal processing unit may be a transparent or regenerative processor. Transparent processing is only to provide flexibility in frequency, bandwidth and signal routing at lower Intermediate Frequency (IF) level for multichannel multibeam transponders. Regenerative processor extracts the information by using a demodulator and then processed. The processed signal is then remodulated before being upconverted to the transmit microwave frequency. The upconverted microwave signal is subsequently amplified by a driver amplifier (DA) and high power amplifier (HPA) before being transmitted by a transmit antenna. A predistortion linearizer is used to improve the linearity of a satellite transmitter which is cascaded at the input of the high power amplifier.

The ground receiver consists of receive antenna followed by a low-noise amplifier. The amplified signal is downconverted to an intermediate frequency (IF) to ease the signal demodulation and processing for information extraction. Thus, a typical microwave communication system consists of various diode and transistor-based microwave circuits and systems such as modulator, demodulator, receiver, IF amplifiers, high power amplifier, linearizer, converters, etc.

1.2 Microwave Active Circuits

In microwave communication systems, various types of amplifiers are used in receive and transmit sections. A microwave amplifier takes the signal of low or intermediate level as its input over a range of frequency band and significantly boosts its power level. In addition to amplification of the desired signal, a practical amplifier introduces noise generated within it and also distorts the signal. Based on the signal power level to be amplified, there are three types of microwave amplifiers: low-noise amplifier (LNA), high gain amplifier and high power amplifier (HPA). Low-noise amplifiers are used as the front end of a receiver where the strength of the desired signal is very low. High gain amplifiers are used in the intermediate stages of a transmitter–receiver section to boost the signal strength to the desired level. The high power amplifiers (HPAs) are used as the final power amplifying unit of a transmitter. At low frequencies, the design of an amplifier involves mainly designing a proper DC bias circuit for maximum power output. However, at microwave frequency range, in addition to the design of appropriate DC bias circuit, it involves the design of input and output matching networks based on the requirements of noise figure, gain and output power over the desired frequency band.

The linearity of the amplifier plays a major role in a communication system. Ideally, throughout the receive–transmit (repeater) chain, linear amplification of modulated signal is required to accurately decode the information. Amplifiers' nonlinearity becomes more severe when higher order modulations such as 8APSK, 16APSK, 16QAM, etc. are used to achieve improved spectrum efficiency. Problem arises when linear amplification is required throughout the chain where power amplifiers are used. To operate the power amplifiers in their linear region, it is required to operate them in power back-off condition which reduces the DC to RF efficiency. The solution is the use of amplifier near its saturation condition and use a linearizer to improve the overall linearity of the transmitter section. There are various types of linearization techniques used in baseband as well as at RF and IF level. At the baseband level, digital linearization technique is used. However, at RF level, diode-based predistortion linearization technique is mostly used due to its low power consumption, broadband performance and compact size.

Various sensitive microwave devices are used in receivers, solid-state power amplifiers, detectors, etc. Limiter circuits are used to protect sensitive devices from intentionally or unintentionally exposure of high power levels. Basically, two types of limiter circuits are used, one is open-loop limiter and another is closed-loop limiter. Limiting device of an open-loop limiter works with self-bias condition. Whereas closed-loop limiter circuit uses a separate detector circuit to detect the RF power level. Then, the detected signal is applied to other device such as voltage-variable amplifier/attenuator to limit the power level in a closed-loop configuration. Schottky diodes, p-i-n diodes as well as various types of Field Effect Transistors (FETs) are the most suitable limiting devices for microwave applications.

Microwave signal power level is an important performance parameter of any communication system. At lower frequency range, signal power is calculated from the measured voltage and current. Direct measurement of power is more convenient and accurate starting from the microwave frequency range to optical spectrum. Thus, as the frequency increases, power level is measured from which voltage and current are calculated. Microwave power detectors are used to detect/measure the power level. Schottky barrier and tunnel diodes are also popularly used as microwave power level detectors. Depending upon the signal structure and applications; average, envelop, peak or pulse powers are measured.

Mixers are used in up- and downconverters to convert/translate signals in frequency domain. They use a stable local oscillator to translate the spectrum containing information to other frequency band. It is fundamentally a multiplier having three ports: a radio frequency (RF) port, a local oscillator (LO) port and an intermediate frequency (IF) port. In practice, the LO signal is CW, whereas the RF and IF signals are modulated carriers. Generally, Schottky barrier diodes and various types of FETs and Bipolar Junction Transistors (BJTs) are commonly used as nonlinear device for realization of microwave mixers.

Frequency multipliers are used to realize microwave frequency source, e.g. local oscillator (LO) of higher frequency from a low-frequency standard. Performance of microwave solid-state oscillator degrades with increasing frequency. To overcome this problem, highly stable oscillators operating at a lower frequency are used followed by frequency multipliers with low conversion loss. The basic realization principle for frequency multiplication is the generation of multiple higher frequencies by distorting (amplitude or phase) a sinusoidal signal and then filtering out the desired higher frequency component. The distortion is realized using nonlinear characteristic of semiconductor devices. Varactor diodes, Schottky barrier diodes (SBD), step recovery diodes (SRD) and bipolar junction transistors (BJTs) or field effect transistors (FETs) are used as nonlinear device.

Microwave and RF amplifiers are used in many broadband applications such as meeting the increasing demand for Internet and multimedia applications. Modern microwave receivers use equalizers to compensate the excess passband negative gain slope due to the reduction in gain of transistors as the frequency increases. Amplitude tilt equalizer provides predetermined gain/loss slope over an operating range of frequency. It is mainly used in cascade with other circuits or systems to improve the overall frequency response. Microwave-variable gain slope equalizers are used for the compensation of gain roll-off of microwave and RF circuits and systems using the provision of slope adjustability.

Modulators are three port networks. One of the ports is for modulating signal and other two ports are RF ports. A conventional modulator is designed by using a complex chain of downconverter, mixers, filters and upconverter. Most frequently used digital modulation techniques are the discrete phase modulation, i.e. M-ary phase shift keying (PSK). BPSK and QPSK are the most popularly used digital phase modulations techniques for wireless and satellite communication systems. Direct carrier modulation at RF/microwave frequency avoids the use of upconverter and downconverter assemblies leading to less hardware complexity and cost. In case of direct modulation techniques, p-i-n diode or FET-based voltage/current-controlled RF resistance are used to modulate the amplitude and phase of the carriers by using the modulating signal as control voltage/current of the device.

Phase shifter provides a known and controllable phase shift when a RF/microwave signal passes through it. Phase shifters are used for the realization of phase array antenna systems, linearizers for high power amplifiers, various types of phase modulators, etc. True time delay (TTD) and phase shifters (PS) are the basic functional blocks for complex signal processing used in variety of applications including arbitrary waveform generation, filtering, antenna beam forming and reconfiguration ability in communication systems.

Attenuators are used to control signal level in the signal path. They provide fixed attenuation or variable/adjustable attenuation depending upon control voltage or current. Fixed attenuators are realized using a combination of RF resistive elements with fixed values. Voltage/current controlled attenuators are realized using combination of voltage/current dependent resistors. At RF/microwave frequency range, forward-biased p-i-n diodes are used as voltage/current dependent RF resistors. Voltage-variable p-i-n diode attenuators are used to control signal level of communication systems by controlling gain of the system. In case of satellite transponders, p-i-n diode based attenuators are used for on-board gain control of a communication transponder by issuing command from ground or automatically by sensing RF power.

RF/microwave switches are used to reconfigure communication systems by diverting signals from one path to another. The switches are also used to realize stepwise variable attenuators, phase shifters and other elements. Realization point of view, there are two types of RF and microwave switches: electromechanical and solid-state switches. Generally, solid-state switches are very compact in size, faster in operation and more reliable than their electromechanical counterparts. However, solid-state switches have higher insertion loss and also exhibit more nonlinearity at higher signal power. Though CMOS and GaAs FETs are also used as switching devices, p-i-n diodes are used widely as solid-state switching device due to high RF power handling capability and low insertion loss at high-frequencies.

Microwave circuits are used for many applications including airborne and space-based systems where temperature-invariant performance is always demanded over a very wide operating temperature range. The success of electronic systems for these applications relies on the ability to design high performance, highly reliable circuits which function in demanding thermal environments. Performance parameters of the microwave diodes and transistors change with temperature. Attenuation and phase shift of p-i-n diode circuits, limiting power level, attenuation and phase shift of Schottky barrier diode circuits and gain, saturated power, nonlinear parameters of transistor-based circuits are all function of temperature. Therefore, microwave active devices need proper bias circuit with temperature compensation mechanism to achieve temperature-insensitive performance [4].

1.3 Microwave Active Devices

Transistors are three-terminal solid-state devices used for the realization of amplifiers, multipliers, mixers, linearizers and other electronic circuits in communication systems. Among the various types of transistors, bipolar junction transistor (BJT) was the first three-terminal device and still being used for many low frequency as well as microwave applications. With the advancement of hetero-junction technology, hetero-junction bipolar transistor (HJBT or HBT) uses a high bandgap semiconductor as emitter material offers better performance in terms of gain at higher microwave frequency and provides sufficient gain over mmwave frequency regions.

With the advancement of microwave transistors fabrication facilities, solid-state amplification is dominated by the use of three-terminal device transistors. Among the microwave transistors, all the low-noise amplifiers for communication systems are dominated by the psuedomorphic and metamorphic High Electron Mobility Transistors (pHEMTs and mHEMTs). High gain with moderate output power amplifiers are also dominated by MEtal Semiconductor FET (MESFET) and Heterojunction FETs (HFETs). With the advancement of high bandgap semiconductor materials such as GaN and SiC, single-ended high power amplification beyond 100 W is achievable from solid-state technology. However, at higher microwave frequency bands and higher output power, travelling wave tube amplifier (TWTA) is the most preferable option and is the only choice in today's scenario.

Currently, Field Effect Transistor (FET) is the most utilized three-terminal semiconductor device used in microwave and mmwave frequency bands. FETs are used as linear, nonlinear as well as control devices. These are used for realization of low-noise amplifier (LNA) and high-gain amplifier as well as high power amplifiers. As nonlinear devices, these are used to realize multipliers, mixers and oscillators. FETs are also used in RF control device applications as switches and attenuators in which they dissipate essentially no power and can easily be designed into broadband circuits. There are various types of field effect transistors (FETs) based on the use of different materials and structures with different doping profiles. FETs can be fabricated from a wide variety of materials like Si, SiGe, GaAs, InP, InGaAs, GaN, SiC, etc. These are unipolar devices since only one type of carrier, mostly electrons, take part for the channel current. Being a unipolar device, FET does not suffer from minority carrier storage delay time & thus have higher cut-off frequency compared to bipolar devices like BJTs. Field effect transistors based on III-V group semiconductors provide superior performance at microwave and mmwave frequency bands for their higher electron velocity at low electric field. MEtal Semiconductor Field Effect Transistors (MESFETs) based on GaAs are the better microwave device compared to junction field effect transistors (JFETs) and Metal Oxide Semiconductor Field Effect Transistors (MOSFETs) with respect to gain, noise figure and power handling capability at microwave and mmwave frequency bands.

Hetero-structure Field Effect Transistor (HFET) are also known as MOdulation-Doped FET (MODFET) and High Electron Mobility Transistor (HEMT) because of the doping variations within the structure and higher electron mobility, respectively. Depending on structure, doping profile and materials, there are different variations of HFETs suitable for various applications. InP-based HEMT provides the highest frequency of operation with the best noise figure. On the other hand, high bandgap semiconductor such as GaN- and SiC-based HFETs provides hundreds of watt microwave power due to their higher breakdown voltage, higher electron mobility at high electric field and high thermal conductivity.

Schottky barrier diode is a metal–semiconductor junction diode where only majority carrier takes part for device operation. The non-existence of minority carrier injection and extraction leads it to a very high speed of operation. High-speed operation capability of the Schottky diode makes it suitable to use widely for microwave and mmwave applications. Lower cut-in voltage of the Schottky diode makes it suitable for very low-voltage high-current operation. The device is used as RF detector, multiplier, mixer, limiter, attenuator, etc. This is also used to generate nonlinear characteristic for linearizer and microwave phase shifter for microwave beam forming network.

A microwave p-i-n diode behaves as current controlled resistance at microwave frequency. The resistance of the diode may vary from 10,000 Ω to less than 1 Ω by control of its bias current. Though all the junction diodes show this type of characteristic, structure and doping profile of p-i-n diodes is designed to achieve wide variable resistance range and linearity with low bias current level. These characteristics of the p-i-n diode made it most suitable for variable attenuators, switches, limiters, phase shifters, modulators, etc.

Step recovery diode (SRD), varactor diode and tunnel diodes are popular devices used for microwave communications. The step recovery diode has ultra-fast switching speed, typically in the range of picosecond, and is used as nonlinear device for multiplier, mixer, etc. Nonlinear behaviour of tunnel diodes is used for low-level RF power detection and multiplier applications. Nonlinear characteristics of varactor diodes are also used to realize multiplier, mixer and linearizer.

1.4 Microwave Circuit Analysis and Measurements

At low frequencies, where circuit dimensions are relatively smaller than the operating wavelength, circuit elements can be considered as lumped components as the phase change along the circuit elements is negligible. Similarly, at low frequencies, the interconnection between two elements is treated as an ideal connection, with no loss, no characteristic impedance and no transmission phase angle. At microwave frequencies, circuit components and interconnecting elements may not conveniently be scaled down with the reducing wavelength. Thus, the circuit analysis cannot be done using conventional circuit voltage and current methodology. Microwave circuits may consist of elements with dimensions such that the voltage amplitude and phase over the length of the device can vary significantly. In this case, the current that flows in the interconnecting wires or through the circuit elements is not the same at different points along the wire or elements. To accurately calculate the behaviour of an element, it is required to consider its length, width and thickness of metal and its proximity to the ground plane, i.e. the element is considered as a distributed element. Similarly, when interconnects become an appreciable fraction of the signal wavelength, these interconnections themselves must be treated as distributed elements or transmission lines.

1.5 Book Outline

The book consists of 20 chapters including this introduction.

Chapters 2 and 3 describe microwave p-i-n diodes, and Schottky barrier diodes respectively. Step recovery, tunnel, backward and varactor diodes are discussed in Chap. 4. These chapters also describe temperature behaviour of the diodes and effect on the performance of the diode-based circuits.

Chapters 5 and 6 describe microwave bipolar and field effect transistors respectively. Chapter 5 also describes hetero-junction bipolar transistors (HBTs) based on GaAs and SiGe semiconductors. Chapter 6 gives a detailed description about metal semiconductor FETs (MESFETs), various types of hetero-junction FETs (HFETs) including pseudomorphic HEMT (pHEMT), metamorphic HEMT (mHEMT) and GaN-based HEMTs.

Chapter 7 describes details about microwave circuit analysis including the concept of S-parameters, ABCD parameters and their inter-relationships through various examples.

The working principle of microwave switches is given in Chap. 8, and about attenuators and their performance dependency on operating temperature and compensation techniques are given in Chap. 9. Chapter 10 describes RF and microwave phase shifters. Chapter 11 describes various RF and microwave modulators. Chapter 12 describes various amplitude tilt active equalizers operating at RF and microwave frequency range. Chapter 13 described various types of microwave detectors for measurement of power level.

The design and analysis of different types of microwave amplifiers is given in Chap. 14. The chapter contains design and analysis of low-noise amplifier (LNA), high gain amplifier and high power amplifier with examples using transistors. This chapter also gives detailed design and analysis of Class-F amplifiers. Chapters 15 and 16 describe design and analysis of various types of microwave limiters and linearizers respectively.

Chapter 17 describes microwave frequency multipliers using various types of diodes and transistors. Chapter 18 provides details on the design and analysis of microwave mixers.

Chapter 19 presents an overview of terrestrial mobile and satellite communication systems with examples of design and development of complete microwave subsystems such as receiver, local oscillator, channel amplifier, and linearized high power amplifiers. Chapter 20 provides various types of multiple choice questions (MCQs) and answers with their explanations.

References

1. Huang X, Guo YJ, Zhang J, Dyadyuk V (2012) A multi-gigabit microwave backhaul. IEEE Commun Mag 50(3):122–129
2. Huang X, Guo YJ, Zhang JA (2014) Multi-gigabit microwave and millimeter-wave communications research at CSIRO. In: 2014 14th International Symposium on Communications and Information Technologies (ISCIT)
3. Wang CX, Haider F, Gao X, You XH, Yang Y, Yuan D, Aggoune H, Haas H, Fletcher S, Hepsaydir E (2014) Cellular architecture and key technologies for 5G wireless communication networks. IEEE Communications Magazine, February 2014
4. Bera SC (2006) Temperature behavior of active circuits at microwave frequency. Ph.D. dissertation, Gujarat University, India

Chapter 2
P-I-N Diode

Abstract This chapter presents the basic working principle of p-i-n diodes including their characteristics, equivalent circuits at different bias conditions and applications in communication systems. Nonlinear behaviours of p-i-n diodes under different bias conditions are presented. Temperature behaviour of p-i-n diodes and concept of equiresistance curves for achieving temperature-invariant RF resistance is also presented.

2.1 Introduction

A microwave p-i-n diode behaves as current-controlled resistance at microwave frequency. The resistance of the diode may vary from 10,000 Ω to less than 1 Ω by controlling its bias current. Though all the junction diodes show this type of characteristic; structure and doping profile of p-i-n diodes are designed to achieve a wide range of variable resistance and linearity with low-bias current level. These characteristics of the p-i-n diodes made them most suitable for realization of variable attenuators, switches, limiters, phase shifters, modulators, etc.

2.2 Basics of P-I-N Diode

A p-i-n diode has a high resistivity nearly Intrinsic (I) layer sandwiched in between a heavily doped p-region and a heavily doped n region as shown in Fig. 2.1 [1]. The thickness of the intrinsic layer is usually in the range of 10–200 μm. The intrinsic layer is either lightly doped p-type or n-type having resistivity of about 1000 ohm-cm. Practically, the lightly doped layer is ν-type or π-type depending upon whether the conductivity is n-type or p-type, respectively.

At zero-bias condition, in case of diodes with purely intrinsic middle layer, there will be no mobile charge carriers in the I (intrinsic) layer, and thus the p-i-n diode exhibits very high resistance. However, in case of ν-type or π-type middle layer,

S. C. Bera, *Microwave Active Devices and Circuits for Communication*, Lecture Notes in Electrical Engineering 533, https://doi.org/10.1007/978-981-13-3004-9_2

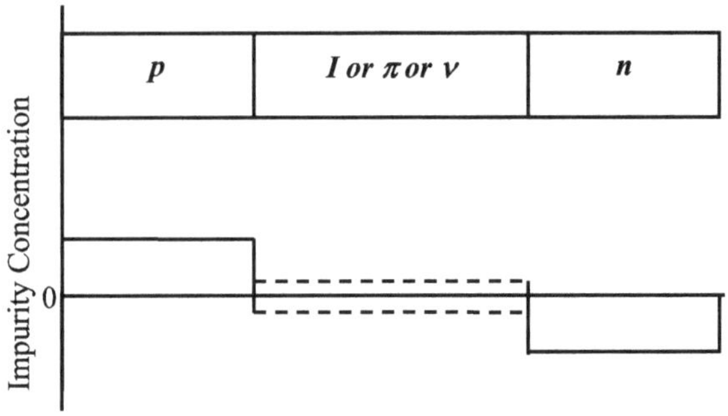

Fig. 2.1 Impurity profile of a p-i-n diode

a very thick depletion region will be formed in the middle layer and a very thin depletion layer will be formed in p- or n-layer, respectively. In this case, at zero-bias condition, the middle layer will not be completely depleted of mobile carriers. A diode exhibits very high resistance when a small reverse-bias voltage, called punch-through voltage, is applied to sweep free of mobile carriers.

When forward bias is applied across a p-i-n diode, carrier injection takes place into the middle layer. Electrons are injected in the middle layer from the n-layer, and holes are injected from the p-layer. Due to the recombination of the carriers in the middle layer, carrier concentration diminishes with the depth. The carrier concentration in the middle layer increases with the increase of forward bias, thus increasing the conductivity, i.e. lowering its resistance. The current-controlled variable RF resistance with high OFF resistance and low ON resistance makes the p-i-n diode suitable for various microwave applications.

2.3 P-I-N Diode Characteristics

The impedance of p-i-n diode is the most important electrical parameter that determines the p-i-n diode-based circuit's performance. In forward-bias condition, the amount of stored charge in the I-layer is forward-bias current (I_d) times the ambipolar lifetime (τ_a). Thus, with the increase of diode forward current, conductivity of the I-layer increases, i.e. intrinsic layer resistance R_i decreases. However, when AC signal is applied with the DC forward bias, then impedance of the p-i-n diode not only depends on the DC forward current but also depends on the AC signal current depending on its frequency.

The conductance of the p-i-n diode is determined by the stored charge (Q_d), i.e. current to the diode. The total diode current $I_d(t)$ is given by [2, 3]

$$I_d(t) = \frac{Q_d}{\tau_a} + \frac{\mathrm{d}Q_d(t)}{\mathrm{d}t} \tag{2.1}$$

The first term is due to the excess carrier distribution that is replaced in every ambipolar carrier life lime τ_a seconds, which happens when the diode biased with constant current. Thus, if τ_a is relatively long and the I-layer is not too thick, the I-layer becomes flooded with carriers at a reasonable forward-bias level and the diode exhibits a low RF resistance. The second term is time-dependent excess charge variation term $\mathrm{d}Q_d/\mathrm{d}t$. This time-varying storage charge is due to the time-varying signal to the diode. The effect of controlling the stored charge in the I-layer by controlling the forward current is known as conductivity modulation. This phenomenon of the p-i-n diode is used in electronically controlled attenuators and amplitude modulator. When a time-varying signal is applied with the constant-current bias, the stored charge is modulated by the presence of time-varying signal. The degree of modulation depends on the frequency of the time-varying modulating signal. When the signal frequency is much higher than the inverse of the ambipolar carrier lifetime and amplitude of the AC current is much less than the applied DC current, then AC modulation of the stored charge in the I-region is negligible. Thus, at this higher frequency, the conductivity modulation due to the AC signal is negligible compared to the DC current effect. At this higher frequency, the I-region Ohmic resistance R_i of a p-i-n diode of area A, width W, and carrier density over I-region $\rho(x)$ is given by [1]

$$R_i = \frac{1}{A}\int_0^W \rho(x)\mathrm{d}x \tag{2.2}$$

Under DC forward current, I_{do} and ambipolar diffusion length L_a, the intrinsic region resistance is given by

$$R_i = \frac{4kT}{qI_{do}}\sinh\left(\frac{W}{2L_a}\right)\left[\tan^{-1}\left(\exp\left(\frac{W}{2L_a}\right)\right) - \frac{\pi}{4}\right] \tag{2.3}$$

Here, q is the electron charge, k is the Boltzmann constant and T is diode temperature. The expression (2.3) is valid for any value of W/L_a. For narrow p-i-n diodes, where $W \ll L_a$, it is assumed that the stored charge is uniformly distributed over the entire I-region and the value of R_i is given by

$$R_i = \frac{W^2}{2\mu_a Q_d} = \frac{W^2}{2\mu_a \tau_a I_{do}} \tag{2.4}$$

Here, μ_a is the ambipolar carrier mobility. This relationship shows that the RF resistance of p-i-n diode depends on device geometry (W), electronic properties of its semiconductor material (μ_a, τ_a) and forward-bias current (I_{do}). Thus, for small-size p-i-n diode ($W \ll L_a$), the diode's intrinsic region resistance R_i changes inversely with the diodes forward-bias current (I_{do}).

Example 2.1 Calculate the intrinsic resistances of a narrow p-i-n diode of I-region width 100 μm at forward current of 10 μA and 10 mA. Consider $\mu_a = 1800\ \text{cm}^2/\text{Vs}$, and $\tau_a = 1\ \mu\text{s}$.

Solution

The intrinsic resistance of a narrow p-i-n diode is given by (2.4)

$$\begin{aligned} R_i &= W^2/(2\mu_a\tau_a I_{do}) \\ &= \frac{\left(100\times10^{-6}\right)^2}{2\times\left(1800\times10^{-4}\right)\times1\times10^{-6}}\times\frac{1000}{I_{do}(\text{in mA})}\ \Omega \\ &= \frac{27.8}{I_{do}(\text{in mA})}\ \Omega \end{aligned}$$

Therefore, R_i at 10 μA = (27.8 × 100) Ω ≈ 2.8 kΩ
And R_i at 10 mA = (27.8/10) Ω ≈ 2.8 Ω.

Practically, minimum resistance of a forward-bias diode is limited by its parasitic series resistance which is typically about 1–2 Ω. On the other side, maximum resistance of the diode is limited by the value of its equivalent shunt capacitance.

Modulation of the stored charge in the I-region takes place under low AC frequency. The time variation of the stored charge is equivalent to reactance across the I-region. This reactance may be capacitive or inductive depending on the size of the p-i-n diodes. In general, for small-size p-i-n diodes, the reactance is capacitive. The value of I-region capacitor (C_i) diminishes with the increase of frequency of the AC signal.

Junctions of a p-i-n diode will not contribute to the RF impedance of the p-i-n diode at high frequency due to the stored charge (Q_d) across the junctions. In equivalent circuit terminology, the stored charge effect is equivalent to a capacitance called diffusion capacitance C_d that is given by [3]

$$C_d = \frac{\partial Q_d}{\partial V_{dj}} = \frac{q}{\eta kT}\tau_a I_d\left(V_{dj}, T\right) \tag{2.5}$$

Here, V_{dj} is voltage across junction of the diode. Thus, the capacitance C_d depends upon the ambipolar carrier lifetime τ_a of the I-region, determined by the structure of the diode, and diode forward current I_d. The diffusion capacitance increases linearly with the increase of I_d.

At low frequencies, there will be time variation of stored charge in the entire I-region. Thus, low-frequency AC signal will modulate the I-region conductivity and junction resistance will be the major contribution to the impedance of p-i-n diode. The small-signal junction resistance will be governed by the slope of the following DC I–V relationship of the diode:

$$I_d\left(V_{dj}, T\right) = I_o\exp\left(\frac{qV_{dj}}{\eta kT} - 1\right) \tag{2.6}$$

Here, I_o is the reverse saturation current of the diode. Thus, small-signal low-frequency AC junction resistance of the forward-biased p-i-n diode is given by

$$R_j(I_d, T) = 1/\frac{\partial I_d}{\partial V_{dj}} = \frac{\eta kT}{qI_d\left(V_{dj}, T\right)} \tag{2.7}$$

At low-frequency operation, the p-i-n diode behaves as simple junction diode. The junction capacitance C_j of the diode of area A and width of the transition region W is given by

$$C_j = \frac{\varepsilon A}{W} \tag{2.8}$$

This junction capacitance will contribute to the low-frequency diode impedance.

The equivalent circuit of a p-i-n diode is shown in Fig. 2.2. Junction resistance R_j and capacitance C_j correspond to p–n junction of the diode. R_i represents resistance of the middle I-layer excluding the swept out region. C_i is I-region capacitance due to the modulation of the stored charges under AC signal. C_d is diffusion capacitance across the junction representing the charge storage due to the diode current given by (2.5). R_s is the ohmic series resistance. L_p and C_p are parasitic inductance and capacitance related to bond wire and packaging of the diode. The value of the equivalent components depends upon size, bias condition and operating AC frequency.

The diffusion capacitor appears in parallel with the junction resistance. Thus, the degree of modulation decreases with the increase of the current as well as frequency. The cut-off frequency f_c is given by

$$f_c = \frac{1}{2\pi\left(C_d + C_j\right)R_j} \tag{2.9}$$

In general, under sufficient forward-bias condition, $C_d \gg C_j$. Thus, putting the values of C_d and R_j from (2.5) and (2.7), respectively, the cut-off frequency becomes

$$f_c = \frac{1}{2\pi\tau_a} \tag{2.10}$$

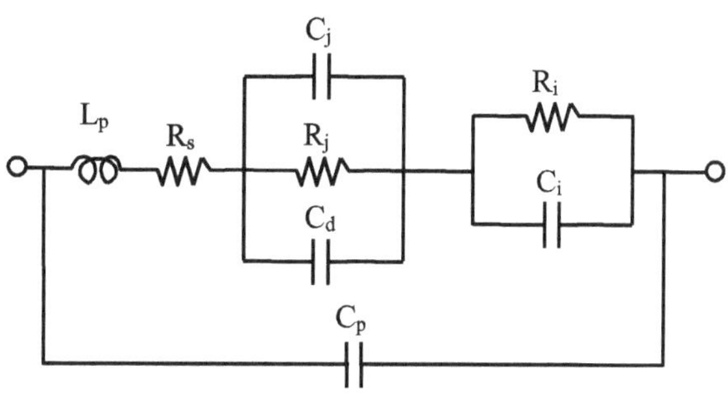

Fig. 2.2 Equivalent circuit of p-i-n diode

At frequencies well below the cut-off frequency f_c, the voltage-variable junction resistance dominates the impedance of a p-i-n diode and thus it behaves as an ordinary p–n junction diode. Therefore, RF signal of frequency well below f_c incident to the diode will be rectified and will generate considerable distortion. However, at frequencies well above $f_c (f \gg f_c)$, effect of junction resistance and the I-region reactance due to the stored charge modulation in the I-region will be negligible and the diode impedance is purely resistive.

Example 2.2 Calculate the cut-off frequency of a p-i-n diode of ambipolar carrier lifetime $\tau_a = 0.01$ μs.

Solution
From (2.10), the cut-off frequency is given by

$$\begin{aligned} f_c &= 1/2\pi\tau_a \\ &= \frac{1}{2 \times \pi \times 0.01 \times 10^{-6}} \text{Hz} \\ &= 15.9 \text{ MHz} \end{aligned}$$

Therefore, this diode can behave as voltage-variable RF resistance at its forward-bias condition without much distortion when the operating frequency $f \gg 15.9$ MHz. Operating frequency $f > 10 f_c$ is considered for using the p-i-n diode as RF resistance without much distortion. Practically, minimum frequency of operation of

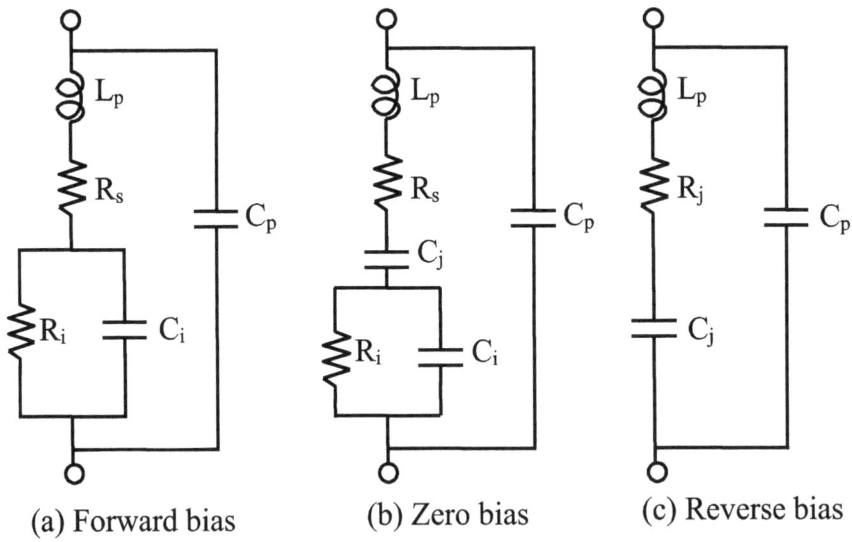

Fig. 2.3 Equivalent circuit of **a** forward biased, **b** zero biased, **c** reversed biased p-i-n diode for $f \gg 1/2\pi\tau_a$

a p-i-n diode as RF resistance not only determines by only f_c. It also depends on operating bias voltage/current of the diode and operating RF power level.

The equivalent circuit of a p-i-n diode is dependent on frequency and bias condition. At frequencies much higher than f_c, the equivalent circuit of a p-i-n diode at different bias conditions is as shown in Fig. 2.3. At forward-bias condition, the effect of junction capacitance and junction resistance can be neglected due to the presence of diffusion capacitance at operation frequency much greater than the cut-off frequency. Thus, the equivalent circuit at forward-bias condition will be as shown in Fig. 2.3a. However, at zero-bias condition, junction resistance R_j is very high, and there will be no presence of diffusion capacitance. Thus, the equivalent circuit of the p-i-n diode at zero-bias condition is as shown in Fig. 2.3b. At reverse-bias condition, if there is no charge carrier in the I-region, then there will be no existence of R_i. Thus, the equivalent circuit at reverse-bias condition is as shown in Fig. 2.3c.

2.4 Nonlinearity of P-I-N Diode

For controlling RF and microwave signals, p-i-n diode-based voltage/current-controlled attenuators, switches and phase shifters are widely used. In forward and reverse-bias condition, p-i-n diode may create signal distortion for its nonlinear high-frequency current–voltage characteristic. The modulation of the I-region stored charge due to the applied AC signal causes time-varying conductivity modulation. This generates harmonics and intermodulation distortion of the diode current output.

In forward-bias condition of p-i-n diode, low-frequency ($f \ll f_c$) RF signal modulates the I-region carriers as the DC current does. Stored charge modulation by the AC signal decreases with the increase of frequency. At high frequencies ($f \gg f_c$), AC signals have less influence on I-region stored charge. However, at microwave frequencies also, there is still a small amount of charge injection by the AC signal. I-region resistance R_i is inversely proportional to the stored charge Q_d as it is clear from (2.4). Therefore, R_i will also vary with the high-frequency AC signal due to the time variation of stored charge causing distortion generation of the AC signal. Stored charge modulation in the I-region depends upon frequency as well as amplitude of the applied AC signal for a particular p-i-n diode. However, width (W) of the I-region of p-i-n diodes also influences the degree of modulation of the stored charge. Larger width p-i-n diodes exhibit less variation of stored charge with increase of the AC signal frequency [4].

Reverse-biased p-i-n diodes are used for RF and microwave switching application. Reverse-biased p-i-n diodes also generate signal distortion. The reason for distortion of reverse-biased p-i-n diode is the finite time-varying capacitance due to the RF and microwave signal. The capacitance of an ideal p-i-n diode is generally assumed as constant for all reverse-bias voltage beyond punch-through voltage. However, in actual p-i-n diodes, reverse-bias capacitance decreases for voltages beyond its punch-through due to the advancement of depletion layer into the end regions with the increase of reverse voltage. Thicker p-i-n diodes exhibit flatter capacitance–voltage relationship than thinner diodes. This results in a lower rate of capacitance change with reverse voltage for thick p-i-n diodes. The magnitude of the reverse-bias distortion is inversely proportional to the slope of the capacitance–voltage characteristic and frequency. This is in contrast to the forward-bias p-i-n diode distortion case where the distortion improves with increasing AC signal frequency.

2.5 Temperature Behaviour of P-I-N Diode

RF resistance, R_i, of p-i-n diodes depends on forward-bias current (I_{do}) and electronic properties of its semiconductor material, i.e. ambipolar carrier mobility (μ_a) and ambipolar carrier lifetime (τ_a) as given by (2.4). The I-region ambipolar mobility μ_a, and the ambipolar carrier lifetime τ_a, are both functions of temperature; therefore, high-frequency resistance of p-i-n diodes also depends on temperature. Forward-bias current I_{do} also changes with temperature if the diode is not driven by a constant current source.

The temperature dependency of mobility is well known and it decreases with nearly square of increasing temperature in the temperature range of interest 223–473 K (−50–220 °C). This temperature dependency can be approximated by [1]

$$\mu(T) = \mu(T_o)\left(\frac{T_o}{T}\right)^n \tag{2.11}$$

Here, $\mu(T_o)$ and $\mu(T)$ are the carrier mobility at temperature T_o and T, respectively and n is a temperature independent constant.

For pure materials, the experimental value of n is 2.42 and 2.2 for n- and p-type Si, respectively; 1.0 and 2.1 for n- and p-type GaAs, respectively; 1.66 and 2.33 for n- and p-type Ge, respectively.

Based on measurements and works on silicon devices, carrier lifetime has been found to increase with increasing temperature. The temperature dependency of carrier lifetime may be modelled as

$$\tau(T) = \tau(T_o)\left(\frac{T}{T_o}\right)^m \tag{2.12}$$

Here, $\tau(T_o)$ and $\tau(T)$ are the carrier lifetime at temperature T_o and T, respectively and m is a temperature independent constant. The value of m depends on variety of factors influencing the carrier lifetime temperature characteristic. The important factors are the diode geometry, i.e. I-region width, diameter and the type of surface passivation material. Depending on these factors, in general, the value of m varies from 0.5 to 1.5 [5, 6]. However, in case of most Si microwave p-i-n diodes, the value of m is near about 1.0 [7].

Therefore, from (2.4), (2.11) and (2.12), temperature dependency of high-frequency resistance of p-i-n diodes can be written as

$$R_i(I_d, T) = \frac{W^2 T^{n-m}}{2\mu_a(T_o)\tau_a(T_o)I_{do}(T)} \tag{2.13}$$

Here, $\mu_a(T_o)$ and $\tau_a(T_o)$ are the ambipolar carrier mobility and ambipolar carrier lifetime, respectively, at temperature T_o, whereas putting the temperature-dependent reverse saturation current I_o to (2.6), the temperature-dependent forward-bias static DC current I_d (T) of p-i-n diodes is given by [1]

$$I_d(V_{dj}, T) = AT^{\nu}\exp\left(\frac{-qE_b}{\eta kT}\right)\left[\exp\left(\frac{qV_{dj}}{\eta kT}\right) - 1\right] \tag{2.14}$$

Here, A is a temperature-independent constant depending on area and electronic properties of material of the diode, ν is a constant ~3 for p-i-n diodes, E_b is the bandgap potential in volt of diode's semiconductor material, V_{dj} is the voltage across the junction of the diode, η is the ideality factor nearly equal to 2.0 due to the recombination of current in p-i-n diodes, q is the electron charge and k is the Boltzmann constant.

Typical current–voltage relationship of p-i-n diodes under forward-bias condition at three different temperatures is shown in Fig. 2.4. Three different types of bias load lines on the same plot are also shown. Temperature dependency of RF resistance of p-i-n diodes is also shown in Fig. 2.5 under fixed-current- and fixed-voltage-bias condition based on Eqs. (2.13) and (2.14) for $n = 2$, $m = 1$, $\nu = 3$

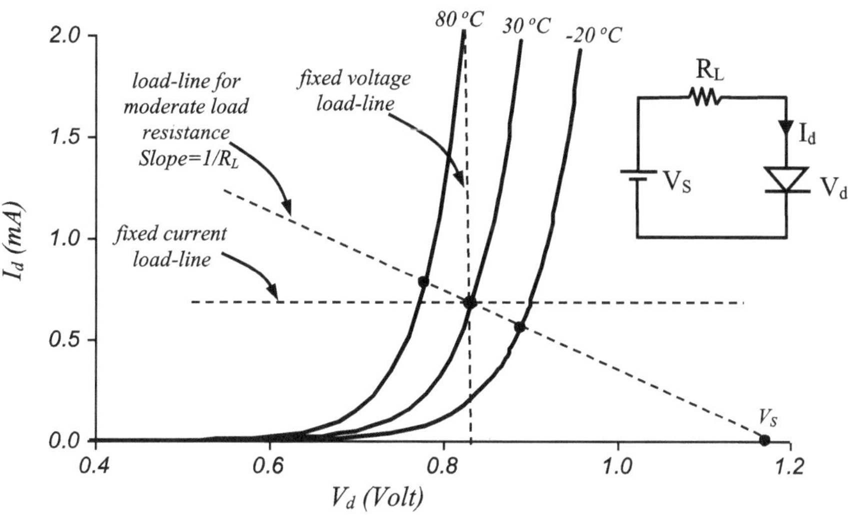

Fig. 2.4 Typical temperature dependency of V–I curve and different bias load line

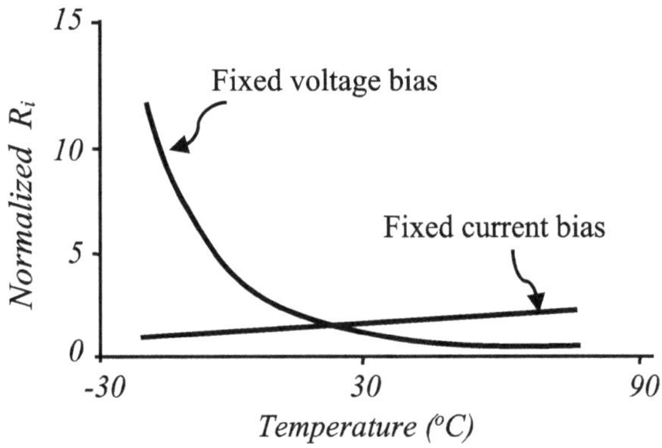

Fig. 2.5 Typical temperature dependency of p-i-n diode RF resistance at fixed-current- and fixed-voltage-bias conditions

and $\eta = 2$ [8]. Operation of diodes under constant-voltage-bias condition never recommended due to its large variation of current with the change of operating temperature. The RF resistance is also very sensitive with the change of temperature under constant-voltage-bias condition, and the RF resistance decreases drastically with the increase of temperature. Diodes are generally known as current-controlled device. Under constant-current-bias condition, i.e. when driver circuit of p-i-n diode provides constant current irrespective of the temperature of the diodes, the RF resistance of the diode increases slowly with the increase of temperature. Thus, with the increase of temperature, the RF resistance of p-i-n diode increases when it is biased by constant-current-bias, and however decreases when it is biased by constant-voltage-bias condition. These are the very interesting characteristics of p-i-n diodes. Therefore, it is possible to achieve temperature-invariant RF resistance of the p-i-n diodes under optimum bias load-line condition which is in between constant-voltage- and constant-current-bias condition.

2.6 Temperature-Invariant RF Resistance of P-I-N Diode

Microwave and RF control circuits based on p-i-n diodes are used in airborne and space-based systems where always temperature-invariant performance demanded over very wide operating temperature range. The success of electronic system for these applications relies on the ability to design high-performance, high-reliable circuits which function in demanding thermal environments. However, high-frequency resistance of p-i-n diodes that determine the performance of the circuits is temperature-dependent as discussed in the previous section. The temperature variation of RF resistance is due to the temperature dependency of carrier mobility and carrier lifetime. The carrier mobility is the property of the diode's semiconductor material. However, temperature characteristic of carrier lifetime influenced by the diode geometry, especially by the I-region area-to-width ratio and the type of material, is used to passivate the diode surface [5, 6]. For larger size silicon p-i-n diodes (n

~ 2) with glass or silicon dioxide passivated surface, the value of m becomes nearly 2 and leads to temperature-insensitive RF resistance. However, the diodes with larger I-region area-to-width ratio are having larger capacitance value across it that leads to limit high-frequency circuit performance.

In case of most of the commercially available microwave p-i-n diodes, the carrier lifetime linearly increases with the increase of temperature, i.e. $m = 1$. Therefore, for most of the practically available microwave p-i-n diodes $(n - m) = \sim 1$, that means, RF resistance linearly increases with the increase of diode temperature under constant current bias. In other words, the well-known constant current biasing will not provide acceptable temperature stability of the p-i-n diode-based circuit performance depending upon the RF resistance of the p-i-n diode such as attenuators and modulators. Therefore, proper temperature compensation mechanism is required to achieve acceptable temperature stability.

There are numerous conventional compensation mechanisms in the circuit design level, which uses temperature sensor to produce a temperature-dependent signal. This signal modifies the current applied to the diode in such a manner that it maintains same RF resistance over the operating temperature. Some of the compensation mechanism is based entirely on analog control signal, while others contain digital control signal acting at discrete temperature steps. However, the successful adjustment of such control circuits is a time-consuming task, often involves trial and error; and the circuits themselves are relatively complex.

Figure 2.6 shows a simple biasing circuit for a p-i-n diode with supply voltage V_S and load resistor R_L. Forward-biased DC equivalent circuit of p-i-n diode is also shown in Fig. 2.6. Here, R_s is the parasitic series resistance, R_i is the I-region resistance and R_j is the DC junction resistance of the diode. Therefore, the voltage–current relationship of the p-i-n diode can be written in terms of diode terminal voltage as

$$I_d(V_d, T) = AT^{\upsilon}\exp\left(\frac{-qE_b}{\eta kT}\right)\left[\exp\left(\frac{q[V_d(T) - I_dR_s - I_dR_i]}{\eta kT}\right) - 1\right] \tag{2.15}$$

For $(n - m) \sim 1$ and neglecting reverse saturation current compared to the forward current, (2.15) can be written as

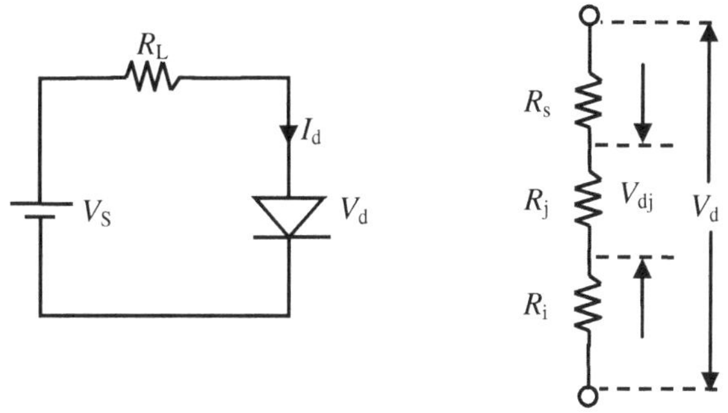

Fig. 2.6 Simple bias circuit and DC equivalent circuit of forward-biased p-i-n diode

$$I_d(V_d, T) = A'T^{\nu} \exp\left(\frac{q[V_d(T) - I_d(V_d, T)R_s - E_b]}{\eta kT}\right) \tag{2.16}$$

Here, A' is a temperature-independent constant. The temperature and current dependency of RF resistance of practical diodes (2.13) can be rewritten in more generic form as

$$R_i(I_d, T) = z\frac{T^{n-m}}{I_{do}(V_d, T)^p} \tag{2.17}$$

Here, p and z are constants. The value of p is near about 1. To achieve temperature-invariant RF resistance of the diodes, the bias network must meet the current requirement to achieve a constant RF resistance, say R_{io} over the operating temperature. Therefore, current–voltage relationship (2.16) is to be satisfied by the bias network to achieve temperature-invariant RF resistance over the temperature range under the constraint imposed on $V_d(T)$ as given by

$$V_d(T) = E_b + \left[z\frac{T^{n-m}}{R_{io}}\right]^{\frac{1}{p}} R_s + \frac{\eta kT}{qp} \ln\left[\frac{z}{A'^p R_{io}} T^{(n-m-\nu p)}\right] \tag{2.18}$$

This parametric equation is the equation of ideal bias-point locus that will satisfy $R_i = R_{io}$ over the diode temperature range. This locus is called *equiresistance* curve for the required RF resistance R_{io} [7, 9, 10].

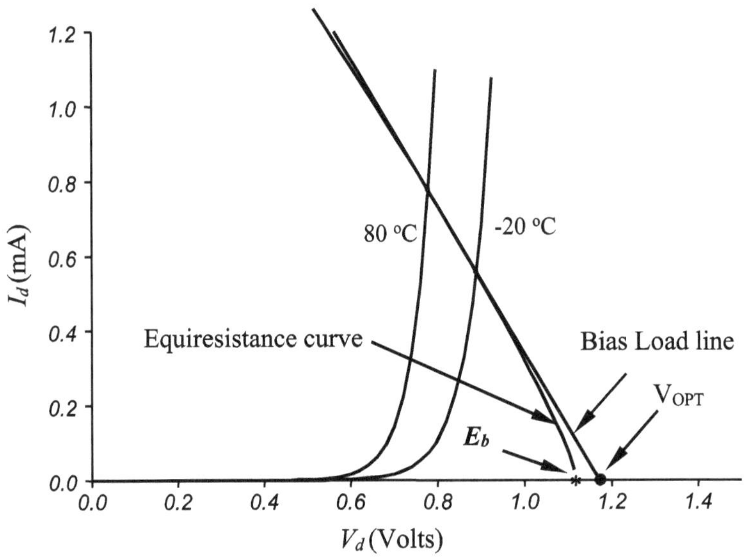

Fig. 2.7 Equiresistance curve and optimum bias load line along with *I-V* characteristic of p-i-n diode

I–V characteristic and corresponding equiresistance curve are shown in Fig. 2.7, for $m = 0.97$, $p = 0.9$, $\eta = 2$ and $n = 2.3$ over the temperature range of −260 to +300 °C. The plot shows that the equiresistance curve is highly linear over the temperature range of interest. Thus, the load line of a simple bias circuit, as shown in Fig. 2.6, can be adjusted to coincide with the equiresistance curve within the temperature range of interest and it will maintain the desired RF resistance. The voltage of the bias network will be voltage axis intercept of the bias load line, and resistance will be the reciprocal of slope of the bias load line. Combining (2.16), (2.17) and (2.18) and eliminating T, another form of equation of the equiresistance curve can be written as

$$V_d(T) = E_b + \frac{\eta k}{q}\left(\frac{R_{io} I_d^p}{z}\right)^{\frac{1}{n-m}} \ln\left[\left(\frac{z}{R_{io}}\right)^{\frac{v}{n-m}} \frac{1}{A I_d^{\frac{vp}{n-m}-1}}\right] \quad (2.19)$$

The equation of tangent to the equiresistance curve of (2.19) at diode bias voltage V_{do} and corresponding current I_{do} at temperature T_o is given by

$$V_d(T) = V_{do} + \left[\frac{p}{n-m}(E_b - V_{do}) + \frac{\eta k T_o}{q}\left(\frac{pv}{n-m} - 1\right)\right] \times \left[1 - \frac{I_d}{I_{do}}\right] \quad (2.20)$$

Taking this tangent to be the optimum bias load line, its voltage axis intercept V_{OPT}, called optimum voltage, is given by

$$V_{\mathrm{OPT}} = V_{do} + \frac{p}{n-m}(E_b - V_{do}) + \frac{\eta k T_o}{q}\left(\frac{pv}{n-m} - 1\right) \quad (2.21)$$

This optimum voltage V_{OPT} will be the open-circuit voltage of the bias network. The voltage axis intercept is seen to be practically independent of the selected bias point (V_{do}, I_{do}), implying that the optimum no-load output voltage of the bias circuit is the same for all resistance values.

For a particular type of diode, the parameter values m, n and p are constants and independent of temperature. Putting typical values of these constants, i.e. $m = 1.0$, $p = 1.0$, $v = 3$, $\eta = 2$ and $n = 2.0$ in (2.21), the optimum open-circuit voltage becomes

$$V_{\mathrm{OPT}} = E_b + \frac{2kT_o}{q} \quad (2.22)$$

Thus, open-circuit voltage is very close to the bandgap potential E_b of the semiconductor material of the p-i-n diode since the last term is of the order of a few millivolts. Parameter E_b is the property of the diode's material; it is independent of process technology and does not vary from one die to another. Moreover, the temperature dependency of bandgap potential (E_b) is also very weak [11] over a practical operating temperature range. Thus, this optimum bias load-line technique for achieving temperature-invariant performance is very stable under different parameter variations of the diodes.

The optimum bias load-line technique for achieving temperature-invariant performance is also applicable for many other junction diodes [10]. For example, RF resistance of Schottky barrier diode and light intensity (brightness) of light-emitting diode can be made temperature-insensitive using the optimum bias load-line technique [12, 13].

Example 2.3 Calculate open-circuit voltage for temperature-invariant intrinsic layer resistance of a silicon p-i-n diode over the temperature range of −20 to 80 °C.

Solution From (2.22), the optimum open-circuit voltage is given by

$$\begin{aligned} V_{\mathrm{OPT}} &= E_b + \frac{2kT_o}{q} \\ &= 1.12 + \frac{2 \times 1.38 \times 10^{-23}}{1.6 \times 10^{-19}} T_o \text{ V} \\ &= 1.12 + 0.0001725 \times T_o \text{ V} \end{aligned}$$

Putting the values of T_o, the optimum voltage (V_{OPT}) is calculated and is shown in Fig. 2.8. It shows that the optimum voltage is practically independent of temperature and is slightly higher than the value of bandgap potential (E_b) of the semiconductor.

Example 2.4 Current-dependent RF resistance, R_i, and current–voltage ($I_d - V_d$) relationship of a p-i-n diode are

$$R_i = \frac{aT}{I_d} \quad \text{and} \quad I_d = b \times \exp\left(\frac{V_d - E_b}{cT}\right)$$

Here, a, b, c are constants. Derive optimum open-circuit voltage for zero temperature coefficient of RF resistance of the diode.

Solution

Temperature coefficient of the p-i-n diode resistance is given by

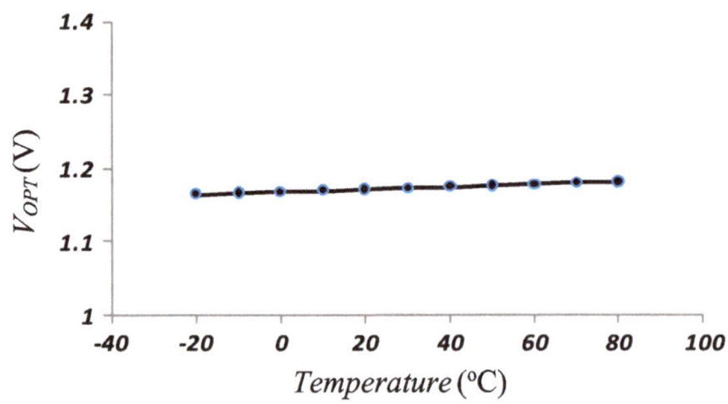

Fig. 2.8 Dependence of optimum open-circuit voltage on temperature

$$\frac{\partial R_i}{\partial T} = \frac{a}{I_d} - \frac{aT}{I_d^2}\frac{\partial I_d}{\partial T} \tag{2.23a}$$

For zero temperature coefficient of the RF resistance, $\partial R_i/\partial T = 0$. Putting this condition to (2.23a), we can write

$$\frac{\partial I_d}{\partial T} = \frac{I_d}{T} \tag{2.23b}$$

Suppose the diode is biased through a bias resistance, R_b and supply voltage V_S. Thus,

$$V_S = I_d R_b + V_d \tag{2.23d}$$

Using (2.23d), the I–V relationship of the p-i-n diode can be written as

$$I_d = b \times \exp\left(\frac{V_S - I_d R_b - E_b}{cT}\right) \tag{2.23e}$$

Differentiating (2.23e) w.r.t. temperature T,

$$\frac{\partial I_d}{\partial T}\left(1 + \frac{I_d R_b}{cT}\right) = -\frac{I_d(V_S - I_d R_b - E_b)}{cT^2} \tag{2.23f}$$

For zero temperature coefficient, suppose $V_S = V_{\text{OPT}}$ and $I_d = I_{do}$. Putting these values to (2.23f) and using (2.23b),

$$\frac{I_{do}}{T}\left(1 + \frac{I_{do} R_b}{cT}\right) = -\frac{I_{do}(V_{\text{OPT}} - I_{do} R_b - E_b)}{cT^2} \tag{2.23f}$$

This leads to the value of optimum supply voltage for zero temperature coefficient of RF resistance of the p-i-n diode as

$$V_{\text{OPT}} = E_b - cT \tag{2.23g}$$

Example 2.5 A microwave circuit as shown in Fig. 2.9 consists of n numbers of batch matched p-i-n diodes in shunt configuration (connected in DC parallel path). Current-dependent RF resistance, R_i, and current–voltage ($I_d - V_d$) relationship of the p-i-n diodes are

$$R_i = \frac{aT}{I_d} \quad \text{and} \quad I_d = bT^3 \times \exp\left(\frac{V_d - E_b}{cT}\right)$$

Here, a, b, c are constants. Derive the optimum open-circuit voltage for zero temperature coefficient of the RF resistance.

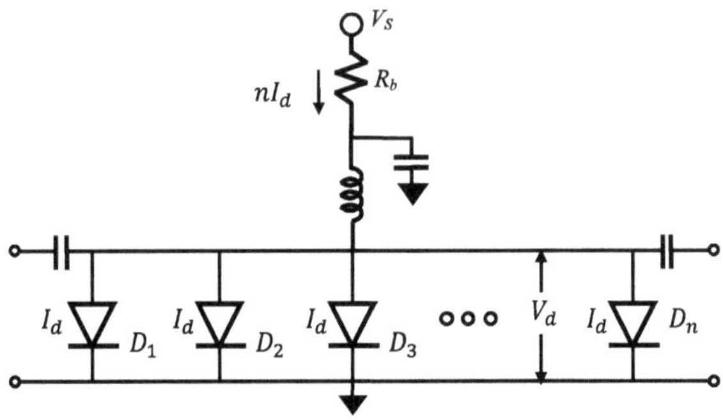

Fig. 2.9 RF circuit using multiple shunt p-i-n diodes

Solution For achieving zero temperature coefficient of RF resistance of the p-i-n diodes,

$$\frac{\partial R_i}{\partial T} = \frac{a}{I_d} - \frac{aT}{I_d^2}\frac{\partial I_d}{\partial T} = 0 \tag{2.24a}$$

$$\text{or,} \quad \frac{\partial I_d}{\partial T} = \frac{I_d}{T} \tag{2.24b}$$

As shown in Fig. 2.9, the diodes are biased through a bias resistance R_b and supply voltage V_S. Here, DC current I_d flows through each diode and voltage across each diode is V_d. Thus,

$$V_S = nI_d R_b + V_d \tag{2.24c}$$

Using (2.24c), the I–V relationship of the p-i-n diode can be written as

$$I_d = bT^3 \times \exp\left(\frac{V_S - nI_d R_b - E_b}{cT}\right) \tag{2.24d}$$

Differentiating (2.24d) w.r.t. temperature T,

$$\frac{\partial I_d}{\partial T}\left(1 + \frac{nI_d R_b}{cT}\right) = \frac{3I_d}{T} - \frac{I_d(V_S - nI_d R_b - E_b)}{cT^2} \tag{2.24e}$$

For zero temperature coefficient, suppose $V_S = V_{\text{OPT}}$ and $I_d = I_{do}$. Putting these values to (2.24e) and using (2.23b),

$$\frac{I_{do}}{T}\left(1 + \frac{nI_{do} R_b}{cT}\right) = \frac{3I_d}{T} - \frac{I_{do}(V_{\text{OPT}} - nI_{do} R_b - E_b)}{cT^2} \tag{2.24f}$$

This leads to the value of optimum supply voltage for zero temperature coefficient of RF resistance as

$$V_{\text{OPT}} = E_b + 2cT \tag{2.24g}$$

This shows that the optimum supply voltage for zero temperature coefficient of RF resistance of diodes is independent on the number of diodes connected in parallel DC path.

Example 2.6 A microwave circuit as shown in Fig. 2.10 consists of n numbers of matched pair p-i-n diodes connected in DC series path. Current-dependent RF resistance, R_i, and current–voltage ($I_d - V_d$) relationship of the p-i-n diodes are

$$R_i = \frac{aT}{I_d} \quad \text{and} \quad I_d = bT^3 \times \exp\left(\frac{V_d - E_b}{cT}\right)$$

Here, a, b, c are constants. Derive the optimum open-circuit voltage for zero temperature coefficient of RF resistance of the diodes.

Solution For achieving zero temperature coefficient of RF resistance of the p-i-n diodes,

$$\frac{\partial R_i}{\partial T} = \frac{a}{I_d} - \frac{aT}{I_d^2}\frac{\partial I_d}{\partial T} = 0 \tag{2.25a}$$

$$\text{or,} \quad \frac{\partial I_d}{\partial T} = \frac{I_d}{T} \tag{2.25b}$$

As shown in Fig. 2.10, the diodes are biased through a bias resistance R_b and supply voltage V_S. Here, DC current I_d flows through each diode, and voltage across each diode is V_d. Thus,

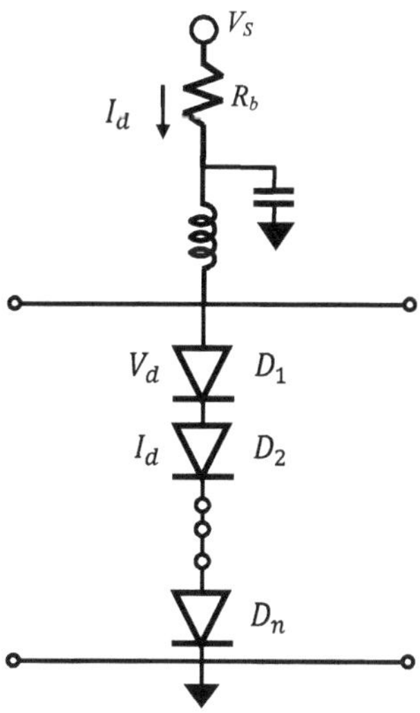

Fig. 2.10 RF circuit diagram using multiple series p-i-n diodes

$$V_S = I_d R_b + n V_d \tag{2.25c}$$

Using (2.25c), the I–V relationship of the p-i-n diode can be written as

$$I_d = bT^3 \times \exp\left(\frac{V_S - I_d R_b - nE_b}{ncT}\right) \tag{2.25d}$$

Differentiating (2.25d) w.r.t. temperature T,

$$\frac{\partial I_d}{\partial T}\left(1 + \frac{I_d R_b}{ncT}\right) = \frac{3I_d}{T} - \frac{I_d(V_S - I_d R_b - nE_b)}{ncT^2} \tag{2.25e}$$

For zero temperature coefficient, suppose $V_S = V_{\mathrm{OPT}}$ and $I_d = I_{do}$. Putting these values to (2.25e) and using (2.25b),

$$\frac{I_{do}}{T}\left(1 + \frac{I_{do} R_b}{ncT}\right) = \frac{3I_{do}}{T} - \frac{I_{do}(V_{\mathrm{OPT}} - I_{do} R_b - nE_b)}{ncT^2} \tag{2.25f}$$

This leads to the value of optimum supply voltage for zero temperature coefficient of RF resistance as

$$V_{\mathrm{OPT}} = n(E_b + 2cT) \tag{2.25g}$$

This shows that the optimum supply voltage for n numbers of series DC path connected diodes is n times the optimum voltage corresponding to a single diode.

References

1. Sze SM (1981) Physics of semiconductor devices, 2nd edn. Wiley Eastern Limited
2. Wilson P (1967) Recombination in silicon p-i-n diodes. Solid State Electron 10:145–154
3. Slatter JAG, Whelan JP (1980) p-i-n diode recovery storage time. Solid State Electron 23:1235–1242
4. Caverly R (2004) Distortion modeling of PIN diode switches and attenuators. In: IEEE MTT-S digest, pp 957–960
5. Caverly RH, Hiller G (1993) Temperature insensitive PIN diode attenuators. Appl Microw 88, Summer 1993
6. Caverly RH, Hiller G (1995) Temperature effect on PIN diode forward bias resistance. Solid State Electron 38(11):1879–1885
7. Bera SC, Bharadhwaj PS (2005) Insight into PIN diode behavior leads to improved control circuit. IEEE Trans Circuits Syst II 52(1):1–4
8. Bera SC, Singh RV, Garg VK, Sharma SB (2007) Optimum bias load-line compensates temperature variation of junction diode's RF resistance. IEEE Trans Microw Theory Tech 55(2):215–221
9. Bera SC, Singh RV, Garg VK (2005) A temperature dependent PIN diode model leads to simple temperature invariant attenuator circuit. Microw J 48(02):104–116
10. Bera SC, Singh RV, Garg VK (2006) Inherent temperature-compensating behavior of junction diodes. Proceedings of international conference on computer and devices for communication

11. Bludau W, Onton A, Heinke W (1974) Temperature dependence of the band gap of silicon. J Appl Phys 45(4):1846–1848
12. Bera SC, Singh RV, Garg VK (2005) Temperature behavior and compensation of light emitting diode. IEEE Photonics Tech Lett 17(11):2286–2288
13. Bera SC, Singh RV, Garg VK (2009) A device for compensating the effect of temperature variation on brightness of light emitting diodes. India Patent, IN235617

Chapter 3
Schottky Diode

Abstract This chapter presents the basic working principle and characteristics of Schottky barrier diodes including its equivalent circuit and applications in communication systems. Temperature behaviour of Schottky barrier diodes and its effect on performance of the diode-based circuits are also presented.

3.1 Introduction

Metal–semiconductor junction diode is known as Schottky barrier diode where only majority carrier takes part for device operation. Non-existence of minority carrier injection and extraction leads it to a very high-speed operating device. High-speed operation capability of the Schottky diode makes it suitable to use widely for microwave and mmwave applications. Lower cut-in voltage of the Schottky diode makes it suitable for very low voltage and high current operation. The device is used as RF detector, multiplier, mixer, limiter, attenuator, phase shifter, etc. This is also used to realize phase shifter and distortion generator for microwave beam forming network and linearizer, respectively.

3.2 Basics of Schottky Diode

Schottky contact is a metal–semiconductor contact that generates potential barrier for carriers to move from one material to another and generates nonlinear voltage–current characteristic similar to p–n junction diode. The semiconductor material may be n- or p-type silicon (Si), Gallium Arsenide (GaAs), Gallium Nitride (GaN), Silicon Carbide (SiC), etc. and metal may be Aluminium (Al), Gold (Au), Tungsten (W), Titanium (Ti), etc. [1–4].

Early detector diodes were point contact diodes, consists of a thin metal whisker pressing on a semiconductor material generally silicon. However, Schottky barrier diodes are fabricated by deposition of a metal layer on a semiconductor. Practically,

S. C. Bera, *Microwave Active Devices and Circuits for Communication*, Lecture Notes in Electrical Engineering 533, https://doi.org/10.1007/978-981-13-3004-9_3

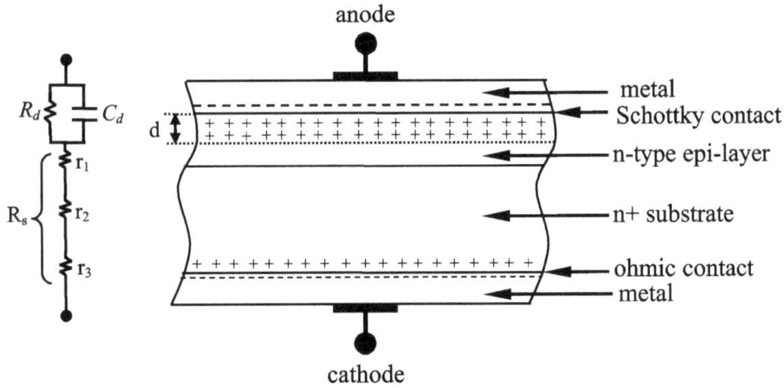

Fig. 3.1 Schematic structure of Schottky barrier diode

n-type semiconductors are used for all microwave Schottky diodes because of superior mobility of electrons than mobility of holes. The basic schematic structure of a Schottky barrier diode is shown in Fig. 3.1 [1, 2]. Metal–semiconductor Schottky junction is formed in between metal and n-type active layer. The n-type active layer is generally grown epitaxially over an n+ substrate. Very low resistance ohmic contact is formed in between n+ substrate and metal as electrode.

The energy band diagram of an n-type Schottky barrier diode when metals and semiconductors are away from each other is shown in Fig. 3.2. Here, E_{Fm} and E_{FS} are Fermi level energies, $q\phi_m$ and $q\phi_s$ are work functions of the metal and semiconductor respectively. E_C and E_V are conduction and valance band-edge energies respectively and $q\chi$ is electron affinity. For the n-type Schottky diode, the work function of metal (ϕ_m) is higher than that of semiconductors (ϕ_s). Electron flows from semiconductor to metal to establish thermodynamic equilibrium when metals and semiconductors come into contact. Due to the charge transfer, Fermi levels align and barrier potential ($\phi_b = \phi_m - \chi$) is developed between metal and n-type semiconductor junction as shown in Fig. 3.3. The Fermi level and band profile of the metal will not change at all due to very high-electron density of the metal side. A depletion region (W) is formed adjacent to the metal in the semiconductor and the depletion region width is given by (3.1) for donor concentration of N_d [2], where ε is dielectric constant of the semiconductor. The depletion layer of width W_1 for Schottky barrier contact and W_2 for Ohmic contact are shown in Fig. 3.3.

$$W = \left[\frac{2\varepsilon V_{bi}}{qN_d}\right]^{1/2} \tag{3.1}$$

The positive charge within the depletion region W due to the uncompensated donor ions matches the negative charge in the metal. It generates the equilibrium contact potential (V_{bi}), which is the difference of work functions of metal and semiconductor and is given by $V_{bi} = (\phi_m - \phi_s)$. The generated contact potential prevents further net electron diffusion in between them. The potential barrier height ϕ_b for

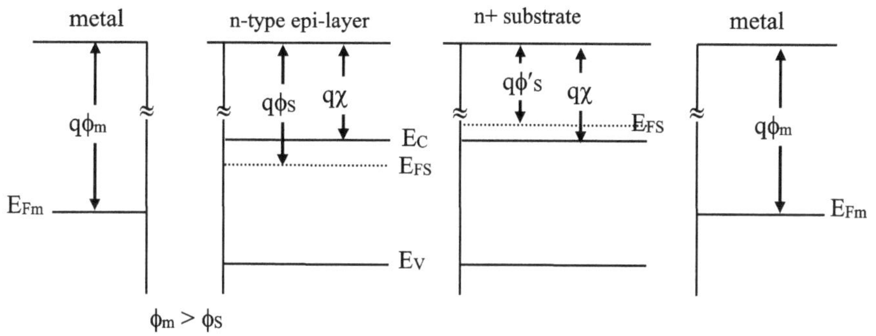

Fig. 3.2 Energy band diagram of Schottky barrier diode when metals & semiconductors are away from each other

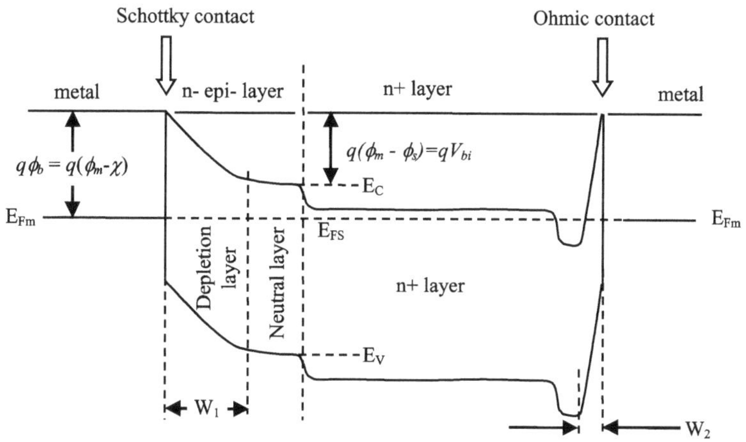

Fig. 3.3 Energy band diagram of Schottky barrier diode when metals and semiconductors are in contact

electron injection from the metal into the n-type semiconductor conduction band is $(\phi_m - \chi)$. Here, $q\chi$ (the electron affinity) is measured from the vacuum level to the semiconductor conduction band edge. The width W of the depletion region is inversely proportional to $\sqrt{N_d}$. Thus, as the doping concentration in the semiconductor increases, the width of the depletion region decreases. This property is used to make ohmic contact with the heavily doped semiconductor and suitable metal as shown in Fig. 3.3. The locally heavy doping in the semiconductor decreases the width of the depletion region (W_2). Thus, probability of tunnelling the electrons through the barrier increases and the junction behaves as ohmic contact, i.e. I–V characteristic of the ohmic junction becomes linear with the same slope under forward and reverse bias condition.

3.3 Schottky Diode Characteristics

The potential difference V_{bi} can decrease or increase due to the application of either forward or reverse bias voltage. The contact potential is reduced from V_{bi} to $(V_{bi} - V)$ with the application of forward voltage V applied across the Schottky barrier junction. Due to reduction of the contact potential, electrons from the semiconductor can diffuse across the depletion region to the metal, this rises forward current from metal to semiconductor through the junction. However, due to the barrier potential ϕ_b, the flow of electrons from metal to semiconductor is negligible. Thus, the forward current is due to the only injection of majority carriers from the semiconductor into the metal. Whereas with the application of reverse voltage V_r, the barrier potential increases from V_{bi} to $(V_{bi} + V_r)$ that reduced to negligible electron flow from semiconductor to metal. In this case, also due to the barrier potential ϕ_b, the flow of electrons from metal to semiconductor is negligible.

At normal operation of Schottky barrier diodes, there is no minority carrier injection. This eliminates the associated storage delay time, which is an important feature of Schottky barrier diodes. Therefore, Schottky barrier diodes are essentially majority carrier devices, although some minority carrier injection occurs at high-current levels. These properties lead to the superior high-frequency properties and switching speed compared to typical *p–n* junctions.

Under thermal equilibrium with no bias condition, the current from the metal to the semiconductor is equal in magnitude and opposite in direction to the current from the semiconductor to the metal. At this condition, the current in one direction say from the semiconductor to the metal ($I_{S \to M}$) is given by Sze [1] as follows:

$$I_{S \to M} = A^* A T^2 \exp\left(\frac{-q\phi_b}{kT}\right) \tag{3.2}$$

Here, A^* is the Richardson constant and A is area of the diode. Under forward bias condition of junction voltage V_{dj}, incorporating ideality factor (η), the diode current is given by

$$I_d = A^* A T^2 \exp\left(\frac{-q\phi_b}{\eta kT}\right)\left[\exp\left(\frac{qV_{dj}}{\eta kT}\right) - 1\right] \tag{3.3}$$

$$= I_o\left[\exp\left(\frac{qV_{dj}}{\eta kT}\right) - 1\right] \tag{3.4}$$

In case of Schottky barrier diodes, the ideality factor $\eta \cong 1.0$–1.2, because of negligible recombination in the depletion region. I_o is defined as the reverse saturation current. The Schottky diode does not have charge storage delay. Thus, its DC transfer characteristic (3.3) is also valid for ac transfer characteristic up to a very high frequency, extended to microwave even up to mmwave frequency range.

Suppose, RF signal of amplitude V_s and angular frequency ω is applied to a finitely DC biased (V_o) Schottky diode. The diode voltage can be represented by (3.5) as follows:

$$V_{dj} = V_o + V_s \cos(\omega t) \tag{3.5}$$

The diode current can be written as

$$I_d = I_o\left[\exp\left(\frac{V_o + V_s\cos(\omega t)}{V_t}\right) - 1\right] \tag{3.6}$$

Here $V_t = (\eta kT/q)$. Representing the exponential term containing AC part by series expansion up to fourth-order term, the diode current can be written as

$$\begin{aligned} I_d &= I_o\left[\exp\left(\frac{V_o}{V_t}\right) \times \exp\left(\frac{V_s\cos(\omega t)}{V_t}\right) - 1\right] \\ &= I_o\left[\exp\left(\frac{V_o}{V_t}\right) - 1\right] \\ &\quad + I_o\exp\left(\frac{V_o}{V_t}\right)\left[\frac{V_s}{V_t}\cos(\omega t) + \frac{V_s^2}{2V_t^2}\cos^2(\omega t) + \frac{V_s^3}{6V_t^3}\cos^3(\omega t)\right. \\ &\quad \left. + \frac{V_s^4}{24V_t^4}\cos^4(\omega t)\right] + \cdots \end{aligned} \tag{3.7}$$

$$\begin{aligned} &= I_o\left[\exp\left(\frac{V_o}{V_t}\right) - 1\right] \Rightarrow \text{dc current due to the bias voltage } V_o \\ &+ I_o\exp\left(\frac{V_o}{V_t}\right) \times \frac{V_s}{V_t}\cos(\omega t) \Rightarrow \text{fundamental ac component} \\ &+ I_o\exp\left(\frac{V_o}{V_t}\right) \times \left[\frac{V_s^2}{4V_t^2} + \frac{V_s^2}{4V_t^2}\cos(2\omega t)\right] \Rightarrow \text{rectified dc + 2nd harmonic} \\ &+ I_o\exp\left(\frac{V_o}{V_t}\right) \times \left[\frac{V_s^3}{8V_t^3}\cos(\omega t) + \frac{V_s^3}{24V_t^3}\cos(3\omega t)\right] \\ &\qquad \Rightarrow \text{fundamental + 3rd harmonic} \\ &+ I_o\exp\left(\frac{V_o}{V_t}\right) \times \left[\frac{V_s^4}{64V_t^4} + \frac{V_s^4}{48V_t^4}\cos(2\omega t) + \frac{V_s^4}{192V_t^4}\cos(4\omega t)\right] \\ &\qquad \Rightarrow \text{dc + 2nd harmonic + 4th harmonic} \\ &+ \cdots \end{aligned} \tag{3.8}$$

The expression (3.8) shows that all the even-order nonlinearity produces DC term and even-order harmonics. Whereas, odd-order nonlinearity produces only odd-order harmonic components. The property of generation of DC term related to RF signal by Schottky diodes is used as power detector. Whereas, the property of generation of higher order harmonics of the applied RF signal by Schottky diodes is utilized as multiplier and mixer.

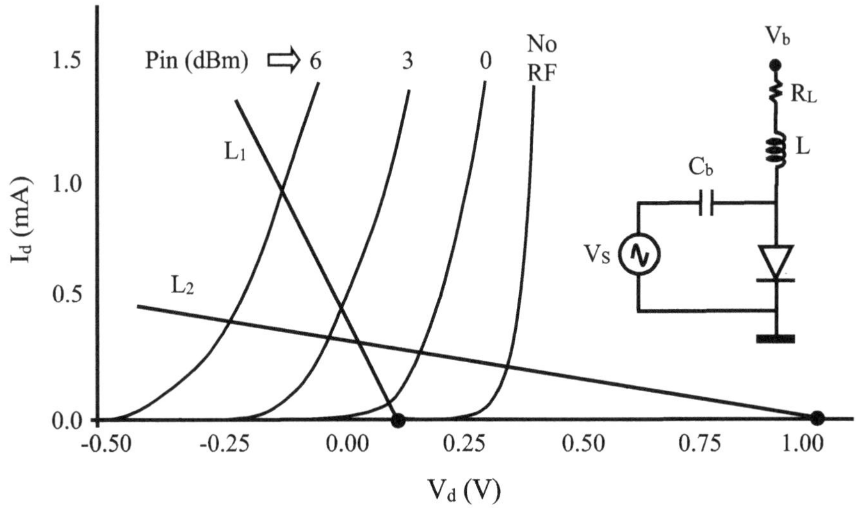

Fig. 3.4 *I–V* characteristic of Schottky diode in presence of RF signal

The expression (3.6) is written under the assumption that the exponential *I–V* relationship is also valid under RF signal operation. This assumption is true if $V_s \ll V_t$. Otherwise, due to the significant rectification of the RF signal exponential, *I–V* characteristic of the Schottky diode will not remain valid. A typical *I–V* characteristic of Schottky diodes in presence of RF signal is shown in Fig. 3.4. The characteristic deviates more from its exponential nature with the increase in RF signal level.

RF resistance of Schottky diodes is given by the inverse slope of the *I–V* curve at the particular operating condition and is given by

$$R_d(I_d) = 1/\frac{\partial I_d}{\partial V_{dj}} \tag{3.9}$$

Under small signal operating condition ($V_s \ll V_t$), the RF resistance of Schottky diodes is given by

$$R_d(I_d) = 1/\frac{\partial I_d}{\partial V_{dj}} = \frac{\eta kT}{q}\frac{1}{(I_d + I_o)} \tag{3.10}$$

However, at large signal operating conditions, the exponential *I–V* relationship is not valid to determine the RF resistance using it. Therefore, the expression (3.10) for RF resistance of the diodes is also not valid under large signal operating conditions. Under large signal RF operation, DC voltage and current across the diode are a function of RF signal level and also depends on bias load line as shown in Fig. 3.4. It is clear from the figure that for bias load line L_1, RF resistance of the diode decreases with the increase of RF signal level. Whereas, in case of bias load line L_2, RF resistance of the diode increases with the increase of RF signal level. Thus, RF

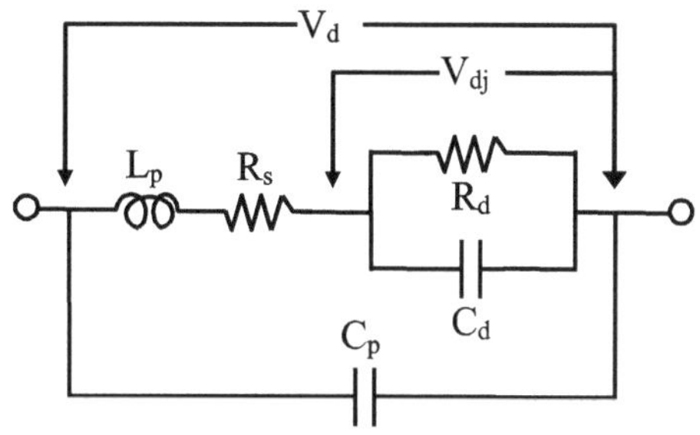

Fig. 3.5 RF equivalent circuit of forward biased Schottky barrier diode

resistance of the diode may increase or decrease with the increase in RF power level depending on the bias load line [5, 6].

RF equivalent circuit of a forward-biased Schottky diode is shown in Fig. 3.5. Here, R_d is the junction resistance of the diode, is a function of DC bias voltage as well as operating RF power of the diode as discussed in the previous paragraphs. The capacitance C_d is the depletion layer capacitance and is nearly independent of operating RF power level. Thus, the capacitor C_d is considered as linear capacitor. R_s is equivalent series resistance of the diode caused by the part of N layer, $N+$ layer and Ohmic contact as shown in Fig. 3.1. The resistor R_s is also considered as linear resistor. L_p and C_p are the parasitic inductance and capacitance related to the packaging of the diode.

3.4 Temperature Behaviour of Schottky Diodes

DC and RF characteristics of the Schottky barrier diodes are very much sensitive to its operating temperature. Temperature dependency of DC and RF parameters of the diodes in a circuit is also determined by the way it is biased. Temperature-dependent I V relationship of the Schottky diodes is given by

$$I_d(V_d(T), T) = A^* A T^2 \exp\left(\frac{-q\phi_b}{\eta k T}\right)\left[\exp\left(\frac{q V_d(T) - I_d(V_d(T), T) R_s}{\eta k T}\right) - 1\right] \tag{3.11}$$

where $V_d(T)$ is the temperature-dependent voltage across the diode and is related to the junction voltage $V_{dj}(T)$ by

$$V_d(T) = V_{dj}(T) - I_d(V_d(T), T) R_s \tag{3.12}$$

A typical I–V characteristic of a Schottky diode is shown in Fig. 3.6 for three different temperatures. Three different bias load lines are also shown in the figure to show the current and voltage variations along the load lines under varying temperature conditions. It shows that the diode current is highly temperature sensitive under

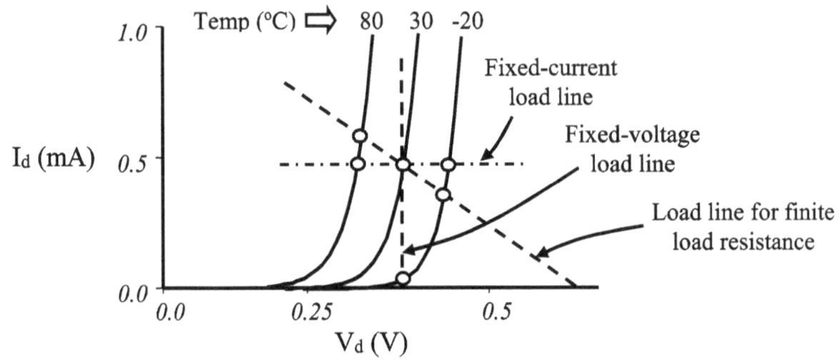

Fig. 3.6 *I–V* characteristic dependency on temperature and load line

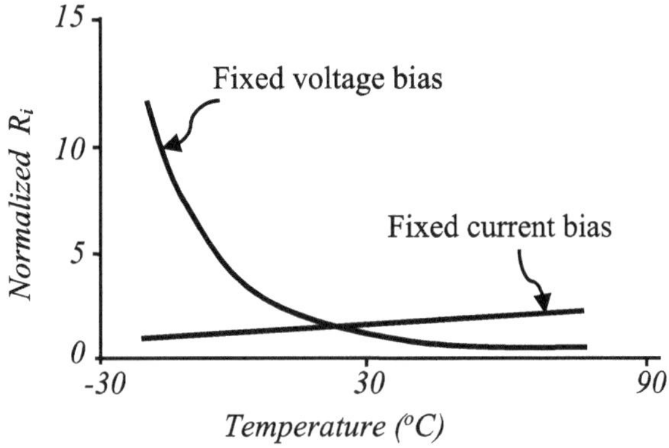

Fig. 3.7 Temperature dependency of Schottky diode RF resistance under different bias conditions

constant-voltage-bias condition. Thus, bias load line not only determines the RF performance of the diode-based circuits it also determines the temperature sensitivity of the circuit parameters.

Under sufficiently forward bias condition, where $I_d \gg I_o$ the temperature-dependent small signal RF resistance R_d can be written from (3.10) as [3]

$$R_d(I_d) = \frac{\eta k}{q} \frac{T}{I_d(V_d(T), T)} \tag{3.13}$$

The variation of the RF resistance of a Schottky diode with temperature, given by (3.13), under fixed-current and fixed-voltage-bias conditions are shown in Fig. 3.7. It shows that, under fixed-current bias, the RF resistance increases linearly with the increase of temperature. Whereas, it decreases exponentially under fixed-voltage-bias-condition.

The practical value of equivalent series resistance R_s is very small compared to the RF resistance R_d of diodes and is defined by (3.13). Thus, in spite of small temperature coefficient of R_s, its temperature effect on overall RF performance is negligible. Similarly, in most of the practical applications, at the operating frequency, the capacitive reactance of the capacitor C_d is very high compared to the diode resistance R_d. Thus, in spite of small temperature coefficient of the capacitance C_d, its temperature effect on overall RF performance can be neglected compared to the temperature effect of R_d. Thus, making temperature invariant RF resistance R_d, determined by (3.13), temperature-independent RF performance of the circuit can be achieved.

3.5 Temperature Invariant RF Resistance

Schottky diode-based microwave circuits such as multiplier, mixer, linearizer, etc. are used in various airborne and space-based systems. In all these applications, always temperature invariant performance is demanded. However, the RF performance of the circuits determined by RF resistance of the diodes is very sensitive to the circuit temperature as discussed in the previous section.

To achieve temperature invariant RF resistance of the diode say $R_d = R_{do}$, and neglecting reverse saturation current compared to the forward current, the I–V relationship will be given by Eq. (3.12) under the constraint imposed on $V_d(T)$ as [5, 6]

$$I_d(V_d(T), T) = A^* A T^2 \exp\left(\frac{q[V_d(T) - I_d(V_d(T), T)]R_s - \phi_b}{\eta k T}\right) \tag{3.14}$$

$$V_d(T) = \phi_b + \frac{z R_s}{R_{do}} T + \frac{\eta k T}{q} \ln\left(\frac{z}{A^* A R_{do} T}\right) \tag{3.15}$$

Here, $z = \eta k/q$. Taking the typical values of the diode parameters as $\phi_b = 0.69$ V and $\eta = 1.2$, the plot of Eq. (3.14) over the temperature range of −273 to 300 °C with the constraint given by Eq. (3.15) is shown in Fig. 3.8. This curve in I–V plane over the diode temperature is called '*equiresistance curve*'. The equiresistance curve intersects the voltage axis at the diode's barrier height voltage (ϕ_b). An important characteristic of the equiresistance curve is that it is a straight line over a very wide range of temperature. Therefore, a bias circuit consists of a constant voltage source with a series connected resistor of predefined value can provide temperature invariant RF resistance of the diode, as shown in Fig. 3.8. The value of the constant voltage and the series connected resistor are such that the bias load line passes along the equiresistance curves over the operating temperature range.

Combining (3.14) and (3.15) and eliminating T, the equation of the equiresistance curve can be written as

$$V_d = \phi_b + I_d R_s + \frac{\eta k}{q}\left(\frac{I_d R_{do}}{z}\right) \ln\left[\left(\frac{z}{R_{do}}\right)^2 \frac{1}{A^* A I_d}\right] \tag{3.16}$$

Differentiating the Eq. (3.16), the slope (M_o) of the equiresistance curve at bias point (V_{do}, I_{do}) corresponding to T_o can be written as

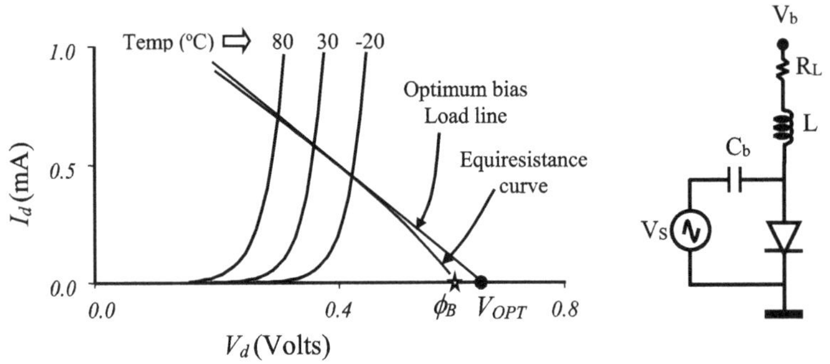

Fig. 3.8 Equiresistance curve and optimum bias load line

$$\frac{1}{M_o} = \frac{\partial V_d}{\partial I_d} = R_s + \frac{\eta k}{q}\frac{T_o}{I_{do}}\left[\frac{q(V_{do} - I_{do}R_s - \phi_b)}{\eta k T_o} - 1\right] \tag{3.17}$$

The optimum bias load line is nothing but the tangent of the equiresistance curve at temperature T_o. Thus, optimum load line at diode bias point V_{do} and I_{do} can be written as

$$V_d = V_{do} + \frac{1}{M_o}(I_d - I_{do}) \tag{3.18}$$

Combining (3.17) and (3.18), the optimum load-line equation can be written as

$$V_d(I_d) = V_{do} + \left[(\phi_b - V_{do} + I_{do}R_s) + \frac{\eta k T_o}{q} - I_{do}R_s\right]\left(1 - \frac{I_d}{I_{do}}\right) \tag{3.19}$$

The optimum open-circuit voltage (V_{OPT}) of the bias circuit is the voltage axis intercept of the load line that can be obtained by putting $I_d = 0$ in the load line Eq. (3.19). Thus, the optimum open-circuit voltage (V_{OPT}) is given by

$$V_{\mathrm{OPT}} = \phi_b + \frac{\eta k T_o}{q} \tag{3.20}$$

The expression (3.20) of the optimum open-circuit voltage shows that V_{OPT} is independent of the diode current or voltage. Therefore, the required open-circuit voltage of the bias circuit for achieving temperature invariant RF resistance is the same for all values of the RF resistance. The second term of the expression (3.20) is about 0.03 V at ambient temperature, thus it can be considered as negligible compared to the metal-to-semiconductor barrier height (ϕ_b) of the diode. Thus, V_{OPT} is determined by the physical parameter (ϕ_b) of the diode and is practically independent on temperature.

Figure 3.9 shows the small signal RF resistance variation with temperature under optimum load-line bias and fixed-current bias for ϕ_b =0.69 V and η =1.2. This

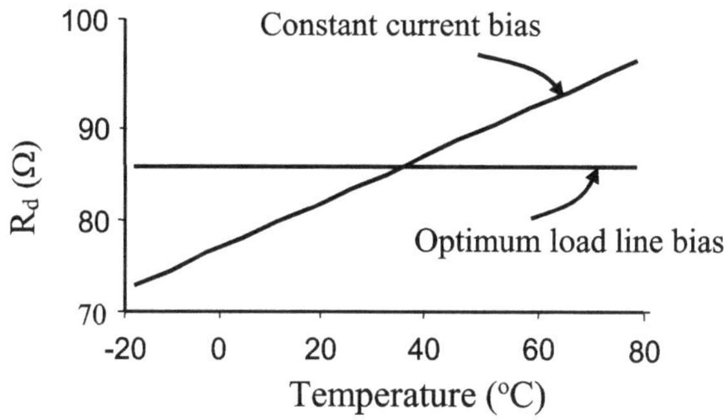

Fig. 3.9 Temperature dependency of small signal RF resistance under constant current and optimum load-line bias conditions

shows that the RF resistance variation decreases from 24% to about 0% over the temperature range of −10 to +60 °C when the diode is biased from conventional constant current bias to optimum open-circuit voltage bias.

Thus, RF resistance of Schottky diodes is (a) highly temperature sensitive under fixed-voltage-biased condition, (b) less sensitive under constant-current-biased condition and (c) minimal sensitivity of about 0% when it is biased by the optimum load line technique.

The optimum open-circuit voltage V_{OPT} as given by (3.20) corresponds to small signal RF resistance. With the increase of RF power level, the I–V characteristic of Schottky diodes deviates from exponential characteristic due to self-biasing effect. Thus, the expression for V_{OPT} also differs from (3.18). Practically, it is observed that V_{OPT} increases with the increase of RF power level under large signal operation.

Example 3.1 Calculate open-circuit voltage for temperature invariant small signal RF resistance of a Silicon Schottky diode. Consider barrier height = 0.69 V and ideality factor = 1.2.

Solution

From (3.20), the optimum open-circuit voltage is given by

$$\begin{aligned} V_{\mathrm{OPT}} &= \phi_b + \frac{\eta k T_o}{q} \\ &= 0.69 + \frac{1.2 \times 1.38 \times 10^{-23}}{1.6 \times 10^{-19}} T_o \text{ V} \\ &= 0.69 + 0.0001035 \times T_o \text{ V} \end{aligned}$$

The second term is temperature dependent but it is very small. Considering temperature $T_o = 300$ K:

$$V_{\mathrm{OPT}} = 0.721 \text{ Volt}$$

Example 3.2 Small signal RF resistance, R_d and current–voltage ($I_d - V_d$) relationship of a Schottky barrier diode under forward bias condition are

$$R_d = \frac{aT}{I_d} \quad \text{and} \quad I_d = bT^2 \times \exp\left(\frac{V_d - \phi_b}{cT}\right)$$

Here, a, b, c are temperature-independent constants. Derive optimum open-circuit voltage for zero temperature coefficient of RF resistance of the diode.

Solution

Temperature coefficient of the small signal RF resistance of the Schottky barrier diode is given by

$$\frac{\partial R_d}{\partial T} = \frac{a}{I_d} - \frac{aT}{I_d^2}\frac{\partial I_d}{\partial T} \tag{3.21a}$$

For zero temperature coefficient of the RF resistance, $\partial R_d/\partial T = 0$. Putting this condition to (3.21a), we can write

$$\frac{\partial I_d}{\partial T} = \frac{I_d}{T} \tag{3.21b}$$

Suppose the diode is biased through a bias resistance, R_b and supply voltage V_S. Thus,

$$V_S = I_d R_b + V_d \tag{3.21d}$$

Using (3.21d), the I–V relationship of the diode can be written as

$$I_d = bT^2 \times \exp\left(\frac{V_S - I_d R_b - \phi_b}{cT}\right) \tag{3.21e}$$

Differentiating (3.21e) with respect to temperature T:

$$\frac{\partial I_d}{\partial T}\left(1 + \frac{I_d R_b}{cT}\right) = \frac{2I_d}{T} - \frac{I_d(V_S - I_d R_b - \phi_b)}{cT^2} \tag{3.21f}$$

For zero temperature coefficient, suppose, $V_S = V_{\text{OPT}}$ and $I_d = I_{do}$. Putting these values to (3.21f) and using (3.21b), we can write

$$\frac{I_{do}}{T}\left(1 + \frac{I_{do} R_b}{cT}\right) = \frac{2I_{do}}{T} - \frac{I_{do}(V_{\text{OPT}} - I_{do} R_b - \phi_b)}{cT^2} \tag{3.21f}$$

This leads to the value of optimum supply voltage for zero temperature coefficient of RF resistance of the Schottky diode as

$$V_{\text{OPT}} = \phi_b + cT \tag{3.21g}$$

References

1. Sze SM (1981) Physics of semiconductor devices, 2nd edn. Wiley Eastern Limited
2. Streetman BG (1986) Solid state electronic devices, 2nd edn. Prentice Hall of India Pvt. Ltd., New Delhi
3. Erikson J, Rorsman N, Zirath H (2001) Microwave silicon carbide Schottky diodes. Electron Lett 37(4):250–252
4. Erikson J, Rorsman N, Zirath H (2003) 4H-silicon carbide Schottky barrier diodes for microwave applications. IEEE Trans Microw Theory Tech 51(3):796–804
5. Bera SC, Singh RV, Garg VK, Sharma SB (2007) Optimum bias load-line compensates temperature variation of Junction diode's RF resistance. IEEE Trans Microw Theory Tech 55(2):215–221
6. Bera SC, Singh RV, Garg VK (2008) Temperature behavior and compensation of Schottky barrier diode. Int J Electron 95(5):457–465

Chapter 4
Special Microwave Diodes

Abstract This chapter presents construction, working principle and characteristic of Step Recovery Diodes (SRDs), tunnel diodes, backward diodes, and varactor diodes. The applications of these diodes in microwave communication systems are also presented.

4.1 Introduction

Step recovery diodes (SRD), tunnel diodes and varactor diodes are popular devices used for microwave communications. The step recovery diodes have very fast switching speed, in the range of picosecond, and used as a nonlinear device for multiplier, mixer, etc. Nonlinear behaviour of tunnel diodes are used for low-level RF power detection and frequency multiplier applications. Nonlinear characteristics of varactor diodes are used to realize multiplier and mixer for receivers and linearizer for high-power amplifiers.

4.2 Step Recovery Diode

When the current through a p–n junction diode is reverse biased from its forward bias, the voltage across the diode does not change from its positive voltage to a negative voltage instantaneously. Thus, the current through the diode also does not stop instantaneously. This happens due to the charge storage effect in the diodes [1–4]. Typical current and voltage transient under a step input for the circuit using p–n junction diode is shown in Fig. 4.1a. The time t_S is storage delay time required for the stored charge in the diode to become zero. The time t_R is the transition time required to build the reverse bias across the diode junction by depletion of the charge carriers.

S. C. Bera, *Microwave Active Devices and Circuits for Communication*, Lecture Notes in Electrical Engineering 533, https://doi.org/10.1007/978-981-13-3004-9_4

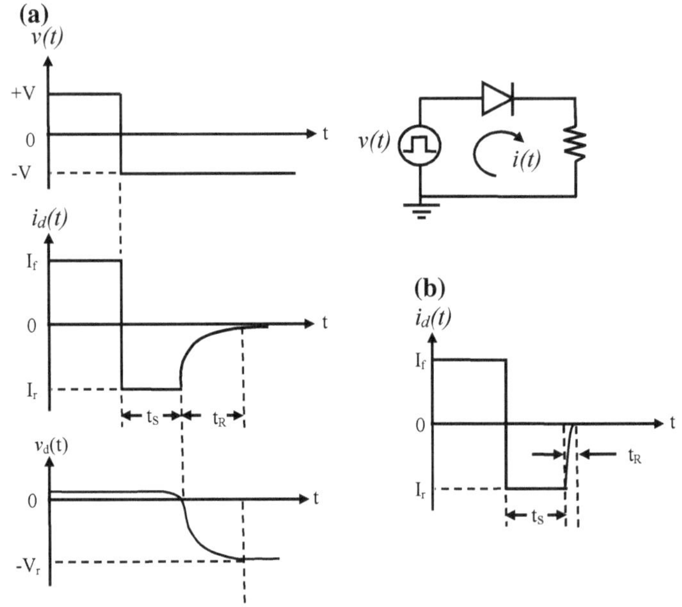

Fig. 4.1 Pulse response of a conventional p–n junction diode

Example 4.1 A p–n junction diode in series with a 100 Ω resistor is forward biased so that a current of 50 mA flows. What will be the current through the circuit just after the voltage across the combination is suddenly reversed to 10 V?

Solution

Initially, the p–n junction diode is in forward biased condition with forward current of 50 mA. Just after providing the reverse voltage of 10 V to the combination, the diode will remain in its forward bias condition due to its charge storage effect with small voltage drop across the diode. Thus, just after reversing the supply voltage, the reverse current (opposite direction of current) will be $10\,\text{V}/100\,\Omega = 100\,\text{mA}$.

Step recovery diodes, also known as 'snap-off-diode' and 'charge storage diode', are used as frequency multipliers and harmonic generators. These diodes are also used for generating pulses having very short rise and fall time for its ultra-fast switching time. For a reverse recovery diode, the fall time t_R should be very small as shown in Fig. 4.1b. Due to the rapid reverse recovery, it generates various higher harmonic frequency components when RF signal is applied to the diode. To generate the very sharp current pulse, a special type of impurity profile is required for step recovery diode [5]. The impurity profile is made such that the generated build-in potential pushes the minority carriers away from the semiconductor junction. The impurity doping profile for a step recovery diode is shown in Fig. 4.2. Here, impurity concentrations within the p-and n-type semiconductors are further away from its p–i and n–i junctions. This impurity profile creates very short transition time (t_R) going

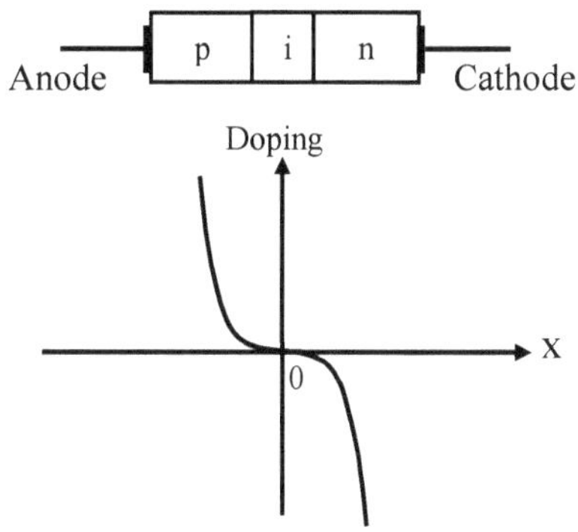

Fig. 4.2 Typical doping profile of SRD

from its conduction to non-conduction phase within very short time in its reverse bias condition.

Under forward bias condition, SRD acts as a normal p–n junction diode. When the SRD is biased from forward to reverse, initially it has a very high conductivity of current like normal p–n junction diode. However, this high-conductivity situation only maintains for a certain period of time. Then, it suddenly stops conducting current. This property of SRD makes it function as generator of higher order harmonics with sufficient magnitudes, when driven by a sinusoidal signal. This property of the diode makes it suitable to use as higher order multiplier and mixer.

Epitaxial p-i-n structure with quite narrow I-layer is used to construct SRDs. Thus, the charges injected under forward bias condition are stored very close to the p–i and n–i junctions. Under reverse bias, the narrow intrinsic layer width ensures very low-dynamic punch-through voltage compared to operating voltage. A significant electric field is generated throughout the entire I-layer. Thus remaining mobile carriers are rapidly swept out, causing very sharp current transition from high- to low-final value of the diode.

4.2.1 Characteristic of SRD

The step recovery diode behaves as an ideal nonlinear capacitor [5]. Ideally, it works as a two-state capacitor with zero switching time between the two states. In one state, it acts as a large (ideally infinite) capacitance under forward charge storage condition. The diffusion capacitance of the I-layer is the forward storage capacitance. In other state, it acts as a small capacitance (the depletion layer capacitance) under its reverse charge storage condition. The capacitor charge consists of electrons and holes stored in the I-layer and are related to the effective minority carrier lifetime τ_a.

I-V characteristic of an SRD under forward bias condition follows the same equation as of simple p–n junction diode, and can be written as

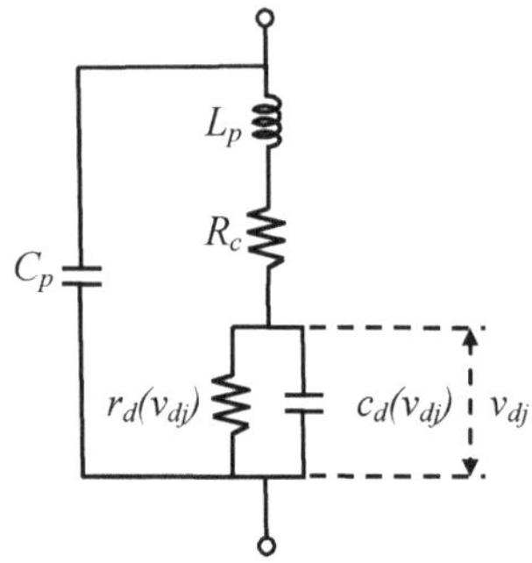

Fig. 4.3 SRD equivalent circuit

$$I_d(V_{dj}, T) = I_o\left[\exp\left(\frac{qV_{dj}}{\eta kT}\right) - 1\right] \quad (4.1)$$

Thus, under forward bias condition, the SRD is modelled as a p–n junction diode and a parallel diffusion capacitance to account the stored charge as given by (4.2) neglecting the depletion charge that is negligible under sufficient forward bias current.

$$c_d(V_{dj}) = \frac{\partial Q_d}{\partial V_{dj}} = \frac{q}{\eta kT}\tau_a I_d(V_{dj}, T) \quad (4.2)$$

Under reverse bias condition, the junction capacitance will be the depletion layer capacitance of the diode of area A and depletion layer width W and is given by (4.3) as follows:

$$c_d(V_{dj}) = \frac{\varepsilon A}{W} \quad (4.3)$$

The most contribution on W is the width of the intrinsic region thus it is nearly independent of reverse bias voltage and is very low.

Under forward bias condition, the diffusion capacitance (c_d) is determined by large amount of free carriers within a volume of depletion width W and area A. Therefore, it is not possible for instantaneous transition from forward capacitance to reverse capacitance. An important parameter of SRD is time constant t_t for switching from its forward capacitance to reverse capacitance. The effect of finite switching time should be taken into account for operating frequencies higher than $1/t_t$. Equivalent circuit of SRD is shown in Fig. 4.3 for the operating frequency range of greater than $1/\tau_a$ and less than $1/t_t$. At frequencies lower than $1/\tau_a$, it is required to consider the effect of carrier recombination and at frequencies higher than $1/t_t$; the effect of finite switching time should be taken into account.

In a semiconductor, the minority carrier lifetime is strongly temperature dependent. The major temperature effect on SRDs is an increase of minority carrier lifetime of about 0.5%/°C. This may not be a problem when diodes are used for generating harmonics for microwave signal generation, as the temperature-dependent optimum bias voltage is used. In general, this temperature dependency is compensated using temperature-dependent resistor for its biasing. However, the problem is severe when an SRD is used for generation of pulses and the position in time of the pulses is important. In this case, a change of minority carrier lifetime causes a change in snap time, this leads to a change in the pulse position in time relative to a driving waveform.

4.3 Tunnel Diodes

Tunnel diode is based on the electron tunnelling effect discovered by Leo Esaki, thus also known as Esaki diode. Tunnel diodes have lower $1/f$ noise, higher rectification current sensitivity, low-temperature variation and broadband performance capability.

Tunnel diode is a highly doped p–n junction diode. A part of its I-V characteristic shows a decrease of current with the increase of voltage. That means its I-V characteristic exhibits negative differential resistance [6]. The depletion region becomes very thin ~100 Å, since the p- and n-regions of the tunnel diodes are highly doped (about 10^{24}–10^{25} m^{-3}). Due to the very thin depletion layer, electrons can tunnel from conduction band of n-region of the diode to the valence band of p-region without change of particle energy. Hence, the flow of current is due to the tunnelling of particle, thus called tunnelling current and the diode is called tunnel diode.

Tunnel diodes are made using highly doped semiconductor, called degenerate semiconductor, where for the n-type semiconductor Fermi level goes above the conduction band and for p-type semiconductor, it goes below valence band. At zero bias, there is no current flow as shown in Fig. 4.4a. Under the application of small forward bias, there will be no conventional forward current due to very high-potential barrier. However, under this small forward bias condition, the current will be due to the tunnelling of electrons from conduction band of the n-region to the valence band in the p-region, as shown in Fig. 4.4b. With the application of certain higher forward bias as shown in Fig. 4.4c, the majority electron's energy in the n-region becomes equal to energy of the p-region's empty states in the valence band. This results in very high-tunnelling current as shown in Fig. 4.4c. With further increase of forward bias, the number of n-region's electrons decrease that are directly opposite to the p-region's valence band empty states. Therefore, decrease in the tunnelling current will start as shown in Fig. 4.4d. This region of the I-V characteristic shows negative differential resistance of the diode. With further increase of applied forward bias voltage, tunnelling current decrease and ultimately drops to zero. However, due to the lowering of potential barrier as shown in Fig. 4.4e, the conventional forward current of the diode starts to increase due to the injection of electron hole through the junction. Whereas under reverse bias condition, current flows in the reverse direction due to the tunnelling of valence band electrons of the p-region directly to the n-region empty states present in the conduction band as shown in Fig. 4.4f with the increase of reverse bias tunnelling current increases which leads to increase of the reverse current.

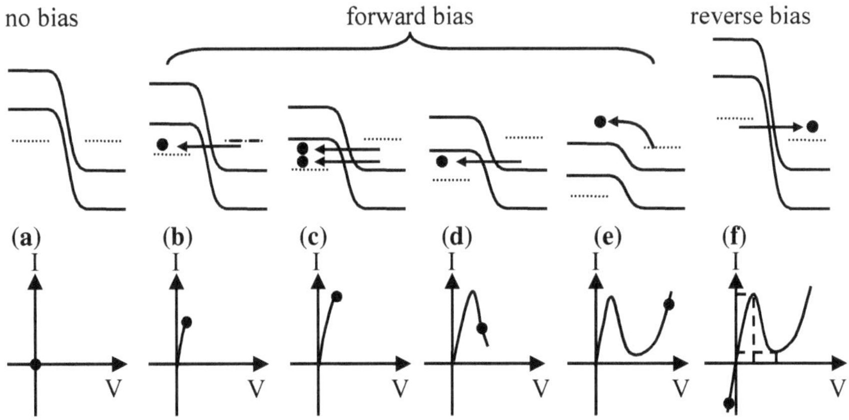

Fig. 4.4 Typical I-V characteristic of tunnel diode at various bias conditions

4.3.1 Characteristic of Tunnel Diode

Figure 4.5a shows I-V characteristic of a tunnel. There are two very distinct features of tunnel diode's I-V characteristics. First, the I-V characteristic is strongly nonlinear and second the characteristic consists of a negative differential resistance region. The negative differential resistance between V_p and V_v is important, and when the diode is used in this region it acts as voltage-controlled negative resistance. The plot of differential resistance versus diode voltage is shown in Fig. 4.5b. It shows that negative differential resistance occurs over the diode voltage of V_p to V_v with a peak near about V_p. The current ratio I_p/I_v is used as figure of merit of a tunnel diode. The total diode current I_d of a tunnel diode is given by [6]

$$I_d = \frac{V_{dj}}{R_o}\exp\left[-\left(\frac{V_{dj}}{V_o}\right)^m\right] + \frac{V_{dj}}{R_V}\exp\left(\frac{V_{dj}-V_V}{V_{ex}}\right) + I_o\left[\exp\left(\frac{qV_{dj}}{\eta kT}\right)-1\right] \quad (4.4)$$

The first term is the tunnel current, where m is a constant of value ranging from 1 to 3, and R_o is the tunnel diode resistance in the ohmic region, i.e. near $V_{dj} = 0\,\text{V}$. Second term of the total current expression is the excess current of the tunnel diode which is due to the additional parasitic tunnelling current. The last term is conventional p–n junction current. The parameters V_o, V_v, R_V and V_{ex} are the curve fitting empirical parameters. The maximum negative differential resistance of the tunnel diode is given by

$$R_{\text{nd}} = R_o\frac{\exp\left(\frac{1+m}{m}\right)}{m} \quad (4.5)$$

The negative resistance of the tunnel diode is used for realization of microwave oscillators and amplifiers. High-frequency equivalent circuit of a tunnel diode is

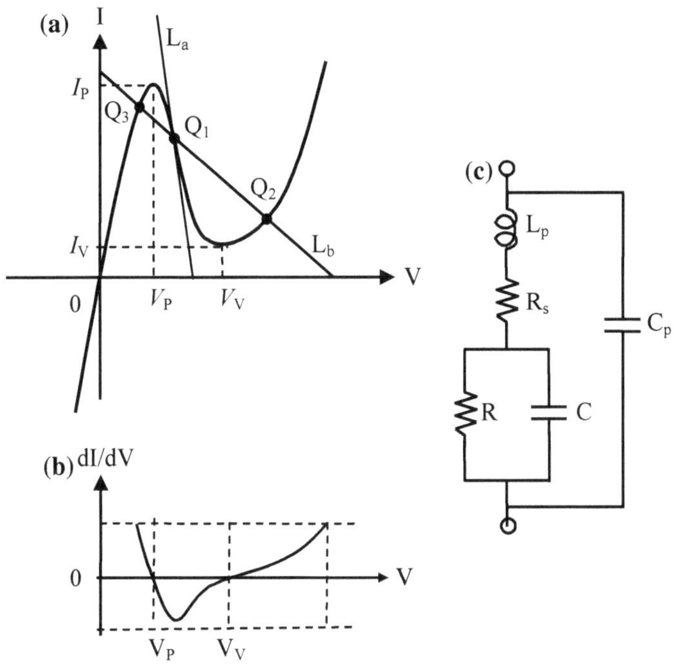

Fig. 4.5 Typical **a** current–voltage characteristic and **b** conductance, **c** equivalent circuit of the tunnel diode

shown in Fig. 4.5c. Here, R is differential resistance and R_s is parasitic resistance due to the lead resistance, ohmic resistance and spreading resistance of the semiconductor. Two typical bias load lines L_a and L_b are shown in the Fig. 4.5a. Quiescent operating point Q_1 corresponding to the bias load line L_a corresponds to a circuit of resistance less than the negative resistance of the diode. It will provide stable or unstable amplification/oscillation depending upon the reactance of the circuit. Whereas, quiescent point Q_1 corresponding to the bias load line L_b is not stable as it may go to the point Q_2 or Q_3 depending upon its previous condition. As the tunnelling current does not depend strongly on operating temperature, the performance of the tunnel diode-based circuits is very stable with change in operating temperature.

As there is no time delay in the tunnelling current as present in the drift and diffusion process, tunnel diodes are capable to operate up to hundreds of GHz range. However, due to the limitations in high-current handling capacity of the tunnel diode and also due to the presence of many other high-speed devices based on advanced material and processes, tunnel diodes are not being used frequently.

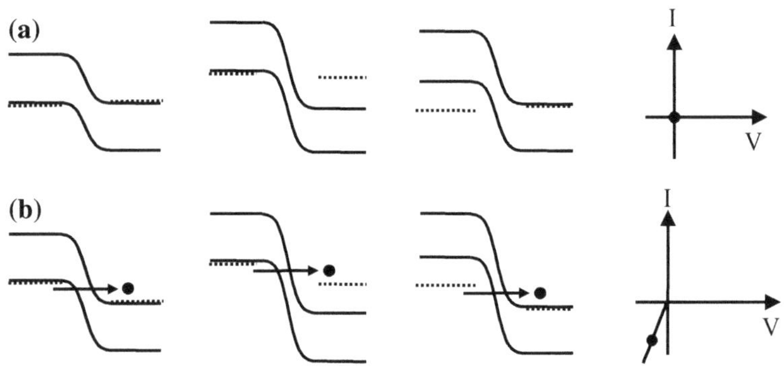

Fig. 4.6 Energy band diagrams of backward diodes with three different doping configurations **a** no bias, **b** reverse bias

4.4 Backward Diode

Backward diode is also a highly doped p–n junction device but it has slightly lower impurity concentration than the tunnel diode. In case of backward diodes, tunnelling current flows only in its reverse bias condition. There is no tunnelling current when a backward diode is forward biased. I-V characteristic of a backward diode has very sharp knee near 0 V. This characteristic enables rectification of very small magnitudes of RF signal and is used as sensitive RF detector.

Energy band diagram of backward diodes for three different combinations of p- and n-region doping concentrations at zero bias and reverse bias conditions are shown in Fig. 4.6. The three different doping combinations in p- and n-region of the backward diodes are such that in thermal equilibrium, (i) the Fermi levels of both p-and n-regions are very close to the band edges, (ii) the Fermi level of p-region is very close to the band edge, however, Fermi level in n-region is above the conduction band edge and (iii) the Fermi level of n-region is very close to the band edge, however, Fermi level in p-region is below the valence band edge.

Energy band diagram and associated current-voltage characteristics of a backward diode are shown in Fig. 4.7. Under forward bias condition of the backward diode, there will be negligible (ideally zero) tunnel current thus it behaves like a conventional p–n junction diode. However, when a small reverse voltage is applied to the backward diode, electron can easily tunnel from the valence band of p-region to the conduction band of n-region and give rise to a tunnelling current. The reverse bias tunnel current (I_R) can be expressed as [6]:

$$I_R = A_1 \exp\left[+\left(\frac{|V_R|}{A_2}\right)^m\right] \tag{4.6}$$

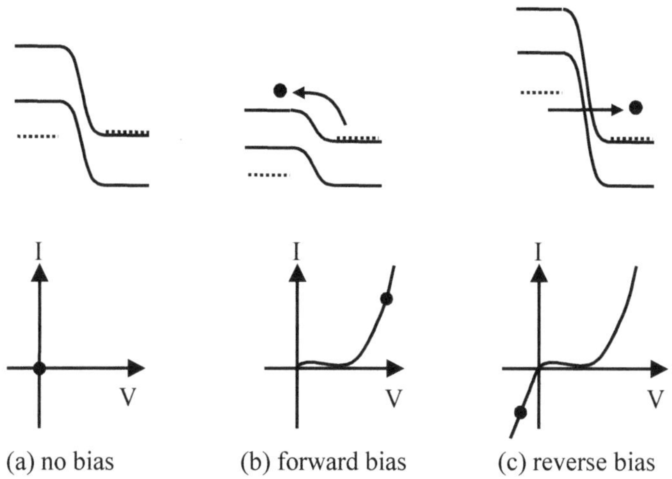

Fig. 4.7 Typical I-V characteristic of backward diode

Fig. 4.8 Typical current–voltage characteristic of backward diode

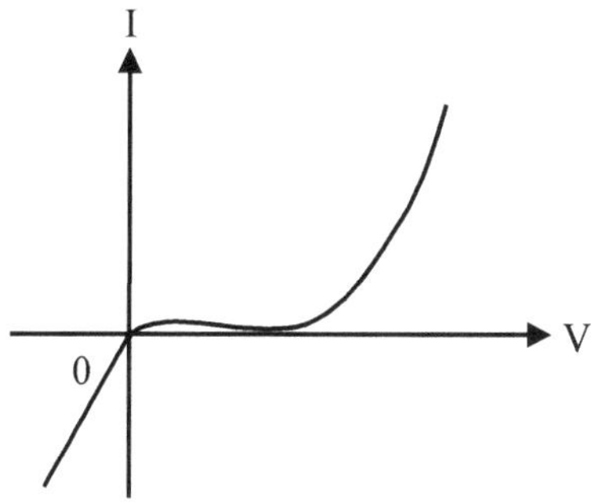

Here, A_1, A_2 and m are all positive constants and are very weak functions of the applied reverse voltage V_R. The expression shows that the reverse current of the backward diode increases exponentially with the applied voltage.

A typical forward- and reverse-biased I-V characteristic of the backward diode is shown in Fig. 4.8. It shows that for small reverse bias, the current is larger than the current under the same amount of forward bias voltage. This enables backward diode to rectify small signals of amplitude smaller than the forward bias cut-in voltage. The reverse bias current is due to the tunnelling of the carriers. Thus, it is fast due to absence of minority carriers charge storage effect.

Tunnelling current is insensitive to the change of temperature. Therefore, the performance of the backward diode is not sensitive to its operating temperature.

4.5 Varactor Diode

Varactor diode is basically a p–n junction or Schottky contact diode usually operated in reverse bias condition to act as voltage variable capacitance [7]. Varactor diode-based electronically tunable elements like capacitive switch banks and continuously variable capacitances are used for voltage variable phase shifters, frequency tunable, i.e. multiband amplifiers, antennas and oscillators. These elements are used for realization of truly generic transceivers that feature adjustable frequency, impedance, and system transfer functions for flexible communication systems by dynamic reconfiguration. The varactor diodes are also used as a nonlinear device for realization of multiplier, mixer, linearizer, etc.

Under reverse biased condition, absolute capacitance and the change of capacitance with the variation of reverse bias voltage are two important parameters for a varactor diode. Absolute capacitance with the other reactance of a circuit determines the frequency of operation whereas the change of capacitance with the change of reverse bias voltage determines the range of frequency or phase adjustability of RF circuit.

Doping profile of a varactor diode determines the absolute capacitance and change of capacitance over the reverse bias voltage. Doping distribution ($N(x)$) along the length (x) of the diode can be expressed as

$$N \propto x^m \tag{4.7}$$

In case of uniform doping distribution, $m = 0$ and for linearly graded doping distribution, $m = 1$. Whereas for hyper-abrupt doping distribution, $m < 0$. The junction capacitance $\left(C_j\right)$ dependency on the applied bias voltage (V) can be expressed as

$$C_j \propto (V + V_{bi})^{-1/(m+2)} \tag{4.8}$$

Here, V_{bi} is the built-in voltage of the diode. Thus, with the increase of reverse bias voltage, the junction capacitance decreases. Larger the value of $1/(m + 2)$, the capacitance variation with the biasing voltage will be larger. This parameter is called sensitivity of a varactor diode. For the hyper-abrupt junction, sensitivity of the varactor diode increases with the increase of $|m|$. A typical hyper-abrupt doping profile is shown in Fig. 4.9. GaAs-based hyper-abrupt Schottky contact layer used to realize a high-frequency performance varactor diode to operate in the higher microwave and mmwave frequency range.

Equivalent circuit of a reverse biased varactor diode is shown in Fig. 4.10. Here, C_j is the bias-dependent junction capacitance. R_p is the parallel equivalent depletion layer resistance corresponding to reverse bias leakage current. Generally, for high-quality diodes, R_p increases with the increase of reverse bias voltage. R_s is the equivalent series resistance of the diode consisting of contact resistance, and resistance of the un-depleted regions of the diode. L_p and C_p are the package inductance and capacitance respectively.

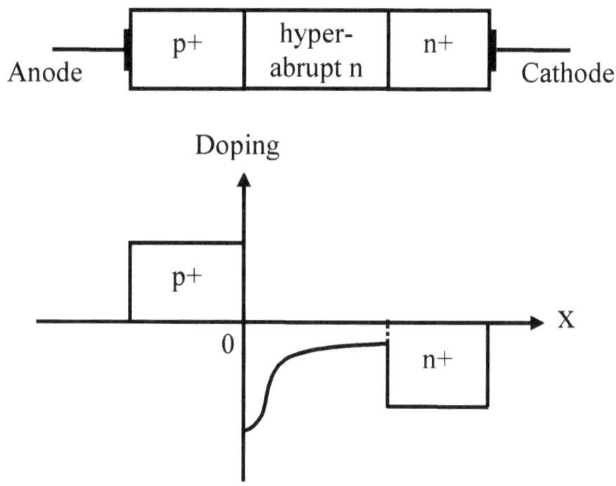

Fig. 4.9 Typical hyper-abrupt doping profile of a varactor diode

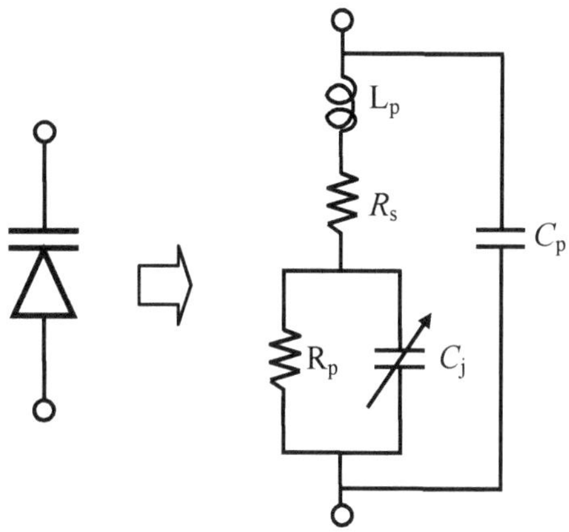

Fig. 4.10 Equivalent circuit of reverse-biased varactor diode

The capacitance of a varactor diode also changes with temperature. With the increase of diode temperature, capacitance increases. Temperature sensitivity of the diode capacitance increases with increase of capacitance sensitivity with the applied voltage.

References

1. Streetman BG (1986) Solid state electronic devices, 2nd edn. Prentice Hall of India Pvt. Ltd., New Delhi
2. Ko Wen-Hsiung (1961) The forward transient behaviour of semiconductor junction diodes. Solid-State Electron 3:59
3. Ko WH (1961) The reverse transient behaviour of semiconductor junction diodes. IRE Trans Elec Dev ED-8:123

4. Kuno HJ (1964) Analysis and characterization of P-N junction diode switching. IEEE Trans Elec Dev 11:8
5. Moll JL, Hamilton SA (1969) Physical modeling of the pulse and harmonic step recovery diode for generation circuits. Proc IEEE 57(7):1250
6. Karlovsky J (1967) The curvature coefficient of germanium tunnel and backward diodes. Solid-State Electron 10:1109
7. Ulriksson B (1979) Continuous varactor-diode phase shifter with optimized frequency response. IEEE Trans Microw Theor Tech 27(7):650–654

Chapter 5
Microwave Bipolar Transistors

Abstract This chapter describes various types of microwave bipolar junction transistors (BJTs). Frequency limitation of BJT and structural modification for suitable operation at microwave frequency band are also discussed. The chapter also describes GaAs- and SiGe-based hetero-junction bipolar transistors (HBTs) suitable for microwave and mmwave applications. Temperature dependency of BJTs and its effects on circuit performance are also discussed.

5.1 Introduction

Transistors are three-terminal solid-state devices used for realization of amplifiers, multipliers, mixers, linearizers and various other active circuits for microwave communication systems. A transistor should have sufficient gain at microwave frequency range for its use as an amplifying device. Thus, the device should have ability of large linear change of voltage/current at its output due to a small change of voltage/current at its input. Various types of amplifiers are realized using transistors for communication systems like low-noise amplifier (LNA), high-gain amplifier and high power amplifiers (HPA). Transistors required for LNA should amplify the signal with the addition of minimum noise, i.e. the device should have low-noise figure at microwave frequency. Power transistors should have higher output current handling capability with high output breakdown voltage, i.e. it should have high output power handling capability, whereas transistor's nonlinear characteristics are used for realization of multiplier, mixer, linearizer, etc. for communication systems.

Depending on material, doping profile, and device structures, there is a wide range of transistors suitable for different applications. Among the various types of transistors, bipolar junction transistor (BJT) was the first three-terminal device and is still being used for many low-frequency as well as microwave applications. With the advancement of hetero-junction technology, hetero-junction bipolar transistor (HJBT or HBT) uses a high bandgap semiconductor as emitter material and hence offers better performance in terms of gain at higher microwave frequency, and provides sufficient gain over mmwave frequency regions.

S. C. Bera, *Microwave Active Devices and Circuits for Communication*, Lecture Notes in Electrical Engineering 533, https://doi.org/10.1007/978-981-13-3004-9_5

5.2 Bipolar Junction Transistor (BJT)

In spite of the existence of various advanced unipolar junction transistors, the first developed three-terminal solid-state device, the bipolar junction transistor (BJT), is still being used an account of its microwave applications for its lower realization cost, lower 1/*f* noise and high packing density using well-established foundry. BJTs are not only used for amplification, but also used as nonlinear devices for realization of multipliers, mixers and linearizers. Basically, the bipolar junction transistor can be considered as two back-to-back p–n junction diodes with appropriate doping profile and structure. A schematic structure of a BJT suitable for microwave applications is shown in Fig. 5.1, and its cross-sectional view is shown in Fig. 5.2. All microwave bipolar transistors are n-p-n type due to the higher mobility of electron than holes. The emitter is heavily n-type doped, shown as n^+, compared to the lightly p-type doping of base. The collector is also lightly n-type doped. Suppose a BJT is working in the forward-active mode, i.e. base–emitter junction is forward-biased and base–collector junction is in reverse-biased condition by applying the bias voltages V_{BE} and V_{CB}, respectively, as shown in Fig. 5.2. Energy band diagrams of the BJT under unbiased and forward-active-bias condition are shown in Fig. 5.3. The basic *I–V* relationships across the junctions are same as p–n junction diode. Working principle of BJT is based on injection of minority carrier electrons into the base from the emitter. The amount of carrier injection is determined by the applied base-to-emitter voltage V_{BE}. The injected minority carriers in the base region diffuse across the base region and are swept to the collector as the base–collector junction is in reversed bias condition. Actually, the base width W_B is much smaller than the diffusion length of the electrons in the base region. Thus, most of the injected electrons cross the base region without recombination. The relationship between the emitter current and base–emitter voltage is same as diode *I–V* relationship and can be written for the common-base-bias condition as [1]

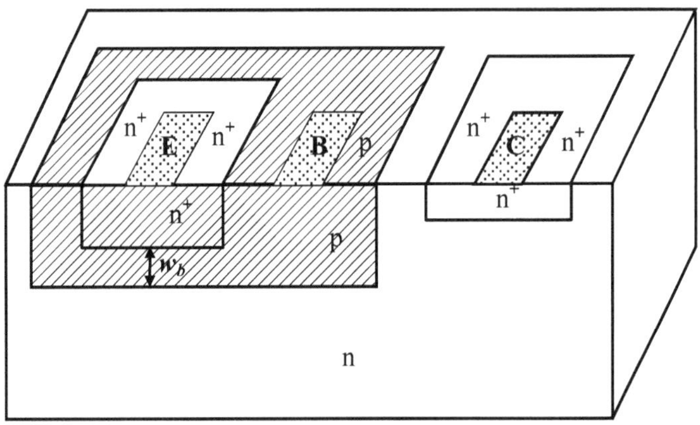

Fig. 5.1 A schematic structure and doping profiles of a BJT

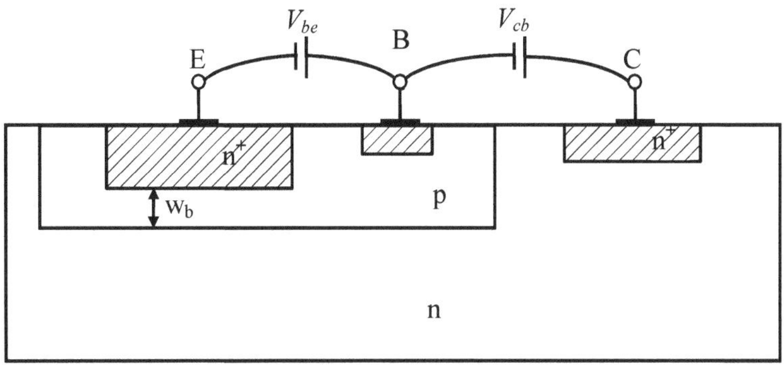

Fig. 5.2 Cross-sectional view of the BJT

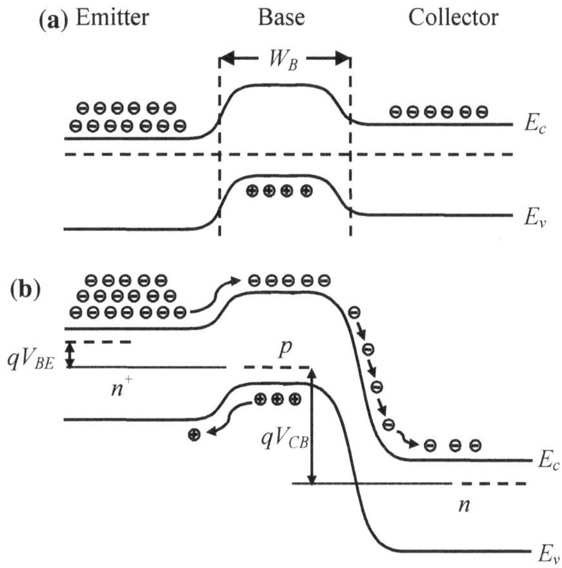

Fig. 5.3 Band diagram of **a** unbiased, **b** forward-active-mode-bias BJT

$$I_E = I_o\left[\exp\left(\frac{q(V_{BE} - I_B R_B - I_E R_E)}{\eta kT}\right) - 1\right] \tag{5.1}$$

Here, I_o is the reverse saturation current, and R_B and R_E are the base and emitter terminal parasitic resistances, respectively. The well-known static common-base current gain α_o is defined as

$$\alpha_o = \frac{\partial I_C}{\partial I_E} = \frac{\partial I_{nE}}{\partial I_E}\frac{\partial I_{nC}}{\partial I_{nE}}\frac{\partial I_C}{\partial I_{nC}} \tag{5.2a}$$

Here, the first term of the common-base current gain $\partial I_{nE}/\partial I_E$ is the emitter efficiency γ; the second term, $\partial I_{nC}/\partial I_{nE}$, is the base transport factor α_t; and the last term is called the collector multiplication factor M. Thus, the common-base current gain can be expressed as

$$\alpha_o = \gamma \alpha_t M \tag{5.2b}$$

The currents of the three terminals are related by

$$I_E = I_C + I_B \tag{5.3}$$

Thus, the static common-emitter current gain β_o, defined as $\partial I_C / \partial I_B$, can be expressed as

$$\beta_o = \frac{\partial I_C}{\partial I_B} = \frac{\alpha_o}{1 + \alpha_o} \tag{5.4}$$

The collector current can be written as

$$I_C = \alpha_o I_E = \alpha_o I_o \left[\exp\left(\frac{q(V_{BE} - I_B R_B - I_E R_E)}{\eta k T} \right) - 1 \right] \tag{5.5}$$

Therefore, an input signal voltage across the base–emitter junction will produce an output current at the collector terminal. In the presence of bias, consider a small change of input voltage, i.e. base-to-emitter voltage, then the change of output current can be determined by

$$g_m = \frac{\partial I_C}{\partial V_{BE}} = \frac{q I_C}{\eta k T} \tag{5.6}$$

This is called transconductance gain g_m of the BJT. Transconductance of BJT is very high due to the exponential relationship of collector current with the base–emitter voltage. The expression suggests that transconductance gain of a BJT is linearly proportional to the collector bias current, I_C.

A typical output characteristic of BJTs in common-emitter configuration is shown in Fig. 5.4. It shows that over the active region, the collector current (I_C) increases with the increase of collector to emitter voltage (V_{CE}). With the increase of V_{CE}, the base width W_B decreases, and this causes an increase in current gain β_o. The increase of current gain with the increase of V_{CE} results in the increase in collector current. The lack of saturation of the collector current in the active region is known as early effect. The voltage axis intercept point with the extrapolated output curves is called the early voltage as denoted by V_A as shown in Fig. 5.4.

Equivalent circuit model of a forward-active-bias BJT in common-emitter configuration, including the terminal parasitic components, is shown in Fig. 5.5 [2]. Here, R_B, R_E, and R_C are the terminal parasitic ohmic resistances in series with base, emitter and collector terminals and C_{EB} and C_{CB} are parasitic capacitances associated with both the p–n junctions. Output of the intrinsic transistor consists of

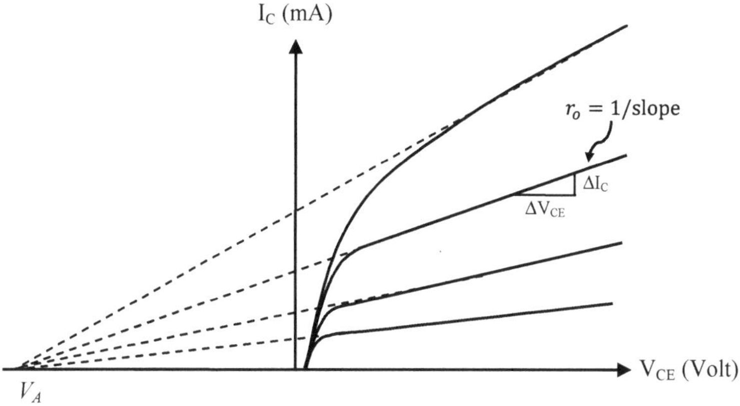

Fig. 5.4 A typical *I–V* characteristic (output characteristic) of common-emitter BJT for different base currents

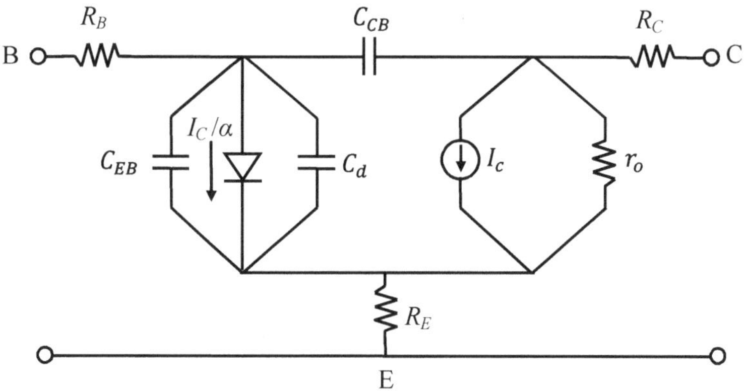

Fig. 5.5 Equivalent circuit of forward-active mode-biased BJT

a current source I_C given by (5.5) with parallel combination of output resistance r_o as given by

$$r_o = \frac{\partial V_{CE}}{\partial I_C} \tag{5.7}$$

The input of the intrinsic transistor is equivalent to a p–n junction diode to model the dynamic resistance of the base-to-emitter junction, and C_d is the diffusion capacitance to model the carrier storage effect.

Example 5.1 Common–base current gain of a bipolar transistor is 0.99. Calculate the common-emitter current gain of the transistor.

Solution

Common–base current gain of the transistor

$$\alpha = 0.99$$

Common-emitter current gain

$$\beta = \frac{\alpha}{1-\alpha} = \frac{0.99}{1-0.99} = 99$$

Example 5.2 A BJT is having common-emitter current gain 100. Considering 10 V supply, V_{BE} =0.7 V and R_E =100 Ω, what is the value of R_B, and R_C to set the quiescent point at I_C = 10 mA and V_{CE} = 8 V.

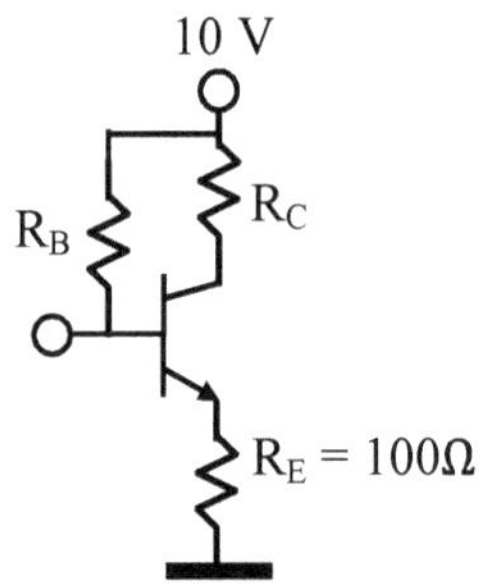

Solution

Here, collector current $I_C = 10\,\text{mA}$

Common-emitter current gain $\beta = 100$

Thus, base current

$$I_B = I_C/\beta = 10/100\,\text{mA} = 0.1\,\text{mA}$$

Therefore, emitter current

$$I_E = I_C + I_B = (10 + 0.1)\,\text{mA} = 10.1\,\text{mA}$$

Thus, voltage drop across the emitter resistance

$$V_E = I_E \times R_E = 10.1\,\text{mA} \times 100\,\Omega = 1.01\,\text{V}$$

Thus, voltage drop across the base resistor R_B

$$V_B = (10 - 0.7 - 1.01)\,\text{V} = 8.29\,\text{V}$$

Thus, base resistor

$$R_B = V_B/I_B = 8.29\,\text{V}/0.1\,\text{mA} = 82.9\,\text{k}\Omega$$

Collector resistor

$$R_C = \frac{(10 - 1.1 - 8)\text{V}}{10\,\text{mA}} = 90\,\Omega$$

5.2.1 *Frequency Limitation of BJT*

At high frequencies, BJTs are used to realize amplifier, multiplier, mixer, linearizer and other circuits. BJT should provide sufficient gain over microwave frequency range to realize high-gain/low-loss circuits. High-frequency performance of a BJT is limited mainly by the minority carrier storage effect in the base region, time delay due to various parasitic capacitances, and other transit time delays along the carrier flow path through the device [2, 3].

Carrier injected from the emitter pass through the base region by diffusion process due to the gradient in distribution of the stored minority carriers in the base and go to the collector. The diffusion process is relatively slower than the carrier transport in the FET channel which is governed by drift of carrier by the applied electric field. The carrier storage in the base region can be modelled by a diffusion capacitance C_d and is given by

$$C_d = \frac{\partial Q_B}{\partial V_{BE}} = \frac{\partial (I_{cB})}{\partial V_{BE}} = g_m \tau_B \tag{5.8}$$

Here, Q_B is the stored charge in the base region and τ_B is the forward transit time of the carriers. The transit time τ_B can be considered as the average time a carrier spends in the base region before going into the collector. The transit time is determined by the structure and material properties of the BJT and is given by [1]

$$\tau_B = \frac{W_B'^2}{\gamma D_n} \tag{5.9}$$

Here, W_B' is the un-depleted width of the base region as shown in Fig. 5.6, D_n is the electron diffusion coefficient and γ is a constant depending on the build-in field. In case of uniformly doped base region, $\gamma = 2$. The transit time τ_B determines the alpha cut-off frequency f_α and can be expressed in terms of the charging time τ_B as

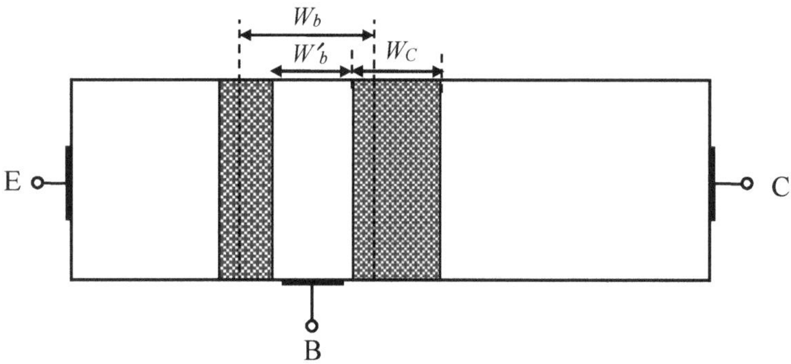

Fig. 5.6 Diagram showing depletion layers of forward-active-biased BJT

$$f_\alpha = \frac{1}{2\pi \tau_B} \tag{5.10}$$

From this it is clear that for microwave BJT, the base region width W_B should be as small as possible. Though minority carrier storage time limit the high-frequency response, BJTs can work very well around the lower microwave frequency range due to the possibility of achieving very small base width (less than 100 nm) in Si technology.

The microwave BJTs, in which τ_B is reduced by minimizing the base width, are limited by the device fabrication technology, and delays due to the various parasitic capacitors which have become the dominant part of its high-frequency performance limitation.

Example 5.3 Calculate alpha cut-off frequency of Si transistor with uniformly doped base region of width 0.25 μm. Consider electron diffusion coefficient for Si, $D_n = 20$ cm^2/s.

Solution

The alpha cut-off frequency is given by

$$f_\propto = \frac{1}{2\pi \tau_B} = \frac{1}{2\pi} \frac{2D_n}{W_B'^2} = \frac{1}{\pi} \frac{D_n}{W_B'^2}$$

Putting $W_B' = 0.25$ μm, $D_n = 20$ cm^2/s.

$$f_\propto = \frac{1}{\pi} \frac{20}{\left(0.25 \times 10^{-4}\right)^2} \text{Hz}$$
$$= 10.2\,\text{GHz}$$

Various parasitic capacitors and resistors will contribute to the total delay time. One of these delay times which determines the higher cut-off frequency of BJT is

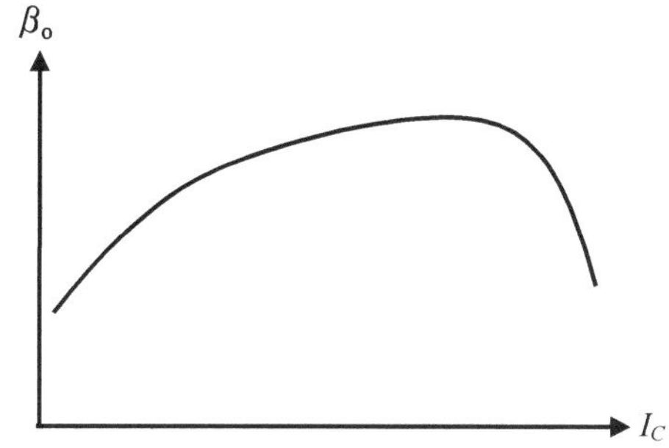

Fig. 5.7 Common-emitter current gain of BJT with its collector current

the emitter charging time, τ_E, which is related to the capacitance of depletion layer and can be expressed as

$$\tau_E = \frac{C_{EB} + C_{CB}}{g_m} \tag{5.11}$$

Thus, to minimize the charging time τ_E, high g_m is required. To achieve this, emitter region should be heavily doped compared to base doping and as per (5.6) the transistor should also be operated at sufficiently high current level. With the increase of collector current, current gain initially increases and attains a maximum value. For further increase of collector current, the injected minority carrier density from emitter to the base approaches the majority carrier density and causes a decrease of emitter efficiency. The decrease in current gain with the increase of collector current is known as Webster effect as shown in Fig. 5.7 [4].

Another factor of delay time which also contributes to the higher cut-off frequency of BJT is the base–collector junction depletion layer delay and can be expressed as

$$\tau_C = \frac{W_C}{v_s} \tag{5.12}$$

where W_C is the width of the collector–base junction depletion layer and v_s is the carrier saturation velocity.

Therefore, the high-frequency cut-off (f_C) of the BJTs can be expressed as

$$f_C = \frac{1}{2\pi(\tau_B + \tau_E + \tau_C)} \tag{5.13}$$

or

$$f_C = \left[2\pi\left(\frac{W_B'^2}{\gamma D_n} + \frac{C_{EB} + C_{CB}}{g_m} + \frac{W_C}{v_s}\right)\right]^{-1} \tag{5.14}$$

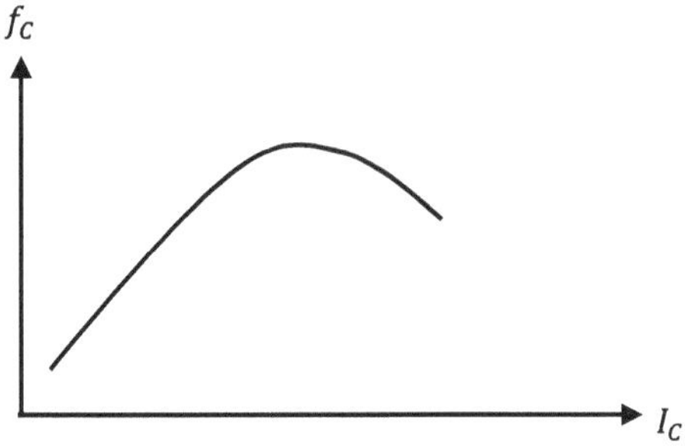

Fig. 5.8 Typical dependency of cut-off frequency on output current

From this expression, it is clear that for achieving higher cut-off frequency (f_C), the BJT should have (a) a very narrow base width W_B, (b) higher transconductance gain g_m and (c) narrow collector region W_C.

Base width W_B is limited by the fabrication technology, and the present Si technology is capable to fabricate submicron base width. Higher value of transconductance can be achieved by operating the transistor at higher current level. However, a maximum current level exists up to which the current gain increase; above this level the current gain decreases and the high-frequency performance drops as shown in Fig. 5.8. Actually, under high current conditions, effective base width increases and extended towards the collector region which degrades the high-frequency performance. This effective increase of base width at very high-level injection of current is known as Kirk effect [5]. To maximize the higher cut-off frequency, collector width should be minimized. However, with decrease of collector width, there will be decrease in collector breakdown voltage, which limits the high-voltage operation for high-power handling.

Another way to increase common-emitter current gain is by increasing the doping density of emitter region compared to the doping density of base. This requirement may be met by lowering the base region doping. However, lower base doping will increase the base resistance that will adversely affect the BJT performance. The other way to meet this requirement is to dope the emitter heavily maintaining base doping lower than emitter to ensure lower base resistance. However, heavy doping of emitter region will lead to bandgap shrinkage called the bandgap narrowing effect [6]. Therefore, for a fixed-base doping, with the increase of emitter doping, though initially the current gain increases, the current gain starts to decrease beyond certain level of doping.

From the above discussion, it is clear that the requirements of heavy emitter doping, lower base doping, smaller base width, smaller collector width, etc. are conflicting requirements and cannot be met by simple structure of BJT. To overcome these conflicting requirements, new materials with advanced fabrication technology are used to develop advanced bipolar devices such as SiGe and GaAs HBTs.

5.2.2 Temperature Behaviour of BJT

Common-emitter configuration of a BJT is the most extensively used configuration for its largest current and power amplification. Almost all the expressions of the BJT parameters as derived in earlier sections contain temperature T [7]. This is due to the fact that the current is governed by the minority carrier density, which is determined by the temperature in addition to other factors. The collector current (I_C) of a BJT is strongly related to the temperature through the exponential temperature dependency of reverse saturation current I_o as expressed in (5.5). Temperature dependency of static characteristic of BJT is shown in Fig. 5.9. The collector current increases with the increase of temperature for a fixed-base-bias current. In addition to these components, the other parasitic components also have temperature dependency. Though common-emitter current gain varies differently with temperature in different types of bipolar transistors, however, in most cases it increases with increasing temperature. Therefore, the common-emitter current gain as well as transconductance gain increases with the increase of temperature which leads to positive temperature coefficient of power gain in BJT-based microwave amplifiers.

In addition to the temperature dependency of gain of BJT-based amplifiers, the higher cut-off frequency is also sensitive to the change of operating temperature. The diffusion coefficient D_n, transconductance gain g_m and saturation velocity v_s of the carrier determine the higher cut-off frequency as expressed in (5.14). The most dominant factor which determines the higher cut-off frequency is the stored charge effect in the base region which is governed by diffusion coefficient. The diffusion coefficient is determined by the Einstein relationship $D_n = \mu kT/q$. As $\mu \propto T^{-2}$ (approximately) in the range of operating temperature for Si, the diffusion coefficient decreases with the increase of temperature and thus decreases the cut-off frequency.

At high-power applications, due to the power dissipation ($I_C \times V_{BE}$) within the BJT, there is thermal heating and rise of the device temperature. With this increase of temperature, there will be further increase of current. With the further increase of

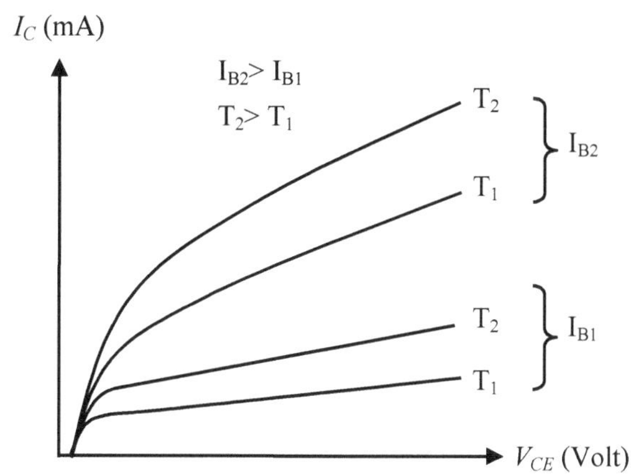

Fig. 5.9 Temperature dependency of static output characteristic of BJT

current, there will be increase of heat dissipation. With this process, the device can burn out, known as thermal runaway, if proper protection is not taken in the circuit design. Thus, the temperature sensitivity of the BJT affects the performance of the circuit. In circuit level, various compensation mechanisms are adopted to minimize the temperature-related performance variations and protection from thermal runaway.

5.3 Hetero-junction Bipolar Transistor (HBT)

Bipolar junction transistors are made with the same material, in general Si, for its base, emitter and collector regions. In case of hetero-junction bipolar transistors (HBTs), different (at least two) materials, i.e. materials having different bandgap energies, are used for different regions [8–19]. Thus, base–emitter and/or base–collector junctions are formed with different bandgap materials; hence, these devices are called hetero-junction device.

In case of homojunction bipolar transistor, it is seen that to achieve higher cut-off frequency, the emitter doping should be much higher than the base region doping and the base width should be as small as possible. This requirement can be met by having a thin base with low doping density. However, a thin base with low doping creates a large base resistance which degrades the device performance. Another way to meet the requirement is doping the emitter region heavily with moderate base region doping. However, heavy doping in the emitter region causes bandgap shrinkage problem. Therefore, though at the lower doping level of base region, current gain increases with the increase of emitter doping, it starts to decrease beyond certain doping level.

From the above discussion, it is clear that these conflicting requirements cannot be met by BJT using a single-material structure. Hetero-junction bipolar transistor (HBT) meets these requirements due to its hetero-junction construction [8]. Schematic structural diagram of an HBT is shown in Fig. 5.10. Emitter of an HBT is made with a wider bandgap material such as GaAlAs, whereas base is made with narrower bandgap material such as GaAs. Aluminium gallium arsenide, $Al_xGa_{1-x}As$, is a ternary semiconductor alloy, where x is mole fraction of aluminium in the alloy and the value is in between 0 and 1. The $Al_xGa_{1-x}As$ has very nearly the same lattice constant as GaAs, and at 300 K its energy bandgap is given by [25]

$$E_g(x) = 1.424 + 1.155x + 0.37x^2 \quad \text{for } x < 0.43 \tag{5.15a}$$

$$= 1.424 + 0.125x + 0.143x^2 \quad \text{for } x > 0.43 \tag{5.15b}$$

Therefore, the energy bandgap of $Al_xGa_{1-x}As$ varies in between 1.424 eV corresponding to GaAs ($x = 0$) and 2.168 eV corresponding to AlAs ($x = 1$). As shown in the diagram, for $x = 0.28$, the bandgap energy of the $Al_{0.28}Ga_{0.72}As$ material is 1.776 eV, about 0.35 eV larger than the bandgap energy of GaAs material. Thus,

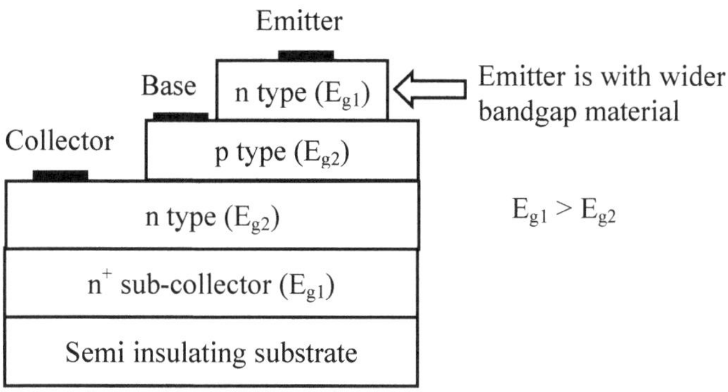

Fig. 5.10 Schematic basic structural diagram of HBT

emitter–base junction becomes hetero-junction whereas collector–base junction is homojunction.

Energy band diagrams of an HBT under no-bias and forward–active-bias conditions are shown in Fig. 5.11. Dotted lines represent abrupt p–n hetero-junction and solid lines represent graded junction. As marked in the figure, ΔE_c and ΔE_v are the conduction band and valence band discontinuities due to the abrupt hetero-junction. In case of graded hetero-junction, the notch in the conduction band discontinuity is very much reduced. The figure shows that the barrier to hole injection from the p-type base to the n-type emitter region is $q\phi_{Bh}$, whereas barrier to electron injection from n-type emitter to p-type base is $q\phi_{Be}$. It is to be noted that the barrier to the holes in both abrupt and graded hetero-junctions are same, whereas it is reduced for electron barrier, which provides improved performance. The dependence of electron current I_n and the hole current I_p on the barrier potential can be expressed as

$$I_n \propto \exp(-q\phi_{Be}/kT) \tag{5.16a}$$

$$I_p \propto \exp(-q\phi_{Bh}/kT) \tag{5.16b}$$

In case of emitter–base homojunction, the potential barrier seen by electrons and holes is the same and only means for controlling the currents are relative doping densities. For hetero-junction, barrier seen by holes is more than the barrier seen by the electrons. As seen in the Fig. 5.11, in graded hetero-junction, the bandgap difference ΔE_g between emitter material and base material appears across the valence band barrier which leads to highly diminishing of hole current contribution compared to the electron current contribution in the total emitter current. The reduction of hole current contribution in total emitter current for HBT with respect to the case for homojunction BJT can be expressed as

$$\frac{I_{pE} \text{ for HBT}}{I_{pE} \text{ for BJT}} = \exp\left(-\frac{\Delta E_g}{kT}\right) \tag{5.17}$$

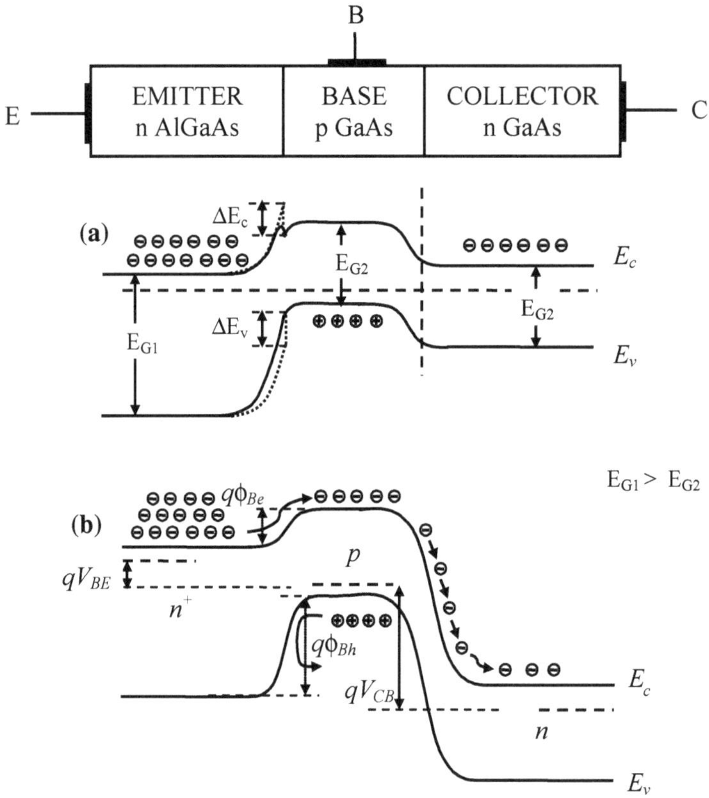

Fig. 5.11 Energy band diagram of **a** unbiased, **b** forward-active-mode-bias HBT

Due to the exponential relationship of the hole current reduction with the difference of bandgap, emitter efficiency $\partial I_{nE}/\partial I_E$ in HBT increases tremendously. Therefore, the advantages for the hetero-junction bipolar transistors (HBTs) compared to the homojunction bipolar transistors (BJTs) are (a) higher emitter efficiency, due to the higher barrier potential in the valence band for the hole injection from base to emitter, and (b) reduced base resistance, because base can be heavily doped without sacrificing the emitter efficiency. Both the advantages ultimately lead to the improved frequency response due to the achievement of higher current gain.

In the high-power wireless communication, HBT technologies are becoming important for their distinct advantages compared to FET technology. HBTs typically exhibit much higher device transconductance, as a result of the exponential relationship between the input voltage and the output current, as opposed to the quadratic relationship for FETs. In an HBT, the turn-on voltage is determined by the intrinsic material properties of the epitaxial layers, whereas for FET, its turn-on voltage is dependent on the actual dimensions of the device, and therefore fluctuates significantly between devices. Furthermore, electron transport occurs in the vertical direction in case of HBTs, whereas laterally in a FET that leads to the overall smaller area for the HBT. A greater power density is thus achieved, and integrated circuits can be designed using less die area for HBT-based circuits. HBTs also provide excellent linearity and significantly lower 1/*f* noise. There are varieties of HBTs based on the different material technologies which are discussed in the following sections.

5.3.1 SiGe HBT

The silicon–germanium hetero-junction bipolar transistors (SiGe HBT) are realized using silicon processing compatible technology [9–12]. The base layer of the SiGe HBT developed using strained and composition-graded SiGe. The SiGe HBTs provide RF performance comparable to GaAs technologies with fabrication cost and reliability similar to the silicon process. This is an Si-based device, and thus it has intrinsically higher thermal conductivity than GaAs HBTs. In comparison to the Si BJTs, it also offers the flexibility of bandgap engineering as well as base and emitter doping adjustment capabilities to achieve better RF performance.

Because of the much matured Si technology with high reliability and low-cost, Si-based devices are more competitive with their III–V group semiconductor-based counterparts, even though the latter show superior high-frequency performance. The smaller bandgap of base layer of $Si_{1-x}Ge_x$ causes an increase in current gain in n-Si/p-$Si_{1-x}Ge_x$/n-Si HBT. Schematic basic structure of an SiGe/Si HBT is shown in Fig. 5.12. Practically graded-base SiGe HBT is realized in which the Ge is graded across the base with the higher Ge content at the collector side. In case of silicon homojunction transistors, the valence and conduction bands are flat, implying that there is no electrostatic field in the base. However, alloy grading of the Ge in the base region of a graded SiGe HBT changes the band structure and introduces a quasi-electric field that drives the electrons across the base. The graded bandgap in the base region of SiGe HBT results in rapid transport of minority carriers through the base region. With the increase of Ge mole fraction x, the cutoff frequency of the HBT increases. Hetero-junction bipolar transistor based on the SiGe/Si technology material also shows its higher cutoff frequency beyond 350 GHz.

The processing compatibility of SiGe and Si-CMOS technology leads to a powerful combination technology called BiCMOS technology for high-frequency mixed-signal circuits [13]. The BiCMOS circuit consists of both bipolar hetero-junction transistors (HBTs) and MOS transistors on a single substrate. This enables combination of high-speed switching and high output driving capabilities of the bipolar junction transistors, and the low-power and high-density characteristics of the CMOS circuit. This aids realization of low-cost very compact integrate transceiver for microwave wireless communication, particularly mobile handsets. This technology enhances output and gate drive capabilities for high-speed logic and power switching circuits, which are required for specialized analog applications.

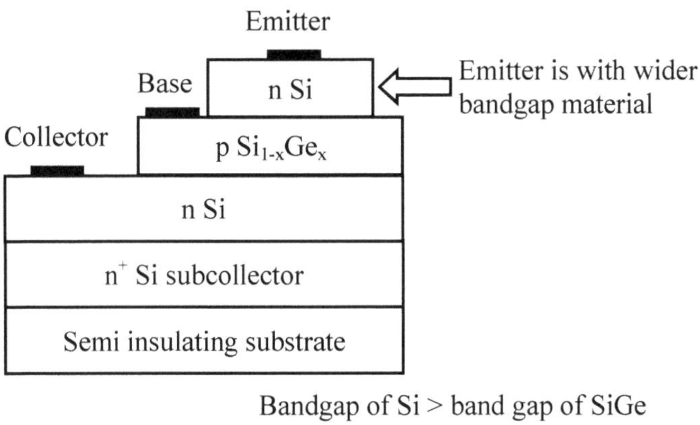

Fig. 5.12 Schematic structural diagram of Si/SiGe-based HBT

5.3.2 *III–V Group Semiconductor HBT*

III–V group semiconductor-based HBTs are attractive for high-speed applications due to their excellent electron transport properties which lead to improved high-frequency performance. Semi-insulating property of GaAs material due to higher bandgap energy makes it suitable for monolithic microwave integrated circuits. Due to its semi-insulating (i.e. high resistivity) property, the parasitic capacitance in between substrate and the other terminals is very less and can have improved high-frequency performance. Schematic structural diagram of three different basic HBTs, i.e. n-AlGaAs/p-GaAs, n-InP/p-InGaAs and n-InAlAs/p-InP are shown in Fig. 5.13 [20–22]. The $In_xGa_{1-x}As$ systems with high indium content have several unique advantages. Their high electron mobility and large peak saturation velocity extend the performance of electronic devices to higher frequency at lower power dissipation. HBT structure based on the InP/InGaAs is shown in Fig. 5.13b, and the schematic HBT structure based on InAlAs/InGaAs is shown in Fig. 5.13c. These are very simple structures; actual devices consist of many other layers with different doping densities for improving the performance of HBTs. The complicated structures are made by growing epitaxial layers using metal beam epitaxy (MBE) and metal–organic chemical vapour deposition (MOCVD) system.

5.3.3 *GaN HBT*

The GaN material system has enormous potential for high-power applications across a range of frequencies. High breakdown field, high electron saturation velocity and high thermal conductivity of GaN make it especially useful for delivering high output power at high frequencies for wireless base stations and satellite communications.

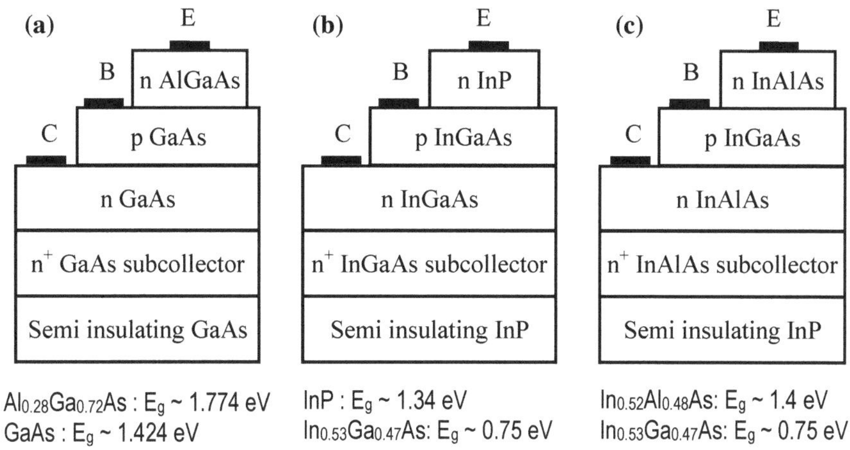

Fig. 5.13 Schematic structural diagram of GaAs/AlGaAs-based HBT

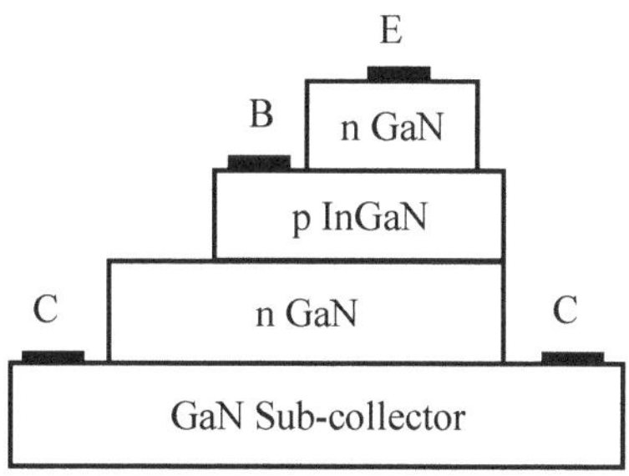

Fig. 5.14 Simplified structural diagram of InGaN/GaN-based HBT

In the high-power wireless communication, GaN-based HBT is becoming one of the most important semiconductor device technologies for its distinct advantages compared to FET technology [23, 24]. A simplified structural diagram of an InGaN/GaN-based HBT is shown in Fig. 5.14. Here, narrower bandgap material InGaN is used as base for HBT action. The energy bandgap of the $In_xGa_{1-x}N$ is given by

$$E_g(x) = 3.427 - 3.86x \quad (x < 0.25) \tag{5.18}$$

Practically graded-base InGaN HBT is realized in which the In is graded across the base with the higher In content at the collector side. The graded bandgap in the base region of InGaN HBT results in rapid transport of minority carriers through the base region leads to improved high-frequency performance.

References

1. Sze SM (1981) Physics of semiconductor devices, 2nd edn, Wiley Eastern Limited
2. Schwierz F, Schippel C (2005) Performance trends of Si-based RF transistors. In: 2005 IEEE conference on electron devices and solid-state circuits, Dec 2005, pp 299–304
3. Johnson E (1965) Physical limitations on frequency and power parameters of transistors. IRE Int Conv Rec 13:27–34
4. Webster WM (1954) On the variation of junction-transistor current amplification factor with emitter current. IRE Proc 42(6):914–920
5. Kirk CT (1962) A theory of transistor cutoff frequency f_T falloff at high current densities. IRE Trans Electron Devices 9(2):164–174
6. Lanyon HPD, Tuft RA (1978) Bandgap narrowing in heavily doped silicon. In: IEEE technical digest on international electron device meet, p 316
7. Arora N, Hauser J, Roulston D (1982) Electron and hole mobilities in silicon as a function of concentration and temperature. IEEE Trans Electron Devices ED-29:292
8. Kroemer H (1957) Theory of a wide-gap emitter for transistors. Proc IRE 45(11):1535–1537
9. Bardin JC (2009) Silicon-germanium heterojunction bipolar transistors for extremely low-noise applications. Ph.D. thesis, California Institute of Technology, Pasadena, California
10. Cressler J, Crabbe E, Comfort J, Stork J, Sun J-C (1993) On the profile design and optimization of epitaxial Si- and SiGe-base bipolar technology for 77 K applications. IEEE Trans Electron Devices 40(3):542–556
11. Krithivasan R, Lu Y, Cressler J, Rieh J-S, Khater M, Ahlgren D, Freeman G (2006) Half terahertz operation of SiGe HBTs. IEEE Electron Device Lett 27(7):567–569
12. Yuan J, Krithivasan R, Cressler J, Khater M, Ahlgren D, Joseph A (2007) On the frequency limits of SiGe HBTs for terahertz applications. In: Bipolar/BiCMOS circuits and technology meeting, 2007, BCTM'07. IEEE, 30 2007–2 Oct 2007, pp 22–25
13. Pruvost S, Delcourt S, Telliez I, Laurens M, Bourzgui N, Danneville F, Monroy A, Dambrine G (2005) Microwave and noise performance of SiGe BiCMOS HBT under cryogenic temperatures. IEEE Electron Device Lett 26:105–108
14. Woods B, Mantooth H, Cressler J (2007) SiGe HBT compact modeling for extreme temperatures. In: 2007 international semiconductor device research symposium, Dec 2007, pp 1–2
15. Tasker P, Fernandez-Barciela M (2002) HBT small signal T and Pi model extraction using a simple, robust and fully analytical procedure. In: IEEEMTT-S international microwave symposium digest, 2002, vol 3, pp 2129–2132
16. Gobert Y, Tasker P, Bachem K (1997) A physical, yet simple, small-signal equivalent circuit for the heterojunction bipolar transistor. IEEE Trans Microw Theory Tech 45(1):149–153
17. Jia S, Yang R, Guo H (2011) SiGe heterojunction bipolar transistor and its applications in microwave communication systems. In: 7th international conference on wireless communications, networking and mobile computing (WiCOM), Sept 2011
18. Escotte L, Phillippe Roux J, Plana R, Graffeuil J, Gruhle A (1995) Noise modeling of microwave heterojunction bipolar transistors. IEEE Trans Electron Devices 42(5):883–889
19. Schamacher H, Erben U, Gruhle A (1994) Low noise performance of SiGe heterojunction bipolar transistors. In: Proceedings of IEEE MTT-S microwave symposium, pp 1167–1170
20. Monier C, Cavus A, Sandhu RS, Lange MD, Chang PC, Sawdai DJ, Gambin VF, Gutierrez Aitken AL (2004) High indium content InAlAs/InGaAs HBT technology for low-power, high-speed applications. In: 205th meeting, The Electrochemical Society Inc.
21. Griffith Z, Lind E, Rodwell M, Fang X-M, Loubychev D, Wu Y, Fastenau J, Liu A (2007) Sub-300 nm InGaAs/InP type-i DHBTs with a 150 nm collector, 30 nm base demonstrating 755 GHz fmax and 416 GHz ft. In: IEEE 19th international conference on indium phosphide and related materials, 2007, IPRM'07, May 2007, pp 403–406

22. Radisic V, Sawdai D, Scott D, Deal WR, Dang L, Li D, Cavus A, Fung A, Samoska L, To R, Gaier T, Lai R (2008) Demonstration of 184 and 255-GHz amplifiers using InP HBT technology. IEEE Microw Wirel Compon Lett 18(4):281283
23. McCarthy LS (1999) AlGaN/GaN heterojunction bipolar transistor. IEEE Electron Device Lett 20(6):277–279
24. McCarthy LS, Smorchkova IP, Xing H, Fini P, Limbo J, Pulfery DL, Speck JS, Rodwel MJW, DenBaars SP, Mishra UK (2001) GaN HBT: toward an RF device. IEEE Trans Electron Devices 48(3):543–551
25. Allford Craig (2016) Resonant Tunnelling in GaAs/AlGaAs Triple Barrier Heterostructures. Ph.D. dissertation, Cardiff University

Chapter 6
Microwave Field Effect Transistors

Abstract This chapter presents structures, working principles and various characteristics of metal–semiconductor FET (MESFET) and different types of heterojunction FETs (HFETs) including pseudo-morphic HEMT (pHEMT), meta-morphic HEMT (mHEMT) and GaN-based HEMTs. Equivalent circuits of the FETs and details of the associated elements are also discussed.

6.1 Introduction

Field effect transistors (FETs) are the most utilized three terminal semiconductor devices in microwave and mmwave frequency bands. They consist of a semiconductor active channel connected to two ohmic contact terminals named source and drain. Another terminal called gate to which voltage is applied to controls current flow through the channel by changing the electric field, and hence the name field effect transistor. There are various types of field effect transistors based on the use of different materials and structures with different doping profiles [1–23]. FETs can be made from a wide variety of materials like Si, SiGe, GaAs, InP, InGaAs, GaN, SiC, etc. These are unipolar devices, hence only one type of carrier, mostly electrons, takes part for the channel current. Being a unipolar device, FET does not suffer from minority carrier storage delay time, and thus enhances the higher cut-off frequency compared to a bipolar device like BJT. FETs are negative temperature coefficient devices, thus with the increase in temperature, current of the device decreases, this prevent damage to FETs from thermal runaway. Field effect transistors based on III–V group semiconductors provide superior performances at microwave and mmwave frequency bands. The superior performances are for their higher electron velocity compared to Si semiconductor as shown in Fig. 6.1 [1]. Electron velocity of GaAs and InGaAs are higher at low electric field, thus these materials are suitable for low power high-frequency performance. Whereas, electron velocities of GaN and SiC materials are higher at higher electric field, thus these are suitable for high power at higher frequency applications.

S. C. Bera, *Microwave Active Devices and Circuits for Communication*, Lecture Notes in Electrical Engineering 533, https://doi.org/10.1007/978-981-13-3004-9_6

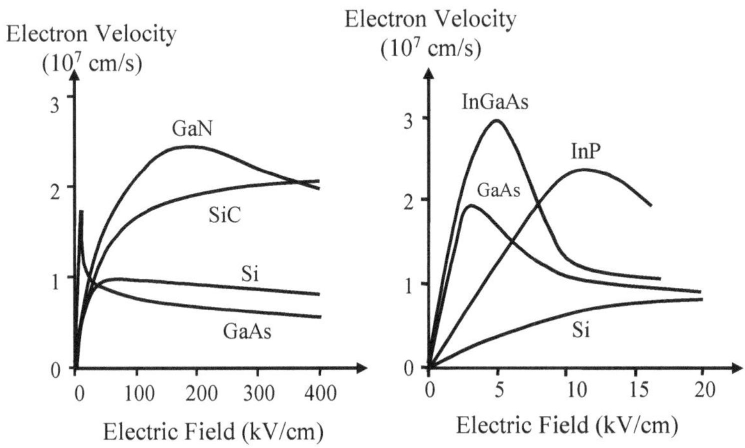

Fig. 6.1 Electron velocity with electric field of different materials

Metal–semiconductor field effect transistors (MESFETs) based on GaAs is the better microwave device compared to junction field effect transistors (JFETs) and metal oxide semiconductor field effect transistors (MOSFETs) with respect to gain, noise figure and power handling capability at microwave and mmwave frequency bands. Schottky barrier gate-channel junction is used in MESFET due to its superior high-frequency performance compared to p–n junction and MOS junction used in JFETs and MOSFETs, respectively.

Hetero-structure field effect transistor (HFET) uses dissimilar bandgap materials in their structures that offer further enhanced performance compared to single material structure MESFET at microwave and mmwave frequency range. HFETs are also known as modulation doped FET (MODFET) and high-electron-mobility Transistor (HEMT) because of the doping variations within the structure and higher electron mobility. Depending on the structure, doping profile and materials, there are different variations of HFETs suitable for various applications. InP-based HEMT provides the highest frequency of operation with the best noise figure. On the other hand, high bandgap semiconductor such as GaN- and SiC-based HFETs provides hundreds of watt microwave power due to their high breakdown voltage and higher thermal conductivity.

FETs are used as linear, and nonlinear devices for various types of amplifiers and control devices. These are used for the realization of low noise amplifier (LNA), high gain amplifier and high power amplifier. GaAs material is more preferable than the Si material for MESFET due to the former's better low field mobility, 8500 cm^2/V-s compared to 1300 cm^2/V-s of Si in addition to many more advantages. Power handling capability of GaAs FETs is about 1 W/mm gate width whereas it is about 10 W/mm for GaN-based power FETs. As nonlinear device, these are used to realize multiplier, mixer and oscillator. FETs are also used as RF control devices to realize switches and attenuators in which they dissipate essentially no extra power from the control signal and can easily be designed into broadband circuits.

6.2 Metal–Semiconductor Field Effect Transistors

Metal–semiconductor field effect transistors (MESFETs) consist of metal–semiconductor Schottky barrier gate contact to control the current flow through the active channel [1–4]. MESFET structure consists of a conducting channel positioned in between a source and a drain contact region. Schematic structural diagram of a MESFET is shown in Fig. 6.2. In between source and drain contact region Schottky contact is placed which controls the carrier flow through the channel from source to drain. The channel current between source to drain is controlled by varying the depletion layer width which modulates the thickness of the conducting channel. The dimensions of the structural diagram are not drawn in proper scale. The length of the gate is very critical for determining the maximum frequency limit. For microwave/millimetre wave frequencies, the gate length L_g is of the order of 0.1 μm. Whereas, the channel depth, a is only about one-tenth of the gate length, i.e., $a \approx 0.1 \times L_g$. Shortening of gate length, maintaining the aspect ratio of gate length to channel depth is one of the most critical technological challenges. High-frequency FETs as well as the fabrication technology are called in terms of the gate length and material. For example, GaAs FETs of 0.25 μm gate length is called 0.25 μ GaAs FET and the technology are known as 0.25 μ GaAs technology. The width (Z) of a FET is responsible for maximum current, i.e., power handling capability. For low noise and low power applications, relatively smaller gate width devices are used. Larger gate width devices are used for higher power applications. To increase the power handling capability, FETs are designed with larger periphery configuring the gate structure with multiple gate fingers.

Invariably n-channel MESFETs are preferable for higher electron mobility than hole mobility. Under the source and drain terminals, n^+ doping is used to make both the contacts ohmic. At zero gate bias, as shown in Fig. 6.2, the active channel is partially depleted. When a negative voltage on the gate terminal is applied, it reverse-biases the gate Schottky diode which will increase the depletion region thus constricting the channel width. The control of the channel width is done by varying the depletion layer width under the gate contact that modulates the thickness of the conducting channel for controlling the current between source and drain. In this way, the source to drain carrier flow is controlled by the Schottky metal gate as shown in Fig. 6.3. Thus, by controlling the gate-bias voltage, MESFETs are used as voltage-controlled RF resistor. Figure 6.4 shows the change in drain current (I_d) with the variation of drain-to-source voltage (V_{ds}) at different fixed gate control voltage (V_g). The current–voltage relationship over this lower range of drain-to-source voltage is linear. Over this region, MESFETs are used as linear RF resistor. As the gate reverse bias voltage is increased the channel current decreases until the channel is pinched

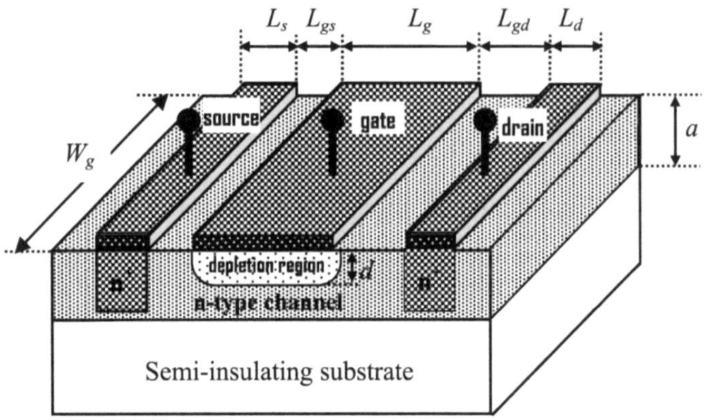

Fig. 6.2 Schematic structural diagram of MESFET

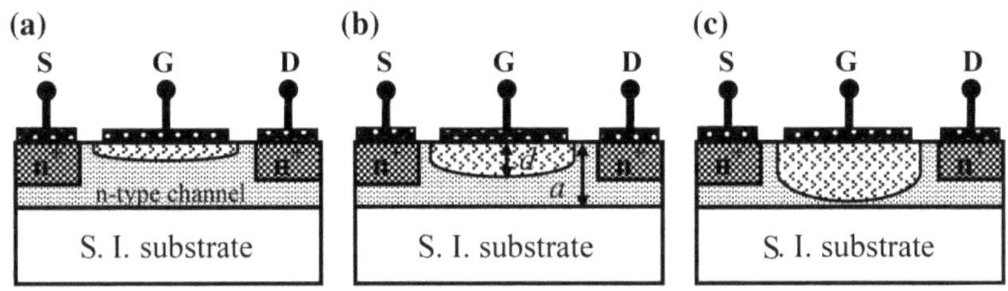

Fig. 6.3 Schematic structural diagram of MESFET with very small drain bias with **a** zero gate bias, **b** moderate reverse gate bias and **c** reverse gate bias up to pinch off

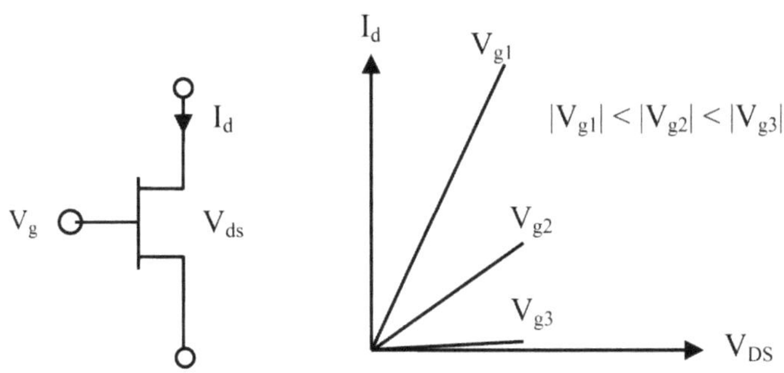

Fig. 6.4 Channel current is linearly related with the drain voltage. With the increase of gate reverse voltage, channel opening decreases thus increases the channel resistance

off and in this condition, there will be no free carrier within the channel under the gate. This situation is shown in Fig. 6.3c.

MESFETs at zero drain bias are used as variable resistor to realize voltage variable amplifier or attenuator (VVA). Gate-bias voltage is used to control the channel resistance, i.e. resistance in between drain and gate terminals which are used as linear

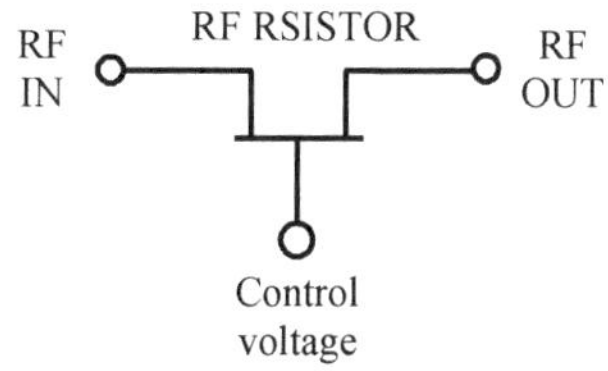

Fig. 6.5 MESFETs are used as voltage variable RF resistor for realization of VVA and switches

RF resistor where the drain and source terminals are used as RF ports as shown in Fig. 6.5. Ideally, the gate is isolated from the channel, thus there will be no current through the gate terminal, thus it dissipates essentially no power of the control signal and can be easily designed into broadband circuits. The RF signal flows from source to drain, and the RF isolated gate is used for controlling the resistance value. The channel resistance (R_c) can be expressed as [1]

$$R_c = \frac{1}{\text{channel conductivity}} \times \frac{\text{channel length}}{\text{channel area}} \tag{6.1a}$$

$$R_c = \frac{1}{qN_d\mu_n} \times \frac{L_g}{W_g\left(a - d\left(V_g\right)\right)} \tag{6.1b}$$

Here, N_d is the channel electron density which is equal to the free carrier density within the channel, μ_n is the low-field electron mobility, L_g is the gate length approximately equal to the channel length, W_g is the gate periphery, a is channel depth and $d(V_g)$ is the depletion layer depth which is a function of the applied gate control voltage. Linearity of the MESFETs as VVA maintained as long as the operating RF power level is very small compared to the applied gate control voltage. Otherwise, depletion layer width will be modulated by the electric field due to the signal amplitude in between source–drain ports.

MESFETs are also used as microwave/mmwave switch. By applying a control voltage to the gate terminal which stops current through the channel, high impedance OFF state is attained. In this condition, the source–drain channel is pinched off, thus offers very high resistance path in between source and drain. The OFF condition of the MESFET equivalently can be considered as a capacitance of value about 0.25 pF per mm of gate width. The ON state occurs when zero control voltage is applied to the gate. At this condition, the channel from source to drain is full of free carriers and represents a very low RF resistance about 2–3 Ω per mm of gate periphery.

At a fixed gate-bias voltage, channel current increases with the increase of drain voltage towards positive value. The increase of channel current is due to the increase of electron velocity with the increase of electric field in between source and drain. The drain current (I_d) is given by

$$I_d = \text{channel area} \times \text{channel conductivity} \times \text{electric field} \tag{6.2a}$$

$$= W_g \times (a - d) \times qN_d\mu_n \frac{\mathrm{d}V}{\mathrm{d}x} \tag{6.2b}$$

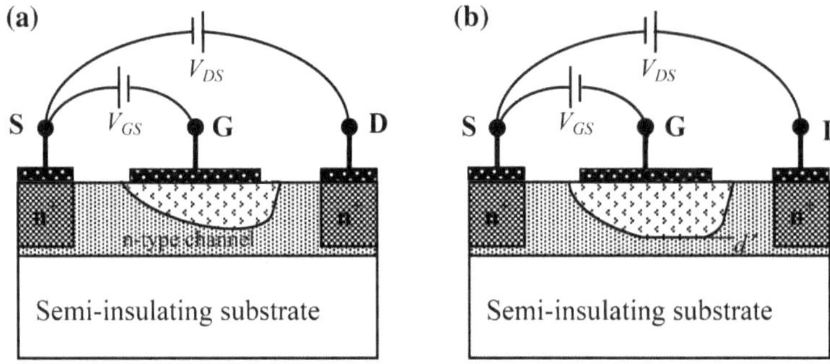

Fig. 6.6 Schematic diagram of MESFET showing depletion layer with non-zero drain bias **a** when $V_{ds} < V_{ds}$ (sat), **b** when $V_{ds} > V_{ds}$ (sat)

$$= W_g \times (a - d) \times qN_d v_n \tag{6.2c}$$

Here, dV/dx is the electric field experienced by the carrier in the direction of carrier flow from the source towards the drain and v_n is the drift velocity of the carriers. With the increase of drain bias voltage, potential difference in between drain and gate will increase, and thus there will be an increase of the depletion layer width towards the drain as shown in Fig. 6.6a. The channel then starts to pinch-off beyond certain value of drain voltage and current starts to saturate. For silicon-like materials, carrier velocity increases monotonically with the increase of applied electric field. However, for materials like GaAs, initially, the carrier velocity increases and attains a peak of about 2×10^7 cm/s at about 3 kV/cm electric field then decreases and finally saturates as shown in Fig. 6.1. Depending upon the gate length, channel pinch-off may coincide with the saturation velocity of the carriers. For very small gate length, carrier saturation may occur before channel pinch-off. Considering the onset of channel pinch-off coincides with the attaining of carrier saturation velocity, the saturated drain current can be written as

$$I_d(\text{sat}) = W_g \times d' \times qN_d v_n(\text{sat}) \tag{6.3}$$

Here, d' is the channel opening near drain at pinch-off condition as shown in Fig. 6.6b and $v_n(\text{sat})$ is the carrier saturation velocity. Beyond onset of pinch-off voltage, with the further increase of drain voltage, there will be lateral extension of depletion region from drain towards source as shown in Fig. 6.6b, which leads to unchanged drain current, i.e. saturation of drain current as shown in Fig. 6.7. There will be sharp increase of drain current with further increase of the drain voltage when it reaches to drain-to-gate breakdown voltage corresponds to drain-to-gate breakdown electric field. The drain-to-gate breakdown voltage, V_{dgB}, mainly depends on the material property, doping concentration and structural configuration. With the increase of gate reverse bias voltage, the drain breakdown voltage will be lesser by the amount of applied gate reverse voltage as shown in Fig. 6.7.

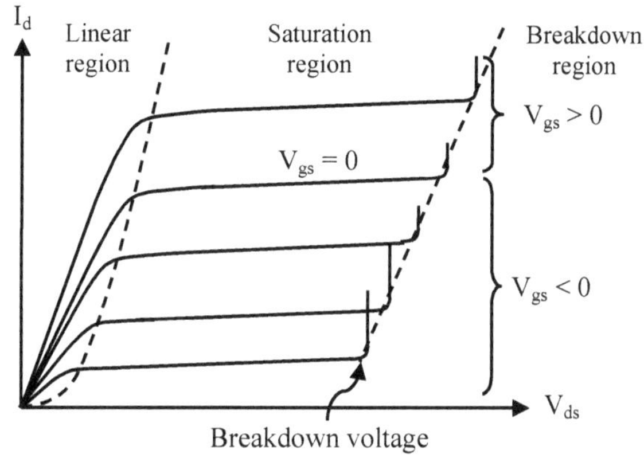

Fig. 6.7 Typical *I–V* characteristics of a n-channel MESFET

6.3 Hetero-Structure Field Effect Transistors (HFETs)

Hetero-structure field effect transistors (HFETs) utilize dissimilar bandgap materials in their structures enabling current carrying electrons to flow within an undoped semiconductor material to offer higher electron mobility compared to single material structure MESFET [8–23]. Thus, HFETs offer further improved performance compared to MESFETs at microwave and mmwave frequency range.

The carriers in the MESFET channel are provided by dopants. Electron mobility in a semiconductor is highly dependent on the dopant concentration as shown in Fig. 6.8. The dopant in the channel causes ionization impurity scattering, which reduces the mobility of the carriers. Thus, electron velocity also decreases with the increase of dopant density as shown in Fig. 6.9 [1]. It shows that the low field electrons drift velocity is more in case of undoped and lightly doped materials compared to heavily doped materials. To increase electron mobility in MESFET, doping concentration in its channel cannot be reduced beyond a certain limit; otherwise carriers responsible for drain current will be reduced. Reduction of doping concentration in MESFET channel ultimately will reduce the drain current and gain of the device. Hetero-structure FET utilizes two dissimilar bandgap materials in their structure with modulation doping to provide carriers in the undoped region, hence the device is called modulation doped field effect transistor (MODFET). Here, channel electrons are provided from the highly doped layer of higher bandgap materials which is spatially separated from the undoped channel of lower bandgap material. Thus, channel carriers do not suffer from the ionized impurity scattering hence achieving high-electron mobility. The HFETs are also called high-electron-mobility transistor (HEMT) because the structure takes the advantage of the higher mobility of the carriers in the undoped layer. In the HFET structure, the carriers flow under the gate like two-dimensional electron gas, thus it is also called two-dimensional-gate FET (TEGFET). Hetero-structure field effect transistors (HFETs) have various advantages compared to the simple MESFETs. HEMTs offer higher cut-off frequency

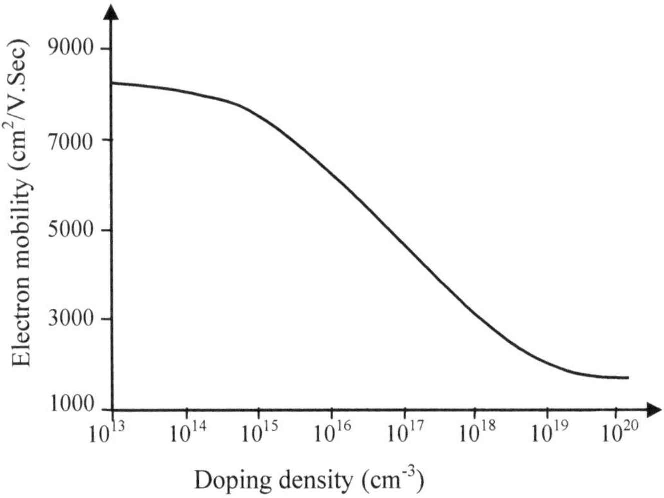

Fig. 6.8 Dependence of electron mobility on doping concentration of GaAs

Fig. 6.9 Dependence of electron velocity on doping concentration of GaAs

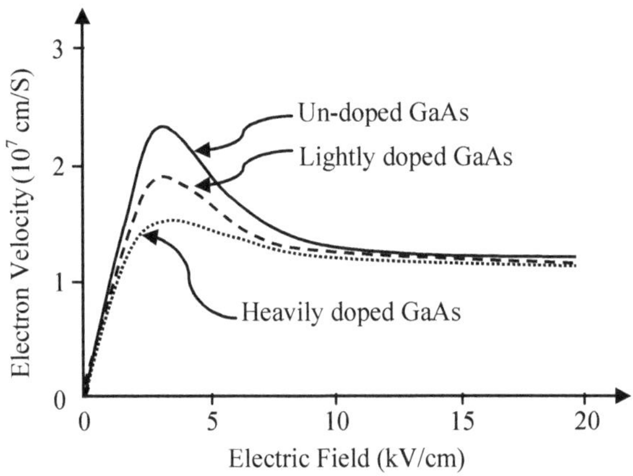

(higher gain) as well as better noise performance due to the reduction of source resistance. Some HFET structures also offer comparatively higher output power at higher microwave/mmwave frequency range. High–electron-mobility transistors have different varieties depending on the doping profiles and use of materials for different layers.

One of the advantage in modulation doped field effect transistor (MODFET) is the use of any superior material as channel since only about 200 Å layer is required in which the carrier transport takes place. Therefore, it enables to use a very high mobility material for the channel. Electron mobility of some semiconductor materials is shown in Table 6.1 with their bandgap energy and lattice constant. InSb and InAs are having very high-electron mobility but these materials cannot be

Table 6.1 Electron mobility and bandgap

Semiconductor material	Electron mobility @ 300 K ($cm^2/V\ s$)	Bandgap @ 300 K (eV)	Lattice constant (Å)
Si	1500	1.12	5.431
GaAs	8500	1.424	5.653
AlAs	200	2.163	5.660
InSb	80,000	2.26	6.479
InAs	33,000	0.36	6.058
InP	4600	1.35	5.8687
$Al_xGa_{1-x}As$	8500–200	$1.424 + 1.247x$	≈same as GaAs
$In_xGa_{1-x}As$	8500–33,000	$1.424 - 1.064x$	5.653–6.058
$In_{0.53}Ga_{0.47}As$	11,000	0.8	≈same as InP

used in MESFETs since processing of these materials is difficult. However, in case of MODFETs, a very narrow layer of these materials can be used as channel. Another aspect of making junction in between two dissimilar bandgap materials is that ideally lattice constants of both the semiconductors should be same. However, an epitaxial layer of a semiconductor whose lattice constant is close but not equal to the lattice constant of the substrate can also be grown with a very thin layer though resulting in a coherent strain. Structures of lattice-matched HEMTs consists of semiconductors with the same lattice constant but different bandgap energy. The hetero-junction structures grown with slightly different lattice constants are known as pseudo-morphic HEMTs or pHEMTs. Heterojunction also formed using two unequal lattice constant materials with graded lattice constants. These devices are called meta-morphic-HEMT (mHEMT). For military, space and commercial applications, at mmwave frequencies, where low noise figure and high gain are required, conventional MESFET technology has been replaced by differnt high-electron-mobility transistors (HEMTs, pHEMTs, mHEMTs). For high-efficiency and high-power amplifications, at microwave and mmwave frequency range, pHEMTs with multiple channels are also being used.

6.3.1 High-Electron-Mobility Transistors (HEMTs)

The high-electron-mobility transistor (HEMT) is a hetero-structure FET where the structure takes the advantage of superior electron mobility in the near-intrinsic/undoped semiconductor materials to achieve better performance primarily in terms of higher gain and better noise figure particularly at microwave and millimetre-wave frequency range.

HEMTs are basically FETs, its structure and working principles are similar to MESFET as discussed in the previous section. The epitaxial layer structures of

HEMTs are different from that of MESFETs [5–23]. In the HEMT structure, compositionally different layers are grown to improve high-frequency performance. Such types of structures are fabricated by epitaxial techniques such as metal–organic chemical vapour deposition (MOCVD) and molecular beam epitaxy (MBE) with the use of recessed gate technology of gate length down to about 0.05 μm. Common materials $Al_xGa_{1-x}As$ and GaAs having different bandgap energies are used as different layers with GaAS as substrate to construct hetero-structure of HEMTs. $Al_xGa_{1-x}As$ has larger bandgap energy than GaAs. The bandgap energy of $Al_xGa_{1-x}As$ increases with the increase of AlAs mole fraction in the $Al_xGa_{1-x}As$ material. Here, lattice constants of both the materials are nearly the same, thus this is also simply referred to as lattice-matched HEMT [5–7]. The epitaxial structure of a basic HEMT of AlGaAs/GaAs is shown in Fig. 6.10 and its energy bandgap diagram is shown in Fig. 6.11. The HEMT structure is fabricated on a semi-insulating GaAs substrate. The HEMT structure consists of an $Al_xGa_{1-x}As$ spacer layer, a n^+ donor $Al_xGa_{1-x}As$ layer, and an n $Al_xGa_{1-x}As$ Schottky contact layer. The highly doped n^+ GaAs layer is used for low resistance ohmic contacts. The formation of two-dimensional electron gas (2DEG) is shown in Fig. 6.11. The higher bandgap energy of $Al_xGa_{1-x}As$ allows free electrons to diffuse from the $Al_xGa_{1-x}As$ to the lower bandgap GaAs near the interface. This forms the two-dimensional electron gas (2DEG). The electrons from the donor atoms in the high barrier material AlGaAs spill over into the low bandgap material GaAs conduction band creating the 2DEG. A potential barrier then confines the electrons to a thin sheet of charge known as the 2DEG. Thus, the 2DEG electron is free to move in the 2D plane of the device but are confined in the vertical directions. The 2DEG structures have very less Coulomb scattering, leads to high-electron-mobility structure compared to doped MESFET channel having lots of ionized donors. The spacer layer serves to separate the 2DEG from any ionized donors in the n^+ donor AlGaAs layer. Here, the source of electrons is the n^+ donor layer and n Schottky layer. Thickness of the Schottky layer of depletion mode devices is chosen to overlap the depletion region at the $Al_xGa_{1-x}As$/2DEG interface. This is to avoid the possibility of conventional MESFET effects that is conduction of electrons through the doped $Al_xGa_{1-x}As$ layer in HEMTs. Higher Schottky barrier height of the metal–semiconductor junction on the AlGaAs is advantageous. However, increase of donor concentration in donor layer of HEMTs decreases breakdown voltage that leads to lowering output power capability. However, structures of power HEMTs with higher breakdown voltages designed and developed using either single or double recess gate technology (as shown in Fig. 6.10) or by reducing the doping in the layer just below the Schottky gate.

The AlGaAs/GaAs HEMT was the first implemented lattice-matched HFET. It shows significant performance improvement, compared to the GaAs MESFETs, in terms of noise figure and gain at microwave frequency due to the 2DEG at undoped GaAs layer. However, amount of the 2D charge density in the GaAs channel is limited due to the limited band discontinuity at the AlGaAs/GaAs interface. Moreover, with availability of semiconductors such as InSb, InAs of higher mobility and advancement of material processing technology, other HEMT structures have been developed and being used. On GaAs and InP substrates $In_xGa_{1-x}As$ layer as active channels

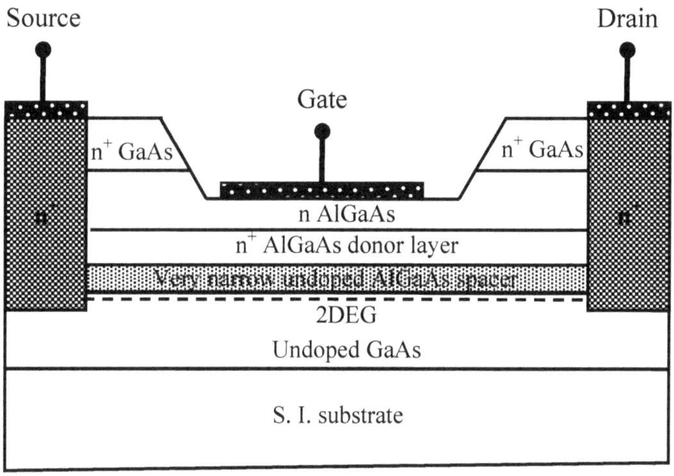

Fig. 6.10 Schematic diagram of AlGaAs/GaAs HEMT structure

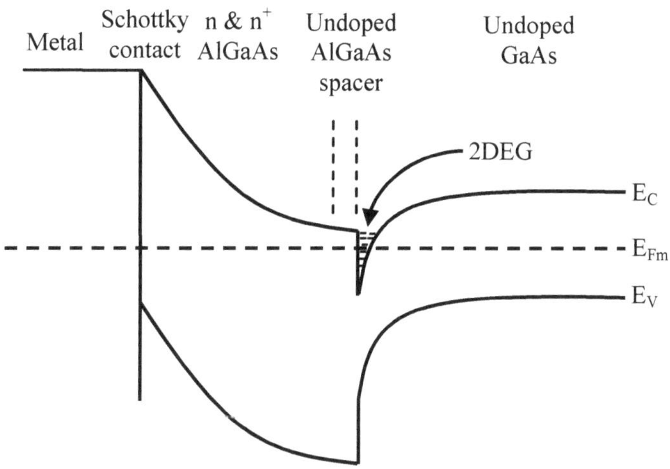

Fig. 6.11 Energy band diagram of HEMT

can be made to achieve better performance for superior mobility of $In_xGa_{1-x}As$. Indium gallium arsenide (InGaAs) is an alloy composed of two III–V semiconducting materials, GaAs and InAs. The lattice constant of InGaAs can be varied by changing the ratio of InAs and GaAs in the alloy. Though $In_{0.53}Ga_{0.47}As$ is lattice constant matched with the InP, it does not matches with the lattice constant of GaAs. Thus, in order to match the lattice constant of InP and to avoid mechanical strain, $In_{0.53}Ga_{0.47}As$ is used and structure as shown in Fig. 6.12 is known as InP HEMT. GaInAs device is normally grown on an indium phosphide (InP) substrate. InP-based HEMTs are advantageous for power HEMTs relative to GaAs-based HEMTs due

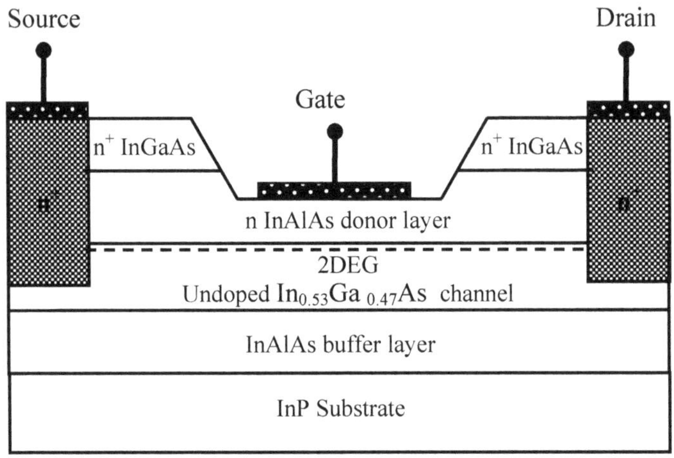

Fig. 6.12 Schematic structural diagram of InP HEMT

to superior thermal conductivity of InP (0.68 W/°C cm) compared to that of GaAs (0.46 W/°C cm). InP HEMTs also have superior microwave performance compared to GaAs HEMTs and its current gain cutoff frequency is in excess of 500 GHz.

Designing circuits using HEMTs are similar to MESFETs including biasing techniques. The Schottky layer of HEMTs depleted with application of a small negative gate bias as in the case of MESFET. With the further increase of negative gate bias, number of free carriers in the 2DEG layer decreases. This is the modulation of the channel carriers (2DEG) due to the applied negative bias. The device provides amplification by the modulation of the carriers until the channel is pinched off. At low drain current, the transconductance of HEMTs is very high because of the conduction of carriers takes place within the undoped channel which is well confined. In addition to this, HEMT provides lower noise figure and higher gain for the higher mobility compared to MESFETs resulting from lower parasitic drain and source resistances.

6.3.2 Pseudo-morphic HEMTs (pHEMTs)

The limitation of the 2D charge density in the channel due to the limited band discontinuity at the AlGaAs/GaAs interface can be solved by reducing the bandgap of the channel material. Further improvement can also be made by using the channel material with higher electron mobility thus higher electron velocity in addition to the lower bandgap material. To take the advantage of matured GaAs processing technology, on GaAs substrate AlGaAs/InGaAs interface is used where InGaAs layer used as channel for its higher electron mobility and lesser bandgap energy compared to GaAs and AlGaAs (Table 6.1).

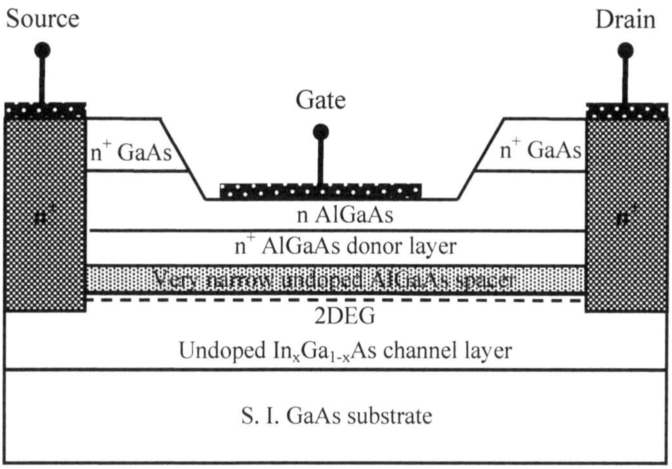

Fig. 6.13 Schematic structural diagram of pHEMT

On GaAs substrates, $In_xGa_{1-x}As$ layer as active channels is made to achieve better performance for superior mobility of $In_xGa_{1-x}As$. Indium gallium arsenide is an alloy composed of GaAs and InAs, the two III–V semiconducting materials. The lattice constant of InGaAs can be changed by changing the ratio of InAs and GaAs in the alloy. Though interatomic spacing, i.e. lattice constant of $In_{0.53}Ga_{0.47}As$ matches with the InP as discussed in case of InP HEMT, it does not matches with the lattice constant of GaAs. Thus, the device formed having interface of two materials of different lattice constant is called pseudo-morphic HEMT (pHEMT) [8, 9]. In this case, a special constraints is adopted to keep the mechanical strain from differences in lattice constants within allowable limit. For the GaAs-based pHEMT using $In_xGa_{1-x}As$, the InAs mole fraction is constrained for $x < 0.3$ and the layer thickness kept very thin, and typically 100–200 Å. A schematic diagram of AlGaAs/InGaAs-based pHEMT structure on GaAs substrate is shown in Fig. 6.13.

Advantages of the AlGaAs/InGaAs pHEMT structure include the enhanced electron mobility ($\approx$13000 cm^2/V-s) and higher saturation velocity of InGaAs. The large conduction band discontinuity in between AlGaAs and InGaAs enables higher 2DEG sheet charge density and improved confinement of carriers in the channel by the InGaAs quantum well. In addition to all these advantages, this pHEMT structure also enjoys the use of more matured GaAs process technology for realization in MMIC.

6.3.3 Meta-morphic HEMTs (mHEMTs)

The AlGaAs/InGaAs-based pHEMT structure on a GaAs substrate enjoys the use of matured and low price fabrication technology (compared to InP processing) in addition to the superior electron transport properties in InGaAs. However, excessive lattice mismatch in between GaAs and $In_xGa_{1-x}As$ does not permit to increase the Indium (In) content in the InGaAs channel to increase the electron mobility in the InGaAs channel. The lattice mismatch limits the In content in $In_xGa_{1-x}As$ below 30% ($x < 0.3$) which limits the electron mobility, thus limits the high-frequency performance in GaAs pHEMT. On the other hand, InAlAs/InGaAs HEMT devices on indium phosphide (InP) substrates have achieved cutoff frequencies of more than 500 GHz due to the higher In content of 53%. Unfortunately, InP substrates remain expensive. Hence, it is attractive to explore other ways to produce such high-performance devices, preferably using MOCVD and GaAs substrates. Another way to use semiconductor materials of different lattice constants is to place a buffer layer of varying lattice constant in between them. This is done in the low strain, i.e. meta-morphic HEMT (mHEMT) [10–13]. This is a structural advancement compared to that of the pHEMT structure. Schematic diagram of an InAlAs/InGaAs mHEMT structure is shown in Fig. 6.14. The buffer layer of InAlAs, with graded concentration of the indium content is used to match its one side lattice constant with GaAs substrate and another side with the $In_xGa_{1-x}As$ channel. The graded indium concentration in the buffer layer gives another advantage that practically any concentration of Indium in the channel can be realized to optimize different parameters of the device for different applications. Practically, for the realization of device of low noise figure, low indium concentration is used, whereas high indium concentration is used to achieve high transconductance gain.

6.4 Microwave GaN HEMTs

The wide bandgap GaN based HEMT is an emerging class of field effect transistor for microwave high-power amplifiers which are going to replace the presently used travelling wave tube amplifiers (TWTA) for high power requirements in broadband microwave communications [14–23]. Properties of GaN semiconductor with other contemporary materials are shown in Table 6.2. Wide bandgap energy of GaN (3.4 eV), high channel charge density ($\approx 10^{13}/cm^2$) of the AlGaN/GaN hetero-structure interface and high thermal conductivity of the GaN (1.7 W/cm-K) channel on the higher thermal conducting SiC (3.5 W/cm-K) substrate makes the structure suitable for handling very high microwave power of the order of 10 W/mm gate width [17–19].

Schematic diagram of basic HEMT of AlGaN/GaN hetero-structure is shown in Fig. 6.15. Due to the generation of tensile strain at the heterointerface of AlGaN/GaN, there will be spontaneous and piezoelectric polarizations. It generates free charge

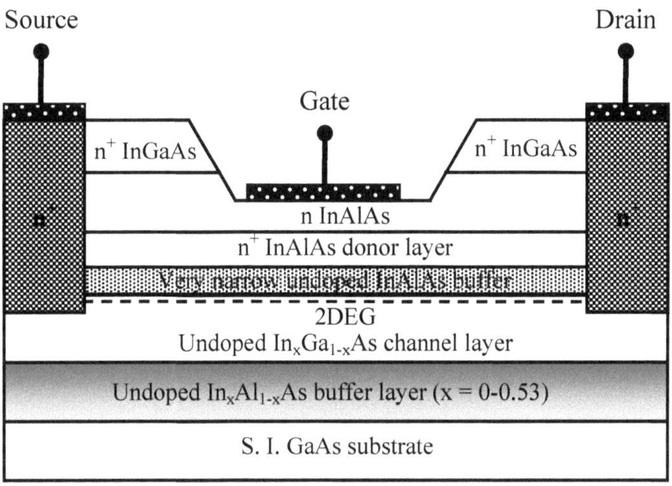

Fig. 6.14 Schematic structural diagram of mHEMT

Table 6.2 Properties of GaN compared to other materials

Material	Electron mobility @ 300 K (cm^2/V s)	Bandgap @ 300 K (eV)	Breakdown field (v/cm)	Electron saturation velocity (cm/s)	Thermal conductivity (W/cm-K)
Si	1350	1.12	6×10^5	1×10^7	1.5
GaAs	8500	1.424	6.5×10^5	1.3×10^7	0.46
SiC	800	2.163	3.5×10^6	2×10^7	3.5
GaN	1600	**3.4**	3.5×10^6	2.5×10^7	**1.7**

density at the interface of AlGaN/GaN epitaxial layer. Due to the difference of bandgap energy of AlGaAs and GaN materials, a quantum well exists in the conduction band discontinuity. The quantum well is of the order of 25 Å, thus the induced charge density behaves as 2DEG. The 2DEG sheet charge density at the interface depends on the Al content in the AlGaN. With the increase of Al content, there will be an increase in the sheet charge density. The achievable 2DEG charge density is of the order of $10^{13}/cm^2$. The origin of the electrons in the 2DEG is different in the polar hetero-structures compared to the conventional HFETs. The channel electrons in conventional HFETs are provided by a doped donor layer, whereas they come from ionized surface donor states in GaN-based HFETs. Therefore, there may not be any intentional doping of the AlGaN layer in the GaN HEMTs. In some AlGaN/GaN hetero-structure, Si impurity is added in the AlGaN which donate electron in the crystal, which then tends to accumulate just beneath the AlGaN/GaN interface which contributes to the sheet 2DEG charge density. Similar to the other

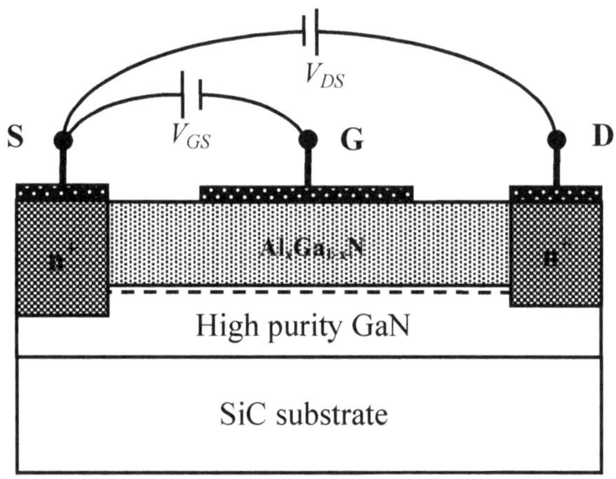

Fig. 6.15 Schematic diagram of GaN HEMT structure

HFETs, in case of GaN HEMTs also, a Schottky barrier gate terminal is used which modulates the sheet charge in the 2DEG.

There is a tremendous increase of the power densities, about 10 W/mm gate width, in the wide bandgap GaN HFETs for its higher breakdown voltage (high voltage operation) and high sheet charge density (high current operation). Furthermore, high voltage (drain supply $\approx$ 50 V) operation makes the higher input and output impedances lead to the possibility of a very broad frequency band matching for maximum output power.

High thermal conductivity of GaN channel and higher thermal conductivity of the SiC substrate makes lower thermal resistance path to dissipate out the heat from the device. However, there is a challenge on packaging techniques to dissipate out the large amount of generated heat in the device for keeping the device's channel temperature within the specified maximum allowable temperature limit due to the continuous device miniaturization. In spite of various efforts, there exists a thermal issue for high power GaN devices for its very high power density and parasitic effects due to the associated packaging that affects the circuit performances reliability and lifetime.

The GaN devices are mainly targeted for high power application at microwave frequency range and single device has the capability of CW output power about 250 W at *S*-band. There is also development of GaN amplifiers at mmwave frequency range and reported operating frequency range goes to beyond 100 GHz.

6.5 Equivalent Circuit of Microwave FETs

To design microwave circuits using FETs, it is important to know its equivalent circuit. Design and analysis of linear circuits such as low noise amplifiers and low power high gain amplifiers need small-signal equivalent circuit, whereas, non-linear

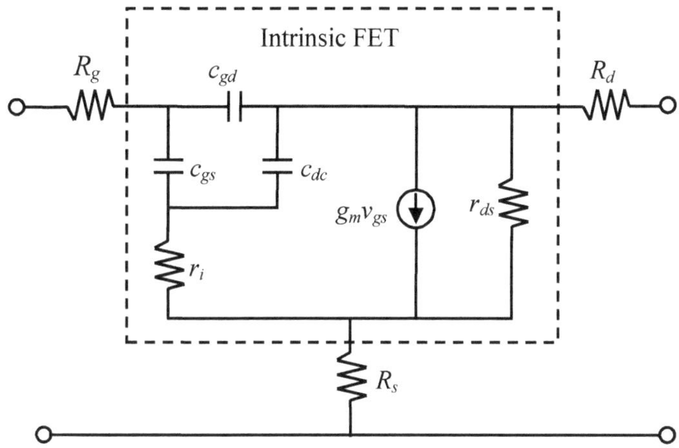

Fig. 6.16 Equivalent circuit of a FET

circuits such as multiplier, mixer, or power amplifiers operating in nonlinear region need large signal equivalent circuit. An equivalent circuit of FET is shown in Fig. 6.16 and the origin of the equivalent circuit elements are shown in Fig. 6.17 [24–40]. This is a generalized equivalent circuit which is applicable for MESFET as well as for different types of HEMTs. However, the element values depend not only on the types of FETs it also changes with the structural dimensions and doping profile of the device. The equivalent circuit elements can be divided into two distinct categories, i.e. extrinsic and intrinsic elements. The first category presents the different terminals of access like the source, drain and gate resistances R_s, R_d and R_g. The intrinsic elements represent the intrinsic characteristics of the FETs. The most important intrinsic elements are the transconductance gain g_m and drain-to-source resistance r_{ds}, its inverse is also known as drain conductance g_d. Both the parameters determine the DC I–V characteristics of the FETs. Performance of a FET based circuit can be analyzed using the equivalent circuit. The important elements of the equivalent circuits are defined and details are discussed in the following sections.

6.5.1 Transconductance Gain (g_m)

The transconductance (g_m) of a device is defined as the slope of its $I_{ds} - V_{gs}$ characteristics with the drain–source voltage (V_{ds}) kept constant. Mathematically it can be expressed as:

$$g_m = \frac{\partial I_{ds}}{\partial V_{gs}}\Big|_{V_{ds=\text{constant}}} \tag{6.4}$$

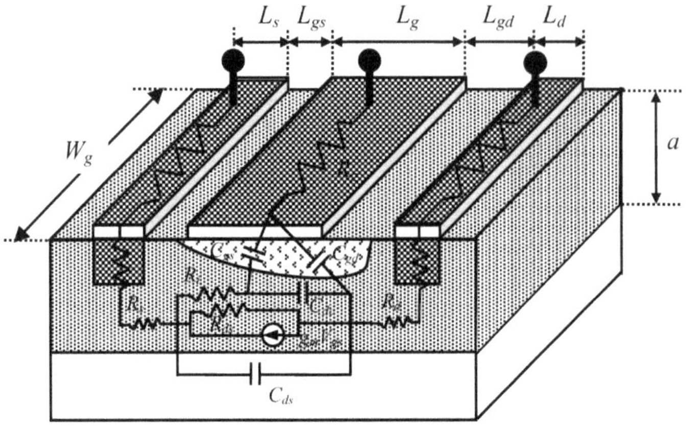

Fig. 6.17 Schematic structure of a FET indicating the origin of its equivalent circuit elements

Typical $I_{ds}-V_{gs}$ characteristics and transconductance dependence on gate voltage of FETs are shown in Fig. 6.18. The transconductance is zero, for the gate-bias voltage below the pinch-off voltage and it increases with the reduction of reverse bias voltage towards zero. Sensitivity of drain current on the change of gate–source voltage determines the transconductance of a FET as shown in the plots. Transconductance variation with the gate-bias voltage for different drain bias voltages is shown in Fig. 6.19. Transconductance is nearly invariant with drain bias voltage when it is above the drain saturation currents. However, over very low drain bias voltage, corresponds to below the drain saturation current, the transconductance becomes very low. With all other characteristics same, a FET with higher transconductance will provide greater gain and superior high frequency performance. The device transconductance is greatly affected by the device dimensions, properties of channel material and doping profile. With the decrease of gate length, transconductance increases. In general, the transconductance is inversely proportional to the gate length of the FETs. The HEMT device always provides better transconductance than GaAs FETs. As with the channel current, FET transconductance is directly proportional to the gate width of the device. For this reason, device comparisons are made by comparing transconductance per unit gate width.

To achieve highest gain from a FET, it is desirable to bias the device at its maximum transconductance point when it is operating under small-signal conditions. However, dependency of transconductance on gate–source and drain–source bias levels introduces non-linearity when the FETs are operated under large signal conditions. For large signal operation, such as in case of high power amplifiers, the transconductance needs to remain constant in the entire operating range of the gate voltage to achieve low distortion in the amplified output signal. Such a transconductance characteristic is shown in Fig. 6.20. This type of characteristics is achieved in case of power FETs by designing it with modified device structure and doping profiles [3].

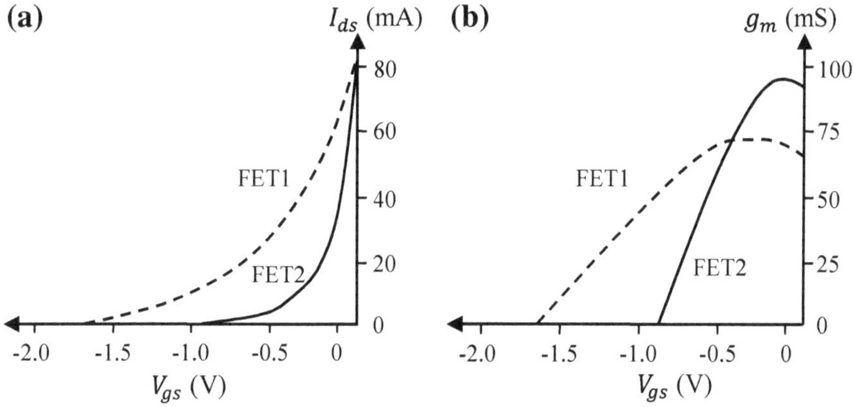

Fig. 6.18 Typical **a** drain current and **b** transconductance variation with gate voltage

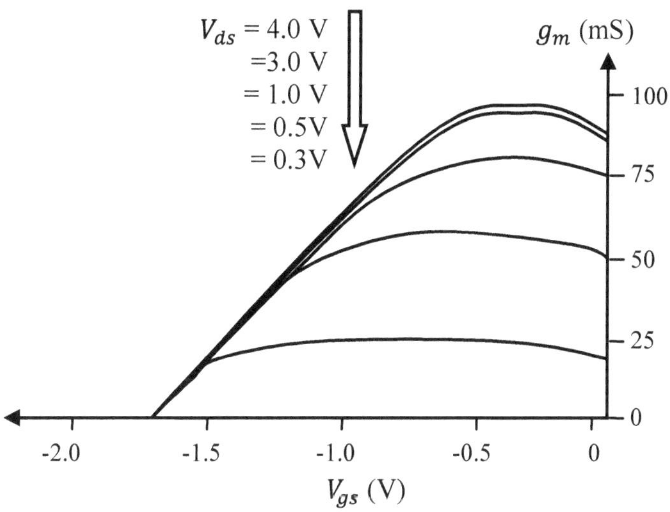

Fig. 6.19 Typical transconductance variation with gate bias voltage for various drain bias voltages

6.5.2 Output Conductance (1/r$_{ds}$)

Output conductance (g_{ds}) is the measure of drain current (I_{ds}) variation with the change of drain–source voltage (V_{ds}) while keeping the gate–source voltage (V_{gs}) constant. Output characteristics of FETs are often more conveniently expressed in terms of output resistance (r_{ds}) which is inverse of the output conductance (g_{ds}). Mathematically it can be written as:

$$g_{ds} = \frac{1}{r_{ds}} = \frac{\partial I_{ds}}{\partial V_{ds}}\Big|_{V_{gs=\text{constant}}} \tag{6.5}$$

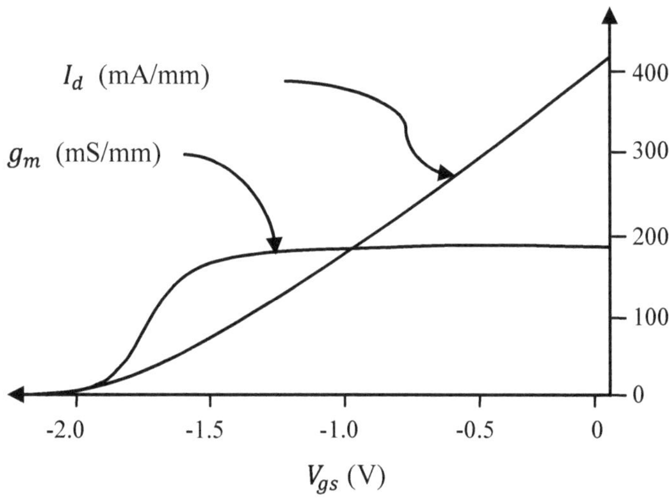

Fig. 6.20 Typical drain current and transconductance variation with gate voltage for a power FET with $V_{ds} = 5$ V

Thus, the output conductance of FETs is determined by the slope of its $I_{ds} - V_{ds}$ characteristics. The output conductance of a FET is important for any circuit-level applications. It plays a significant role in determining the output matching network for a FET based circuit. In general, a FET with low value of output conductance, i.e. higher value of output resistance is desirable. This implies that the FETs $I_{ds} - V_{ds}$ characteristic should have ideally zero slopes. A typical output resistance characteristic with the change of drain supply voltage for different values of gate-bias voltages is shown in Fig. 6.21. At low drain–source bias levels, the output resistance is low and then increases with the increase of V_{ds} towards drain current saturation. This behaviour is valid for all the gate–source bias levels except when it reaches near the pinch-off levels as shown in Fig. 6.21. At the pinch-off bias levels, the active channel shows high value of resistance which is nearly independent of drain–source bias voltage. From this characteristic, it is also clear that the FET at its lower level of drain–source bias level acts as a voltage-controlled microwave resistor controllable by the gate–source voltage. This operating region is also used as non-linear resistance for the realization of microwave multiplier and mixer circuits.

The output resistance of a FET is affected by the device dimensions, material properties and doping profile. It decreases with the increase of device periphery as the channel current increases. Power FETs are having very large gate periphery, thus its output resistance is very small. For a typical case of 0.5 μm gate length GaAs MESFET, the output resistance is about 150 Ω/mm gate width. Thus, for a power FET of output power of about 7.5 W, the gate periphery will be about 10 mm (considering output power capability = 0.75 W/mm), leads to the output resistance of about 15 Ω. Thus, designing an amplifier using this power FET it is required to transfer this 15 Ω output resistance to the load resistance of 50 Ω. Larger the

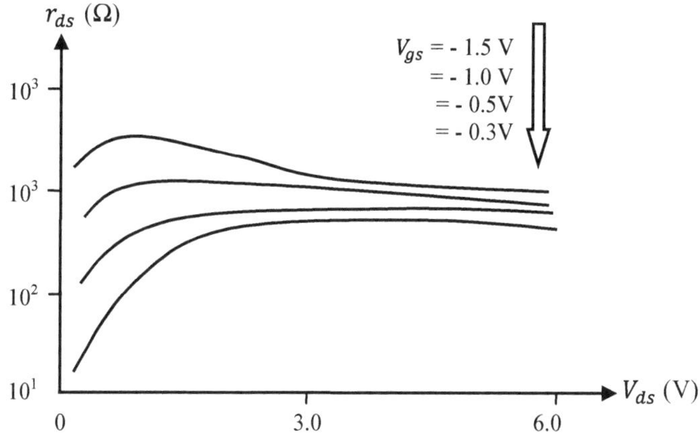

Fig. 6.21 Typical output resistance variation with drain voltage

difference in between the output resistance and the load resistance, there will be difficulty in designing the matching network over a broad frequency range. For further higher power MESFET chip, there will be further reduction of output resistance and narrower bandwidth of the power amplifier. GAN material based HEMTs are having higher output resistance per gate width. This is one of the reasons that GaN power amplifiers provide broadband output power performance.

6.5.3 Gate–Source and Gate–Drain Capacitances (C_{gs} and C_{gd})

Amount of charge stored under the gate terminal is governed by the applied terminal voltages. In normal operation of FETs, reverse bias of the gate–drain region is more than the reverse bias of gate–source region. Thus at the drain end, the gate depletion region is deeper compared to that at the source end. Also, the depletion region is closer to the drain end compared to the source end. Due to the redistribution of stored charge under the gate terminal by the change of gate–drain and gate–source reverse voltages, there are two capacitances known as gate–source capacitance (C_{gs}) and gate–drain capacitance (C_{gd}).

The gate–source capacitance is defined as when the gate–drain voltage remains constant, the rate of change of free charge under the gate terminal with respect to the change of gate–source voltage. Mathematically, it can be written as

$$C_{gs} = \frac{\partial Q_g}{\partial V_{gs}}\Big|_{V_{gd=\text{constant}}} \tag{6.6a}$$

Here, Q_g is the free charge under the gate terminal. In general, FETs are operated at common source configurations where gate and drain voltages are applied with respect to the source terminal. The controllable voltages are gate–source and drain-to-source voltages. Therefore, the gate–source capacitance is also defined with respect to the constant drain–source voltage as below

$$C_{gs} = \frac{\partial Q_g}{\partial V_{gs}}\Big|_{V_{ds}=\text{constant}} \tag{6.6b}$$

Capacitance values, as per the two different definitions, differ slightly. For microwave circuit applications, the gate–source capacitance has an important role. Input impedance as well as the microwave frequency response depends on this capacitance. Neglecting the contribution of other capacitances, the input impedance of a FET is mainly governed by the gate–source capacitive reactance in series with a few ohms of resistance. At a higher frequency of operation the input capacitance acts as a short circuit. The lower the C_{gs} value, higher the frequency at which the input impedance becomes short. At such frequencies, the FETs will not be able to provide sufficient gain. Thus, it is desirable to have lower gate–source capacitance value to achieve high-frequency performance. The gate–source capacitance is also depends on the device structure as well as channel doping. Higher the channel doping, higher is the value of C_{gs}. With the increase of gate periphery, the capacitance value increases. The capacitance value decreases with the decrease of gate–strip length. Thus, the value of C_{gs} is proportional to both the gate width and gate length. Actually, the gate–source capacitance is not only due to the charge under the gate strip, it is also due to the fringing charge in the gate–drain and gate–source spaces. The gate–source capacitance is also a nonlinear component. Its value depends on the gate and drain voltage levels. The dependence of gate–source capacitance on the terminal voltages is shown in Fig. 6.22. For a fixed drain–source voltage, the gate–source capacitance increases monotonically with the increase of gate–source voltage towards zero value as it is expected due to the decrease of depletion layer thickness with the increase of gate–source voltage. The C_{gs} value also increases with the increase of drain–source voltage this is due to the extension of the depletion layer towards drain and source regions, thus effectively it increases the area of the gate–source capacitor.

The gate–drain capacitance (C_{gd}) is also due to the charge stored under the gate terminal. The gate–drain capacitance is defined as the rate of change of free charge under the gate terminal with respect to the gate–drain voltage when the gate–source voltage remains constant. Mathematically, it can be written as

$$C_{gd} = \frac{\partial Q_g}{\partial V_{gd}}\Big|_{V_{gs}=\text{constant}} \tag{6.7}$$

In circuit applications, the capacitance C_{gd} is responsible for the reverse isolation. It determines the stability of a FET based amplifier. The smaller the gate–drain capacitance value higher will be the output to input isolation and greater will be the stability of the circuit. In case of oscillator circuit, the C_{gd} will provide the

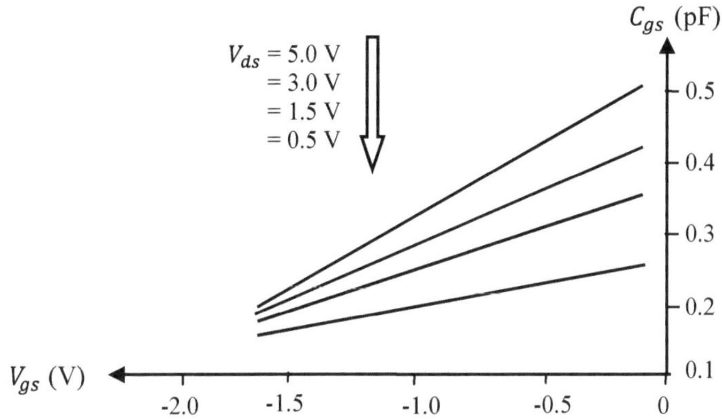

Fig. 6.22 Typical gate–source capacitance variation with gate voltage

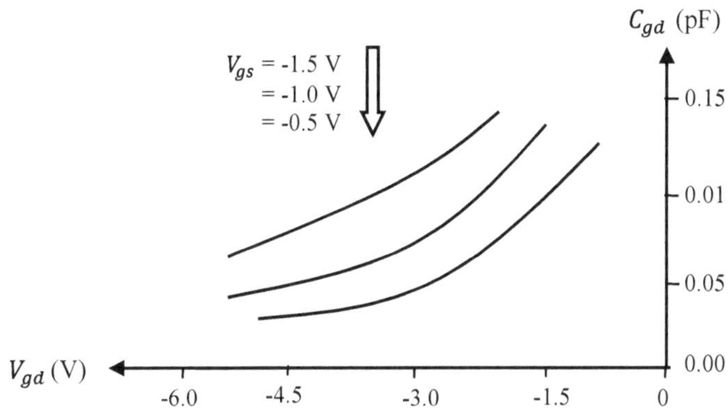

Fig. 6.23 Typical gate–drain capacitance variation with the gate–drain voltage

positive feedback for oscillation. As in the case of C_{gs}, the gate–drain capacitance also increases with the increase of gate length and width. It also increases with the increase of channel doping density. In general, for a particular FET, the value of C_{gd} is always smaller than the value of C_{gs}. The dependence of C_{gd} on the terminal voltages is shown in Fig. 6.23.

6.5.4 Charging Resistance ($\mathbf{R_i}$)

The charging resistance R_i represents the time required to charge the gate depletion region capacitance C_{gs}. The value of R_i is about few ohms with a strong dependency on gate–source as well as drain–source voltages. In general, it increases with the increase of gate as well as drain reverse bias voltages. Practically, the contribution

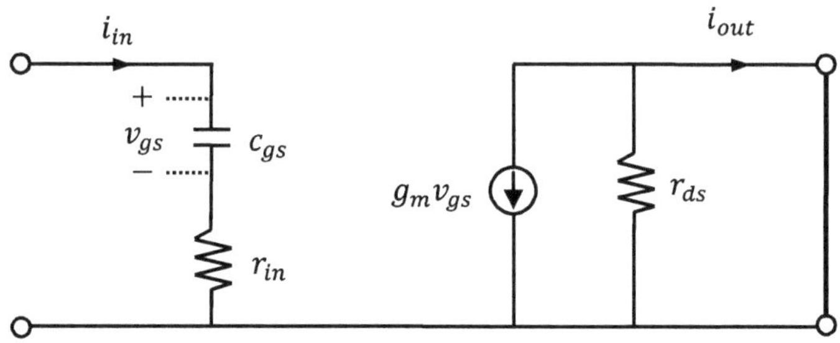

Fig. 6.24 Equivalent circuit of a FET for Example 6.1

of R_i to signal distortion is negligible compared to the contributions due to the other strongly nonlinear elements such as g_m, r_{ds} and C_{gs}. However, bias dependency of R_i is important to consider its contribution in input resistance of the FETs.

Example 6.1 Compute short circuit current gain of a FET whose equivalent circuit is as shown in Fig. 6.24. Also, compute the frequency at which its short circuit current gain becomes unity.

Solution

From Fig. 6.24, the voltage across the gate to source capacitor is given by

$$v_{gs}(f) = \frac{i_{in}(f)}{j2\pi f c_{gs}}$$

Here, i_{in} is the input current at frequency f. The short circuit output current i_{out} is given by

$$i_{\text{out}}(f) = -g_m v_{gs}(f)$$

Therefore, the short circuit current gain β is given by

$$\beta(f) = \frac{i_{\text{out}}(f)}{i_{\text{in}}(f)} = \frac{-g_m v_{gs}(f)}{j2\pi f c_{gs} v_{gs(f)}} = \frac{j g_m}{2\pi f c_{gs}}$$

$$\text{or,} \qquad |\beta(f)| = \frac{g_m}{2\pi f c_{gs}}$$

Suppose, the frequency for which the short-circuit current gain becomes unity is f_T. Therefore, we can write

$$|\beta(f_T)| = 1 = \frac{g_m}{2\pi f_T c_{gs}}$$

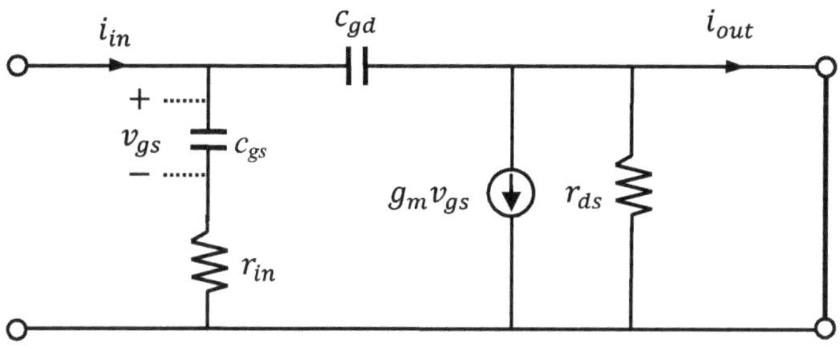

Fig. 6.25 Equivalent circuit of a FET for Example 6.2

Therefore, the short circuit current gain, $$f_T = \frac{g_m}{2\pi c_{gs}} \tag{6.8}$$

Example 6.2 Compute the frequency for unity short circuit current gain of a FET whose equivalent circuit is as shown in Fig. 6.25. Calculate its value for $g_m =$ 0.1 mho, $c_{gs} = 0.25$ pF, $c_{gd} = 0.01$ pF, $r_{ds} = 400\ \Omega$, $r_{\text{in}} = 2.5\ \Omega$.

Solution

From Fig. 6.25, the voltage across the gate to source capacitor is given by

$$\begin{aligned} v_{gs}(f) &= \frac{i_{\text{in}}(f)/(j2\pi f c_{gd})}{r_i + 1/(j2\pi f c_{gs}) + 1/(j2\pi f c_{gd})} \times \frac{1}{j2\pi f c_{gs}} \\ &= \frac{i_{\text{in}}(f)}{j2\pi f(c_{gs} + c_{gd}) - 4\pi^2 f^2 r_{\text{in}} c_{gs} c_{gd}} \end{aligned}$$

Here, i_{in} is the input current at frequency f. The short circuit output current $i_{\text{out}}(f)$ is given by

$$\begin{aligned} i_{\text{out}}(f) &= -g_m v_{gs}(f) + \frac{i_{\text{in}}(f) \times (r_{\text{in}} + 1/(j2\pi f c_{gs}))}{r_{\text{in}} + 1/(j2\pi f c_{gs}) + 1/(j2\pi f c_{gd})} \\ &= \frac{-g_m i_{\text{in}}(f)}{j2\pi f(c_{gs} + c_{gd}) - 4\pi^2 f^2 r_{\text{in}} c_{gs} c_{gd}} + \frac{i_{\text{in}}(f) \times (j2\pi f r_{\text{in}} c_{gs} + 1) c_{gd}}{j2\pi f r_{\text{in}} c_{gs} c_{gd} + c_{gs} + c_{gd}} \end{aligned}$$

Therefore, the short circuit current gain $\beta(f)$ is given by

$$\beta(f) = \frac{i_{\text{out}}(f)}{i_{\text{in}}(f)} = \frac{-(g_m + 4\pi^2 f^2 r_{\text{in}} c_{gs} c_{gd}) + j2\pi f c_{gd}}{-4\pi^2 f^2 r_{\text{in}} c_{gs} c_{gd} + j2\pi f(c_{gs} + c_{gd})}$$

$$\text{or,} \quad |\beta(f)| = \left(\frac{(g_m + 4\pi^2 f^2 r_{\text{in}} c_{gs} c_{gd})^2 + (2\pi f c_{gd})^2}{(4\pi^2 f^2 r_{\text{in}} c_{gs} c_{gd})^2 + (2\pi f(c_{gs} + c_{gd}))^2} \right)^{1/2} \tag{6.9}$$

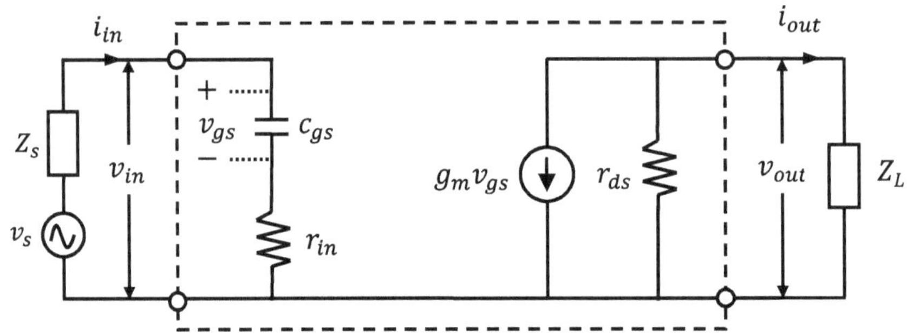

Fig. 6.26 Equivalent circuit of a FET for Example 6.3

In practice, $r_{\text{in}} \ll 2\pi f c_{gs}$, and $g_m \gg 2\pi f c_{gd}$. Therefore, the output short-circuit current gain can be written as

$$|\beta(f)| = \frac{g_m}{2\pi f\left(c_{gs} + c_{gd}\right)} \tag{6.10}$$

Suppose, the frequency for which the short-circuit current gain becomes unity is f_T. Therefore, we can write

$$|\beta(f_T)| = 1 = \frac{g_m}{2\pi f_T\left(c_{gs} + c_{gd}\right)}$$

$$\text{Therefore, the short circuit current gain,}\quad f_T = \frac{g_m}{2\pi\left(c_{gs} + c_{gd}\right)}$$

Putting the values for $g_m = 0.1$ mho, $c_{gs} = 0.25$ pF, $c_{gd} = 0.01$ pF, the frequency where the short current gain becomes unity is

$$f_T = \frac{0.1 \times 1000}{2\pi(0.25 + 0.01)} \text{GHz} = 61.2 \text{ GHz}$$

Putting $|\beta(f_T)| = 1$ in (6.9), and using $g_m = 0.1$ mho, $c_{gs} = 0.25$ pF, $c_{gd} = 0.01$ pF, $r_{ds} = 400\ \Omega$, $r_{\text{in}} = 2.5\ \Omega$, one can calculate the frequency for unity short circuit current gain without any approximation, it is about 61.83 GHz.

Example 6.3 Derive the maximum frequency of oscillation (f_{max}) which is the frequency for which the maximum available power gain of the device becomes unity of a FET. The equivalent circuit of the FET is as shown in Fig. 6.26 inside the dotted area. Also, derive relationship in between the maximum frequency of operation (also called maximum frequency of oscillation), f_{max} and frequency for unity short circuit current gain, f_T.

Solution

The input signal v_s of impedance Z_s is applied at the input of the FET and the output is terminated with impedance Z_L as shown in Fig. 6.26, the voltage across the gate to source capacitor is given by

$$v_{gs}(f) = \frac{v_s(f)}{r_{\text{in}} + 1/\left(j2\pi f c_{gs}\right) + Z_s} \times \frac{1}{j2\pi f c_{gs}} \tag{6.11a}$$

For complex conjugate matching of input port, the source impedance will be

$$Z_s = r_{\text{in}} - 1/\left(j2\pi f c_{gs}\right) \tag{6.11b}$$

Putting this value of Z_s, the voltage across c_{gs} is given by

$$v_{gs}(f) = \frac{v_s(f)}{j4\pi f c_{gs} r_{\text{in}}} \tag{6.11c}$$

The current through the output load impedance is given by

$$i_{\text{out}}(f) = \frac{-v_s(f) g_m}{j4\pi f c_{gs} r_{\text{in}}} \times \frac{r_{ds}}{r_{ds} + Z_L} \tag{6.12a}$$

Thus, the output signal voltage is given by

$$v_{\text{out}}(f) = \frac{-v_s(f) g_m}{j4\pi f c_{gs} r_{\text{in}}} \times \frac{r_{ds} Z_L}{r_{ds} + Z_L} \tag{6.12b}$$

For complex conjugate matching at the output port:

$$Z_L = r_{ds} \tag{6.12c}$$

Putting $Z_L = r_{ds}$, the output signal voltage under complex conjugate matched condition is given by

$$v_{\text{out}}(f) = \frac{-v_s(f) g_m r_{ds}}{j8\pi f c_{gs} r_{\text{in}}} \tag{6.12d}$$

Under complex conjugate matching at the input port, the input signal voltage is given by

$$v_{in}(f) = \frac{v_s(f)}{2} \tag{6.13}$$

Thus, the maximum available power gain of the FET is given by

$$P_{avs}(f) = \frac{\left(v_{\text{out}}(f)\right)^2 / r_{ds}}{\left(v_{\text{in}}(f)\right)^2 / r_{\text{in}}} = \left(\frac{v_{\text{out}}(f)}{v_{\text{in}}(f)}\right)^2 \times \frac{r_{\text{in}}}{r_{ds}} \tag{6.14}$$

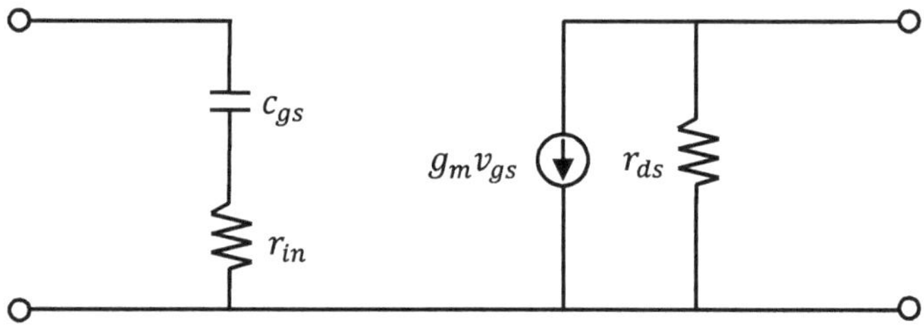

Fig. 6.27 Equivalent circuit of a FET for Example 6.1

Putting $v_{\text{out}}(f)$ and $v_{\text{in}}(f)$ from (6.12d) and (6.13), respectively, in (6.14), the maximum available power gain is given by

$$P_{avs}(f) = \left|\frac{-g_m r_{ds}}{j4\pi f c_{gs} r_{\text{in}}}\right|^2 \times \frac{r_i}{r_{ds}} = \left(\frac{g_m}{4\pi f c_{gs}}\right)^2 \times \frac{r_{ds}}{r_{\text{in}}} \tag{6.15}$$

As per the definition of the maximum frequency of operation: $P_{avs}(f) = 1$, for $f = f\text{max}$. Thus, from (6.15), we can write

$$\left(\frac{g_m}{4\pi f\text{max}\, c_{gs}}\right)^2 \times \frac{r_{ds}}{r_{\text{in}}} = 1 \tag{6.15}$$

$$\text{or,}\quad f\text{max} = \frac{g_m}{4\pi c_{gs}} \times \sqrt{\frac{r_{ds}}{r_{\text{in}}}} \tag{6.16}$$

From, the previous example we have seen

$$f_T = \frac{g_m}{2\pi c_{gs}} \tag{6.18}$$

Thus, the maximum frequency of operation of the FET is given by

$$\text{or,}\quad f\text{max} = \frac{f_T}{2} \times \sqrt{\frac{r_{ds}}{r_{\text{in}}}} \tag{6.19}$$

Example 6.4 Calculate the frequency, f_T for unity short circuit current gain and maximum frequency of operation, $f\text{max}$ of a FET whose equivalent circuit is as shown in Fig. 6.27, for $g_m = 0.15$ mho, $c_{gs} = 0.65$ pF, $r_{ds} = 500\ \Omega$, $r_{in} = 3.5\ \Omega$.

Solution

The frequency for unity short circuit current gain is given by

$$f_T = \frac{g_m}{2\pi c_{gs}}$$

Putting, $g_m = 0.15$ mho and $c_{gs} = 0.30$ pF:

$$f_T = \frac{0.15 \times 1000}{2\pi \times 0.65}\text{GHz} = 36.73\ \text{GHz}$$

The maximum frequency of operation is given by

$$f\text{max} = \frac{f_T}{2} \times \sqrt{\frac{r_{ds}}{r_{\text{in}}}}$$

Putting, $f_T = 36.73$ GHz, $r_{ds} = 500\ \Omega$ and $r_{\text{in}} = 3.5\ \Omega$:

$$f\text{max} = \frac{36.73}{2} \times \sqrt{\frac{500}{3.5}} = 219.5\ \text{GHz}$$

6.6 Maximum Frequency of Operation

Maximum frequency of operation for a FET is mainly determined by transit time of carriers through the channel [40]. The transit time is the time required for carriers to travel from source to drain. At low-field condition, when the mobility is constant, the transit time is given by

$$\tau = \frac{L_g}{\mu E_x} \tag{6.20a}$$

Here, L_g is the gate length and E_x is the electric field along the channel. Suppose the drain supply voltage is V_D with respect to the source terminal. Therefore,

$$E_x = \frac{V_d}{L_g} \tag{6.20b}$$

Putting E_x from (6.20b) to (6.20a):

$$\tau = \frac{L_g^2}{\mu V_d} \tag{6.20c}$$

Whereas, at saturated velocity condition, the transit time is given by

$$\tau = \frac{L_g}{v_s} \tag{6.21}$$

Here, v_s is the saturated velocity of the carriers. It can also be shown that the ratio of transconductance g_m and gate to source capacitance c_{gs} is equal to v_s/L_g. Therefore, the frequency for unity short circuit current gain can also be written as

$$f_T = \frac{v_s}{2\pi L_g} \tag{6.22}$$

Example 6.5 Calculate the frequency, f_T for a GaAs MESFET of gate length 1 μm. Consider the electrons are moving with their saturated velocity of 1.3×10^7 cm/s.

Solution

Under the condition of saturation velocity of carriers, f_T is given by (6.22)

$$\begin{aligned} f_T &= \frac{v_s}{2\pi L_g} \\ &= \frac{1.3 \times 10^7}{2\pi \times 10^{-4}} \text{Hz} \\ &= 20.7\ \text{GHz} \end{aligned}$$

References

1. Sze SM (1981) Physics of semiconductor devices, 2nd edn. Wiley Eastern Limited
2. Statz H, Newman P, Smith I, Pucel R, Haus H (1987) GaAs FET device and circuit simulation in spice. IEEE Trans Electron Devices ED-34(2):160–169
3. Tkachenko AY, Bartle D, Wei C-J (1999) Correlation between ungated recess width and linearity of GaAs MESFETs. Microw J 1–8
4. Pal D, Noll A (2014) Improvement of transconductance flatness of GaAs MESFETs. In: CS mantech conference, pp 219–222, Denver, Colorado, USA, 19th–22nd May 2014
5. Lai R, Mei X, Deal W, Yoshida W, Kim Y, Liu P, Lee J, Uyeda J, Radisic V, Lange M, Gaier T, Samoska L, Fung A (2007) Sub 50 nm InP HEMT device with fmax greater than 1-THz. In: IEEE international electron devices meeting, IEDM 2007, pp 609–611, Dec 2007
6. Grundbacher R, Lai R, Barsky M, Tsai R, Gaier T, Weinreb S, Dawson D, Bautista J, Davis J, Erickson N, Block T, Oki A (2002) 0.1 m InP HEMT c from X-band to W-band. In: 14th Indium phosphide and related materials conference, IPRM, pp 455–458
7. Greenberg D, Del Alamo L, Bhat R (1995) Impact ionization and transport in InAlAs/n+-InP HFET. IEEE Trans Electron Devices 42(9):1574–1582
8. Lin YC, Chang EY, Yamaguchi H, Wu WC, Chang CY (2007) A δ-doped InGaP/InGaAs pHEMT with different doping profiles for device linearity improvement. IEEE Trans Electron Devices 54(7):1617–1625
9. Yang MT, Chan YJ (1996) Device linearity comparisons between doped channel and modulation doped designs in Pseudomorphic $Al_{0.3}Ga_{0.7}As/In_{0.2}Ga_{0.8}As$ heterostructures. IEEE Trans Electron Devices 43(8):1174–1180
10. Leuther A, Tessmann A, Massler H, Losch R, Schlechtweg M, Mikulla M, Ambacher O (2008) 35 nm metamorphic HEMT MMIC technology. In: 20th international conference on indium phosphide and related materials, MoA3.3, May 2008
11. Tessmann A, Leuther A, Massler H, Kuri M, Loesch R (2008) A metamorphic 220–320 GHz HEMT amplifier MMIC. In: IEEE compound semiconductor integrated circuit symposium, CSICS 2008, IEEE, Monterey, California, Piscataway, NJ, 12–15 Oct 2008, pp 31–34
12. Hülsmann A, Leuther A, Kallfass I, Weber R, Tessmann A, Schlechtweg M, Ambacher O (2009) Advanced mHEMT technologies for space applications. 20th international symposium on space terahertz technology, Charlottesville, 20–22 April 2009
13. Leuther A, Tessmann A, Kallfass I, Losch R, Seelmann-Eggebert M, Wadefalk N, Schafer F, Puyol JDG, Schlechtweg M, Mikulla M, Ambacher O (2009) Metamorphic HEMT technology for low-noise applications. In: 21st IEEE international conference on indium phosphide & related materials, IPRM, pp 188–191, 411

14. Pengelly RS, Wood SM, Milligan JW, Sheppard ST, Pribble WL (2012) A review of GaN on SiC high electron-mobility power transistors and MMICs. IEEE Trans Microw Theory Tech 60(6)
15. Kazutaka I, Norihiko U, Seigo S (2011) High power and high efficiency GaN-HEMT for microwave communication applications. In: IEEE IMWS-IWPT proceedings
16. Ambacher O, Smart J, Shealy JR, Weimann NG, Chu K, Murphy M, Schaff WJ, Eastman LF, Dimitrov R, Wittmer L, Stutzman M, Rieger W, Hilsenbeck J (1999) Two-dimensional electron gases induced by spontaneous and piezoelectric polarization charges in N- and Ga-face AlGaN/GaN heterostructures. J Appl Phys 85(6):3222–3333
17. Wu YF et al (1996) Very high breakdown voltage and large transconductance realized on GaN heterojunction field-effect transistors. Appl Phys Lett 69:1438–1440
18. Wu YF, Kapolnek D, Ibbetson JP, Parikh P, Keller BP, Mishra UK (2001) Very-high power density AlGaN/GaNHEMTs. IEEE Trans Electron Devices 48:586–590
19. Mitani E, Aojima M, Maekawa A, Sano S (2007) An 800-W AlGaN/GaN HEMT for S-band high-power application. In: CS mantech on-line Digest
20. Chung Jinwook W, Hoke William E, Chumbes Eduardo M, Palacios Tomas (2010) AlGaN/GaN HEMT With 300-GHz f_{max}. IEEE Electron Device Lett 31(3):195–197
21. Palacios T, Chakraborty A, Rajan S, Poblenz C, Keller S, DenBaars SP, Speck JS, Mishra UK (2005) High-power AlGaN/GaN HEMTs for Ka-band applications. IEEE Electron Device Lett 26(11):781–783
22. Quay R, Musser M, van Raay F, Maier T, Mikulla M (2009) Managing power density of high-power GaN devices. In: IEEE MTT-S international microwave symposium workshop Digest, Boston, MA, pp 71–86
23. Wu Y-F, Kapolnek D, Ibbetson JP, Parikh P, Keller BP, Mishra UK (2001) Very-high power density AlGaN/GaN HEMTs. IEEE Trans Electron Devices 48:586–590
24. Curtice W, Ettenberg M (1985) A nonlinear GaAs FET model for use in the design of output circuits for power amplifiers. IEEE Trans Microw Theory Tech 33(12):1383–1393
25. Minasian RA (1977) Simplified GaAs MESFET model to 10 GHz. Electron Lett 13(18):549–551
26. Kondoh H (1986) An accurate fet modelling from measured S-parameters. In: IEEE MTT-S international microwave symposium digest, pp 377–380, June 1986
27. Fukui H (1979) Determination of the basic device parameters of a GaAs MESFET. Bell Syst Tech J 58(3):771–795
28. Chen T, Kumar M (1988) Novel GaAs FET modeling technique for MMICs. In: IEEE GaAs IC Symposium Digest, pp 49–52
29. Dambrine G et al (1988) A new method for determining the FET small-signal equivalent circuit. IEEE Trans Microw Theory Tech 36(7):1151–1159
30. Berroth M, Bosch R (1991) High-frequency equivalent circuit of GaAs FET's for large-signal applications. IEEE Trans Microw Theory Tech 39(2):224–229
31. Ooi B et al (1997) A novel approach for determining the GaAs MESFET small-signal equivalent-circuit elements. IEEE Trans Microw Theory Tech 45(12):2084–2088
32. Angelov I, Zirath H, Rorsman N (1992) A new empirical nonlinear model for HEMT and MESFET devices. IEEE Trans Microw Theory Tech 40(12):2258–2266
33. Jeon K, Kwon Y, Hong S (1997) A frequency dispersion model of GaAs MESFET for large-signal applications. IEEE Microw Guided Wave Lett 7(3):78–80
34. Lazaro A, Pradell L, O'Callaghan J (1999) FET noise-parameter determination using a novel technique based on 50-Ω noise-figure measurements. IEEE Trans Microw Theory Tech 47(3):315–324
35. Rorsman N, Garcia M, Karlsson C, Zirath H (1996) Accurate small-signal modeling of HFET's for millimeter-wave applications. IEEE Trans Microw Theory Tech 44:432–437
36. Dambrine G, Cappy A, Heliodore F, Playez E (1988) A new method for determining the FET small-signal equivalent circuit. IEEE Trans Microw Theory Tech 36:1151–1159
37. Curtice WR, Ettenberg M (1985) A nonlinear GaAs FET model for use in the design of output circuits for power amplifiers. IEEE Trans Microw Theory Tech MTT-33(12):1383–1393

38. Kompa G, Lin F (1990) FET modelling using an analytical extraction method based on broadband S-parameter measurement. In: 20th European microwave conference proceedings, Budapist, Hungary, pp 778–783, September 1990
39. Kompa G, Novotny M (1992) Highly consistent FET model parameter extraction based on broadband S-parameter measurements. IEEE MTT-S international microwave symposium digest, pp. 293–296
40. Higashiwaki M, Mimura T, Matsui T (2008) AlGaN/GaN heterostructure field-effect transistors on 4H-SiC substrates with current-gain cutoff frequency of 190 GHz. Appl Phys Express 1:021103

Chapter 7
Microwave Circuit Analysis

Abstract This chapter describes detailed transmission line theory and analysis including concept of *S*-parameters, *ABCD* parameters and their interrelationships through various examples. It gives Smith chart analysis and describes various transmission lines like waveguide, coaxial, stripline, microstrip and coplanar transmission lines. This chapter also presents detailed analysis about power transfer theory describing available power gain, transducer power gain and operating power gain for microwave networks.

7.1 Introduction

Objective of the microwave circuit analysis is to provide a complete description of terminal behaviour of the circuit as of low-frequency circuit analysis. Microwave circuit analysis needs the knowledge about transmission line theory, microwave networks, network analysis and impedance matching.

At low frequencies where circuit dimensions are relatively smaller than the operating wavelength, the circuit elements can be considered as lumped components as the phase change along the circuit elements is negligible. Similarly, at low frequencies, the interconnection between two elements is treated as an ideal connection with no loss, no characteristic impedance and no transmission phase angle. At microwave frequencies, dimension of circuit components and interconnecting elements may not be possible to conveniently made smaller compared with the operating wavelength. Thus, the circuit analysis technique cannot be done using conventional circuit voltage and current. Microwave circuits may consist of elements with dimensions such that the voltage amplitude and phase over the length of the device can vary significantly. In this case, the current that flows in the interconnecting wires or through the circuit elements is not the same at different points along the wire or elements. To accurately calculate the behaviour of an element, it is required to consider its length, width and thickness of metal and its proximity to the ground plane, i.e. the element is considered as a distributed element. Similarly, when interconnections become an

S. C. Bera, *Microwave Active Devices and Circuits for Communication*, Lecture Notes in Electrical Engineering 533, https://doi.org/10.1007/978-981-13-3004-9_7

appreciable fraction of the signal wavelength, these interconnections themselves must be treated as distributed elements or transmission lines.

Though all electronic circuits are nonlinear in character; however, some circuits can be treated as linear circuits depending upon their operating conditions. Amount of nonlinearity not only depends on particular device but it also depends on bias condition, signal level and operating frequency. A circuit that behaves as nonlinear circuit at some operating signal level may behave as a linear circuit under comparatively lower signal power level. In general, microwave switches, attenuators, phase shifters, modulators, low noise and IF amplifiers operate at their linear operating regions, whereas high power amplifiers, linearizers and limiters shall be treated as nonlinear circuits.

7.2 Transmission Line Theory and Analysis

Transmission lines are used to guide and carry electromagnetic waves from one point of a system to another. At the higher operating frequency, where length of the transmission lines used as interconnecting elements is comparable or more than the wavelength, the approximations useful for calculations at lower frequencies are no longer accurate. In this case, the voltage and current along the transmission line at a given time may not be the same at all points. Thus, the length of the interconnecting line is important when the signal includes frequency components with corresponding wavelengths comparable to or less than the length of the line.

For analysis, a transmission line can be represented by a network with circuit components distributed throughout the line as shown in Fig. 7.1a. Here, R in Ω/m is resistance, L in H/m is inductance in both conductors, G in S/m is conductance of the dielectric media and C in F/m is capacitance between the conductors, all are per unit length. The capacitance C and inductance L represent electromagnetic field that passes along the transmission line, whereas the resistance R and the conductance G represent ohmic loss along the line and dielectric loss across the transmission lines, respectively.

Kirchhoff's voltage and current laws can be applied to this line section to derive the propagation constant γ and the characteristic impedance Z_o of the line [1]. Applying Kirchhoff's voltage law to the circuit shown in Fig. 7.1a,

$$v(l,t) - R\Delta l\, i(l,t) - L\Delta l \frac{\partial i(l,t)}{\partial t} - v(l+\Delta l, t) = 0 \tag{7.1a}$$

or

$$-\frac{v(l+\Delta l, t) - v(l,t)}{\Delta l} = R\, i(l,t) + L\frac{\partial i(l,t)}{\partial t} \tag{7.1b}$$

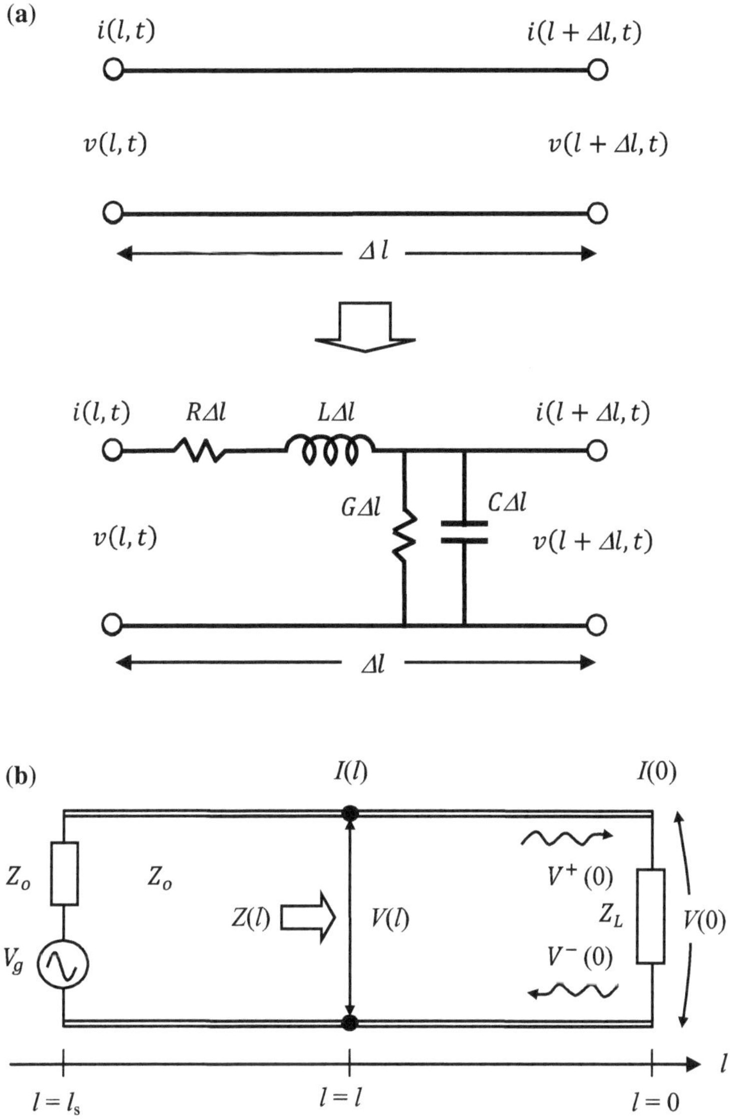

Fig. 7.1 **a** Equivalent circuit of a transmission line of very small length Δl, **b** EM wave travelled in a transmission line of characteristic impedance Z_o terminated by a load impedance Z_L

For $\Delta l \rightarrow 0$

$$-\frac{\partial v(l,t)}{\partial l} = Ri(l,t) + L\frac{\partial i(l,t)}{\partial t} \tag{7.1c}$$

Similarly, applying Kirchhoff's current law to the circuit shown in Fig. 7.1a,

$$i(l,t) - G\Delta l\, v(l+\Delta l,t) - C\Delta l \frac{\partial v(l+\Delta l,t)}{\partial t} - i(l+\Delta l,t) = 0 \tag{7.2a}$$

For $\Delta l \to 0$

$$-\frac{\partial i(l,t)}{\partial l} = Gv(l,t) + C\frac{\partial v(l,t)}{\partial t} \tag{7.2b}$$

The first-order differential equations (7.1c) and (7.2b) are the generalized transmission line equations valid for any voltage and current waveforms along the transmission line [1]. Suppose, the voltage and current are time-varying sinusoidal functions represented by

$$v(l,t) = Re\left(V(l)\mathrm{e}^{j\omega t}\right) \tag{7.3a}$$

$$i(l,t) = Re\left(I(l)\mathrm{e}^{j\omega t}\right) \tag{7.3b}$$

Using (7.3a) and (7.3b), the generalized transmission line equations can be written as

$$-\frac{\mathrm{d}V(l)}{\mathrm{d}l} = (R + j\omega L)I(l) \tag{7.4a}$$

$$-\frac{\mathrm{d}I(l)}{\mathrm{d}l} = (G + j\omega C)V(l) \tag{7.4b}$$

Differentiating (7.4a) and using (7.4b), we get,

$$\frac{\mathrm{d}^2V(l)}{\mathrm{d}l^2} = (R + j\omega L)(G + j\omega C)V(l) \tag{7.5a}$$

Similarly, differentiating (7.4b) and using (7.4a), we get,

$$-\frac{\mathrm{d}^2I(l)}{\mathrm{d}l^2} = (R + j\omega L)(G + j\omega C)I(l) \tag{7.5b}$$

Defining, the propagation constant,

$$\gamma = \alpha + j\beta = \sqrt{(R + j\omega L)(G + j\omega C)} \tag{7.6a}$$

For lossless transmission lines, $R = 0$ and $G = 0$. Putting these values, for lossless transmission lines, the propagation constant can be expressed as

$$\gamma = j\beta = j\omega\sqrt{LC} \tag{7.6b}$$

Thus, the phase velocity v_p is given by

$$v_p = \omega/\beta = 1/\sqrt{LC} \tag{7.6c}$$

(7.5a) and (7.5b) can be written as

$$\frac{d^2V(l)}{dl^2} = \gamma^2 V(l) \tag{7.7a}$$

and

$$-\frac{d^2I(l)}{dl^2} = \gamma^2 I(l) \tag{7.7b}$$

(7.7a) and (7.7b) are the ordinary differential equations. The solutions can be written as

$$V(l) = V^+e^{-\gamma l} + V^-e^{+\gamma l} \tag{7.8a}$$

$$I(l) = I^+e^{-\gamma l} - I^-e^{+\gamma l} \tag{7.8b}$$

The two terms in each solution represent travelling waves along positive l and negative l directions respectively. Here, V^+ (I^+) and V^- (I^-) are the amplitudes of voltage (current) waves travelling in positive and negative l drection respectively. The characteristic impedance Z_o is the ratio of voltage and current for an infinitely long transmission line. In case of an infinitely long transmission line, there will be no reflecting waves. Thus, using (7.4a) and (7.8a), the characteristic impedance can be written as

$$Z_o = \frac{V^+}{I^+} = \frac{(R + j\omega L)}{\gamma} = \sqrt{\frac{R + j\omega L}{G + j\omega C}} \tag{7.9}$$

For lossless transmission lines, $R = 0$ and $G = 0$. Putting these values, for lossless transmission lines, the characteristic impedance can be expressed as

$$Z_o = \sqrt{L/C} \tag{7.10}$$

When a signal is fed to a transmission line of infinite length, there will be no reflection of the signal. The voltage and current ratios at any location and time on the line give the same impedance as if the signal generator is terminated by an impedance of Z_o. This impedance is called characteristic impedance of the transmission line. When a transmission line of finite length is terminated by its characteristic impedance (Z_o), then there also will be no reflection of the signal and the line is called matched transmission line. In case of lossless transmission line, the characteristic impedance is purely resistive and is denoted as R_o.

In practice, a transmission line may be terminated by any load, i.e. other than its characteristic impedance. In this case, when a wave travelling from the generator reaches the load impedance, the wave (V^+) is incident on it and some of the power gets absorbed in the load. Some wave $\left(V^-\right)$ is reflected from the load as backward wave, as shown in Fig. 7.1b. The load $Z_L$is connected in the line and l is along the transmission line. Origin of the axis is chosen at the load point, i.e. at l =0. As the generator impedance is matched to the characteristic impedance of the line (Z_o), the reflected wave that reaches by the generator is absorbed by the generator impedance. At steady-state condition, under sinusoidal excitation of angular frequency ω, the voltage and current waves along the lossless transmission line at l is given by

$$v(l,t) = Re\left[V(l)e^{i\omega t}\right] \tag{7.11}$$

$$i(l,t) = Re\left[I(l)e^{i\omega t}\right] \tag{7.12}$$

where $V(l)$ and $I(l)$ can be expressed as

$$V(l) = V^+e^{-j\beta l} + V^-e^{+j\beta l} \tag{7.13}$$

$$I(l) = \frac{V^+}{Z_o}e^{-j\beta l} - \frac{V^-}{Z_o}e^{+j\beta l} \tag{7.14}$$

Here, β is the phase constant and is given by $\beta = \omega/c = 2\pi/\lambda$. Ratio of the reflected wave amplitude and incident wave amplitude is called reflection coefficient (Γ) and is given by

$$\Gamma = \frac{V^-}{V^+} \tag{7.15}$$

Thus, in terms of the reflection coefficient, the voltage and current waves along the lossless transmission line at l is given by

$$V(l) = V^+e^{-j\beta l}\left(1 + \Gamma e^{+j2\beta l}\right) \tag{7.16}$$

$$I(l) = \frac{V^+}{Z_o}e^{-j\beta l}\left(1 - \Gamma e^{+j2\beta l}\right) \tag{7.17}$$

At the load end, $V(0) = Z_L I(0)$. Therefore,

$$Z_L = Z_o\frac{1+\Gamma}{1-\Gamma} \tag{7.18}$$

Thus, the reflection coefficient for a load impedance of Z_L of the transmission line of characteristic impedance Z_o is given by

$$\Gamma = \frac{Z_L - Z_o}{Z_L + Z_o} \tag{7.19}$$

In microwave frequency, it is very convenient to express load impedance by reflection coefficient involving incident and reflected waves instead of expressing it by the ratio of voltage and current as it is done for low-frequency circuits.

The reflection coefficient including its phase term can be written as

$$\Gamma(l) = \frac{V^-(z)}{V^+(z)} = \frac{V^- e^{+j\beta l}}{V^+ e^{-j\beta l}} = \Gamma e^{+j2\beta l} \tag{7.20}$$

Therefore, under sinusoidal excitation of angular frequency ω, the steady-state voltage and current waveforms can be written as

$$v(l,t) = Re\left[V^+ e^{-j\beta l}(1+\Gamma(l))e^{j\omega t}\right] \tag{7.21}$$

$$i(l,t) = Re\left[\frac{V^+}{Z_o} e^{-j\beta l}(1-\Gamma(l))e^{j\omega t}\right] \tag{7.22}$$

These expressions give the amplitude variation of voltage and current waveforms along the transmission line. These variations are due to the generation of standing wave for the load impedance other than its characteristic impedance.

Figures 7.2 and 7.3 show the normalized amplitude of voltage and current waves along the transmission line for different values of load impedances. It is clear from the figures that amplitude of the voltage waveform changes along the transmission line due to the generation of standing wave. Comparing (7.12) with (7.11), it is also clear that the position of the voltage maximum will be the position of the current minimum along the transmission line. For example, in case of open load condition, there will be maximum voltage amplitude and minimum current amplitude at the load end, whereas there will be minimum voltage amplitude and maximum current amplitude at the shorted load end.

The ratio of the maximum voltage amplitude ($V_{\max}$) to the minimum voltage amplitude $\left(V_{\min}\right)$ is defined as voltage standing wave ratio (VSWR), i.e.

$$\text{VSWR} = \frac{V_{\max}}{V_{\min}} = \frac{1+|\Gamma|}{1-|\Gamma|} \tag{7.23}$$

Therefore, when transmission lines are not terminated by its characteristic impedance, variation of voltage and current amplitudes leads to the variation of impedance along the transmission line. The input impedance $Z_{\text{in}}(l)$ of a transmission line is the impedance looking from one end of the transmission line and is given by

$$Z_{\text{in}}(l) = \frac{V(l)}{I(l)} = Z_o \frac{1+\Gamma(l)}{1-\Gamma(l)} \tag{7.24}$$

Putting the value of $\Gamma(l)$ from (7.20), the input impedance becomes

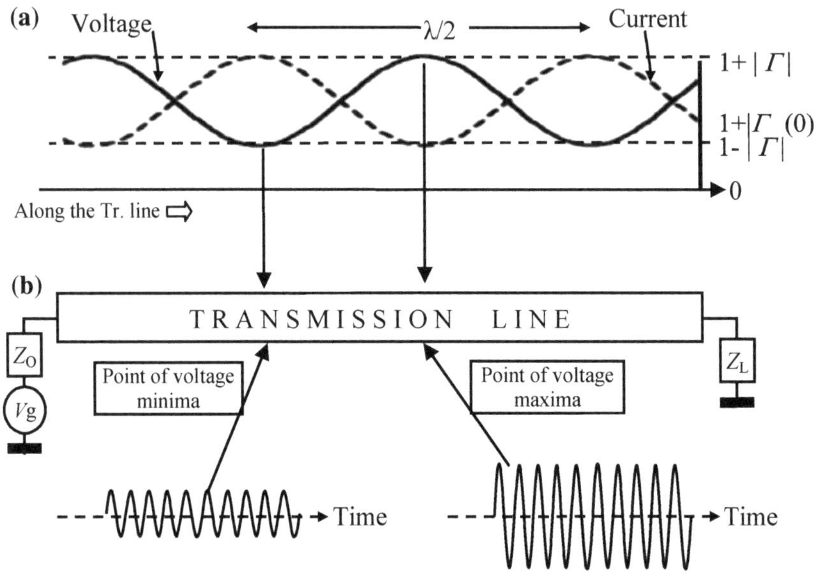

Fig. 7.2 Normalized voltage and current wave along the transmission line with load impedance $Z_L \neq Z_o$ **a** amplitude along the transmission line, **b** time domain voltage waveform at minima and maxima on the transmission line

$$Z_{\text{in}}(l) = Z_o \frac{1 + \Gamma e^{+j2\beta l}}{1 - \Gamma e^{+j2\beta l}} = Z_o \frac{Z_L - jZ_o \tan(\beta l)}{Z_o - jZ_L \tan(\beta l)} \tag{7.25}$$

Therefore, the input impedance of a transmission line at a distance l from the load end towards the generator (i.e. at the source end) terminated with Z_L can be derived by putting $l = -l$ in (7.25) as given by

$$Z_{\text{in}} = Z_o \frac{Z_L + jZ_o \tan(\beta l)}{Z_o + jZ_L \tan(\beta l)} \tag{7.26}$$

This is the most widely used formula for microwave circuit design and analysis using transmission lines.

From (7.25), the impedance of a transmission line at a distance l towards the generator from the load impedance of reflection coefficient Γ_L can be written as

$$Z_{\text{in}} = Z_{\text{in}}(-l) = Z_o \frac{1 + \Gamma_L e^{-j2\beta l}}{1 - \Gamma_L e^{+j2\beta l}} = Z_o \frac{1 + |\Gamma_L| e^{j(\phi - 2\beta l)}}{1 - |\Gamma_L| e^{j(\phi - 2\beta l)}} \tag{7.27}$$

Here, ϕ is the phase of the reflection coefficient at the load end (i.e. load reflection coefficient). This (7.27) gives the variation of impedance along a transmission line when we move from the load end towards generator. Suppose l_{max} is the distance

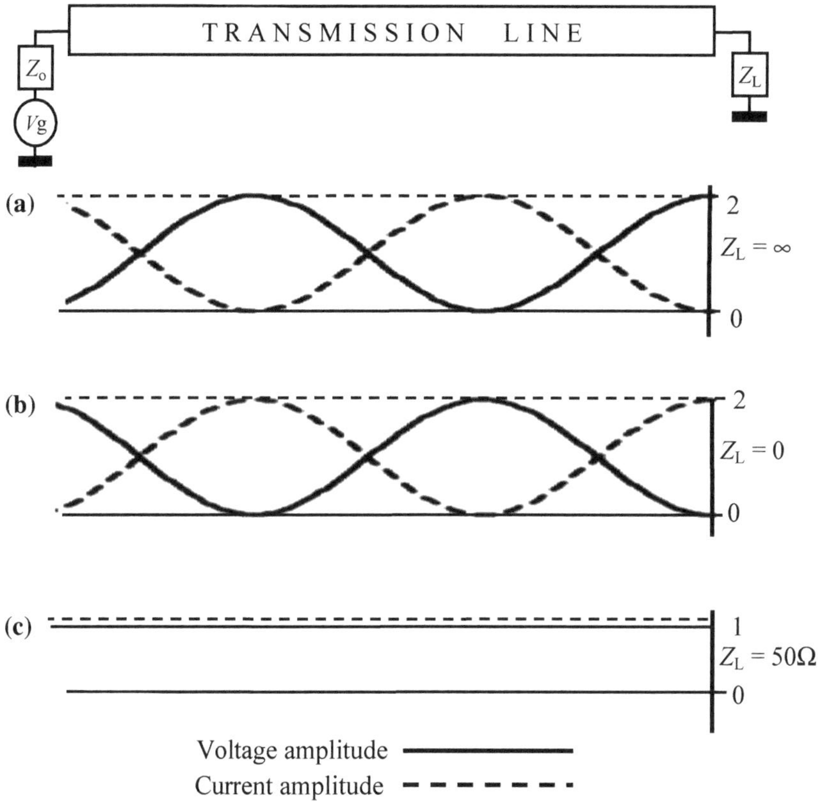

Fig. 7.3 Normalized voltage wave along the transmission line with load impedance **a** Z_L = open, **b** Z_L = short and **c** $Z_L = Z_o$

from the load end towards generator where the impedance is maximum, i.e. voltage maximum and current minimum point. Then,

$$2\beta l_{\max} = \phi \tag{7.28}$$

Therefore, the voltage minima point $l_{\min}$ from the load end will be

$$2\beta l_{\min} = \phi + \pi \tag{7.29}$$

From (7.27), the impedance at the voltage maxima point is given by:

$$Z_{\max} = Z_o \frac{1 + |\Gamma_L|}{1 - |\Gamma_L|} \tag{7.30a}$$

$$= Z_o \times \text{VSWR} \tag{7.30b}$$

This is a pure resistance. Similarly, the impedance at the voltage minimum point can be written as

$$Z_{\min} = Z_o \frac{1 - |\Gamma_L|}{1 + |\Gamma_L|} \tag{7.31a}$$

$$= Z_o/\text{VSWR} \tag{7.31b}$$

7.3 Microwave Transmission Lines

Low insertion loss, negligible signal distortion, high transmitted power capability, wide operation frequency band, low electromagnetic interference (EMI) and proper electromagnetic compatibility (EMC) are the requirements for any transmission line. In addition to these, easy and low-cost production, and easy integration of passive and active devices & components are also demanded from a good transmission line.

Waveguides and coaxial lines transmit electromagnetic waves totally confined by metallic walls, so both satisfy all the above-mentioned requirements except easy integration of the components, whereas open transmission lines such as microstrip lines, striplines, coplanar waveguides, etc., where waves propagate along the line and its close vicinity have less power transmitting capability, higher insertion loss, poorer EMI/EMC and higher coupling to neighbouring circuitry.

7.3.1 Losses in Transmission Lines

The propagation constant for a transmission line is given by (7.6a)

$$\begin{aligned}\gamma &= \alpha + j\beta = \sqrt{(R + j\omega L)(G + j\omega C)}\\ &= j\omega\sqrt{LC}\left(1 + \frac{R}{j\omega L}\right)^{1/2}\left(1 + \frac{G}{j\omega C}\right)^{1/2}\\ &= j\omega\sqrt{LC}\left(1 + \frac{R}{j2\omega L} - \frac{R^2}{j^2 8\omega^2 L^2} + \cdots\right)\left(1 + \frac{G}{j2\omega C} - \frac{G^2}{j^2 8\omega^2 C^2} + \cdots\right)\end{aligned}$$

For a low-loss transmission line, neglecting higher order terms,

$$\gamma = \alpha + j\beta = j\omega\sqrt{LC}\left[1 + \frac{1}{j2\omega}\left(\frac{R}{L} + \frac{G}{C}\right) + \left(\frac{R^2}{8\omega^2 L^2} + \frac{G^2}{8\omega^2 C^2}\right) - \frac{RG}{4\omega^2 LC}\right]$$

Equating the real terms from both the sides,

$$\alpha = \sqrt{LC}\left[\frac{1}{2}\left(\frac{R}{L} + \frac{G}{C}\right)\right] \tag{7.32a}$$

Putting, $Z_o = 1/Y_o = \sqrt{L/C}$ in (7.32a),

$$\alpha = \frac{R}{2Z_o} + \frac{G}{2Y_o} = \alpha_c + \alpha_d \tag{7.32b}$$

The first term (α_c) of (7.32b) is related to the series equivalent resistance of the transmission line, this corresponds to the conductor or resistive loss of the line. Resistive losses within a transmission line arise from the resistance of the conductors. When an electromagnetic wave flows through a conductor, amplitude of the field decreases with the depth of the conductor. Thus, most of the currents flow through skin of the conductor. The skin depth (δ) is defined as the distance at which the field falls $1/e$ of its value at its surface where the wave enters. The skin depth (δ) is given by

$$\delta = \sqrt{\frac{2}{\mu_o \omega \sigma}}, \quad \sigma \text{ is conductivity of the conductor} \tag{7.33}$$

Thus, the actual area through which the current flows through the conductor is limited by the skin effect. As frequency $\omega = 2\pi f$ increases, area through which current flows decreases and thus the resistive loss increases.

Example 7.1 Conductivity of silver is 3×10^7 S/m. Calculate its skin depth at 10 GHz.

Solution

$$\mu_o = 4\pi \times 10^{-7} \text{ H/m}$$
$$\sigma = 3 \times 10^7 \text{ S/m}$$
$$\omega = 2\pi \times 10^{10} \text{ Hz}$$

From (7.33), the skin depth is given by

$$\delta = \sqrt{\frac{2}{\mu_o \omega \sigma}} = \sqrt{\frac{2}{(4\pi \times 10^{-7})(2\pi \times 10^{10})(3 \times 10^7)}} \text{ m}$$
$$= 0.92 \times 10^{-6} \text{ m}$$
$$= 0.92\ \mu\text{m}$$

This example shows that at microwave frequency, the skin depth for a good conductor is very small. Thus, the difference microwave performance (loss) between a component using pure good conductor and a component using bad conductor plated with good conductor would be expected to be negligible. This enables the use of silver- or gold-plated low-cost microwave components such as silver- or gold-plated waveguides and planar transmission lines instead of using fully gold- or silver-based costly component, without compromising their RF performance.

The second term (α_d) of (7.32b) is related to the shunt conductance (G) of the transmission line, and this corresponds to the loss within the dielectric support material of the transmission line. Within the dielectric material, displacement current cannot leads the field strength by $\pi/2$ due to the polarization delay. This leads to the appearance of an active displacement current component which is in phase with the field strength. Under this condition, the permittivity can be expressed as

$$\varepsilon_r = \varepsilon_r' - j\varepsilon_r'' \tag{7.34}$$

This leads to the power dissipation within the dielectric material under alternate electromagnetic field. The power dissipation per unit volume within the dielectric material is given by

$$P_d = \omega\varepsilon_o\varepsilon_r' E^2 \tan(\delta) \tag{7.35}$$

where $\delta = \varepsilon_r''/\varepsilon_r'$ and E is the electric field. This (7.35) shows that the dielectric loss increases linearly with frequency, whereas due to the decrease of skin depth, the resistive loss increases with square root of the frequency. Therefore, at higher operating frequency, the dielectric loss dominates over resistive loss.

Radiation is other source of transmission line losses of electromagnetic energy when it propagates through a transmission line. Though it is negligible at lower range of microwave frequency, it becomes significant at higher microwave frequency range. Transmission line losses due to radiation are generally of not much concern for waveguide, coaxial line, stripline and grounded coplanar waveguide due to their closed configurations. In all these configurations, the signal path is surrounded by some form of grounded conductors, whereas in microstrip line configurations, the signal line is open to environment. Thus, microstrip line configurations are more prone to radiation loss and there are several EMI-related issues related to the radiation.

Mismatch loss is another source of transmission losses of electromagnetic energy. When a transmission line is terminated to a load of impedance different from its characteristic impedance, then some of the incident power is reflected back from the load. The amount of power that is not delivered to the load due to the impedance mismatch is the mismatch loss. The mismatch loss is defined as the power delivered to the load to the power available from the transmission line network. With reference to Fig. 7.4, the mismatch loss is given by

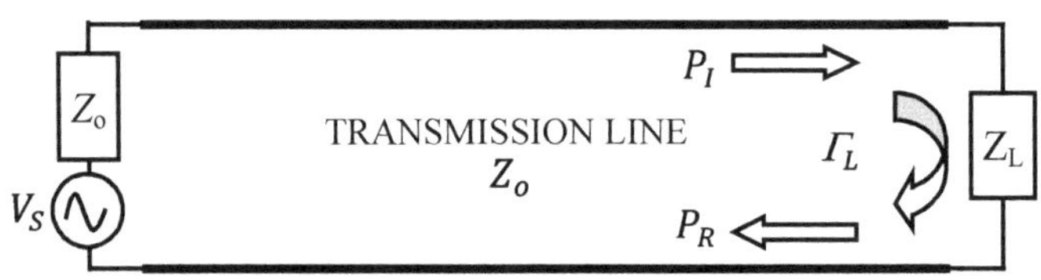

Fig. 7.4 Mismatch loss in transmission line

$$L_{\text{MISMATCH}} = \frac{P_I - P_R}{P_I} \tag{7.36a}$$

$$= 1 - \frac{P_R}{P_I} \tag{7.36b}$$

Here, P_I is the power incident to the load and P_R is the power reflected from the load. The (7.36b) can be written as

$$L_{\text{MISMATCH}} = 1 - \Gamma_L^2 \tag{7.36c}$$

Here, Γ_L is the load reflection coefficient and is given by $\Gamma_L = P_R/P_I$. Thus, in decibel, the mismatch loss is given by

$$(L_{\text{MISMATCH}})_{\text{dB}} = 10\log(1 - \Gamma_L^2) \tag{7.36d}$$

7.3.2 Coaxial Transmission Lines

A coaxial transmission line consists of a centre conductor with another conductor around it and the space in between the conductors is filled with dielectric material as shown in Fig. 7.5. The principal mode of transmission that propagates along coaxial transmission lines is TEM since it is a two-wire transmission line. The useful frequency range of a coaxial transmission line is restricted to the principal mode of operation which is below the cut-off frequency of first higher order mode. Maximum power handling of a coaxial line is set by the dielectric breakdown and heat dissipation within the line due to attenuation losses contributed by resistive and dielectric losses. The inductance and capacitance per unit length of a coaxial transmission line can be calculated considering the entire current flow through the surface of the conductor as follows.

Suppose $2a$ and $2b$ are the diameters of inner and outer conductors of a coaxial transmission line as shown in Fig. 7.5. The electric and magnetic field distributions of TEM wave propagating through a coaxial line is as shown in Fig. 7.6a. The electric field is in the radial direction and varies as $1/r$ and has no variation in the ϕ direction, whereas the magnetic field is in the ϕ direction and varies as $1/r$ in the radial direction. Suppose current I flows through the line. Therefore, the magnetic flux B everywhere is tangent to a circle centred on the centre conductor as shown in Fig. 7.6b. Therefore, at a distance x from the centre of the centre conductor, we can write

$$\oint B.dl = 2\pi x B \tag{7.37a}$$

Using Ampere's circuital law, we can write

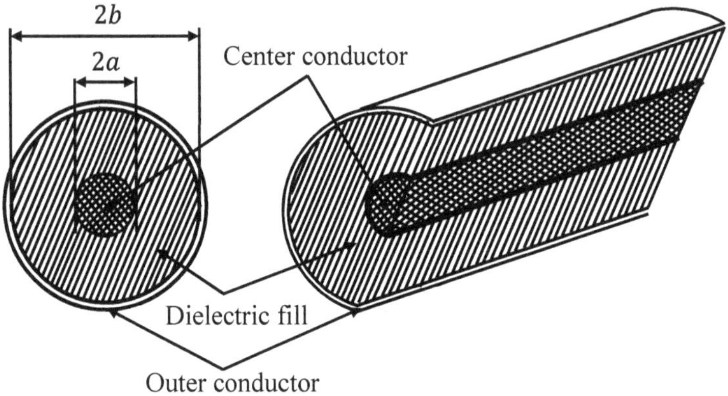

Fig. 7.5 Cross section of coaxial transmission line

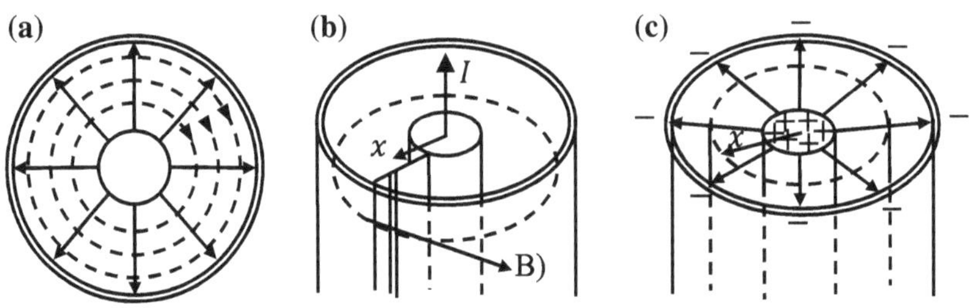

Fig. 7.6 **a** Field distribution for TEM wave of a coaxial line, **b** diagram to determine *L* and **c** diagram to determine *C*

$$\oint B.dl = 2\pi x B = \mu_o \mu_r I \tag{7.37b}$$

Here, μ_o is permiability in vacuum and μ_r is relative permeability of the medium. Thus,

$$B = \frac{\mu_o \mu_r I}{2\pi x} \tag{7.37c}$$

Thus, magnetic flux dΦ at a distance x from the centre of the conductors through a strip of thickness dx and length l is given by

$$\mathrm{d}\Phi = Bl\mathrm{d}x = \frac{\mu_o \mu_r I}{2\pi x} l\mathrm{d}x \tag{7.37d}$$

Suppose inductance per unit length is L; therefore, inductance of the line of length l is Ll. Thus, total magnetic flux is given by

$$\Phi = \int_a^b \frac{\mu_o \mu_r I}{2\pi x} l \mathrm{d}x = \frac{\mu_o \mu_r I l}{2\pi} \ln\left(\frac{b}{a}\right) = LlI \tag{7.37e}$$

Thus, inductance per unit length is given by

$$L = \frac{\mu_o \mu_r}{2\pi} \ln\left(\frac{b}{a}\right) \tag{7.37f}$$

The capacitance per unit length for a coaxial line can be obtained by evaluating the voltage difference between the conductors for a given charge on each. Suppose ξ is the charge density per unit length of the conductor and E is the electric field at a distance x as shown in Fig. 7.6b. The flux through the Gaussian surface, as shown in Fig. 7.6b, is $E \times 2\pi x l$. The total charge along the length l is $l\xi$. By applying Gauss' law,

$$E \times 2\pi x l = l\xi/\varepsilon_r \varepsilon_o \tag{7.38a}$$

Thus, electric field at a distance x is given by

$$E = \frac{\xi}{2\pi \varepsilon_r \varepsilon_o} \frac{1}{x} \tag{7.38b}$$

The voltage between the cylinders can be obtained by integrating the electric field along a radial line as shown below:

$$\Delta V = \frac{\xi}{2\pi \varepsilon_r \varepsilon_o} \int_a^b \frac{1}{x} \mathrm{d}x = \frac{\xi}{2\pi \varepsilon_r \varepsilon_o} \ln\left(\frac{b}{a}\right) \tag{7.38c}$$

Suppose equivalent capacitance per unit length is C. Therefore, the total capacitance Cl is given by

$$Cl = \frac{\xi l}{\Delta V} = \xi l \Big/ \left[\frac{\xi}{2\pi \varepsilon_r \varepsilon_o} \ln\left(\frac{b}{a}\right)\right] \tag{7.38d}$$

or capacitance per unit length is given by

$$C = \frac{2\pi \varepsilon_r \varepsilon_o}{\ln\left(\frac{b}{a}\right)} \tag{7.38e}$$

Thus, for a lossless coaxial transmission line, the characteristic impedance can be written from (7.10), (7.37f) and (7.38e) as

$$Z_o = \sqrt{\frac{L}{C}} = \frac{1}{2\pi} \sqrt{\frac{\mu_o \mu_r}{\varepsilon_o \varepsilon_r}} \ln\left(\frac{b}{a}\right) \tag{7.39a}$$

Putting $\mu_o = 4\pi \times 10^{-7}$ H/m and $\varepsilon_o = 8.854 \times 10^{-12}$ F/m

$$Z_o = \frac{377}{2\pi}\sqrt{\frac{\mu_r}{\varepsilon_r}}\ln\left(\frac{b}{a}\right)\ \Omega \tag{7.39b}$$

$$Z_o = 138\sqrt{\frac{\mu_r}{\varepsilon_r}}\log_{10}\left(\frac{b}{a}\right)\ \Omega \tag{7.39c}$$

Example 7.2 Inner and outer diameters of an air–dielectric coaxial cable are 0.05 and 0.15 in., respectively. Calculate its characteristic impedance.

Solution

Characteristic impedance is given by (7.39c)

$$Z_o = 138\sqrt{\frac{\mu_r}{\varepsilon_r}}\log_{10}\left(\frac{b}{a}\right)\ \Omega$$

Here, $a = 0.05$ in., $b = 0.15$ in., $\mu_r = \varepsilon_r = 1$. Thus,

$$\begin{aligned} Z_o &= 138 \times \log_{10}\left(\frac{0.15}{0.05}\right)\ \Omega \\ &= 138 \times 0.477\ \Omega \\ &= 65.8\ \Omega \end{aligned}$$

In a coaxial transmission line, the space between the two conductors can support higher order modes, which is similar to waveguide mode of propagation. The approximate cut-off wavelength (λ_c) for the first higher order waveguide mode is given by

$$\lambda_c = \pi(a + b)\sqrt{\varepsilon_r} \tag{7.40a}$$

Thus, the first higher order cut-off frequency (f_c) is given by

$$f_c = \frac{c}{\pi(a + b)\sqrt{\varepsilon_r}}, \quad c \text{ is speed of light in vacuum} \tag{7.40b}$$

$$= \frac{c/a}{\pi(1 + b/a)\sqrt{\varepsilon_r}} \tag{7.40c}$$

A coaxial line is useful only below its cut-off frequency of the first higher order mode. The cut-off frequency is determined by the size (a) of the coaxial transmission line for a given dielectric constant (ε_r) and characteristic impedance (b/a) of the line.

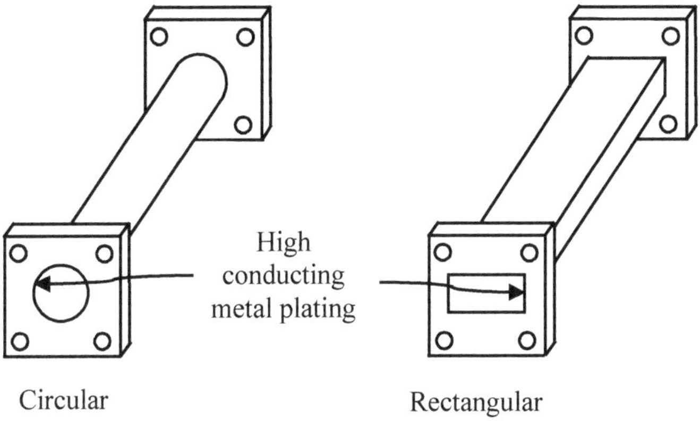

Fig. 7.7 Waveguide structures with circular and rectangular cross section

7.3.3 Waveguides

Literally, a transmission line which guides electromagnetic (EM) waves permitting its propagation along it, confining its energy within it or in vicinity of it can be considered as waveguide. Under this definition, coaxial line, twin-wire line, microstrip line, fibre optic cables, hollow metal pipe, all are waveguides. However, in microwave engineering, a hollow metal (conductor) pipe of rectangular, circular or elliptic cross section used to guide and carry EM waves from one place to another with minimum loss of energy is known as waveguide. Structure of such waveguides is shown in Fig. 7.7. The waveguides are superior transmission lines in terms of lower attenuation and higher power handling capability compared to coaxial, microstrip and stripline transmission lines. For these superior performances, waveguides are popularly used for high-power and low-loss applications in microwave and mmwave frequency range.

Waveguides can be analysed using the following Maxwell's electromagnetic equations in terms of electric $(\overline{E})$ and magnetic $(\overline{H})$ fields inside the waveguides [1, 2].

$$\nabla \times \overline{H} = \overline{J} + \frac{\partial \overline{D}}{\partial t} \quad (7.41a)$$

$$\nabla \times \overline{E} = -\frac{\partial \overline{B}}{\partial t} \quad (7.41b)$$

$$\nabla \cdot \overline{D} = \rho \quad (7.41c)$$

$$\nabla \cdot \overline{B} = 0 \quad (7.41d)$$

Here, $\overline{D} = \varepsilon \overline{E}$ and $\overline{B} = \mu \overline{H}$ are electric and magnetic flux densities, respectively, constants ε and μ are permittivity and permeability, respectively, of the medium through which EM wave propagates, $\overline{J}$ and ρ are the electric current density and electric charge density, respectively. In case of waveguide, considering perfect dielectric medium containing no charge and no conduction currents, we can write, $\overline{J} = 0$ and $\rho = 0$ in (7.41).

Taking curl ($\nabla\times$) of (7.41b) and substituting (7.41a), under the condition of $\overline{J} = 0$, we can write,

$$\nabla \times \left(\nabla \times \overline{E}\right) = -\nabla \times \left(\frac{\partial \overline{B}}{\partial t}\right) = -\mu\varepsilon \frac{\partial^2 \overline{E}}{\partial t^2} \tag{7.42}$$

Using the vector identity, $\nabla \times \left(\nabla \times \overline{E}\right) = \nabla\nabla \cdot \overline{E} - \nabla^2 \overline{E}$, (7.42) can be written as

$$\nabla\nabla \cdot \overline{E} - \nabla^2 \overline{E} = -\mu\varepsilon \frac{\partial^2 \overline{E}}{\partial t^2} \tag{7.43}$$

Here, $\rho = 0$. Thus, from (7.41c), $\nabla \cdot \overline{D} = 0$ implies $\nabla \cdot \overline{E} = 0$. Thus, (7.43) can be written as

$$\nabla^2 \overline{E} = \mu\varepsilon \frac{\partial^2 \overline{E}}{\partial t^2} \tag{7.44a}$$

This is the wave equation for the electric field $\overline{E}$. Similar wave equation for the magnetic field can be derived and written as

$$\nabla^2 \overline{H} = \mu\varepsilon \frac{\partial^2 \overline{H}}{\partial t^2} \tag{7.44b}$$

The time-varying electric and magnetic fields can be expressed as $\overline{E}(r, t) = E(r)\mathrm{e}^{j\omega t}$ and $\overline{H}(r, t) = H(r)\mathrm{e}^{j\omega t}$. Thus, wave equations can be expressed as

$$\nabla^2 \overline{E} = -\omega^2 \mu\varepsilon \overline{E} \tag{7.45a}$$

$$\nabla^2 \overline{H} = -\omega^2 \mu\varepsilon \overline{H} \tag{7.45b}$$

Analysis of a waveguide can be done using these wave equations putting appropriate boundary conditions corresponding to the waveguide. The boundary conditions for the waveguide walls are the tangential component of electric field $\overline{E}$ should be zero and the normal component of the magnetic field $\overline{H}$ should be zero. Consider a rectangular waveguide in a coordinate system as shown in Fig. 7.8, and in general always $a > b$. Suppose the waveguide is invariant in the z direction and the wave is propagating in the positive z direction as $\mathrm{e}^{-\gamma z}$ (in $-z$ direction can also be considered), here, $\gamma = \alpha + j\beta$. Equation (7.41b) can be written as

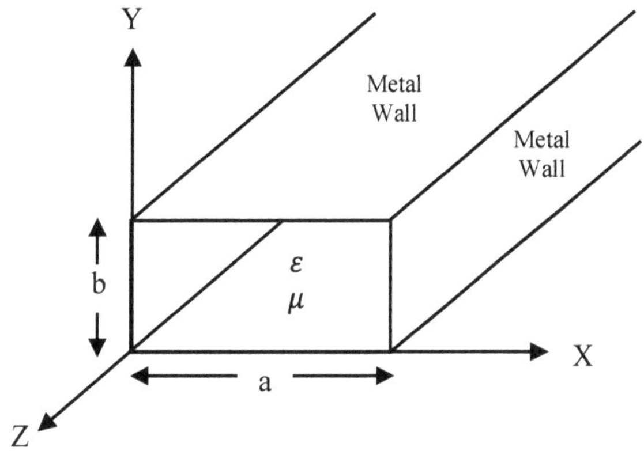

Fig. 7.8 Cross section of a rectangular waveguide of infinite length

$$\begin{bmatrix} \hat{x} & \hat{y} & \hat{z} \\ \frac{\partial}{\partial x} & \frac{\partial}{\partial y} & \frac{\partial}{\partial z} \\ E_x & E_y & E_z \end{bmatrix} = -j\omega\mu\left(\hat{x}H_x + \hat{y}H_y + \hat{z}H_z\right) \tag{7.46}$$

Equating the x, y, z components from left and right sides,

$$\frac{\partial E_z}{\partial y} - \frac{\partial E_y}{\partial z} = -j\omega\mu H_x \tag{7.47a}$$

$$\frac{\partial E_z}{\partial x} - \frac{\partial E_x}{\partial z} = -j\omega\mu H_y \tag{7.47b}$$

$$\frac{\partial E_y}{\partial x} - \frac{\partial E_x}{\partial y} = -j\omega\mu H_z \tag{7.47c}$$

Considering spatial variation of the wave in z direction as $\mathrm{e}^{-j\omega z}$:

$$\frac{\partial\left(\mathrm{e}^{-\gamma z}\right)}{\partial z} = -\gamma \mathrm{e}^{-\gamma z}$$

Equations (7.47) can be simplified to

$$\frac{\partial E_z}{\partial y} + \gamma E_y = -j\omega\mu H_x \tag{7.48a}$$

$$\frac{\partial E_z}{\partial x} + \gamma E_x = j\omega\mu H_y \tag{7.48b}$$

$$\frac{\partial E_y}{\partial x} - \frac{\partial E_x}{\partial y} = -j\omega\mu H_z \tag{7.48c}$$

Similarly, following expressions can be derived from Eq. (7.41a)

$$\frac{\partial H_z}{\partial y} + \gamma H_y = j\omega\varepsilon E_x \tag{7.49a}$$

$$\frac{\partial H_z}{\partial x} + \gamma H_x = -j\omega\varepsilon E_y \tag{7.49b}$$

$$\frac{\partial H_y}{\partial x} - \frac{\partial H_x}{\partial y} = j\omega\varepsilon E_z \tag{7.49c}$$

Combining (7.48a) and (7.49b) and putting

$$k_c^2 = \gamma^2 + \omega^2\mu\varepsilon \tag{7.50}$$

The x component of magnetic field (H_x) can be expressed as

$$H_x = -\frac{\gamma}{k_c^2}\frac{\partial H_z}{\partial x} + j\frac{\omega\varepsilon}{k_c^2}\frac{\partial E_z}{\partial y} \tag{7.51}$$

Similarly, combining (7.48b) and (7.49a) y component of magnetic field (H_y) can be expressed as

$$H_y = -\frac{\gamma}{k_c^2}\frac{\partial H_z}{\partial y} - j\frac{\omega\varepsilon}{k_c^2}\frac{\partial E_z}{\partial x} \tag{7.52a}$$

Combining (7.49a) and (7.48b) x component of electric field (E_x) can be expressed as

$$E_x = -\frac{\gamma}{k_c^2}\frac{\partial E_z}{\partial x} - j\frac{\omega\mu}{k_c^2}\frac{\partial H_z}{\partial y} \tag{7.52b}$$

Similarly, combining (7.49b) and (7.48a) y component of electric field (E_y) can be expressed as

$$E_y = -\frac{\gamma}{k_c^2}\frac{\partial E_z}{\partial y} + j\frac{\omega\mu}{k_c^2}\frac{\partial H_z}{\partial x} \tag{7.52c}$$

Equations (7.52a) to (7.52c) show very interesting and important characteristics of waves transmitting through waveguide. These equations show that the x and y component of magnetic fields (H_x, H_y) and electric fields (E_x, E_y) can be expressed in terms of only z component of electric field (E_z) and magnetic field (H_z). In short, all the transverse components can be determined from only the axial components. This finding implies that, if E_z and H_z both are zero, all the fields within the waveguide vanish. To support wave propagation through a waveguide, either an electric field along the direction of propagation (E_z) or a magnetic field along the direction of propagation (H_z) must exist.

Depending on the existence of either electric field or magnetic field along the direction of propagation, there are two modes of propagation in a waveguide. One is transverse magnetic (TM) mode of wave propagation, for which magnetic field along the wave propagation is zero, i.e. $H_z = 0$. Another is transverse electric (TE) mode of wave propagation, for which electric field along the wave propagation is zero, i.e. $E_z = 0$. In general, the propagation of EM wave through a waveguide may be a combination of both these types of waves.

7.3.3.1 Transverse Magnetic (TM) Wave

Characteristics of the transverse magnetic (TM) waves propagating along a rectangular waveguide can be derived by putting appropriate boundary conditions in the wave equations.

For TM wave, magnetic field component along the direction of propagation is zero. Thus, $H_z = 0$. The wave equation for the electric field (7.44a) along the direction of propagation (z) can be written as

$$\frac{\partial^2 E_z}{\partial x^2} + \frac{\partial^2 E_z}{\partial y^2} + \gamma^2 E_z = -\omega^2 \mu \varepsilon E_z \tag{7.53a}$$

This equation is to be solved under the following boundary conditions, for the rectangular waveguide as shown in Fig. 7.8:

$$E_z = 0, \quad \text{at} \quad x = 0, x = a, y = 0 \ \text{ and } \ y = b \tag{7.53b}$$

The solution of the partial differential equation (7.53a) can be written as

$$E_z = (C_1 \cos Ax + C_2 \sin Ax)(C_3 \cos By + C_4 \sin By)\mathrm{e}^{-\gamma z} \tag{7.54}$$

Here, $k_c^2 = A^2 + B^2$, the constants C_1, C_2, C_3, C_4, A and B can be selected to satisfy the boundary conditions (7.53b) as follows:

$$C_1 = 0, \quad C_3 = 0, \quad A = \frac{m\pi}{a} \ \text{ and } \ B = \frac{n\pi}{b} \tag{7.55}$$

Putting the values of constants from (7.55) to Eq. (7.54), the electric vector of TM wave along the propagation can be written as replacing $C_2 C_4 = C$,

$$E_z = C \sin\left(\frac{m\pi}{a}x\right) \sin\left(\frac{n\pi}{b}y\right)\mathrm{e}^{-\gamma z} \tag{7.56a}$$

Other electric and magnetic field components can be written from (7.52a) to (7.52c) as follows:

$$E_x = -C\frac{\gamma}{k_c^2}\frac{m\pi}{a} \cos\left(\frac{m\pi}{a}x\right) \sin\left(\frac{n\pi}{b}y\right)\mathrm{e}^{-\gamma z} \tag{7.56b}$$

$$E_y = -C\frac{\gamma}{k_c^2}\frac{n\pi}{b} \sin\left(\frac{m\pi}{a}x\right) \cos\left(\frac{n\pi}{b}y\right)\mathrm{e}^{-\gamma z} \tag{7.56c}$$

$$H_x = jC\frac{\omega\varepsilon}{k_c^2}\frac{n\pi}{b}\sin\left(\frac{m\pi}{a}x\right)\cos\left(\frac{n\pi}{b}y\right)e^{-\gamma z} \tag{7.56d}$$

$$H_y = -jC\frac{\omega\varepsilon}{k_c^2}\frac{m\pi}{a}\cos\left(\frac{m\pi}{a}x\right)\sin\left(\frac{n\pi}{b}y\right)e^{-\gamma z} \tag{7.56e}$$

The set of equations from (7.56a) to (7.56e) determines the electric and magnetic field variations of transverse magnetic (TM) wave in a rectangular waveguide propagating in z direction.

The constants m and n can have any integer value. Thus, multiple solutions for the electric and magnetic fields of the TM wave propagation in a waveguide are possible. These are called mode of propagation. These transverse magnetic modes of propagation are designated as TM_{mn} modes. From (7.56a), for either $m = 0$ or $n = 0$, the electric field vector of the wave along the direction of propagation becomes zero, i.e. $E_z = 0$. This implies that there is no existence of TM waves for either $m = 0$ or $n = 0$. Minimum possible values for m and n are 1 and the mode is designated as TM_{11}.

7.3.3.2 Transverse Electric (TE) Wave

Characteristics of the transverse electric (TE) waves propagating along a rectangular waveguide can be derived using the wave equations by putting appropriate boundary conditions.

For TE waves, the electric field component along the direction of propagation is zero. Thus, $E_z = 0$. The wave equation for the magnetic field (7.44b) along the direction of propagation (z) can be written as

$$\frac{\partial^2 H_z}{\partial x^2} + \frac{\partial^2 H_z}{\partial y^2} + \gamma^2 H_z = -\omega^2\mu\varepsilon H_z \tag{7.57a}$$

This equation is to be solved under the following boundary conditions for the rectangular waveguide as shown in Fig. 7.8:

$$\begin{aligned} \frac{\partial H_z}{\partial x} &= 0, \quad \text{at} \quad x = 0, \quad x = a \text{ and} \\ \frac{\partial H_z}{\partial y} &= 0 \quad \text{at} \quad y = 0, \quad y = b \end{aligned} \tag{7.57b}$$

The solution of the partial differential equation (7.57a) can be written as

$$E_z = (C_1\cos Ax + C_2\sin Ax)(C_3\cos By + C_4\sin By)e^{-\gamma z} \tag{7.58}$$

Here, $k_c^2 = A^2 + B^2$, the constants C_1, C_2, C_3, C_4, A and B can be selected to satisfy the boundary conditions (7.57b) as follows:

$$C_2 = 0, \quad C_4 = 0, \quad A = \frac{m\pi}{a} \quad \text{and} \quad B = \frac{n\pi}{b} \tag{7.59}$$

Putting the values of constants from (7.59) to Eq. (7.58), the magnetic field vector of TE wave along the propagation can be written as, replacing $C_1C_3 = C$:

$$H_z = C\cos\left(\frac{m\pi}{a}x\right)\cos\left(\frac{n\pi}{b}y\right)e^{-\gamma z} \tag{7.60a}$$

Other electric and magnetic field components can be written from (7.52a) to (7.52d) as follows:

$$E_x = jC\frac{\omega\mu}{k_c^2}\frac{n\pi}{b}\cos\left(\frac{m\pi}{a}x\right)\sin\left(\frac{n\pi}{b}y\right)e^{-\gamma z} \tag{7.60b}$$

$$E_y = -jC\frac{\omega\mu}{k_c^2}\frac{m\pi}{a}\sin\left(\frac{m\pi}{a}x\right)\cos\left(\frac{n\pi}{b}y\right)e^{-\gamma z} \tag{7.60c}$$

$$H_x = C\frac{\gamma}{k_c^2}\frac{m\pi}{a}\sin\left(\frac{m\pi}{a}x\right)\cos\left(\frac{n\pi}{b}y\right)e^{-\gamma z} \tag{7.60d}$$

$$H_y = C\frac{\gamma}{k_c^2}\frac{n\pi}{b}\cos\left(\frac{m\pi}{a}x\right)\sin\left(\frac{n\pi}{b}y\right)e^{-\gamma z} \tag{7.60e}$$

The set of equations from (7.60a) to (7.60e) determines the electric and magnetic field variations of transverse electric (TE) wave in a rectangular waveguide propagating in z direction. The constants m and n can have any integer value. Thus, multiple solutions for the electric and magnetic fields of the TE wave propagation in a waveguide are possible. From (7.60a), it is clear that for TE_{mn} waves, it is possible to make either m or n equal to zero (however, both m and n cannot be zero simultaneously) without resulting all fields to vanish. Therefore, possible modes for TE_{mn} are TE_{10}, TE_{01}, TE_{11}, TE_{12}, TE_{21}, etc.

7.3.4 Cut-Off Frequency and Guide Wavelength

Putting $k_c^2 = A^2 + B^2$ in (7.50),

$$\gamma^2 = A^2 + B^2 - \omega^2\mu\varepsilon \tag{7.61}$$

Values of A and B are the same for TM_{mn} mode (7.55) and TE_{mn} mode (7.59). Putting these values of A and B, and $\gamma = \alpha + j\beta$ in (7.61),

$$\gamma = \alpha + j\beta = \sqrt{\left(\frac{m\pi}{a}\right)^2 + \left(\frac{n\pi}{b}\right)^2 - \omega^2\mu\varepsilon} \tag{7.62}$$

Equation (7.62) determines the propagation constant for TM_{mn} and TE_{mn} modes in a rectangular waveguide. At low frequencies, where

$$\left(\frac{m\pi}{a}\right)^2 + \left(\frac{n\pi}{b}\right)^2 > \omega^2\mu\varepsilon$$

γ is real, implies β, the propagation phase constant is zero, i.e. $\gamma = \alpha$, where α is the attenuation constant. In this condition, there is no transmission of EM waves at these frequencies. At the frequencies, where

$$\left(\frac{m\pi}{a}\right)^2 + \left(\frac{n\pi}{b}\right)^2 < \omega^2\mu\varepsilon$$

γ is imaginary. It implies α, the attenuation constant is zero. The propagation constant $\gamma = j\beta$. At these frequencies, the EM wave transmission will be without any attenuation (in case of no conductor loss and no dielectric loss in the waveguide).

The frequency below which the EM wave propagation does not take place is called cut-off frequency. The cut-off frequency ω_c is given by

$$\omega_c = \frac{1}{\sqrt{\mu\varepsilon}}\sqrt{\left(\frac{m\pi}{a}\right)^2 + \left(\frac{n\pi}{b}\right)^2} \tag{7.63}$$

The frequencies above the cut-off, the propagation constant is imaginary and is given by

$$j\beta = j\sqrt{\omega^2\mu\varepsilon - \left(\frac{m\pi}{a}\right)^2 + \left(\frac{n\pi}{b}\right)^2} \tag{7.64a}$$

Putting cut-off frequency from (7.63), the phase constant can be written as

$$\beta = \sqrt{\omega^2\mu\varepsilon - \omega_c^2\mu\varepsilon} \tag{7.64b}$$

Therefore, the phase velocity $v_p = (\omega/\beta)$, can be written as

$$v_p = (\omega/\beta) = \frac{1}{\sqrt{\mu\varepsilon}}\frac{1}{\sqrt{1-\omega_c^2/\omega^2}} = \frac{c}{\sqrt{\mu_r\varepsilon_r}\sqrt{1-\omega_c^2/\omega^2}} \tag{7.65}$$

where $\mu = \mu_o\mu_r$, $\varepsilon = \varepsilon_o\varepsilon_r$ and $c = 1/\sqrt{\mu_o\varepsilon_o}$ is the wave velocity of EM waves in vacuum (unguided).

The expression (7.65) of phase velocity in waveguide indicates that the propagation velocity is greater than the velocity (c) of EM wave in free space. With the increase of frequency (ω) above the cut-off frequency (ω_c), the phase velocity (v_p) decreases from very large value towards the velocity (c) in free space.

Wavelength of the wave propagating within the waveguide is called guide wavelength λ_g and can be written from (7.64) as

$$\lambda_g = \frac{2\pi}{\beta} = \frac{2\pi}{\sqrt{\omega^2\mu\varepsilon - \omega_c^2\mu\varepsilon}} \tag{7.66a}$$

$$= \frac{2\pi}{\sqrt{\omega^2\mu\varepsilon - \omega_c^2\mu\varepsilon}} \tag{7.66b}$$

$$= \frac{c}{f\sqrt{\mu_r\varepsilon_r}}\frac{1}{\sqrt{1-(f_c/f)^2}} \tag{7.66c}$$

$$= \frac{\lambda_o}{\sqrt{\mu_r\varepsilon_r}\sqrt{1-(\lambda_o/\lambda_c)^2}} \tag{7.66d}$$

where $\lambda_o = c/f$ is wavelength in vacuum for the frequency f.

Example 7.3 Derive expression of cut-off frequency, guide wavelength and phase velocity of TM_{11} mode in a rectangular waveguide of dimensions a and b.

Solution

Putting $m = n = 1$ in (7.63), the cut-off frequency of TM_{11} mode for a rectangular waveguide of dimension (a, b) can be written as follows:

$$f_{c(TM_{11})} = \frac{1}{2\pi\sqrt{\mu\varepsilon}}\sqrt{\left(\frac{\pi}{a}\right)^2 + \left(\frac{\pi}{b}\right)^2} \tag{7.67a}$$

The guide wavelength of TM_{11} mode for a rectangular waveguide of dimension (a, b) can be written from (7.66d) as

$$\lambda_g = \frac{c}{f\sqrt{\mu_r\varepsilon_r}}\frac{1}{\sqrt{1-\left(f_{c(TM_{11})}/f\right)^2}} \tag{7.67b}$$

Here, f is the operating frequency. The phase velocity can be written from (7.65) as

$$v_p = \frac{1}{\sqrt{\mu_r\varepsilon_r}}\frac{c}{\sqrt{1-\left(f_{c(TM_{11})}/f\right)^2}} \tag{7.67c}$$

Example 7.4 Derive expression of cut-off frequency, guide wavelength and phase velocity of TE_{10} mode in a rectangular waveguide of dimensions a and b.

Solution

Putting $m = 1$ and $n = 0$ in (7.63), the cut-off frequency of TE_{10} mode for a rectangular waveguide of dimension (a, b) can be written as follows (Fig. 7.9):

$$f_{c(TE_{10})} = \frac{1}{2\pi\sqrt{\mu\varepsilon}}\frac{\pi}{a} = \frac{1}{\sqrt{\mu_r\varepsilon_r}}\frac{c}{2a} \tag{7.68a}$$

Comparing (7.68a) with (7.67a),

$$f_{c(TE_{10})} < f_{c(TM_{11})}$$

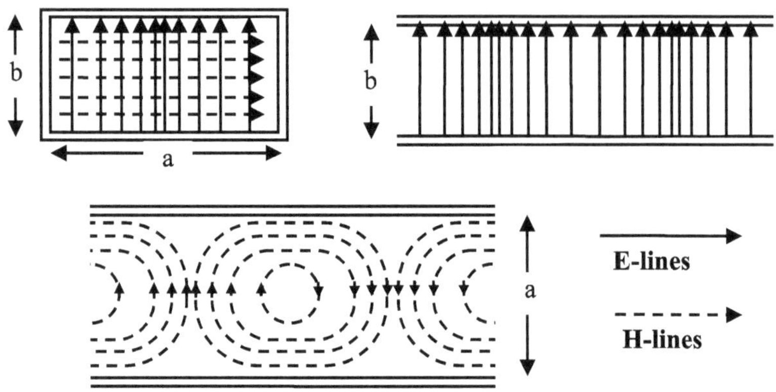

Fig. 7.9 E- and H-field lines of TE_{10} in rectangular waveguide cross section

This implies that the cut-off frequency of TE_{10} is lower than the cut-off frequency of TM_{11} for a rectangular waveguide. Considering $a > b$, it can be concluded that cut-off frequency of TE_{10} is the lowest among all TE_{mn} and TM_{mn} modes in a rectangular waveguide. Thus, TE_{10} is the lowest order mode in a rectangular waveguide and is called dominant mode.

The guide wavelength of TE_{10} mode for a rectangular waveguide of dimension (a, b) can be written as

$$\lambda_g = \frac{c}{f\sqrt{\mu_r \varepsilon_r}} \frac{1}{\sqrt{1 - \left(f_{c(TE_{10})}/f\right)^2}} \tag{7.68b}$$

Here, f is the operating frequency. The phase velocity can be written from (7.65) as

$$v_p = \frac{1}{\sqrt{\mu_r \varepsilon_r}} \frac{c}{\sqrt{1 - \left(f_{c(TE_{10})}/f\right)^2}} \tag{7.68c}$$

Example 7.5 Derive expressions of electric and magnetic fields for TE_{10} waves propagating in a rectangular waveguide of dimensions a and b, and also diagrammatically show the electric and magnetic field lines.

Solution

Putting $m = 1$ and $n = 0$ in (7.60a) to (7.60e), and replacing $k_c^2 = A^2 + B^2 = (\pi/a)^2$, the electric and magnetic field components can be written for TE_{10} mode in a rectangular waveguide as

$$E_x = 0 \tag{7.69a}$$

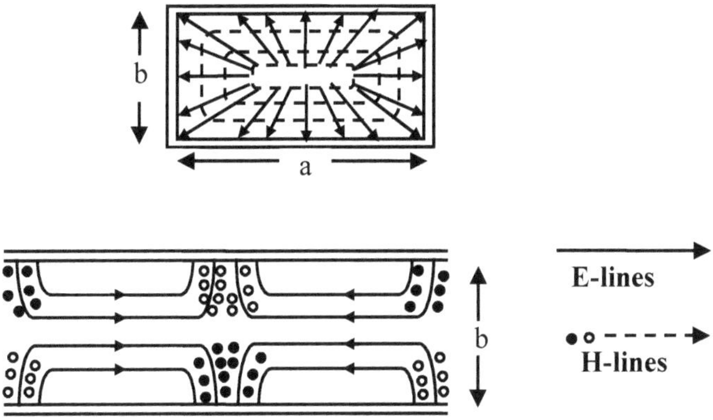

Fig. 7.10 E- and H-field lines of TM_{11} in rectangular waveguide cross section

$$E_y = -jC\frac{\omega\mu a}{\pi}\sin\left(\frac{\pi}{a}x\right)e^{-\gamma z} \tag{7.69b}$$

$$E_z = 0 \tag{7.69c}$$

$$H_x = C\frac{\gamma a}{\pi}\sin\left(\frac{\pi}{a}x\right)e^{-\gamma z} \tag{7.69d}$$

$$H_y = 0 \tag{7.69e}$$

$$H_z = C\cos\left(\frac{\pi}{a}x\right)e^{-\gamma z} \tag{7.69f}$$

Based on these expressions, the electric and magnetic field lines are shown in different cross sections of a rectangular waveguide in Fig. 7.9.

Example 7.6 Derive expressions of electric and magnetic fields for TM_{11} waves propagating in a rectangular waveguide of dimensions a and b. Also diagrammatically show the electric and magnetic field lines.

Solution
Putting $m = 1$ and $n = 1$ in (7.56a) to (7.56e), the electric and magnetic field components for TM_{110} mode can be written in a rectangular waveguide as

$$\lambda E_x = -C\frac{\gamma}{k_c^2}\frac{\pi}{a}\cos\left(\frac{\pi}{a}x\right)\sin\left(\frac{\pi}{b}y\right)e^{-\gamma z} \tag{7.70a}$$

$$E_y = -C\frac{\gamma}{k_c^2}\frac{\pi}{b}\sin\left(\frac{\pi}{a}x\right)\cos\left(\frac{\pi}{b}y\right)e^{-\gamma z} \tag{7.70b}$$

$$E_z = C\sin\left(\frac{\pi}{a}x\right)\sin\left(\frac{\pi}{b}y\right)e^{-\gamma z} \tag{7.70c}$$

$$H_x = jC\frac{\omega\varepsilon}{k_c^2}\frac{\pi}{b}\sin\left(\frac{\pi}{a}x\right)\cos\left(\frac{\pi}{b}y\right)e^{-\gamma z} \tag{7.70d}$$

$$H_y = -jC\frac{\omega\varepsilon}{k_c^2}\frac{\pi}{a}\cos\left(\frac{\pi}{a}x\right)\sin\left(\frac{\pi}{b}y\right)e^{-\gamma z} \tag{7.70e}$$

$$H_z = 0 \tag{7.70f}$$

Here, $k_c^2 = A^2 + B^2 = (\pi/a)^2 + (\pi/b)^2$.

Based on these expressions, the electric and magnetic field lines are shown in different cross sections of a rectangular waveguide in Fig. 7.10.

Example 7.7 Calculate the cut-off frequencies for different $TE_{10}, TE_{11}, TE_{20}, TE_{21}$ and $TM_{11}, TM_{21}, TM_{12}$ modes in a rectangular waveguide of dimension $a = 10$ mm $b = 5$ mm, and filled with dielectric of $\mu_r = 1, \varepsilon_r = 4$.

Solution

The cut-off frequencies for a rectangular waveguide is given by (7.63)

$$\omega_{c(mn)} = \frac{1}{\sqrt{\mu\varepsilon}}\sqrt{\left(\frac{m\pi}{a}\right)^2 + \left(\frac{n\pi}{b}\right)^2}$$
$$= \frac{c\pi}{\sqrt{\mu_r\varepsilon_r}}\sqrt{\left(\frac{m}{a}\right)^2 + \left(\frac{n}{b}\right)^2}$$

Therefore,

$$f_{c(mn)} = \frac{c}{2\sqrt{\mu_r\varepsilon_r}}\sqrt{\left(\frac{m}{a}\right)^2 + \left(\frac{n}{b}\right)^2} \tag{7.71}$$

Putting $\mu_r = 1$, $\varepsilon_r = 4$, $a = 10$ mm $= 1$ cm, $b = 5$ mm $= 0.5$ cm and $c = 3 \times 10^{10}$ cm/s in (7.71),

$$f_{c(mn)} = \frac{3 \times 10^{10}}{2\sqrt{1 \times 4}}\sqrt{\left(\frac{m}{1}\right)^2 + \left(\frac{n}{0.5}\right)^2}\ \text{Hz}$$
$$= \frac{30}{4}\sqrt{m^2 + 4n^2}\ \text{GHz}$$
$$= 7.5 \times \sqrt{m^2 + 4n^2}\ \text{GHz}$$

Putting the values of m, n cut-off frequencies of different modes are shown in the following Table:

Mode	Cut-off frequency (GHz)
TE_{10}	7.5
TE_{01}, TE_{20}	15.0
TE_{11}, TM_{11}	16.77
TE_{21}, TM_{21}	21.21
TE_{12}, TM_{12}	30.92

Example 7.8 Determine all the possible propagating modes in a rectangular waveguide of dimension $a = 10$ mm $b = 5$ mm and filled with dielectric of $\mu_r = 1, \varepsilon_r = 6.0$ operating at 10 GHz.

Solution
The cut-off frequencies for a rectangular waveguide are given by (7.71)

$$f_{c(mn)} = \frac{c}{2\sqrt{\mu_r \varepsilon_r}} \sqrt{\left(\frac{m}{a}\right)^2 + \left(\frac{n}{b}\right)^2}$$

Putting $\mu_r = 1$, $\varepsilon_r = 9.9$, $a = 10$ mm $= 1$ cm, $b = 5$ mm $= 0.5$ cm and $c = 3 \times 10^{10}$ cm/s,

$$\begin{aligned} f_{c(mn)} &= \frac{3 \times 10^{10}}{2\sqrt{1 \times 9.9}} \sqrt{\left(\frac{m}{1}\right)^2 + \left(\frac{n}{0.5}\right)^2} \text{ Hz} \\ &= 4.767 \times \sqrt{m^2 + 4n^2} \text{ GHz} \end{aligned}$$

Putting the values of m, n cut-off frequencies of different modes are shown in the following Table:

Mode	Cut-off frequency (GHz)
TE_{10}	6.123
TE_{01}, TE_{20}	12.247
TE_{11}, TM_{11}	13.693
TE_{21}, TM_{21}	17.321

The table shows that the only mode which has cut-off frequency below the operating frequency (10 GHz) is TE_{10}. Therefore, only TE_{10} mode can propagate through the waveguide at operating frequency 10 GHz.

7.3.5 Planar Transmission Lines

Planar transmission lines have structures that allow controlling its characteristic impedance by controlling its dimension in a single plane. Planar transmission lines consist of multiconductor lines comprising a solid dielectric substrate having single or multilayers of metallization. In the structure, signal and ground currents flow through separate conductors. These structures are very compact and suitable for realization of microwave integrated circuits (MICs) using simpler fabrication process. There are various types of planar transmission line structures. Among various configurations stripline, microstrip line and coplanar waveguide are commonly used for realization of microwave integrated circuits.

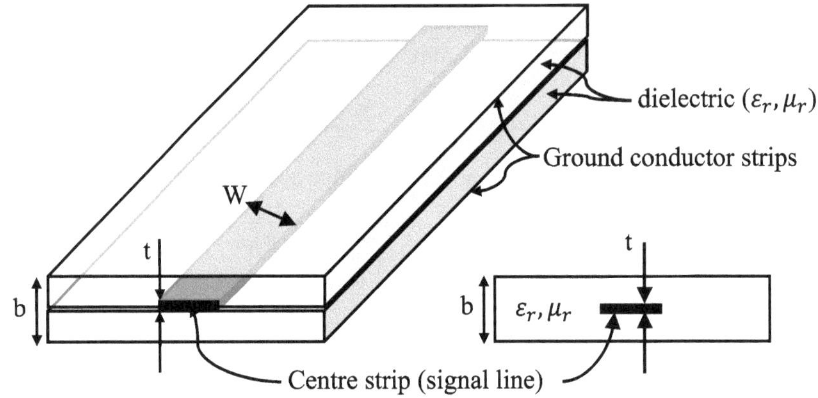

Fig. 7.11 Stripline configuration

7.3.5.1 Stripline

Stripline configuration consists of three flat conductor strips; the centre strip is the signal line that runs in between two ground strips filled with suitable dielectric material. Sketch of stripline configuration is shown in Fig. 7.11. Width (W), thickness (t) of the centre strip, height of the stripline (b) and dielectric constant of the filled dielectric material determine the characteristic impedance of the line.

Stripline supports propagation of TEM mode because the region between the two ground conductor strips contains a single medium. Thus, the phase velocity and the characteristic impedance of a stripline do not vary with frequency. Electric and magnetic field distributions in stripline are shown in Fig. 7.12. Here, electromagnetic field is strongly confined near the centre strip and within the top and bottom ground planes. In a stripline, the return current path for a high-frequency signal will remain directly above and below the signal strip on the ground planes. The high-frequency signal thus remains entirely inside the substrate, minimizing emissions and providing natural shielding against any incoming radiation. The minimum emission leads to negligible radiation and thus the stripline transmission lines have two types of losses, conductor loss and dielectric loss. The dielectric loss is proportional to frequency and dominates at higher operating frequencies. The ohmic skin losses in the strip conductor and ground planes depend on the conductivity of the metal conductors and surface roughness of the conductors.

The characteristic impedance Z_o of the stripline depends on the dielectric constant and cross-sectional geometry of the structure. The characteristic impedance is very sensitive to the ratio of centre strip width W to the dielectric thickness b. However, it is relatively insensitive to the thickness of the centre conductor strip. The approximate characteristic impedance Z_o of a stripline can be expressed in the following simple form:

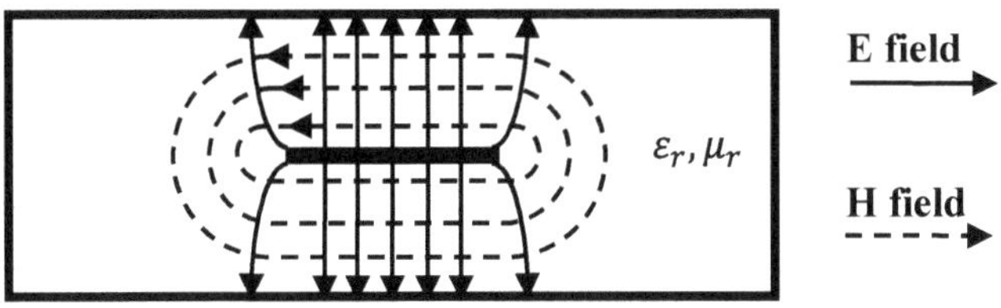

Fig. 7.12 *E*- and *H*-fields distribution in stripline configuration

$$Z_o = \frac{30\pi}{\sqrt{\varepsilon_r}} \frac{b}{W_e + 0.441b} (\text{in } \Omega) \tag{7.72}$$

Here, W_e is the effective width of the centre strip conductor and is given by

$$\begin{aligned} \frac{W_e}{b} &= \frac{W}{b} \quad \text{for} \quad \frac{W}{b} > 0.35 \\ &= \frac{W}{b} - \left(0.35 - \frac{W}{b}\right)^2 \quad \text{for} \quad \frac{W}{b} < 0.35 \end{aligned}$$

It shows that the characteristic impedance Z_o of the stripline decreases with the increase of width W of the strip for fixed substrate height b.

7.3.5.2 Microstrip

The microstrip line is the most widely used planar transmission line for RF and microwave circuits. It consists of a single dielectric substrate with ground plane on one side and a strip on the other side as shown in Fig. 7.13. The widespread use of microstrip line is due to its planar nature, ease of fabrication using various simple processes, easy integration of surface-mount devices (SMD) due to easy access of the top surface, good heat sinking and good mechanical support. Width (W), thickness (t) of the upper strip, height (h) and dielectric constant (ε_r) of the dielectric material determine the characteristic impedance of the line. Microstrip configuration is a strip over a ground plane and thus it tends to radiate due to openness of the upper strip as the spacing between the ground plane and the strip increases. However, a substrate thickness of a few percent of a wavelength (or less) minimizes radiation without restricting width of the strip.

Electric and magnetic field distributions in a microstrip configuration are shown in Fig. 7.14. Here, electromagnetic fields extend in the space above the dielectric. Thus, this configuration becomes a mixed dielectric transmission structure (air and dielectric). To ensure that EM fields are confined near the strip and mostly within the dielectric, it is necessary to use high dielectric constant material as substrate. Another advantage of using high dielectric constant substrate is the reduced circuit dimension due to the reduction of phase velocity.

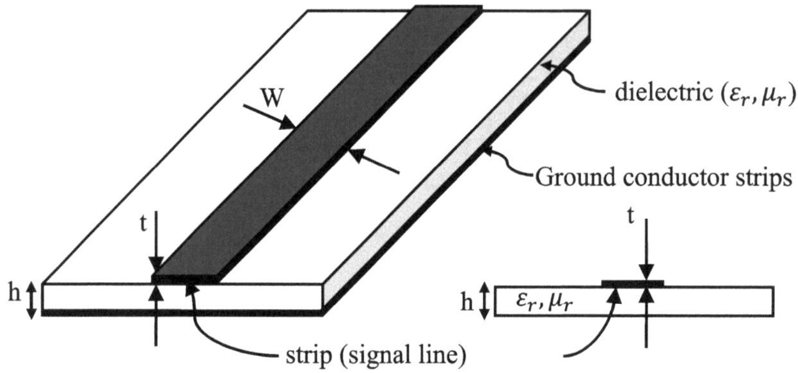

Fig. 7.13 Microstrip line configuration

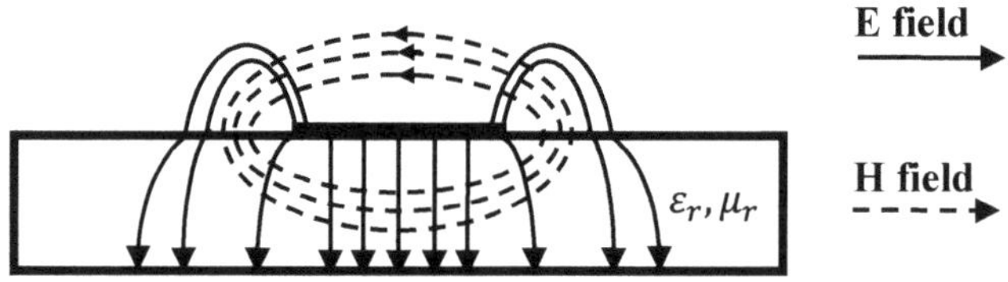

Fig. 7.14 *E*- and *H*-fields distribution in microstrip line configuration

In contrast to stripline, the two-media nature (substrate discontinuity due to the air above upper strip) of microstrip causes its dominant mode to be hybrid (not TEM) called quasi TEM. This fact can be ascertained from Maxwell's equations as follows.

The tangential component of the electric field is continuous at the air–dielectric interface. Thus,

$$(E_{\text{dielectric}})_x = (E_{\text{air}})_x \tag{7.73a}$$

Here, x is tangential to the dielectric–air interface and normal to the conductor strip. Since $\varepsilon_{\text{dielectric}} \neq \varepsilon_{\text{air}}$, the tangential component of displacement density D becomes discontinuous, i.e

$$(D_{\text{dielectric}})_x \neq (D_{\text{air}})_x \tag{7.73b}$$

Using Maxwell's $(\nabla \times H)$ equation and combining with (7.73b), we can write,

$$\left[(\nabla \times H)_x\right]_{\text{dielectric}} \neq \left[(\nabla \times H)_x\right]_{\text{air}} \tag{7.74}$$

If we assume that the mode of EM wave propagation is pure TEM, then $H_z = 0$. Putting this in (7.74), we can write,

$$\left[\frac{\partial H_y}{\partial z}\right]_{\text{dielectric}} \neq \left[\frac{\partial H_y}{\partial z}\right]_{\text{air}} \tag{7.75}$$

This leads to,

$$\left[H_y\right]_{\text{dielectric}} \neq \left[H_y\right]_{\text{air}} \tag{7.76}$$

This inequality violates the required field continuity condition for the normal components of magnetic field. This implies that the assumption $H_z = 0$ is not true. This leads to the conclusion that a pure TEM mode cannot be supported by a microstrip line configuration. However, in most of the practical configurations, major portion of the electromagnetic energy is confined within the dielectric material below the strip and the electric flux crossing the air–dielectric boundary is small. Due to the quasi-TEM mode of propagation, the phase velocity, characteristic impedance and field variation in the guide cross section all become mildly frequency dependent.

Taking the effect of quasi-TEM mode of propagation, concept of effective dielectric constant ε_{eff} is used for simple analysis in design of microstrip line. Practically, part of the electric field is in air ($\varepsilon_r = 1$) and other is in the dielectric (ε_r). Considering the effective dielectric constant ε_{eff}, it is assumed that most of the electric fields are constrained within the substrate, but a fraction of the total energy exists within the air above the strip. The value of ε_{eff} is somewhere in between 1 and ε_r, which is used to calculate capacitance per unit length of the line. With the increase of strip width or decrease of substrate thickness more and more electric lines of forces will be within the dielectric. Thus, ε_{eff} increases with the increase in the strip width and decrease in substrate thickness.

With increasing frequency, the effective dielectric constant gradually increases towards that of the substrate, so that the phase velocity gradually decreases. This is true even with a non-dispersive substrate material. The ε_{eff} changes with the free-space wavelength, λ_o. The dispersion becomes more pronounced with the decreasing ratio of strip width to substrate thickness, (W/h). Dispersion is less pronounced as the strip width becomes relatively wider, and in this condition, the effective dielectric constant, $\varepsilon_{\text{eff}} = \varepsilon_r$.

The effective dielectric constant is given by

$$\varepsilon_{\text{eff}} = \frac{\varepsilon_r + 1}{2} + \frac{\varepsilon_r - 1}{2}\left[\frac{1}{\sqrt{1 + \frac{12h}{W}}} + 0.04\left(1 - \frac{W}{h}\right)^2\right] \quad \text{for} \quad \frac{W}{h} < 1 \tag{7.77}$$

$$\text{Else, } \varepsilon_{\text{eff}} = \frac{\varepsilon_r + 1}{2} + \frac{\varepsilon_r - 1}{2}\left[\frac{1}{\sqrt{1 + \frac{12h}{W}}}\right]$$

whereas characteristic impedance of microstrip lines can be expressed as

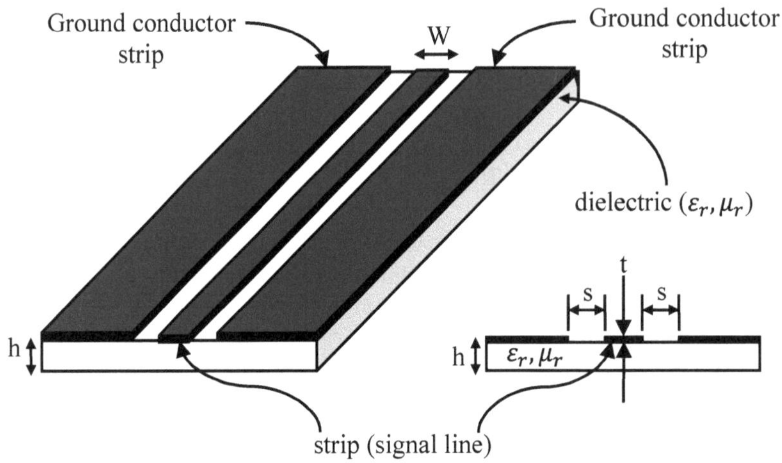

Fig. 7.15 Coplanar waveguide configuration

$$
\begin{aligned}
Z_o &= \frac{60}{\sqrt{\varepsilon_{\text{eff}}}} \ln\left(\frac{8h}{W} + \frac{W}{4h}\right) \Omega \quad \text{when} \quad \frac{W}{h} < 1 \\
&= \frac{120\pi}{\sqrt{\varepsilon_{\text{eff}}}} \frac{1}{\left[\frac{W}{h} + 1.393 + \frac{2}{3}\ln\left(\frac{W}{h} + 1.444\right)\right]} \Omega \quad \text{when} \quad \frac{W}{h} \geq 1
\end{aligned} \tag{7.78}
$$

Practically, the microstrip impedance (Z_o) with finite width ground plane is equal to the impedance value with infinite width ground plane, if the ground width is at least greater than three times of strip width (W).

The sources of attenuation of microstrip lines are due to finite conductivity of the strip conductor, finite resistivity of the substrate and its dumping phenomena, i.e. polarization effect, radiation effects and magnetic loss in case of magnetic substrates such as ferrites.

Microstrip lines with wider strip width have lower current density and thus lower ohmic loss. One way to lower the conductor loss of microstrip line is to suspend the substrate over the air, which increases the width of the strip due to the lower effective dielectric constant. In addition, the insertion loss of the microstrip is reduced because air essentially has no/less dielectric loss compared to standard substrates. However, in this case, the separation between the signal and ground paths increase, which leads to an increase of radiation loss. The radiation loss depends on the dielectric constant, substrate thickness and the circuit geometry. The higher the dielectric constant, the higher the concentration of energy in the substrate region that leads to lesser radiation loss. Another benefit of having a higher dielectric constant is the reduction of circuit size. To minimize radiation loss in a circuit, the number of discontinuities, such as bends and T-junctions, should be made as small as possible. Dielectric loss can be reduced using substrates with a low dielectric loss tangent.

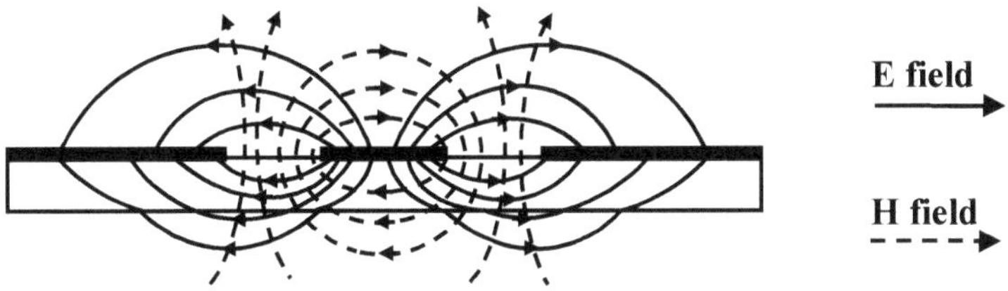

Fig. 7.16 *E*- and *H*-fields distribution in CPW configuration

Power handling capacity of a microstrip line is limited by heating caused by ohmic and dielectric losses and by dielectric breakdown. An increase in temperature due to conductor and dielectric losses limits the average power handling of the microstrip line, whereas the substrate breakdown in between the strip conductor and the ground plane limits the peak-power-handling capability.

Metallic enclosure is used for microwave active as well as passive microstrip circuits such as amplifiers, mixers, filters, etc. to protect them from environment and to provide EM shielding. The presence of conducting (metallic) top and side walls affect both the characteristic impedance, Z_o and the effective dielectric constant, ε_{eff}. To reduce the effect of metallic enclosure for practical circuits, the height of the cover from the circuits should be more than ten times the substrate thickness, and the distance to walls should be more than six times the substrate thickness.

7.3.5.3 Coplanar Waveguide (CPW)

Coplanar waveguide (CPW) consists of a centre strip separated by a narrow gap from two ground planes on either side as shown in Fig. 7.15. Characteristic impedance of a CPW configuration is determined by the dimensions of the centre strip, the gap, the thickness and permittivity of the dielectric substrate. The gap in the CPW configuration is usually very small, about half of the dielectric substrate thickness, supports electric fields primarily concentrating the fields in the substrate area to minimize radiation. CPW can support propagation of EM waves down to zero frequency. The mode of EM wave propagation is quasi TEM for the presence of mixed dielectric medium. Typical electric and magnetic field distribution in a coplanar waveguide is shown in Fig. 7.16.

Two ground planes of the CPW ideally to be maintained at the same potential to prevent propagation of unwanted modes. In case of different potentials, the CPW mode will become uneven, with a higher field in one gap than the other resulting in combined propagation of desired coplanar mode and undesired parasitic slotline mode. To suppress the propagation of undesired slotline mode, air bridges between ground planes have to be applied as shown in Fig. 7.17. Practically, the spacing between the bond wires must be one-quarter wavelength apart or less.

Coplanar waveguides are widely used for high-frequency MMIC applications, for good access to grounding points and lower dispersion characteristic. It provides

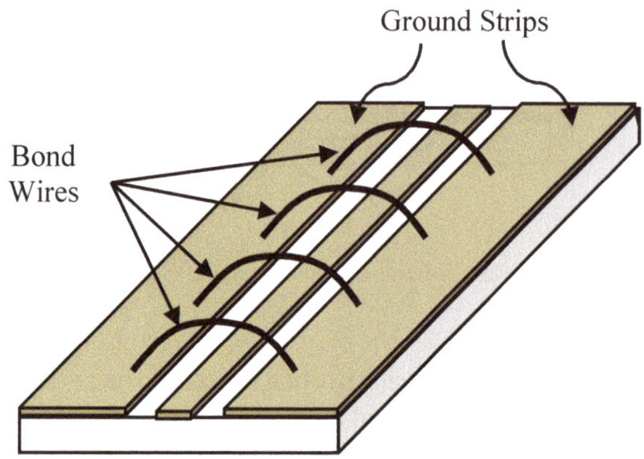

Fig. 7.17 Coplanar waveguide with bond wires

broadband performance as it does not need via holes for shunt and series-connected elements.

7.3.5.4 Substrate Integrated Waveguide (SIW)

Metallic waveguides are having low-loss and high-power handling capability suitable for high-*Q* and high-power component realization and no EM radiation. However, they are bulky, voluminous, expensive and non-planar, so their integration with planar components and packaged circuits is impractical. Whereas planar transmission lines such as microstrip and coplanar waveguide are convenient for integration of surface-mount devices and realization of compact components and packaged circuits but suffers from higher insertion loss and EM radiation. Alternative option is the substrate integrated waveguide (SIW) which shows almost all advantages of metallic waveguides and planar transmission lines [3–7].

Schematic diagram of surface integrated waveguide (SIW) is shown in Fig. 7.18. Series of filled via holes are used for vertical walls. The vertical walls have a mesh structure and are able to reflect all components of electric field. Upper and lower conductive layers are the main conductive layers. The two periodic rows of metallic vias are connecting the top and bottom ground planes of a dielectric substrate. The SIW structure can be manufactured by standard lamination techniques. SIW structure is a promising alternative to conventional all-metal waveguides for the design of microwave/mmwave components and systems. Using SIW structure, relatively low-loss and high-*Q* waveguide structure can be realized as highly integrated planar microwave/mmwave components with compact size, low cost with very low EM radiation.

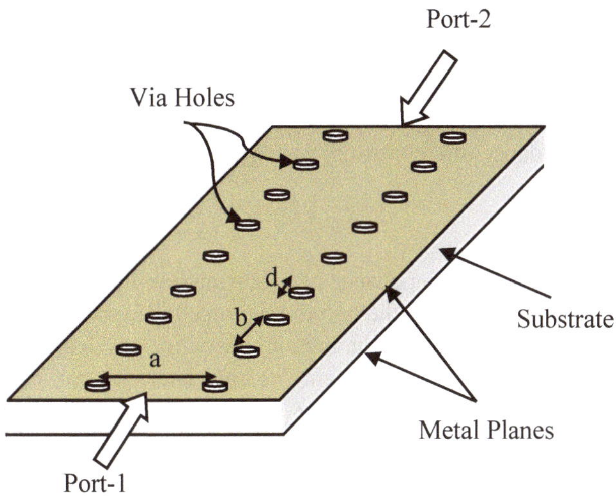

Fig. 7.18 Schematic of substrate integrated waveguide (SIW) structure

7.4 Transmission Line Elements

Transmission line elements with particular termination and particular length are having interesting features and thus frequently used as circuit elements. Some of the examples are given below:

(1) *Short-circuited transmission line* (Fig. 7.19): Putting $Z_L = 0$ in (7.26), the input impedance of a short-circuited transmission line of length l will be

$$Z_{\text{in}}(\text{short}) = jZ_o \tan(\beta l) \tag{7.79}$$

It is inductive for $\beta l < \pi/2$, i.e. $l < \lambda/4$ and capacitive for $\pi/2 < \beta l < \pi$, i.e. $\lambda/4 < l < \lambda/2$. Therefore, a short-circuited transmission line can be used as a capacitive or inductive element in a circuit.

(2) *Open-circuited transmission line* (Fig. 7.20): Putting $Z_L = \propto$ in (7.26), the input impedance of an open-circuited transmission line of length l will be

$$Z_{\text{in}}(\text{open}) = -jZ_o \cot(\beta l) \tag{7.80}$$

Fig. 7.19 Short-circuited transmission line

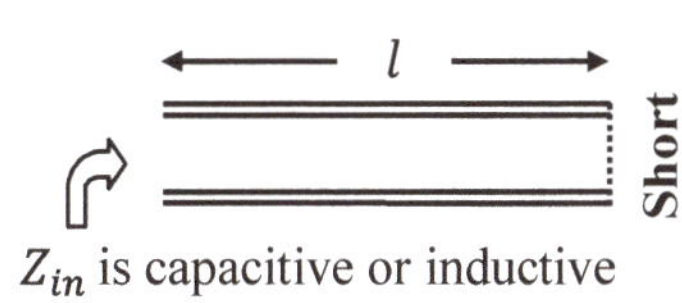

It is capacitive for $\beta l < \pi/2$, i.e. $l < \lambda/4$ and inductive for $\pi/2 < \beta l < \pi$, i.e. $\lambda/4 < l < \lambda/2$. Therefore, an open-circuited transmission line can also be used as a capacitive or an inductive element in a circuit.

(3) *Half-wavelength transmission line section* (Fig. 7.21): Putting $l = \lambda/2$ in (7.26), the input impedance of the transmission line,

$$Z_{\text{in}}(\lambda/2) = Z_L \tag{7.81}$$

It implies that the input impedance of a transmission line of half wavelength is the load impedance. Due to the one-to-one impedance transformer property of the half-wavelength transmission line, it is used for physical spacing between two parts of a circuit without changing its impedance at a particular frequency.

(4) *Quarter-wavelength transmission line section* (Fig. 7.22): Putting $l = \lambda/4$ in (7.26), the input impedance of the transmission line,

$$Z_{\text{in}}(\lambda/4) = \frac{Z_o^2}{Z_L} \tag{7.82}$$

It implies that the input impedance is resistive if the load impedance is resistive but with different value. This property of the quarter-wavelength transmission line is used as impedance transformer to transform resistive impedance to another resistive impedance of different value. For $Z_L = 0$, the input impedance, $Z_{\text{in}} = \propto$. This implies that the input impedance of a quarter-wavelength short-circuited transmission line tends to infinity. This property is used to realize an RF choke for providing DC bias to a microwave circuit. Similarly, for $Z_L = \propto$, the input impedance, $Z_{\text{in}} = 0$. This implies that the input impedance of a quarter-wavelength open-circuited transmission line tends to zero. This property is used to provide short-circuit termination for a particular frequency. Thus, open-circuit $\lambda/4$ stub transmission line can act as band-stop filter.

(5) *One-eighth wavelength transmission line section* (Fig. 7.23): Putting $l = \lambda/8$ in (7.26), the input impedance of the transmission line,

$$Z_{\text{in}}\left(\frac{\lambda}{8}\right) = \frac{Z_o(Z_L + jZ_o)}{Z_o + jZ_L} \tag{7.83}$$

It implies that if Z_L is real, then $|Z_{\text{in}}| = Z_o$, i.e. independent of the value of load impedance Z_L at a particular frequency. However, other than this particular frequency, the input impedance is dependent on the load impedance. This

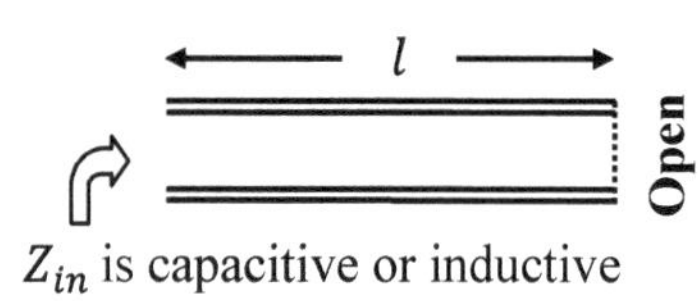

Fig. 7.20 Open circuitted transmission line

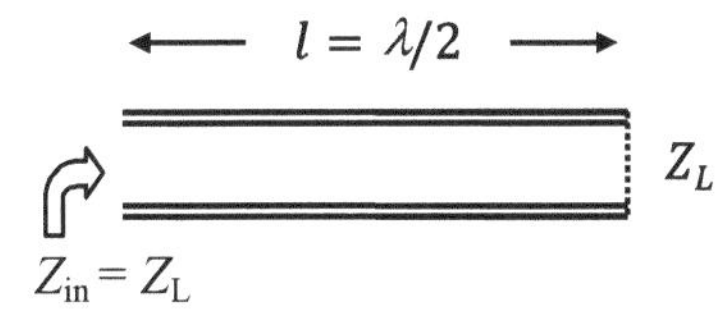

Fig. 7.21 Half-wavelength transmission line

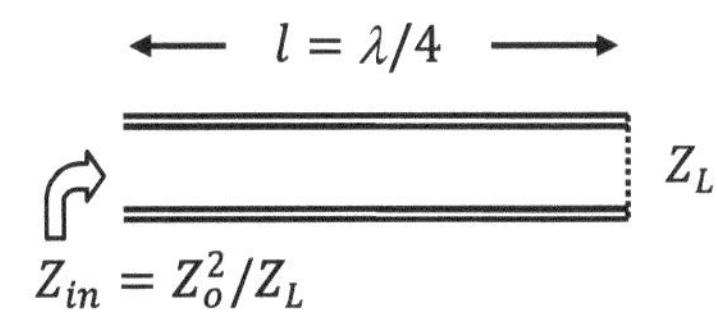

Fig. 7.22 Quarter-wavelength transmission line

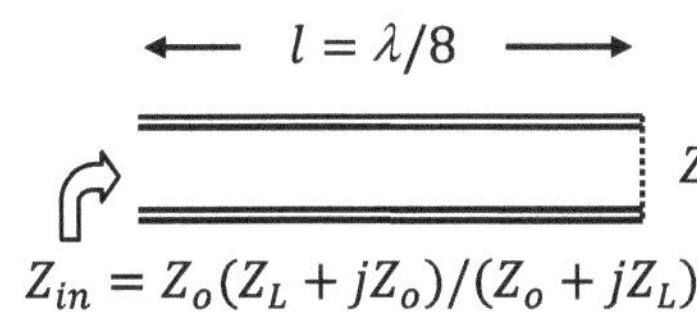

Fig. 7.23 One-eighth wavelength transmission line

property is used to realize amplitude tilt equalizer. In a particular case, when Z_L is complex and $Z_o = |Z_L|$, then the input impedance Z_{in} becomes real. This property of the one-eighth wavelength transmission line section is used to transform complex impedance to real impedance satisfying $Z_o = |Z_L|$.

7.5 Smith Chart Analysis

Smith chart is one of the most useful graphical tools for high-frequency circuit design and analysis. The Smith chart is a graphical representation of complex impedance/admittance on a complex reflection coefficient plane. It is a graph showing both the normalized impedance and corresponding reflection coefficient in a single plane. Smith chart is a simple and very convenient tool for transmission line and circuit analysis. It is also a practical tool to design impedance matching circuit at microwave frequency.

Complex impedance Z of a transmission line can be written as

$$Z(l) = R + jX \tag{7.84}$$

Suppose the characteristic impedance of the line is Z_o. The normalized line impedance $z(l)$ with respect to the reference impedance Z_o can be written as

$$z(l) = (R + jX)/Z_o = r + jx \tag{7.85}$$

The reflection coefficient $\Gamma(l)$ of the line for a system of reference impedance Z_o can be written as

$$Z(l) = Z_o \frac{1 + \Gamma(l)}{1 - \Gamma(l)} \tag{7.86}$$

or

$$z(l) = \frac{Z(l)}{Z_o} = \frac{1 + \Gamma(l)}{1 - \Gamma(l)} \tag{7.87}$$

Suppose

$$\Gamma(l) = \Gamma_r + j\Gamma_{im} \tag{7.88}$$

Combining (7.85), (7.87) and (7.88),

$$r + jx = \frac{1 + \Gamma_r + j\Gamma_{im}}{1 - \Gamma_r - j\Gamma_{im}} \tag{7.89}$$

$$= \frac{1 - \Gamma_r^2 - \Gamma_{im}^2 + j2\Gamma_{im}}{(1 - \Gamma_r)^2 + \Gamma_{im}^2} \tag{7.90}$$

Equating real and imaginary parts separately,

$$r = \frac{1 - \Gamma_r^2 - \Gamma_{im}^2}{(1 - \Gamma_r)^2 + \Gamma_{im}^2} \tag{7.91a}$$

$$x = \frac{2\Gamma_{im}}{(1 - \Gamma_r)^2 + \Gamma_{im}^2} \tag{7.91b}$$

(7.91a) can be written as

$$\left(\Gamma_r - \frac{r}{1 + r}\right)^2 + \Gamma_{im}^2 = \left(\frac{1}{1 + r}\right)^2 \tag{7.92}$$

The result (7.92) for the real part indicates that on the complex plane with coordinates (Γ_r, Γ_{im}), all the possible impedances with a given normalized resistance r are found on a circle with

$$\text{Center:} \quad \left(\frac{r}{1 + r}, 0\right) \tag{7.93a}$$

and

$$\text{Radius:} \quad \left(\frac{1}{1 + r}\right) \tag{7.93b}$$

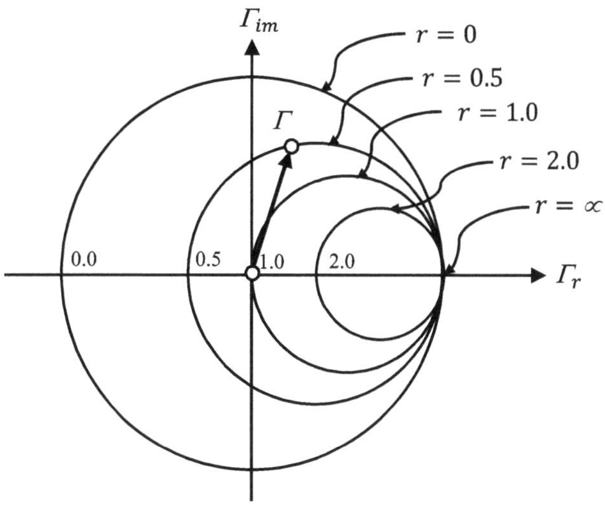

Fig. 7.24 Constant resistance (r) circles on complex Γ plane

Figure 7.24 shows the plots of (7.92) for different constant values of normalized resistance r. These circles are the constant resistance circles. There are many other circles which can be drawn for different values of the constant resistance r. Boundary circle corresponds to $r = 0$. With the increase of resistance values, circles move towards right and the large value circles collapse towards $r = \propto$. As the normalized resistance r varies from 0 to $\propto$, a family of circles completely contained inside the domain of the reflection coefficient $|\Gamma| \leq 1$.

Similarly, (7.91b) can be written as

$$(\Gamma_r - 1)^2 + \left(\Gamma_{im} - \frac{1}{x}\right)^2 = \left(\frac{1}{x}\right)^2 \tag{7.94}$$

The result (7.94) for the imaginary part indicates that on the complex plane with coordinates (Γ_r, Γ_{im}), all the possible impedances with a given normalized reactance x are found on a circle with

$$\text{Center: } \left(1, \frac{1}{x}\right) \tag{7.95a}$$

and

$$\text{Radius: } \left(\frac{1}{x}\right) \tag{7.95b}$$

Figure 7.25 shows the plots of (7.95) for different constant values of normalized reactance x. These are constant reactance arcs. The positive imaginary components

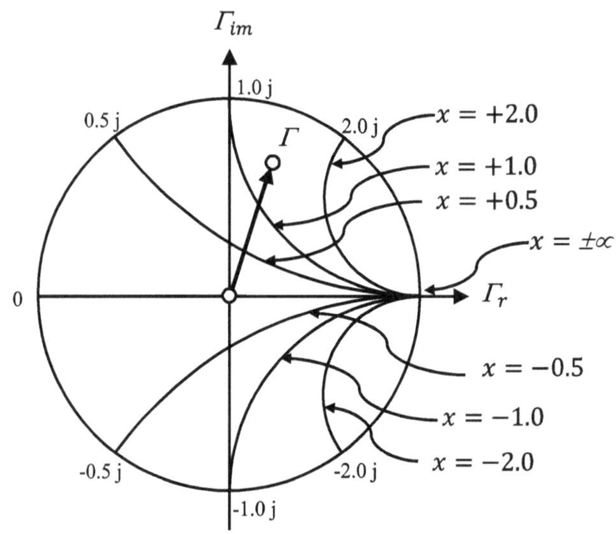

Fig. 7.25 Constant reactance (x) circles on complex Γ plane

indicate that the impedance is inductive and negative imaginary values indicate that the impedance is capacitive. There are many other such arcs which can be drawn for different values of reactances. The horizontal axis itself is a constant reactance arc of infinite radius for $x = 0$. As in case of constant resistance circles, the smaller reactance values have arcs of larger circles. As the reactance increases, the radius of the arcs becomes smaller and the arcs collapse towards $x = \propto$. These circles are the constant reactance circles. As the normalized reactance x varies from $-\propto$ to $+\propto$, a family of arcs is completely contained inside the domain of the reflection coefficient $|\Gamma| \leq 1$. Plot of families of constant resistance circles and reactance arcs on a single reflection coefficient plane gives the impedance Smith chart.

Smith chart containing plot of admittance on Γ plane is also very useful particularly for matching circuit design using shunt elements. Normalized admittance $y(l)$ is given by

$$y(l) = \frac{Y(l)}{Y_o} \tag{7.96a}$$

$$= \frac{Z_o}{Z(l)} \tag{7.96b}$$

using (7.86),

$$y(l) = \frac{Z_o}{Z(l)} = \frac{1 - \Gamma(l)}{1 + \Gamma(l)} \tag{7.96c}$$

From (7.20), the reflection coefficient $\Gamma(l)$ of a lossless transmission line can be written as

$$\Gamma(l) = |\Gamma| e^{+j2\beta l} \tag{7.97a}$$

Therefore,

$$\begin{aligned} \Gamma(l + \lambda/4) &= |\Gamma| e^{+j2\beta(l+\lambda/4)} \\ &= |\Gamma| e^{+j2\beta l} \times e^{+j\pi} \\ &= -\Gamma(l) \end{aligned} \tag{7.97b}$$

Thus, (7.96c) can be written as

$$y(l) = \frac{1 - \Gamma(l)}{1 + \Gamma(l)} = \frac{1 + \Gamma(l + \lambda/4)}{1 - \Gamma(l + \lambda/4)} \tag{7.98}$$

Comparing (7.98) with (7.87):

$$y(l) = z(l + \lambda/4) \tag{7.99}$$

Thus, rotating all the resistance circles and reactance arcs of the Smith chart by 180° around the complex Γ plane provide a mapping of complex admittance onto the complex Γ plane. Figure 7.26 shows the impedance Z and admittance Y Smith charts. The distance from origin to any other point on the Smith chart gives magnitude of the reflection coefficient $|\Gamma|$. The angle between $x = 0$ line and the line connecting the point to the origin gives the phase angle of the reflection coefficient $\angle\Gamma$. Smith chart containing impedance and admittance chart on a single reflection coefficient plane is shown in Fig. 7.27. Here, the horizontal axis is a constant reactance circle of value $r = 0$. Thus, all impedances on the horizontal axis are pure resistances. The upper half of the chart corresponds to inductive impedances of all positive constant reactances and negative susceptance arcs. The lower half of the chart corresponds to capacitive impedances of all negative constant reactances and positive susceptance arcs.

Reflection coefficient $\Gamma(l)$ of a lossless transmission line at a distance l from a load towards generator is given by $\Gamma(-l) = |\Gamma| e^{-j2\beta l}$. The magnitude of $\Gamma(-l)$ remains constant with the change of transmission line length though its phase changes. Therefore, moving through a lossless transmission line, magnitude of reflection coefficient remains constant. This result gives very important set of constant reflection coefficient circles on Γ plane. The voltage standing wave ratio VSWR is given by (7.23)

$$\text{VSWR} = \frac{1 + |\Gamma|}{1 - |\Gamma|} \tag{7.100}$$

From (7.100), it is clear that the VSWR remains constant with the change of transmission line length. Figure 7.28 shows constant VSWR circles (dark dashed)

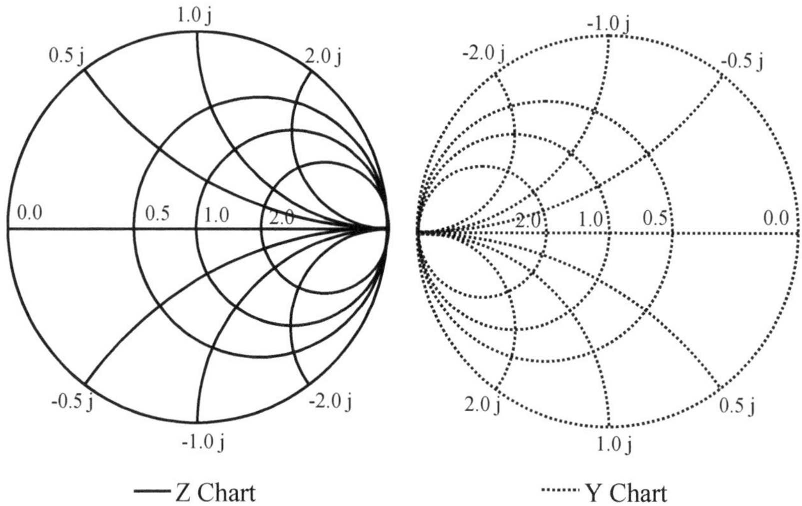

Fig. 7.26 Smith chart: impedance and admittance plots on complex reflection coefficient plane

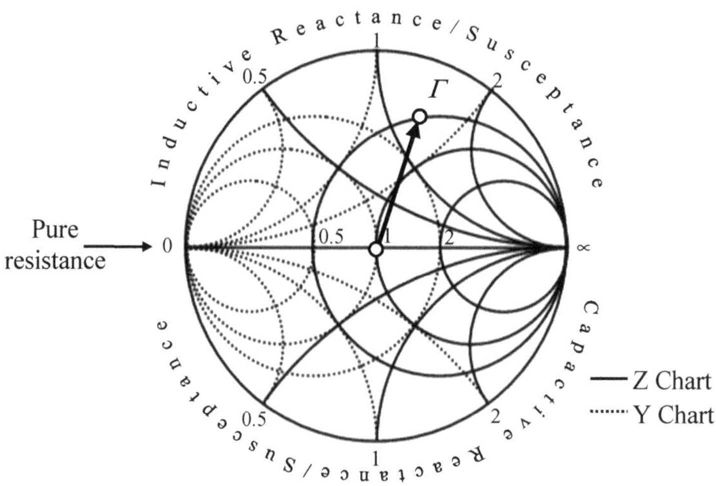

Fig. 7.27 Smith chart: impedance and admittance plots on a single complex reflection coefficient plane

on complex reflection coefficient plane. Here, it is shown that a load impedance Z_L changes to Z_{IN} through a series transmission line of length l along a constant VSWR circle while moving towards generator.

Example 7.9 Compute transferred input impedance of a load $(25 + 50j)\Omega$ transferred through a series-connected transmission line of characteristic impedance 50 Ω and length (a) $\lambda/8$ and (b) $\lambda/4$.

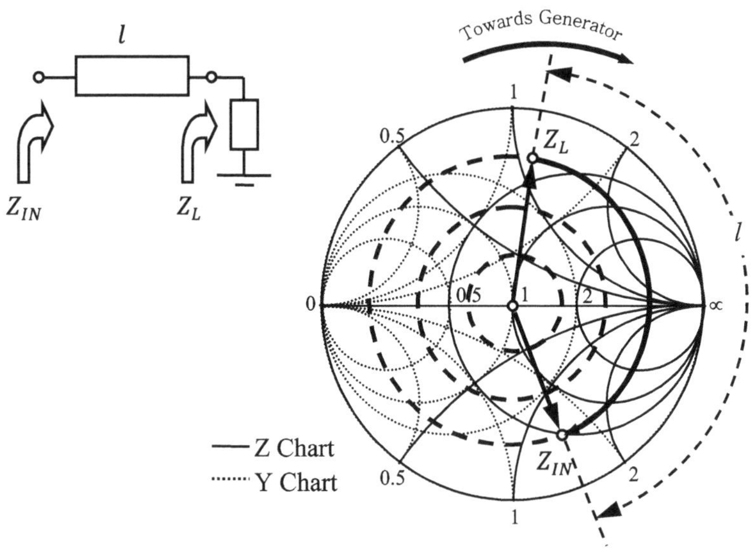

Fig. 7.28 Constant VSWR circles

Solution

The problem is shown schematically in Fig. 7.29. Here, we have to determine Z_{IN} for the two cases.

The normalized load impedance is given by

$$z_L = \frac{(25 + 50j)}{50} = 0.5 + 1j$$

The normalized load z_L is plotted marked as A on the smith chart in Fig. 7.29. The load reflection coefficient for this load is shown as vector Γ_L. A constant VSWR circle is drawn (dashed circle) corresponding to this reflection coefficient. Input impedance will change along the constant VSWR circle.

(a) The length $\lambda/8$ corresponds to 90° on the Smith chart. Moving the point A by 90° towards generator (clockwise direction), it reaches at the point B which corresponds to normalized impedance of $(4 - 1j)$, i.e. $(200 - 50j)\Omega$.
(b) The length $\lambda/4$ corresponds to 180° on the Smith chart. Moving the point A by 180° towards generator (clockwise direction), it reaches at the point C which corresponds to normalized impedance of $(0.4 - 0.8j)$, i.e. $(20 - 40j)\Omega$. From the plot, it is clear that this point corresponds to normalized admittance of $(0.5 + 1j)$.

Example 7.10 Compute transferred input impedance of a pure resistive load of 100 Ω, when it is shunted by an (a) open-circuited transmission line of length $\lambda/8$ and

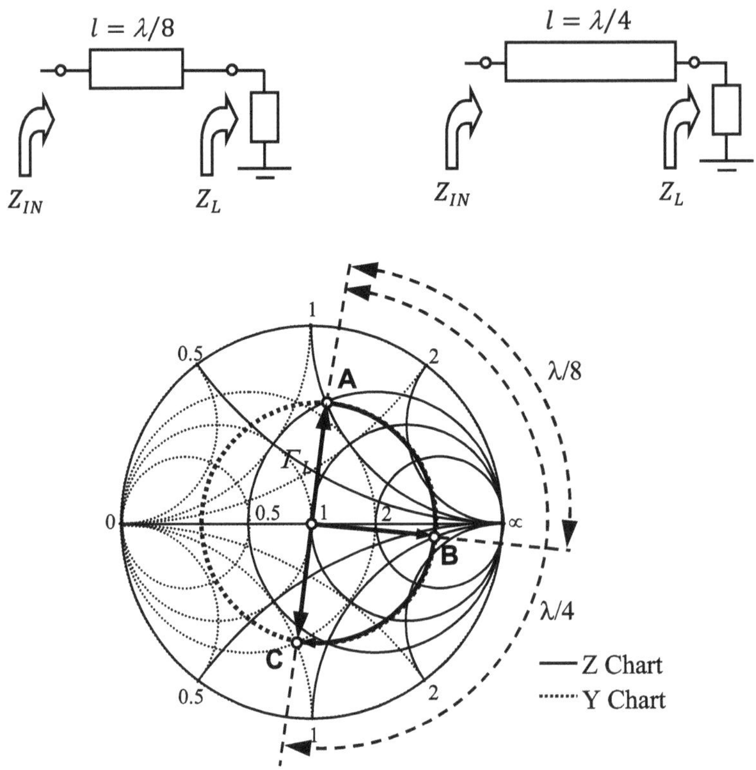

Fig. 7.29 Impedance transfer for series transmission line

(b) short-circuited transmission line of length $\lambda/8$. The characteristic impedances of the shunt lines are 50 Ω.

Solution
The normalized load impedance is given by

$$z_L = \frac{(100 + 0j)}{50} = 2 + 0j$$

The normalized load z_L is plotted marked as A on the Smith chart in Fig. 7.30. Connecting an open- or short-circuit stub parallel to a load will change its reactive part keeping conductance constant. Thus, the input impedance will change along the constant conductance circle with the change of stub length.

(a) As shown in Fig. 7.30, $\lambda/8$ open-circuit stub corresponds to normalized impedance of $-1j$, and thus the normalized susceptance of the $\lambda/8$ open-circuit line is $+1j$. Therefore, moving along the constant conductance (1/2 = 0.5) circle, the point B is reached. The admittance corresponding to the point B is

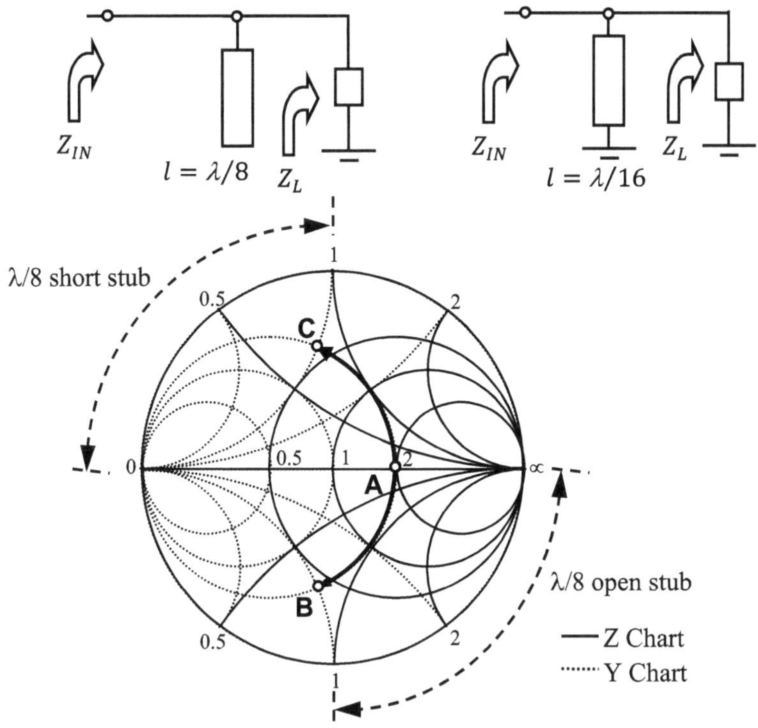

Fig. 7.30 Impedance transfer for series transmission line

(0.5 + 1j). This corresponds to normalized impedance of (0.4 – 0.8j). Thus, the transformed impedance is (20 – 40j) Ω.

(b) As shown in Fig. 7.30, $\lambda/8$ short-circuit stub corresponds to normalized impedance of +1j, and thus normalized susceptance of the $\lambda/8$ open-circuit line is −1j. Therefore, moving along the constant conductance (1/2 = 0.5) circle, the point C is reached. The admittance corresponding to the point C is (0.5 – 1j). This is corresponding to normalized impedance of (0.4 + 0.8j). Thus, the transformed impedance is (20 + 40j)Ω.

Lumped elements in different configurations are also used as impedance transformer. Figure 7.31 shows the transfer of impedance z_L on a Smith chart by the use of resistance, capacitance and inductance as lumped elements in different configurations.

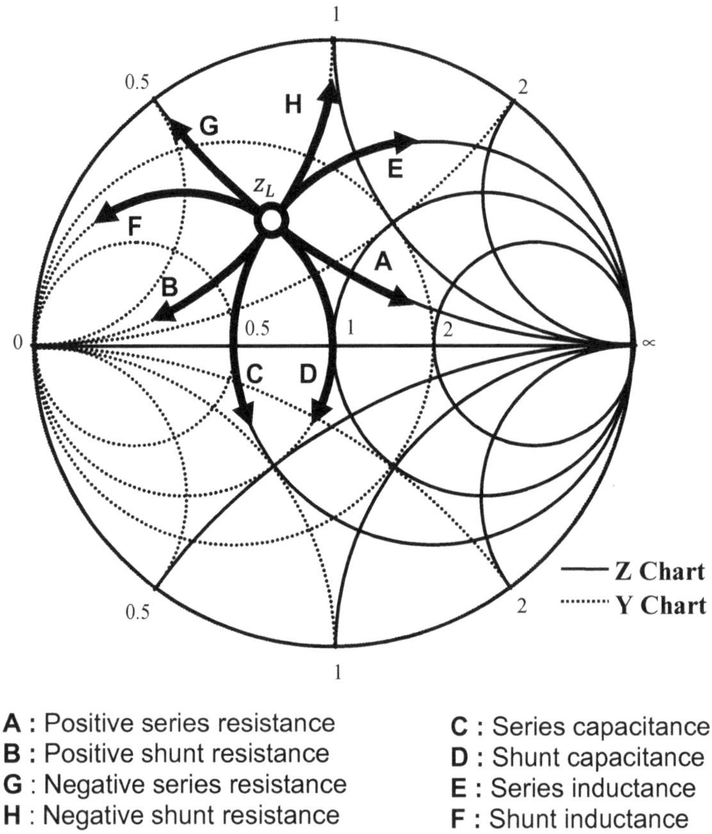

Fig. 7.31 Impedance transfer using lumped elements

7.6 Network Theory of Circuits and Transmission Lines

Several microwave devices and elements are connected together by transmission lines to form a microwave network. A large complex network may be divided into subnetworks usually 1- or 2-port networks for the purpose of analysis and design before being interconnected to form the complex network. Network analysis approach enables to know the port behaviours in terms of voltages, currents and impedances of a network without knowing inside details about it. Therefore, the network analysis is very useful for the analysis of microwave systems, instead of working with Maxwell equations. Knowing the network parameters related to the ports enable us to treat the network as a black box when embedded within a large network.

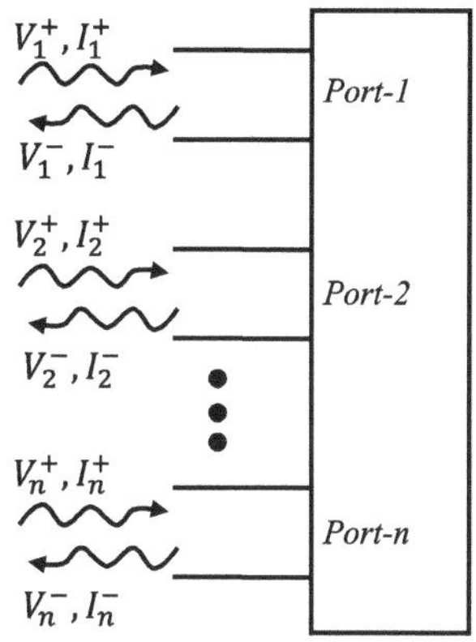

Fig. 7.32 Schematic of an n-port network

7.6.1 S-Parameter Network Representation

In case of TEM waves, like waves in coaxial lines, unique current and voltage can be defined; however, it is not possible to uniquely define for non-TEM waves, like waves in rectangular waveguide. Moreover, there is no suitable instrument to directly measure voltages and currents in complex value at microwave frequencies. Therefore, microwave networks are not conveniently represented by impedance or admittance matrices. For microwave networks, a more appropriate representation is scattering parameter (S-parameter) matrix representation where directly measurable reflected, transmitted and incident waves are used. Scattering matrix relates the incident, transmitted and reflected waves at the ports which can be measured directly with a network analyzer. Knowing the S-parameters these can be converted to any other matrix parameters which are suitable for particular analysis of the microwave network depending upon applications [8–12].

Consider an n-port network as shown in Fig. 7.32. Here, V_1^+, V_2^+, ..., V_n^+ are the amplitudes of the voltage wave and I_1^+, I_2^+, ..., I_n^+ are the amplitudes of the current wave incident on port-1, port-2, ..., port-n, respectively, whereas V_1^-, V_2^-, ..., V_n^- are the amplitudes of the voltage wave and I_1^-, I_2^-, ..., I_n^- are the amplitudes of the current wave reflected from the port-1, port-2, ..., port-n, respectively. The scattering matrix $[S]$ which relates the reflected and incident voltage waves is written as $\left[V^-\right] = [S]\left[V^+\right]$

or

$$\begin{bmatrix} V_1^- \\ V_2^- \\ \vdots \\ V_n^- \end{bmatrix} = \begin{bmatrix} S_{11} & S_{12} & \cdots & S_{1n} \\ S_{21} & S_{22} & \cdots & S_{2n} \\ & \vdots & & \\ S_{n1} & S_{n2} & \cdots & S_{nn} \end{bmatrix} \begin{bmatrix} V_1^+ \\ V_2^+ \\ \vdots \\ V_n^+ \end{bmatrix} \tag{7.101}$$

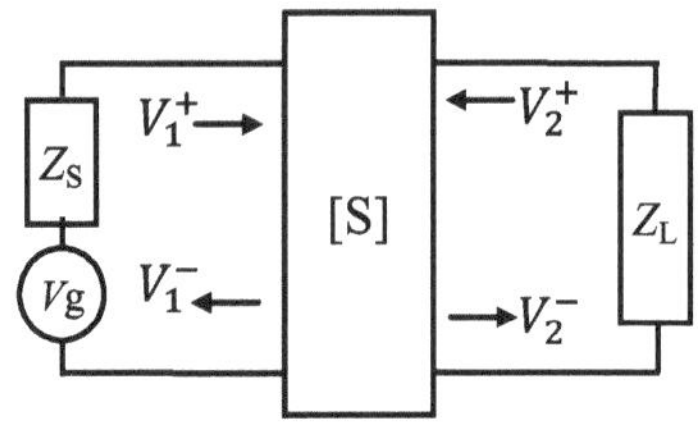

Fig. 7.33 Schematic of a 2-port network

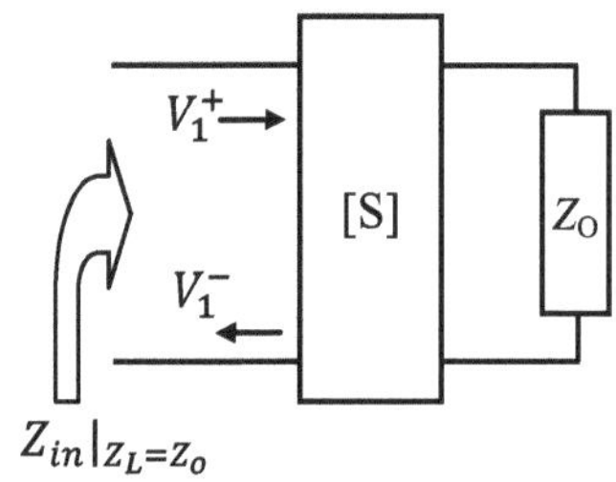

Fig. 7.34 2-port network with matched output port

Thus, each element of the scattering matrix is defined as

$$S_{ij} = \frac{V_i^-}{V_j^+}, \text{ with the condition: } V_k^+ = 0 \quad \text{for} \quad k \neq j \tag{7.102}$$

This means, there is no incident wave other than the port-j. Therefore, the transmission line on these ports must be terminated by a matched load.

When a 2-port network is driven by a voltage source V_g at its port-1 and port-2 is terminated by load of impedance Z_L as shown in Fig. 7.33, then the incident and reflected voltage waves are related by,

$$\begin{bmatrix} V_1^- \\ V_2^- \end{bmatrix} = \begin{bmatrix} S_{11} & S_{12} \\ S_{21} & S_{22} \end{bmatrix} \begin{bmatrix} V_1^+ \\ V_2^+ \end{bmatrix} \tag{7.103}$$

or

$$V_1^- = S_{11} V_1^+ + S_{12} V_2^+ \tag{7.104}$$
$$V_2^- = S_{21} V_1^+ + S_{22} V_2^+ \tag{7.105}$$

Thus, from Fig. 7.34, the S-parameters will be determined by,

$$S_{11} = \frac{V_1^-}{V_1^+}, \quad \text{when} \quad V_2^+ = 0 \tag{7.106}$$

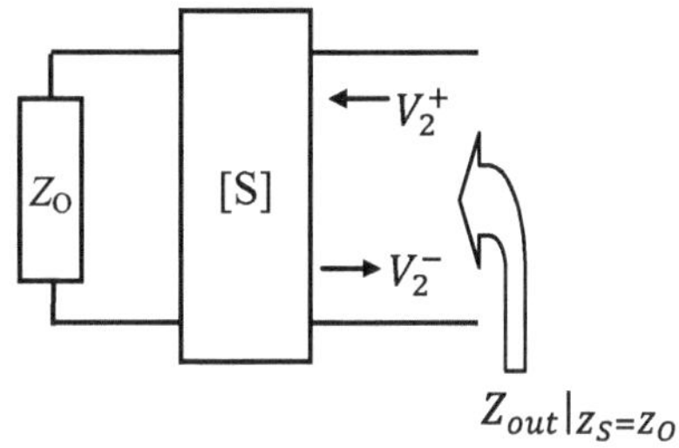

Fig. 7.35 2-port network with matched input port

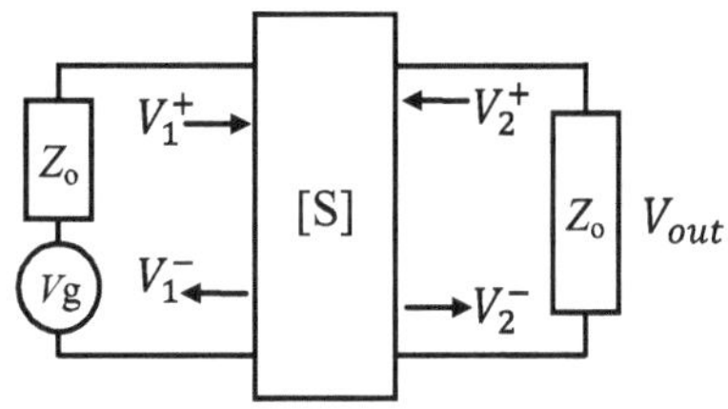

Fig. 7.36 A 2-port network with Z_o at input and output

Comparing with (7.15), it is clear that S_{11} is nothing but the reflection coefficient of the input port when the output port is terminated by its characteristic impedance. Thus, from (7.19) S_{11} can be written as

$$S_{11} = \Gamma_{\text{in}}|_{Z_L=Z_O} = \frac{Z_{\text{in}}|_{Z_L=Z_o} - Z_O}{Z_{\text{in}}|_{Z_L=Z_o} + Z_O} \tag{7.107}$$

In similar way from Fig. 7.35, S_{22} can be determined as

$$S_{22} = \frac{V_2^-}{V_2^+}, \quad \text{when} \quad V_1^+ = 0 \tag{7.108}$$

Thus, S_{22} is nothing but the reflection coefficient of the output port when the input port of the network is terminated by its characteristic impedance Z_o as shown in Fig. 7.36. Thus, from (7.19), S_{22} can be written as

$$S_{22} = \Gamma_{\text{out}}|_{Z_S=Z_O} = \frac{Z_{\text{out}}|_{Z_S=Z_O} - Z_O}{Z_{\text{out}}|_{Z_S=Z_O} + Z_O} \tag{7.109}$$

From (7.105), S_{21} is given by

$$S_{21} = \frac{V_2^-}{V_1^+}, \quad \text{when} \quad V_2^+ = 0 \tag{7.110}$$

Therefore, S_{21} is the ratio of voltage wave reached to the output matched load to the incident voltage wave at the input port. S_{21} is called forward transmission when the output port is terminated by its characteristic impedance Z_o. The incident voltage wave V_1^+ at the input port is given by

$$V_1^+ = \frac{V_g Z_o}{Z_o + Z_o} = \frac{V_g}{2} \tag{7.111}$$

Since the output port is terminated to its characteristic impedance Z_o, the voltage wave V_2^- is the output voltage V_{out} across the output termination Z_o. Thus, the forward transmission coefficient is given by

$$S_{21} = \frac{2V_{out}}{V_g} \tag{7.112}$$

In a similar way, the reverse transmission coefficient S_{12} is given by

$$S_{12} = \frac{V_1^-}{V_2^+}, \quad \text{when} \quad V_1^+ = 0 \tag{7.113}$$

Thus, the reverse transmission S_{12} is the ratio of voltage wave reached at the input terminated by Z_o to the incident voltage wave at the output port from a source of impedance Z_o. Thus, S-parameters of a network do not depend on the source or load impedance; however, it depends upon the characteristic impedance of the ports in which it is connected.

In a general way, the voltage and current at port-n can be written by

$$V_n = V_n^+ + V_n^- \tag{7.114}$$

$$I_n = (V_n^+ - V_n^-)/Z_{on} \tag{7.115}$$

Here, Z_{on} is the characteristic impedance of the nth port. Considering Z_{on} as real, the power delivered to the nth port is given by

$$P_n = \frac{1}{2} Re\left[V_n I_n^*\right] \tag{7.116}$$

$$= \frac{1}{2Z_{on}} Re\left[\left|V_n^+\right|^2 - \left|V_n^-\right|^2\right] \tag{7.117}$$

Presence of Z_{on} term implies that the power delivered to the port is dependent on the characteristic impedance of the line connected to the port. To make power delivered to a port independent of the characteristic impedance of that port, a new set of incident and reflected wave is defined. Suppose the new set of incident wave (a_n) and reflected wave (b_n) for port-n is defined as

$$a_n = \frac{V_n^+}{\sqrt{Z_{on}}} \quad \text{and} \quad b_n = \frac{V_n^-}{\sqrt{Z_{on}}} \tag{7.118}$$

Thus, the voltage and current at the nth port is given by

$$V_n = V_n^+ + V_n^- = \sqrt{Z_{on}}(a_n + b_n) \tag{7.119}$$

$$I_n = \frac{1}{Z_{on}}\left(V_n^+ - V_n^-\right) = \frac{1}{\sqrt{Z_{on}}}(a_n - b_n) \tag{7.120}$$

Now, the power delivered to the nth port is given by

$$P_n = \frac{1}{2}Re\left[V_n I_n^*\right] \tag{7.121}$$

$$= \frac{1}{2}Re\left[(a_n + b_n)\left(a_n^* - b_n^*\right)\right] \tag{7.122}$$

$$= \frac{1}{2}Re\left[|a_n|^2 - |b_n|^2 + a_n^* b_n - a_n b_n^*\right] \tag{7.123}$$

or

$$P_n = \frac{1}{2}Re\left[|a_n|^2 - |b_n|^2\right] \tag{7.124}$$

This shows that the power delivered to the port is independent of the characteristic impedance (Z_{on}) of the line connected to that port. The generalized scattering matrix $[S]$ is the relations in between the normalized incident wave (a_n) and the normalized reflected wave (b_n) and is given by

$$[b] = [S][a] \tag{7.125}$$

or

$$\begin{bmatrix} b_1 \\ b_2 \\ \vdots \\ b_n \end{bmatrix} = \begin{bmatrix} S_{11} & S_{12} & \cdots & S_{1n} \\ S_{21} & S_{22} & \cdots & S_{2n} \\ & \vdots & & \\ S_{n1} & S_{n2} & \cdots & S_{nn} \end{bmatrix} \begin{bmatrix} a_1 \\ a_2 \\ \vdots \\ a_n \end{bmatrix} \tag{7.126}$$

where each element of the S-parameter is defined as

$$S_{ij} = \frac{b_i}{a_j}, \text{ with the condition: } a_k = 0 \quad \text{for} \quad k \neq j \tag{7.127}$$

The elements of the generalized scattering matrix can be determined by putting the values of a_j and b_i from (7.118).

$$S_{ij} = \frac{V_i^-}{\sqrt{Z_{oi}}} . \frac{\sqrt{Z_{oi}}}{V_j^+}, \text{ with the condition: } V_k^+ = 0 \quad \text{for} \quad k \neq j \tag{7.128}$$

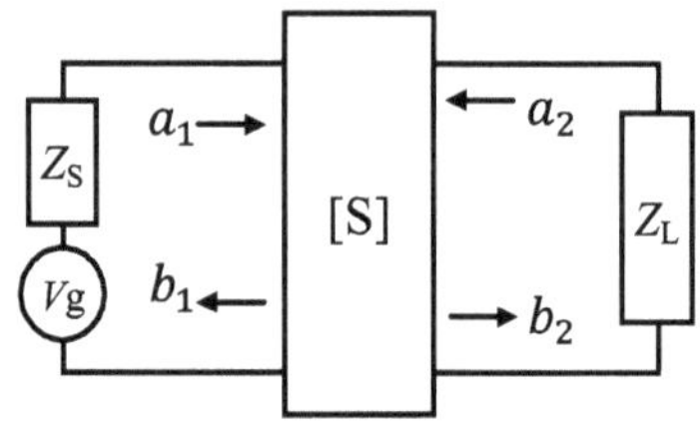

Fig. 7.37 A generalized 2-port S-parameter representation

and

$$S_{ii} = \frac{V_i^-}{V_i^+}, \text{ with the condition: } V_k^+ = 0 \quad \text{for} \quad k \neq i \tag{7.129}$$

In case of 2-port network as shown in Fig. 7.37, the generalized S-parameter can be written as

$$b_1 = S_{11}a_1 + S_{12}a_2 \tag{7.130}$$

$$b_2 = S_{21}a_1 + S_{22}a_2 \tag{7.131}$$

In matrix form,

$$\begin{bmatrix} b_1 \\ b_2 \end{bmatrix} = \begin{bmatrix} S_{11} & S_{12} \\ S_{21} & S_{22} \end{bmatrix} \begin{bmatrix} a_1 \\ a_2 \end{bmatrix} \tag{7.132}$$

In the similar way, the S-parameters can be determined as follows:

$$S_{11} = \frac{b_1}{a_1}, \quad \text{when} \quad a_2 = 0 \tag{7.133}$$

Putting the values of a_1 and b_1 from (7.118)

$$S_{11} = \frac{V_1^-}{\sqrt{Z_{o1}}} \cdot \frac{\sqrt{Z_{o1}}}{V_1^+} = \frac{V_1^-}{V_1^+}, \quad \text{when} \quad V_2^+ = 0 \tag{7.134}$$

This is the same as (7.106) and is the input port reflection coefficient when the output port is matched terminated.

Similarly, other parameters can be determined as

$$S_{12} = \frac{b_1}{a_2} = \frac{V_1^-}{\sqrt{Z_{o1}}} \frac{\sqrt{Z_{o2}}}{V_2^+}, \quad \text{when} \quad V_1^+ = 0 \tag{7.135}$$

$$S_{21} = \frac{b_2}{a_1} = \frac{V_2^-}{\sqrt{Z_{o2}}} \frac{\sqrt{Z_{o1}}}{V_1^+}, \quad \text{when} \quad V_2^+ = 0 \tag{7.136}$$

$$S_{22} = \frac{b_2}{a_2} = \frac{V_2^-}{\sqrt{Z_{o2}}} \cdot \frac{\sqrt{Z_{o2}}}{V_2^+} = \frac{V_2^-}{V_2^+}, \quad \text{when} \quad V_1^+ = 0 \tag{7.137}$$

7.6.2 S-Parameters for Matched Reciprocal Lossless Networks

The scattering matrices for the matched, reciprocal and lossless networks have special properties. The input impedance at each port of a matched network is equal to the reference impedance Z_o when all other ports are terminated by matched loads. Therefore, the reflection coefficient of each port is zero, and thus there will be no reflected signal from the port if a signal is incident only on that port. Therefore, for a matched device/network,

$$S_{ii} = \frac{V_i^-}{V_i^+} = 0, \quad \text{for all } i \tag{7.138}$$

This means that a matched device has a scattering matrix where all diagonal elements, S_{ii} are zero.

A reciprocal network does not contain any active device or non-reciprocal medium such as plasmas or ferrites. For a reciprocal network, the scattering matrix is symmetric. Therefore, the condition for a reciprocal network is

$$S_{ij} = S_{ji} \tag{7.139}$$

For example, a reciprocal network will have $S_{21} = S_{12}$ or, $S_{23} = S_{32}$, etc. Therefore, for a reciprocal network, the S-parameter matrix satisfies

$$[S]^{\mathrm{T}} = [S] \tag{7.140}$$

For a lossless network, the net power delivered to the network must come out from it. Thus, for a lossless network there will be no absorption of power by it, i.e. there will be no power dissipation and generation of heat within the network. With reference to an n-port network as shown in Fig. 7.32, the power incident to the port-n is given by

$$P_n^+ = \frac{1}{2}\left[a_n a_n^*\right] \tag{7.141a}$$

The total power incident to the network through all the n ports is given by

$$P^+ = \sum_{n=1}^{n} \frac{1}{2}\left[a_n a_n^*\right] \tag{7.141b}$$

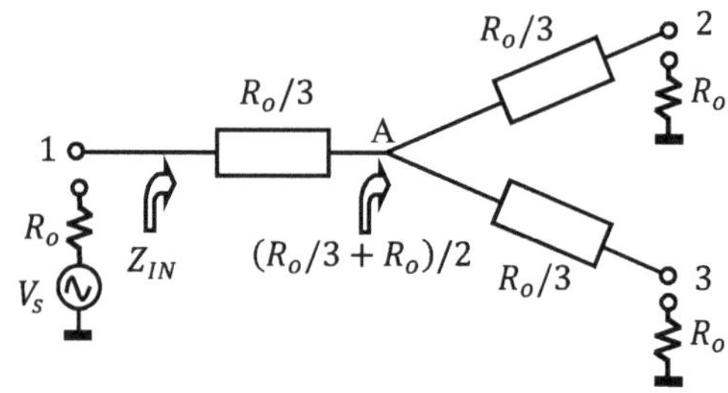

Fig. 7.38 Schematic diagram of two-way resistive power divider

$$= \frac{1}{2}[a_n]^H[a_n] \tag{7.141c}$$

Here, operator H indicates the conjugate transpose.

Similarly, the exiting power, i.e. power coming out from the port-n is given by

$$P_n^- = \frac{1}{2}\left[b_n b_n^*\right] \tag{7.142a}$$

The total reflected power from the network through all the n ports is given by

$$P^- = \sum_{n=1}^{n} \frac{1}{2}\left[b_n b_n^*\right] \tag{7.142b}$$

$$= \frac{1}{2}[b_n]^H[b_n] \tag{7.142c}$$

Therefore, the resultant power delivered through the port-n to the network is given by

$$\Delta P = P^+ - P^- = \frac{1}{2}\left[[a_n]^H[a_n] - [b_n]^H[b_n]\right] \tag{7.143a}$$

Putting $[b_n] = [S][a_n]$ in (7.143a)

$$\Delta P = \frac{1}{2}\left[[a_n]^H[a_n] - [S]^H[a_n]^H[S][a_n]\right]$$

$$= \frac{1}{2}[a_n]^H\left[[I] - [S]^H[S]\right][a_n] \tag{7.143b}$$

Here, $[I]$ is the identity matrix. For a lossless network, the total power delivered to the network, i.e. the power absorbed by the device must be zero, i.e. $\Delta P = 0$.

Thus, for lossless network from (7.143b)

$$[I] - [S]^H[S] = 0$$

or

$$[S]^H[S] = [I] \tag{7.144}$$

A matrix that satisfies $[S]^H[S] = [I]$ is a special matrix known as unitary matrix. Thus, S-parameter matrix of a lossless network is a unitary matrix. The columns of a unitary matrix form an orthogonal set. That means each column of the scattering matrix will have a magnitude equal to one while the inner product of dissimilar columns must be zero. Thus, for a lossless network, the S-parameters satisfies

$$\sum_{m=1}^{n} \lfloor S_{mn} \rfloor^2 = 1 \quad \text{for all } n \tag{7.145a}$$

and

$$\sum_{n=1}^{n} S_{ni} S_{nj}^* = 0 \quad \text{for all } i \neq j \tag{7.145b}$$

For example, the relationship of the S-parameters for a lossless 2-port network will be

$$\lfloor S_{11} \rfloor^2 + \lfloor S_{21} \rfloor^2 = 1$$
$$\lfloor S_{12} \rfloor^2 + \lfloor S_{22} \rfloor^2 = 1$$

and

$$S_{11} S_{12}^* + S_{21} S_{22}^* = 0$$
$$S_{12} S_{11}^* + S_{22} S_{21}^* = 0$$

Example 7.11 A 2-port network has the following scattering matrix:

$$[S] = \begin{bmatrix} 0.467\angle - 144.01^\circ & 0.884\angle - 142.53^\circ \\ 0.884\angle - 142.53^\circ & 0.467\angle 38.95^\circ \end{bmatrix}$$

Determine whether the network is reciprocal, matched and lossless.

Solution

Here, $S_{12} = 0.884\angle - 142.53^\circ = S_{21} = 0.884\angle - 142.53^\circ$. Thus, S-parameter matrix is symmetric. So, the network is reciprocal. $S_{11} = 0.467\angle - 144.01^\circ \neq 0$ also $S_{22} = 0.467\angle 38.95^\circ \neq 0$. Thus, the ports of the network are not matched.

To check for lossless, we can check unitary property of the matrix,

$$|S_{11}|^2 + |S_{21}|^2 = 0.467^2 + 0.884^2 = 1$$
$$|S_{12}|^2 + |S_{22}|^2 = 0.884^2 + 0.467^2 = 1$$
$$S_{11} S_{12}^* + S_{21} S_{22}^* = 0.467\angle - 144.01^\circ \times 0.884\angle 142.53^\circ$$

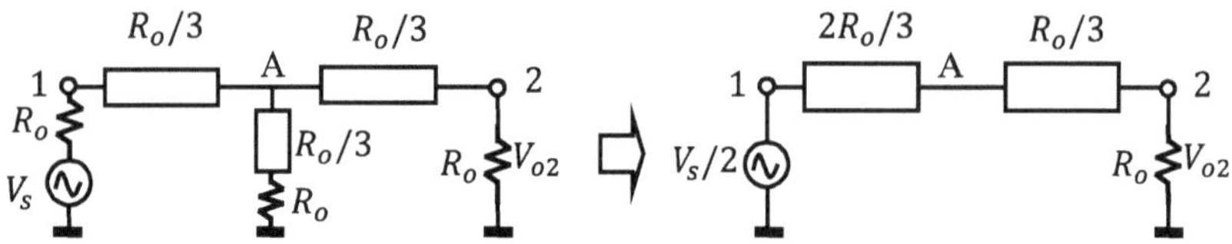

Fig. 7.39 Equivalent circuit for signal V_s at port-1 to calculate V_o

$$+0.884\angle-142.53^\circ \times 0.467\angle-38.95^\circ = 0$$

Similarly, $S_{11}^* S_{12} + S_{21}^* S_{22} = 0$

Thus, the matrix elements satisfy all the conditions for a unitary matrix. Thus, the network is lossless.

Example 7.12 Compute S-parameter of a 3-port two-way resistive power divider as shown in Fig. 7.38.

Solution

Suppose a signal V_s is applied to the port-1 through a source resistance R_o and the other two ports are terminated by resistance R_o as shown in Fig. 7.38. The impedance at the junction seeing towards the port-2 and port-3 is the parallel combination of two resistors of values $(R_o/3 + R_o)$. Therefore, the input impedance, Z_{IN} at port-1 is given by

$$\begin{aligned} Z_{IN} &= \left(\frac{R_o}{3}\right) + \left(\frac{R_o}{3} + R_o\right) \| \left(\frac{R_o}{3} + R_o\right) \\ &= R_o \end{aligned} \tag{7.146}$$

Therefore, the port-1 is a matched port. Since the structure is symmetric, other two ports are also matched. Thus, $S_{11} = S_{22} = S_{33} = 0$.

To calculate output voltage wave under the application of signal V_s at port-1, the equivalent circuit is shown in Fig. 7.39. The signal voltage V_{o2} at the port-2 is given by

$$\begin{aligned} V_{o2} &= \left(\frac{V_s/2}{2R_o/3 + R_o/3 + R_o}\right) R_o \\ &= \frac{V_s}{4} \end{aligned} \tag{7.147}$$

Thus, the transmission coefficient S_{21} is given by

$$S_{21} = \frac{V_{o2}}{V_s/2} = \frac{V_s/4}{V_s/2} = \frac{1}{2} \tag{7.148}$$

Since the network is symmetric in structure, we can write

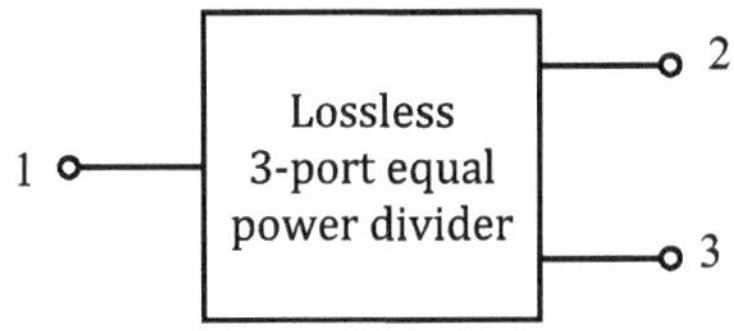

$$[S] = \begin{bmatrix} S_{11} & S_{12} & S_{13} \\ S_{21} & S_{22} & S_{23} \\ S_{31} & S_{32} & S_{33} \end{bmatrix}$$

Fig. 7.40 Block schematic of two-way lossless equal power divider

$$S_{21} = S_{12} = S_{31} = S_{32} = S_{23} = S_{32} = 1/2 \tag{7.149}$$

Therefore, the S-parameter matrix of the resistive power splitter is given by

$$[S] = \begin{bmatrix} S_{11}\ S_{12}\ S_{13} \\ S_{21}\ S_{22}\ S_{23} \\ S_{31}\ S_{32}\ S_{33} \end{bmatrix} = \frac{1}{2}\begin{bmatrix} 0\ 1\ 1 \\ 1\ 0\ 1 \\ 1\ 1\ 0 \end{bmatrix} \tag{7.150}$$

From the S-parameter matrix, it is clear that the network is matched and reciprocal.

Example 7.13 Compute S-parameter matrix of a lossless two-way equal power divider as shown in Fig. 7.40.

Solution
The two-way power divider does not contain any active device and non-reciprocal components. Thus, the network is reciprocal, we can write

$$S_{12} = S_{21}, \quad S_{13} = S_{31} \quad \text{and} \quad S_{23} = S_{32} \tag{7.151}$$

Suppose the port-1 is the input port and it is matched to the reference impedance. Thus, $S_{11} = 0$. For equal power division to port-2 and port-3, we can write (without considering the phase difference if any)

$$S_{21} = S_{31} = 1/\sqrt{2} \tag{7.152}$$

Thus, the S-parameter matrix can be written as

$$[S] = \begin{bmatrix} 0 & 1/\sqrt{2} & 1/\sqrt{2} \\ 1/\sqrt{2} & S_{22} & S_{23} \\ 1/\sqrt{2} & S_{23} & S_{33} \end{bmatrix} \tag{7.153}$$

Since the network is lossless, the matrix is unitary. Thus, applying (7.145a) we can write

$$1/2 + |S_{22}|^2 + |S_{23}|^2 = 1 \tag{7.154a}$$

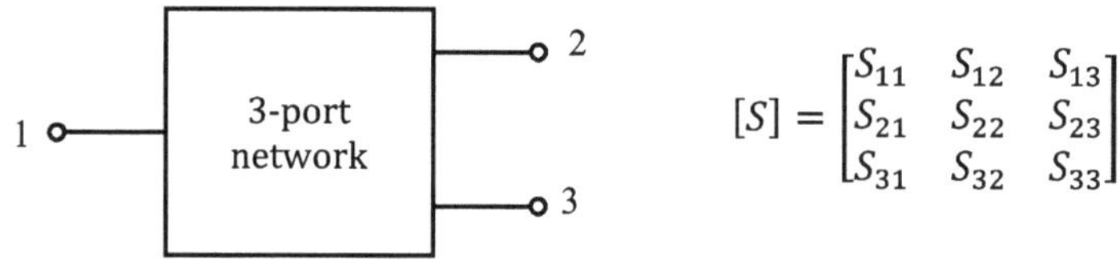

Fig. 7.41 Schematic diagram of two-way resistive power divider

and

$$1/2 + |S_{23}|^2 + |S_{33}|^2 = 1 \tag{7.154b}$$

From (7.154a) and (7.154b), we can write

$$S_{22} = S_{33} \tag{7.155}$$

Now, applying (7.145b), we can write

$$0 + 1/\sqrt{2}S_{22} + 1/\sqrt{2}S_{23} = 0 \tag{7.156a}$$

and

$$1/2 + S_{22}S_{23}^* + S_{23}S_{33}^* = 0 \tag{7.156b}$$

Using (7.155), (7.156a) and (7.156b), we can write

$$S_{22} = S_{33} = -S_{23} = 1/2$$

Therefore, the S-parameter matrix of the lossless 2-way power splitter is

$$[S] = \begin{bmatrix} 0 & 1/\sqrt{2} & 1/\sqrt{2} \\ 1/\sqrt{2} & 1/2 & -1/2 \\ 1/\sqrt{2} & -1/2 & 1/2 \end{bmatrix} \tag{7.157}$$

Here, S_{22} and S_{33} are non-zero. Thus, for the 3-port lossless network, the other ports are not matched.

Example 7.14 Is it possible to have a 3-port matched, reciprocal and simultaneously lossless network?

Solution

Suppose all the 3-ports of the network shown in Fig. 7.41 are matched and the network is reciprocal. Then, we can write

$$S_{11} = S_{22} = S_{33} = 0 \tag{7.158}$$

and

$$S_{12} = S_{21}, \quad S_{13} = S_{31}, \quad S_{23} = S_{32}$$

Thus, the S-parameter matrix of a matched reciprocal 3-port network can be written as

$$[S] = \begin{bmatrix} 0 & S_{12} & S_{13} \\ S_{12} & 0 & S_{23} \\ S_{13} & S_{23} & 0 \end{bmatrix} \tag{7.159}$$

Now, if the network is simultaneously lossless. The following conditions for unitary matrix also should be satisfied.

Since the network is lossless, the matrix is unitary. Thus, applying (7.145a), we can write

$$|S_{12}|^2 + |S_{13}|^2 = 1 \tag{7.160a}$$
$$|S_{12}|^2 + |S_{23}|^2 = 1 \tag{7.160b}$$
$$|S_{13}|^2 + |S_{23}|^2 = 1 \tag{7.160c}$$

and applying (7.145b),

$$S_{12}S_{13}^* = 0 \tag{7.161a}$$
$$S_{12}S_{23}^* = 0 \tag{7.161b}$$
$$S_{13}S_{23}^* = 0 \tag{7.161c}$$

To satisfy (7.161), at least two parameters among S_{12}, S_{13} and S_{23} must be zero but this is contradictory to Eq. (7.160). Therefore, it is not possible to have a 3-port network with all ports matched, reciprocal and lossless.

Example 7.15 Compute S-parameter of a series network of impedance Z_a as shown in Fig. 7.42.

Solution

The S-parameters of the series network can be determined as follows:

$$\begin{aligned} S_{11} &= \frac{b_1}{a_1}, \quad \text{when} \quad a_2 = 0 \\ &= \frac{V_1^-}{V_1^+}, \quad \text{when} \quad V_2^+ = 0 \\ &= Z_a/(Z_a + 2Z_o) \end{aligned}$$

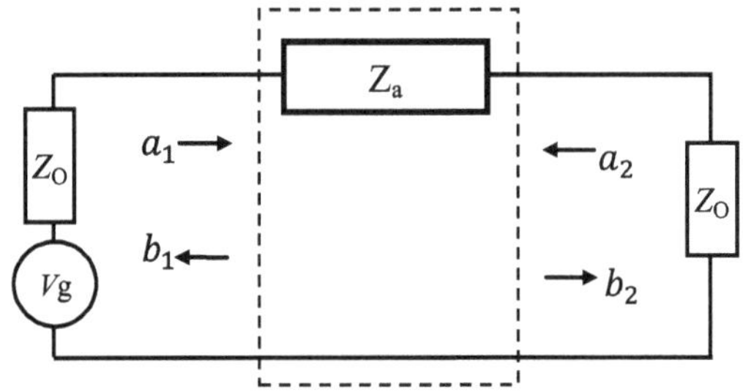

Fig. 7.42 *S*-parameter determination of a 2-port series network

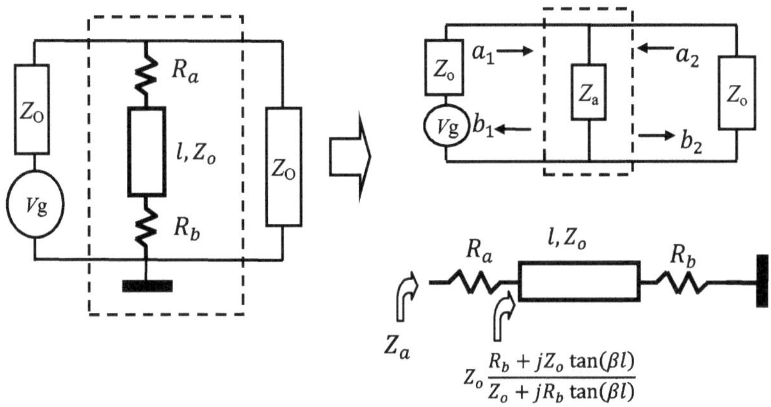

Fig. 7.43 *S*-parameter determination of a 2-port shunt network

It is a symmetric network, and therefore,

$$S_{11} = S_{22} = (Z_a/(Z_a + 2Z_o). \tag{7.162a}$$

Similarly, from (7.136) and using (7.112), S_{21} can be determined as

$$\begin{aligned} S_{21} &= \frac{b_2}{a_1} = \frac{V_2^-}{\sqrt{Z_o}} \frac{\sqrt{Z_o}}{V_1^+} = \frac{V_2^-}{V_1^+}, \quad \text{when} \quad V_2^+ = 0 \\ &= 2Z_o/(Z_a + 2Z_o) \end{aligned} \tag{7.162b}$$

Thus, in matrix form, *S*-parameter of the series network can be written as

$$\begin{bmatrix} S_{11} & S_{12} \\ S_{21} & S_{22} \end{bmatrix} = \begin{bmatrix} \frac{Z_a}{2Z_o+Z_a} & \frac{2Z_o}{2Z_o+Z_a} \\ \frac{2Z_o}{2Z_o+Z_a} & \frac{Z_a}{2Z_o+Z_a} \end{bmatrix} \tag{7.162c}$$

Example 7.16 Compute *S*-parameter of a shunt network which consists of series combination of a transmission line (Z_o) of length l in between two resistors R_a and R_b as shown in Fig. 7.43.

Solution

The shunt network consists of a transmission line of length l and two resistors R_a and R_b connected in series as shown in Fig. 7.43. Using (7.26), the input impedance Z_a of the shunt network can be written as

$$Z_a = R_a + Z_o \frac{R_b + jZ_o \tan(\beta l)}{Z_o + jR_b \tan(\beta l)} \tag{7.163a}$$

Here, Z_o is the characteristic impedance of the shunt transmission line and port terminations. The S-parameters of the shunt network can be determined as follows:

$$\begin{aligned} S_{11} &= \frac{b_1}{a_1}, \quad \text{when} \quad a_2 = 0 \\ &= \frac{V_1^-}{V_1^+}, \quad \text{when} \quad V_2^+ = 0 \\ &= -Z_o/(Z_o + 2Z_a) \end{aligned}$$

It is a symmetric network, and therefore,

$$S_{11} = S_{22} = -Z_o/(Z_o + 2Z_a). \tag{7.163b}$$

Similarly, from (7.136) and using (7.112), S_{21} can be determined as

$$\begin{aligned} S_{21} &= \frac{b_2}{a_1} = \frac{V_2^-}{\sqrt{Z_o}} \frac{\sqrt{Z_o}}{V_1^+} = \frac{V_2^-}{V_1^+}, \quad \text{when} \quad V_2^+ = 0 \\ &= 2Z_a/(Z_o + 2Z_a) = S_{12} \end{aligned} \tag{7.163c}$$

Thus, in matrix form, S-parameter of the shunt network can be written as

$$\begin{bmatrix} S_{11} & S_{12} \\ S_{21} & S_{22} \end{bmatrix} = \begin{bmatrix} \frac{-Z_o}{Z_o+2Z_a} & \frac{2Z_a}{Z_o+2Z_a} \\ \frac{2Z_a}{Z_o+2Z_a} & \frac{-Z_o}{Z_o+2Z_a} \end{bmatrix} \tag{7.163d}$$

In most of the practical cases, ports of a microwave network are terminated by the same characteristic impedance (mostly $Z_o = 50\ \Omega$). However, generalized S-parameters can be used wherever the ports of a network are terminated with different characteristic impedances (say $Z_o = 25\ \Omega$, $70\ \Omega$, etc.). From the above discussion, it is clear that in case of S-parameter representations, parameters to be measured are incident, reflected and transmitted waves.

At this point, the advantages of using S-parameter matrix in microwave circuits can be summarized as

- At microwave frequency; incident, reflected and transmitted waves, all are directly measurable quantities, required to determine S-parameters.

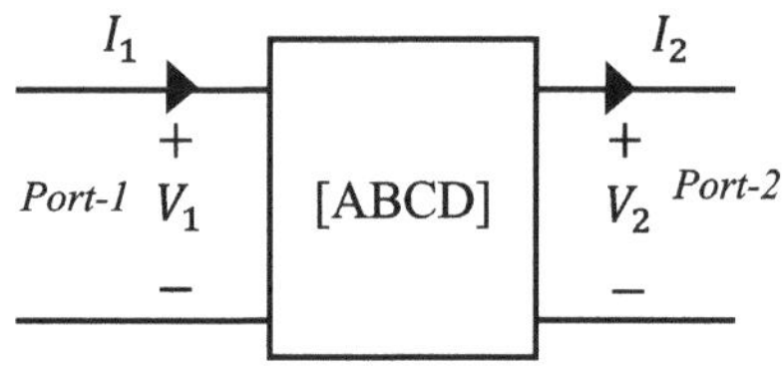

Fig. 7.44 *ABCD* parameter representation of a 2-port network

- No suitable instrument is available to measure voltage and current in their complex form at microwave frequency.
- Matched load required for *S*-parameters determination are available in broadband microwave applications, whereas it is difficult to make open and short circuits over broadband. Practical microwave short often behaves as inductive element and open behaves as capacitive element.
- Termination of active devices with Z_o (mostly = 50 Ω) causes no oscillation to active devices, whereas short and open at the active device and circuit ports may result in oscillation, i.e. may not remain stable.

7.6.3 ABCD Parameter for Network Representation

The most practical way to represent a microwave network is by *S*-parameters since it is determined from measurable quantities. However, *Z*, *Y* and *ABCD* parameter representation of microwave networks sometimes leads to convenient and easy way for analysis. Most of the microwave circuits and systems consist of cascaded connection of several 2-port networks. To analyse these types of circuits, it is convenient to define 2-port parameter representation as transmission parameters or *ABCD* parameters. The transmission parameter matrix of circuit or system formed by several cascaded 2-port networks is the product of the transmission matrices of the individual 2-port networks.

The transmission matrix relates input voltage and current to the output voltage and current of a given 2-port network as shown in Fig. 7.44 as follows:

$$\begin{aligned} V_1 &= AV_2 + BI_2 \\ I_1 &= CV_2 + DI_2 \end{aligned} \tag{7.164}$$

or in matrix form,

$$\begin{bmatrix} V_1 \\ I_1 \end{bmatrix} = \begin{bmatrix} A & B \\ C & D \end{bmatrix} \begin{bmatrix} V_2 \\ I_2 \end{bmatrix} \tag{7.165a}$$

where the matrix elements can be determined as follows:

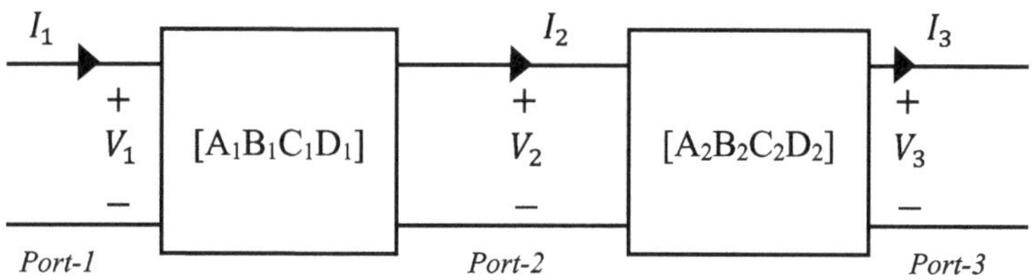

Fig. 7.45 *ABCD* parameters of cascaded two 2-port networks

$$A = \frac{V_1}{V_2}, \text{ open-circuit voltage ratio,} \quad \text{when} \quad I_2 = 0 \tag{7.165b}$$

$$B = \frac{V_1}{I_2}, \text{ short-circuit transfer impedance,} \quad \text{when} \quad V_2 = 0, \tag{7.165c}$$

$$C = \frac{I_1}{V_2}, \text{ open-circuit transfer admittance,} \quad \text{when} \quad I_2 = 0 \tag{7.165d}$$

$$D = \frac{I_1}{I_2}, \text{ short-circuit current ratio,} \quad \text{when} \quad V_2 = 0 \tag{7.165e}$$

Suppose two 2-port networks of parameters $[A_1B_1C_1D_1]$ and $[A_2B_2C_2D_2]$ are connected in cascade as shown in Fig. 7.45. The two networks can be represented by

$$\begin{bmatrix} V_1 \\ I_1 \end{bmatrix} = \begin{bmatrix} A_1 & B_1 \\ C_1 & D_1 \end{bmatrix} \begin{bmatrix} V_2 \\ I_2 \end{bmatrix} \tag{7.166}$$

$$\begin{bmatrix} V_2 \\ I_2 \end{bmatrix} = \begin{bmatrix} A_2 & B_2 \\ C_2 & D_2 \end{bmatrix} \begin{bmatrix} V_3 \\ I_3 \end{bmatrix} \tag{7.167}$$

Combining two matrixes (7.166) and (7.167), the relationship in between port-1 and port-3 can be written as

$$\begin{bmatrix} V_1 \\ I_1 \end{bmatrix} = \begin{bmatrix} A_1 & B_1 \\ C_1 & D_1 \end{bmatrix} \begin{bmatrix} A_2 & B_2 \\ C_2 & D_2 \end{bmatrix} \begin{bmatrix} V_3 \\ I_3 \end{bmatrix} \tag{7.168}$$

From (7.168), it is clear that the *ABCD* parameter of cascaded network is just matrix multiplication of the individual networks.

Example 7.17 A π network is shown in Fig. 7.46. It consists of a series transmission line of length l and characteristic impedance Z_o in between two shunt elements Z_a and Z_b, respectively. Determine *ABCD* parameter of this π network.

Solution

The complete network can be considered as combination of three simple networks (1, 2 and 3) connected in cascade configuration as shown in Fig. 7.46.

The network-1 can be represented as

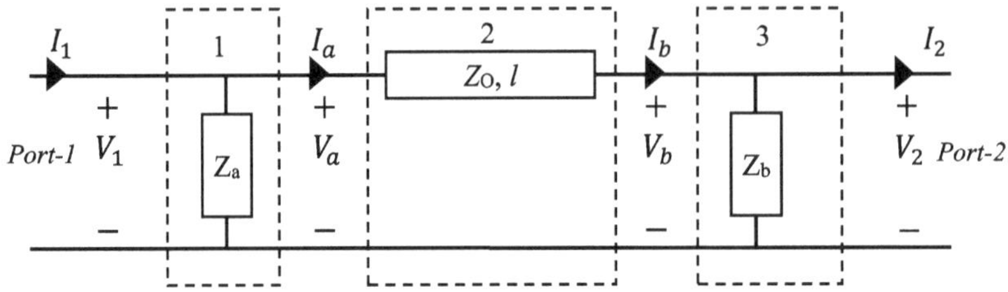

Fig. 7.46 A 2-port π network

$$\begin{bmatrix} V_1 \\ I_1 \end{bmatrix} = \begin{bmatrix} 1 & 0 \\ Y_a & 1 \end{bmatrix} \begin{bmatrix} V_a \\ I_a \end{bmatrix} \tag{7.169a}$$

Here, $Y_a = 1/Z_a$. The network-2 can be represented as

$$V_a = V_b e^{j\beta l} = V_b \cos(\beta l) + jV_b \sin(\beta l) = V_b \cos(\beta l) + jI_b Z_o \sin(\beta l)$$
$$I_a = I_b e^{j\beta l} = I_b \cos(\beta l) + jI_b \sin(\beta l) = I_b \cos(\beta l) + jV_b Y_o \sin(\beta l)$$

Here, $Y_o = 1/Z_o$. Thus,

$$\begin{bmatrix} V_a \\ I_a \end{bmatrix} = \begin{bmatrix} \cos(\beta l) & jZ_o \sin(\beta l) \\ jY_o \sin(\beta l) & \cos(\beta l) \end{bmatrix} \begin{bmatrix} V_b \\ I_b \end{bmatrix} \tag{7.169b}$$

Similarly, network-3 can be represented as

$$\begin{bmatrix} V_b \\ I_b \end{bmatrix} = \begin{bmatrix} 1 & 0 \\ Y_b & 1 \end{bmatrix} \begin{bmatrix} V_2 \\ I_2 \end{bmatrix} \tag{7.169c}$$

Here, $Y_b = 1/Z_b$. Therefore, the complete network can be represented by the matrix multiplication of these three *ABCD* matrices as

$$\begin{bmatrix} V_1 \\ I_1 \end{bmatrix} = \begin{bmatrix} 1 & 0 \\ Y_a & 1 \end{bmatrix} \begin{bmatrix} \cos(\beta l) & jZ_o \sin(\beta l) \\ jY_o \sin(\beta l) & \cos(\beta l) \end{bmatrix} \begin{bmatrix} 1 & 0 \\ Y_b & 1 \end{bmatrix} \tag{7.169d}$$

Thus, measurable *S*-parameters can be converted to *ABCD* or other parameters for convenience of analysis depending upon the way the 2-port subnetworks are interconnected to form a larger network.

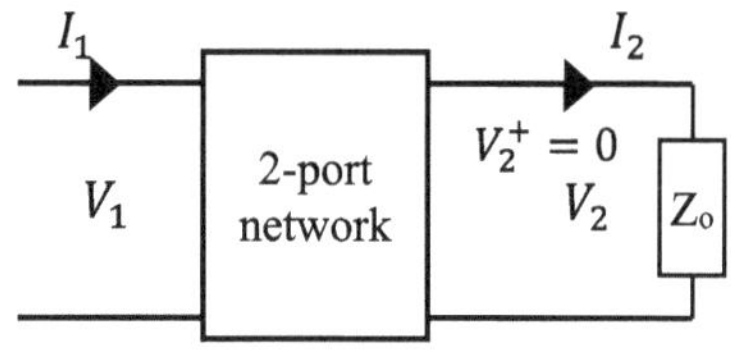

Fig. 7.47 A two-port network with matched output termination

7.6.4 *Conversion in Between* ABCD *and* S-*Parameters*

The conversion relationship in between 2-port S-parameters and $ABCD$ parameters can be derived in the following way:

$$\begin{bmatrix} b_1 \\ b_2 \end{bmatrix} = \begin{bmatrix} S_{11} & S_{12} \\ S_{21} & S_{22} \end{bmatrix} \begin{bmatrix} a_1 \\ a_2 \end{bmatrix} \quad \text{and} \quad \begin{bmatrix} V_1 \\ I_1 \end{bmatrix} = \begin{bmatrix} A & B \\ C & D \end{bmatrix} \begin{bmatrix} V_2 \\ I_2 \end{bmatrix}$$

From (7.114) and (7.115),

$$V_1 = V_1^+ + V_1^- \quad \text{and} \quad I_1 = \left(V_1^+ - V_1^-\right)/Z_o$$

or

$$V_1^+ = (V_1 + Z_o I_1)/2 \quad \text{and} \quad V_1^- = (V_1 - Z_o I_1)/2$$

$$S_{11} = \left.\frac{b_1}{a_1}\right|_{a_2=0} = \left.\frac{V_1^-}{V_1^+}\right|_{V_2^+=0} = \left.\frac{AV_2 + BI_2 - Z_o CV_2 - Z_o DI_2}{AV_2 + BI_2 + Z_o CV_2 + Z_o DI_2}\right|_{V_2^+=0}$$

From Fig. 7.47, for $V_2^+ = 0$,

$$V_2 = I_2 Z_o \quad \text{and} \quad I_2 = V_2/Z_o = V_2 Y_o$$

$$S_{11} - \frac{A + BY_o - CZ_o - D}{A + BY_o + CZ_o + D} \tag{7.170a}$$

Similarly,

$$S_{21} = \left.\frac{b_2}{a_1}\right|_{a_2=0} = \left.\frac{V_2^-}{V_1^+}\right|_{V_2^+=0} = \frac{V_2}{(V_1 + Z_o I_1)/2} = \left.\frac{2V_2}{AV_2 + BI_2 + Z_o CV_2 + Z_o DI_2}\right|_{V_2^+=0}$$

or

$$S_{21} = \frac{2}{A + BY_o + CZ_o + D} \tag{7.170b}$$

Relationship of S_{12} and S_{22} with $ABCD$ parameters can be determined as follows:

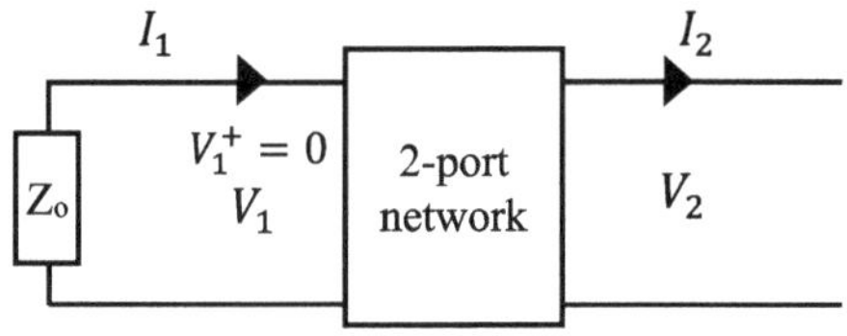

Fig. 7.48 A two-port network with matched input termination

$$S_{12} = \left.\frac{b_1}{a_2}\right|_{a_1=0} = \left.\frac{V_1^-}{V_2^+}\right|_{V_1^+=0} \quad \text{where,}$$

$$V_2 = V_2^+ + V_2^- \quad \text{and} \quad -I_2 = \left(V_2^+ - V_2^-\right)/Z_o$$

Thus,

$$S_{12} = \frac{V_1}{(V_2 - Z_o I_2)/2}$$

The *ABCD* matrix can also be written as

$$\begin{bmatrix} V_2 \\ I_2 \end{bmatrix} = \frac{1}{\Delta}\begin{bmatrix} D & -B \\ -C & A \end{bmatrix}\begin{bmatrix} V_1 \\ I_1 \end{bmatrix}, \quad \text{where} \quad \Delta = AD - BC$$

Putting V_2 and I_2:

$$S_{12} = \frac{2\Delta}{A + BY_o + CZ_o + D} \tag{7.170c}$$

Similarly,

$$S_{22} = \left.\frac{b_2}{a_2}\right|_{a_1=0} = \left.\frac{V_2^-}{V_2^+}\right|_{V_1^+=0}$$

or

$$S_{22} = \frac{V_2 + Z_o I_2}{V_2 - Z_o I_2}$$

From Fig. 7.48, the condition $V_1^+ = 0$, gives $V_1 = -I_1 Z_o$, and $I_1 = -V_1 Y_o$. Thus,

$$S_{22} = \frac{-A + BY_o - CZ_o + D}{A + BY_o + CZ_o + D} \tag{7.170d}$$

In a similar way, *ABCD* parameters can be expressed by *S*-parameters as follows:

$$A = \frac{(1 + S_{11})(1 - S_{22}) + S_{12}S_{21}}{2S_{21}} \tag{7.170e}$$

$$B = Z_o \frac{(1 + S_{11})(1 + S_{22}) - S_{12}S_{21}}{2S_{21}} \tag{7.170f}$$

$$C = Y_o \frac{(1 - S_{11})(1 - S_{22}) - S_{12}S_{21}}{2S_{21}} \tag{7.170g}$$

$$D = \frac{(1 - S_{11})(1 + S_{22}) + S_{12}S_{21}}{2S_{21}} \tag{7.170h}$$

Scattering matrix relates the port incident, transmitted and reflected waves which are directly measurable with a network analyzer. Once, the S-parameters of a network are known, these can be converted to $ABCD$ parameters using the derived relationship or to any other suitable parameters such as Z or Y parameters for the analysis of the microwave network.

7.7 Power Transfer in Microwave Networks

At microwave frequency range, for a mismatched load (load other than the characteristic impedance Z_o of the line), impedance looking down a transmission line varies continuously along the line. Thus, it is convenient to express microwave power transfer in terms of reflection coefficient of the load instead of impedance [13, 14].

7.7.1 *Power Transfer from Source to a Load*

Suppose a microwave generator of voltage V_S with impedance Z_S is directly connected to a load of impedance Z_L as shown in Fig. 7.49, the load can be represented by its reflection coefficient Γ_L (amplitude including phase) with respect to the reference impedance Z_o as

$$\Gamma_L = \frac{V_L^-}{V_L^+} = \frac{Z_L - Z_o}{Z_L + Z_o} \tag{7.171a}$$

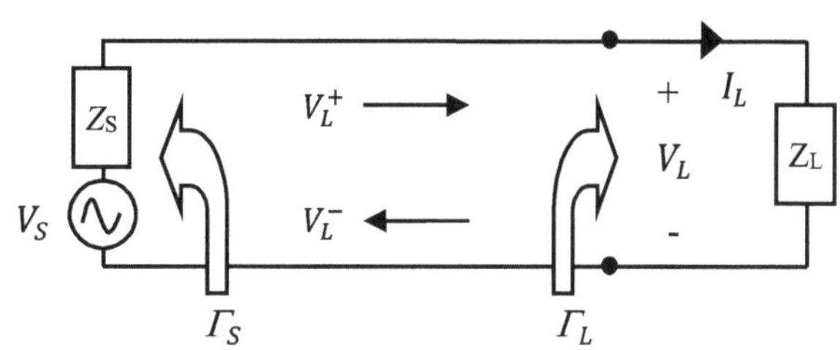

Fig. 7.49 Power transfer from source of impedance Z_S to a load Z_L

Similarly, source reflection coefficient with respect to the reference impedance Z_o is

$$\Gamma_S = \frac{Z_S - Z_o}{Z_S + Z_o} \tag{7.171b}$$

The voltage V_L across the load is given by

$$V_L = V_S \frac{Z_L}{Z_S + Z_L} = V_L^+ + V_L^- = V_L^+(1 + \Gamma_L) \tag{7.171c}$$

or

$$V_L^+ = \frac{V_S}{1 + \Gamma_L} \frac{Z_L}{Z_S + Z_L} = \frac{V_S}{2} \frac{1 - \Gamma_S}{1 - \Gamma_S \Gamma_L} \tag{7.171d}$$

Similarly,

$$V_L^- = \frac{V_S}{1 + \Gamma_L} \frac{\Gamma_L}{\Gamma_S + \Gamma_L} = \frac{V_S}{2} \frac{1 - \Gamma_S}{1 - \Gamma_S \Gamma_L} \Gamma_L \tag{7.171e}$$

The incident power to the load is given by

$$P_{incL} = \frac{1}{2} \frac{\left|V_L^+\right|^2}{Z_o} \tag{7.172a}$$

or

$$P_{incL} = \frac{1}{8} \frac{|V_S|^2}{Z_o} \frac{|1 - \Gamma_S|^2}{|1 - \Gamma_S \Gamma_L|^2} \tag{7.172b}$$

And the reflected power from the load is given by

$$P_{refL} = \frac{1}{2} \frac{\left|V_L^-\right|^2}{Z_o} \tag{7.173a}$$

or

$$P_{refL} = \frac{1}{8} \frac{|V_S|^2}{Z_o} \frac{|1 - \Gamma_S|^2}{|1 - \Gamma_S \Gamma_L|^2} |\Gamma_L|^2 \tag{7.173b}$$

Therefore, the power reached to the impedance Z_L is given by

$$P_{inL} = P_{incL} - P_{refL} \tag{7.174a}$$

or

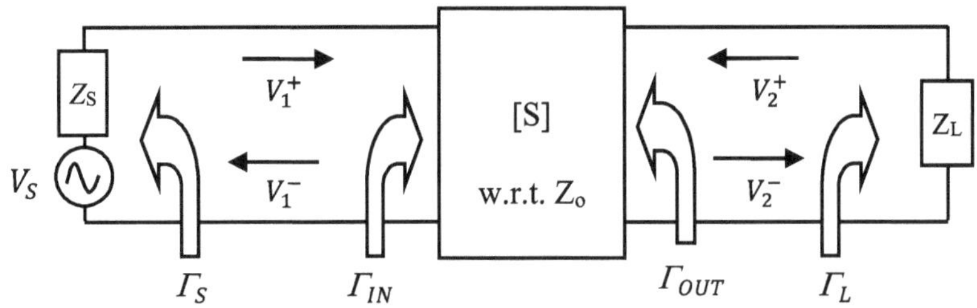

Fig. 7.50 Power transfer to a load Z_L

$$P_{inL} = P_{incL}\left(1 - \frac{P_{refL}}{P_{incL}}\right) \tag{7.174b}$$

or

$$P_{inL} = P_{incL}\left(1 - |\Gamma_L|^2\right) \tag{7.174c}$$

or

$$P_{inL} = \frac{1}{8}\frac{|V_S|^2}{Z_O}\frac{|1-\Gamma_S|^2}{|1-\Gamma_S\Gamma_L|^2}\left(1 - |\Gamma_L|^2\right) \tag{7.174d}$$

Equations (7.172b), (7.173b) and (7.174d) are the incident, reflected and delivered power to the load of impedance Z_L from a source of impedance Z_S in terms of their reflection coefficients.

In case of complex-conjugate-matched load for the given source impedance, i.e. for $Z_L = Z_S^*$, the power delivered to the load becomes

$$\text{or,}\quad P_{inL}|_{Z_L=Z_S^*} = \frac{1}{8}\frac{|V_S|^2}{Z_O}\frac{|1-\Gamma_S|^2}{1-|\Gamma_S|^2} \tag{7.174e}$$

$$\text{or,}\quad P_{inL}|_{Z_L=Z_S^*} = \frac{1}{8}\frac{|V_S|^2}{Re(Z_S)} \tag{7.174f}$$

This is nothing but the maximum power available from the generator. Thus, under matched load condition, generator delivered maximum power to the load.

7.7.2 Power Transfer to and from a 2-Port Network

A 2-port network is connected in between a source of impedance Z_S and a load of impedance Z_L as shown in Fig. 7.50. The network is represented by its S-parameters with respect to the reference impedance Z_o.

As defined in the previous section, the refection coefficients of the load and source with respect to Z_o can be written as

$$\Gamma_L = \frac{V_2^+}{V_2^-} = \frac{Z_L - Z_o}{Z_L + Z_o} \tag{7.175a}$$

$$\Gamma_S = \frac{V_1^+}{V_1^-} = \frac{Z_S - Z_o}{Z_S + Z_o} \tag{7.175b}$$

Thus, S-parameters of the network can be written as

$$V_1^- = S_{11} V_1^+ + S_{12} \Gamma_L V_2^- \tag{7.176a}$$

and

$$V_2^- = S_{21} V_1^+ + S_{22} \Gamma_L V_2^- \tag{7.176b}$$

Input reflection coefficient Γ_{IN} and output reflection coefficient Γ_{OUT} can be written as

$$\Gamma_{\mathrm{IN}} = \frac{V_1^-}{V_1^+} = S_{11} + \frac{\Gamma_L S_{12} S_{21}}{1 - \Gamma_L S_{22}} \tag{7.177a}$$

$$\Gamma_{\mathrm{OUT}} = \frac{V_2^-}{V_2^+} = S_{22} + \frac{\Gamma_S S_{12} S_{21}}{1 - \Gamma_S S_{11}} \tag{7.177b}$$

The incident voltage V_1^+ can be derived in the similar way as in previous section and can be written as

$$V_1^+ = \frac{V_S}{2} \frac{1 - \Gamma_S}{1 - \Gamma_S \Gamma_{\mathrm{IN}}} \tag{7.177c}$$

Power input to the network is given by

$$P_{\mathrm{IN}} = \frac{1}{2} \frac{\left|V_1^+\right|^2}{Z_o} \left(1 - |\Gamma_{\mathrm{IN}}|^2\right) \tag{7.177d}$$

or

$$P_{\mathrm{IN}} = \frac{1}{8} \frac{|V_S|^2}{Z_o} \frac{|1 - \Gamma_S|^2}{|1 - \Gamma_S \Gamma_{\mathrm{IN}}|^2} \left(1 - |\Gamma_{\mathrm{IN}}|^2\right) \tag{7.177e}$$

Power delivered to the load is given by

$$P_L = \frac{1}{2} \frac{\left|V_2^-\right|^2}{Z_o} \left(1 - |\Gamma_L|^2\right) \tag{7.177f}$$

From (7.176b), V_2^- can be expressed as

$$V_2^- = \frac{S_{21}V_1^+}{1 - S_{22}\Gamma_L} \tag{7.177g}$$

Putting this value of V_2^- in (7.177f) and using (7.177c),

$$P_L = \frac{1}{8}\frac{|V_S|^2}{Z_o}|S_{21}|^2\frac{(1 - |\Gamma_L|^2)|1 - \Gamma_S|^2}{|1 - S_{22}\Gamma_L|^2|1 - \Gamma_S\Gamma_{IN}|^2} \tag{7.177h}$$

Maximum power available from the source is given by

$$P_{\text{AVS}} = P_{\text{IN}}|_{\Gamma_{\text{IN}}=\Gamma_S^*} = \frac{1}{8}\frac{|V_S|^2}{Z_o}\frac{|1 - \Gamma_S|^2(1 - |\Gamma_S|^2)}{\left|1 - |\Gamma_S|^2\right|^2} \tag{7.178a}$$

or

$$P_{\text{AVS}} = \frac{1}{8}\frac{|V_S|^2}{Z_o}\frac{|1 - \Gamma_S|^2}{1 - |\Gamma_S|^2} \tag{7.178b}$$

or

$$P_{\text{AVS}} = \frac{1}{8}\frac{|V_S|^2}{Re(Z_S)} \tag{7.179}$$

Now, maximum power available from the network P_{AVN} can be determined by putting the conditions,

$$Z_L = Z_{\text{OUT}}^*, \quad \text{i.e.} \quad \Gamma_L = \Gamma_{\text{OUT}}^* \tag{7.180a}$$

Thus,

$$P_{\text{AVN}} = \frac{1}{8}\frac{|V_S|^2}{Z_o}|S_{21}|^2\frac{|1 - \Gamma_S|^2}{(1 - |\Gamma_{\text{OUT}}|^2)|1 - S_{11}\Gamma_S|^2} \tag{7.180b}$$

Transducer gain G_T is defined as the ratio of power delivered to the load to power available from the source. This can be represented by

$$G_T = \frac{P_L}{P_{AVS}} = \frac{1 - |\Gamma_S|^2}{|1 - \Gamma_S\Gamma_{\text{IN}}|^2}|S_{21}|^2\frac{1 - |\Gamma_L|^2}{|1 - S_{22}\Gamma_L|^2} \tag{7.181a}$$

Thus, the transducer gain G_T depends upon both the source impedance Γ_S and the load impedance Γ_L.

In case of load and source impedance of Z_o as shown in Fig. 7.51, $\Gamma_S = \Gamma_L = 0$. Putting these values in (7.181a), the transducer gain can be represented by

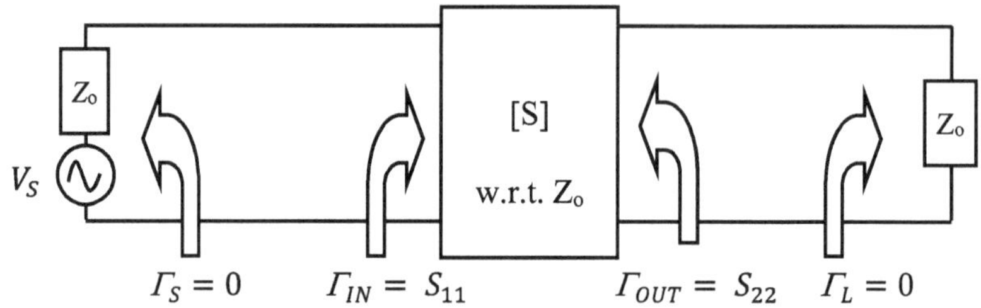

Fig. 7.51 The 2-port network with source and load impedance $= Z_o$

$$G_T = \frac{P_L}{P_{\text{AVS}}} = |S_{21}|^2 \tag{7.181b}$$

Under this condition, the transducer gain in dB is given by

$$G_T \text{ (in dB)} = 10\log\left(\frac{P_L}{P_{AVS}}\right) = 10\log\left(|S_{21}|^2\right) = 20\log(|S_{21}|) \tag{7.181c}$$

Operating power gain G_P of a network is defined as the ratio of power delivered to the load P_L to power input P_{IN} to the network. Thus, for a 2-port network, from (7.177h) and (7.177f), G_P can be written as

$$G_P = \frac{P_L}{P_{\text{IN}}} = \frac{1}{\left(1-|\Gamma_{\text{IN}}|^2\right)}|S_{21}|^2\frac{1-|\Gamma_L|^2}{|1-S_{22}\Gamma_L|^2} \tag{7.182}$$

The operating power gain depends on the load impedance Γ_L; however, it is independent of the source impedance Γ_S and output impedance Γ_{OUT}.

Available power gain G_A of a network is defined as the ratio of power available from the network P_{AVN} to power available from the source P_{AVS}. Thus, for a 2-port network, from (7.180b) and (7.178b), G_A can be written as

$$G_A = \frac{P_{\text{AVN}}}{P_{\text{AVS}}} = \frac{1-|\Gamma_S|^2}{|1-S_{11}\Gamma_S|^2}|S_{21}|^2\frac{1}{\left(1-|\Gamma_{\text{OUT}}|^2\right)} \tag{7.183}$$

The available power gain depends upon the input matching condition, Γ_S, because actual power transferred to the network (device) is not necessarily the same as available power from the source which happens only for conjugate match at input. However, available power gain G_A is independent of the input impedance Γ_{IN} and load impedance Γ_L.

Example 7.18 A 2-port network has the following scattering matrix:

$$[S] = \begin{bmatrix} 0\angle 0^\circ & 1/3\angle 0^\circ \\ 1/3\angle 0^\circ & 0\angle 0^\circ \end{bmatrix}$$

Compute the input return loss when the output of the network is terminated by a (a) matched load, (b) electrical short and (c) electrical open.

Solution

Input reflection coefficient of a 2-port network is given by (7.177a)

$$\Gamma_{\mathrm{IN}} = S_{11} + \frac{\Gamma_L S_{12} S_{21}}{1 - \Gamma_L S_{22}}$$

(a) In case of matched load, the load reflection coefficient $\Gamma_L = 0$. Thus, input reflection coefficient $\Gamma_{\mathrm{IN}} = 0$. Thus, input return loss is infinite in dB.

(b) In case of shorted load, $Z_L = 0$, the load reflection coefficient,

$$\Gamma_L = \frac{Z_L - Z_o}{Z_L + Z_o} = \frac{0 - Z_o}{0 + Z_o} = -1$$

Thus, input reflection coefficient,

$$\Gamma_{\mathrm{IN}} = 0\angle 0^\circ + \frac{-1 \times 1/3\angle 0^\circ \times 1/3\angle 0^\circ}{1 + 0} = -1/9$$

Thus, input return loss (RL) is given by

$$\mathrm{RL} = -20\log(|\Gamma_{\mathrm{IN}}|) = 20\log(1/9) = 19.08\ \mathrm{dB}$$

(b) In case of open load, $Z_L = \propto$. The load reflection coefficient,

$$\Gamma_L = \frac{Z_L - Z_o}{Z_L + Z_o} = 1$$

Thus, input reflection coefficient,

$$\Gamma_{\mathrm{IN}} = 0\angle 0^\circ + \frac{1 \times 1/3\angle 0^\circ \times 1/3\angle 0^\circ}{1 - 0} = 1/9$$

Thus, input return loss is infinite in dB,

$$\mathrm{RL} = -20\log(|\Gamma_{\mathrm{IN}}|) = 20\log(1/9) = 19.08\ \mathrm{dB}$$

Example 7.19 Compute S-parameter matrix of a 2-port network which consists of a 50 Ω series resistor in a 50 Ω system. Also, derive the port return losses and insertion loss of the network.

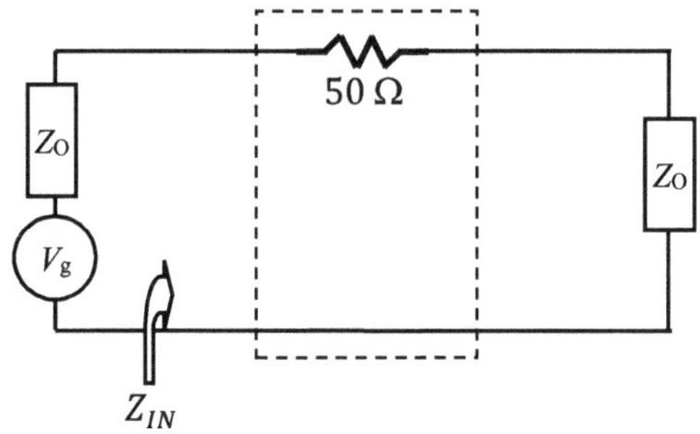

Fig. 7.52 S-parameter determination of a 2-port series network

Solution

The network is shown in Fig. 7.52. S_{11} is the input reflection coefficient when output of the network is terminated by the reference impedance 50 Ω and is given by

$$S_{11} = \frac{Z_{\text{IN}} - Z_o}{Z_{\text{IN}} + Z_o}$$

Here, $Z_{\text{IN}} = 50 + Z_o = 100\ \Omega$. Thus,

$$S_{11} = \frac{100 - 50}{100 + 50} = \frac{1}{3}$$

The network is symmetric. Thus, $S_{11} = S_{22} = 1/3$.

To compute S_{21}, input port of the network is driven by a voltage source v_g with source impedance $Z_o = 50\ \Omega$ and the output is terminated by load impedance $Z_o = 50\ \Omega$.

The load voltage v_o is given by

$$v_o = \frac{v_g \times Z_o}{Z_o + 50 + Z_o} = \frac{v_g}{3}$$

Thus,

$$S_{21} = \frac{v_o}{v_g/2} = \frac{2}{3}$$

The network is symmetric. Thus, $S_{12} = S_{21} = 2/3$.

Therefore, the S-parameter matrix of the network is given by

$$\begin{bmatrix} S_{11} & S_{12} \\ S_{21} & S_{22} \end{bmatrix} = \begin{bmatrix} 1/3\angle 0^\circ & 2/3\angle 0^\circ \\ 2/3\angle 0^\circ & 1/3\angle 0^\circ \end{bmatrix}$$

The same result can also be obtained using (7.162c), putting $Z_a = 50\ \Omega$ and $Z_o = 50\ \Omega$ as:

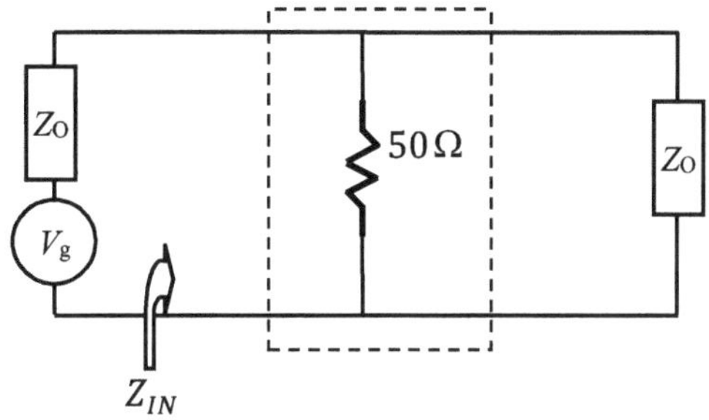

Fig. 7.53 S-parameter determination of a 2-port series network

$$\begin{bmatrix} S_{11} & S_{12} \\ S_{21} & S_{22} \end{bmatrix} = \begin{bmatrix} \frac{Z_a}{2Z_o+Z_a} & \frac{2Z_o}{2Z_o+Z_a} \\ \frac{2Z_o}{2Z_o+Z_a} & \frac{Z_a}{2Z_o+Z_a} \end{bmatrix} = \begin{bmatrix} 1/3\angle 0^\circ & 2/3\angle 0^\circ \\ 2/3\angle 0^\circ & 1/3\angle 0^\circ \end{bmatrix}$$

The input and output return losses of the network (50 Ω series resistor) in a 50 Ω system are

$$\mathrm{RL} = -20\log|S_{11}| = -20\log|S_{22}| = -20\log|1/3| = 9.54 \text{ dB}$$

The insertion loss of the network is

$$\mathrm{IL} = -20\log|S_{21}| = -20\log|2/3| = 3.52 \text{ dB}$$

Example 7.20 Compute S-parameter matrix of a 2-port network which consists of a 50 Ω shunt resistor in a 50 Ω system. Also, derive the port return losses and insertion loss of the network.

Solution

The network is shown in Fig. 7.53. S_{11} is the input reflection coefficient when output of the network is terminated by the reference impedance 50 Ω and is given by

$$S_{11} = \frac{Z_{\mathrm{IN}} - Z_o}{Z_{\mathrm{IN}} + Z_o}$$

Here, $Z_{\mathrm{IN}} = 50||Z_o = 25\ \Omega$. Thus,

$$S_{11} = \frac{25 - 50}{25 + 50} = -\frac{1}{3}$$

The network is symmetric. Thus, $S_{11} = S_{22} = -1/3$.

To compute S_{21}, input port of the network is driven by a voltage source v_g with source impedance $Z_o = 50\ \Omega$ and the output is terminated by load impedance $Z_o = 50\ \Omega$.

The load voltage v_o is given by

$$v_o = \frac{v_g \times 50||Z_o}{Z_o + 50||Z_o} = \frac{v_g}{3}$$

Thus,

$$S_{21} = \frac{v_o}{v_g/2} = \frac{2}{3}$$

The network is symmetric. Thus, $S_{12} = S_{21} = 2/3$.
Therefore, the S-parameter matrix of the network is given by

$$\begin{bmatrix} S_{11} & S_{12} \\ S_{21} & S_{22} \end{bmatrix} = \begin{bmatrix} 1/3\angle 180^\circ & 2/3\angle 0^\circ \\ 2/3\angle 0^\circ & 1/3\angle 180^\circ \end{bmatrix}$$

The same result can also be obtained using (7.163d), putting $Z_a = 50\ \Omega$ and $Z_o = 50\ \Omega$ as

$$\begin{bmatrix} S_{11} & S_{12} \\ S_{21} & S_{22} \end{bmatrix} = \begin{bmatrix} \frac{-Z_o}{Z_o+2Z_a} & \frac{2Z_a}{Z_o+2Z_a} \\ \frac{2Z_a}{Z_o+2Z_a} & \frac{-Z_o}{Z_o+2Z_a} \end{bmatrix} = \begin{bmatrix} 1/3\angle 180^\circ & 2/3\angle 0^\circ \\ 2/3\angle 0^\circ & 1/3\angle 180^\circ \end{bmatrix}$$

The input and output return losses of the network (50 Ω series resistor) in a 50 Ω system are

$$\mathrm{RL} = -20\log|S_{11}| = -20\log|S_{22}| = -20\log|1/3| = 9.54\ \mathrm{dB}$$

The insertion loss of the network is

$$\mathrm{IL} = -20\log|S_{21}| = -20\log|2/3| = 3.52\ \mathrm{dB}$$

Example 7.21 The S-parameter matrix of a 3-port network is as follows:

$$[S] = \begin{bmatrix} 0 & 0 & 1\angle 0^\circ \\ 1\angle 0^\circ & 0 & 0 \\ 0 & 1\angle 0^\circ & 0 \end{bmatrix} \tag{7.184}$$

The 3-port network is converted to a 2-port network by terminating its port-3 with a matched load as shown in Fig. 7.55. Compute the S-parameter matrix of the converted 2-port network. Identify both the networks.

Solution

The S-parameters of the 3-port network are related to the different signal levels as

$$b_1 = 0 \times a_1 + 0 \times a_2 + 1 \times a_3 \tag{7.185a}$$

$$b_2 = 1 \times a_1 + 0 \times a_2 + 0 \times a_3 \tag{7.185b}$$

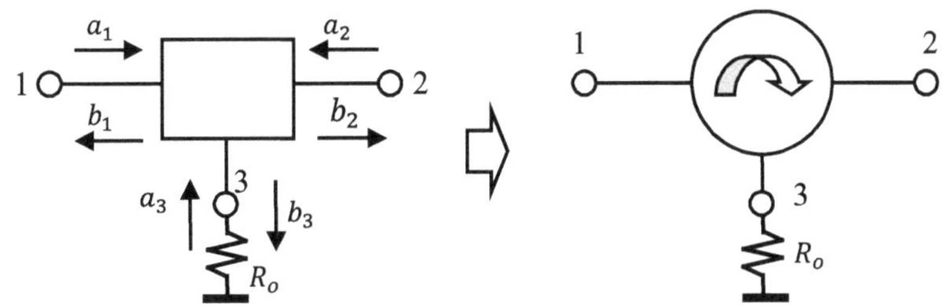

Fig. 7.54 3-port network with third port matched termination

Fig. 7.55 3-port network with third port shorted termination

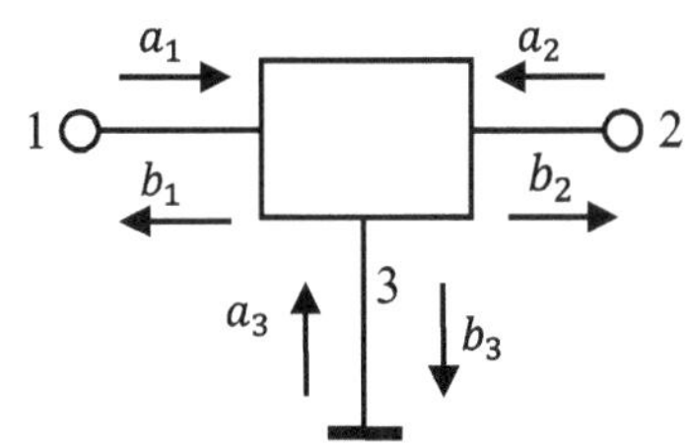

$$b_3 = 0 \times a_1 + 1 \times a_2 + 0 \times a_3 \tag{7.185c}$$

The third port is matched terminated, and thus,

$$a_3 = 0 \tag{7.186}$$

Thus, the 3-port network becomes a 2-port network. Combining (7.185) and (7.186), we can write,

$$b_1 = 0 \times a_1 + 0 \times a_2 \tag{7.187a}$$

$$b_2 = 1 \times a_1 + 0 \times a_2 \tag{7.187b}$$

Thus, the S-parameter matrix of the converted 2-port network,

$$[S] = \begin{bmatrix} 0 & 0 \\ 1\angle 0^\circ & 0 \end{bmatrix} \tag{7.188}$$

The 3-port S-parameter matrix given in (7.184) is for a non-reciprocal device called circulator as shown in Fig. 7.54. S-parameter matrix of the network with third port matched termination is given by (7.188). This is the S-parameter matrix of isolators. Thus, a circulator with its one port matched termination works as an isolator (Fig. 7.54).

Example 7.22 The S-parameter matrix of a 3-port network is as follows:

$$[S] = \begin{bmatrix} 0 & 0 & 1\angle 0^\circ \\ 1\angle 0^\circ & 0 & 0 \\ 0 & 1\angle 0^\circ & 0 \end{bmatrix} \quad (7.189)$$

The 3-port network is converted to a 2-port network by terminating its port-3 with a shorted load as shown in Fig. 7.55. Compute the S-parameter matrix of the converted 2-port network.

Solution
The S-parameters of the 3-port network are related to the different signal levels as

$$b_1 = 0 \times a_1 + 0 \times a_2 + 1 \times a_3 \quad (7.190a)$$
$$b_2 = 1 \times a_1 + 0 \times a_2 + 0 \times a_3 \quad (7.190b)$$
$$b_3 = 0 \times a_1 + 1 \times a_2 + 0 \times a_3 \quad (7.190c)$$

The third port is terminated by an electrical short, and thus,

$$a_3 = -b_3 \quad (7.191a)$$

Using (7.190c) and (7.191a),

$$a_3 = -a_2 \quad (7.191b)$$

Putting $a_3 = -a_2$ into (7.190a) and (7.190b), we can write

$$b_1 = 0 \times a_1 + 0 \times a_2 + 1 \times (-a_2) \quad (7.192a)$$
$$b_2 = 1 \times a_1 + 0 \times a_2 + 0 \times (-a_2) \quad (7.192b)$$

or

$$b_1 = 0 \times a_1 + (-1)1 \times a_2 \quad (7.192c)$$
$$b_2 = 1 \times a_1 + 0 \times a_2 \quad (7.192d)$$

Thus, the S-parameter matrix of the converted 2-port network is

$$[S] = \begin{bmatrix} 0 & 1\angle 180^\circ \\ 1\angle 0^\circ & 0 \end{bmatrix} \quad (7.193)$$

Example 7.23 A 50 Ω transmission line is terminated by a load impedance of $(100 + j100)$ Ω. Calculate the magnitude of load reflection coefficient.

Solution
Magnitude of reflection coefficient of load impedance Z_L is given by

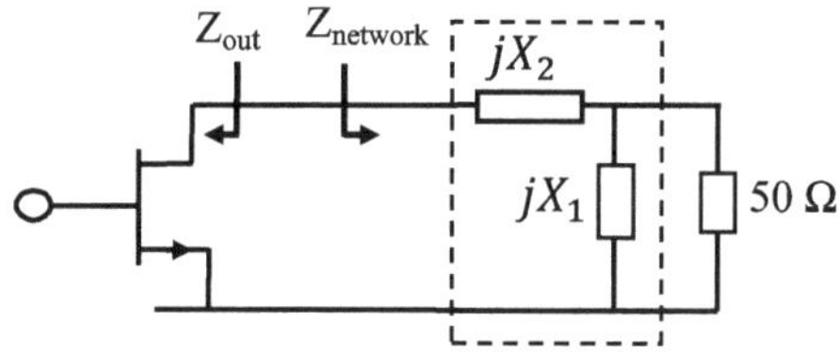

Fig. 7.56 Circuit diagram of Example 7.24

$$|S_{11}| = \left|\frac{Z_L - Z_o}{Z_L + Z_o}\right|$$

Putting $Z_L = (100 + j100)\,\Omega$ and $Z_o = 50\,\Omega$,

$$\begin{aligned}|S_{11}| &= \left|\frac{100 + j100 - 50}{100 + j100 + 50}\right| \\ &= \left|\frac{1 + j2}{3 + j2}\right| \\ &= \sqrt{\frac{5}{13}}\end{aligned}$$

Example 7.24 A unilateral transistor has output impedance $Z_{out} = (10 - j10)\,\Omega$. Calculate the value of the series and shunt components of the matching network as shown in Fig. 7.56 for complex conjugate match at the output of the device to 50 Ω load.

Solution

The impedance of the network is

$$Z_{nework} = jX_2 + jX_1||50$$

For complex conjugate matching at the output of the device,

$$Z_{nework} = Z^*_{out} = 10 + j10$$

Therefore,

$$jX_2 + jX_1||50 = 10 + j10$$

or

$$-X_1X_2 + j50(X_1 + X_2) = j10(X_1 + 50) + (500 - 10X_1)$$

On equating real and imaginary parts equal for both the sides separately

$$X_1 = +25$$

$$X_2 = -10 \quad \text{for} \quad X_1 = +25$$
$$X_2 = +30 \quad \text{for} \quad X_1 = -25$$

Thus, the two sets of series and shunt components are

$$jX_1 = +j25, \quad jX_2 = -j10$$

and

$$jX_1 = -j25, \quad jX_2 = +j30$$

Example 7.25 Input resistance of a short-circuited lossless transmission line is 25 Ω. The input impedance becomes 100 Ω when its other end is kept open. What will be the characteristic impedance of the transmission line?

Solution

Input impedance of a lossless transmission line of characteristic impedance Z_o when terminated with load impedance Z_L is given by (7.26)

$$Z_{\text{in}} = Z_o \frac{Z_L + jZ_o \tan(\beta l)}{Z_o + jZ_L \tan(\beta l)}$$

Therefore, input impedance of a shorted ($Z_L = \infty$) ($Z_L =$) line is given by

$$Z_{\text{inShort}} = jZ_o \tan(\beta l) \tag{7.194}$$

And the input impedance of an open-circuited ($Z_L =$) line is given by

$$Z_{\text{inOpen}} = -jZ_o \cot(\beta l) \tag{7.195}$$

Multiplying (7.194) and (7.195), it can be written as

$$Z_O = \sqrt{Z_{\text{inShort}} \times Z_{\text{inOpen}}} \tag{7.196}$$

Putting $Z_{\text{inShort}} = 25$ and $Z_{\text{inOpen}} = 100\ \Omega$ in (7.196), the characteristic resistance of the line is

$$Z_O = \sqrt{25 \times 100}\ \Omega$$
$$= 50\ \Omega$$

Example 7.26 The *S*-parameters of a transistor at 14 GHz are as follows:

$$S_{11} = 0.498\angle 121°, \quad S_{12} = 0.1137\angle -49°$$
$$S_{21} = 3.080\angle -30°, \quad S_{22} = 0.225\angle -150°$$

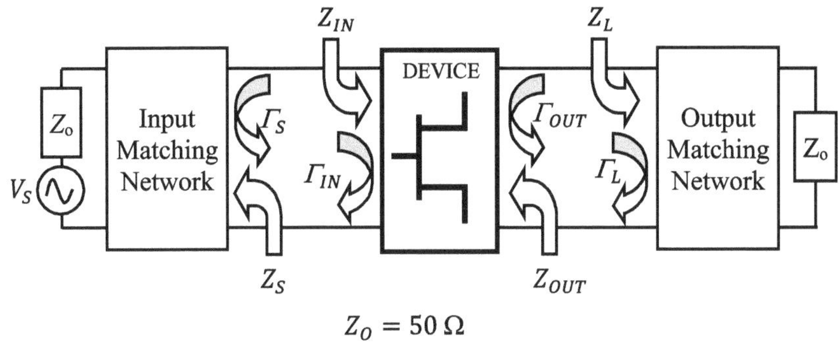

Fig. 7.57 Microwave amplifier with its associated matching networks

Calculate (a) transducer power gain, (b) available power gain and (c) operating power gain for the amplifier with input and output matching networks of reflection coefficients (Fig. 7.57)

$$\Gamma_S = 0.455\angle - 160°, \quad \Gamma_L = 0.58\angle 175°$$

Solution

With reference to the Fig. 7.57, The input reflection coefficient (7.177a)

$$\begin{aligned}\Gamma_{\mathrm{IN}} &= S_{11} + \frac{\Gamma_L S_{12} S_{21}}{1 - \Gamma_L S_{22}} \\ &= 0.498\angle 121° + \frac{0.58\angle 175° \times 0.1137\angle - 49° \times 3.080\angle - 30°}{1 - 0.58\angle 175° \times 0.225\angle - 150°} \\ &= -0.256 + j0.427 + \frac{-0.0212 + j0.2020}{0.8817 - j0.0552} \\ &= (-0.256 + j0.427) + (-0.0383 + j0.2267) \\ &= -0.2947 + j6536 = 0.717\angle - 65.726°\end{aligned}$$

The output reflection coefficient (7.177b)

$$\begin{aligned}\Gamma_{\mathrm{OUT}} &= S_{22} + \frac{\Gamma_S S_{12} S_{21}}{1 - \Gamma_S S_{11}} \\ &= 0.225\angle - 150° + \frac{0.455\angle - 160° \times 0.1137\angle - 49° \times 3.080\angle - 30°}{1 - 0.455\angle - 160° \times 0.498\angle 121°} \\ &= -0.195 - j0.113 + \frac{-0.082 + j0.137}{0.824 + j0.143} \\ &= (-0.195 - j0.113) + (-0.069 + j0.178) \\ &= -0.264 + j0.065 = 0.272\angle - 13.885°\end{aligned}$$

Transducer gain G_T is defined as the ratio of power delivered to the load to power available from the source and is given by (7.181a)

$$G_T = \frac{P_L}{P_{\text{AVS}}} = \frac{1-|\Gamma_S|^2}{|1-\Gamma_S\Gamma_{\text{IN}}|^2}|S_{21}|^2\frac{1-|\Gamma_L|^2}{|1-S_{22}\Gamma_L|^2}$$

$$= \frac{1-|0.455\angle-160°|^2}{|1-0.455\angle-160° \times 0.717\angle-65.726°|^2}$$
$$\times |3.080\angle-30°|^2$$
$$\times \frac{1-|0.58\angle175°|^2}{|1-0.225\angle-150° \times 0.58\angle175°|^2}$$
$$= \frac{0.798}{0.651} \times 9.486 \times \frac{0.664}{0.780}$$
$$= 1.218 \times 9.486 \times 0.850$$
$$= 9.83 = 9.92 \text{ dB}$$

Operating power gain G_P of a network is defined as the ratio of power delivered to the load P_L to power input P_{IN} to the network and is given by (7.182)

$$G_P = \frac{P_L}{P_{\text{IN}}} = \frac{1}{\left(1-|\Gamma_{\text{IN}}|^2\right)}|S_{21}|^2\frac{1-|\Gamma_L|^2}{|1-S_{22}\Gamma_L|^2}$$

$$= \frac{1}{\left(1-|0.717\angle-65.726°|^2\right)}$$
$$\times |3.080\angle-30°|^2$$
$$\times \frac{1-|0.58\angle175°|^2}{|1-0.225\angle-150° \times 0.58\angle175°|^2}$$
$$= \frac{1}{0.486} \times 9.486 \times \frac{0.664}{0.780}$$
$$= 2.058 \times 9.486 \times 0.850$$
$$= 16.60 = 12.2 \text{ dB}$$

Available power gain G_A of a network is defined as the ratio of power available from the network P_{AVN} to power available from the source P_{AVS} and is given by (7.183)

$$G_A = \frac{P_{\text{AVN}}}{P_{\text{AVS}}} = \frac{1-|\Gamma_S|^2}{|1-S_{11}\Gamma_S|^2}|S_{21}|^2\frac{1}{\left(1-|\Gamma_{\text{OUT}}|^2\right)}$$

$$= \frac{1-|0.455\angle-160°|^2}{|1-0.498\angle121° \times 0.455\angle-160°|^2}$$
$$\times |3.080\angle-30°|^2$$

$$\times \frac{1}{\left(1 - |0.272\angle - 13.885^\circ|^2\right)}$$
$$= \frac{0.793}{0.699} \times 9.486 \times \frac{1}{0.926}$$
$$= 1.134 \times 9.486 \times 1.080$$
$$= 11.62 = 10.65 \text{ dB}$$

This example shows different gains of the amplifier for the same matching condition, i.e. for $\Gamma_S = 0.455\angle - 160^\circ$ and $\Gamma_L = 0.58\angle 175^\circ$. Practically, gain of an amplifier is measured by the ratio of power delivered to the load to the power available from the source. In this case, the measured power gain is nothing but the transducer power gain, G_T which is 9.92 dB in this example.

Example 7.27 The S-parameters of a transistor at 14 GHz are as follows:

$$S_{11} = 0.653\angle - 160^\circ, \quad S_{12} = 0.000$$
$$S_{21} = 2.565\angle 30^\circ, \quad S_{22} = 0.452\angle - 90^\circ$$

Calculate (a) transducer power gain, (b) available power gain and (c) operating power gain for the amplifier with input and output matching networks of reflection coefficients

$$\Gamma_S = 0.653\angle 160^\circ, \quad \Gamma_L = 0.452\angle 90^\circ$$

Solution

The input reflection coefficient (7.177a)

$$\Gamma_{\text{IN}} = S_{11} + \frac{\Gamma_L S_{12} S_{21}}{1 - \Gamma_L S_{22}}$$

Here, $S_{12} = 0.000$, therefore:

$$\Gamma_{\text{IN}} = S_{11} = 0.653\angle - 160^\circ$$

The output reflection coefficient (7.177b)

$$\Gamma_{\text{OUT}} = S_{22} + \frac{\Gamma_S S_{12} S_{21}}{1 - \Gamma_S S_{11}}$$

Here, $S_{12} = 0.000$, therefore:

$$\Gamma_{\text{OUT}} = S_{22} = 0.452\angle - 90^\circ$$

Transducer gain G_T is defined as the ratio of power delivered to the load to power available from the source and is given by (7.181a)

$$
\begin{aligned}
G_T = \frac{P_L}{P_{\text{AVS}}} &= \frac{1-|\Gamma_S|^2}{|1-\Gamma_S\Gamma_{\text{IN}}|^2}|S_{21}|^2\frac{1-|\Gamma_L|^2}{|1-S_{22}\Gamma_L|^2} \\
&= \frac{1-|0.653\angle 160°|^2}{|1-0.653\angle 160° \times 0.653\angle -160°|^2} \\
&\quad \times |2.565\angle 30°|^2 \\
&\quad \times \frac{1-|0.452\angle 90°|^2}{|1-0.452\angle -90° \times 0.452\angle 90°|^2} \\
&= \frac{0.574}{0.329} \times 6.579 \times \frac{0.796}{0.633} \\
&= 1.743 \times 6.579 \times 1.257 \\
&= 14.42 = 11.59\ \text{dB}
\end{aligned}
$$

Operating power gain G_P of a network is defined as the ratio of power delivered to the load P_L to power input P_{IN} to the network and is given by (7.182)

$$
\begin{aligned}
G_P = \frac{P_L}{P_{\text{IN}}} &= \frac{1}{\left(1-|\Gamma_{\text{IN}}|^2\right)}|S_{21}|^2\frac{1-|\Gamma_L|^2}{|1-S_{22}\Gamma_L|^2} \\
&= \frac{1}{\left(1-|0.653\angle -160°|^2\right)} \\
&\quad \times |2.565\angle 30°|^2 \\
&\quad \times \frac{1-|0.452\angle 90°|^2}{|1-0.452\angle -90° \times 0.452\angle 90°|^2} \\
&= \frac{1}{0.574} \times 6.579 \times \frac{0.796}{0.633} \\
&= 1.743 \times 6.579 \times 1.257 \\
&= 14.42 = 11.59\ \text{dB} = G_T!!
\end{aligned}
$$

Available power gain G_A of a network is defined as the ratio of power available from the network P_{AVN} to power available from the source P_{AVS} and is given by (7.183)

$$
\begin{aligned}
G_A = \frac{P_{\text{AVN}}}{P_{\text{AVS}}} &= \frac{1-|\Gamma_S|^2}{|1-S_{11}\Gamma_S|^2}|S_{21}|^2\frac{1}{\left(1-|\Gamma_{OUT}|^2\right)} \\
&= \frac{1-|0.653\angle 160°|^2}{|1-0.653\angle 160° \times 0.653\angle -160°|^2} \\
&\quad \times |2.565\angle 30°|^2 \\
&\quad \times \frac{1}{\left(1-|0.452\angle -90°|^2\right)}
\end{aligned}
$$

$$= \frac{0.574}{0.329} \times 6.579 \times \frac{1}{0.796}$$
$$= 1.743 \times 6.579 \times 1.257$$
$$= 14.42 = 11.59 \text{ dB} = G_T!! = G_P!!!$$

This example shows that the same gain value of 11.59 dB for transducer, operating and available power gain. In this case, it can be noticed that the device is unilateral, i.e. $S_{12} = 0.000$ and the source and load reflection coefficients $\Gamma_S = 0.653\angle 160° = S_{11}^*$ and $\Gamma_L = 0.452\angle 90° = S_{22}^*$, respectively, make the amplifier perfectly (complex conjugate) matched at its both ports. Actually, any amplifier with input and output ports perfectly matched provides same gain values corresponding to transducer, operating and available power gain.

References

1. Gupta KC (1983) Microwaves. Wiley Eastern Limited
2. Marks RB, Williams DF (1992) A general waveguide circuit theory. J Res Nat Inst Stand Technol 97:533–562
3. Uchimura H, Takenoshita T, Fujii M (1998) Development of a laminated waveguide. IEEE Trans Micro Theory Tech 46(12):2438–2443
4. Deslandes D, Wu K (2003) Single-substrate integration technique of planar circuits and waveguide filters. IEEE Trans Micro Theory Techs 53(9):2968–2977
5. Wu K, Zhang YL, Hong W (2005) Novel substrate integrated waveguide cavity filter with defected ground structure. IEEE Trans Micro Theory Tech 53(4)
6. Bozzi M, Georgiadis A, Wu K (2011) Review of substrate-integrated waveguide circuits and antennas. IET Microw Antennas Propag 5:909–920
7. Belenguer A, Esteban H, Boria V (2014) Novel empty substrate integrated waveguide for high-performance microwave integrated circuits. IEEE Trans Microw Theory Tech 62:832–839
8. Kurokawa K (1965) Power waves and the scattering matrix. IEEE Trans Microw Theory Tech 13(3):194–202
9. Tippet JC, Speciale RA (1982) A rigorous technique for measuring the scattering matrix of a multiport device with a 2-port network analyzer. IEEE Trans Microw Theory Tech 30:661–666
10. S parameters… circuit analysis and design. Hewlett Packard Application Note, 95A, Feb 1973
11. Frickey DA (1994) Conversions between S, Z, Y, H, ABCD, and T parameters which are valid for complex source and load impedances. IEEE Trans Microwave Theory Tech 42:205–211
12. Marks RB, Williams DF (1995) Comments on conversions between S, Z, Y, H, A BCD, and T parameters which are valid for complex source and load impedances. IEEE Trans Microwave Theory Tech 43:914–915
13. Stanculescu M, Iordache L, Iordache M, Niculae D, Bucata V (2017) Using S parameters in wireless power transfer analysis. In: The 10th international symposium on advanced topic in electrical engineering, pp 107–112, March 23–25, 2017
14. Sah AK, Pant DP (2014) Analysis and optimization of wireless power transfer link. In: Proceedings of international conference on technology and innovation management and IOE graduate conference, pp 82–93, Oct 10–11, 2014

Chapter 8
Microwave Switches

Abstract This chapter presents design and analysis of various types of microwave switches using p-i-n diodes and transistors. It describes series-, shunt-, TEE- and π-type switches with their merits and demerits. Various applications of active switches in microwave communications are also presented.

8.1 Introduction

Microwave switches are used in most modern systems including satellite communications, cell phone communications, software-defined radios, radars, and many other applications. The key features of RF switches are the combination of low insertion loss at ON state, high isolation at OFF state, high power handling capability, high linearity, low DC power consumption and compact in size enabling easy integration with other circuitry. The functions of RF/microwave switches are to reconfigure systems by diverting signals from one direction to another. The switches are also used to realize step wise variable attenuators, phase shifters and other elements. From realization point of view, there are two types of microwave switches, electromechanical and solid-state switches. Generally, solid-state switches are very compact in size, faster in operation and more reliable than electromechanical. However, solid-state switches have higher insertion loss and also exhibit relatively more nonlinearity at higher level of signal power. The p-i-n diodes, various FETs and HEMTs are used as solid-state switching device. Among them, p-i-n diodes are widely used for high-frequency applications as solid-state switching device for its high RF power handling capability and comparatively lower insertion loss. MESFETs, and various HEMTs are used for low-power microwave/RF switch for easy realization in MMIC technology integrated with other circuitry. Extremely high two-dimensional electron gas (2-DEG) of GaN HEMTs results in very low ON resistance, and due to its high breakdown fields it is capable to operate at higher operating temperatures which makes it suitable for realization of high power switches.

S. C. Bera, *Microwave Active Devices and Circuits for Communication*, Lecture Notes in Electrical Engineering 533, https://doi.org/10.1007/978-981-13-3004-9_8

8.2 Switch Circuits Based on P-I-N Diode

Low ON resistance at forward bias and high OFF resistance at reverse bias of a p-i-n diode enables it to be used as a switching device [1–5]. Si- and GaAs-based p-i-n diodes are mostly used for switching applications. However, SiC based p-i-n diode-based switches are also developed for high power applications for their higher power handling capability [4, 5]. The function of a switch is either to allow a signal to a specific signal path or to block the signal from following this path. In case of ideal switch, circuit paths are either closed by providing zero impedance or opened by an ideal open circuit with infinite impedance. RF/microwave switches based on p-i-n diodes utilize DC voltage/current dependent variable impedance of the p-i-n diode to direct the signal flow. Practically, RF/microwave p-i-n diodes cannot provide ideally short or open impedance. Thus, at ON state, always there is some amount of incident signal absorbed by the switch and some amount also will be reflected by the switch. This reduction of signal power leads to non-zero value of insertion loss (IL) which is the ratio of the output signal power to the input signal power. Similarly, at OFF state of a switch, there is always some small amount of signal passing through the switch. This leads to finite value of isolation. At OFF state, isolation of a switch is also defined as the ratio of output power to the input power.

Thus, in case of source and load impedance of Z_o, the insertion loss and isolation of a switch in dB is given by

$$20 \log_{10}(|S_{21}|) \tag{8.1}$$

Here, S_{21} is the forward transmission coefficient of the switch at ON and OFF state for insertion loss and isolation, respectively. Sometimes, isolation of a switch is defined as the ratio of insertion loss at OFF to ON state, i.e. loss at OFF state with respect to loss at ON state.

RF/microwave switches are implemented in two ways: reflective-type and absorptive-type as shown in Figs. 8.1 and 8.2 respectively. In ON state, both types of switches connect the signal path to the load through a very low-impedance path. However, in OFF state they work differently. In case of reflective switch, at OFF state, two extreme impedances, i.e. either short or open impedances are created across the signal path to reflect back most of the signals, ideally, full signal to its source. Due to the maximum mismatch in both the extreme impedance cases, signals are reflected back to provide desired isolation. In case of absorptive-type switch at OFF state, switch connects the signal to a matched dummy load of impedance Z_o. Therefore, at OFF state, the signals will be absorbed by the dummy load and there will be no reflected power to the source.

As discussed in Chap. 2 that the p-i-n diode can be used as voltage-/current-controlled RF/microwave impedance. Figure 8.3 shows the schematic circuit diagram of switches using p-i-n diode in series or shunt configuration as variable impedance. Under positive control voltage V_C, the p-i-n diode operates in forward-bias condition and acts as low impedance, whereas under negative voltage V_C, the

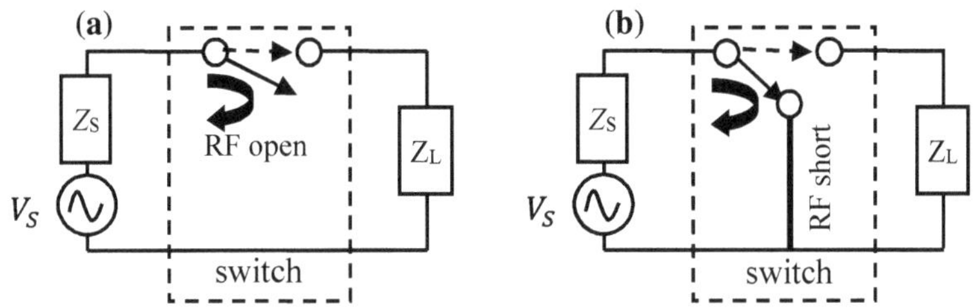

Fig. 8.1 Reflective switch is in OFF state **a** open in the RF path, **b** short across the RF path

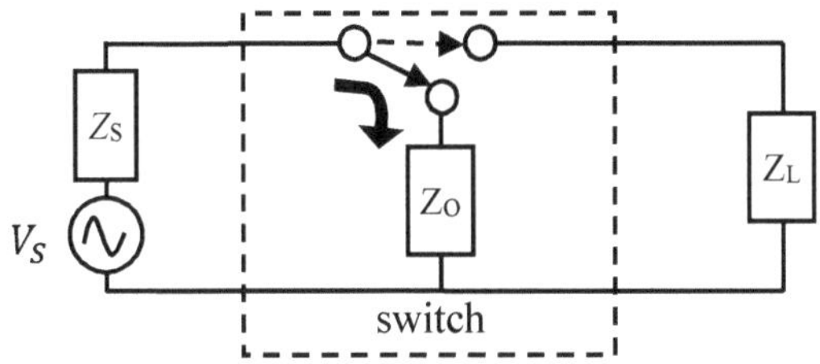

Fig. 8.2 Absorptive RF/microwave switch

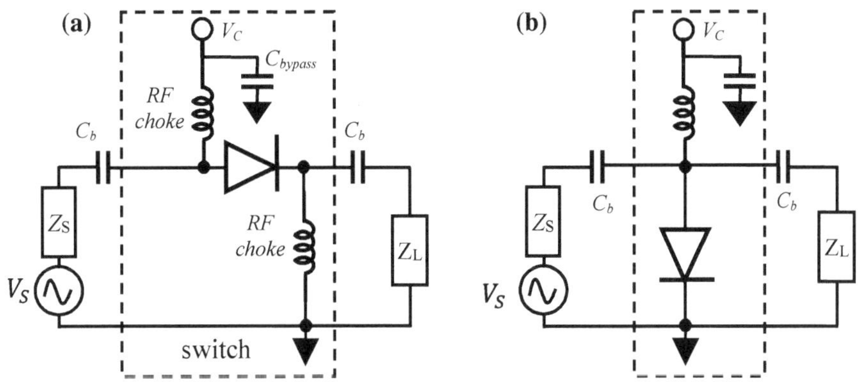

Fig. 8.3 Reflective switches configuration **a** series diode **b** shunt diode

p-i-n diode operates in reverse-bias condition and acts as high impedance. Thus, in case of low-impedance state, p-i-n diode operates in forward-bias condition and in high-impedance state the p-i-n diode operates in reverse-bias condition. The bias voltage/current is provided to the diode through RF choke and bypass capacitor, the details of which will be discussed in the driver circuit section. DC blocking capacitors (C_b) are used to block DC voltage towards generator and load without affecting RF performance.

Under forward-bias condition, the equivalent RF circuit of a p-i-n diode is shown in Fig. 8.4. At low-impedance state, the intrinsic layer resistance R_i is very low compared to the capacitive reactance (i.e. $R_i \ll 1/2\pi f C_i$). The RF/microwave frequency range

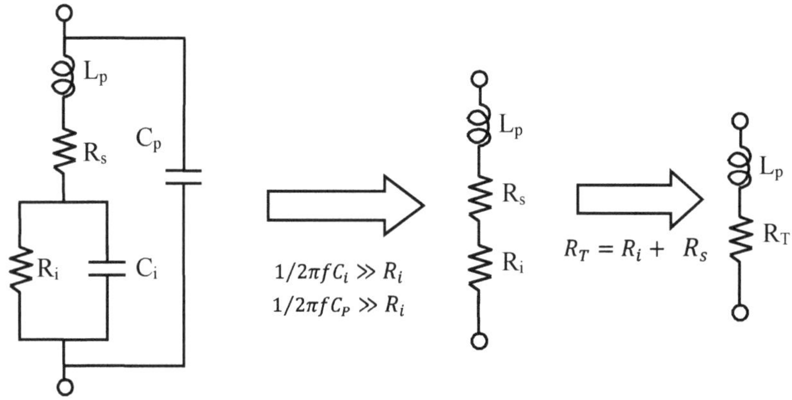

Fig. 8.4 Equivalent circuit of p-i-n diode at low-impedance state (forward-bias condition)

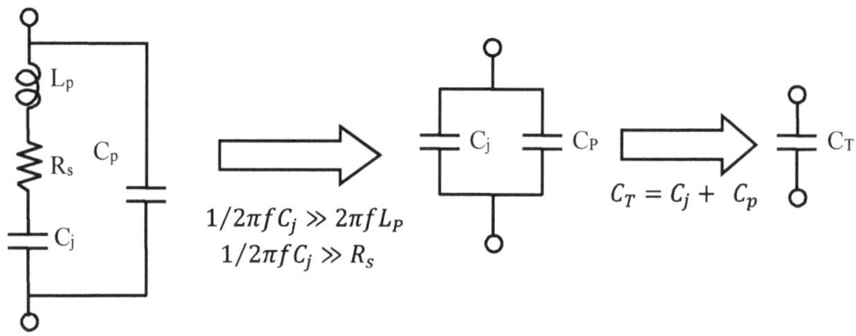

Fig. 8.5 Equivalent circuit of p-i-n diode at high-impedance state (reverse-bias condition)

where the effect of package capacitance is negligible compared to the highly forward-biased resistance $R_i + R_s$, the equivalent low impedance (Z_{TL}) of the diode can be considered as series combination of $(R_i + R_s) = R_T$ and L_p as shown in Fig. 8.4.

Under reverse-bias condition, the equivalent RF circuit of a p-i-n diode is shown in Fig. 8.5. At microwave frequency range where the capacitive reactance, i.e. $1/2\pi fC_j \gg R_s$ and also $1/2\pi fC_j \gg 2\pi fL_P$, the equivalent RF circuit of the p-i-n diode can be considered as a capacitance $C_T = C_j + C_p$ and the high impedance of the diode at OFF condition can be written as $Z_{TH} = 1/j2\pi fC_T$.

8.3 Series Switch Configuration

The series-configured switch circuit with equivalent circuit element of p-i-n diodes can be represented as shown in Fig. 8.6a, and b at ON and OFF state, respectively.

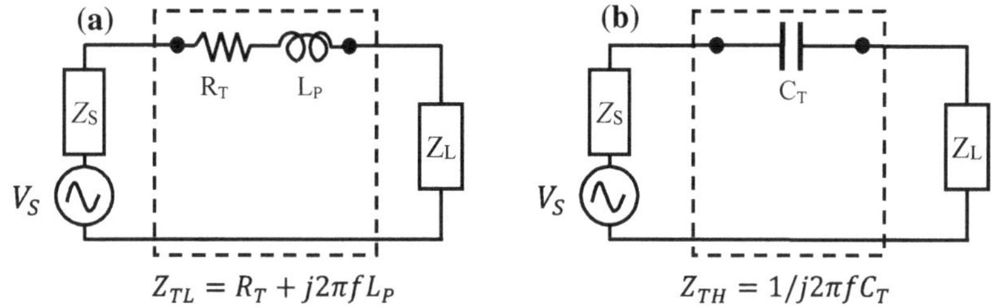

Fig. 8.6 Series switch configuration **a** ON state **b** OFF state

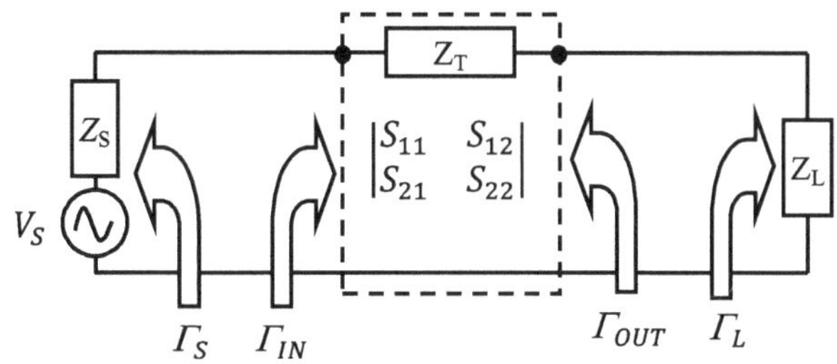

Fig. 8.7 Series switch configuration

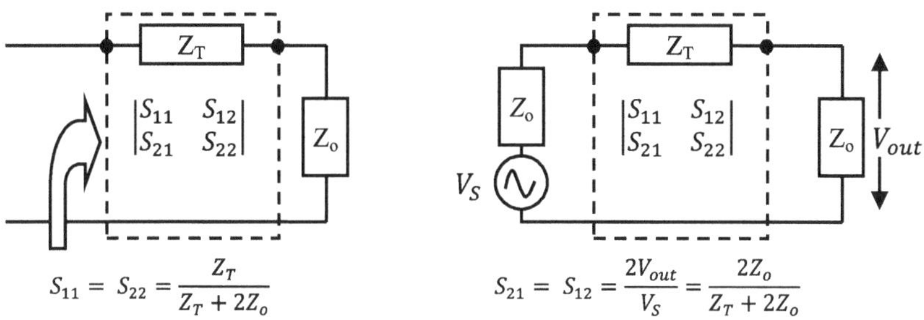

Fig. 8.8 Series switch configuration for S-parameter determination

Therefore, insertion loss, isolation and return loss of the switching circuit when it is connected in between a source of impedance Z_S and a load of impedance Z_L as shown in Fig. 8.7 can be derived as follows:

Here,

$$Z_T = Z_{TL} = R_T + j2\pi f L_P \text{ at low-impedance state} \quad (8.2a)$$

$$= Z_{TH} = \frac{1}{j2\pi f C_T} \text{ at high-impedance state} \quad (8.2b)$$

For the network of series impedance Z_T, the $[S]$ parameter of the series switch with respect to Z_o as shown in Fig. 8.8 is given by (using (7.162c))

$$\begin{bmatrix} S_{11} & S_{12} \\ S_{21} & S_{22} \end{bmatrix} = \begin{bmatrix} \frac{Z_T}{Z_T+2Z_o} & \frac{2Z_o}{Z_T+2Z_o} \\ \frac{2Z_o}{Z_T+2Z_o} & \frac{Z_T}{Z_T+2Z_o} \end{bmatrix} \tag{8.3}$$

The source and load reflection coefficients with respect to the reference impedance Z_o are given by

$$\Gamma_S = \frac{Z_S - Z_o}{Z_S + Z_o}, \quad \Gamma_L = \frac{Z_L - Z_o}{Z_L + Z_o} \tag{8.4a}$$

The input and output reflection coefficients of the switch can be written from (7.1774a) and (7.1774b) for the load and source impedances Z_L and Z_S, respectively, as

$$\Gamma_{\text{IN}} = S_{11} + \frac{\Gamma_L S_{12} S_{21}}{1 - \Gamma_L S_{22}} = \frac{Z_T + Z_L - Z_o}{Z_T + Z_L + Z_o} \tag{8.4b}$$

$$\Gamma_{\text{OUT}} = S_{22} + \frac{\Gamma_S S_{12} S_{21}}{1 - \Gamma_S S_{11}} = \frac{Z_T + Z_S - Z_o}{Z_T + Z_S + Z_o} \tag{8.4c}$$

For the series-configured switch, the ratio of power available from the source (P_{avs}) to the power delivered to the load (P_L) can be written as (using (7.181a))

$$\frac{P_{avs}}{P_L} = \left[\frac{|1 - \Gamma_S \Gamma_{\text{IN}}|^2}{1 - |\Gamma_S|^2}\right]\left[\frac{1}{|S_{21}|^2}\right]\left[\frac{|1 - S_{22}\Gamma_L|^2}{1 - |\Gamma_L|^2}\right] \tag{8.5}$$

Thus, putting the values of reflection coefficients to this equation, insertion loss and isolation of the series-connected switch can be determined. The first and last terms are losses due to the input and output mismatches, respectively. The middle term is the loss of the switch when both of its ports are terminated with reference impedance Z_o and is given from (8.3) as

$$\frac{1}{|S_{21}|^2} = \left|\frac{2Z_o + Z_T}{2Z_o}\right|^2 \tag{8.6}$$

Suppose, the load and source impedances are equal and resistive, i.e. R_o then at ON state, putting $Z_T = R_T + j2\pi f L_P = R_T + jX_L$, the insertion loss of the series switch will be

$$\text{IL} = \frac{1}{|S_{21}|^2} = \left|\frac{2R_o + R_T + jX_L}{2R_o}\right|^2 \tag{8.7a}$$

or

$$\text{IL(in dB)} = 10\log_{10}\left[\left(1 + \frac{R_T}{2R_o}\right)^2 + \left(\frac{X_L}{2R_o}\right)^2\right] \tag{8.7b}$$

Similarly, at OFF state, putting $Z_T = 1/(j2\pi fC_T) = -jX_C$, the loss of the series switch, i.e. isolation is

$$\text{Isolation} = \frac{1}{|S_{21}|^2} = \left|\frac{2R_o - jX_C}{2R_o}\right|^2 \tag{8.8a}$$

or

$$\text{Isolation (in dB)} = 10\ \log_{10}\left[1 + \left(\frac{1}{4\pi fC_T R_o}\right)^2\right] \tag{8.8b}$$

Example 8.1 Determine the insertion loss and isolation of a single p-i-n diode-based series switch operating at 4 GHz with the following parameters: $R_s = 2\ \Omega$, $R_i = 2\ \Omega$, $L_p = 0.1$ nH, $C_j = 0.02$ pF, $C_p = 0.01$ pF.

Solution

Here, $R_T = (R_i + R_s) = 4\ \Omega$, $C_T = C_j + C_p = 0.03$ pF and $L_p = 0.1$ nH.

To calculate insertion loss using (8.7a)

$$R_T = 4\ \Omega,$$
$$X_L = 2\pi fL_p = 2\pi \times 4 \times 0.1 = 2.513\ \Omega$$

From (8.7a),

$$\text{IL} = \frac{1}{|S_{21}|^2} = \left|\frac{2R_o + R_T + jX_L}{2R_o}\right|^2 = \left|\frac{2 \times 50 + 4 + j2.513}{2 \times 50}\right|^2 = 1.0822$$

or

$$\text{IL (in dB)} = 10\log_{10}[1.082] = 0.343\ \text{dB}$$

To calculate isolation using (8.8a),

$$X_C = 1/2\pi fC_T = 1000/2\pi \times 4 \times 0.03 = 1326.29\ \Omega$$

From (8.8a),

$$\text{Isolation} = \frac{1}{|S_{21}|^2} = \left|\frac{2R_o - jX_C}{2R_o}\right|^2 = \left|1 - \frac{j1326.29}{2 \times 50}\right|^2 = 176.905$$

or

$$\text{Isolation (in dB)} = 10\log_{10}[176.905] = 22.48\ \text{dB}$$

Thus, for a 50 Ω system, the ON-state insertion loss is 0.343 dB and OFF-state isolation is 22.48 dB.

8.4 Shunt Switch Configuration

The shunt-configured switch with p-i-n diode as control impedance can be represented by the circuit diagrams as shown in Fig. 8.9a, b at its ON and OFF states, respectively. Suppose the shunt impedance of the diode is Z_T as shown in Fig. 8.10. In this case, the shunt impedance of the switch can be written as

$$Z_T = Z_{TH} = \frac{1}{j2\pi f C_T} = -jX_C \quad \text{at ON state} \tag{8.9a}$$

$$= Z_{TL} = R_T + j2\pi f L_P = R_T + jX_L \text{ at OFF state} \tag{8.9b}$$

From (7.163d), using Fig. 8.11, the [S] parameter matrix of the shunt switch with respect to Z_o can be written as

$$\begin{bmatrix} S_{11} & S_{12} \\ S_{21} & S_{22} \end{bmatrix} = \begin{bmatrix} \frac{Z_T||Z_o - Z_o}{Z_T||Z_o + Z_o} & \frac{2Z_T}{2Z_T + Z_o} \\ \frac{2Z_T}{2Z_T + Z_o} & \frac{Z_T||Z_o - Z_o}{Z_T||Z_o + Z_o} \end{bmatrix} \tag{8.10}$$

The source and load reflection coefficients with respect to the reference impedance Z_o are given by

$$\Gamma_S = \frac{Z_S - Z_o}{Z_S + Z_o}, \quad \Gamma_L = \frac{Z_L - Z_o}{Z_L + Z_o} \tag{8.11}$$

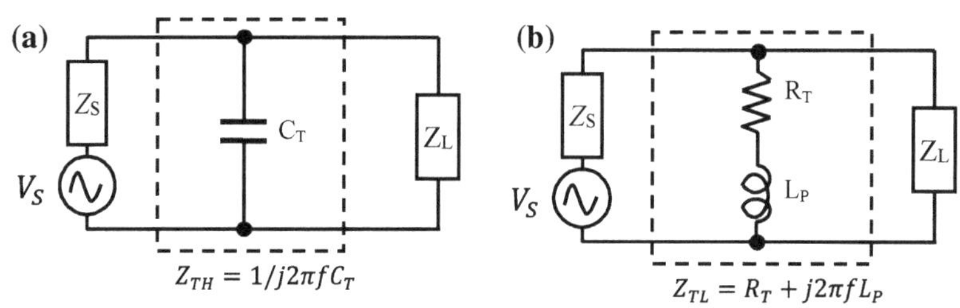

Fig. 8.9 Shunt switches configuration **a** ON state **b** OFF state

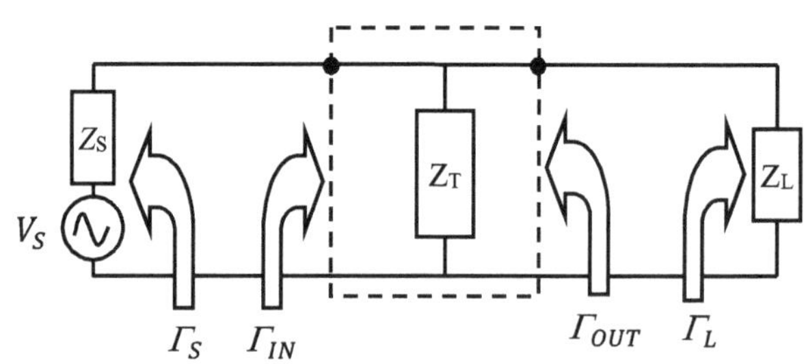

Fig. 8.10 Shunt switch configuration for reflection coefficient determination

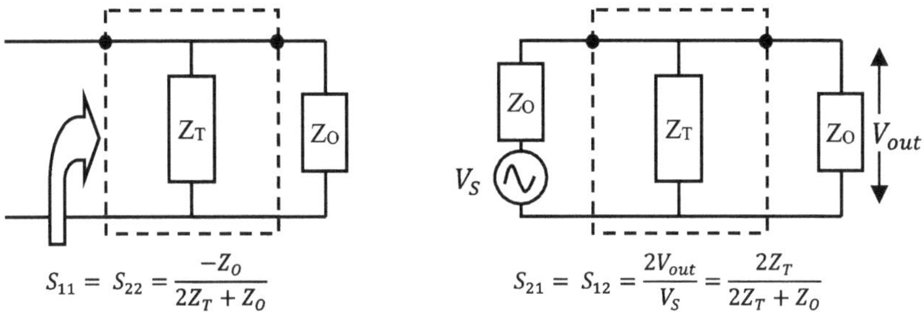

Fig. 8.11 Shunt switch configuration for S-parameter determination

The input and output reflection coefficients of the switch for the load and source impedances Z_L and Z_S, respectively, are

$$\Gamma_{\mathrm{IN}} = S_{11} + \frac{\Gamma_L S_{12} S_{21}}{1 - \Gamma_L S_{22}} = \frac{Z_T||Z_L - Z_o}{Z_T||Z_L + Z_o} \tag{8.12a}$$

$$\Gamma_{\mathrm{OUT}} = S_{22} + \frac{\Gamma_S S_{12} S_{21}}{1 - \Gamma_S S_{11}} = \frac{Z_T||Z_S - Z_o}{Z_T||Z_S + Z_o} \tag{8.12b}$$

For the shunt-configured switch, the ratio of power available from the source (P_{avs}) to the power delivered to the load (P_L) can be written as (using (7.181a))

$$\frac{P_{avs}}{P_L} = \left[\frac{|1 - \Gamma_S \Gamma_{\mathrm{IN}}|^2}{1 - |\Gamma_S|^2}\right]\left[\frac{1}{|S_{21}|^2}\right]\left[\frac{|1 - S_{22}\Gamma_L|^2}{1 - |\Gamma_L|^2}\right] \tag{8.13}$$

Thus, putting the values of reflection coefficients from (8.12a), insertion loss and isolation of the shunt-connected switch can be determined. The first and last terms are losses due to the input and output impedance mismatches, respectively. The middle term is the loss of the switch when both the ports are terminated with the reference impedance Z_o and is given by

$$\frac{1}{|S_{21}|^2} = \left|\frac{2Z_T + Z_o}{2Z_T}\right|^2 = \left|1 + \frac{Z_o}{2Z_T}\right|^2 \tag{8.14a}$$

Suppose, the source and load impedances are resistive, i.e. $Z_o = R_o$ then putting ON-state Z_T value, the insertion loss of the shunt switch can be written as

$$\mathrm{IL} = \frac{1}{|S_{21}|^2} = \left|\frac{2Z_T + Z_o}{2Z_T}\right|^2 = \left|1 + \frac{Z_o}{2Z_T}\right|^2 = \left|1 - \frac{Z_o}{j2X_C}\right|^2 \tag{8.14b}$$

$$= \frac{1}{|S_{21}|^2} = 1 + (\pi f C_T R_o)^2 \tag{8.14c}$$

or

$$\text{IL (in dB)} = 10\log_{10}\left[1 + (\pi f C_T R_o)^2\right] \tag{8.15}$$

Similarly, putting OFF-state Z_T value, loss of the shunt switch, i.e. isolation will be

$$\text{Isolation} = \frac{1}{|S_{21}|^2} = \left|1 + \frac{R_o}{2(R_T + jX_L)}\right|^2 \tag{8.16a}$$

$$= \frac{(2R_T + R_o)^2 + (2X_L)^2}{(2R_T)^2 + (2X_L)^2} \tag{8.16b}$$

or

$$\text{Isolation (in dB)} = 10\log_{10}\left[\frac{(2R_T + R_o)^2 + (2X_L)^2}{(2R_T)^2 + (2X_L)^2}\right] \tag{8.16c}$$

Example 8.2 Determine the insertion loss and isolation of a single shunt p-i-n diode-based switch operating at 4 GHz for the following parameters: $R_s = 2\ \Omega$, $R_i = 2\ \Omega$, $L_p = 0.1$ nH, $C_j = 0.02$ pF, $C_p = 0.01$ pF.

Solution

Here, $R_T = (R_i + R_s) = 4\ \Omega$, $C_T = C_j + C_p = 0.03$ pF and $L_p = 0.1$ nH.

To calculate insertion loss of the shunt diode switch using (8.14b),

$$X_C = \frac{1}{2\pi f C_T} = \frac{1000}{2 \times \pi \times 4 \times 0.03} = 1326.29\ \Omega$$

From (8.14b),

$$\text{IL} = \frac{1}{|S_{21}|^2} = \left|1 - \frac{Z_o}{j2X_C}\right|^2 = \left|1 - \frac{50}{j2 \times 1326.29}\right|^2 = 1.0000355$$

or

$$\text{IL (in dB)} = 10\log_{10}[1.000355] = 0.002\ \text{dB}$$

To calculate isolation using (8.16a),

$$X_L = 2\pi f L_p = 2\pi \times 4 \times 0.1 = 2.513\ \Omega$$

From (8.8a),

$$\text{Isolation} = \frac{1}{|S_{21}|^2} = \left|1 + \frac{R_o}{2(R_T + jX_L)}\right|^2 = \left|1 + \frac{50}{2(4 + j2.513)}\right|^2 = 37.97$$

or

$$\text{Isolation (in dB)} = 10\log_{10}[37.97] = 15.79\ \text{dB}$$

Thus, for a 50 Ω system, the ON-state insertion loss is 0.0002 dB and OFF-state isolation is 15.79 dB.

Examples 8.1 and 8.2 show that switch with series p-i-n diode configuration provides better isolation, whereas switch with shunt diode configuration provides better insertion loss.

8.5 Compound Switch Configuration

Examples of series and shunt switches with same diode parameters show that shunt switch provides better insertion loss compared to series switch configuration. However, series switch configuration provides better isolation compared to shunt switch configuration. Shunt switch provides lower insertion loss due to the absence of any switch elements in series with the signal path.

At RF and microwave frequencies, practically, it is very difficult to achieve very high isolation using a single p-i-n diode either in shunt or series configuration. The causes of this limitation are not only the parasitic reactance of the diode but also it is due to the radiation effects in the transmission medium and inadequate shielding. To overcome this limitation, multiple p-i-n diodes can be used in a single switch to increase the isolation with the penalty of more insertion loss. All p-i-n diodes may be connected only in series or in parallel. Also, there are switch designs called compound switch configuration that uses combinations of series and shunt diodes to achieve improved isolation performance.

Simple series–shunt and TEE configurations are two basic configurations for compound switch. In the insertion loss (ON) state for a compound switch, the series diode is operated at forward bias and the shunt diodes are operated at zero- or reverse-bias conditions, whereas in isolation (OFF) state opposite bias is provided to series and shunt diodes.

8.5.1 Series–Shunt Configuration

The schematic circuit diagram of a series–shunt configuration switch is shown in Fig. 8.12. The equivalent RF circuit of the series–shunt-configured switch is shown in Fig. 8.13. Suppose, the impedance of the series- and shunt-connected p-i-n diodes are Z_{T1} and Z_{T2}, respectively, and the switch is connected in between a source and a load of impedance Z_o. With reference to the equivalent circuit as shown in Fig. 8.13, $[S]$ parameter matrix of the switch can be determined following the similar approach as it is done in case of series and shunt switch configuration.

With reference to Fig. 8.13, the input and output reflection coefficient of the switch can be written as

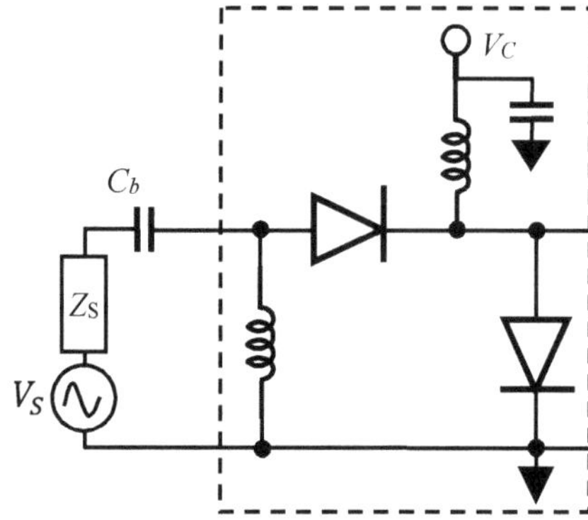

Fig. 8.12 Schematic circuit diagram of series–shunt compound switch configuration

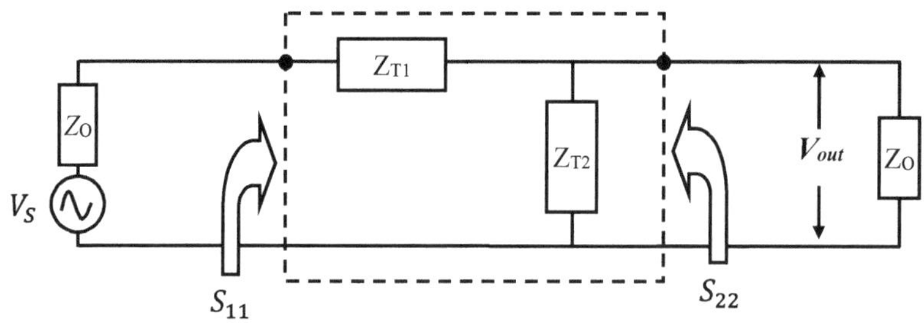

Fig. 8.13 Equivalent RF circuit of the series–shunt switch configuration

$$S_{11} = \frac{Z_{T1} + Z_{T2}||Z_o - Z_o}{Z_{T1} + Z_{T2}||Z_o + Z_o} \tag{8.17a}$$

$$S_{22} = \frac{Z_{T2}||(Z_{T1} + Z_o) - Z_o}{Z_{T2}||(Z_{T1} + Z_o) + Z_o} \tag{8.17b}$$

and

$$S_{21} = \frac{2V_{\text{out}}}{V_S} = \frac{2Z_{T2}||Z_o}{Z_o + Z_{T1} + Z_{T2}||Z_o} \tag{8.17c}$$

Similarly, connecting the source V_S with impedance at the port-2 of the switch and a load at the port-1, the reverse transmission coefficient S_{12} can be written as

$$S_{12} = \frac{2Z_oZ_{T2}}{[Z_o + Z_{T2}||(Z_{T1} + Z_o)][Z_{T1} + Z_{T2} + Z_o]} \tag{8.17d}$$

It can be shown that $S_{12} = S_{21}$, because the network is reciprocal. From (8.17c) the insertion loss and isolation of the series–shunt switch can be calculated as follows:

$$\frac{1}{S_{21}} = \frac{Z_o + Z_{T1} + Z_{T2}||Z_o}{2Z_{T2}||Z_o} \tag{8.18a}$$

or

$$\frac{1}{S_{21}} = 1 + \frac{Z_o + Z_{T1}}{2Z_{T2}} + \frac{Z_{T1}}{2Z_o} \tag{8.18b}$$

Insertion loss can be calculated by putting the values of Z_{T1} and Z_{T2} corresponding to the low impedance and high impedance of the p-i-n diodes, respectively, as

$$Z_{T1} = Z_{TL} = R_T + j2\pi f L_P = R_T + jX_L \quad \text{p-i-n diode ON}$$
$$Z_{T2} = Z_{TH} = \frac{1}{j2\pi f C_T} = -jX_C \quad \text{p-i-n diode OFF}$$

Thus,

$$\frac{1}{S_{21}} = \left[1 + \frac{R_T}{2Z_o} - \frac{X_L}{X_C}\right] + j\left[\frac{X_L}{2Z_o} + \frac{Z_o + R_T}{2X_C}\right] \tag{8.19a}$$

Therefore,

$$\text{IL} = \frac{1}{|S_{21}|^2} = \left(1 + \frac{R_T}{2Z_o} - \frac{X_L}{X_C}\right)^2 + \left(\frac{X_L}{2Z_o} + \frac{Z_o + R_T}{2X_C}\right)^2 \tag{8.19b}$$

$$\text{IL (in dB)} = 10\log_{10}\left[\left(1 + \frac{R_T}{2Z_o} - \frac{X_L}{X_C}\right)^2 + \left(\frac{X_L}{2Z_o} + \frac{Z_o + R_T}{2X_C}\right)^2\right] \tag{8.19c}$$

Isolation of the series–shunt switch can be calculated by putting the values of Z_{T1} and Z_{T2} corresponding to the high and low impedance of the p-i-n diodes, respectively, in (8.18b) as

$$Z_{T2} = Z_{TL} = R_T + j2\pi f L_P = R_T + jX_L \quad \text{p-i-n diode ON}$$
$$Z_{T1} = Z_{TH} = \frac{1}{j2\pi f C_T} = -jX_C \quad \text{p-i-n diode OFF}$$

Thus,

$$\frac{1}{S_{21}} = \left[1 + \frac{Z_o R_T - X_C X_L}{2\left(R_T^2 + Z_L^2\right)}\right] + j\left[\frac{X_C}{2Z_o} + \frac{Z_o X_L + R_T X_C}{2\left(R_T^2 + X_L^2\right)}\right] \tag{8.20a}$$

Therefore,

$$\text{Isolation} = \frac{1}{|S_{21}|^2} = \left(1 + \frac{Z_o R_T - X_C X_L}{2\left(R_T^2 + X_L^2\right)}\right)^2 + \left(\frac{X_C}{2Z_o} + \frac{Z_o X_L + R_T X_C}{2\left(R_T^2 + X_L^2\right)}\right)^2 \tag{8.20b}$$

$$\text{Isolation(in dB)} \Rightarrow 10\log_{10}\left[\left(1+\frac{Z_oR_T - X_CX_L}{2(R_T^2+X_L^2)}\right)^2 + \left(\frac{X_C}{2Z_o}+\frac{Z_oX_L+R_TX_C}{2(R_T^2+X_L^2)}\right)^2\right] \tag{8.20c}$$

Example 8.3 Determine the insertion loss and isolation of the switch as shown in Fig. 8.12 using series and shunt p-i-n diodes operating at 4 GHz for the following parameters: $R_s = 2\ \Omega$, $R_i = 2\ \Omega$, $L_p = 0.1$ nH $C_j = 0.02$ pF, $C_p = 0.01$ pF.

Solution

Here, $R_T = (R_i + R_s) = 4\ \Omega$, $C_T = C_j + C_p = 0.03\text{pF and}L_p = 0.1\text{nH}$.

To calculate insertion loss using (8.19b),

$$\begin{aligned}
R_T &= 4\ \Omega,\\
X_L &= 2\pi fL_p = 2\pi \times 4 \times 0.1 = 2.513\ \Omega\\
X_C &= 1/2\pi fC_T = 1000/2\pi \times 4 \times 0.03 = 1326.29\ \Omega
\end{aligned}$$

From (8.19b),

$$\begin{aligned}
\text{IL} = \frac{1}{|S_{21}|^2} &= \left[1+\frac{R_T}{2Z_o}-\frac{X_L}{X_C}\right]^2 + \left[\frac{X_L}{2Z_o}+\frac{Z_o+R_T}{2X_C}\right]^2\\
&= \left[1+\frac{4}{2\times 50}-\frac{2.513}{1326.29}\right]^2 + \left[\frac{2.513}{2\times 50}+\frac{50+4}{2\times 1326.29}\right]^2 = 1.0797
\end{aligned}$$

or

$$\text{IL (in dB)} = 10\log_{10}[1.0797] = 0.333\,\text{dB}$$

To calculate isolation using (8.20b),

$$\begin{aligned}
\text{Isolation} = \frac{1}{|S_{21}|^2} &= \left(1+\frac{Z_oR_T - X_CX_L}{2(R_T^2+X_L^2)}\right)^2 + \left(\frac{X_C}{2Z_o}+\frac{Z_oX_L+R_TX_C}{2(R_T^2+X_L^2)}\right)^2\\
&= \left(1+\frac{200-1326.29\times 2.513}{2(4^2+2.513^2)}\right)^2 + \left(\frac{1326.29}{2\times 50}+\frac{50\times 2.513+4\times 1326.29}{2(4^2+2.513^2)}\right)^2\\
&= 4788.92 + 18208.820.0021 = 22997.75
\end{aligned}$$

or

$$\text{Isolation (in dB)} = 10\log_{10}[22997.75] = 43.617\,\text{dB}$$

For a 50 Ω system, the ON-state insertion loss is 0.333 dB and OFF-state isolation is 43.617 dB.

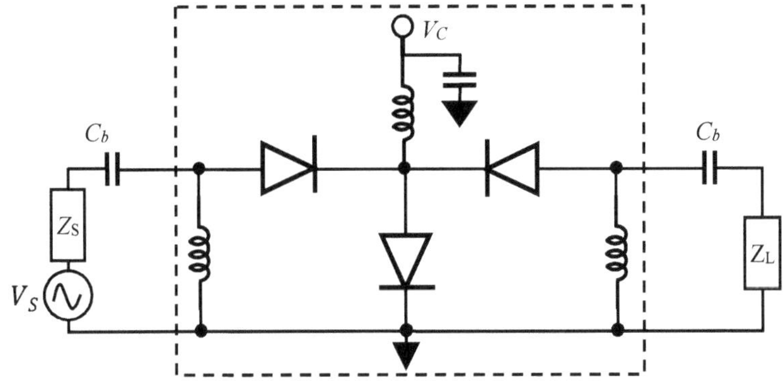

Fig. 8.14 TEE-type switch configuration

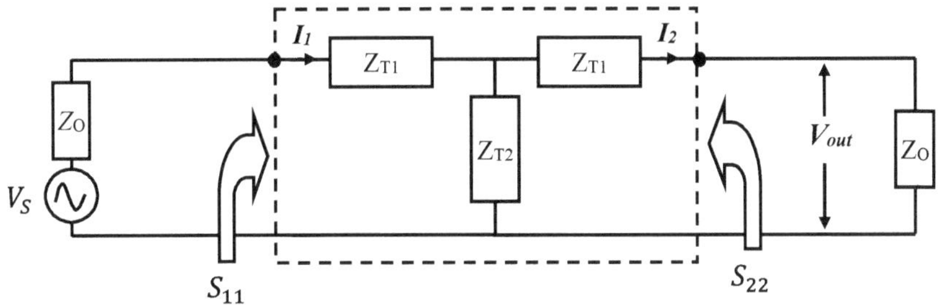

Fig. 8.15 TEE-type switch configuration for reflection coefficient derivation

8.5.2 TEE Configuration

The schematic circuit diagram of a TEE configuration switch is shown in Fig. 8.14. The equivalent RF circuit of the TEE-configured switch is shown in Fig. 8.15. Suppose, the impedances of the series- and shunt-connected p-i-n diodes are Z_{T1} and Z_{T2}, respectively, and the switch is connected in between a source and a load of impedance of Z_o. With reference to the equivalent circuit as shown in Fig. 8.15, [S] parameter matrix of the switch can be determined following the previous approach.

With reference to Fig. 8.15, the input and output reflection coefficient of the switch can be written as

$$S_{11} = S_{22} = \frac{Z_{T1} + Z_{T2}||(Z_{T1} + Z_o) - Z_o}{Z_{T1} + Z_{T2}||(Z_{T1} + Z_o) + Z_o} \tag{8.21}$$

and

$$\begin{aligned} S_{21} = S_{12} &= \frac{2V_{\text{out}}}{V_S} = \frac{2Z_o I_2}{V_S} \\ &= \frac{2Z_o Z_{T2} I_1}{V_S(Z_{T1} + Z_{T2} + Z_o)} \\ &= \frac{2Z_o Z_{T2}}{(Z_o + Z_{T1} + Z_{T2})(Z_o + Z_{T1} + Z_{T2}||(Z_{T1} + Z_o))} \qquad (8.22a) \\ &= \frac{2Z_o Z_{T2}}{2Z_o Z_{T1} + 2Z_o Z_{T2} + 2Z_{T1} Z_{T2} + Z_{T1}^2 + Z_o^2} \qquad (8.22b) \end{aligned}$$

Thus, insertion loss and isolation can be determined as follows:

$$\frac{1}{S_{21}} = \frac{2Z_o Z_{T1} + 2Z_o Z_{T2} + 2Z_{T1} Z_{T2} + Z_{T1}^2 + Z_o^2}{2Z_o Z_{T2}} \qquad (8.22c)$$

or

$$\frac{1}{S_{21}} = 1 + \frac{Z_{T1}}{Z_{T2}} + \frac{Z_{T1}}{Z_o} + \frac{Z_{T1}^2}{2Z_o Z_{T2}} + \frac{Z_o}{2Z_{T2}} \qquad (8.22d)$$

Insertion loss of the TEE switch can be calculated by putting the values of Z_{T1} and Z_{T2} corresponding to the low impedance and high impedance of the p-i-n diodes, respectively, as

$$\begin{aligned} Z_{T1} &= Z_{TL} = R_T + j2\pi f L_P = R_T + jX_L \quad \text{p-i-n diode ON} \\ Z_{T2} &= Z_{TH} = \frac{1}{j2\pi f C_T} = -jX_C \quad \text{p-i-n diode OFF} \end{aligned}$$

Thus,

$$\frac{1}{S_{21}} = \left[1 + \frac{R_T}{Z_o} - \frac{X_L(R_T + Z_o)}{Z_o X_C}\right] + j\left[\frac{X_L}{Z_o} + \frac{Z_o R_T + R_T^2 - X_L^2 + Z_o^2}{2Z_o X_C}\right] \qquad (8.23a)$$

Therefore,

$$\text{IL} = \left(1 + \frac{R_T}{Z_o} - \frac{X_L(R_T + Z_o)}{Z_o X_C}\right)^2 + \left(\frac{X_L}{Z_o} + \frac{Z_o R_T + R_T^2 - X_L^2 + Z_o^2}{2Z_o X_C}\right)^2 \qquad (8.23b)$$

or

$$\begin{aligned} \text{IL(in dB)} = 10\log_{10}\Bigg[&\left(1 + \frac{R_T}{Z_o} - \frac{X_L(R_T + Z_o)}{Z_o X_C}\right)^2 \\ &+ \left(\frac{X_L}{Z_o} + \frac{Z_o R_T + R_T^2 - X_L^2 + Z_o^2}{2Z_o X_C}\right)^2\Bigg] \qquad (8.23c) \end{aligned}$$

Isolation of the TEE switch can be calculated by putting the values of Z_{T1} and Z_{T2} corresponding to the high and low impedance of the p-i-n diodes, respectively, in (8.22b) as

$$Z_{T2} = Z_{TL} = R_T + j2\pi fL_P = R_T + jX_L \text{ p-i-n diode ON}$$

$$Z_{T1} = Z_{TH} = \frac{1}{j2\pi fC_T} = -jX_C \text{ p-i-n diode OFF}$$

Thus,

$$\frac{1}{S_{21}} = \left[1 + \frac{Z_o^2 R_T - R_T X_C^2 - 2Z_o X_L X_C}{2Z_o\left(R_T^2 + X_L^2\right)}\right] - j\left[\frac{X_C}{Z_o} + \frac{2Z_o R_T X_C - X_C^2 X_L - Z_o^2 X_L}{2Z_o\left(R_T^2 + X_L^2\right)}\right] \tag{8.24a}$$

Therefore,

$$\text{Isolation} = \frac{1}{|S_{21}|^2} = \left(1 + \frac{Z_o^2 R_T - R_T X_C^2 - 2Z_o X_L X_C}{2Z_o\left(R_T^2 + X_L^2\right)}\right)^2 + \left(\frac{X_C}{Z_o} + \frac{2Z_o R_T X_C - X_C^2 X_L - Z_o^2 X_L}{2Z_o\left(R_T^2 + X_L^2\right)}\right)^2 \tag{8.24b}$$

$$\text{Isolation (in dB)} \Rightarrow 10\log_{10}\left[\left(1 + \frac{Z_o^2 R_T - R_T X_C^2 - 2Z_o X_L X_C}{2Z_o\left(R_T^2 + X_L^2\right)}\right)^2 + \left(\frac{X_C}{Z_o} + \frac{2Z_o R_T X_C - X_C^2 X_L - Z_o^2 X_L}{2Z_o\left(R_T^2 + X_L^2\right)}\right)^2\right] \tag{8.24c}$$

Similarly, connecting the source V_S with impedance at the port-2 of the switch and a load at the port-1, the reverse transmission coefficient S_{12} can be determined. In this case, the reverse transmission coefficient S_{12} is the same as forward transmission coefficient S_{21} as the network is reciprocal (does not consist of any active element and non-reciprocal element).

Example 8.4 Determine the insertion loss and isolation of a TEE-type switch as shown in Fig. 8.14 using series and shunt p-i-n diodes operating at 4 GHz for the following parameters of the diodes (Figs. 8.4 and 8.5): $R_s = 2\ \Omega$, $R_i = 2\ \Omega$, $L_p = 0.1$ nH, $C_j = 0.02$ pF, $C_p = 0.01$ pF.

Solution

Here, $R_T = (R_i + R_s) = 4\ \Omega$, $C_T = C_j + C_p = 0.03$ pF and $L_p = 0.1$ nH.

Therefore,

$$\begin{aligned}
R_T &= 4\ \Omega,\\
X_L &= 2\pi f L_p = 2\pi \times 4 \times 0.1 = 2.513\ \Omega\\
X_C &= 1/2\pi f C_T = 1000/2\pi \times 4 \times 0.03 = 1326.29\ \Omega
\end{aligned}$$

From (8.23b),

$$\begin{aligned}
\text{IL} &= \left(1 + \frac{R_T}{Z_o} - \frac{X_L(R_T + Z_o)}{Z_o X_C}\right)^2 + \left(\frac{X_L}{Z_o} + \frac{Z_o R_T + R_T^2 - X_L^2 + Z_o^2}{2 Z_o X_C}\right)^2\\
&= \left[1 + \frac{4}{50} - \frac{2.513(4+50)}{50 \times 1326.29}\right]^2 + \left[\frac{2.513}{50} + \frac{50 \times 4^2 - 2.513^2 + 50^2}{2 \times 50 \times 1326.29}\right]^2\\
&= [1.07795]^2 + [0.07509]^2 = 1.1676
\end{aligned}$$

or

$$\text{IL (in dB)} = 10\log_{10}[1.1676] = 0.673\,\text{dB}$$

To calculate isolation using (8.24b),

$$\begin{aligned}
\text{Isolation} &= \left(1 + \frac{Z_o^2 R_T - R_T X_C^2 - 2 Z_o X_L X_C}{2 Z_o \left(R_T^2 + X_L^2\right)}\right)^2\\
&\quad + \left(\frac{X_C}{Z_o} + \frac{2 Z_o R_T X_C - X_C^2 X_L - Z_o^2 X_L}{2 Z_o \left(R_T^2 + X_L^2\right)}\right)^2\\
&= \left(1 + \frac{50^2 \times 4 - 4 \times 1326.29^2 - 2 \times 50 \times 2.513 \times 1326.29}{2 \times 50\left(4^2 + 2.513^2\right)}\right)^2\\
&\quad + \left(\frac{1326.29}{50} + \frac{2 \times 50 \times 4 \times 1326.29 - 1326.29^2 \times 2.513 - 50^2 \times 2.513}{2 \times 50\left(4^2 + 2.513^2\right)}\right)^2\\
&= 2957004.87 + 10868818.84 = 13825823.7
\end{aligned}$$

Thus,

$$\text{Isolation (in dB)} = 10\log_{10}[13825823.7] = 71.4\,\text{dB}$$

For a 50 Ω system, the ON-state insertion loss is 0.673 dB and OFF-state isolation is 71.4 dB.

8.6 Compound Switch Analysis Using ABCD Parameter

It is more convenient to use ABCD parameters for the analysis of compound switch circuits. Any compound switch can be considered as multiple simple networks connected in cascade. Thus, it is convenient to determine the ABCD matrix of the individual networks and then multiplying them to determine ABCD parameters of the compound switch from which S-parameters of the complete switch can be determined.

With reference to Fig. 8.16, the input and output reflection coefficient of the switch can be written as follows:

The TEE switch is divided into three simple networks. Input voltage V_1 and input current I_1 of the network-1, as shown in Fig. 8.16, is related by

$$V_1 = A_1 V_2 + B_1 I_2$$

$$I_1 = C_1 V_2 + D_1 I_2$$

The network-1 consists of only one element Z_{T1} connected in the series line. Thus, the parameters can be determined as

$$A_1 = \left.\frac{V_1}{V_2}\right|_{I_2=0} = 1, \quad B_1 = \left.\frac{V_1}{I_2}\right|_{V_2=0} = Z_{T1}$$

$$C_1 = \left.\frac{I_1}{V_2}\right|_{I_2=0} = 0, \quad D_1 = \left.\frac{I_1}{I_2}\right|_{V_2=0} = 1$$

Input voltage V_2 and input current I_2 of the network-2, as shown in Fig. 8.16, is related by

$$V_2 = A_2 V_3 + B_2 I_3$$

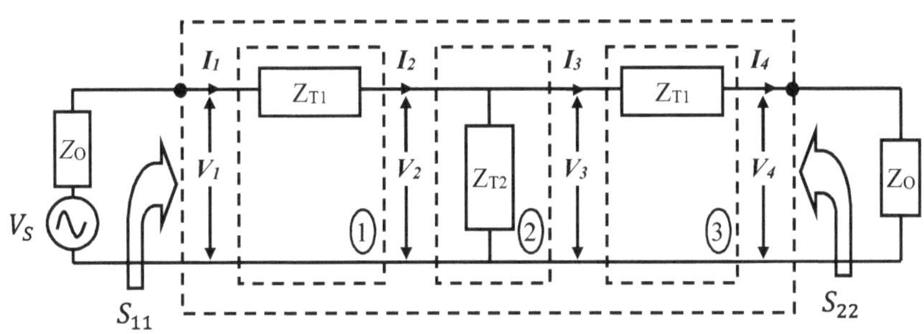

Fig. 8.16 TEE-type switch configuration for reflection coefficient derivation

$$I_2 = C_2 V_3 + D_2 I_3$$

Whereas, the network-2 consists of only one element Z_{T2} connected in shunt with the line. Thus, the parameters can be determined as:

$$A_2 = \left.\frac{V_2}{V_3}\right|_{I_3=0} = 1, \quad B_2 = \left.\frac{V_2}{I_3}\right|_{V_3=0} = 0$$

$$C_2 = \left.\frac{I_2}{V_3}\right|_{I_3=0} = 1/Z_{T2}, \quad D_2 = \left.\frac{I_2}{I_3}\right|_{V_3=0} = 1$$

Therefore, the ABCD matrix of the TEE configuration switch can be expressed as

$$\begin{bmatrix} A & B \\ C & D \end{bmatrix} = \begin{bmatrix} 1 & Z_{T1} \\ 0 & 1 \end{bmatrix} \begin{bmatrix} 1 & 0 \\ 1/Z_{T2} & 1 \end{bmatrix} \begin{bmatrix} 1 & Z_{T1} \\ 0 & 1 \end{bmatrix} \tag{8.25a}$$

$$\begin{bmatrix} A & B \\ C & D \end{bmatrix} = \begin{bmatrix} 1 + Z_{T1}/Z_{T2} & Z_{T1} \\ 1/Z_{T2} & 1 \end{bmatrix} \begin{bmatrix} 1 & Z_{T1} \\ 0 & 1 \end{bmatrix} \tag{8.25b}$$

$$\begin{bmatrix} A & B \\ C & D \end{bmatrix} = \begin{bmatrix} 1 + \frac{Z_{T1}}{Z_{T2}} & 2Z_{T1} + \frac{Z_{T1}^2}{Z_{T2}} \\ \frac{1}{Z_{T2}} & 1 + \frac{Z_{T1}}{Z_{T2}} \end{bmatrix} \tag{8.25c}$$

The $[S]$ parameter matrix of the switch can be determined using the relationship between S-parameters and ABCD parameters as determined in Chap. 7. The forward transmission coefficient S_{21} of the switch can be determined using the following equation:

$$S_{21} = \frac{2}{A + BY_o + CZ_o + D}$$

Putting the values of ABCD parameters from (6.25c):

$$\begin{aligned} S_{21} &= \frac{2}{1 + \frac{Z_{T1}}{Z_{T2}} + \frac{2Z_{T1}}{Z_o} + \frac{Z_{T1}^2}{Z_o Z_{T2}} + \frac{Z_o}{Z_{T2}} + 1 + \frac{Z_{T1}}{Z_{T2}}} \\ &= \frac{2Z_o Z_{T2}}{2Z_o Z_{T1} + 2Z_o Z_{T2} + 2Z_{T1} Z_{T2} + Z_{T1}^2 + Z_o^2} \end{aligned} \tag{8.26}$$

This is the same result as derived in (8.22c). Other elements of the S-parameters matrix can be derived in the similar way.

8.7 Switch Circuits Based on FETs

Switches based on p-i-n diodes require higher DC current at ON condition and also behave as nonlinear element at lower microwave frequency. Therefore, p-i-n diode-based switches are not suitable over lower microwave frequency of operation and applications where sufficient DC current is not available. This is the reason FETs are used in place of p-i-n diodes as switching device for mobile applications.

All types of field effect transistors (MESFETs, HEMTs and pHEMTs) at zero drain bias are used as variable resistors to construct microwave switches [14–16]. Among varieties of FETs, mostly n-channel depletion mode FETs are used as variable RF resistor, which has very low and very high source-to-drain resistance values in absence and presence of gate bias voltage, respectively [6–10]. As discussed in Chap. 6, with the application of negative voltage at gate terminal, the produced electric field below the gate causes the conduction channel to narrow, which increases the resistance in between source to drain. Ideally, the gate is isolated from the channel, thus there will be no current through the gate terminal, and it dissipates essentially no DC power and can easily be designed as broadband switches. Thus, MESFETs at zero drain bias are used as variable RF resistance to realize microwave switches by applying control voltage to its gate terminal. Realization of MMIC-based microwave systems using MESFETs and various HEMTs switches integrated with other circuits are more convenient than using p-i-n diode-based circuits. Therefore, for low-power applications, FET-based switches are more convenient.

MESFETs and various HEMTs are used for low-power microwave/RF switch for easy realization in MMIC technology integrated with other circuitry. Extremely high two-dimensional electron gas (2-DEG) of GaN HEMTs results in very low ON resistance and due to its high breakdown fields it is capable to operate at higher operating temperatures which makes it suitable for realization of high power switches.

Schematic diagram of a MESFET is shown in Fig. 8.17. The channel resistance can be written as (6.1b)

$$R_c = \frac{1}{qN_d\mu_n} \times \frac{L_g}{W\left(a - d\left(V_g\right)\right)} \tag{8.27}$$

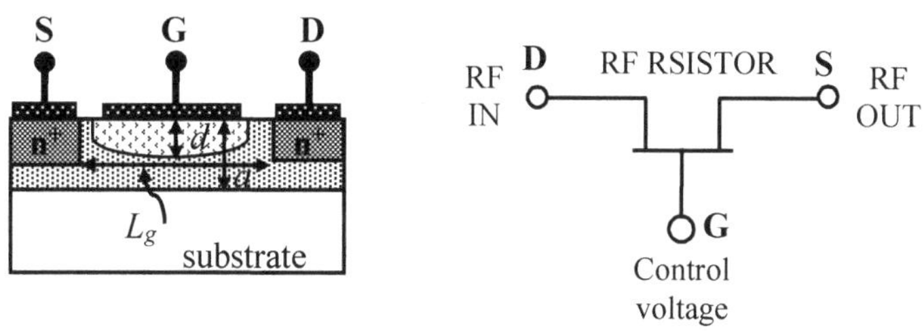

Fig. 8.17 MESFET used as voltage-variable linear RF resistor for microwave switch

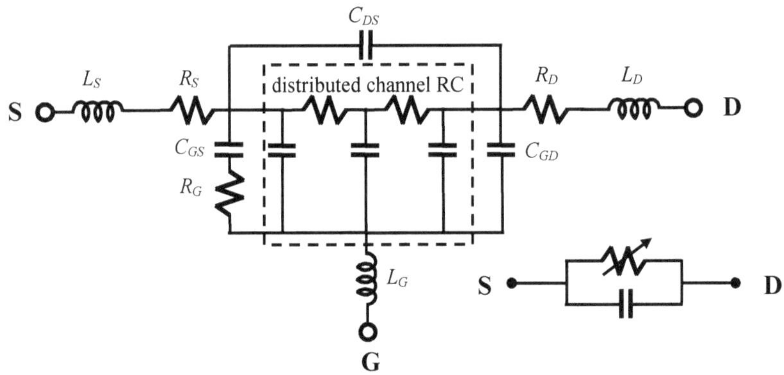

Fig. 8.18 Equivalent circuit of a zero-gate-bias FET

Here, N_d is the channel electron density which is equal to the free carrier density within the channel, μ_n is the low field electron mobility, L_g is the gate length approximately equal to the channel length, W is the gate periphery, a is channel depth and $d(V_g)$ is the depletion layer depth which is function of the applied gate control voltage. Equivalent circuit model of a zero-gate-biased MESFET is shown in Fig. 8.18. The voltage-variable channel resistance can be considered as distributed RC network, where the resistances and capacitances are voltage variable. Lumped elements are included in each terminal to take into account the parasitics of the extrinsic MESFET. This FET model is used to analyse switches using CAD tool. However, for simplicity the equivalent circuit of a FET, at zero drain-bias condition, can be considered as combination of resistors and capacitors which are represented as a parallel RC circuit with an effective variable resistance and an effective capacitance as shown in Fig. 8.18. However, these parasitic capacitances have a significant influence on the switch performance at higher microwave frequencies.

For switch circuit realization, FET's are arranged in both series and shunt configurations. For compound switch configuration, series FETs provide a through-path for the ON state, while the shunt FETs provide isolation for the OFF state. For the operation of series FETs and shunt FETs associated with each switch state requires opposite or complementary conduction states, and therefore opposite or complementary gate biases.

Circuit diagram for MESFET-based switches are the same as p-i-n diode-based circuits, only p-i-n diodes are replaced by FETs. Schematic circuit diagrams of TEE- and π-type switches are shown in Fig. 8.19 and 8.20, respectively. At ON-state condition, the series FETs should operate in low-impedance state and parallel transistors should operate in high-impedance state. Thus, in case of TEE- and π-type switch configurations as shown in Figs. 8.19 and 8.20, the control voltage V_{G1} will be high negative voltage and V_{G2} will be zero-bias voltage at ON-state condition and it will be reversed in OFF-state condition.

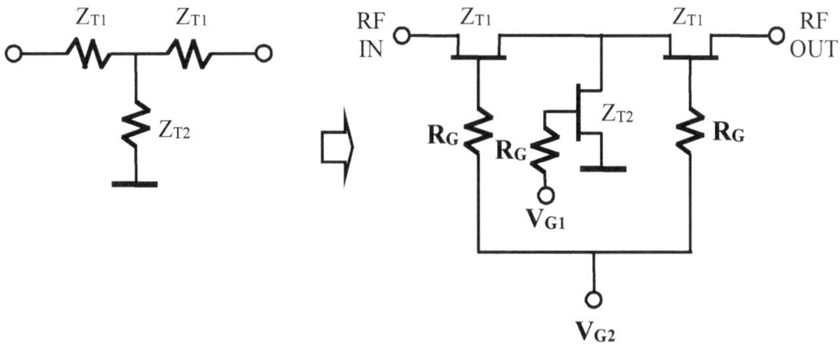

Fig. 8.19 FET-based TEE-type switch circuit diagram

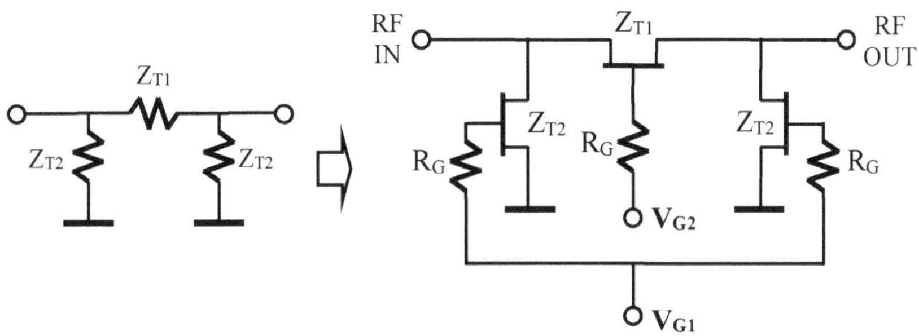

Fig. 8.20 FET-based π-type switch circuit diagram

8.8 Applications of RF/Microwave Switches

High-performance communication systems need in-build reconfiguration ability to achieve optimum performance with change of applications. Routing of RF/microwave signals is required to add more communication functionality. In case of satellite communication payloads, more complex redundancy measures are required to make the system more reliable to ensure long life. Space-based communication systems also need reconfigurable antennas that can support multiple frequency bands or reconfigurable service area. To achieve all these flexibilities, it is required to change the signal path, signal level or signal phase. Microwave switches are the essential building blocks to realize all these flexibilities. Adaptive power amplifiers with tuneable frequency bands by switching matching networks can also be used to maximize efficiency of a communication system over different bands of frequencies as per requirement. Stepwise variable attenuators and phase shifters are developed using microwave switches. RF/microwave switches are also used in automated measurement systems to control signal paths. With the help of switch matrix, an instrument can measure multiple DUTs. Multiple tests with different instruments

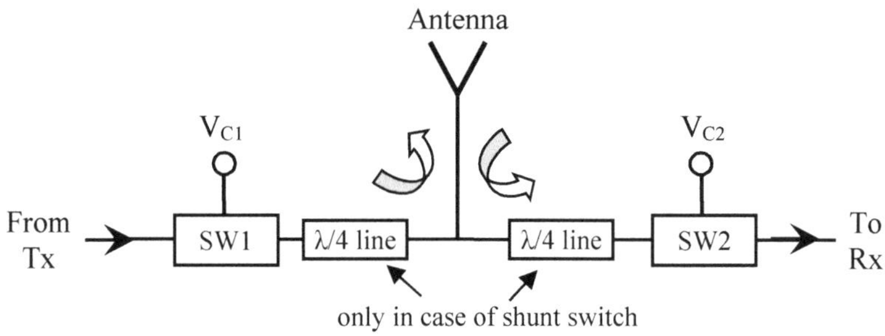

Fig. 8.21 Single antenna used to transmit as well as receive the signals

can also be run on the same DUT or multiple instruments can test multiple DUTs very efficiently.

Other than to provide reconfiguration ability to communication systems, the switches are also used to protect a subsystem just by disconnecting it, to prevent overstress. For example, in case of radar systems microwave switch protects the receiver from high-level transmitted power where a single antenna is used as transmit and receive antenna. Switches used in the transceiver application are known as transmit–receive (TR) switch. This TR switch connects antenna to transmitter in the transmit state and to receiver in the receive state. Schematic diagram of the system using single antenna for transmit and receive is shown in Fig. 8.21. In transmit state, switch-1 (SW1) is ON and switch-2 (SW2) is OFF, thus transmitter gets connected to the antenna and receiver is disconnected, whereas in receive state, SW2 is ON and SW1 is in OFF condition; therefore, the antenna gets connected to the receiver and transmitter is disconnected. Switches in their OFF state should provide isolation by providing high impedance to the antenna; otherwise, it will change the impedance matching condition of the antenna. To satisfy this condition, at least one series diode should be in the switch configuration. However, shunt-configured switches can also be used by connecting one-half wavelength transmission lines in between antenna and switches. Instead of using two separate switches, single switch with different configurations can also be used as shown in Fig. 8.22. This is a single-pole-double-throw (SPDT) switch that acts as TR switch. Under sufficiently positive-bias voltage V_{C1} and negative or zero voltage V_{C2}, the transmitter will be connected to the antenna and the receiver will be isolated. In the other state, V_{C2} will be sufficiently positive-bias voltage and V_{C1} will be zero or negative voltage, thus antenna will be connected to the receiver and transmitter will be isolated.

FET-based SPDT switch configuration is shown in Fig. 8.23. Under sufficiently negative control voltage to V_{G1} and zero control voltage to V_{G2}, low-loss signal path will be established in between RF1 and RF2 and there will be high isolation in between RF1 and RF3 path. To establish low-loss signal path in between RF1 and RF3, high negative control voltage is to be applied to V_{G2} and zero control voltage to V_{G1}. This will provide high isolation in between RF1 and RF2 path.

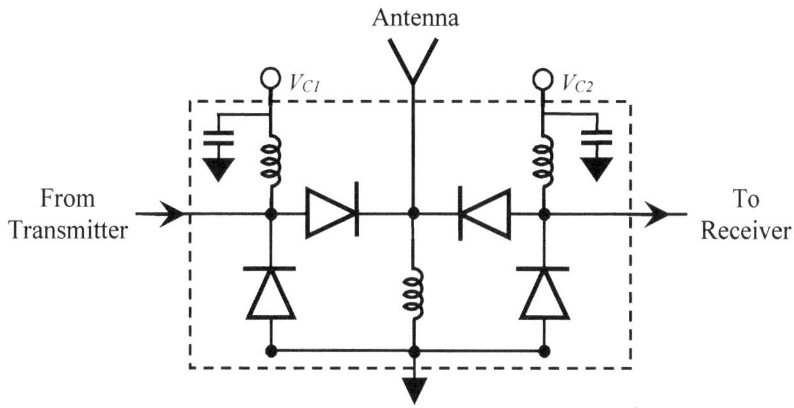

Fig. 8.22 TR switch with series–shunt configuration

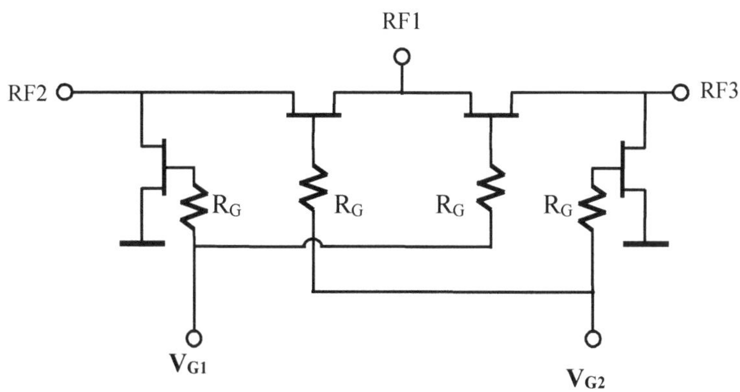

Fig. 8.23 FET-based reflective-type SPDT Switch

The SPDT switch diagrams as shown in Figs. 8.22 and 8.23, respectively, are both reflective types. That means the return loss of the OFF-state arm is poor. Terminated switches can also be realized using p-i-n diodes and FETs by putting a resistor in series with the shunt devices to achieve reference resistance R_o (50 Ω) of the shunt path at ON condition. Schematic circuit diagram of a terminated SPDT switch using FETs is shown in Fig. 8.24. The combined resistance of shunt FET at ON condition and R_S should be equal to the reference impedance R_o (=50 Ω).

Example 8.5 Calculate OFF-state arm isolation and return loss of a switch as shown in Fig. 8.23. Consider ON-state and OFF-state impedances of the device are purely resistive and values are 10 and 2000 Ω, respectively.

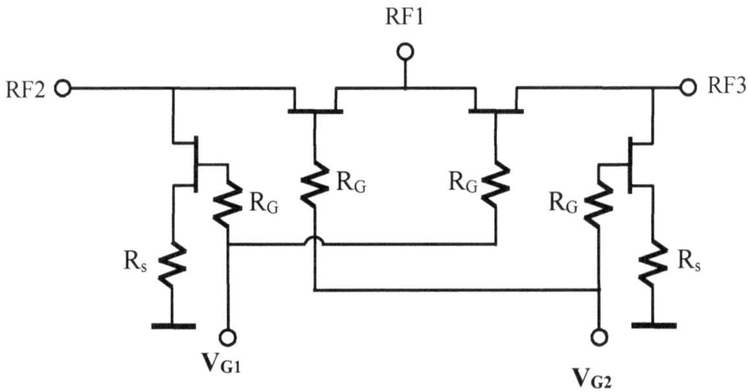

Fig. 8.24 FET-based terminated type SPDT Switch

Solution

With reference to the circuit of Fig. 8.23, suppose the signal path RF1 to RF2 is ON and the path RF1 to RF3 is OFF. At this condition the equivalent circuit will be as shown in Fig. 8.25.

When a signal of voltage v_s and reference impedance 50 Ω is applied to terminal RF1 then the signal voltage v_{out} at the RF3 port across a load of impedance 50 Ω will be

$$v_{out} = 0.002212\, v_s$$

Thus, transmission coefficient from RF1 to RF3 port will be

$$S_{31} = \frac{2v_{out}}{v_s} = 0.004425$$

Thus the OFF-state isolation will be

$$20\log(S_{31}) = 20\ \log(0.004425) = -47.08\,\text{dB}$$

With reference to Fig. 8.25c, the impedance seen by the RF3 port is given by

$$(58.78||50 + 2000)||10 = 9.951\ \Omega$$

Thus, the return loss of the RF3 port is

$$\text{RL(port RF3)} = 20\log\frac{|9.951 - 50|}{|9.951 + 50|}\text{dB} = -3.50\,\text{dB}$$

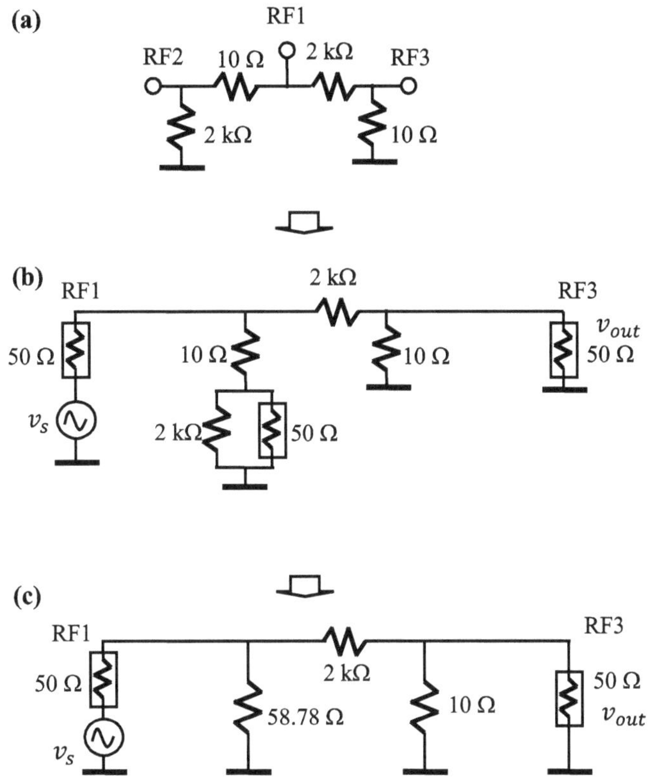

Fig. 8.25 Equivalent circuits for the switch of Example 8.5

Example 8.6 Calculate OFF-state isolation of a terminated switch as shown in Fig. 8.24. Consider ON-state and OFF-state impedances of the device are purely resistive and values are 10 and 2000 Ω, respectively.

Solution

With reference to the circuit of Fig. 8.24, suppose the signal path RF1 to RF2 is ON and the path RF1 to RF3 is OFF. At this condition, the equivalent circuit will be as shown in Fig. 8.26. Here, the value of R_s will be 40 Ω.

When a signal of voltage v_s and reference impedance 50 Ω is applied to terminal RF1, then the signal voltage v_{out} at the RF3 port across a load of impedance 50 Ω will be

$$v_{out} = 0.006584 v_s$$

Thus, transmission coefficient from RF1 to RF3 port will be

$$S_{31} = \frac{2v_{out}}{v_s} = 0.013169$$

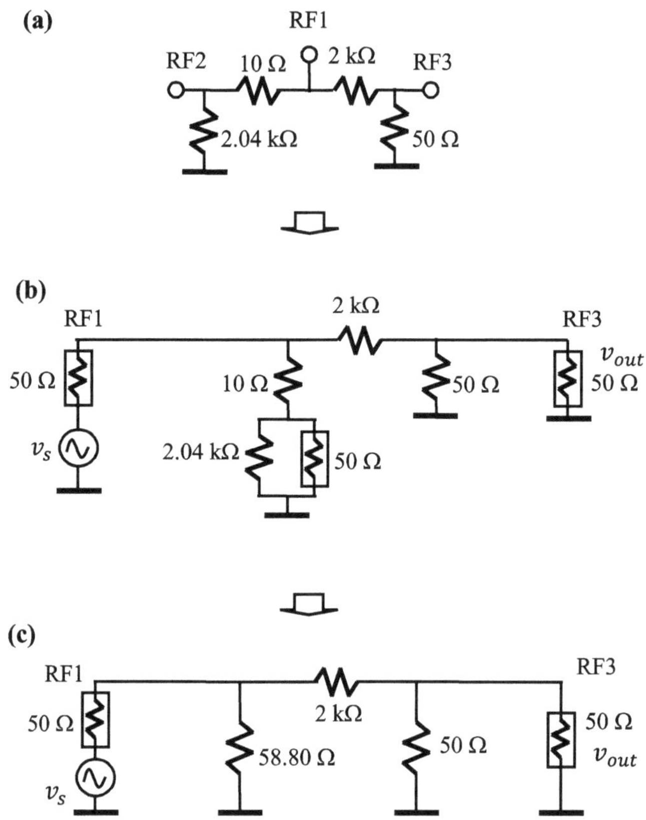

Fig. 8.26 Equivalent circuits for the switch of Example 8.6

Thus, the OFF-state isolation will be

$$20\log(S_{31}) = 20\log(0.013169) = -37.61 \text{ dB}$$

With reference to Fig. 8.26c, the impedance seen by the RF3 port is given by

$$(58.80||50 + 2000)||50 = 48.796\,\Omega$$

Thus, the return loss of the RF3 port is

$$\text{RL (port RF3)} = 20\log\frac{|48.796 - 50|}{|48.796 + 50|}\text{dB} = -38.28\,\text{dB}$$

From the Examples 8.5 and 8.6, it is clear that in case of reflective-type switches, return loss of the isolated port is poorer compared to terminated switch using same number of shunt/series devices. To achieve better isolation in the OFF-state arm,

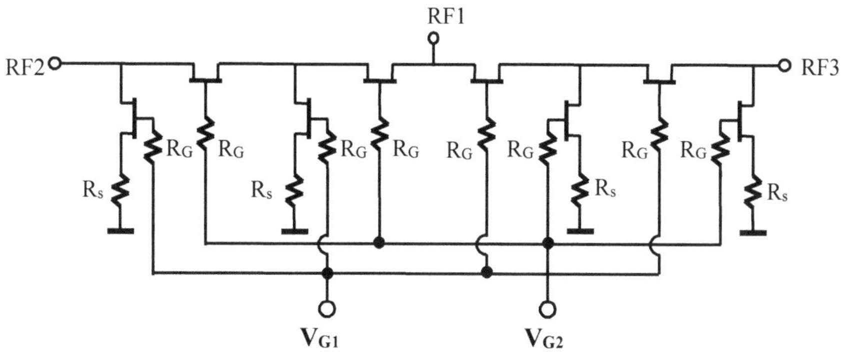

Fig. 8.27 FET-based terminated SPDT switch with better OFF-state isolation

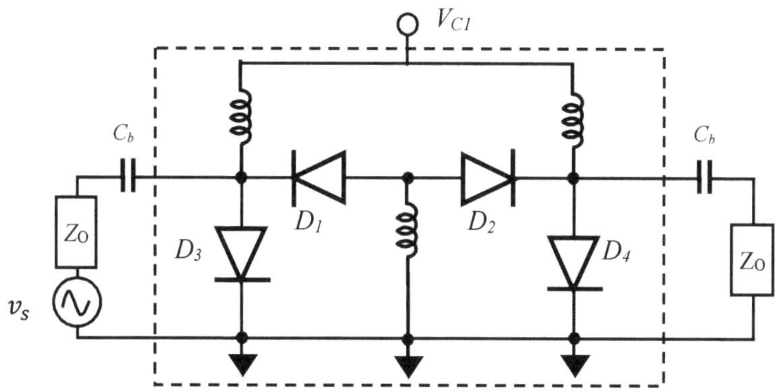

Fig. 8.28 Schematic circuit diagram of π type switch using series-shunt p-i-n diodes

more number of series and shunt devices need to be added in the switch configurations as shown in Fig. 8.27. In this case, extra isolation is achievable at the expense of increased insertion loss in the ON-state arm.

Example 8.7 Determine the insertion loss and isolation of a π-type switch using series and shunt p-i-n diodes as shown in Fig. 8.28 operating at 10 GHz for the following parameters of the diodes: $R_s = 2\ \Omega$, $R_i = 2\ \Omega$, $L_p = 0.1$ nH, $C_j = 0.02$ pF and $C_p = 0.01$ pF.

Solution

Equivalent circuit diagram of the π-type switch of Fig. 8.28 is shown in Fig. 8.29 considering impedances of the series and shunt diodes are Z_{T1} and Z_{T2} respectively. The transmission coefficient of the switch is given by

$$S_{21} = \frac{2V_{\text{out}}}{V_S} = \frac{2Z_{T2} \times Z_{T2}||Z_o}{(Z_o + Z_{T2})(Z_{T2}||Z_o + 2Z_{T1} + Z_{T2}||Z_o)} \tag{8.28}$$

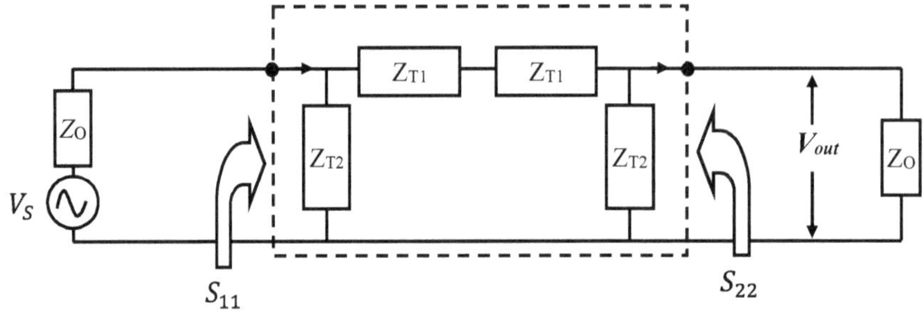

Fig. 8.29 Equivalent circuit diagram of π type switch

Therefore, insertion loss is given by

$$\frac{1}{S_{21}} = \frac{(Z_o + Z_{T2})(Z_o Z_{T2} + Z_o Z_{T1} + Z_{T1} Z_{T2})}{Z_o Z_{T2}^2}$$

or

$$\frac{1}{S_{21}} = 1 + \frac{2Z_{T1}}{Z_{T2}} + \frac{Z_o}{Z_{T2}} + \frac{Z_{T1}}{Z_o} + \frac{Z_{T1} Z_o}{Z_{T2}^2} \tag{8.29}$$

Insertion loss of the π switch can be calculated by putting the values of Z_{T1} and Z_{T2} correspond to the low impedance and high impedance of the p-i-n diodes, respectively, as

$$Z_{T1} = Z_{TL} = R_T + j2\pi f L_P = R_T + jX_L \quad \text{p-i-n diode ON}$$

$$Z_{T2} = Z_{TH} = \frac{1}{j2\pi f C_T} = -jX_C \quad \text{p-i-n diode OFF}$$

Putting these values in (8.29), insertion loss of the switch at ON condition is given by

$$\frac{1}{S_{21}} = \left[1 + \frac{R_T}{Z_o} - \frac{2X_L}{X_C} - \frac{Z_o R_T}{X_C^2}\right] + j\left[\frac{X_L}{Z_o} + \frac{Z_o}{X_C} + \frac{2R_T}{X_C} - \frac{Z_o X_L}{X_C^2}\right]$$

Therefore,

$$\mathrm{IL} = \left(1 + \frac{R_T}{Z_o} - \frac{2X_L}{X_C} - \frac{Z_o R_T}{X_C^2}\right)^2 + \left(\frac{X_L}{Z_o} + \frac{Z_o}{X_C} + \frac{2R_T}{X_C} - \frac{Z_o X_L}{X_C^2}\right)^2 \tag{8.30}$$

Here, $R_T = (R_i + R_s) = 4\,\Omega$, $C_T = C_j + C_p = 0.03\,\text{pF}$ and $L_p = 0.1\,\text{nH}$.

Putting $Z_O = 50\,\Omega$, $R_T = 4\,\Omega$, $X_L = 2\pi f L_p = 2\pi \times 10 \times 0.1 = 6.283\,\Omega$ and $X_C = 1000/(2\pi \times 10 \times 0.03) = 530.516\,\Omega$

$$\text{IL} = \left(1 + \frac{4}{50} - \frac{2 \times 6.283}{530.516} - \frac{50 \times 4}{530.516^2}\right)^2 + \left(\frac{6.283}{50} + \frac{50}{530.516} + \frac{2 \times 4}{530.516} - \frac{50 \times 6.283}{530.516^2}\right)^2$$

$$\text{IL} = 1.1143 + 0.0547 = 1.169$$

or

$$\text{IL (in dB)} = 10\log_{10}(1.169) = 0.678\,\text{dB}$$

At OFF condition the isolation of the π switch can be calculated by putting the values of Z_{T1} and Z_{T2} corresponding to the high impedance and low impedances of the p-i-n diodes, respectively, as

$$Z_{T1} = Z_{TH} = \frac{1}{j2\pi fC_T} = -jX_C \quad \text{p-i-n diode OFF}$$
$$Z_{T2} = Z_{TL} = R_T + j2\pi fL_P = R_T + jX_L \quad \text{p-i-n diode ON}$$

Putting these values in (8.29), isolation of the switch at OFF condition is given by

$$\frac{1}{S_{21}} = \left(1 + \frac{Z_oR_T - 2X_CX_L}{R_T^2 + X_L^2} - \frac{2Z_oR_TX_CX_L}{\left(R_T^2 - X_L^2\right)^2 + (2R_TX_L)^2}\right) + j\left(-\frac{2R_TX_C + Z_oX_L}{R_T^2 + X_L^2} - \frac{X_L}{Z_o} - \frac{Z_oX_C\left(R_T^2 - X_L^2\right)}{\left(R_T^2 - X_L^2\right)^2 + (2R_TX_L)^2}\right)$$

Therefore,

$$\text{Isolation} = \left(1 + \frac{Z_oR_T - 2X_CX_L}{R_T^2 + X_L^2} - \frac{2Z_oR_TX_CX_L}{\left(R_T^2 - X_L^2\right)^2 + (2R_TX_L)^2}\right)^2 + \left(-\frac{2R_TX_C + Z_oX_L}{R_T^2 + X_L^2} - \frac{X_L}{Z_o} - \frac{Z_oX_C\left(R_T^2 - X_L^2\right)}{\left(R_T^2 - X_L^2\right)^2 + (2R_TX_L)^2}\right)^2 \quad (8.31)$$

Putting $Z_O = 50\,\Omega$, $R_T = 4\,\Omega$, $X_L = 6.283\,\Omega$ and $X_C = 530.516\,\Omega$

$$\text{Isolation} = (1 - 116.56 - 433.2)^2 + (-82.16 - 0.126 + 202.344)^2$$
$$= 301141.9 + 14413.12 = 315555.05$$

or

$$\text{Isolation (in dB)} = 10\log_{10}(318402.66) = 55.0\,\text{dB}$$

The ON-state insertion loss is 0.678 dB and OFF-state isolation is 55 dB.

Example 8.8 Determine the insertion loss and isolation of a π-type switch as shown in Fig. 8.28 using series and shunt p-i-n diodes operating at 10 GHz for the following parameters: $R_s = 2\ \Omega$, $R_i = 2\ \Omega$, $L_p = 0.2$ nH, $C_j = 0.04$ pF and $C_p = 0.02$ pF.

Solution

For a π-type switch as shown in Fig. 8.28 using series and shunt p-i-n diodes, the insertion loss at ON condition and isolation in OFF condition is given by (8.30) and (8.31), respectively, as

$$\text{IL} = \left(1 + \frac{R_T}{Z_O} - \frac{2X_L}{X_C} - \frac{Z_o R_T}{X_C^2}\right)^2 + \left(\frac{X_L}{Z_O} + \frac{Z_o}{X_C} + \frac{2R_T}{X_C} - \frac{Z_o X_L}{X_C^2}\right)^2$$

$$\text{Isolation} = \left(1 + \frac{Z_o R_T - 2X_C X_L}{R_T^2 + X_L^2} - \frac{2Z_o R_T X_C X_L}{\left(R_T^2 - X_L^2\right)^2 + (2R_T X_L)^2}\right)^2 + \left(-\frac{2R_T X_C + Z_o X_L}{R_T^2 + X_L^2} - \frac{X_L}{Z_L} - \frac{Z_o X_C\left(R_T^2 - X_L^2\right)}{\left(R_T^2 - X_L^2\right)^2 + (2R_T X_L)^2}\right)^2$$

Here, $R_T = (R_i + R_s) = 4\ \Omega$, and $C_T = C_j + C_p = 0.06$ pF and $L_p = 0.2$ nH.

Putting $Z_o = 50\,\Omega$, $R_T = 4\,\Omega$, $X_L = 2\pi f L_p = 2\pi \times 10 \times 0.2 = 12.566\,\Omega$ and $X_C = 1000/(2\pi \times 10 \times 0.06) = 265.258\,\Omega$, the insertion loss is given by

$$\text{IL} = \left(1 + \frac{4}{50} - \frac{2 \times 12.566}{265.258} - \frac{50 \times 4}{265.258^2}\right)^2 + \left(\frac{12.566}{50} + \frac{50}{265.258} + \frac{2 \times 4}{265.258} - \frac{50 \times 12.566}{265.258^2}\right)^2$$

$$\text{IL} = 0.9651 + 0.213 = 1.178$$

or

$$\text{IL (in dB)} = 10\log_{10}(1.178) = 0.710\,\text{dB}$$

$$\text{Isolation} = (1 - 37.183 - 44.083)^2 + (-15.81 - 0.251 + 62.23)^2 = 6442.67 + 2131.062 = 8573.73$$

or

$$\text{Isolation (in dB)} = 10\log_{10}(8573.73) = 39.33\,\text{dB}$$

The ON-state insertion loss is 0.710 dB and OFF-state isolation is 39.33 dB.

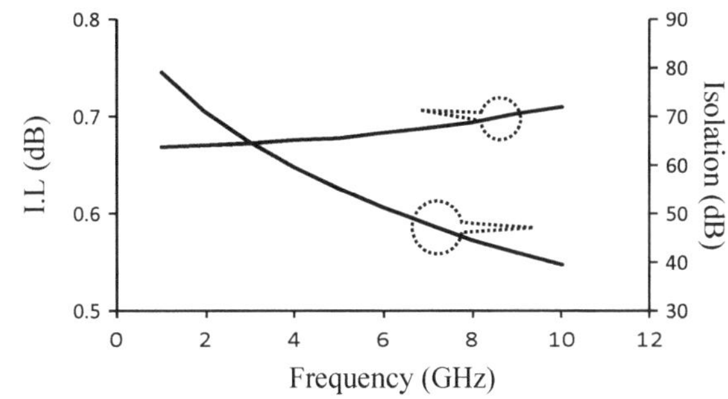

Fig. 8.30 Insertion loss and isolation of π type switch over frequency

Example 8.9 A π type switch as shown in Fig. 8.28 using series and shunt p-i-n diodes for the following parameters: $R_s = 2\ \Omega$, $R_i = 2\ \Omega$, $L_p = 0.2$ nH, $C_j = 0.04$ pF and $C_p = 0.02$ pF. Plot the insertion loss and isolation of the attenuator over the frequency range of 1–10 GHz.

Solution

For a π type switch as shown in Fig. 8.28 using series and shunt p-i-n diodes, the insertion loss at ON condition and isolation in OFF condition is given by (8.30) and (8.31), respectively, as

$$\mathrm{IL} = \left(1 + \frac{R_T}{Z_o} - \frac{2X_L}{X_C} - \frac{Z_o R_T}{X_C^2}\right)^2 + \left(\frac{X_L}{Z_o} + \frac{Z_o}{X_C} + \frac{2R_T}{X_C} - \frac{Z_o X_L}{X_C^2}\right)^2$$

$$\begin{aligned}\mathrm{Isolation} = &\left(1 + \frac{Z_o R_T - 2X_C X_L}{R_T^2 + X_L^2} - \frac{2Z_o R_T X_C X_L}{\left(R_T^2 - X_L^2\right)^2 + (2R_T X_L)^2}\right)^2 \\ &+ \left(-\frac{2R_T X_C + Z_o X_L}{R_T^2 + X_L^2} - \frac{X_L}{Z_o} - \frac{Z_o X_C\left(R_T^2 - X_L^2\right)}{\left(R_T^2 - X_L^2\right)^2 + (2R_T X_L)^2}\right)^2\end{aligned}$$

Here, $R_T = (R_i + R_s) = 4\ \Omega$, and $C_T = C_j + C_p = 0.06$ pF and $L_p = 0.2$ nH.

Putting $Z_o = 50\,\Omega$, $R_T = 4\,\Omega$, $X_L = 2\pi f L_p = 2\pi \times 10 \times 0.2 = 12.566\,\Omega$ and $X_C = 1000/(2\pi \times 10 \times 0.06) = 265.258\,\Omega$, the insertion loss and isolation is calculated and shown in Fig. 8.30.

This plot shows that ON-state insertion loss increases and OFF-state isolation decreases with the increase of frequency. This is due to the finite values of parasitic capacitance and inductance of the diode.

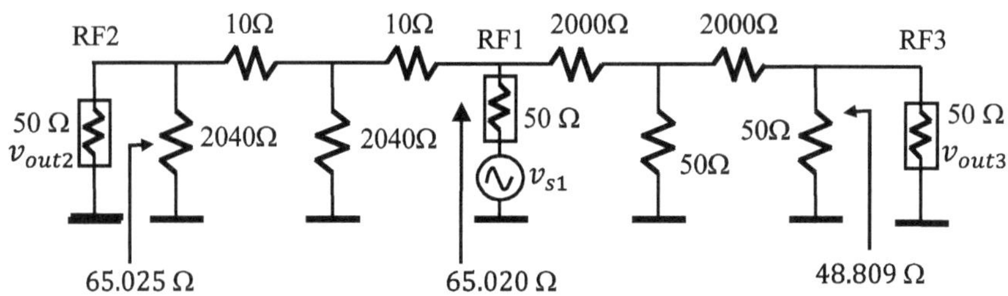

Fig. 8.31 RF equivalent circuit diagram of SPDT switch of Example 8.10

Example 8.10 Determine S-parameter matrix of the FET-based SPDT switch as shown in Fig. 8.27. Consider RF1 to RF2 path ON and RF1 to RF3 path open. Assume ON-state and OFF-state impedances of the FETs are purely resistive of values 10 and 2000 Ω, respectively, and $R_s = 40\ \Omega$. Also, calculate port return losses, insertion loss and isolations.

Solution

RF equivalent circuit diagram of the SPDT switch is shown in Fig. 8.31 at the condition RF1 to RF2 path ON and RF1 to RF3 path open.

From this equivalent circuit, the input impedance at port-1 can be written as

$$\begin{aligned} Z_{\text{IN1}} &= (((50||2040) + 10)||2040 + 10)||(((50||50) + 2000)||50 + 2000) \\ &= 65.020\ \Omega \end{aligned}$$

Therefore,

$$S_{11} = \frac{65.020 - 50}{65.020 + 50} = 0.13059\angle 0^{\circ}$$

In a similar way,

$$S_{22} = \frac{65.025 - 50}{65.025 + 50} = 0.13062\angle 0^{\circ}$$

$$S_{33} = \frac{48.809 - 50}{48.809 + 50} = 0.01206\angle 180^{\circ}$$

Suppose a voltage signal v_{s1} with source impedance 50 Ω is connected to port-1 and other two terminals are terminated with load impedances 50 Ω, the output voltage signal at port-2 will be

$$v_{\text{out2}} = 0.3993\, v_{s1}$$

Thus,

$$S_{21} = \frac{2v_{\text{out2}}}{v_{s1}} = 0.79863\angle 0^{\circ}$$

Similarly, other transmission coefficient parameters can be calculated and the S-parameter matrix of the SPDT switch is

$$[S] = \begin{bmatrix} 0.13059\angle 0^{\circ} & 0.79863\angle 0^{\circ} & 0.00033\angle 0^{\circ} \\ 0.79863\angle 0^{\circ} & 0.13062\angle 0^{\circ} & 0.00023\angle 0^{\circ} \\ 0.00033\angle 0^{\circ} & 0.00023\angle 0^{\circ} & 0.01206\angle 180^{\circ} \end{bmatrix}$$

From the S-parameter matrix, the port return losses are
Return loss of port-1 = 20 log(0.13059) = 17.680 dB
Return loss of port-2 = 20 log(0.13062) = 17.682 dB
Return loss of port-3 = 20 log(0.01206) = 38.377 dB
Insertion loss and isolations are
Insertion loss from port-1 to port-2 = 20 log(0.79863) = 1.953 dB
Isolation in between port-1 to port-3 = 20 log(0.00033) = 69.566 dB
Isolation in between port-2 to port-3 = 20 log(0.00023) = 72.585 dB

References

1. Caulton M, Rosen A, Stabile A, Gombar A (1982) Pin diodes for low frequency high power switching application. IEEE Trans Microw Theory Tech MTT-30:875–886
2. Caverly RH, Hiller G (1990) Establishing the minimum reverse bias for a p-i-n diode in a high-power switch. IEEE Trans Microw Theory Tech 38:1938–1943
3. Drozdovskaia L (2001) RF and microwave frequency properties of a reverse-biased thick switching p-i-n diode. IEEE Trans Microw Theory Tech 49(8):1370–1373
4. Bludov AV, Boltovets MS, Vassilevski KV, Zorenko AV, Zekentes K, Krivutsa VA, Kritskaya TV, Lebedev AA (2004) Microwave switching, based on 4H-SiC p-i-n diodes. Tech Phys Lett 30(2):123–125
5. Camara N, Zekentes K, Romanov LP, Kirillov AV, Boltovets MS, Vassilevski KV, Haddad G (2006) Microwave p-i-n diodes and switches based on 4H-SiC. IEEE Electron Device Lett 27(2):108–110
6. Gopinath A, Rankin JB (1985) GaAs FET RF switches. IEEE Trans Electron Devices ED-32:1272–1278
7. Makioka S, Anda Y, Miyatsuji K, Ueda D (2001) Super self-aligned GaAs RF switch IC with 0.25 dB extremely low insertion loss for mobile communication systems. IEEE Trans Electron Devices 48(8):1510–1514
8. Yang Z, Hu X, Yang J, Simin G, Shur M, Gaska R (2009) Maximum powers of low-loss series-shunt FET RF switches. Solid-State Electron 53:117–119
9. Campbell C, Dumka D (2010) Wideband high power GaN on SiC SPDT switch MMICs. IEEE MTT-S, IMS 2010, Anaheim, CA, USA, pp 145–148
10. Prasad A, Fager C, Thorsell M, Andersson CM, Yhland K (2014) Symmetrical large-signal modeling of microwave switch FETs. IEEE Trans Microw Theory Tech 62(8):1590–1598

11. Simin G, Hu X, Ilinskaya N, Kumar A, Koudymov A, Zhang J, Khan MA, Gaska R, Shur MS (2000) A 7.5 kW/mm current switch using AlGaN/GaN metal–oxide–semiconductor heterostructure field effect transistors on SiC substrates. Electron Lett 36:2043–2044
12. Ishida H, Hirose Y, Murata T, Kanda A, Ikeda Y, Matsuno T, Inoue K, Uemoto Y, Tanaka T, Egawa T, Ueda D (2003) A high power Tx/Rx switch IC using AIGaN/GaN HFETs. In: IEDM technical digest, pp 583–586
13. Koudymov A, Rai S, Adivarahan V, Gaevski M, Yang J, Simin G, Khan MA (2004) Monolithically integrated high-power broad-band RF switch based on III-N insulated gate transistors. IEEE Microw Wireless Compon Lett 14(12):560–562
14. Kaper V, Thompson R, Prunty T, Shealy JR (2004) Monolithic AlGaN/GaN HEMT SPDT switch. In: Proceedings of 12th GAAS symposium, Amsterdam, The Netherlands, pp 84–86
15. Sattu A, Yang J, Shur M, Gaska R, Simin G (2010) AlGaN/GaN Microwave switch with hybrid slow and fast gate design. IEEE Electron Device Lett 31(12):10389–1391
16. Sattu A, Billingsley D, Deng J, Yang J, Simin G, Shur M, Gaska R (2011) Low-loss AlInN/GaN microwave switch. Electronics Lett 47(15)

Chapter 9
Microwave Attenuators

Abstract This chapter presents design and analysis of diode- and transistor-based microwave attenuator circuits including their applications, performance variation and compensation with operating temperature. Complete design details of series-, shunt-, π-, TEE- and hybrid-configured attenuators with their merits and demerits are given. Detailed design and analysis of driver circuits based on equiresistance curves of p-i-n diodes for achieving temperature-insensitive attenuation of the attenuator circuits is also presented.

9.1 Introduction

An attenuator provides known amount of attenuation when it is connected in a signal path having known source and load impedances. These are used to control signal level in the signal path. It may provide a fixed or variable/adjustable attenuation depending upon control voltage or current. Fixed attenuators are realized using combination of RF-resistive elements with fixed values, whereas voltage/current-controlled attenuators are realized using combination of voltage/current-dependent resistors. At RF/microwave frequency range, forward-biased p-i-n diodes, BJTs or FETs are used as voltage/current-dependent RF resistors [1–13]. Voltage-variable p-i-n diode attenuators are used to control signal level of communication systems by controlling gain of the system [1–7]. In case of satellite transponders, p-i-n diode or FET-based attenuators are used for on-board gain control of a communication transponder by issuing command from ground or automatically by sensing RF power level.

Important performance parameters of a microwave attenuator are operating frequency, attenuation range, insertion loss, port return losses, power handling capability and DC power requirements. RF switches discussed in the previous section are also attenuators with maximum attenuation at its OFF state and minimum attenuation at its ON state. OFF-state reflective switch offers maximum attenuation by reflecting the signal which is undesirable as function of attenuator. Variable attenuator provides stepwise or continuous attenuation from its minimum to maximum attenuation range. Known amount of attenuation and port matching over its entire attenuation range are

S. C. Bera, *Microwave Active Devices and Circuits for Communication*, Lecture Notes in Electrical Engineering 533, https://doi.org/10.1007/978-981-13-3004-9_9

important parameters. When input and output ports of an attenuator are matched, then attenuation offered by the attenuator is entirely due to the transmission loss. In case of mismatch, some amount of the loss will be due to the reflection of the signal.

Variable attenuators using p-i-n diodes are mostly used due to their simplicity in circuit and lower cost compared to transistor-based circuits. However, the RF resistance of p-i-n diodes is very sensitive to its operating temperature [1–7]. Thus, attenuation of a p-i-n diode-based attenuator is also a function of temperature, which is undesirable. Some specially fabricated p-i-n diodes having zero temperature coefficient of RF resistance can be used in an attenuator when biased at a constant forward-bias current [1, 2]. However, this requirement cannot be met without sacrificing other aspects of the p-i-n diode's performance. Conventionally, analog as well as digital compensation circuits with external temperature sensor is used to compensate the attenuation. Another technique based on p-i-n diodes intrinsic property called "optimum bias load-line technique" is used to achieve temperature-compensated continuous variable as well as stepwise variable attenuation [3–7].

GaAs MESFETs and HEMTs at zero-drain-bias condition are used as variable resistors to construct microwave attenuators. The attenuation of transistor-based attenuators is highly temperature-sensitive and is preferable only where continuously variable attenuation is required. Stepwise variable attenuators are realized by switching fixed microwave attenuators (also called PADs) using transistors as microwave switch. In this case, attenuation values remain nearly constant over a very wide range of temperature since the attenuation values are mainly governed by the fixed PADs. These are most suitable for realization in monolithic microwave integrated circuit (MMIC) configuration.

9.2 Diode-Based Attenuator Circuits

RF resistance of p-i-n diode is function of its DC bias condition. By controlling the bias current/voltage, its intrinsic layer resistance (R_i) can be varied from few ohms to several thousand ohms. The RF equivalent circuit of a p-i-n diode is shown in Fig. 9.1 over the operating frequency for which $f \gg 1/2\pi\tau_a$, where τ_a is carrier lifetime. Thus, RF impedance of the p-i-n diode can be written as

$$Z_T = (R_i||C_i + R_s + j2\pi f L_p)||\left(1/j2\pi f C_p\right) \quad (9.1)$$

Over the operating frequency range and bias condition where $R_i \gg 2\pi f L_p$ and also $R_i \ll 1/2\pi f C_p$, the RF equivalent circuit of the forward-biased p-i-n diode can be represented as shown in Fig. 9.1b. In this case, the impedance Z_T of the p-i-n diode can be written as

$$Z_T = (R_i||C_i) + R_s \quad (9.2)$$

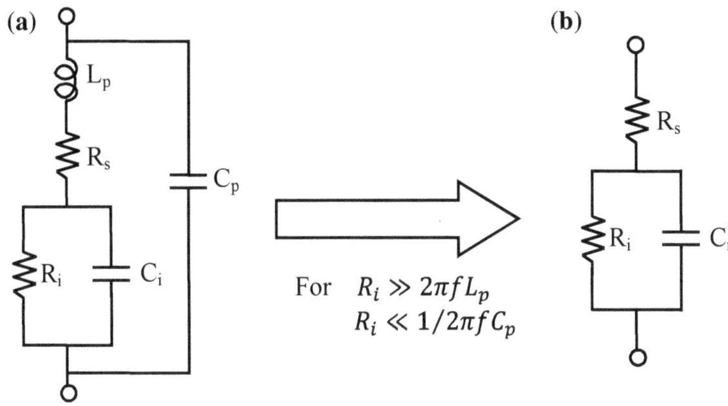

Fig. 9.1 Forward-biased equivalent circuit of p-i-n diode for $f \gg 1/2\pi\tau_a$

As discussed in Chap. 2, under forward-bias condition, the intrinsic layer resistance R_i is current-controlled resistance, whereas other component values are practically independent of its DC bias. Thus, with the change of bias voltage/current resistance, R_i of the forward-biased p-i-n diode acts as variable resistance. Using this variable resistance, voltage/current variable microwave attenuators are realized.

9.3 Series- or Shunt-Connected Element

A simple voltage/current-dependent variable attenuator is a circuit where a p-i-n diode is connected in series or shunt in the signal path as shown in Fig. 9.2. RF equivalent circuits of the series-diode and shunt-diode attenuator circuits are shown in Fig. 9.3a, b respectively. The circuits are same as series and shunt switch circuits; however, here diodes are operated over its wide forward-bias range instead of only high forward- and reverse-bias condition as in switch. The S-parameter matrix of the series diode attenuator circuit can be written as (using 7.162c)

$$\begin{bmatrix} S_{11} & S_{12} \\ S_{21} & S_{22} \end{bmatrix} = \begin{bmatrix} \frac{Z_T}{Z_T+2Z_o} & \frac{2Z_o}{Z_T+2Z_o} \\ \frac{2Z_o}{Z_T+2Z_o} & \frac{Z_T}{Z_T+2Z_o} \end{bmatrix} \tag{9.3}$$

Thus, attenuation of the series diode attenuator is given by

$$\text{Attenuation(in dB)} = 20\log_{10}\left(\frac{1}{|S_{21}|}\right) = 20\log_{10}\left(\left|1 + \frac{Z_T}{2Z_o}\right|\right) \tag{9.4a}$$

The input and output return losses of the series diode attenuator are given by

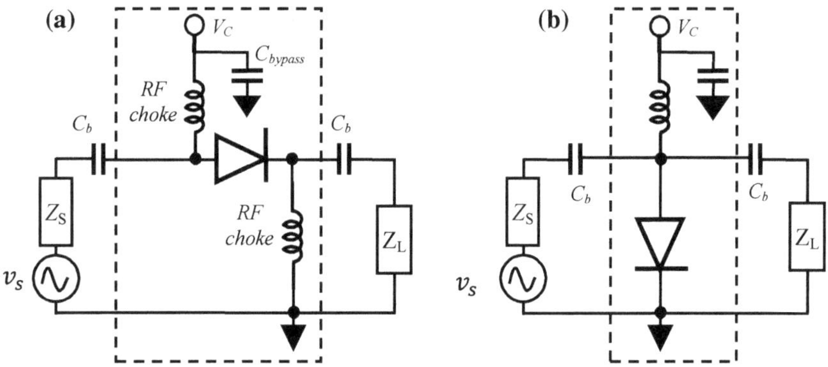

Fig. 9.2 Simple attenuator configuration **a** series diode and **b** shunt diode

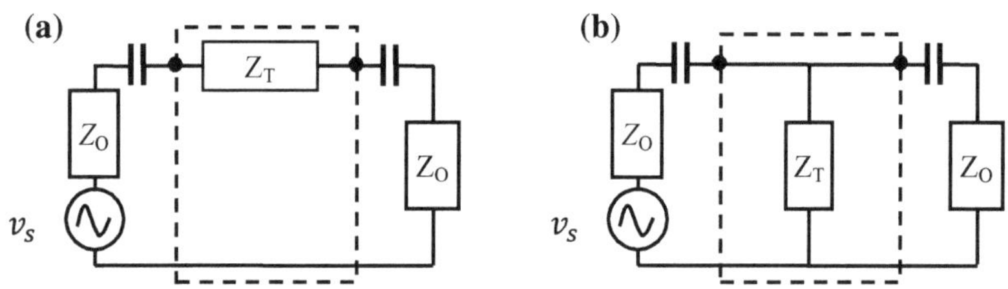

Fig. 9.3 RF equivalent circuit of attenuators with **a** series diode **b** shunt diode

$$\mathrm{RL(in\ dB)} = 20\log_{10}\left(\frac{1}{S_{11}}\right) = 20\log_{10}\left(1 + \frac{2Z_o}{Z_T}\right) \tag{9.4b}$$

The variation of attenuation and return loss with the change of diode resistance R_i is shown in Fig. 9.4 for typical values of other parameters of the diode. It shows that with the increase of attenuation, return loss of the series attenuator degrades drastically. This is the reason for not using attenuators with single series diode configuration. At higher level of attenuation, major amount of attenuation is due to reflection of power from the series element which is highly undesirable for practical systems. Reflected power due to the mismatch not only affects the preceding system but also results in undesirable signal ripple in the frequency response due to generation of standing wave.

For shunt-connected diode, the S-parameter matrix is given by (using 7.163d)

$$\begin{bmatrix} S_{11} & S_{12} \\ S_{21} & S_{22} \end{bmatrix} = \begin{bmatrix} \frac{-Z_o}{2Z_T+Z_o} & \frac{2Z_T}{2Z_T+Z_o} \\ \frac{2Z_T}{2Z_T+Z_o} & \frac{-Z_o}{2Z_T+Z_o} \end{bmatrix} \tag{9.5}$$

Therefore, the attenuation of the shunt-connected diode attenuator is given by

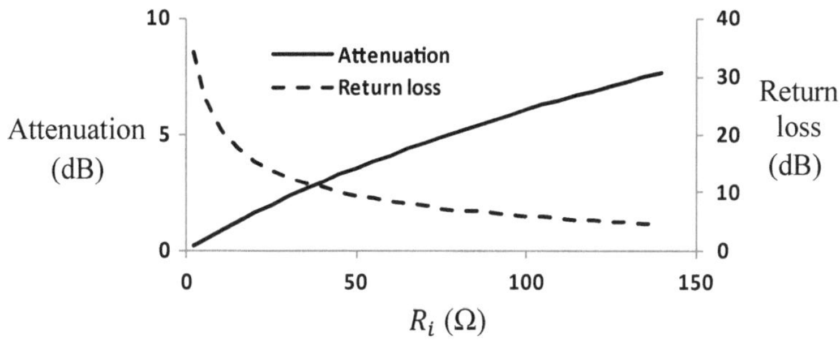

Fig. 9.4 Series diode attenuator characteristic with variation of R_i

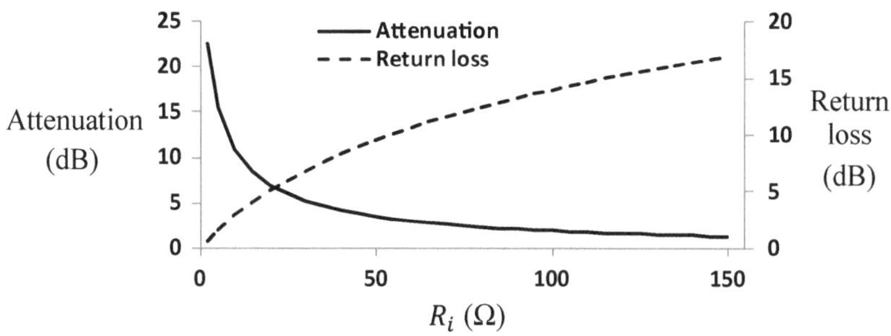

Fig. 9.5 Shunt diode attenuator characteristic with variation of R_i

$$\text{Attenuation(in dB)} = 20\log_{10}\left(\frac{1}{|S_{21}|}\right) = 20\log_{10}\left(\left|1 + \frac{Z_o}{2Z_T}\right|\right) \tag{9.6a}$$

Similarly, the return losses (both input and output) of the shunt-connected diode attenuator are given by

$$\text{RL(in dB)} = 20\log_{10}\left(\frac{1}{S_{11}}\right) = 20\log_{10}\left(\left|1 - \frac{2Z_T}{Z_o}\right|\right) \tag{9.6b}$$

The plot of attenuation and return losses of the shunt-connected diode attenuator is shown in Fig. 9.5. In this case also with the increase of attenuation, return loss is decreasing. Therefore, this configuration of attenuators is also not suitable for practical applications. Practically, it is always preferable to use input- and output-matched attenuators, i.e. with good input and output return losses.

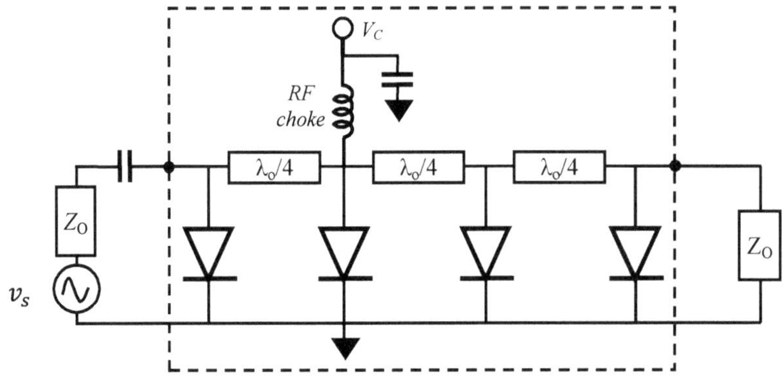

Fig. 9.6 Attenuator circuit using multiple shunt p-i-n diodes

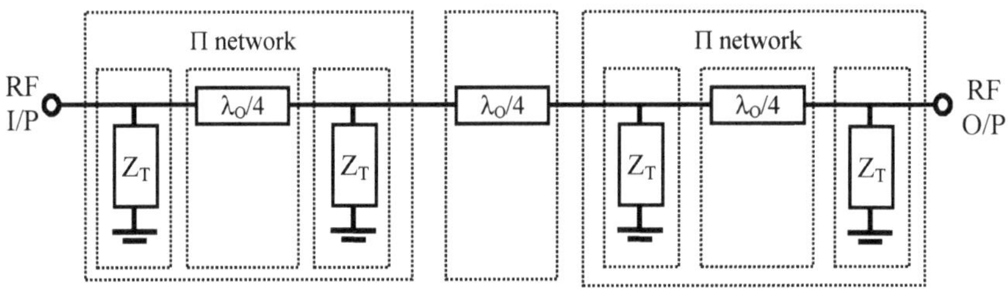

Fig. 9.7 Equivalent circuit of multiple shunt-connected p-i-n diode attenuator

9.4 Multiple Shunt-Connected Element

One way to achieve better input/output return losses is the use of multiple numbers of shunt diodes with separation of $\lambda_o/4$ transmission line length that corresponds to the operating frequency of f_o as shown in Fig. 9.6. Return loss of this configuration will be better due to the cancellation of signals reflected from the multiple diodes due to their mutual phase difference of 180°. *S*-parameter elements of the attenuator can be determined by matrix multiplication of *ABCD* parameter matrix of the individual element of the RF equivalent circuit as shown in Fig. 9.7. The *ABCD* parameter matrix of the attenuator is given by

$$\begin{bmatrix} A & B \\ C & D \end{bmatrix} = \begin{bmatrix} A_d & B_d \\ C_d & D_d \end{bmatrix}\begin{bmatrix} A_l & B_l \\ C_l & D_l \end{bmatrix}\begin{bmatrix} A_d & B_d \\ C_d & D_d \end{bmatrix}\begin{bmatrix} A_l & B_l \\ C_l & D_l \end{bmatrix}\begin{bmatrix} A_d & B_d \\ C_d & D_d \end{bmatrix}\begin{bmatrix} A_l & B_l \\ C_l & D_l \end{bmatrix}\begin{bmatrix} A_d & B_d \\ C_d & D_d \end{bmatrix} \tag{9.7a}$$

Here, Z_T is the forward-biased p-i-n diode impedance and is given by (9.2). The *ABCD* parameter matrix of the shunt diode is given by

$$\begin{bmatrix} A_d & B_d \\ C_d & D_d \end{bmatrix} = \begin{bmatrix} 1 & 0 \\ 1/Z_T & 1 \end{bmatrix} \tag{9.7b}$$

And *ABCD* parameter matrix of a series transmission line of length l is given by

$$\begin{bmatrix} A_l & B_l \\ C_l & D_l \end{bmatrix} = \begin{bmatrix} \cos\left(\frac{\pi\lambda_o}{2\lambda}\right) & jZ_o \sin\left(\frac{\pi\lambda_o}{2\lambda}\right) \\ j\frac{1}{Z_o}\sin\left(\frac{\pi\lambda_o}{2\lambda}\right) & \cos\left(\frac{\pi\lambda_o}{2\lambda}\right) \end{bmatrix} \tag{9.7c}$$

Here, λ is the wavelength corresponding to any operating frequency f. Thus, *ABCD* parameter of the $\lambda_o/4$ transmission line at operating frequency f_o can be determined by putting $\lambda = \lambda_o$ and can be written as

$$\begin{bmatrix} A_{l_o} & B_{l_o} \\ C_{l_o} & D_{l_o} \end{bmatrix} = \begin{bmatrix} 0 & jZ_o \\ j\frac{1}{Z_o} & 0 \end{bmatrix} \tag{9.7d}$$

Thus, *ABCD* matrix of the attenuator can be written as

$$\begin{bmatrix} A & B \\ C & D \end{bmatrix} = \begin{bmatrix} j\frac{Z_o}{Z_T} & jZ_o \\ j\frac{Z_o^2+Z_T^2}{Z_o Z_T^2} & j\frac{Z_o}{Z_T} \end{bmatrix} \begin{bmatrix} 0 & jZ_o \\ j\frac{1}{Z_o} & 0 \end{bmatrix} \begin{bmatrix} j\frac{Z_o}{Z_T} & jZ_o \\ j\frac{Z_o^2+Z_T^2}{Z_o Z_T^2} & j\frac{Z_o}{Z_T} \end{bmatrix} \tag{9.7e}$$

Here, first and last terms are *ABCD* parameter matrix of the Π network containing two shunt elements of impedance Z_T and one series element of $\lambda_o/4$ length transmission line as shown in Fig. 9.7.

The equation (9.7e) can be written as

$$\begin{bmatrix} A & B \\ C & D \end{bmatrix} = \begin{bmatrix} -j\frac{Z_o Z_T^2 + Z_o\left(Z_o^2+Z_T^2\right)}{Z_T^3} & -j\frac{Z_o\left(Z_o^2+Z_T^2\right)}{Z_T^2} \\ -j\frac{Z_o^2 Z_T^2 + \left(Z_o^2+Z_T^2\right)^2}{Z_o Z_T^4} & -j\frac{Z_o Z_T^2 + Z_o\left(Z_o^2+Z_T^2\right)}{Z_T^3} \end{bmatrix} \tag{9.7f}$$

From this *ABCD* parameter matrix, *S*-parameters of the attenuator can be determined as follows (reciprocal matrix with symmetric structure):

$$S_{21} = S_{12} = 2/(A + B/Z_o + CZ_o + D) \tag{9.8a}$$

and

$$S_{11} = S_{22} = (B/Z_o - CZ_o)/(A + B/Z_o + CZ_o + D) \tag{9.8b}$$

Putting the values of A, B, C and D, we can write

$$S_{21} = S_{21} = j\frac{2Z_T^4}{4Z_oZ_T^3 + 4Z_o^2Z_T^2 + 2Z_o^3Z_T + 2Z_T^4 + Z_o^4} \tag{9.9a}$$

$$S_{11} = S_{22} = \frac{Z_o^2\left(Z_o^2 + Z_T^2\right)}{4Z_oZ_T^3 + 4Z_o^2Z_T^2 + 2Z_o^3Z_T + 2Z_T^4 + Z_o^4} \tag{9.9b}$$

Therefore, the attenuation at frequency f_o of the four-element shunt-connected attenuator is given by

$$\begin{aligned}\text{Attenuation (in dB)} &= 20\log_{10}(1/|S_{21}|)\\ &= 20\log_{10}\left(\left|\frac{4Z_oZ_T^3 + 4Z_o^2Z_T^2 + 2Z_o^3Z_T + 2Z_T^4 + Z_o^4}{2Z_T^4}\right|\right)\end{aligned} \tag{9.10a}$$

Similarly, the return losses at frequency f_o of the four shunt-connected diode attenuators are given by

$$\begin{aligned}\text{RL(in dB)} &= 20\log_{10}\left(\frac{1}{S_{11}}\right) = 20\log_{10}\left(\frac{1}{S_{22}}\right)\\ &= 20\log_{10}\left(\left|\frac{4Z_oZ_T^3 + 4Z_o^2Z_T^2 + 2Z_o^3Z_T + 2Z_T^4 + Z_o^4}{Z_o^2\left(Z_o^2 + Z_T^2\right)}\right|\right)\end{aligned} \tag{9.10b}$$

The plot of attenuation and return losses at frequency f_o for four shunt-connected diode attenuators is shown in Fig. 9.8. It shows that the return losses of the attenuator are better than that of the attenuator with single shunt-connected diode attenuator for any value of attenuation. This can be further improved by connecting more numbers of shunt-connected diodes with separation of $\lambda_o/4$ wavelength. Figure 9.9 shows the attenuation and return losses of an attenuator with eight shunt-connected diodes with separation of $\lambda_o/4$ wavelength. In this case, the return losses are better than that of attenuator with four shunt-connected diodes. Figure 9.10 shows the return losses for the four and eight shunt-connected diode attenuators with the attenuation values. From this, it is clear that to realize higher value of attenuation with good return losses, more number of diodes need to be used in case of shunt-configured attenuators. A photograph of such an attenuator using eight shunt-connected diodes with separation of $\lambda_o/4$ wavelength is shown in Fig. 9.11.

In case of multiple shunt-connected diode-based attenuators, the elements are connected with separation of quarter wavelength transmission line corresponding to a particular frequency f_o. Therefore, at other frequencies, the separation will be different from corresponding quarter wavelength transmission line. Thus, effective cancellation of the reflected wave will go on decreasing with the frequency away from f_o. To achieve port matching over a broad frequency band, matched attenuator circuits are designed.

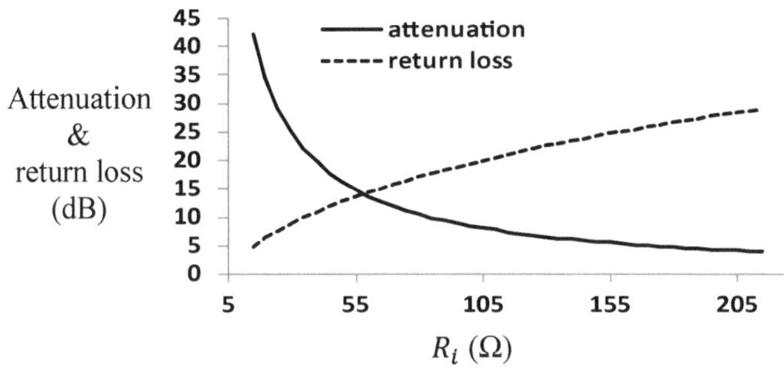

Fig. 9.8 Attenuation and return losses of the four-element shunt-connected diode attenuator at frequency f_o

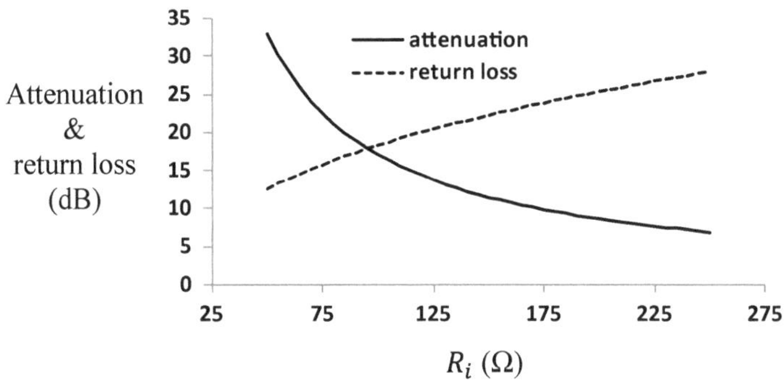

Fig. 9.9 Attenuation and return losses of the shunt-connected eight-diode attenuator at frequency f_o

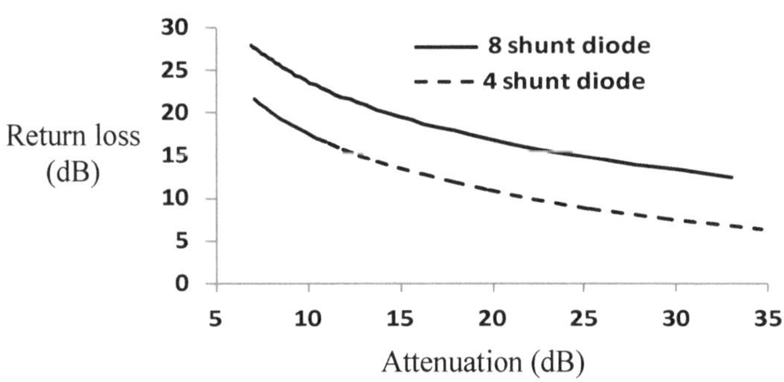

Fig. 9.10 Dependence of return losses with attenuation and number of diodes

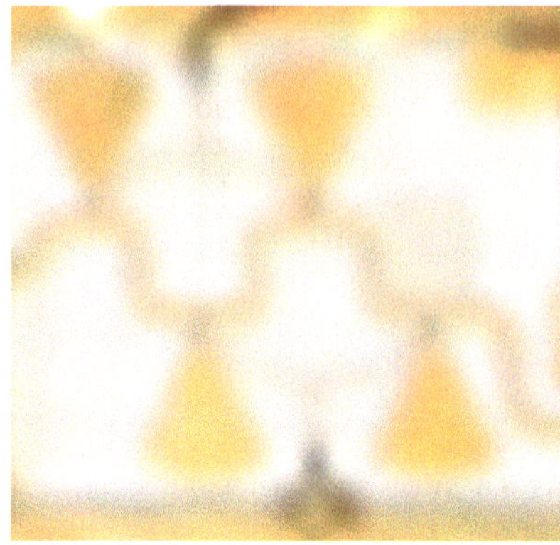

Fig. 9.11 Photograph of an attenuator with eight shunt-connected p-i-n diodes with separation of quarter-wavelength microstrip lines

9.5 Matched Attenuator Circuits

Matched attenuators ideally have matched input and output ports for all the values of attenuation. Among various configurations, generally TEE- and π-type configurations are used.

9.5.1 *TEE Attenuator*

Schematic circuit diagram of the diode-based TEE configuration attenuator is shown in Fig. 9.12. Here, DC control voltage V_{C1} is for biasing the two series-connected diodes and V_{C2} is for biasing the shunt diode. The bias configuration ensures same amount of current to both the series diodes. Also the configuration ensures independent control of currents for series and shunt diodes. Considering same characteristic of both the series diodes, the RF impedance of both the series diodes will be same. The RF equivalent circuit of the TEE-configured attenuator is shown in Fig. 9.13.

The conditions for the perfectly matching of input and output ports are

$$Z_{\text{IN}} = Z_o \quad \text{and} \quad Z_{\text{OUT}} = Z_o \tag{9.11a}$$

Both the conditions lead to

$$Z_o = Z_{T1} + Z_{T2}||(Z_{T1} + Z_o) \tag{9.11b}$$

or

$$Z_o^2 = Z_{T1}^2 + 2Z_{T1}Z_{T2} \tag{9.11c}$$

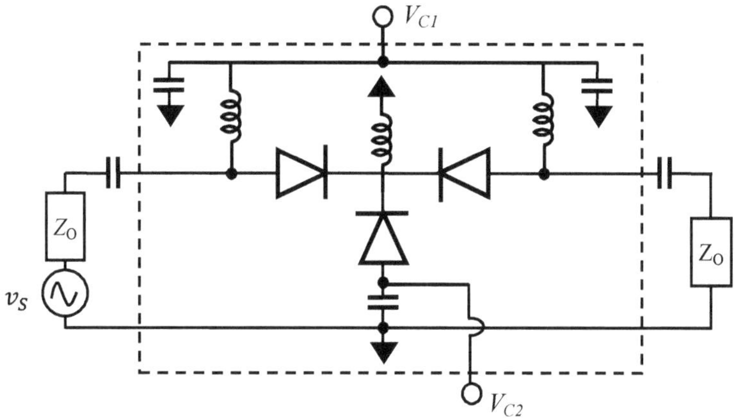

Fig. 9.12 Schematic circuit diagram of TEE attenuator

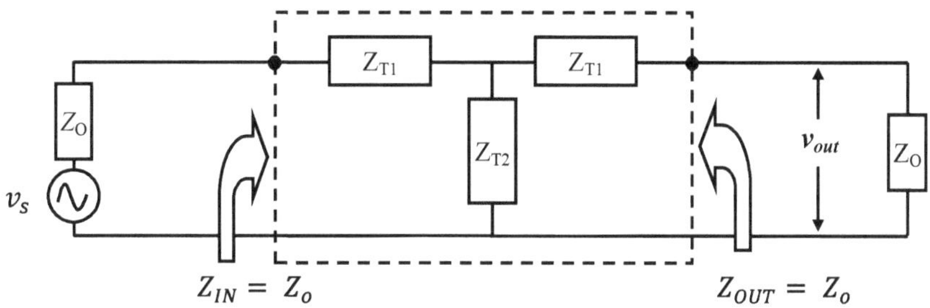

Fig. 9.13 Shunt switch configuration for reflection coefficient determination

This is the required relationship between Z_{T1} and Z_{T2} for achieving perfect input and output impedance matching.

Transmission coefficient of the TEE attenuator is given by

$$S_{21} = \frac{2v_{\text{out}}}{v_s} = \frac{2Z_o Z_{T2}}{(Z_o + Z_{T1})(Z_o + Z_{T1} + 2Z_{T2})} \tag{9.12a}$$

Combining (9.11c) and (9.12a),

$$S_{21} = \frac{2v_{\text{out}}}{v_s} = \frac{Z_{T2}}{(Z_o + Z_{T1} + Z_{T2})} \tag{9.12b}$$

$$\begin{aligned} \text{Attenuation(in dB)} &= 20\log_{10}\left(\frac{1}{|S_{21}|}\right) \\ &= 20\log_{10}\left(\left|\frac{Z_o + Z_{T1} + Z_{T2}}{Z_{T2}}\right|\right) \end{aligned} \tag{9.13}$$

Considering all the impedances are resistive, assume,

$$Z_{T1} = R_1, \quad Z_{T2} = R_2, \quad \text{and} \quad Z_o = R_o$$

Then, attenuation N is given by

$$N = \frac{P_{avs}}{P_L} = \frac{1}{|S_{21}|^2}$$

Therefore, using (9.12b) attenuation can be expressed as

$$\sqrt{N} = \frac{R_o + R_1 + R_2}{R_2} \tag{9.14a}$$

where

$$R_o^2 = R_1^2 + 2R_1R_2 \tag{9.14b}$$

Combining (9.14a) and (9.14b), we can determine the values of R_1 and R_2 in terms of required attenuation as

$$R_1 = R_o \frac{\left(\sqrt{N} - 1\right)}{\left(\sqrt{N} + 1\right)} \tag{9.15a}$$

and

$$R_2 = R_o \frac{2\sqrt{N}}{(N - 1)} \tag{9.15b}$$

Figure 9.14 shows the requirement of resistances for the TEE attenuator for different attenuation values satisfying perfect impedance matching of input and output ports. This suggests that bias voltage/current requirements for series and shunt diodes are different. In Fig. 9.12, a simple bias circuit is shown to provide independent bias to the series- and shunt-connected diodes.

Example 9.1 Calculate resistance values of a perfectly matched TEE-type resistive attenuator of attenuation 3-dB in a 50 Ω system, and also write its S-parameter matrix.

Solution

Attenuation N is given by

$$N = \frac{P_{avs}}{P_L}$$

Here, attenuation in dB is given by

$$10 \times \log(N) = 3 \text{ dB}$$

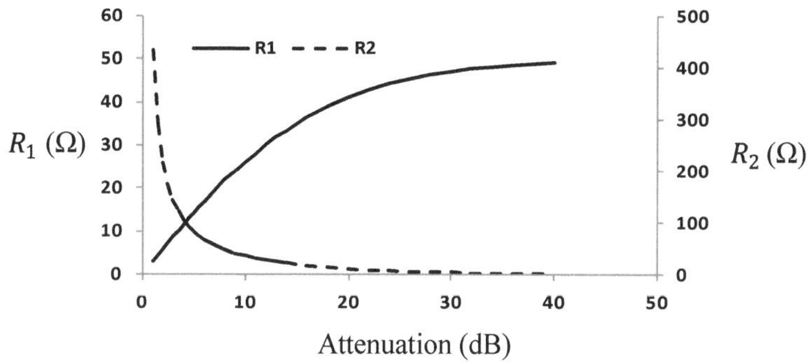

Fig. 9.14 Values of R_1 and R_2 for different attenuation values with perfect port matching

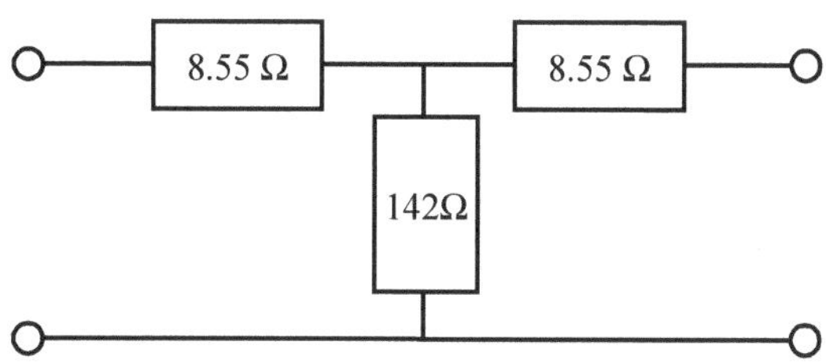

Fig. 9.15 3-dB TEE-type attenuator circuit

Thus,

$$N = 10^{3/10}$$

$$R_o = 50$$

For a TEE-type attenuator, from (9.15a) and (9.15b)

$$R_1 = R_o \frac{\left(\sqrt{N} - 1\right)}{\left(\sqrt{N} + 1\right)} = 50 \times \frac{\left(10^{3/20} - 1\right)}{\left(10^{3/20} + 1\right)} = 8.55\ \Omega$$

and

$$R_2 = R_o \frac{2\sqrt{N}}{(N - 1)} = 50 \times \frac{2 \times 10^{3/20}}{\left(10^{3/10} - 1\right)} = 141.9\ \Omega$$

Thus, the resistance values of the 3-dB TEE-type attenuator are 8.55 and 141.9 Ω as shown in Fig. 9.15.

For the matched attenuator $S_{11} = S_{22} = 0$

$$S_{21} = S_{12} = 1/\sqrt{N} = 1/10^{3/20} = 0.7079$$

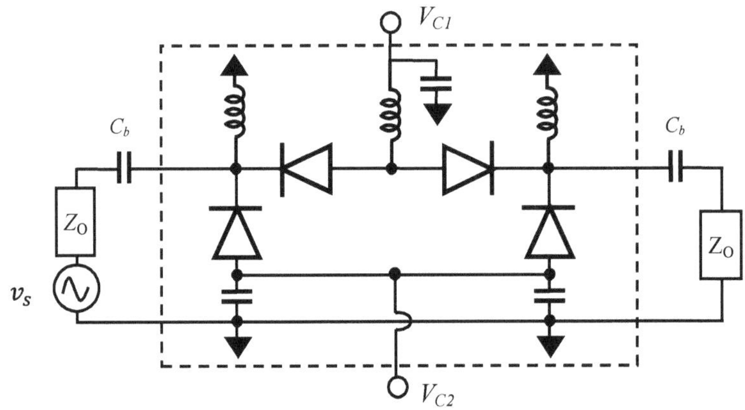

Fig. 9.16 Schematic circuit diagram of π attenuator

Thus, the S-parameters matrix is

$$\begin{bmatrix} S_{11} & S_{12} \\ S_{21} & S_{22} \end{bmatrix} = \begin{bmatrix} 0 & 0.7079 \\ 0.7079 & 0 \end{bmatrix}$$

9.5.2 π *Attenuator*

Schematic circuit diagram of a diode-based π configuration attenuator is shown in Fig. 9.16. Here, DC control voltage V_{C1} is for biasing the two series-connected diodes and V_{C2} is for biasing of the shunt diodes. The bias configuration ensures independent current controllers for the series- and shunt-connected diodes. However, V_{C1} provides same current for both the series diodes and V_{C2} provides same current to both the shunt diodes. Considering Z_{T1} is the impedance of each series pair diode and Z_{T2} is the impedance of each shunt pair diode, the RF equivalent circuit of the --π configured attenuator is shown in Fig. 9.17. The conditions for the perfectly matching of input and output ports are

$$Z_{\mathrm{IN}} = Z_o \quad \text{and} \quad Z_{\mathrm{OUT}} = Z_o \tag{9.16a}$$

Both the conditions lead to

$$Z_o = Z_{T2}||(Z_{T1} + Z_{T2}||Z_o) \tag{9.16b}$$

or

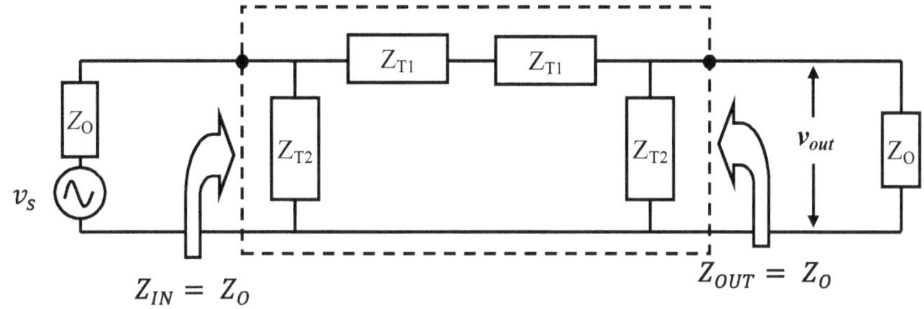

Fig. 9.17 Equivalent RF circuit diagram of the π attenuator

$$Z_o^2 = \frac{Z_{T1}Z_{T2}^2}{Z_{T1} + Z_{T2}} \tag{9.16c}$$

This is the relationship in between Z_{T1} and Z_{T2} for achieving perfect input and output impedance matching.

Transmission coefficient of the π attenuator is given by

$$S_{21} = \frac{2v_{\text{out}}}{v_s} = \frac{Z_{T2}^2}{2Z_oZ_{T2} + 2Z_oZ_{T1} + 2Z_{T1}Z_{T2} + Z_{T2}^2} \quad (9.17) \qquad (9.17\text{a})$$

$$\begin{aligned} \text{Attenuation(in dB)} &= 20\log_{10}\left(\frac{1}{|S_{21}|}\right) \\ &= 20\log_{10}\left(\left|\frac{2Z_oZ_{T2} + 2Z_oZ_{T1} + 2Z_{T1}Z_{T2} + Z_{T2}^2}{Z_{T2}^2}\right|\right) \end{aligned} \tag{9.18a}$$

Considering all the impedances resistive, i.e.

$$Z_{T1} = R_1, \quad Z_{T2} = R_2, \quad \text{and} \quad Z_o = R_o$$

Then, attenuation N is given by

$$N = \frac{P_{avs}}{P_L} = \frac{1}{|S_{21}|^2}$$

Therefore, attenuation can be expressed as

$$\sqrt{N} = \frac{2R_oR_2 + 2R_oR_1 + 2R_1R_2 + R_2^2}{R_2^2} \tag{9.18b}$$

Combining (9.16c) and (9.18b), we can determine the value of R_1 and R_2 in terms of required attenuation as

$$R_1 = R_o\frac{(N-1)}{4\sqrt{N}} \tag{9.19a}$$

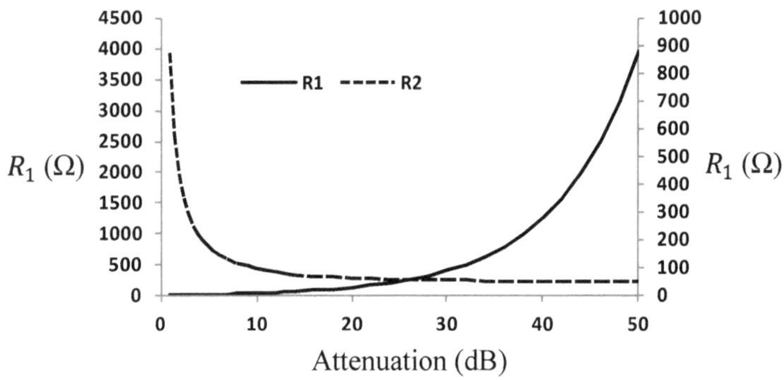

Fig. 9.18 Values of R_1 and R_2 for different attenuation values with perfect port matching for PI-type attenuator

and

$$R_2 = R_o \frac{\sqrt{N} + 1}{\sqrt{N} - 1} \tag{9.19b}$$

Figure 9.18 shows the requirement of resistance values of the π attenuator for different attenuation values satisfying perfect impedance matching of input and output ports. This suggests that bias voltage/current requirements for series and shunt diodes are different. Figure 9.16 shows the bias circuit to provide independent bias voltage/currents to the series- and shunt-connected diodes.

Example 9.2 Calculate resistance values of a perfectly matched π-type resistive attenuator of attenuation 6-dB in a 50 Ω system, and also write its S-parameter matrix.

Solution
Attenuation N is given by

$$N = \frac{P_{avs}}{P_L}$$

Here, attenuation in dB is given by

$$10 \times \log(N) = 6 \text{ dB}$$

Thus,

$$N = 10^{6/10}$$

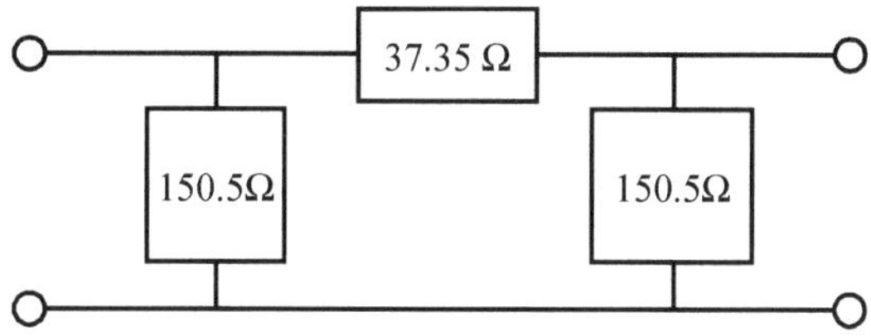

Fig. 9.19 6-dB π-type attenuator circuit

$$R_o = 50$$

For a π-type attenuator, from (9.19a) and (9.19b)

$$R_1 = R_o \frac{(N-1)}{4\sqrt{N}} = 50 \times \frac{\left(10^{6/10} - 1\right)}{4 \times 10^{3/20}} = 18.68\ \Omega$$

and

$$R_2 = R_o \frac{\sqrt{N}+1}{\sqrt{N}-1} = 50 \times \frac{\left(10^{6/20} + 1\right)}{\left(10^{6/20} - 1\right)} = 150.5\ \Omega$$

Thus, the resistance values of the 3-dB π-type attenuator are $R_1 = 18.68$ and $R_2 = 150.5\ \Omega$ as shown in Fig. 9.19.

For the matched attenuator $S_{11} = S_{22} = 0$

$$S_{21} = S_{12} = 1/\sqrt{N} = 1/10^{6/20} = 0.5$$

Thus, the S-parameter matrix is

$$\begin{bmatrix} S_{11} & S_{12} \\ S_{21} & S_{22} \end{bmatrix} = \begin{bmatrix} 0.0 & 0.5 \\ 0.5 & 0.0 \end{bmatrix}$$

9.5.3 *Quadrature Hybrid Matched Attenuator*

The TEE- and π-type attenuators provide attenuation with ideally perfect input and output impedance matching. To achieve this, bias circuit should provide current to the diodes in such a way that it should maintain the required relationship in between the series and shunt impedances. Practically, it is very difficult to meet this bias requirement by using a simple driver circuit. Attenuator configuration using 3-dB 90° hybrid coupler solves this problem. In this case, two p-i-n diodes ideally with same bias current can be used to achieve variable attenuation meeting the requirement of port impedance matching.

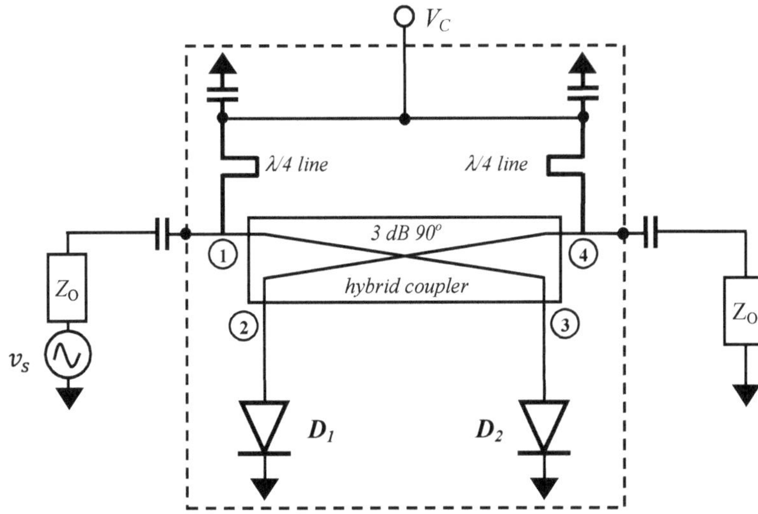

Fig. 9.20 Quadrature hybrid matched attenuator configuration with shunt DC path

Schematic circuit diagram of a p-i-n diode attenuator using 3-dB quadrature hybrid is shown in Fig. 9.20. Here, control voltage V_C is used to provide bias current to both the diodes D_1 and D_2. Both the diodes are connected in parallel as far as DC equivalent circuit is concerned. Therefore, voltage applied by the bias circuit will be same for both the diodes. Practically, it is very unlikely that the DC characteristic, i.e. I–V characteristic of both the diodes, will be same. In such a situation, application of same voltage cannot ensure equal currents through both the diodes leading to different impedances. To ensure same currents through both the diodes, it is preferable to connect both the diodes in DC series path as shown in Fig. 9.21.

The 3-dB quadrature hybrid has four ports with equal power sharing and 90° phase difference in between the two output ports. When a signal fed to input port-1, then half of the signal power will reach to the port-2 and remaining half power will reach to the port-3. The phase difference between the two output signals will be 90°. Fourth port is the isolated port with respect to the port-1. Therefore, there will be no power at the port-4 when power is incident on the port-1. RF equivalent circuit of the quadrature hybrid configuration p-i-n diode attenuator is shown in Fig. 9.22. Here, Z_{T1} and Z_{T2} are the impedances of diodes D_1 and D_2, respectively. Suppose an RF source of amplitude v_s is connected to the port-1 of the hybrid through resistive impedance Z_O and load impedance Z_O is connected at the port-4. Thus, incident voltage amplitude at port-1 will be

$$V_1^+ = \frac{v_s}{2} \tag{9.20a}$$

As per the characteristic of the 3-dB 90° hybrid coupler, incident voltage amplitude to the diode D_1 and D_2 will respectively be

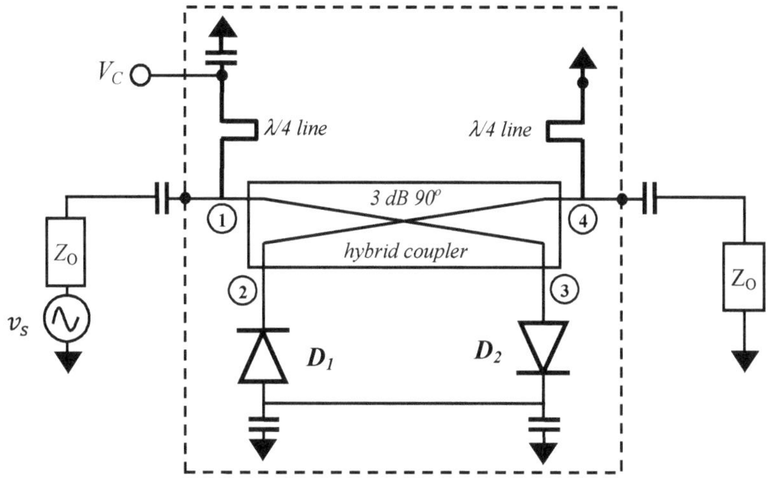

Fig. 9.21 Quadrature hybrid matched attenuator configuration with series DC path

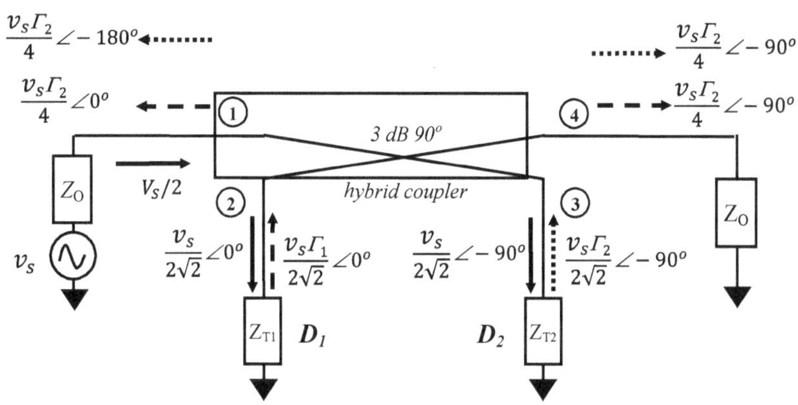

Fig. 9.22 Quadrature hybrid matched attenuator configuration

$$\frac{v_s}{2\sqrt{2}}\angle\theta^\circ \text{ and } \frac{v_s}{2\sqrt{2}}\angle(\theta-90)^\circ \tag{9.20b}$$

Here, θ is the insertion phase of the signal from port-1 to port-2. The reflection coefficients of the diodes with respect to the impedance Z_o are given by

$$\Gamma_1 = \frac{Z_{T1}-Z_o}{Z_{T1}+Z_o} \text{ and } \Gamma_2 = \frac{Z_{T2}-Z_o}{Z_{T2}+Z_o} \tag{9.20c}$$

The reflected signal amplitude from the diodes D_1 and D_2 will be

$$\frac{v_s\Gamma_1}{2\sqrt{2}}\angle\theta^\circ \text{ and } \frac{v_s\Gamma_2}{2\sqrt{2}}\angle(\theta-90)^\circ \tag{9.20d}$$

The reflected signals from both the diodes will be the incident signal for the port-2 and port-3. The signal amplitude at port-1 and port-4 due to the incident signal on port-2 will be

$$\frac{v_s \Gamma_1}{4} \angle 2\theta^\circ \quad \text{and} \quad \frac{v_s \Gamma_1}{4} \angle (2\theta - 90)^\circ \tag{9.20e}$$

Similarly, the signal amplitude at port-1 and port-4 due to the incident signal on port-3 will be

$$\frac{v_s \Gamma_2}{4} \angle (2\theta - 180)^\circ \quad \text{and} \quad \frac{v_s \Gamma_2}{4} \angle (2\theta - 90)^\circ \quad \text{respectively} \tag{9.20f}$$

Therefore, signal amplitude reflected from port-1 and transmitted to port-4 will respectively be

$$\frac{v_s \Gamma_1}{4} \angle 2\theta^\circ + \frac{v_s \Gamma_2}{4} \angle (2\theta - 180)^\circ \tag{9.20g}$$

and

$$\frac{v_s \Gamma_1}{4} \angle (2\theta - 90)^\circ + \frac{v_s \Gamma_2}{4} \angle (2\theta - 90)^\circ \tag{9.20h}$$

Thus, the reflection coefficient of this symmetric structured attenuator is given by

$$S_{11} = S_{22} = \frac{\Gamma_1 \angle 2\theta^\circ - \Gamma_2 \angle 2\theta^\circ}{2} \tag{9.21a}$$

Therefore, return losses of the attenuator are given by

$$\text{RL(in dB)} = 20 \log_{10} \left(\frac{2}{|\Gamma_1 - \Gamma_2|} \right) \tag{9.21b}$$

The reflected signal from the port-1, i.e. from the input port, will be non-zero if both the diodes are not matched, i.e. if $\Gamma_1 \neq \Gamma_2$. However, for the matched pair of diodes, $\Gamma_1 = \Gamma_2$ for same bias condition and return loss will be zero, i.e. perfectly matched.

The transmission coefficient of the attenuator can be calculated as follows:

$$S_{21} = S_{12} = \frac{\Gamma_1 \angle (2\theta - 90)^\circ + \Gamma_2 \angle (2\theta - 90)^\circ}{2} \tag{9.22a}$$

Therefore, the attenuation of the attenuator can be written as

$$\text{Attenuation(in dB)} = 20 \log_{10} \left(\frac{2}{|\Gamma_1 + \Gamma_2|} \right) \tag{9.22b}$$

Fig. 9.23 Photograph of a multistage quadrature hybrid attenuator

In case of perfectly matched pair of p-i-n diodes, i.e. for $\Gamma_1 = \Gamma_2 = \Gamma$, the attenuation of the attenuator can be expressed as

$$\text{Attenuation(in dB)} = 20\log_{10}\left(\frac{1}{|\Gamma|}\right) \quad (9.22c) \qquad (9.22b)$$

Photograph of a 3°dB 90° Lange coupler-based attenuator realized on a 20–mil-thick alumina substrate is shown in Fig. 9.23.

Frequency bandwidth of the quadrature hybrid attenuator is limited by the bandwidth of the coupler. At microwave frequency bands, Lange coupler is used for realization of p-i-n diode attenuator that can operate over an octave bandwidth. TEE- and π-type attenuators are basically very broadband attenuators due to the absence of any frequency-limited elements like quarter wave transmission line or coupler. Thus, frequency bandwidths of a TEE and π attenuators are limited by its bias circuit and parasitics of p-i-n diode itself.

9.6 Driver Circuit for P-I-N Diode Attenuators

Microwave and RF variable attenuators based on p-i-n diodes are used in airborne and space-based systems where temperature-insensitive attenuation is always demanded over very wide operating temperature range. However, RF resistance of p-i-n diodes that mainly determine the attenuation of the circuit is temperature dependent as follows (2.17):

$$R_i(T, I_d) = z\frac{T^{n-m}}{[I_{do}(T, V_d)]^p} \qquad (9.23a)$$

Therefore, function of a driver circuit is not only to provide just DC voltage and current to the diodes without affecting RF performance but also to regulate the bias

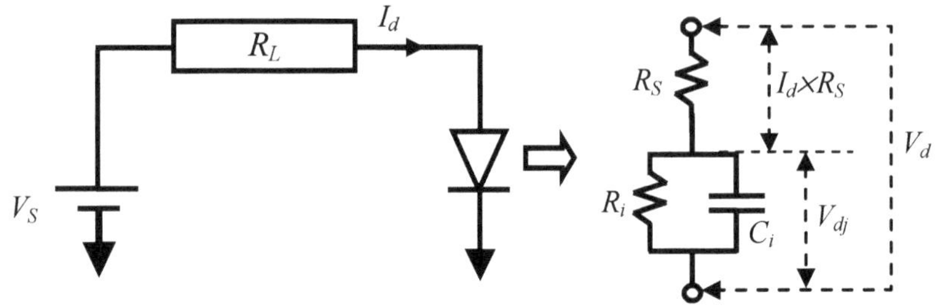

Fig. 9.24 A simple bias circuit of p-i-n diode

voltage/current over the operating temperature range to ensure temperature-invariant attenuation.

From (9.23a), the temperature coefficient of RF resistance can be written as

$$T_{cR_i} = \frac{\partial R_i/\partial T}{R_i} = \frac{n-m}{T} - \frac{p}{I_d}\frac{\partial I_d}{\partial T} \tag{9.23b}$$

Driver circuit in its simplest form consists of a DC supply voltage V_S and a current limiting series resistor R_L as shown in Fig. 9.24. Thus, diode current I_d is given by

$$I_d = \frac{V_S - V_{dj}}{R_L + R_s} \tag{9.23c}$$

Combining (9.23b) and (9.23c), the temperature coefficient of the RF resistance is given by

$$T_{cR_i} = -\frac{n-m}{T} - \frac{p}{\left(V_S - V_{dj}\right)}\frac{\partial V_{dj}}{\partial T} \tag{9.23d}$$

Under sufficiently forward-bias condition, I–V relationship of a p-i-n diode can be written as (2.16)

$$I_d(V_d, T) = A'T^{\nu}\exp\left(\frac{q[V_d(T) - I_d(V_d, T) \times R_S - E_b]}{\eta kT}\right) \tag{9.23e}$$

Combining (9.23d) and (9.23e), temperature coefficient of RF resistance of p-i-n diodes can be expressed as

$$T_{cR_i} = \frac{V_S - V_{dj} + \frac{p}{n-m}\left(E_b - V_{dj}\right) + \frac{p\nu+m-n}{n-m}\frac{\eta kT}{q}}{\frac{T}{n-m}\left[\left(V_S - V_{djO}\right) + \frac{\eta kT}{q}\right]} \tag{9.24}$$

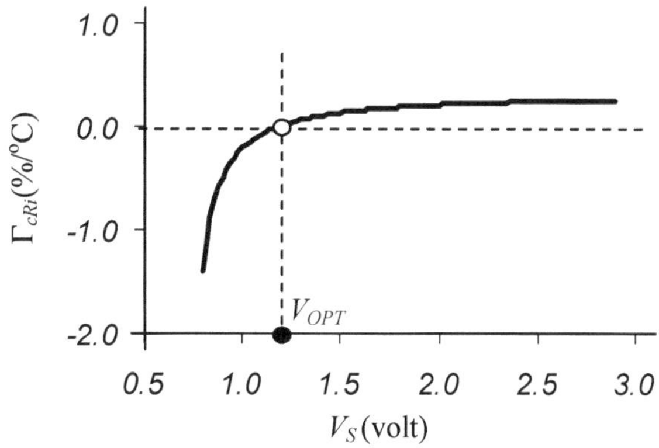

Fig. 9.25 Temperature coefficient of rf resistance of p-i-n diode with different supply voltages

The plot of T_{cR_i} as a function of supply voltage V_S, keeping V_{dj} constant, is shown in Fig. 9.25 for typical parameter values of a p-i-n diode. From the plot and (9.24), the following can be concluded:

1. T_{cR_i} is zero at a particular value of supply voltage V_S. The supply voltage for which the temperature coefficient of p-i-n diode RF resistance becomes zero is called '*Optimum Supply Voltage*' V_{OPT}.
2. T_{cR_i} is positive for V_S more than V_{OPT}. When the diode is driven by a high supply voltage (with high current limiting resistor), i.e. under constant-current-bias condition, T_{cR_i} becomes constant and is given by $(n - m)/T$.
3. T_{cR_i} is negative for V_S less than V_{OPT}. Temperature coefficient of the RF resistance becomes very high when the supply voltage V_S goes towards V_{dj}, corresponding to the constant-voltage-bias condition (i.e. with $R_L = 0$).

Based on this supply voltage-dependent temperature coefficient of a p-i-n diode RF resistance, temperature-insensitive attenuation can be achieved from a simple bias circuit shown in Fig. 9.24 with $V_S = V_{\mathrm{OPT}}$. The optimum supply voltage is obtained from (9.24) by setting $T_{cR_i} = 0$ at $V_S = V_{\mathrm{OPT}}$. Thus, V_{OPT} is given by

$$V_{\mathrm{OPT}} = V_{\mathrm{do}} + \frac{p}{n-m}(E_b - V_{\mathrm{do}}) + \frac{\eta k T_o}{q}\left(\frac{pv}{n-m} - 1\right) \quad (9.25a) \qquad (9.25)$$

This is the same expression as obtained (2.21) in Chap. 2. When the driver circuit is designed with the optimum supply voltage, the bias load lines follow the equiresistance curve of the diodes and maintain temperature-invariant attenuation.

Figure 9.26 shows equiresistance curves and the corresponding optimum bias load lines of a p-i-n diode at different bias currents. Here, all the load lines for different currents corresponding to different attenuation levels, intercept the voltage axis at nearly the same point at t 1.15 V; this is nothing but the bandgap potential in volts of the p-i-n diode semiconductor, i.e. silicon. This is the basis for the design of p-i-n diode attenuator driver circuit for achieving temperature-invariant attenuation.

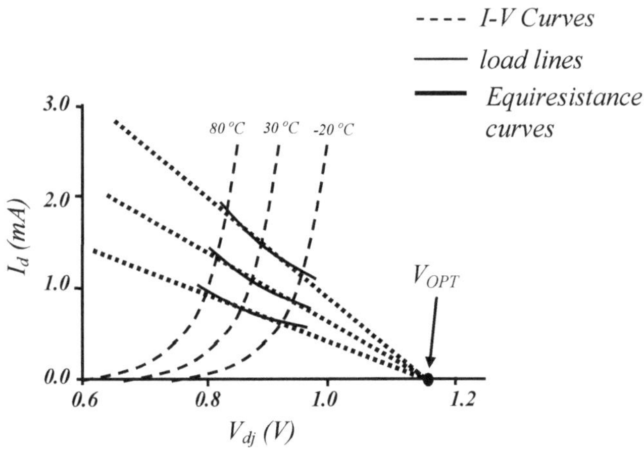

Fig. 9.26 Typical equiresistance curves with load lines of a p-i-n diode

Based on the straight line property of p-i-n diode equiresistance curves, suggested p-i-n diode attenuator driver circuit is shown in Fig. 9.24 with $V_S = V_{\text{OPT}}$ and the bias resistance $R_L = R_O$ given by

$$R_O = \frac{V_{\text{OPT}} - V_{\text{dO}}}{I_{\text{dO}}} \quad (9.25b) \tag{9.25}$$

Here, I_{dO} and V_{dO} are diode current and voltage corresponding to a particular RF resistance R_{dO}, i.e. attenuation. In case of attenuator circuits where N number of diodes are connected in DC series path, the optimum supply voltage will be $N \times V_{\text{OPT}}$.

Setting the diode current $I_d = (V_S - V_d)/R_L$, and differentiating RF resistance R_i with respect to V_S, the variation of RF resistance of p-i-n diode becomes

$$\frac{\partial R_i}{R_i} = -p \frac{\partial V_S}{V_S - V_d} \tag{9.26}$$

In the driver circuit, $V_S = V_{\text{OPT}}$ is slightly greater than the voltage drop ($V_d = V_{\text{dj}} + I_d \times R_s$) across the diode. Thus, it is required to use very stable voltage source for the driver circuit. Figure 9.27 shows a driver circuit including a stable voltage source. Here, V_{OPT} is derived using potential divider constituted by resistor R_a and R_b. Unity gain buffer is used to eliminate the loading effect of the potential divider.

Digital driver circuits are needed where stepwise variable attenuation is required. A digital control circuit for temperature-invariant stepwise variable performance of diode-based circuits is shown in Fig. 9.28. The optimum supply voltage (V_{OPT}) is provided by a regulated voltage supply with the associated potential divider. Output of the buffer circuit is connected to a bank of predetermined resistors (R_1 to R_n) through an analog multiplexer. Depending upon the digital data to the analog multiplexer, one

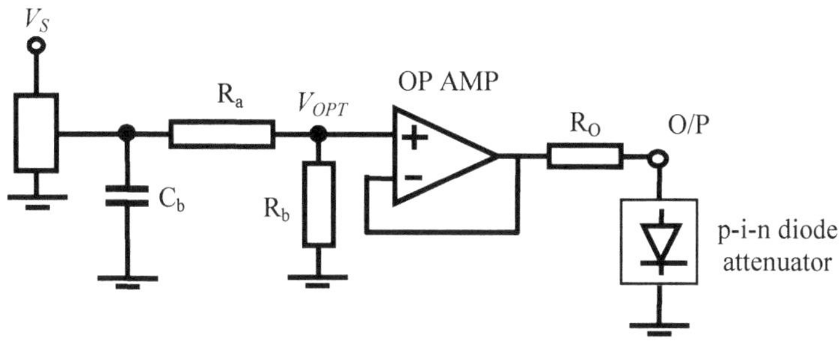

Fig. 9.27 Attenuator driver circuit with optimum load-line bias

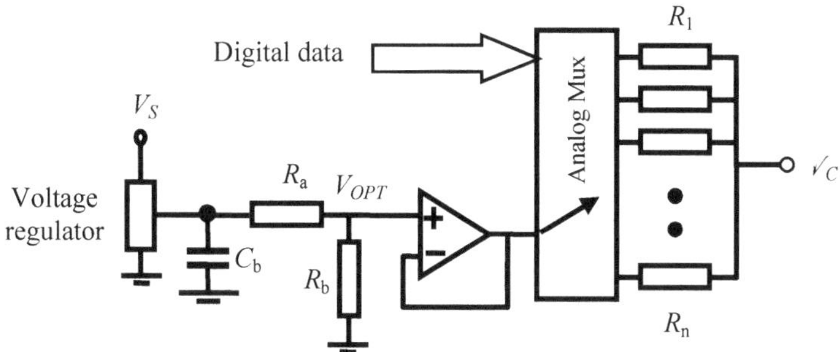

Fig. 9.28 Attenuator driver circuit for stepwise variable attenuation

resistor at a time will be selected from resistors (R_1 to R_n) to provide temperature-invariant stepwise variable attenuation.

9.7 Effect of Nonideal Components in Driver Circuits

Deviation in V_{OPT}, due to the setting tolerance of V_S, will not only deviate the initial settings of the attenuation but temperature coefficient of the attenuation will also change as given by (9.26). Thus, setting of V_{OPT} value is a critical aspect of the driver circuits. Variation of RF resistance due to the tolerance of the bias resistor is given by

$$\frac{\partial R_i}{R_i} = -p\frac{\partial R_L}{R_L} \tag{9.27}$$

Thus, the variation of RF resistance is in the same magnitude as the tolerance of the bias resistor (for $p \sim 1$). Temperature coefficient of the RF resistance is determined only by the setting of the open-circuit voltage (V_{OPT}) of the driver circuit.

Therefore, tolerance of the bias resistor (R_L) only contributes to the initial setting of the attenuation value. It will not affect the temperature coefficient of the attenuation.

Temperature coefficient of V_S and R_L will affect the attenuation values. In case of the digital driver circuit shown in Fig. 9.28, finite ON resistance of the analog multiplexer is a part of the bias resistor having resistance of few tens of Ohms (of 74HC4051) to few hundreds of Ohms (of CD4051). Moreover, ON resistance of the analog multiplexer has reasonably high-temperature coefficient. Thus, temperature coefficient of the bias resistance and supply voltage must be taken into account when control circuits are designed. By modifying V_{OPT} and R_O, it is possible to eliminate or minimize the effect of temperature coefficient of the voltage source and bias resistors [7].

Suppose modified optimum voltage and optimum resistor at ambient temperature are V_{OPTM} and R_{OM}, respectively. Then, we can write

$$V_{\mathrm{OPT}} = V_d + I_d R_O \tag{9.28a}$$

$$V_{\mathrm{OPTM}} = V_d + I_d R_{\mathrm{OM}} \tag{9.28b}$$

Here, V_d, I_d, V_{OPTM} and R_{OM} all have finite temperature coefficients but V_{OPT} and R_O are the temperature-invariant optimum voltage and bias resistance. Thus, differentiating (9.28a) and (9.28b) with respect to T, we can write

$$\frac{\partial V_{\mathrm{OPT}}}{\partial T} = \frac{\partial V_d}{\partial T} + R_O \frac{\partial I_d}{\partial T} = 0 \tag{9.28c}$$

and

$$\frac{\partial V_{\mathrm{OPTM}}}{\partial T} = \frac{\partial V_d}{\partial T} + R_{\mathrm{OM}} \frac{\partial I_d}{\partial T} + I_d \frac{\partial R_{\mathrm{OM}}}{\partial T} \tag{9.28d}$$

Combining (9.28c) and (9.28d) and eliminating I_d from (9.28a) and (9.28b), we can write

$$V_{\mathrm{OPTM}} = \frac{V_{\mathrm{OPT}}\left(\frac{\partial V_d}{\partial T} - \frac{\partial V_{\mathrm{OPTM}}}{\partial T}\right) + V_d\left(\frac{\partial V_{\mathrm{OPTM}}}{\partial T} - \alpha(V_{\mathrm{OPT}} - V_d)\right)}{\frac{\partial V_d}{\partial T} - \alpha(V_{\mathrm{OPT}} - V_d)} \tag{9.29a}$$

Here, the temperature coefficient of the bias resistor is α and is given by $\alpha = (\partial R_{\mathrm{OM}}/\partial T)/R_{\mathrm{OM}}$.

In case of finite temperature coefficient of the bias resistor but temperature-invariant optimum supply voltage,

$$\partial V_{\mathrm{OPTM}}/\partial T = 0 \tag{9.29b}$$

In this condition, the modified optimum voltage can be written as

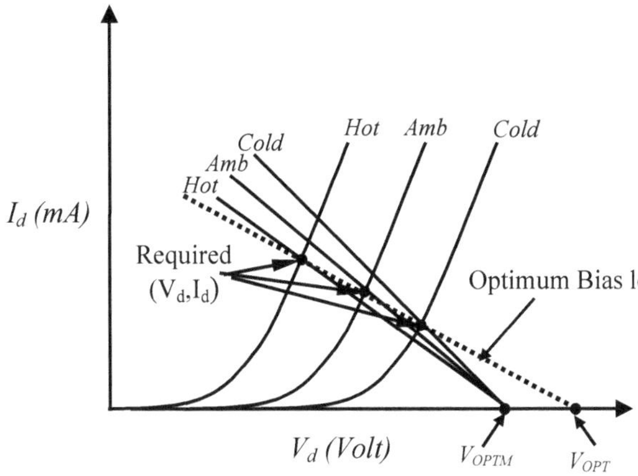

Fig. 9.29 Graphical representation of V_{OPTM} for positive temperature coefficient of bias resistor

$$V_{\mathrm{OPTM}} = \frac{V_{\mathrm{OPT}} \frac{\partial V_d}{\partial T} - \alpha V_d (V_{\mathrm{OPT}} - V_d)}{\frac{\partial V_d}{\partial T} - \alpha (V_{\mathrm{OPT}} - V_d)} \tag{9.29c}$$

Figure 9.29 shows graphical presentations of how the modified optimum voltage provides temperature-invariant performance under positive temperature coefficient of the bias resistor. The slope of the load lines decreases with the increase of temperature due to the positive temperature coefficient of the bias resistor. Here, it is shown that V_{OPTM} with positive temperature coefficient bias resistor tracks the required diode voltage and current (V_d, I_d) over the operating temperature range.

In case of finite temperature coefficient of the supply voltage but with the temperature-invariant bias resistor, $\alpha = 0$ the modified optimum voltage can be written as

$$V_{\mathrm{OPTM}} = \frac{V_{\mathrm{OPT}} \left(\frac{\partial V_d}{\partial T} - \frac{\partial V_{\mathrm{OPTM}}}{\partial T} \right) + V_d \frac{\partial V_{\mathrm{OPTM}}}{\partial T}}{\frac{\partial V_d}{\partial T}} \tag{9.30}$$

Figure 9.30 shows the graphical presentations of how V_{OPTM} provides temperature-invariant attenuation under the positive temperature coefficient of the supply voltage. Here, it is shown that V_{OPTM} with positive temperature coefficient tracks the required V_d and I_d for temperature-invariant attenuation of the attenuator. Here, all the load lines will remain parallel since $\alpha = 0$ but V_{OPTM} increases with the increase of temperature for its positive temperature coefficient.

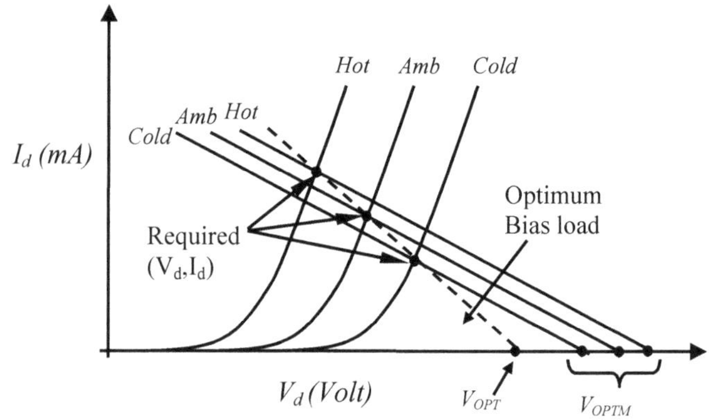

Fig. 9.30 Graphical representation of V_{OPTM} for positive temperature coefficient of bias voltage

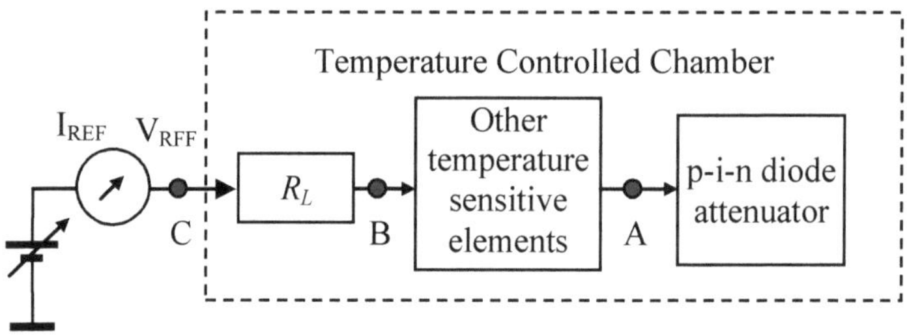

Fig. 9.31 Schematic diagram of a measurement setup to include temperature coefficient of temperature-sensitive elements

Under forward-bias condition, though temperature effect of C_d and R_s can be considered negligible compared to the temperature effect of R_d, however, absolute value of R_s should be taken into account to determine bias resistor (R_O) of the driver circuits.

9.8 Experimental Determination of V_{OPTM}

Figure 9.31 shows schematic block diagram of a scheme used to determine the modified optimum voltage including the effect of temperature coefficient of the bias resistor. To determine the optimum supply voltage of the driver circuit, RF circuit of the attenuator including other temperature-sensitive elements such as multiplexer and bias resistors are kept within the temperature-controlled chamber as shown in Fig. 9.31. The voltage and current of the circuit are measured at point 'C' instead of at the diode terminal point 'A'. The value of the resistor R_L is kept approximately equal to the actual value required for the circuit.

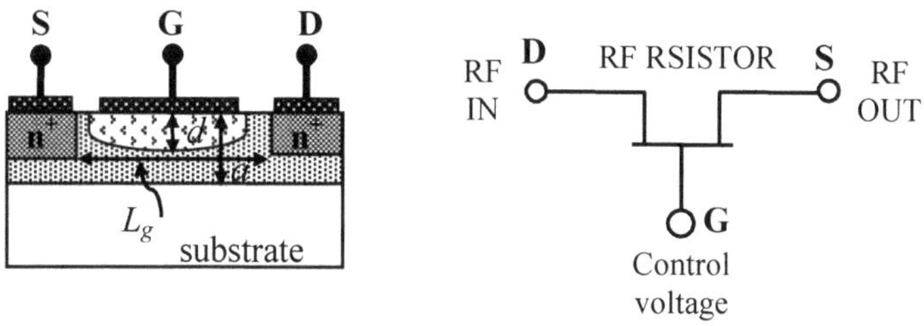

Fig. 9.32 MESFETs are used as voltage-variable linear RF resistor for attenuator

In case of p-i-n diodes, the equiresistance curves are nearly straight lines and intercept the voltage axis at same points for entire operating current condition. Therefore, three-point (at three temperatures) measurement will give accurate V_{OPT}.

To determine V_{OPTM}, three sets of voltages and currents can be measured at three different temperatures of the chamber, adjusting voltage/current to the circuit for same attenuation value. The optimum load line will be the best-fitted straight line passing all these three points. The intercept of the load line with the voltage axis is the modified optimum open-circuit voltage V_{OPTM} of the bias network.

9.9 FET-Based Attenuators

GaAs MESFETs at zero drain bias are used as variable resistors to construct microwave attenuators. Mostly, n-channel depletion mode FET is used as variable RF resistor, which has low source-to-drain resistance in the absence of gate-bias voltage. With the application of negative voltage at gate terminal, the produced electric field below the gate causes the conduction channel to narrow, which increases the resistance in between source and drain. Thus, MESFETs at zero drain bias are used as variable RF resistance to realize voltage-variable attenuator (VVA). Gate-bias voltage is used to control the channel resistance, i.e. resistance between drain and source terminals which is used as linear RF resistor where the drain and source terminals are used as RF ports as shown in Fig. 9.32. Ideally, the gate is isolated from the channel, and thus there will be no current through the gate terminal; thus, it dissipates essentially no DC power and can easily be designed into broadband circuits. The RF signal flows from source to drain, while the RF isolated gate is used for controlling the resistance value. The channel resistance (R_C) can be written as (9.31)

$$R_c = \frac{1}{qN_d\mu_n} \times \frac{L_g}{W\left(a - d\left(V_g\right)\right)} \tag{9.31}$$

Here, N_d is the channel electron density which is equal to the free carrier density within the channel, μ_n is the low field electron mobility, L_g is the gate length

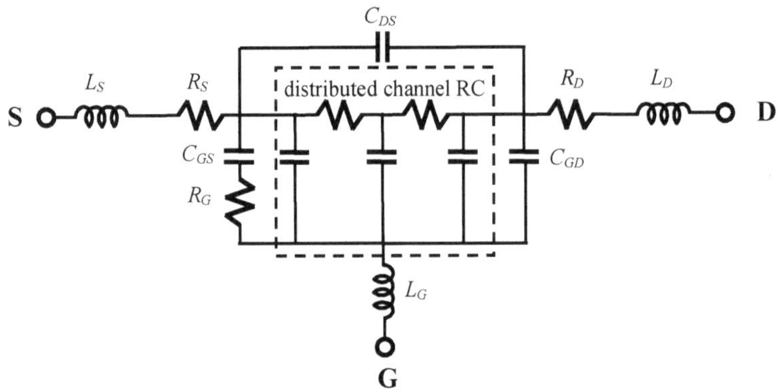

Fig. 9.33 Equivalent circuit of a zero-gate-bias FET

approximately equal to the channel length, W is the gate periphery, a is the channel depth and $d(V_g)$ is the depletion layer depth which is function of the applied gate control voltage. Linearity of the MESFETs as VVA is maintained as long as the operating RF power level is very small compared to the applied gate control voltage. Otherwise, depletion layer width will be modulated by the electric field due to the RF signal amplitude between source–drain ports. Equivalent circuit model of a zero-gate-biased MESFET is shown in Fig. 9.33. The voltage-variable channel resistance can be considered as distributed RC network, where the resistances and capacitances are voltage variable. Lumped elements are included in each terminal to take into account the parasitics of the extrinsic MESFET. This FET model may be used to design attenuator using CAD tool. However, at linear operating condition where the drain-to-source resistance is proportional to the applied gate voltage, the equivalent circuit of FETs can be represented as a parallel RC circuit with an effective resistance and an effective capacitance. FET attenuator relies primarily on the variation of drain-to-source resistance by varying the applied gate voltage. However, these parasitic capacitances have a significant influence on the attenuator performance at high frequencies.

Circuit schematic of a TEE-type attenuator using GaAs MESFETs is shown in Fig. 9.34. RF resistance (Z_{T1}) of the two series-connected FETs are controlled by same control voltage V_{G1} through gate-bias resistors R_G, whereas RF resistance Z_{T2} of the shunt-connected FET is controlled by control voltage V_{G2}. Here, the control voltages V_{G1} and V_{G2} are dependent on each other for achieving the impedance matching condition given by (9.11c) and attenuation given by (9.13) as a function of the FET impedances. Circuit schematic of a π-type attenuator using GaAs MESFETs is shown in Fig. 9.35. RF resistance Z_{T1} of the two shunt-connected FETs are controlled by the same control voltage V_{G2}, whereas RF resistance Z_{T1} of the series-connected FET is controlled by control voltage V_{G1}. Here, the control voltages V_{G1} and V_{G2} are also dependent on each other for achieving the impedance matching

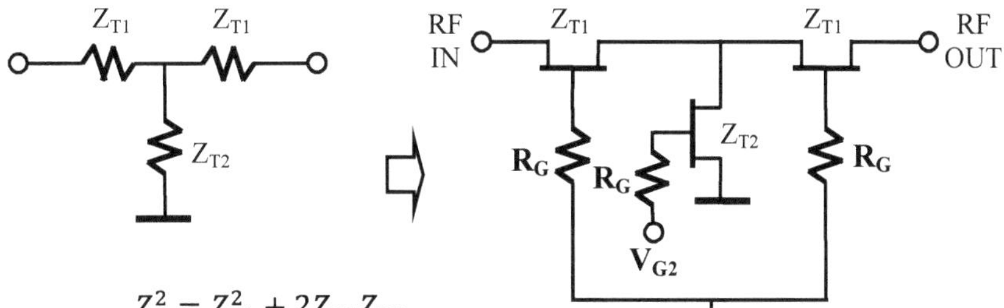

$$Z_o^2 = Z_{T1}^2 + 2Z_{T1}Z_{T2}$$

$$Attn = 20log_{10}\left(\left|\frac{Z_o + Z_{T1} + Z_{T2}}{Z_{T2}}\right|\right)$$

Fig. 9.34 FET-based TEE-type variable attenuator circuit diagram

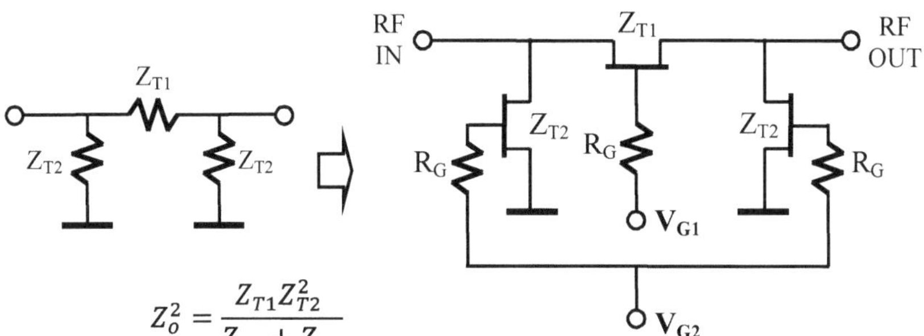

$$Z_o^2 = \frac{Z_{T1}Z_{T2}^2}{Z_{T1} + Z_{T2}}$$

$$Attn = 20log_{10}\left(\left|\frac{2Z_OZ_{T2} + 2Z_OZ_{T1} + 2Z_{T1}Z_{T2} + Z_{T2}^2}{Z_{T2}^2}\right|\right)$$

Fig. 9.35 FET-based π-type variable attenuator circuit diagram

condition given by (9.16c) and attenuation given by (9.18a) as a function of the FET impedances.

These FET-based TEE- and π-type attenuators can be used as continuously variable analog attenuator or as stepwise variable attenuator just by providing the gate-bias voltages as analog or discrete stepwise control voltage. Stepwise variable digital attenuators can also be realized by switching fixed attenuators (PADs) where FETs or p-i-n diodes are used as switch. Schematic diagram of such type of attenuator is shown in Fig. 9.36. The fixed attenuators are TEE- or π-type attenuators using fixed RF resistors of proper resistance values for different attenuations. Switches are used to select a fixed attenuation path or through path. Thus, the above configuration can provide stepwise attenuation values ideally from 0 to 16 dB in steps of 1 dB by using 4-bit digital control voltage. Practically, the through paths and switches have finite insertion loss. Therefore, the 0-dB attenuation case will be practically insertion loss of the attenuator when all the through paths are selected. Attenuation values of

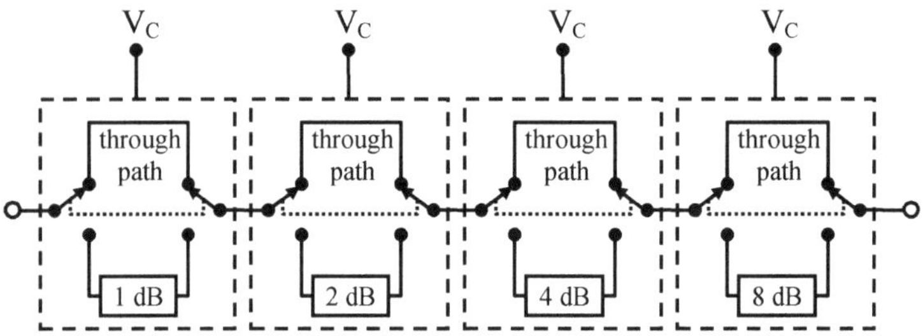

Fig. 9.36 Digital attenuator using fixed attenuators and FET as switch

the fixed attenuators should be selected taking into account the insertion loss of the through paths and switches.

Example 9.3 S-parameter matrix in a 50 Ω system of a 2-port attenuator is as follows:

$$[S] = \begin{bmatrix} 0.243\angle -125.93^\circ & 0.706\angle -15.74^\circ \\ 0.706\angle -15.74^\circ & 0.243\angle -125.93^\circ \end{bmatrix}$$

Find the input return losses of the attenuator when the output port is terminated with a load of (a) 50 Ω, (b) 100 Ω and (c) 10 Ω.

Solution

(a) For output load termination of 50 Ω, the input return loss of the attenuator in dB is

$$\text{RL(in dB)} = 20\log\left(\frac{1}{|S_{11}|}\right) = 20\log\left(\frac{1}{0.243}\right) = 12.288\ \text{dB}$$

(b) For output load termination of 100 Ω, the load reflection coefficient is

$$\Gamma_L = \frac{100-50}{100+50} = \frac{1}{3}\angle 0^\circ$$

For load reflection coefficient Γ_L, the input reflection coefficient Γ_{IN} is given by

$$\Gamma_{\text{IN}} = S_{11} + \frac{S_{12}S_{21}\Gamma_L}{1 - S_{22}\Gamma_L}$$

Putting the S-parameter values and Γ_L, the input return loss in dB is given by

$$\Gamma_{\text{IN}} = S_{11} + \frac{S_{12}S_{21}\Gamma_L}{1 - S_{22}\Gamma_L}$$

$$\Gamma_{\text{IN}} = 0.243\angle-125.93^\circ + \frac{0.706\angle-15.74^\circ \times 0.706\angle-15.74^\circ \times 0.333\angle 0^\circ}{1 - 0.243\angle-125.93^\circ \times 0.333\angle 0^\circ}$$

$$\Gamma_{\text{IN}} = 0.288\angle-92.53^\circ$$

Therefore, the input return loss in dB is

$$20\log\frac{1}{|\Gamma_L|} = 20\log\frac{1}{0.288} = 10.824\text{ dB}$$

(c) For output load termination of 10 Ω, the load reflection coefficient is

$$\Gamma_L = \frac{10-50}{10+50} = -\frac{2}{3}\angle 0^\circ$$

For load reflection coefficient Γ_L, the input reflection coefficient Γ_{IN} is given by

$$\Gamma_{\text{IN}} = S_{11} + \frac{S_{12}S_{21}\Gamma_L}{1 - S_{22}\Gamma_L}$$

Putting the S-parameter values and Γ_L, the input return loss in dB is given by

$$\Gamma_{\text{IN}} = S_{11} + \frac{S_{12}S_{21}\Gamma_L}{1 - S_{22}\Gamma_L}$$

$$\Gamma_{\text{IN}} = 0.243\angle-125.93^\circ + \frac{0.706\angle-15.74^\circ \times 0.706\angle-15.74^\circ \times 0.667\angle 0^\circ}{1\angle - 0.243\angle-125.93^\circ \times 0.667\angle 0^\circ}$$

$$\Gamma_{\text{IN}} = 0.480\angle 6.396^\circ$$

Therefore, the input return loss in dB is

$$20\log\frac{1}{|\Gamma_L|} = 20\log\frac{1}{0.480} = 6.384\text{ dB}$$

Example 9.4 Calculate S-parameter matrix of the π network as shown in Fig. 9.37a in a 50 Ω system, and also compute attenuation and return losses of the network.

Solution

The network is symmetrical. Thus, the input and output reflection coefficients for a 50 Ω system are given by

$$S_{11} = S_{22} = \frac{(100||50+50)||100-50}{(100||50+50)||100+50} = -0.048 = 0.048\angle 180^\circ$$

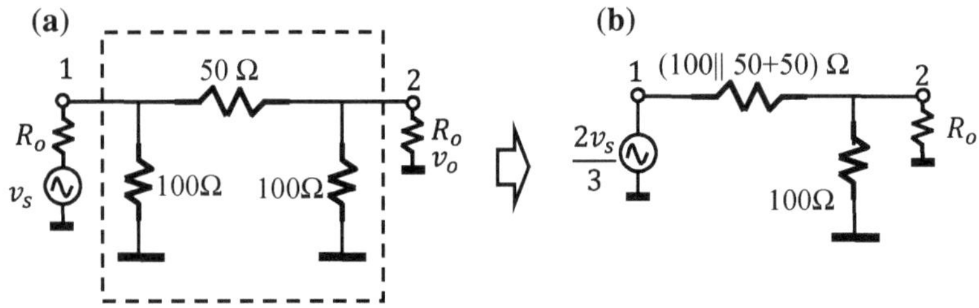

Fig. 9.37 A resistive π-type network

The forward and reverse transmission coefficients are given by

$$S_{12} = S_{21} = \frac{v_o}{v_s/2}$$

Equivalent circuit is shown in Fig. 9.37b to calculate the output voltage. The output voltage is given by

$$v_o = \frac{2v_s}{3}\frac{100||50}{(100||50+50)+100||50}$$

Thus,

$$S_{12} = S_{21} = \frac{v_o}{v_s/2} = \frac{4}{3}\frac{100||50}{(100||50+50)+100||50} = 0.381\angle 0^\circ$$

Thus S-parameter matrix of the network is

$$\begin{bmatrix} S_{11} & S_{12} \\ S_{21} & S_{22} \end{bmatrix} = \begin{bmatrix} 0.048\angle 180^\circ & 0.381\angle 0^\circ \\ 0.381\angle 0^\circ & 0.048\angle 180^\circ \end{bmatrix}$$

The attenuation of the network is

$$20\log|S_{21}| = 20\log(0.381) = 8.383\text{ dB}$$

Input and output return losses are

$$-20\log|S_{11}| = -20\log|S_{11}| = -20\log(0.048) = 26.44\text{ dB}$$

Example 9.5 Find the resistance values of a π- and TEE-type resistive matched attenuator for maximum attenuation.

Solution

For a π-type resistive matched attenuator, the values of resistors are given by (9.19a) and (9.19b) as

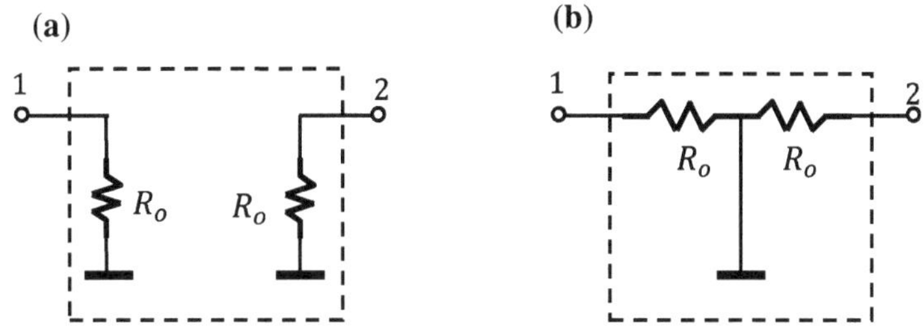

Fig. 9.38 π- and TEE-matched attenuator circuits for maximum attenuation

$$R_1 = R_o \frac{(N-1)}{4\sqrt{N}} \quad \text{and} \quad R_2 = R_o \frac{\left(\sqrt{N}+1\right)}{\left(\sqrt{N}-1\right)}$$

Here, $N = 1/|S_{21}|^2$ is the attenuation. For maximum attenuation, N is very high. Thus,

$$R_1 = R_o \frac{(1-1/N)}{4/\sqrt{N}} = \propto \quad \text{and} \quad R_2 = R_o \frac{\left(1+1/\sqrt{N}\right)}{\left(1-1/\sqrt{N}\right)} = R_o$$

The π-type resistive attenuator circuit for maximum attenuation is shown in Fig. 9.38a.

For a TEE-type resistive matched attenuator, the values of resistors are given by (9.19a) and (9.19b) as

$$R_1 = R_o \frac{\left(\sqrt{N}-1\right)}{\left(\sqrt{N}+1\right)} \quad \text{and} \quad R_2 = R_o \frac{2\sqrt{N}}{(N-1)}$$

Here, $N = 1/|S_{21}|^2$ is the attenuation. N is very high for maximum attenuation. Thus,

$$R_1 = R_o \frac{\left(1-1/\sqrt{N}\right)}{\left(1+1/\sqrt{N}\right)} = R_o \quad \text{and} \quad R_2 = R_o \frac{2/\sqrt{N}}{(1-1/N)} = 0$$

The TEE-type resistive attenuator circuit for maximum attenuation is shown in Fig. 9.38b.

Example 9.6 Find the attenuation and return losses of a quadrature hybrid attenuator at zero-bias condition in case any one of the two p-i-n diodes becomes (a) RF shorted and (b) RF open (assume diode behaves as perfect RF open at no-bias condition).

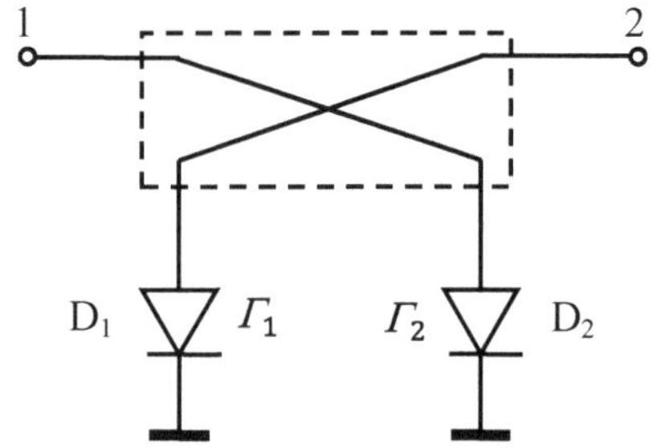

Fig. 9.39 Quadrature hybrid matched attenuator configuration

Solution

Attenuation and return losses of the 3-dB quadrature hybrid matched attenuator as shown in Fig. 9.39 are given by (9.21b, 9.22b)

$$\text{RL(in dB)} = 20\log_{10}\left(\frac{2}{|\Gamma_1 - \Gamma_2|}\right)$$

$$\text{Attenuation(in dB)} = 20\log_{10}\left(\frac{2}{|\Gamma_1 + \Gamma_2|}\right)$$

Here, Γ_1 and Γ_2 are the reflection coefficients of the diodes D_1 and D_2, respectively.

(a) Suppose diode D_1 becomes RF shorted, and thus, $\Gamma_1 = -1$. Under no DC bias condition, impedance of the diode D_2 is very high and can be considered as RF open. Thus, $\Gamma_2 = +1$.

$$\text{Return losses(in dB)} = 20\log_{10}\left(\frac{2}{|-1-1|}\right) = 0\text{ dB}$$

$$\text{Attenuation(in dB)} = 20\log_{10}\left(\frac{2}{|-1+1|}\right) = \propto$$

This result shows that full RF power will be reflected to the input port and it will provide maximum attenuation.

(b) Suppose diode D_1 becomes RF open, and thus, $\Gamma_1 = +1$. Under no DC bias condition, impedance of the diode D_2 is very high and can be considered as RF open. Thus, $\Gamma_2 = +1$.

$$\text{Return losses(in dB)} = 20\log_{10}\left(\frac{2}{|+1-1|}\right) = \propto$$

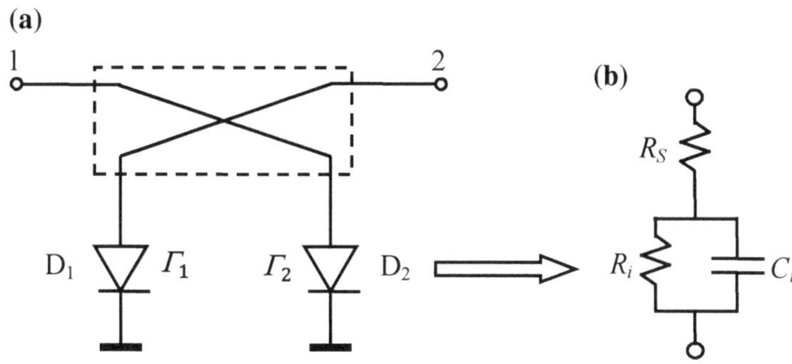

Fig. 9.40 Quadrature hybrid matched attenuator configuration

$$\text{Attenuation(in dB)} = 20\log_{10}\left(\frac{2}{|+1+1|}\right) = 0\text{ dB}$$

This result shows that in case one diode is RF open, loss of the attenuator is zero at no-bias condition.

Example 9.7 Compute and plot attenuation of a quadrature hybrid balanced attenuator as shown in Fig. 9.40a at operating frequency 10 GHz with respect to intrinsic resistance R_i of the diode over the range of 20–100 Ω. Consider equivalent circuit of the p-i-n diodes as shown in Fig. 9.40b with $R_S = 5\ \Omega$, $C_i = 0.10$ pF.

Solution

The RF impedance of the diodes is given by

$$Z_T = (R_i||C_i) + R_s$$

Diodes are batch matched and thus

$$\Gamma_1 = \Gamma_2 = \Gamma = \frac{(R_i||C_i) + R_s - R_o}{(R_i||C_i) + R_s + R_o}$$

or

$$|\Gamma| = \left[\frac{(R_i + R_s - R_o)^2 + (2\pi fC_iR_i(R_s - R_o))^2}{(R_i + R_s + R_o)^2 + (2\pi fC_iR_i(R_s + R_o))^2}\right]^{1/2}$$

Putting $R_S = 5\ \Omega$, $R_o = 50\ \Omega$, $f = 10\,\text{GHz}$

$$|\Gamma| = \left[\frac{(R_i - 45)^2 + (0.28274R_i)^2}{(R_i + 55)^2 + (0.34557R_i)^2}\right]^{1/2}$$

Thus, attenuation of the attenuator in dB is given by

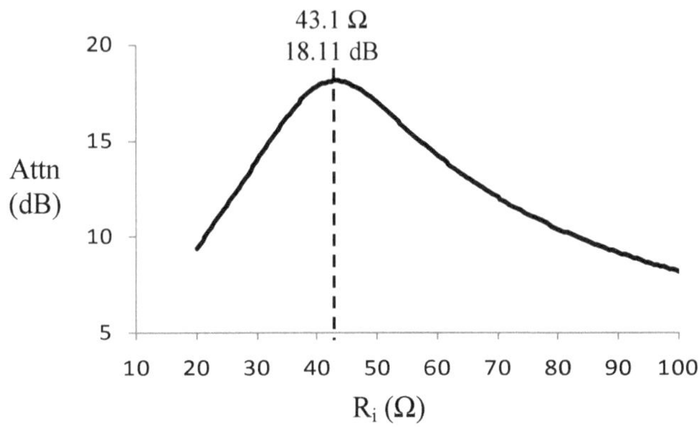

Fig. 9.41 Quadrature hybrid matched attenuator configuration

or,

$$\text{Attenuation (in dB)} = 20\log_{10}\left(\frac{1}{|\Gamma|}\right) = 10\log_{10}\left[\frac{(R_i+55)^2+(0.34557R_i)^2}{(R_i-45)^2+(0.28274R_i)^2}\right]$$

Figure 9.41 shows the plot of attenuation in dB with the variation of resistance R_i.

Here, it is to be noted that (a) the maximum attenuation is only 18.11 dB. The lower value of maximum attenuation is due to the finite capacitive reactance of the diode. (b) The maximum attenuation is at $R_i = 43.1\ \Omega$ instead of 50 Ω. This is due to the effects of diode series resistance of 5 Ω and capacitive reactance. (c) However, the return losses of the attenuator are ideally infinite in spite of the parasitic series resistance and capacitive reactance of the diodes. This is due to the balanced attenuator configuration with matched pair diodes.

Example 9.8 Find the attainable maximum attenuation of the balanced attenuator in Example 9.6 by optimizing the bias-dependent resistor value R_i. What will be the value of R_i at this condition? What will be the attenuation of the attenuator if one diode becomes failed as open at the maximum attenuation bias condition?

Solution

In Fig. 9.41, it is shown that the maximum attainable attenuation is 18.11 dB. At the maximum attenuation condition, $R_i = 43.1\ \Omega$.

In case of diode D_1 failed as open, the reflection coefficient of the diode is

$$\Gamma_1 = +1$$

For the diode D_2, the reflection coefficient is

$$\Gamma_2 = \frac{(R_i||C_i) + R_s - R_o}{(R_i||C_i) + R_s + R_o}$$

Therefore, the attenuation of the attenuator is

$$\text{Attenuation(in dB)} = 20\log_{10}\left(\frac{2}{|1+\Gamma_2|}\right)$$

or

$$|1+\Gamma_2| = \left[\frac{(2R_i + 2R_s)^2 + (4\pi\, f C_i R_i R_s)^2}{(R_i + R_s + R_o)^2 + (2\pi\, f C_i R_i (R_s + R_o))^2}\right]^{1/2}$$

$$\text{Attenuation (in dB)} = 20\log_{10}\left[2\left(\frac{(R_i + 55)^2 + (0.34557\,R_i)^2}{(2R_i + 10)^2 + (0.06283\,R_i)^2}\right)^{1/2}\right]$$

Putting $R_i = 43\ \Omega$ corresponding to maximum attenuation bias condition:

$$\text{Attenuation (in dB)} = 6.286\ \text{dB}$$

Example 9.9 An attenuator consists of a p-i-n diode. Current-dependent RF resistance R_d and diode current–voltage I_d–V_d relationship of the p-i-n diode are

$$R_d = \frac{aT}{I_d} \quad \text{and} \quad I_d = bT^2 \times \exp\left(\frac{V_d - E_b}{cT}\right)$$

Here, a, b, c are constants. Derive the optimum open-circuit voltage for zero temperature coefficient of attenuation of the attenuator.

Solution

For achieving zero temperature coefficient of attenuation of the attenuator using the p-i-n diodes, the RF resistance of the diodes should remain temperature-independent. Thus,

$$\frac{\partial R_d}{\partial T} = \frac{a}{I_d} - \frac{aT}{I_d^2}\frac{\partial I_d}{\partial T} = 0 \qquad (9.32a)$$

or

$$\frac{\partial I_d}{\partial T} = \frac{I_d}{T} \qquad (9.32b)$$

Suppose the diode is biased through a bias resistance R_b and supply voltage V_S.

$$V_S = I_d R_b + V_d \qquad (9.32c)$$

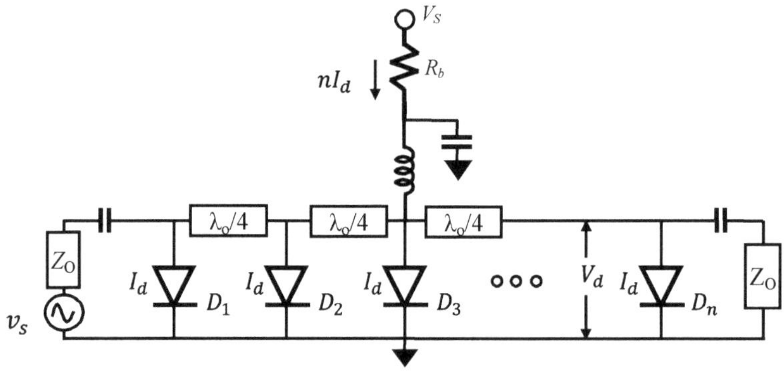

Fig. 9.42 Attenuator circuit using multiple shunt p-i-n diodes

Using (9.32c), the I–V relationship of the p-i-n diode can be written as

$$I_d = bT^2 \times \exp\left(\frac{V_S - I_d R_b - E_b}{cT}\right) \tag{9.32d}$$

Differentiating (9.32d) w.r.t. temperature T,

$$\frac{\partial I_d}{\partial T}\left(1 + \frac{I_d R_b}{cT}\right) = \frac{2I_d}{T} - \frac{I_d(V_S - I_d R_b - E_b)}{cT^2} \tag{9.32e}$$

Suppose for zero temperature coefficient $V_S = V_{\mathrm{OPT}}$ and $I_d = I_{do}$. Putting these values to (9.32e) and using (9.32b),

$$\frac{I_{do}}{T}\left(1 + \frac{I_{do} R_b}{cT}\right) = \frac{2I_{do}}{T} - \frac{I_{do}(V_{\mathrm{OPT}} - I_{do} R_b - E_b)}{cT^2} \tag{9.32f}$$

This leads to the value of optimum supply voltage for zero temperature coefficient of RF resistance as

$$V_{\mathrm{OPT}} = E_b + cT \tag{9.32g}$$

Example 9.10 An attenuator as shown in Fig. 9.42 consists of n-numbers of batch matched p-i-n diodes in shunt configuration. Current-dependent RF resistance, R_d, and current–voltage ($I_d - V_d$) relationship of the p-i-n diodes are

$$R_d = \frac{aT}{I_d} \quad \text{and} \quad I_d = bT^2 \times \exp\left(\frac{V_d - E_b}{cT}\right)$$

Here, a, b, c are constants. Derive the optimum open-circuit voltage for zero temperature coefficient of attenuation of the attenuator.

Solution

For achieving zero temperature coefficient of attenuation of the attenuator using the p-i-n diodes, the RF resistance of the diodes should remain temperature-independent. Thus,

$$\frac{\partial R_d}{\partial T} = \frac{a}{I_d} - \frac{aT}{I_d^2}\frac{\partial I_d}{\partial T} = 0 \tag{9.33a}$$

or

$$\frac{\partial I_d}{\partial T} = \frac{I_d}{T} \tag{9.33b}$$

As shown in Fig. 9.42, the diodes are biased through a bias resistance R_b and supply voltage V_S. Here, DC current I_d flows through each diode, and voltage across each diode is V_d. Thus,

$$V_S = nI_d R_b + V_d \tag{9.33c}$$

Using (9.33c), the I–V relationship of the p-i-n diode can be written as

$$I_d = bT^2 \times \exp\left(\frac{V_S - nI_d R_b - E_b}{cT}\right) \tag{9.33d}$$

Differentiating (9.33d) w.r.t. T,

$$\frac{\partial I_d}{\partial T}\left(1 + \frac{nI_d R_b}{cT}\right) = \frac{2I_d}{T} - \frac{I_d(V_S - nI_d R_b - E_b)}{cT^2} \tag{9.33e}$$

Suppose, for zero temperature coefficient, $V_S = V_{\mathrm{OPT}}$ and $I_d = I_{do}$. Putting these values to (9.33e) and using (9.33b),

$$\frac{I_{do}}{T}\left(1 + \frac{nI_{do} R_b}{cT}\right) = \frac{2I_d}{T} - \frac{I_{do}(V_{\mathrm{OPT}} - nI_{do} R_b - E_b)}{cT^2} \tag{9.33f}$$

This leads to the value of optimum supply voltage for zero temperature coefficient of RF resistance as

$$V_{\mathrm{OPT}} = E_b + cT \tag{9.33g}$$

This shows that the optimum supply voltage is independent of the number of diodes connected in parallel DC path.

Example 9.11 An attenuator as shown in Fig. 9.43 consists of two matched pair p-i-n diodes connected in DC series path. Current-dependent RF resistance, R_d, and current–voltage ($I_d - V_d$) relationship of the p-i-n diodes are

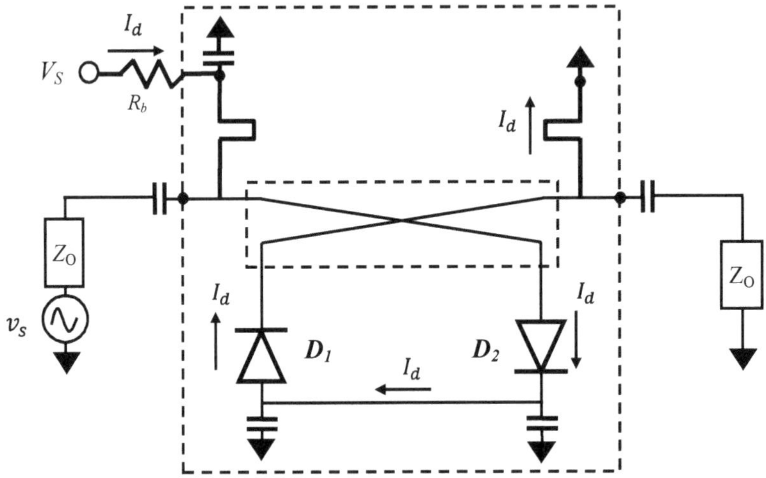

Fig. 9.43 Attenuator circuit using multiple shunt p-i-n diodes

$$R_d = \frac{aT}{I_d} \quad \text{and} \quad I_d = bT^2 \times \exp\left(\frac{V_d - E_b}{cT}\right)$$

Here, a, b, c are constants. Derive the optimum open-circuit voltage for zero temperature coefficient of attenuation of the attenuator.

Solution

For achieving zero temperature coefficient of attenuation of the attenuator using the p-i-n diodes, the RF resistance of the diodes should remain temperature-independent. Thus,

$$\frac{\partial R_d}{\partial T} = \frac{a}{I_d} - \frac{aT}{I_d^2}\frac{\partial I_d}{\partial T} = 0 \tag{9.34a}$$

or

$$\frac{\partial I_d}{\partial T} = \frac{I_d}{T} \tag{9.34b}$$

As shown in Fig. 9.43, the diodes are biased through a bias resistance R_b and supply voltage V_S. Here, DC current I_d flows through each diode and voltage across each diode is V_d. Thus,

$$V_S = I_d R_b + 2V_d \tag{9.34c}$$

Using (9.34c), the I–V relationship of the p-i-n diode can be written as

$$I_d = bT^2 \times \exp\left(\frac{V_S - I_d R_b - 2E_b}{2cT}\right) \tag{9.34d}$$

Differentiating (9.34d) w.r.t. T,

$$\frac{\partial I_d}{\partial T}\left(1 + \frac{I_d R_b}{2cT}\right) = \frac{2I_d}{T} - \frac{I_d(V_S - I_d R_b - 2E_b)}{2cT^2} \tag{9.34e}$$

Suppose, for zero temperature coefficient, $V_S = V_{\mathrm{OPT}}$ and $I_d = I_{do}$. Putting these values to (9.34e) and using (9.34b),

$$\frac{I_{do}}{T}\left(1 + \frac{I_{do} R_b}{2cT}\right) = \frac{2I_{do}}{T} - \frac{I_{do}(V_{\mathrm{OPT}} - I_{do} R_b - 2E_b)}{2cT^2} \tag{9.34f}$$

This leads to the value of optimum supply voltage for zero temperature coefficient of RF resistance as

$$V_{\mathrm{OPT}} = 2(E_b + cT) \tag{9.34g}$$

This shows that the optimum supply voltage for two DC series-connected diodes is two times of the optimum supply voltage of one diode. In general, optimum supply voltage for n-numbers of series-connected diodes is n-times of the optimum voltage required for one diode.

References

1. Caverly RH, Hiller G (1993) Temperature insensitive PIN diode attenuators. Appl Microw Summer 88
2. Caverly RH, Hiller G (1995) Temperature effect on PIN diode forward bias resistance. Solid State Electron 38(11):1879–1885
3. Bera SC, Bharadhwaj PS (2005) Insight into PIN diode behavior leads to improved control circuit. IEEE Trans Circ Syst-II 52(1):1–4
4. Bera SC, Singh RV, Garg VK (2005) A temperature dependent PIN diode model leads to simple temperature invariant attenuator circuit. Microw J 48(02):104–116
5. Bera SC, Singh RV, Garg VK, Sharma SB (2007) Optimum bias load-line compensates temperature variation of junction diode's RF resistance. IEEE Trans Microw Theor Tech 55(2):215–221
6. Bera SC, Bharadhwaj PS, Singh RV (2009) Control circuit for diode based RF circuits. EP1652020, 2008; IN229296, 2009; US7538612, 2009; JP4851792, 2011
7. Bera SC, Singh RV, Garg VK (2009) Driver circuits for temperature-invariant performance of junction diodes. IET Circ Dev Syst 3(4):143–152
8. Tajima Y, Tsukii T, Mozzi R, Tong E, Hanes L, Wrona B (1982) GaAs monolithic wideband (2–18 GHz) variable attenuators. IEEE MTT-S Dig 479–481
9. Mondal JP, Milnes AG, Oakes JG (1985) Theoretical analysis for microwave T- and π-type attenuator circuits using MESFETs. Int J Electron 58(2):231–247
10. Lizama G, Andrade T, Benton R (1987) 1–6 GHz GaAs MMIC linear attenuator with integral drivers. IEEE Microw mm-wave Monolithic Circ Symp Digest 105–107
11. Bachert PS (1996) FET attenuator, 0–1 GHz. Appl Microw Wirel 57–64

12. Dindo S, Meierer R, North R (1997) Computer-aided design of MMIC variable attenuators. Microw J 134–140
13. Daoud SM, Shastry PN (2005) A novel wideband MMIC voltage controlled attenuator with a bandpass filter topology. 13th GaAs Symp 629–632

Chapter 10
Microwave Phase Shifters

Abstract This chapter presents design, analysis and applications of various microwave phase shifters based on varactor diodes, p-i-n diodes and switched delay lines. Design and analysis of continuous variable and constant insertion loss of four-quadrant microwave phase shifters based on p-i-n diodes are described. The chapter also describes control circuit for temperature-insensitive stepwise variable phase shifters.

10.1 Introduction

Phase shifter provides known amount of controllable phase shift when RF/microwave signal passes through it. Variable phase shifter is a linear two-port device, which changes the phase of the signal in response to an external analog or digital command [1–10]. It is realized in its simplest form by connecting a reactive element in series or shunt with a transmission line [3, 4]. Phase shifters are used for the realization of phased array antenna systems, linearizers for high power amplifiers, various types of phase modulators, etc. True time delay (TTD) lines and phase shifters (PS) are the basic functional blocks for complex signal processing used in variety of applications including arbitrary waveform generation, filtering, antenna beamforming and reconfiguration ability in communication systems. Important performance parameters for a phase shifter are frequency range, total phase shift, accuracy, resolution, bandwidth, insertion loss, port return losses, PM-to-AM conversion, power handling, switching speed, etc.

10.2 Phase Shift and Time Delay

Under sinusoidal excitation of angular frequency ω, steady-state voltage and current wave along a transmission line at l is given by (7.3, 7.8)

S. C. Bera, *Microwave Active Devices and Circuits for Communication*, Lecture Notes in Electrical Engineering 533, https://doi.org/10.1007/978-981-13-3004-9_10

$$v(l,t) = Re\left[(V^{+}e^{-\gamma l} + V^{-}e^{+\gamma l})e^{i\omega t}\right] \tag{10.1a}$$

$$i(l,t) = Re\left[\left(\frac{V^{+}}{Z_o}e^{-\gamma l} - \frac{V^{-}}{Z_o}e^{+\gamma l}\right)e^{i\omega t}\right] \tag{10.1b}$$

where γ is the propagation constant and is a complex number, which can be expressed as

$$\gamma = \alpha + j\beta \tag{10.2}$$

Here, β is the phase constant and is given by $\beta = \omega/c = 2\pi/\lambda_g$. The real part α is known as attenuation constant and expressed in neper/unit length, and the imaginary part β is the phase constant and expressed in radian/unit length. The attenuation constant α determines how the signal amplitude reduces as it propagates along the transmission line. The phase constant β determines the difference of signal phase in between the input end and the point of interest. A signal experiences a phase shift of 360° (2π radians) when it travels through a transmission line of length β. Phase shift may also be seen as time delay. As a sinusoidal signal travels through a transmission line, it experiences time delay and associated phase shift as shown in Fig. 10.1. Suppose a signal of frequency $f(= \omega/2\pi)$ travelled through a distance x with propagation velocity v_p, then time delay will be

$$\text{Time delay} = \frac{x}{v_p} = \frac{x}{f\lambda_g} = \frac{\beta x}{\omega} \tag{10.3}$$

Therefore, RF/microwave phase shifters can be realized by switching several transmission lines of different physical lengths. Phase shift of 360° (2π radians) is equivalent to delay of one time period, $T = 1/f$. Time delay can be expressed in terms of phase shift ($\Delta\Phi$) as

$$\text{Time delay} = \frac{\Delta\Phi}{2\pi f} \quad \Delta\Phi \text{ in radian} \tag{10.4}$$

Phase shift $\Delta\Phi$ experienced by a signal travelling through a distance Δl can be expressed as

$$\Delta\Phi = 2\pi\frac{\Delta l}{\lambda_g} \tag{10.5}$$

Thus by diverting a signal from one path to another of different lengths, phase shift can be obtained. In this case, the phase shift is dependent on guide wavelength λ_g. Therefore, phase shift value obtained by changing the line length deviates linearly from the intended value as the signal frequency deviates in either direction from

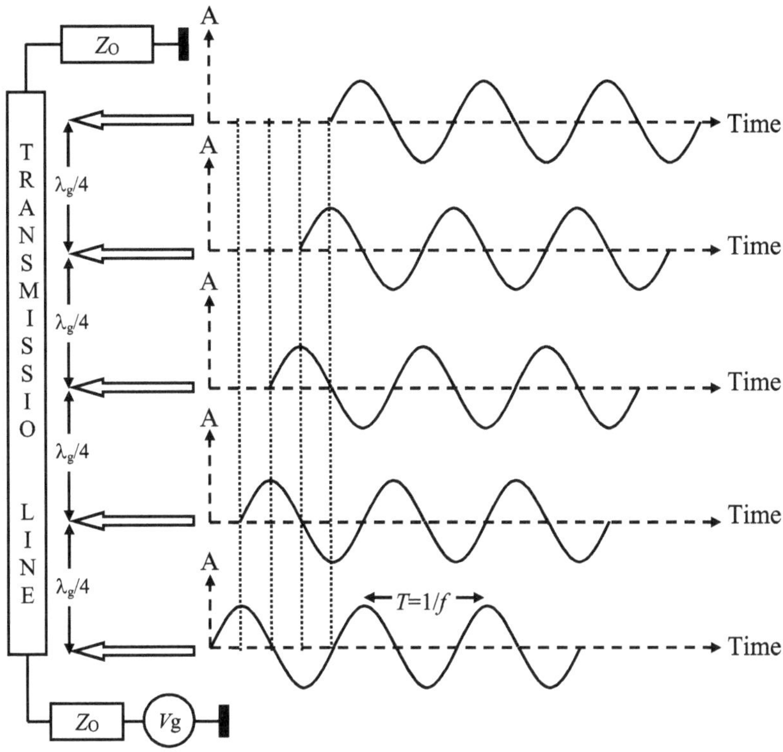

Fig. 10.1 Phase shift/delay experienced by a sinusoidal wave travelling through a transmission line

the nominal centre frequency. However, time delay will remain same as the signal frequency changes from the centre frequency. Therefore, switching transmission lines, constant phase shift cannot be obtained over a wide frequency band. Some other techniques are used for broadband phase shifter that will be discussed in the following sections.

10.3 Types of Phase Shifter

There are two distinct types of phase shifters with two distinct characteristics as shown in Fig. 10.2. In both the cases, insertion phase in each setting is linearly proportional to the frequency. In the first case, Fig. 10.2a, insertion phase slope changes from one setting to another which implies that time delay is different from one setting to another. Therefore, phase shifters of this category are also known as time shifter. In the second case, Fig. 10.2b, insertion phase slope is constant for all the phase setting, i.e. phase with respect to frequency characteristics which are parallel,

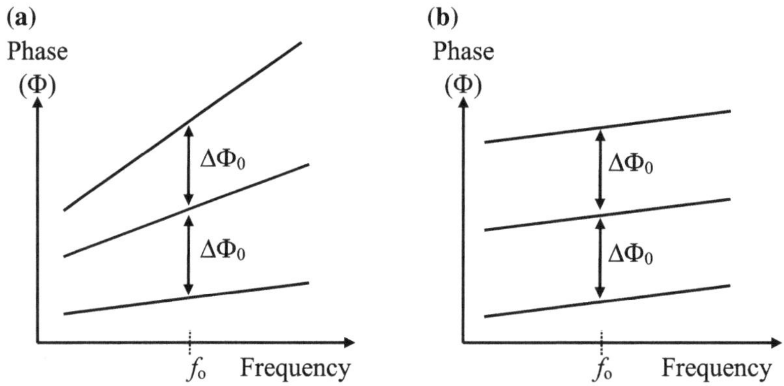

Fig. 10.2 Phase shift with frequency, relative phase shift **a** changes with frequency, **b** constant over the frequency

and implies that time delay remains constant from one setting to another. Therefore, this type of phase shifter changes the phase of the incoming signal maintaining constant time delay.

The first types of phase shifters are generally realized by switching different lengths of transmission line. Thus, these types of phase shifters are inherently wide bandwidth, limited only by the limitation of switching circuit's bandwidth. On the other hand, special networks are required for the second types of phase shifters to obtain the desired phase characteristic. These types of phase shifters are inherently narrowband in nature. Wider bandwidths are obtainable but at a higher complexity of the networks.

At a single frequency, both types of phase shifters can provide identical phase shift. When the frequency is varied, difference between these two types of phase shifters becomes apparent. Therefore, for very narrowband applications, both types of phase shifters can perform the same function. However, for broadband applications, types of phase shifters are to be properly chosen depending upon the applications.

Depending on the control mechanism, phase shifters can be categorized into two types, analog and digital phase shifters. Analog phase shifters are voltage-variable devices offering continuously variable phase shift. The phase control is achieved by analog control voltage. In general, the control voltage varies the capacitance of varactor diodes to change the reactance of the circuit for achieving variable phase shift of the RF/microwave signal. Digital phase shifters are realized by using p-i-n diodes or FETs. The diodes and FETs are used as variable impedance or high-frequency switches to achieve desired phase shifter performance.

10.4 Realization of Phase Shifters

Depending on the characteristics of the phase shifters and control signals, different techniques are used to realize it. The subsequent sections will provide the design and realization details of various phase shifters.

10.4.1 Switched Transmission Line Phase Shifter

Switched transmission line phase shifter design is the most straightforward because it uses the time delay difference by switching the signal path in between two transmission lines to provide the desired phase shift. The switching elements used for the digital phase shifters may be mechanical or electronic. In case of electronic digital phase shifters, p-i-n diodes, field effect transistors (FET), or microelectromechanical systems (MEMS) are used. Among all the switching devices, p-i-n diodes are commonly used in phase shifters for their relatively simple bias circuit, high-speed switching and acceptable lower loss. As discussed in the previous chapter, the change of RF resistance from few ohms to several kilo-ohms of a p-i-n diode from its forward bias to reverse bias makes it suitable as RF switching device.

The standard switched-line phase shifter switches transmission line segments of different electrical lengths to obtain predetermined amount of phase shift. Thus, the phase shift is dependent only on the lengths of the transmission lines used. One of the two transmission lines is used as reference line, and the other is used as a delay line. The most important characteristic of the switched transmission line phase shifter is that the phase shift is approximately a linear function of frequency, which means phase shift is not constant over the operating frequency range; however, time shift is constant over the frequency range from one phase shift setting to another. Here, the phase shift created is dependent only by the length of the transmission lines, making the phase shifter very stable over time and temperature. The temperature-dependent RF resistances of the p-i-n diodes affect the insertion loss of the phase shifter but it does not affect the phase shift.

10.4.1.1 Design of Switched-Line Phase Shifter

Figure 10.3 shows basic configuration of the switched-line phase shifter with RF source of impedance Z_o and also loads of impedance Z_0. Here, two single-pole double-throw (SPDT) switches are used to select one of the two transmission lines at a time, either reference line of length L_R or delay line of length L_1. The switches are mechanical or p-i-n diode/transistor/MEM-based electronic RF switches. Practically, isolation of the RF switches should be better very high (practically >25 dB) to avoid phase errors. The insertion loss of the phase shifter is the combination of losses due to the switches and transmission line. Insertion loss vari-

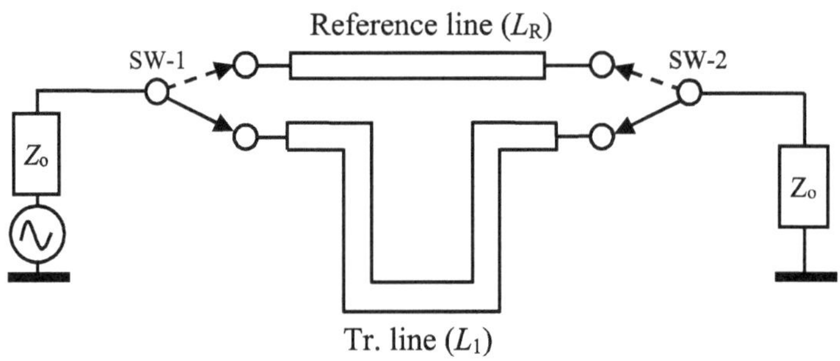

Fig. 10.3 Basic configuration of switched-line phase shifter

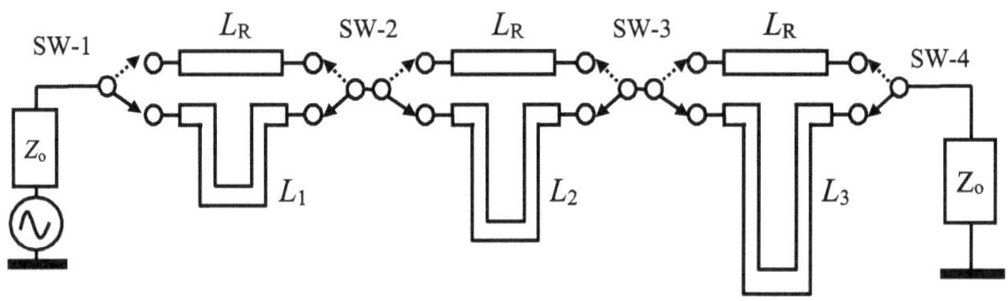

Fig. 10.4 Switched transmission line digital phase shifter

ation from one phase shift to another is equal to insertion loss difference in between the reference line and delay line.

The phase shift of the switched transmission line phase shifter is given by (10.5) and can be written as

$$\Delta\Phi(f) = 2\pi \frac{(L_1 - L_R)}{\lambda_g} = 2\pi f \frac{(L_1 - L_R)}{v_p} \tag{10.6}$$

Here, λ_g is the guide wavelength at a particular operating $f = v_p/\lambda_g$. From (10.6), it is clear that the phase shift value deviates from the intended value linearly as the frequency of the signal deviates in either direction from the nominal (centre) frequency. To obtain phase shift of 180°, 90° and 45°, the required path length L_1 is a half, quarter and one-eight wavelength longer than the reference line length L_R, respectively.

Schematic configuration of a switched transmission line 3-bit digital phase shifter is shown in Fig. 10.4. Here, three switched-line phase shifters are connected in cascade. Using proper combination of ON/OFF states of four switches (SW-1 to SW-4), a 3-bit phase shifter can provide discrete phase shifts from 0° to 360° in steps of 360°/8 = 45°. By connecting more number of switched-line phase shifter in cascade configuration, step size of the phase shifter can be reduced.

10.4.2 Varactor Diode-Based Analog Phase Shifters

Phase shift of an analog phase shifter provides predetermined continuously variable phase shift depending on the applied analog control voltage. Thus, analog phase shifter provides very high-resolution variable phase shift. Most commonly, varactor diodes are used as control element for achieving continuously variable phase shift. Capacitance of reverse-biased varactor diodes is used as variable reactance for analog phase shifters [3, 4].

Configuration of an analog phase shifter using varactor diode as variable reactance is shown in Fig. 10.5. Two output ports (port-2 and port-3) of a 3-dB 90° hybrid coupler are terminated with the varactor diodes to ground. Bandwidth of these types of phase shifter depends on the bandwidth of the quadrature coupler itself. The coupler can be a Lange coupler, branch line hybrid coupler, or a rat-race coupler on microstrip or stripline configuration. The coupler splits the input signal at port-1 to two equal amplitudes with a phase difference of 90° and incident on the port-2 and port-3. The incident signals are reflected from the diodes and combined at the isolated port (port-4) which is the output port of the phase shifter. Considering identical characteristic of both the diodes (matched pair), the magnitudes and angles of reflected signals due to the reflections from the diodes are equal. These two reflected signals will combine at the output port and cancel at the input port. Thus, the transmission coefficient of the phase shifter is given by (using 9.22a)

$$S_{21} = \frac{\Gamma_d \angle(2\theta - 90)^\circ + \Gamma_d \angle(2\theta - 90)^\circ}{2} \tag{10.7a}$$

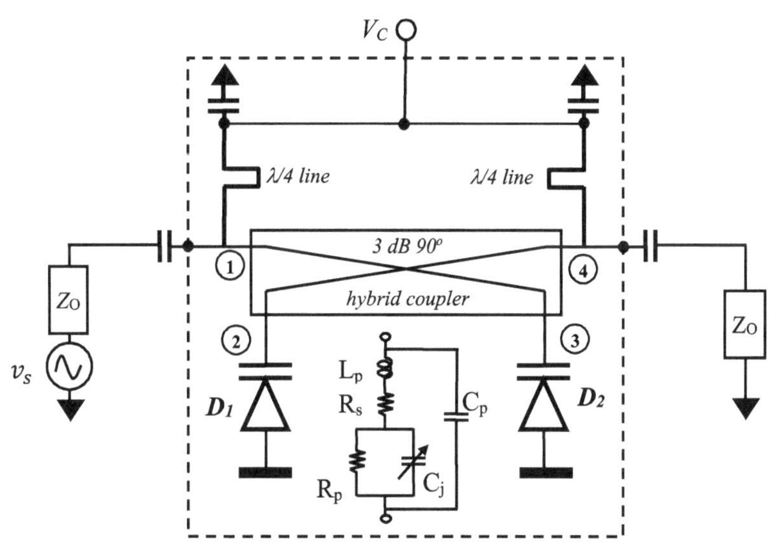

Fig. 10.5 Analog phase shifter using varactor diodes

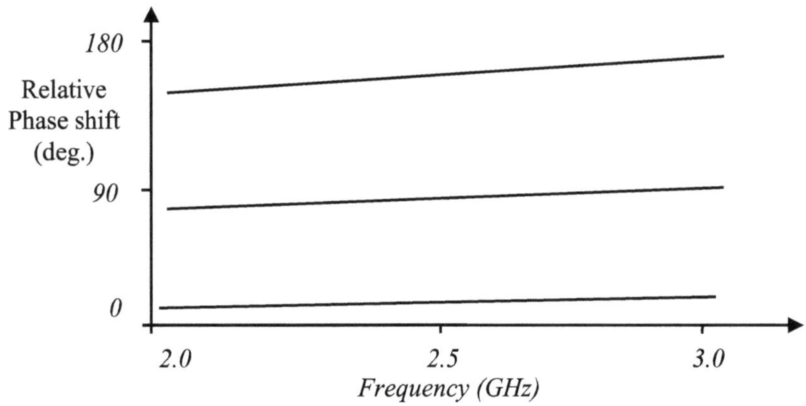

Fig. 10.6 phase characteristic of the varactor diode-based voltage-variable phase shifter for three different settings of the capacitance

$$= \Gamma_d \angle (2\theta - 90)^\circ \tag{10.7b}$$

Here, θ is the insertion phase of the signal in between port-1 and port-2. Γ_d is the reflection coefficient of the diodes with respect to reference impedance Z_o and is given by

$$\Gamma_d = \frac{Z_d - Z_o}{Z_d + Z_o} \tag{10.8a}$$

Here, Z_d is the varactor diode impedance as shown in Fig. 10.5. Neglecting the effect of package inductance L_p, capacitance C_p and reverse-bias resistance R_p, the equivalent diode impedance is simplified to a series combination of voltage-variable capacitance C_i and resistance R_s. Thus, reflection coefficient of the diode can be written as

$$\Gamma_d = \frac{(R_s - Z_o) - jX_c}{(R_s + Z_o) - jX_c} \tag{10.8b}$$

Here, $X_c = 1/(2\pi f C_j)$. Therefore, the insertion phase of the phase shifter is given by

$$\angle S_{21} = \tan^{-1}\left(\frac{X_c}{Z_o - R_s}\right) - \tan^{-1}\left(\frac{-X_c}{R_s + Z_o}\right) + \angle(2\theta - 90)^\circ \tag{10.9}$$

The capacitance C_j can be controlled by changing the reverse-bias voltage of the varactor diodes. Typical phase shift relative to the insertion phase of the reflection-type phase shifter is shown in Fig. 10.6. It provides a voltage-variable phase shift with range from 0° close to 180°.

10.4.3 Four-Quadrant Continuously Variable Phase Shifter

Vector modulator-based continuously variable four-quadrant phase shifter can be designed using p-i-n diodes [5–10]. Bias-controlled RF resistance of p-i-n diodes is used to realize the phase shifter. Schematic block diagram of a phase shifter is shown in Fig. 10.7. Here, input signal is divided into two equal amplitudes with phase difference of 90° by a 3-dB 90° power divider. Both the signals are selectively attenuated and phase shifted by an amount of 0° or 180° using voltage/current-controlled variable attenuator and finally combined by an in-phase power combiner. As shown in the vector diagram in Fig. 10.8 that by continuously varying the amplitude of both the vectors simultaneously and switching the phase of the vectors by 180°, four quadrants, i.e. 360° phase shift can be achieved. As shown in the vector diagram, the output signal (v_o) will be the vector addition of the two controlled signals v_1 and v_2. To maintain output signal level constant over the full phase shift range, both the vectors need amplitude control (av_1 and bv_2). Thus, signal level at the output is given by

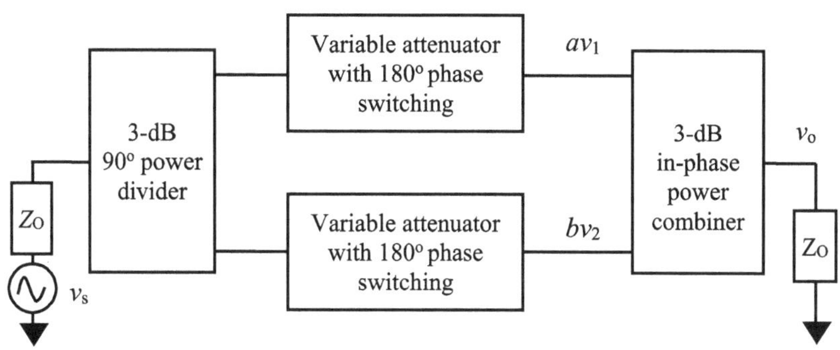

Fig. 10.7 Block diagram of vector modulator phase shifter

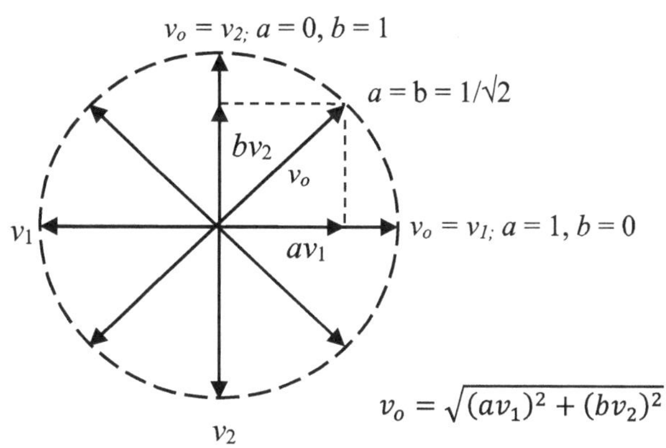

Fig. 10.8 Vector modulator diagram for the four-quadrant phase shifter

$$v_o = \sqrt{(av_1)^2 + (bv_2)^2} \tag{10.10}$$

Ideally, to achieve constant attenuation over the entire phase shift of 0°–360°, the loss is 6 dB. 3 dB loss is due to the combination of two equally divided signals with 90° phase shift, and another 3 dB loss is due to the selective control of both the divided signals over the full phase shift range to achieve constant amplitude of the combined signal at the output.

10.4.3.1 Variable Attenuator with 180° Phase Switching

The variable attenuator with 180° phase switching capability can be achieved from a p-i-n diode-based reflection-type attenuator using 3-dB 90° coupler. The attenuator is same as discussed in Chap. 9 and is shown in Fig. 10.9. Considering both the diodes are having matched characteristic, the transmission coefficient of the attenuator is given by (10.7b)

$$S_{21} = \Gamma_{pd}\angle(2\theta - 90)^\circ \tag{10.11a}$$

Here, θ is the insertion phase of the signal in between input and coupled ports and Γ_{pd} is the reflection coefficient of the ground-connected p-i-n diode with respect to impedance Z_o and is given by

$$\Gamma_{pd} = \frac{Z_{pd} - Z_o}{Z_{pd} + Z_o} \tag{10.11b}$$

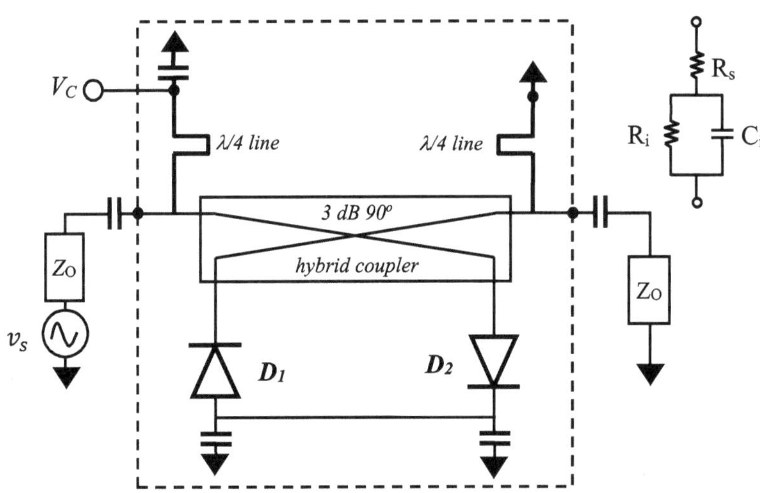

Fig. 10.9 Quadrature hybrid matched attenuator configuration with series DC path

Here, Z_{pd} is the p-i-n diode's impedance neglecting the package inductance L_p, capacitance C_p as shown in Fig. 10.9. Thus, reflection coefficient of the grounded diode can be written as

$$\Gamma_{pd} = \frac{(R_i||C_i + R_s) - Z_o}{(R_i||C_i + R_s) + Z_o}$$

or

$$\Gamma_{pd} = \frac{(R_i + R_s - Z_o) - j2\pi f C_i R_i (Z_o - R_s)}{(R_i + R_s + Z_o) + j2\pi f C_i R_i (Z_o + R_s)} \tag{10.11c}$$

Combining (10.11a) and (10.11c), amplitude and phase of the transmission coefficient can be written as

$$|S_{21}| = \frac{(R_i + R_s - Z_o)^2 + (2\pi f C_i R_i (Z_o - R_s))^2}{(R_i + R_s + Z_o)^2 + (2\pi f C_i R_i (Z_o + R_s))^2} \tag{10.12a}$$

$$\angle S_{21} = -\tan^{-1}\left[\frac{4\pi f C_i R_i^2 Z_o}{(R_i + R_s + Z_o)(R_i + R_s - Z_o) - (2\pi f C_i R_i)^2 (Z_o^2 - R_s^2)}\right] \tag{10.12b}$$

Amplitude (in dB) and phase characteristics of the attenuator based on (10.12a) and (10.12b) are shown in Fig. 10.10, over the diode's resistance R_i =5 Ω–2 kΩ considering f =4 GHz, R_s =3 Ω and C_i =0.02 pF. It shows that phase of the signal is switched by 180° when the diode resistance crosses the value equal to the characteristic impedance Z_o = 50 Ω. The phase deviation from 180° is due to the capacitive reactance of the diode. This can be eliminated by implementing inductance in parallel with the diode. The value of the inductance will be such that parallel resonance frequency will be the operating frequency of the attenuator. The reflection-type p-i-n diode attenuator using 3-dB 90° coupler is used to achieve high level of attenuation with phase reversal characteristic at its maximum attenuation level.

10.4.3.2 Four-Quadrant Phase Shifter

Schematic RF circuit diagram of the four-quadrant phase shifter is shown in Fig. 10.11. The most important circuit module is the variable attenuator with all possible attenuation range and with the provision of 180° phase switching. To realize this phase shifter, p-i-n diode based on two variable attenuators with 180° phase switching is used. Input 3-dB 90° coupler can be implemented by Lange coupler to achieve amplitude and phase performance over broad frequency band. Output 3-dB

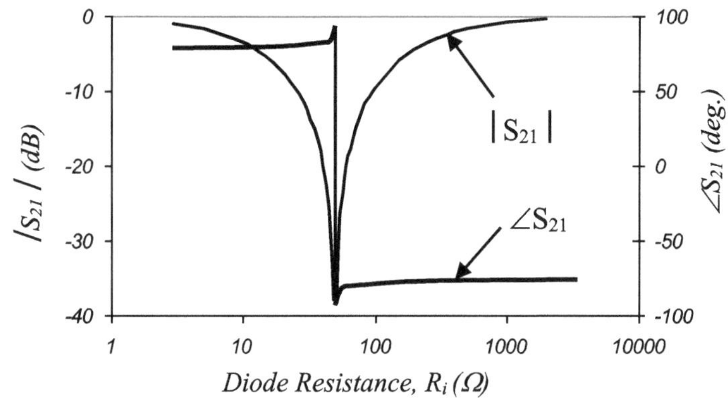

Fig. 10.10 Amplitude and phase characteristics of the attenuator with bias-dependent diode resistance

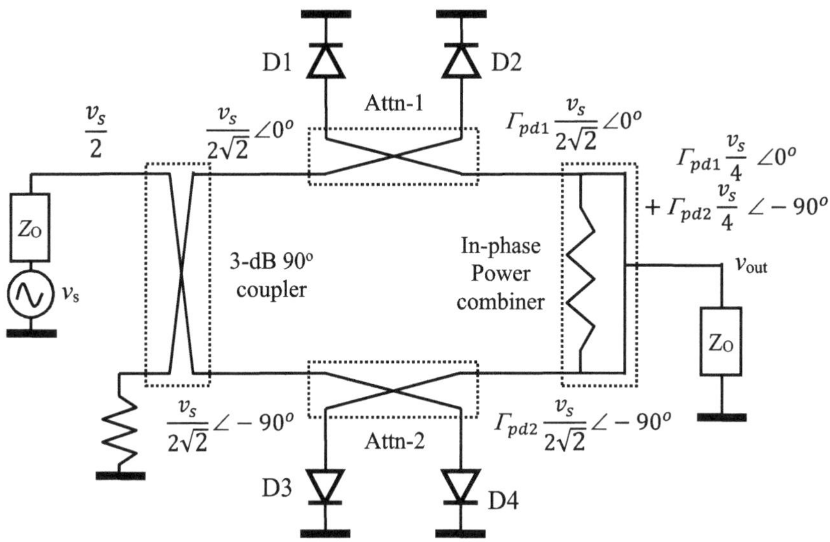

Fig. 10.11 Schematic circuit diagram of the four-quadrant phase shifter using p-i-n diodes

in-phase power combiner can be realized using Wilkinson-type power divider. Here, input signal is divided into equal amplitudes in between the direct and coupled ports, with a 90° phase difference. These two divided signals are selectively attenuated and phase shifted by the variable attenuators and finally added by the Wilkinson power divider. From Fig. 10.11, eliminating the common phase shift in both the signal arms, the transmission coefficient of the phase shifter can be written as

$$S_{21} = \frac{\Gamma_{\mathrm{pd1}} \frac{v_s}{4} + \Gamma_{\mathrm{pd2}} \frac{v_s}{4} \angle -90^{\circ}}{\frac{v_s}{2}} \tag{10.13a}$$

or

$$S_{21} = \frac{\Gamma_{pd1}}{2} + \frac{\Gamma_{pd2}}{2}\angle - 90° \tag{10.13b}$$

where Γ_{pd1} and Γ_{pd2} are the reflection coefficient given by (10.11c) of the grounded p-i-n diodes in attn-1 and attn-2, respectively, as shown in Fig. 10.11.

Suppose

$$\Gamma_{pd1} = \rho_{pd1}\angle\theta_{pd1} \quad \text{and} \quad \Gamma_{pd2} = \rho_{pd2}\angle\theta_{pd2} \tag{10.13c}$$

Then,

$$|S_{21}| = \left[\left(\frac{\rho_{pd1}}{2}\right)^2 + \left(\frac{\rho_{pd2}}{2}\right)^2 + \frac{1}{2}\rho_{pd1}\rho_{pd2}\cos\theta\right]^{\frac{1}{2}} \tag{10.14a}$$

and

$$\angle S_{21} = \tan^{-1}\left(\frac{\rho_{pd2}\sin\theta}{\rho_{pd1} + \rho_{pd2}\cos\theta}\right) \tag{10.14b}$$

where θ is the phase angle in between the two signals combined at the output port of the phase shifter and is given by

$$\theta = \theta_{pd2} - \theta_{pd1} - 90° \tag{10.14c}$$

Ideally, the phase angles θ_{pd1} and θ_{pd2} will be 0 or 180°. Therefore, angle θ between the two signal vectors coming to the output of the phase shifter will be always 90°. Thus, in ideal case putting $\theta = 90°$ in (10.14a), the minimum insertion loss is 3 dB. This loss is due to the 90° phase difference in between the two signals combined at the output. The maximum loss will happen when one of the attenuators will give maximum attenuation. Thus putting ρ_{pd1} or $\rho_{\mathrm{pd1}} = 0$ in (10.14a), the maximum attenuation becomes 6 dB. Therefore, to maintain constant amplitude for the entire phase shift range of 360°, proper biasing to the attenuators is required to keep fixed insertion loss of the phase shifter ideally at 6 dB.

The photograph of a S-band four-quadrant phase shifter is shown in Fig. 10.12. The circuit elements are realized with MIC technology on 20-mil-thick alumina substrate (ε_{r} =9.9) in microstrip line configuration. Lange couplers are used for both the attenuators and input 3 dB 90° coupler. Wilkinson power combiner is used for the output combiner. Beam-lead p-i-n diodes are used for its negligible parasitic inductive and capacitive reactances at S-band frequency. Biasing scheme of the phase shifter is shown in Fig. 10.12. Resistors R_1 and R_2 are used to set optimum open-circuit voltage (V_{OPT}) to achieve temperature-invariant attenuation of the attenuators [11, 12]. Variable resistors R_{b1} and R_{b2} are used to control the bias of both the attenuators independently. Thus, adjusting the values of the resistors R_{b1} and R_{b2},

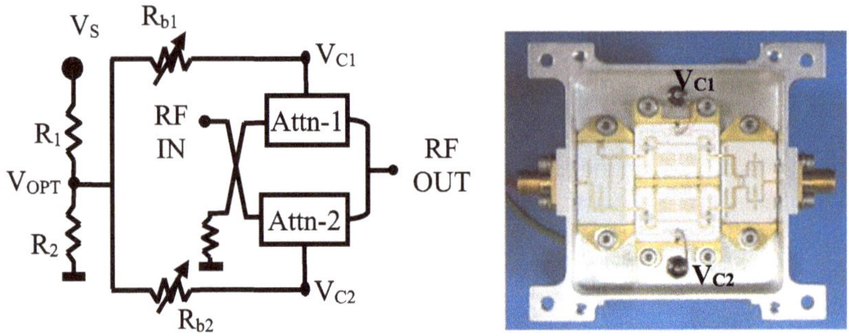

Fig. 10.12 Schematic phase shifter circuit with biasing scheme and photograph of a four-quadrant continuously variable phase shifter

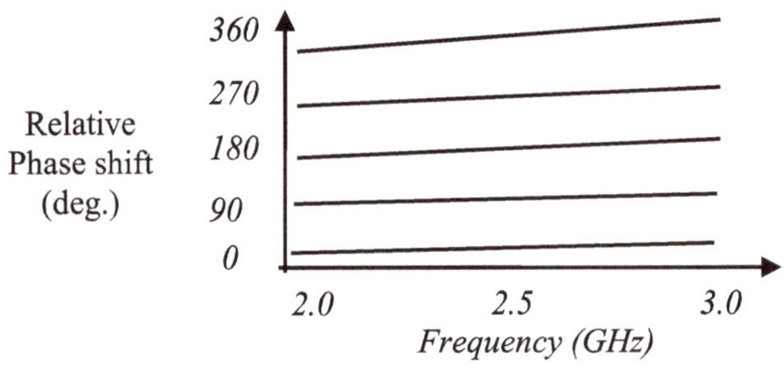

Fig. 10.13 Phase characteristic of the four-quadrant phase shifter for various settings of attenuations of attn-1 and attn-2

any phase from 0 to 360° can be obtained as shown in Fig. 7.13 by maintaining the insertion loss constant. Without any bias to the p-i-n diodes, practically the insertion loss of the phase shifter is about 4.5 dB, and extra 1.5 dB is due to the loss in power divider, combiner and attenuator circuits. In practice, it is possible to achieve continuously variable phase shift with constant loss of 7.5 dB.

10.4.4 Four-Quadrant Digital Phase Shifter

Four-quadrant digitally controlled phase shifter can be realized using the vector modulator configuration discussed in the previous section. Here, the bias voltages to the diodes of both the attenuators are controlled digitally. Figure 10.14 shows the digital bias control circuit for the phase shifter. Biases of the two attenuators are controlled simultaneously by selecting the bias resistors using digitally controlled analog switch (multiplexer) circuits. V_{OPT1} and V_{OPT2} are the optimum open-circuit

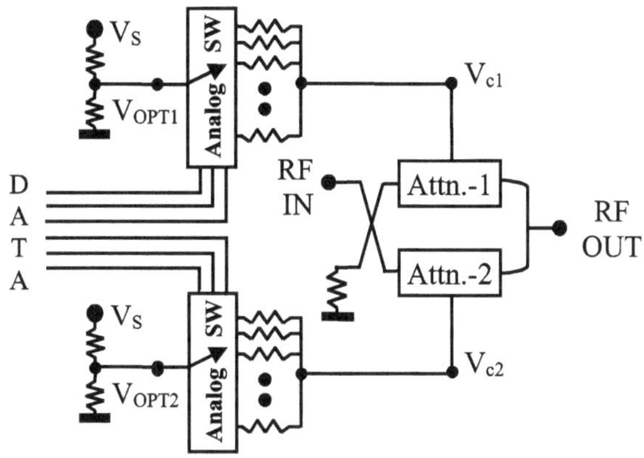

Fig. 10.14 Four-quadrant digital phase shifter using vector modulator configuration

voltages of the p-i-n diode-based attenuator-1 and attenuator-2, respectively, to obtain temperature-invariant attenuation of the attenuators that is required for achieving temperature-invariant phase shift.

10.5 Applications of RF/Microwave Phase Shifter

RF and microwave phase shifters have many applications in communication systems such as phase discriminators, linearization of high power amplifiers, beamforming networks and phased array antennas.

Generally, the loaded line phase shifters are popularly used in phased array antennas. In case of electronically steered antenna arrays, the phase shifters are used to set proper phase of the signals from individual antenna elements to steer the antenna beam in the desired direction without physically rotating the antenna. Each radiating element requires a phase shifter that applies the necessary phase shift to steer the antenna beam. The phase of each element is so adjusted that individual signals line up in the desired direction. In this case, the phase shift settings should not change with frequency to keep the beam in the desired direction over the range of operating frequency band.

Array beamforming techniques are used to form simultaneously available multiple beams. Array beamforming is used to achieve low side lobe level with high gain and controlled beam width. In case of analog array beamforming system, systematically cable lengths to the elements are varied to control the signal phase. In this case of array beamforming technique, both phase and amplitude need to be controlled by the individual signals to achieve adjustment of side lobe level and to steer nulls.

References

1. Garver RV (1969) 360° varactor linear phase modulator. IEEE Trans Microwave Theory Tech 17(3):137–146
2. Upshur JI, Geller BD (1990) Low-loss 360° X-band analog phase shifter. In: IEEE MTT-S international microwave symposium digest, 8–10 May 1990
3. Ulriksson B (1979) Continuous varactor-diode phase shifter with optimum frequency response. IEEE Transactions on Microwave Theory and Techniques, vol. MTT-27, pp 650–654, July 1979
4. Niehenke EC, DiMarco VV, Friedberg A, (1985) Linear analog Hyperabrupt varactor diode phase shifters. IEEE MTT-S Digest, 657–660
5. Yoo TW, Song JH, Park MS (1997) 360° reflection-type analogue phase shifter implemented with a single 90° branch-line coupler. Electron Lett 33(3):224–226
6. Liew YH, Joe J, Leong MS (1999) A Novel 360° analog phase shifter with linear voltage phase relationship. Microwave Conf Asia Pac 1:17–20
7. Bera SC, Raval BM, Singh RV, Garg VK (2000) Electronically controlled continuously variable four quadrant microwave phase shifter. In: 31st mid term symposium on emerging wireless communication technology and systems. IETE, 8–9 Apr 2000
8. Lin C-S, Chang S-F, Hsiao W-C (2008) A full-360° reflection-type phase shifter with constant insertion loss. IEEE Microwave Wirel Compon Lett 18(2):106–108
9. Lin C-S, Chang S-F, Chang C-C, Shu Y-H (2007) Design of a reflection-type phase shifter with wide relative phase shift and constant insertion loss. Microwave Theory Tech IEEE Trans 55:1862–1868
10. Xiao MX, Cheung SW, Yuk TI (2009) A wideband 360° analog phase shifter design. In: Proceedings of 22nd IEEE Canadian conference on electrical and computer engineering, pp 524–527
11. Bera SC, Bharadhwaj PS, Singh RV (2011) Control circuit for diode based RF circuits, EP1652020, 2008; IN229296, 2009; US7538612, 2009; JP4851792
12. Bera SC, Singh RV, Garg VK (2009) Driver circuits for temperature-invariant performance of junction diodes. IET Circuits Devices Syst 3(4):143–152

Chapter 11
Microwave Modulators

Abstract Direct carrier modulation at RF/microwave frequency avoids the use of upconverter and downconverter assemblies reducing hardware complexity and cost. This chapter presents design, analysis and applications of various direct microwave carrier modulators. Here, the design and analysis of p-i-n diode and transistor-based amplitude modulator, bi-phase-balanced modulator, I–Q vector modulator and other phase modulators are discussed.

11.1 Introduction

Modulators are three-port networks. One of the ports is for modulating signal and the other two ports are RF ports. One of the RF ports is for the un-modulated carrier, i.e. local oscillator (LO) input and other is for the RF-modulated output. A complex chain of downconverter, mixer, filter, and upconverter is used to design a conventional modulator. Direct carrier modulation at RF/microwave frequency avoids the use of upconverter and downconverter assemblies leading to less hardware complexity, cost and compact realization in MMIC technology [1, 2]. In case of direct modulation techniques, p-i-n diode or FET-based voltage/current-controlled RF resistance are used to modulate the amplitude and phase of the carriers by using modulating signal as control voltage/current of the device. Variable RF resistance of p-i-n diodes and 180° phase reversal of reflection-type attenuators around 50 Ω makes the p-i-n diodes suitable for various types of modulators. Most frequently used digital modulation techniques are the discrete phase modulation, i.e. M-array phase shift keying (PSK). BPSK and QPSK are the most popularly used digital phase modulations techniques for wireless and satellite communication systems. In case of BPSK and QPSK modulators, p-i-n diodes are used as a switch with 180° phase switching characteristic by using a 3-dB hybrid coupler. However, in the case of M-array PSK modulators, p-i-n diodes are used as a variable attenuator with 180° phase switching characteristic in reflection type configuration. I–Q modulator is the most important element for the realization of broadband direct carrier digital modulators. Broadband microwave I–Q

S. C. Bera, *Microwave Active Devices and Circuits for Communication*, Lecture Notes in Electrical Engineering 533, https://doi.org/10.1007/978-981-13-3004-9_11

modulator modules typically cover an instantaneous bandwidth of up to 2-octaves, e.g. 1–4 GHz. Such modules are usually based on diodes or microwave transistors for amplitude and phase control.

11.2 Amplitude Modulators

Amplitude modulation (AM) has been the most historical and popular method for transmitting information. It modifies the amplitude of a carrier signal in accordance with a modulating signal. The simplest way to realize amplitude modulator is to control insertion loss of a voltage/current-controlled attenuator using the modulating signal as the control voltage/current. A basic circuit diagram of a p-i-n-diode-based simple direct amplitude modulator circuit is shown in Fig. 11.1. Analog or digital modulating signal is applied to the modulating input of the modulator. Modulating input terminal of the modulator is isolated from the RF signal path by using high impedance line and a low impedance radial stub of length $\lambda/4$ corresponds to the RF frequency. Level of the modulating signal is so adjusted that at its maximum value the RF resistance of the diode remains more than 50 Ω to avoid the phase and attenuation characteristic reversal. Another way to avoid phase and attenuation reversal is the use of 50 Ω RF resistors in series with the p-i-n diodes at the coupled (port-2) and direct (port-3) ports of the coupler. The series 50 Ω resistor will ensure the impedance of both the ports more than 50 Ω for any level of modulating signal. A DC bias for the p-i-n diodes is required to obtain a specific attenuation level over which time-varying modulating signal is superimposed. One important requirement for the amplitude modulators is linearity. The modulator must translate the modulating signal with the introduction of minimal distortion. One drawback of this simple amplitude modulator circuit is that the amplitude of the modulated RF carrier is not linearly proportional to the control voltage. Various linearization techniques are used to minimize the AM signal distortion.

The p-i-n-diode-based amplitude modulator, as shown in Fig. 11.1, can be used as a pulse amplitude modulation circuit, i.e. as a switching circuit for RF power levels. Pulse modulation is achieved by applying forward bias up to its maximum attenuation level. After attaining the maximum attenuation level, the RF power can be pulsed “ON” by applying a high positive constant bias to drive the p-i-n diodes to its low impedance state. At the end of the required pulse width, the DC bias voltage is switched corresponding to the maximum attenuation level.

Another way of implementing the amplitude modulator is by summing the modulating signal into an automatic level control (ALC) loop. This type of simple modulator using p-i-n diode is used for controlling RF signal level in a closed-loop ALC systems. In the ALC system, RF signal is sampled, detected and amplified to generate a control signal that is applied to the control input of the modulator to provide constant RF level output. ALC-based amplitude modulation scheme offers better linearity limited by the available ALC dynamic range, which is mainly governed by the dynamic range of the detector. In this case, the maximum modulating signal

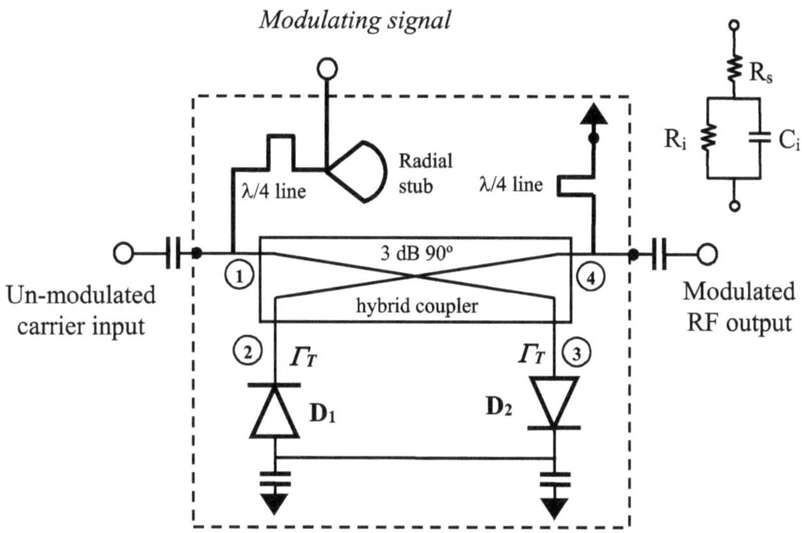

Fig. 11.1 Schematic diagram of p-i-n diode based direct amplitude modulator

frequency is lower compared to the open-loop alternatives due to the settling time of the closed-loop ALC system. Where, the settling time of an ALC system is governed by the loop stability requirement.

Considering the coupler as an ideal coupler and zero insertion phases from its input to the coupled port, the S-parameter of the modulator of Fig. 11.1 is given by

$$[S] = \begin{bmatrix} 0 & -j\Gamma_T \\ -j\Gamma_T & 0 \end{bmatrix} \tag{11.1}$$

Here, Γ_T is the voltage reflection coefficient of the diode used as reflection termination as shown in Fig. 11.1. From the equivalent circuit of the p-i-n diodes, the magnitude of the transmission coefficient of the modulator can be written as (10.12a)

$$|S_{21}| = |\Gamma_T| = \frac{(R_i + R_s - Z_o)^2 + (2\pi f C_i R_i (Z_o - R_s))^2}{(R_i + R_s + Z_o)^2 + (2\pi f C_i R_i (Z_o + R_s))^2} \tag{11.2}$$

Maximum frequency of operation will be governed by the effect of the parasitic capacitance C_i. For higher microwave and mmwave frequency range, devices such as MESFET, pHEMT and HBTs capable to work at higher frequency are preferable.

11.3 Phase Modulators

Most frequently used digital modulation techniques are the binary phase-shift keying (BPSK), different variances of quadrature phase-shift keying (QPSK, OQPSK, DQPSK), M-array phase-shift keying (M-PSK) and M-array quadrature amplitude modulations (M-QAM). Among all the modulation techniques, BPSK and QPSK are the most frequently used digital phase modulation techniques for satellite communication systems. Due to the advancement of high-efficiency low distortion power amplifiers technology, higher order modulation techniques such as 8APSK, 16APSK are also being used in satellite communication. Microwave transistors such as MESFET, pHEMT, HBT or p-i-n diodes are used for the realization of direct microwave phase modulators [3–7]. Transistors are preferable to realize phase modulators at higher microwave and mmwave frequency range for ease in realization using MMIC technology [8–14]; whereas, p-i-n diodes are used at a comparatively lower microwave frequency range for its simpler operation. Bi-phase modulator is the basic element to realize BPSK, QPSK and other higher order phase modulators.

11.3.1 Bi-Phase Modulators

Bi-phase modulator is the basic phase modulator used to realize higher order phase modulators. This is also directly used as binary phase-shift keying (BPSK) modulator. Bi-phase modulators are also used to realize vector modulator, which is used in communication systems. For example, in a linearizer for power amplifiers, a vector modulator is preferred to generate simultaneous amplitude and phase nonlinearities to compensate power amplifiers nonlinearities, instead of using a variable attenuator cascaded with a variable phase shifter. A vector modulator is also widely used as I–Q modulator.

For bi-phase modulator, a reflection-type variable resistance configuration is used to control the attenuation level. The circuit diagram of a bi-phase modulator using p-i-n diodes is the same as shown in Fig. 11.1. This is an analog reflection-type attenuator (RTA), where p-i-n diodes are used as variable RF resistance. For higher microwave frequency range, Schottky diodes are also used as a control device. For an ideal bi-phase modulator using reflection-type attenuator, RF impedance should be pure resistance and vary from zero to infinity to achieve all possible signal amplitudes with 0° and 180° phase shift. However, due to the p-i-n diodes parasitic series resistance (R_s) and parallel capacitance (C_i), bi-phase modulators performance deviate with the increase of frequency. The amplitude of the modulator is governed by (11.2), whereas the phase of the signal is given by (10.12b)

$$\angle S_{21} = -\tan^{-1}\left[\frac{4\pi f C_i R_i^2 Z_o}{(R_i + R_s + Z_o)(R_i + R_s - Z_o) - (2\pi f C_i R_i)^2\left(Z_o^2 - R_s^2\right)}\right] \tag{11.3}$$

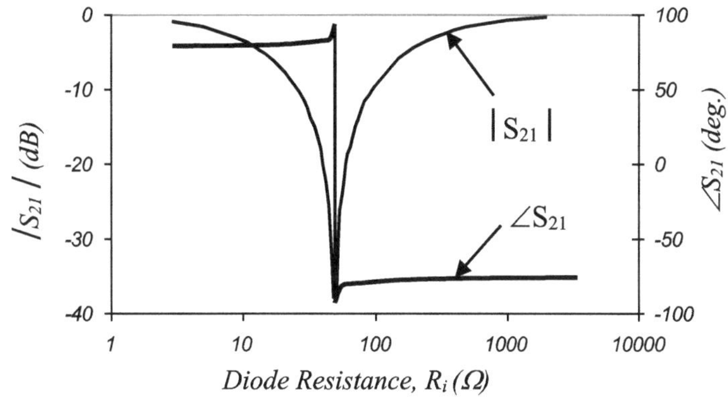

Fig. 11.2 Typical amplitude and phase characteristic of p-i-n-diode-based bi-phase modulator

Taking typical values of $R_s = 5\ \Omega$ and $C_i = 0.2$ pF for a p-i-n diode, the amplitude and phase characteristic of the bi-phase modulator is shown in Fig. 11.2 at 4 GHz. The plot shows that the insertion phase change deviates from its ideal value of 180° for the RF resistance around 50 Ω (just lower and higher than 50 Ω). Also, the insertion loss is finite, i.e. deviates from the ideal value of 0 dB. One way to minimize the phase deviation from its ideal value of 180° is the use of external inductance in parallel with the diode, which minimizes the effect of parasitic capacitance. However, it is possible only over a very narrow band of frequency. Another drawback of using p-i-n diode at its low RF resistance is the requirement of high forward current. This leads to the requirement of more DC power from the control signal.

Because of the near-zero DC control power requirement and better performance at higher microwave frequency, transistors are preferable than p-i-n diodes for the realization of bi-phase modulator. Another advantage of using transistors for the modulator is that it can be realized in a very small size using MMIC technology. Schematic circuit diagram of a bi-phase modulator using transistor is shown in Fig. 11.3. In this case also 3-dB 90° coupler is used to realize transistor-based RTA and the working principle is the same as the p-i-n-diode-based bi-phase modulator. The control signal is applied to the gate terminal of the transistors through high-value bias resistors. Thus, the requirement of current from the control signal is very less for the entire range of RF resistance. The cold FETs, i.e. FETs with no bias voltages are used as variable resistors [3, 4]. The FET's dynamic range of variable resistors is smaller and parasitic elements are larger compared to that of diodes. Therefore, practically amplitude and phase errors are introduced by the parasitic elements (feed inductance, junction capacitances, etc.) that deviate from the required performance of a modulator. To minimize the effect of parasitic components, the proper value of inductance in series with the gate or drain and a resistance in parallel with the transistor is used.

A simple bi-phase modulator allows passing the carrier signal to the modulated output port, known as carrier leakage. Thus, some of the transmitted power is shared

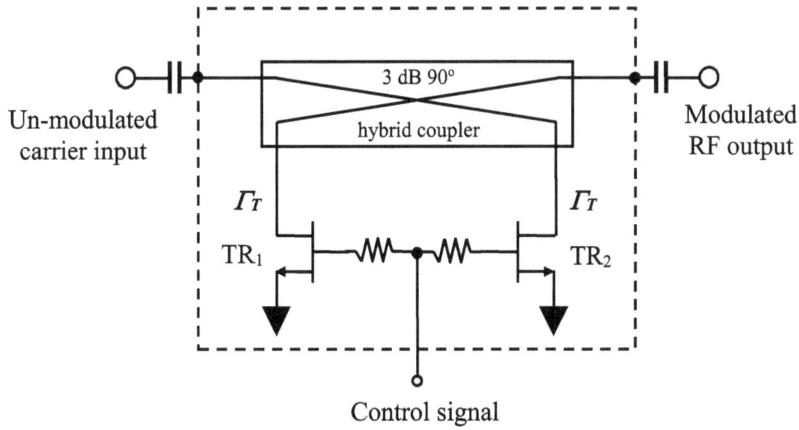

Fig. 11.3 Bi-phase modulator using transistors

by the un-modulated carrier which is undesirable. To alleviate the problem due to the parasitic of the devices and to suppress the un-modulated carrier transmission from the output of the modulator, balanced modulator configuration is used.

11.3.2 Bi-Phase-Balanced Modulators

The balanced structure is used for bi-phase modulators to overcome the problem of amplitude and phase deviation due to the parasitic effects of p-i-n diode and transistors. The balanced or push–pull configuration also eliminates the carrier leakage problem i.e. removes the carrier signal from the modulated output signal in addition to the elimination of the amplitude and phase errors caused by the parasitic of the used active devices. The balanced structure gives good constellation symmetry due to the use of the second attenuator operated in anti-phase.

Schematic circuit diagrams for the balanced bi-phase modulators using p-i-n diodes and transistors are shown in Figs. 11.4 and 11.5, respectively. Here, two branches are used with 180° phase difference using two 3-dB 90° couplers. Complementary baseband control signals (C and $\overline{C}$) are applied to the control inputs of the branches. Therefore, if transistors or p-i-n diodes of one arm provides more than 50 Ω RF resistance then the diodes/transistors of the other arm will provide less than 50 Ω RF resistance for achieving ideally equal amplitude but opposite phase signals due to the reflection from the devices.

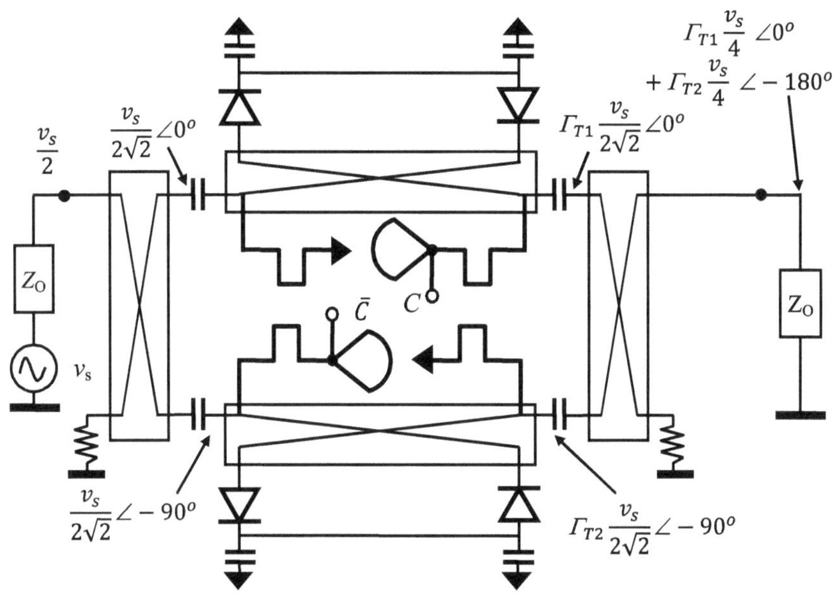

Fig. 11.4 Balanced bi-phase modulator using p-i-n diodes

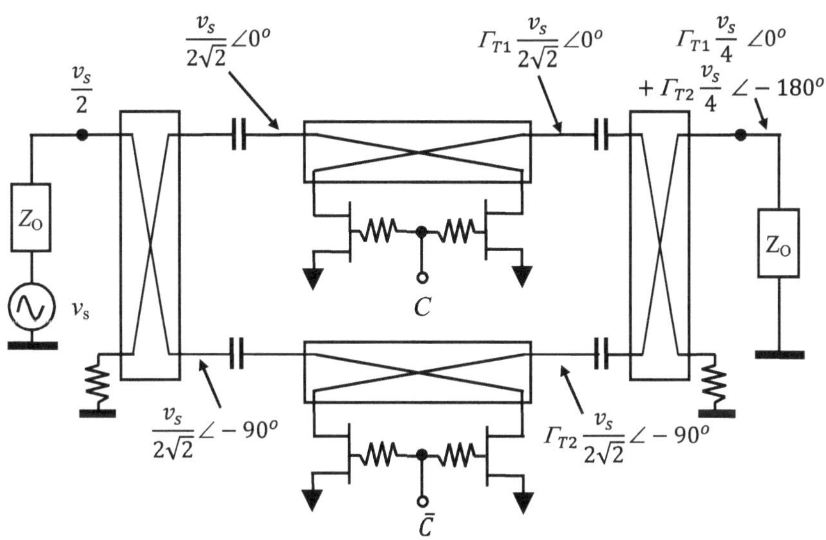

Fig. 11.5 Balanced bi-phase modulator using transistors

The transmission coefficient of the balanced bi-phase modulator is given by

$$S_{21} = \frac{\Gamma_{T1}\frac{v_s}{4} + \Gamma_{T2}\frac{v_s}{4}\angle -180^\circ}{\frac{v_s}{2}} \tag{11.4a}$$

$$\text{or,}\quad S_{21} = \frac{1}{2}(\Gamma_{T1} - \Gamma_{T2}) \tag{11.4b}$$

Here Γ_{T1} and Γ_{T2} are the reflection coefficients of the attenuators of direct and complementary arms, respectively. Suppose, Γ_C and $\Gamma_{\overline{C}}$ are the reflection coefficients of the attenuators at signal level C and complementary signal level $\overline{C}$, respectively. Therefore, if the direct path attenuator is having signal C and the complementary path is having $\overline{C}$ then the transmission coefficient is

$$S_{21} = \frac{1}{2}\left(\Gamma_C - \Gamma_{\overline{C}}\right) \tag{11.4c}$$

$$= \Gamma_{\text{bpm}}\ (\text{say}) \tag{11.4d}$$

Whereas, if $\overline{C}$ and C are applied to the direct path attenuator and complementary path, respectively, then the transmission coefficient of the modulator is given by

$$S_{21} = \frac{1}{2}\left(\Gamma_{\overline{C}} - \Gamma_C\right) = -\frac{1}{2}\left(\Gamma_C - \Gamma_{\overline{C}}\right) \tag{11.4e}$$

$$= -\Gamma_{\text{bpm}} \tag{11.4f}$$

From (11.4d) and (11.4f), it is clear that there is 180° phase difference in between the two states and this phase difference is independent of the amount of phase and amplitude deviations from its ideal values. The 180° phase difference is coming from the 90° phase difference from the input coupler and another 90° phase difference coming from the output coupler. The phase change of 180° is solely determined by the input and output couplers. Many broadband coupler technologies are available. Therefore, broadband bi-phase modulators can be realized using this configuration. Another advantage of the balanced bi-phase modulator is that this configuration provides carrier suppression at the output, which is not possible from the simple bi-phase modulator.

In comparison with the reflection-type attenuators of Figs. 11.1 and 11.3, the balanced bi-phase modulator needs three additional 3-dB 90° couplers, two additional active device (p-i-n diodes or transistors) and an additional complementary baseband control signal $\left(\overline{C}\right)$. A larger area for the circuit is required to realize the transistor-based configuration in MMIC due to increase in circuit complexity.

11.4 I–Q Vector Modulators

In communication systems, a vector modulator is an important element. In digital communication applications, to generate I–Q modulated signals a vector modulator is widely used. In case of microwave linearizer for power amplifier, a vector modulator is used to generate simultaneously required amplitude and phase distortion to compensate for power amplifier nonlinearity. Therefore, vector modulators can replace the use of a variable attenuator cascaded with a variable phase shifter for linearizer realization. Vector modulators are also used in phased-array antenna applications for an adaptive beam-forming network.

Traditionally, vector modulators are designed by two main configurations for microwave and mmwave applications. In the first approach, the RF power is divided into two equal amplitudes and 90° phase shift. Then the signal is combined with a 3-dB in-phase power combiner after passing through two bi-phase modulators. This configuration has inherent 3-dB insertion loss due to the addition of two signals with 90° phase difference.

In the second approach, a voltage/current-controlled attenuator cascaded with a 360° voltage/current-controlled variable phase shifter is used. In this configuration, there is no inherent insertion loss. However, this configuration needs a variable attenuator with a constant phase and a variable phase shifter with constant insertion loss, which is very difficult to design without complicated amplitude and phase control techniques. In terms of realization, the first configuration is more attractive especially in microwave and mmwave applications. The system is composed of a bi-phase variable attenuator, which is already discussed in the previous section.

Block diagram of IQ modulator circuit using two bi-phase (BPSK) balanced modulators is shown in Fig. 11.6. Here, the RF signal is divided into two equal amplitude outputs, with a 90° phase difference in between them using a 3-dB quadrature hybrid. The in-phase or 0° channel is designated as the I channel and the quadrature or 90° channel is designated as the Q channel. Each signal passes through a bi-phase modulator which controls the signal amplitude and selects 0° or 180° phase state for both I and Q paths. The phase state of the bi-phase modulators determines the phase quadrant of the output signal. The amplitude and phase of the resultant vector are determined by the magnitude of each of the signals which are combined at the output. Thus, by varying the amplitude of both the paths and selection of the either 0° or 180° phase state of the bi-phase modulator, the resultant can be placed anywhere within the bounded dotted square area shown in Fig. 11.7.

The transmission coefficient (S_{21}) of the I–Q vector modulator can be written as

$$S_{21} = \frac{\Gamma_{\text{bpm1}}\frac{V_s}{4} + \Gamma_{\text{bpm2}}\frac{V_s}{4}\angle -90^\circ}{\frac{V_s}{2}} \quad (11.5\text{a})$$

$$\text{or,}\ S_{21} = \frac{\Gamma_{\text{bpm1}}}{2} + \frac{\Gamma_{\text{bpm2}}}{2}\angle -90^\circ \quad (11.5\text{b})$$

Two bi-phase modulators are used for the vector modulator. In its simplest configuration, the bi-phase modulators can be designed using single 3-dB 90° coupler with p-i-n diode or transistor as shown in Figs. 11.1 and 11.3, respectively. Therefore, overall three numbers of 3-dB 90° couplers and one Wilkinson-type equal power combiner is required to implement the vector modulator. The configuration is exactly the same as the 360° phase shifter as discussed in the previous chapter. However, to implement the vector modulator using balanced bi-phase modulators, eight active devices, nine hybrid couplers, one Wilkinson power combiner, and four baseband signal channels are required. Thus, to realize the vector modulator using balanced

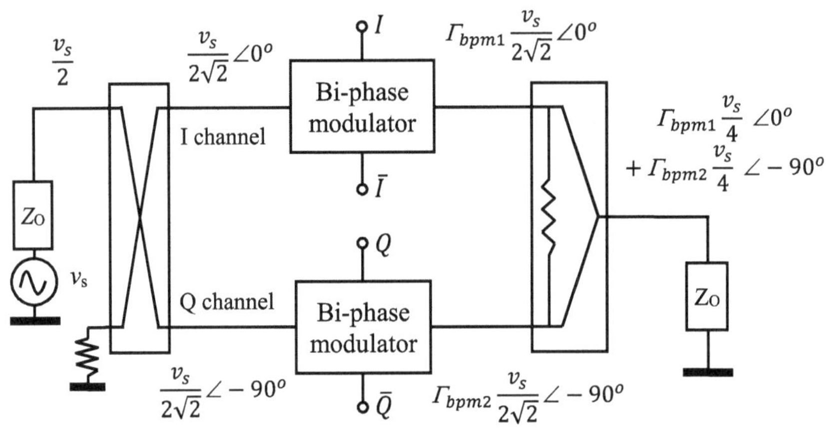

Fig. 11.6 I–Q modulator circuit using two bi-phase (BPSK) balanced modulator

Fig. 11.7 Vector signal diagram of I-Q modulator

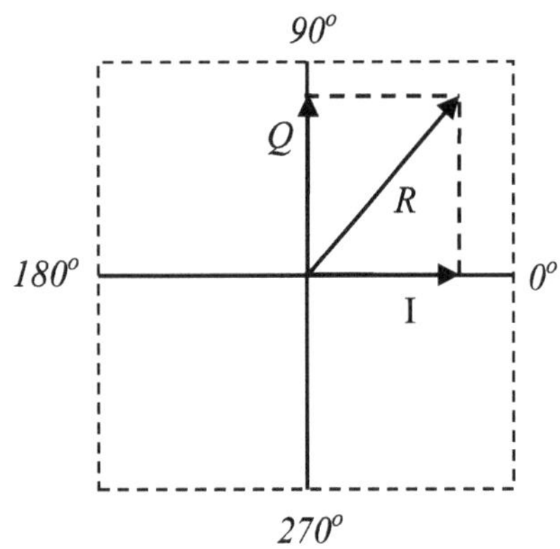

bi-phase modulator a large chip area is needed in monolithic microwave integrated circuit (MMIC) technology. The 3-dB quadrature hybrid is an important element to determine the operating frequency range of an I–Q vector modulator. It should have a very low quadrature phase error, i.e. phase difference between the two output deviations should be very small from its ideal value of 90°. To meet this requirement over a broad frequency band, Lange coupler is preferable. For the in-phase output combiner, a Wilkinson power combiner is used for its better phase and amplitude balance performance. In addition to the couplers performances, I and Q channel's transmission line length must also be phase matched.

Table 11.1 Signal levels for QPSK modulator

States	I	$\overline{I}$	Q	$\overline{Q}$
(0, 0)	Low	High	Low	High
(0, 1)	Low	High	High	Low
(1, 0)	High	Low	Low	High
(1, 1)	High	Low	High	Low

11.5 PSK Modulators

Balanced bi-phase modulator as shown in Figs. 11.4 and 11.5 using diodes and transistors, respectively, can be used as BPSK modulator. For BPSK modulation, there are same amplitudes with two phase states 0 and 180°. The two states are represented as state-0 (C is low and $\overline{C}$ is high) and state-1 (C is high and $\overline{C}$ is low).

Here digital modulating signals are applied to the control input C and complementary signal input $\overline{C}$. When C is in the high state and $\overline{C}$ is in the low state, the phase delay of the signal path will be 180° compared with the condition when C is low and $\overline{C}$ is high. It is required to generate complementary signal level from the modulating signal. Thus, the bi-phase balanced modulator acts as BPSK modulator when the modulating signal applied to its control input C and signal level complementary to the modulating signal applied to the input $\overline{C}$.

The I/Q vector modulator circuit as shown in Fig. 11.6 is the basic functional block for realization of multilevel digital modulation schemes, for example QPSK, M-APSK and M-QAM. There are four phase states with the same amplitude in the case of QPSK modulation. The states are represented as states of (0, 0), (0, 1), (1, 0) and (1, 1). The signal levels to the four control inputs are shown in Table 11.1.

In all the four states, the amplitude of the modulated signal remains the same only if the signal phase of the four states corresponds to four quadrants.

11.6 QAM Modulators

Higher order modulations such as multilevel QAM schemes are required in high data rate communication systems to utilize spectrum more effectively. QAM modulators are also realized using the I/Q vector modulator circuit as shown in Fig. 11.6 [2]. In this case, each bi-phase modulator operates in different amplitude settings each with 0 and 180° phase shift. For example, two amplitude settings are required for 16-QAM and eight amplitude settings are required for 256-QAM with each 0° and 180° phase offset. Therefore, driver circuits generate control voltages from digital bits to control phase as well as the amplitude of the bi-phase modulator. Any deviation in the amplitude of the bi-phase modulators, due to temperature variation or inherent imbalance of the modulator, leads to the imperfection of constellation. To solve the problem of temperature variation of amplitude in case of p-i-n diode bi-phase modulator, optimum bias load line technique can be used for the driver circuit realization.

References

1. Boveda A, Orilgoso F, Alonso JI (1993) A 0.7–3 GHz GaAs QPSK/QAM direct modulator. IEEE J Solid State Circuits 28:1340–1349
2. Ashitiani AE, Nam S, d'Espona A, Lucyszyn S, Robertson ID (1998) Direct multilevel carrier modulation using millimeter-wave balanced vector modulators. IEEE Trans Microw Theory Tech 46:2611–2619
3. Lodhi T, Edgar DL, McLelland H, Ferguson S, Elgaid K, Stanley CR, Thayne IG (2000) A 77 GHz coplanar waveguide MMIC BPSK vector modulator realized using InP technology. In: IEEE gallium arsenide integrated circuit symposium digest, pp 183–186
4. Chongcheawchamnan M, Bunnjaweht S, Kpogla D, Lee D, Robertson ID (2002) Microwave I-Q vector modulator using a simple technique for compensation of FET parasitics. IEEE Trans Microw Theory Tech 50:1642–1646
5. Chang H-Y, Wang H, Wang Y-C, Chao P-C, Chen C-H, Niu D-C (2002) A 45–90 GHz BPSK modulator using HBT technology. In: Proceedings of Asia–Pacific microwave conference, Kyoto, Japan, vol 1, pp 60–62
6. Chang H, Huang T, Wang H, Wang Y, Chao P, Chen C (2004) Broad-band HBT BPSK and IQ modulator MMICs and millimeter-wave vector signal characterization. IEEE Trans Microw Theory Tech 52(3):908–919
7. Alavi MS, Viswesaran A, Staszewski RB, de Vreede LCN, Long JR, Akhnoukh A (2011) A 2 GHz digital I/Q modulator in 65 nm CMOS. In: Proceedings of IEEE Asian solid-state circuits conference, pp 277–280
8. Devlin LM, Minnis BJ (1990) A versatile vector modulator design for MMIC. In: IEEE MTT-S international microwave symposium digest, pp 519–522
9. Norris GB, Boire DC, St. Onge G, Wutke C, Barratt C, Coughlin W, Chickanosky J (1990) A fully monolithic 4 to 18 GHz digital vector modulator. In: IEEE MTT-S international symposium digest, pp 789–792
10. Telliez I, Couturier A-M, Rumelhard C, Versnaeyen C, Champion P, Fayol D (1991) A compact monolithic microwave demodulator-modulator for 64-QAM digital radio links. IEEE Trans Microw Theory Tech 39:1947–1954
11. Gokdemir T, Nam S, Ashtiani AE, Robertson ID, Karacaoglu U (1998) Millimeter wave monolithic-balanced BPSK modulator using a miniaturized backward-wave coupler. In: IEEE MTT-S international symposium digest, pp 877–880
12. Desrosiers R, Cowles J, Hornbuckle C, Gutierrez-Aitken A, Becker J (1998) Monolithic 14 GHz widerband InP HBT BPSK modulator. In: Proceedings of gallium arsenide integrated circuit symposium, pp 135–138
13. Nam S, Shala N, Ang KS, Ashtiani AE, Gokdemir T, Robertson ID, Marsh SP (1999) Monolithic millimeter-wave balanced bi-phase amplitude modulator in GaAs/InGaP HBT technology. In: IEEE MTT-S international microwave symposium digest, vol 1, pp 243–2461
14. Ashtiani AE, Gokdemir T, Vilches A, Hu Z, Robertson ID, Marsh SP (2000) Monolithic GaAs/InGaP HBT balanced vector modulator for millimeter-wave wireless systems. In: IEEE radio frequency integrated circuits symposium digest, pp 187–190

Chapter 12
Amplitude Tilt Microwave Equalizers

Abstract Equalizers are used in cascade with other circuits and systems to improve the overall gain flatness over its operating frequency range. This chapter presents design and analysis of fixed and adjustable amplitude tilt microwave equalizers. Passive components, e.g. resistor-, capacitor- and inductor-based fixed slope equalizers and p-i-n diode-based adjustable slope equalizers are discussed. Versatile equalizers with adjustable positive, negative slope and adjustable insertion loss are also presented to compensate gain variation over frequency and temperature range of microwave systems.

12.1 Introduction

Amplitude tilt equalizer provides predetermined gain/loss ($|S_{21}|$) slope over the operating frequency range [1–8]. It is mainly used in cascade with other circuits or systems to improve the overall gain response over its operating frequency range. Variable gain slope microwave equalizers are used for compensation of gain roll-off of microwave and RF circuits and systems using the provision of slope adjustability. To meet the increasing demand of broadband applications such as internet and multimedia, microwave and RF broadband amplifiers are required. Due to the typically falling-off gain of transistors with the increase of frequency, modern microwave receivers uses equalizers to compensate for the excess passband negative gain slope. In this case, equalizer with positive gain slope is needed to cascade with the transistor-based amplifiers for achieving broadband gain performance. Wideband travelling wave tube amplifiers (TWTAs) have parabolic gain response with maximum gain at the centre frequency and fall-off gradually at the band edges, sometimes by 11 dB or more. In this case, the solution is to use parabolic gain equalizer with maximum loss at the centre frequency to get an overall flat response over the required frequency band. Broadband pre-distortion linearizers require equalizers with very high positive and also very high negative gain slopes to achieve required amplitude and phase distortion over a wide operating frequency band.

S. C. Bera, *Microwave Active Devices and Circuits for Communication*, Lecture Notes in Electrical Engineering 533, https://doi.org/10.1007/978-981-13-3004-9_12

12.2 Different Configurations

There are several configurations of gain slope equalizer circuits which are based on lumped resistors, capacitors and/or inductors connected as series and shunt elements as shown in Fig. 12.1a [1, 2]. In case of multiple shunt networks configuration as shown in Fig. 12.1b, the shunt networks are connected quarter-wavelength apart to achieve better input and output port return losses. Both the equalizers provide positive gain slopes with the increase of frequency. However, these equalizers are not much suitable at microwave frequency range due to parasitic effects of the used lumped elements. Substrate integrated waveguide (SIW) and photonic bandgap (PBG)-based positive gain slope equalizers are also used. However, these are having disadvantages of packaging due to their definite conductor pattern on the ground plane.

Most suitable microwave gain slope equalizers are based on combination of lumped and distributed components. In these cases, multiple shunt elements containing one or more fixed resistors and an open- or short-circuited transmission lines are used to realize the equalizer as shown in Fig. 12.2. Here, shunt elements are connected with separation of quarter-wavelength transmission line for better ports return losses.

Example 12.1 Compute S-parameter of a shunt network consists of a resistance R_a shunted by a transmission line (Z_o) of length l as shown in Fig. 12.3. Also plot transmission coefficient and reflection coefficient in dB of the network over the

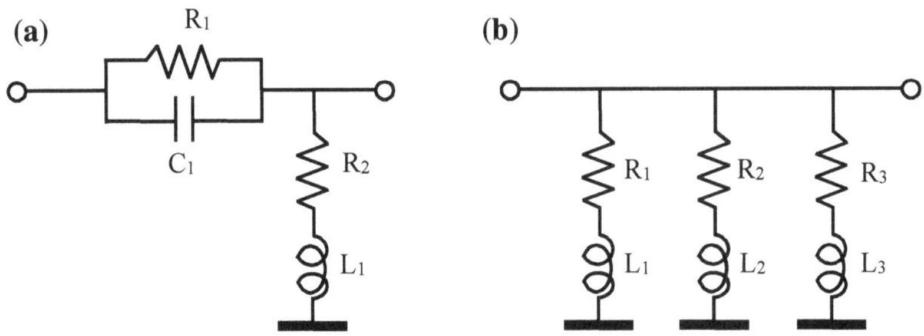

Fig. 12.1 Configurations of amplitude tilt equalizers using lumped components

Fig. 12.2 Configurations of amplitude tilt equalizers using transmission line elements

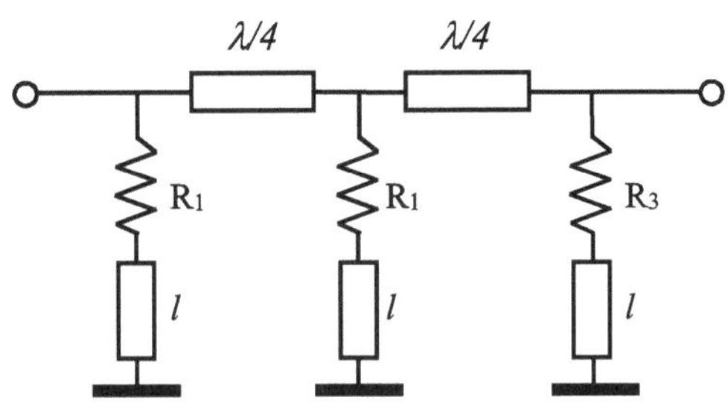

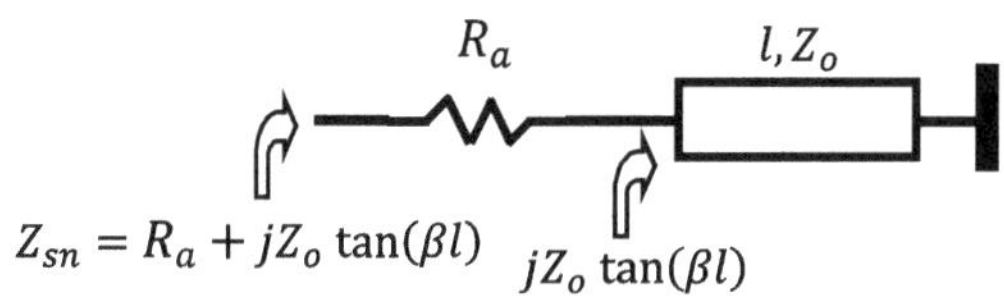

Fig. 12.3 The shunt network consists of a resistance and a transmission line of Example 12.1

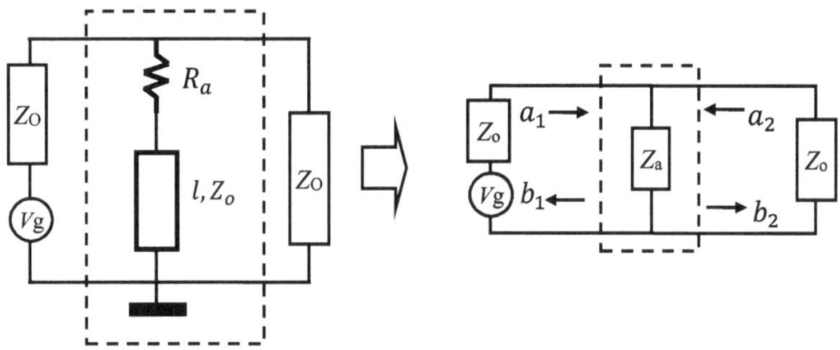

Fig. 12.4 The two port network consists of the shunt network of Example 12.1

frequency range of 1–20 GHz for $R_a = 30\ \Omega$ and $l = \lambda_o/8$, where λ_o is free space wavelength at 10 GHz.

Solution

The shunt network consists of a transmission line of length l and a resistor R_a as shown in Fig. 12.3. Using (7.79), putting $Z_L = 0$ the input impedance Z_a of the shunt network can be written as

$$Z_{sn} = R_a + jZ_o \tan\left(\frac{2\pi}{\lambda} l\right)$$

Here, Z_o is the characteristic impedance of the shunt transmission line. The S-parameters of the shunt network can be determined as follows from (7.134) using Fig. 12.4:

$$\begin{aligned} S_{11} &= \frac{b_1}{a_1}, \text{ when } a_2 = 0 \\ &= \frac{V_1^-}{V_1^+}, \text{ when } V_2^+ = 0 \\ &= -Z_o/(Z_o + 2Z_{sn}) \end{aligned}$$

It is a symmetric network, therefore:

$$S_{11} = S_{22} = -Z_o/(Z_o + 2Z_{sn})$$

Similarly from (7.79) and using (7.55) and (7.136), S_{21} can be determined as

$$S_{21} = \frac{b_2}{a_1} = \frac{V_2^-}{\sqrt{Z_o}} \frac{\sqrt{Z_o}}{V_1^+} = \frac{V_2^-}{V_1^+}, \text{ when } V_2^+ = 0$$
$$= 2Z_{sn}/(Z_o + 2Z_{sn}) = S_{12}$$

Thus, in matrix form, S-parameter of the shunt network can be written as

$$\begin{bmatrix} S_{11} & S_{12} \\ S_{21} & S_{22} \end{bmatrix} = \begin{bmatrix} \frac{-Z_o}{Z_o+2Z_{sn}} & \frac{2Z_{sn}}{Z_o+2Z_{sn}} \\ \frac{2Z_{sn}}{Z_o+2Z_{sn}} & \frac{-Z_o}{Z_o+2Z_{sn}} \end{bmatrix}, \text{ where } Z_{sn} = R_a + jZ_o \tan\left(\frac{2\pi}{\lambda} l\right)$$

The transmission coefficient of the shunt network is given by

$$S_{21} = \frac{2Z_{sn}}{Z_o + 2Z_{sn}} = \frac{2R_a + j2Z_o \tan\left(\frac{2\pi}{\lambda} l\right)}{Z_o + 2R_a + j2Z_o \tan\left(\frac{2\pi}{\lambda} l\right)}$$

Transmission coefficient in dB can be written as

$$20 \times \log(|S_{21}|) = 10 \times \log\left(\frac{4R_a^2 + 4Z_o^2 \tan^2\left(\frac{2\pi}{\lambda} l\right)}{(Z_o + 2R_a)^2 + 4Z_o^2 \tan^2\left(\frac{2\pi}{\lambda} l\right)}\right)$$

Putting $R_a = 30\ \Omega$, $Z_o = 50\ \Omega$, $l = \lambda_o/8$, the transmission coefficient can be written as

$$\text{Transmission coefficient} = 10 \times \log\left(\frac{3600 + 10{,}000 \tan^2\left(\frac{\pi}{40} f_{\text{GHz}}\right)}{12{,}100 + 10{,}000 \tan^2\left(\frac{\pi}{40} f_{\text{GHz}}\right)}\right) \text{dB}$$
$$= 10 \times \log\left(\frac{36 + 100 \tan^2\left(\frac{\pi}{40} f_{\text{GHz}}\right)}{121 + 100 \tan^2\left(\frac{\pi}{40} f_{\text{GHz}}\right)}\right) \text{dB}$$

Similarly, the reflection coefficient in dB can be written as

$$\text{Reflection coefficient} = 10 \times \log\left(\frac{2500}{12{,}100 + 10{,}000 \tan^2\left(\frac{\pi}{40} f_{\text{GHz}}\right)}\right) \text{dB}$$
$$= 10 \times \log\left(\frac{25}{121 + 100 \tan^2\left(\frac{\pi}{40} f_{\text{GHz}}\right)}\right) \text{dB}$$

The variation of transmission and reflection coefficients over the frequency range of 1–20 GHz is plotted in Fig. 12.5 using the above expressions. This network provides decrease of insertion loss with the increase of frequency.

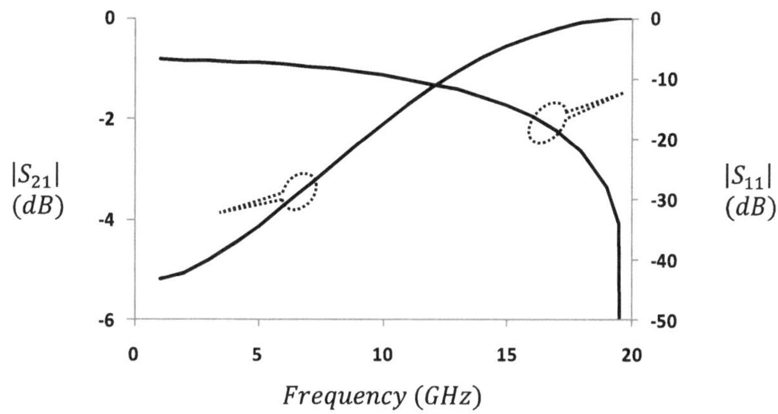

Fig. 12.5 Transmission and reflection coefficients of Example 12.1

12.3 Equalizer with Adjustable Amplitude Slope

Frequency-dependent microwave impedances are used in shunt configuration to a main transmission line to generate gain/loss slope over operating frequency range. Practically, most useful microwave equalizers are configured for generating various types of gain slopes by adjusting one or more circuit elements to provide monotonous negative and positive, parabolic and multi-peak gain slopes over the operating frequency range [3–6]. For active equalizers, p-i-n diodes are used to vary the impedances of the networks to achieve various types of gain slopes over the operating frequency range. A short-circuited shunt network consists of a transmission line in between two resistors R_a and R_b are used as shown in Fig. 12.6. The length of the transmission line is integer numbers (n) of one-eighth wavelength corresponding to the centre frequency (f_o) of operation. In case of adjustable gain slope requirements, the resistors R_a and/or R_b is synthesized using p-i-n diodes. The input impedance Z_{sn} of the shunt network of Fig. 12.6 with transmission line length of $n\lambda_o/8$ is given by (7.79):

$$Z_{sn} = R_a + Z_o \frac{R_b + jZ_o \tan\left(\frac{n\pi\lambda_o}{4\lambda}\right)}{Z_o + jR_b \tan\left(\frac{n\pi\lambda_o}{4\lambda}\right)} \tag{12.1}$$

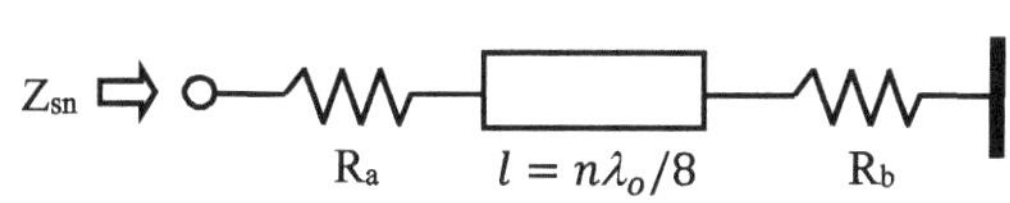

Fig. 12.6 A short-circuited network consists of a transmission line of length $n\lambda_o/8$ in between two resistors

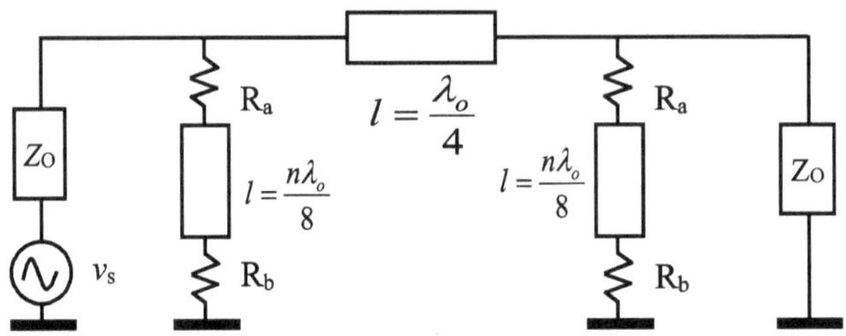

Fig. 12.7 Equalizer with two shunt networks with separation of $\lambda_o/4$

Here, Z_o is the shunt transmission line's characteristic impedance, n is an integer (1, 2, 3, ...), λ_o and λ are the guide wavelengths that correspond to the operating centre frequency f_o and any other operating frequency f, respectively.

For $R_b = Z_o$, the value of the input impedance Z_{sn} is $R_a + Z_o$. This implies that the input impedance Z_{sn} is independent on frequency. However, for $R_b \neq Z_o$, depending upon the value of R_b, the shunt network's input impedance may increase or decrease with the frequency.

In practice, multiple numbers of shunt networks are used in a separation of $\lambda_o/4$ length transmission line for realization of equalizer to achieve more gain slope and better port return losses over the interested frequency band. The schematic circuit diagram of an equalizer with two identical shunt networks is shown in Fig. 12.7. The ABCD parameter of the equalizer of Fig. 12.7 is given by the matrix multiplication of the three elements: the first shunt network, the series transmission line of length $\lambda_o/4$ and the second shunt network. Therefore, the ABCD parameter of the equalizer with two shunt elements separated by the $\lambda_o/4$ transmission line is given by

$$\begin{bmatrix} A & C \\ B & D \end{bmatrix} = \begin{bmatrix} 1 & 0 \\ \frac{1}{Z_{sn}} & 1 \end{bmatrix} \begin{bmatrix} \cos\left(\frac{\pi\lambda_o}{2\lambda}\right) & jZ_o \sin\left(\frac{\pi\lambda_o}{2\lambda}\right) \\ j\frac{1}{Z_o}\sin\left(\frac{\pi\lambda_o}{2\lambda}\right) & \cos\left(\frac{\pi\lambda_o}{2\lambda}\right) \end{bmatrix} \begin{bmatrix} 1 & 0 \\ \frac{1}{Z_{sn}} & 1 \end{bmatrix} \tag{12.2}$$

Thus, transforming ABCD parameter matrix (12.2), the scattering parameters of the equalizer can be written as

$$S_{11} = S_{22} = \frac{-\frac{2Z_o}{Z_{sn}}\cos\left(\frac{\pi\lambda_o}{2\lambda}\right) - j\frac{Z_o^2}{Z_{sn}^2}\sin\left(\frac{\pi\lambda_o}{2\lambda}\right)}{\left(2 + \frac{2Z_o}{Z_{sn}}\right)\cos\left(\frac{\pi\lambda_o}{2\lambda}\right) + j\left(2 + \frac{2Z_o}{Z_{sn}} + \frac{Z_o^2}{Z_{sn}^2}\right)\sin\left(\frac{\pi\lambda_o}{2\lambda}\right)} \tag{12.3}$$

$$S_{21} = S_{12} = \frac{2}{\left(2 + \frac{2Z_o}{Z_{sn}}\right)\cos\left(\frac{\pi\lambda_o}{2\lambda}\right) + j\left(2 + \frac{2Z_o}{Z_{sn}} + \frac{R_o^2}{Z_{sn}^2}\right)\sin\left(\frac{\pi\lambda_o}{2\lambda}\right)} \tag{12.4}$$

Figure 12.8 shows S_{21} response of the equalizer for $n = 1$ over the frequency range of 5–15 GHz for a fixed value of the resistor $R_a = 50\ \Omega$ and different values of the resistance $R_b = 15$, 50 and 150 Ω. For $n = 1$, the length of the shunt transmission line is $\lambda_o/8$ and the achieved gain slope monotonously increases ($R_b < 50\ \Omega$),

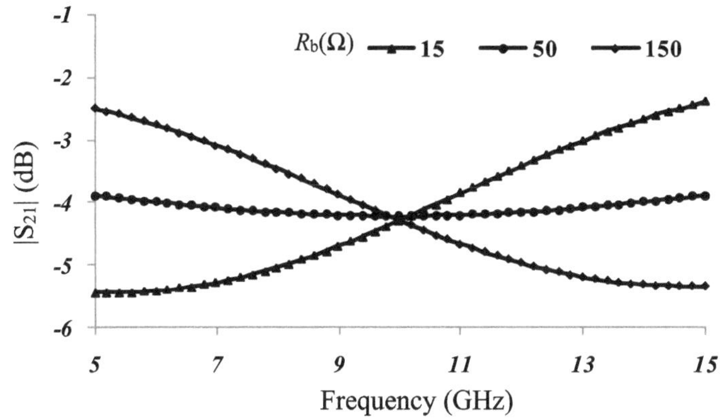

Fig. 12.8 Frequency response of the equalizer for $n = 1$ and $R_a = 50\ \Omega$. Positive gain slope for $R_b = 15$, zero gain slope for $R_b = 50\ \Omega$ and negative gain slope for $R_b = 150\ \Omega$

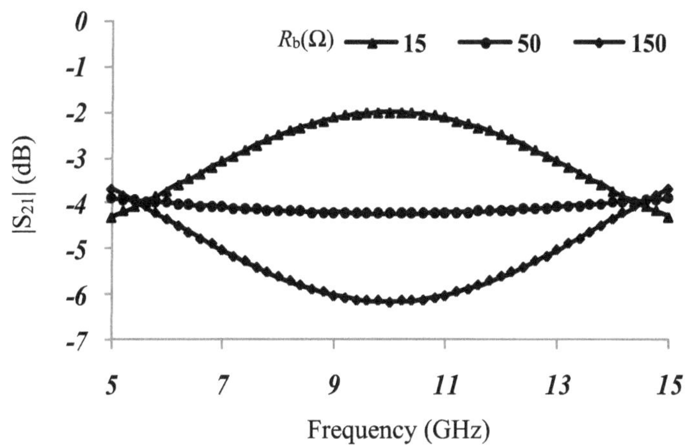

Fig. 12.9 Frequency response ($|S_{21}|$) of the equalizer for $n = 2$, and $R_a = 50\ \Omega$. Maximum at centre frequency for $R_b = 15\ \Omega$, zero gain slope for $R_b = 50\ \Omega$ and minimum at centre frequency for $R_b = 150\ \Omega$

decreases ($R_b > 50\ \Omega$) or remains constant ($R_b = 50\ \Omega$) over the frequency range. In case of solid-state systems such as receivers, driver amplifiers and power amplifiers, gain decreases with the increase of frequency due to inherent characteristic of the solid-state devices. To compensate the gain slope of these systems, the equalizer particularly with positive gain slope ($R_b < 50\ \Omega$) is used.

Frequency response (S_{21}) for $n = 2$, i.e. for the length of the shunt transmission line of $\lambda_o/4$ is shown in Fig. 12.9. It shows that the equalizer provides parabolic gain slope over frequency range of 5–15 GHz. To compensate parabolic gain variation, this type of equalizers can be used. For example, gains of broadband TWTAs are higher at its centre frequency and decreases with the increase and decrease of frequency.

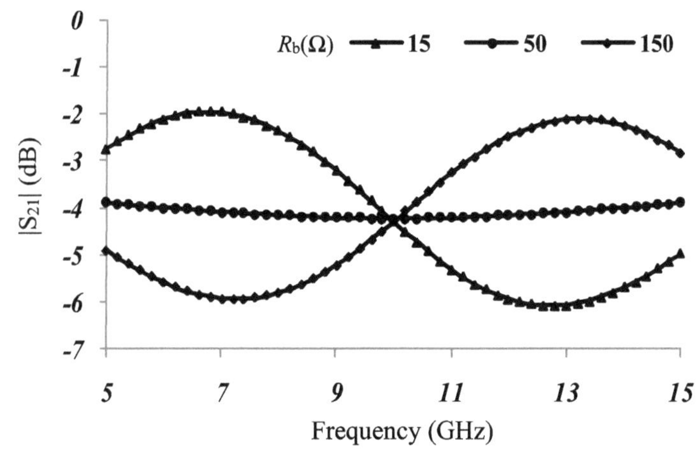

Fig. 12.10 Frequency response of the equalizer for $n = 3$, $R_a = 50\ \Omega$. Maximum is near lower frequency for $R_b = 15\ \Omega$, zero gain slope for $R_b = 50\ \Omega$ and minimum is near higher frequency for $R_b = 150\ \Omega$

Here, the equalizer with shunt transmission line of length $\lambda_o/4$ and with $R_b > 50\ \Omega$ is suitable for the gain slope compensation.

Gain response with multiple peaks is required to equalize systems where several circuits with different gain versus frequency responses are cascaded to achieve system performance. Figures 12.10 and 12.11 show the other two types of gain slope characteristics which correspond to $n = 3$ and 4, i.e. for the length of shunt transmission line of $3\lambda_o/8$ and $\lambda_o/2$, respectively. Thus, with the increase of n, i.e. length of the transmission line of the shunt network, more and more number of peaks in the gain response is achievable. It also shows that with the increase of length, same gain slope is achieved over less frequency range that is required to equalize gain slope of relatively narrow band systems.

To optimize gain flatness of various circuits and systems having different gain slopes, slope adjustability of a gain slope equalizer is very much required. In practice, at the design stage, it may not be known the requirement of amount as well as whether positive or negative gain slope is required. Thus, equalizer with provision of adjustable positive and negative gain slopes is preferable to adjust the gain slopes during production phase. In these cases, resistor R_b is synthesized using p-i-n diodes. Whereas, the cases where either adjustable positive or adjustable negative gain slopes are required, p-i-n diodes are used to realize variable resistors R_a.

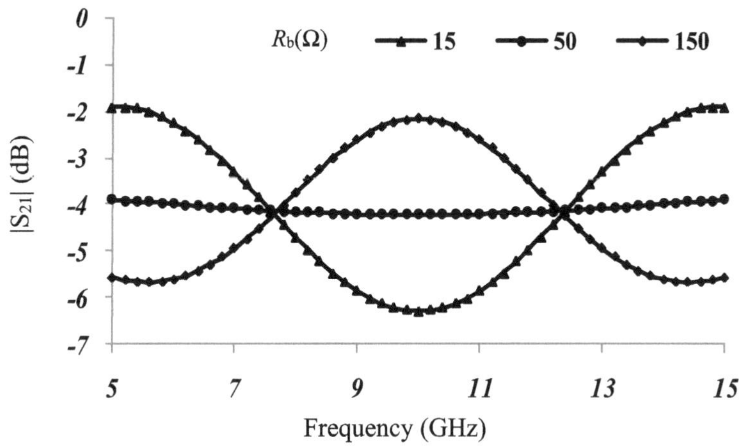

Fig. 12.11 Frequency response of the equalizer for $n = 4$, $R_a = 50\ \Omega$ and $R_b = 15$, 50 and 150 Ω

12.4 Equalizer with Adjustable Parabolic Gain Slope for Broadband MPM

Microwave power module (MPM) uses short length TWT to achieve higher output power with lesser gain. Remaining gain is realized by cascading solid-state amplifiers (SSAs) at its input to make the overall system with smaller size and lesser mass. Broadband TWTs suffer from parabolic gain slope, sometimes nearly 11 dB with maximum gain around centre frequency. To achieve flat gain response over the operating frequency range, gain slope equalizer is incorporated in the solid-state amplifier cascaded with the TWT. An equalizer circuit having parabolic gain slope with maximum loss at the centre frequency is required to serve this purpose. Circuit diagram of such an equalizer is shown in Fig. 12.12. Here, the resistor R_b is fixed but the resistor R_a is variable. The variable resistor R_a is synthesized using p-i-n diode to achieve electronically adjustable gain slope. RF equivalent circuit of the p-i-n diode is parallel combination of RF resistor r_d and parasitic capacitor c_d as shown in Fig. 12.12. Figure 12.13 shows frequency response of the equalizer for $n = 4$, i.e. length of the shunt transmission line of $\lambda_o/2$. The traces are for different values of p-i-n diode resistance which corresponds to different bias currents of the diodes, taking $C_d = 0.017$ pF. The plots show that the parabolic gain slope up to 10 dB with minimum loss of about 2 dB at the edges is achievable with the change of diode current by maintaining return losses better than 10 dB for all the settings. To achieve more gain slope and/or better return losses, more number of identical shunt networks with the separation of $\lambda_o/4$ transmission line length may be used.

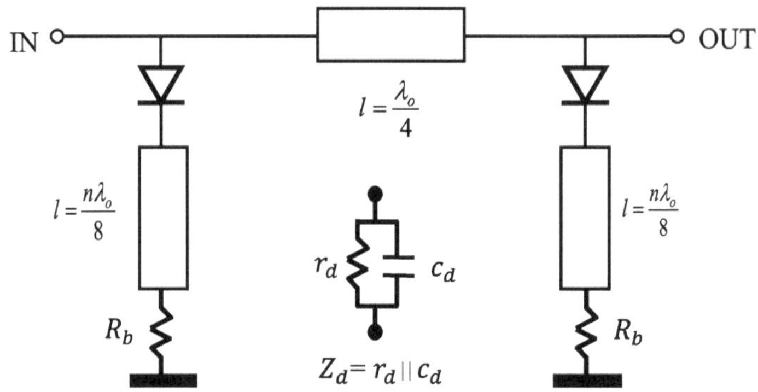

Fig. 12.12 Equalizer circuit with p-i-n diodes as variable resistor to achieve adjustable either positive or negative gain slope over frequency

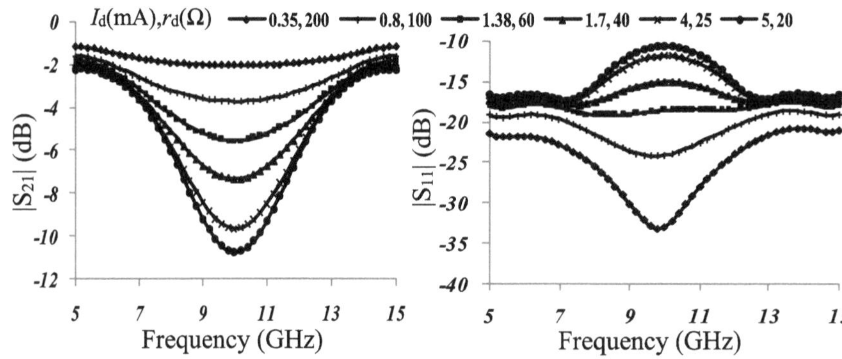

Fig. 12.13 Equalizer circuit with p-i-n diodes as variable resistor to achieve adjustable parabolic gain slope over frequency for $R_a = 15\ \Omega$

12.5 Equalizer with Adjustable Positive Gain Slopes for Solid-State Circuits

Broadband solid-state RF and microwave systems such as receivers, power amplifiers, etc. suffer from negative gain slope over operating frequency range due to the inherent behaviour of the used solid-state devices. For satellite communication, the allocated bandwidth for each frequency band is not very high. For example, the full Ku-band downlink frequency band is from 10.7 to 12.75 GHz, i.e. about 2 GHz bandwidth. However, the gain slope requirement may be very large for the Ku-band subsystems which use many circuits in cascade configuration.

Circuit diagram of an equalizer operating over the frequency range of 10.7–12.8 GHz, to improve the gain flatness of solid-state subsystem, will be the same as shown in Fig. 12.12 with different values of the circuit components. The design goal of the equalizer is to provide positive gain slope over this frequency range

Fig. 12.14 Photograph of a Ku-band amplitude tilt equalizer

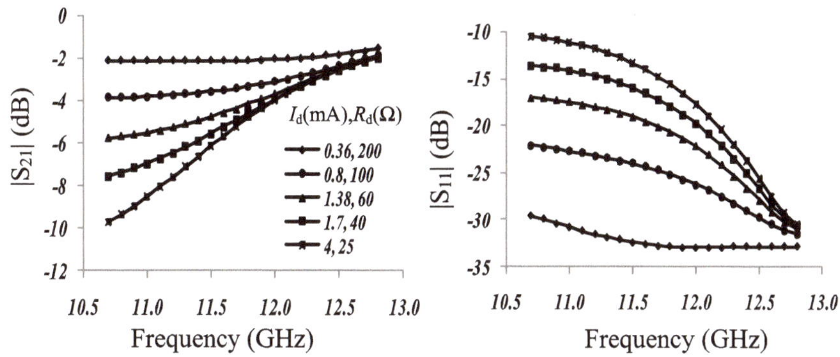

Fig. 12.15 Frequency response of the equalizer for $n = 7$ and $R_b = 250\ \Omega$

with minimum insertion loss at the highest frequency of operation, i.e. at 12.8 GHz, where the gain of the subsystem is minimum. To meet this requirement, the length of the shunt transmission line of the equalizer is selected as $7\lambda_o/8$ ($n = 7$), λ_o is corresponding to the centre frequency of 11.75 GHz. Two identical shunt elements are used with the separation of $\lambda_o/4$ to achieve better input and output port matching. The equalizer is designed in microstrip configuration on a 10 mil-thick alumina substrate ($\varepsilon_r = 9.9$). Photograph of the equalizer is shown in Fig. 12.14. Thin film chip resistors are used for R_b to perform at Ku-band frequency. Bended shunt transmission lines are used to make the circuit on a small size substrate of 0.5 mm × 0.5 mm. Beam lead p-i-n diodes are used for their negligible parasitic inductance and very small parasitic capacitance (only 0.017 pF). High-impedance $\lambda_o/4$ line with radial stub is used to provide bias of the p-i-n diodes.

Frequency response of the equalizer is shown in Fig. 12.15. Equalizer provides variable gain slope response with the variation of circuit current I_d corresponding to diode resistance 25, 40, 60, 100 and 200 Ω for the fixed resistance values of R_b = 250 Ω. Return losses of the circuit are better than 10 dB over the entire bandwidth for all the gain slope settings.

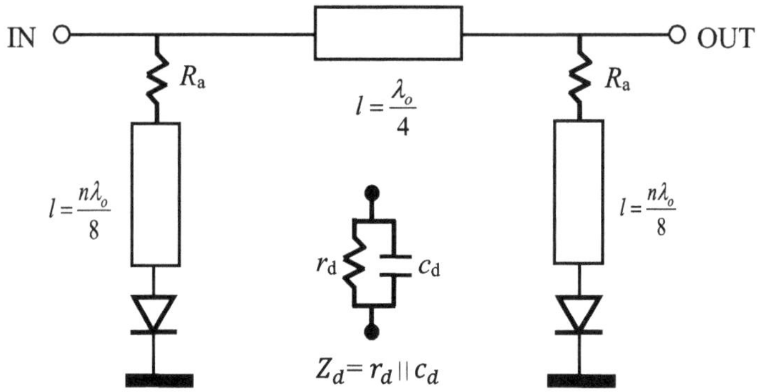

Fig. 12.16 Equalizer circuit with p-i-n diodes as variable resistor

12.6 Equalizer with Adjustable Positive as Well as Negative Gain Slopes

In case of complex microwave systems and subsystems, it is required to incorporate an equalizer which provides adjustable positive as well as negative gain slope over the frequency of operation. Circuit diagram of such an equalizer is shown in Fig. 12.16.

Here, the resistors R_a are fixed but the resistors R_b are variable. The variable resistors R_b are synthesized using p-i-n diodes to achieve electronically adjustable gain slope. Figure 12.17 shows frequency response of the equalizer for $n = 7$, i.e. length of the shunt transmission line of $7\lambda_o/8$ and $R_a = 40\ \Omega$. The traces are for different values of p-i-n diode resistance corresponding to different bias current of the diodes, taking $C_d = 0.017$ pF. The traces show that the gain slopes of the equalizer are positive as well as negative depending on the diode bias current. For all the gain slope settings, the port return losses are better than 12 dB. In this case also, to increase the maximum gain slope with better port impedance matching, i.e. return loss, more numbers of shunt networks are to be used. Another way to increase the maximum gain slope is the use of low value of resistors R_a. In this case, penalty will be the poor port return losses.

12.7 Versatile Equalizer with Variable Gain Slope and Insertion Loss

The gain of microwave circuits and systems are function of operating frequency as well as its operating temperature. To compensate the gain variation over the frequency as well as temperature, a versatile variable gain slope equalizer with the provision of adjustable gain (i.e. insertion loss in this case) is required [7, 8]. To achieve adjustable positive and negative gain slopes with the provision of adjustable insertion loss, both

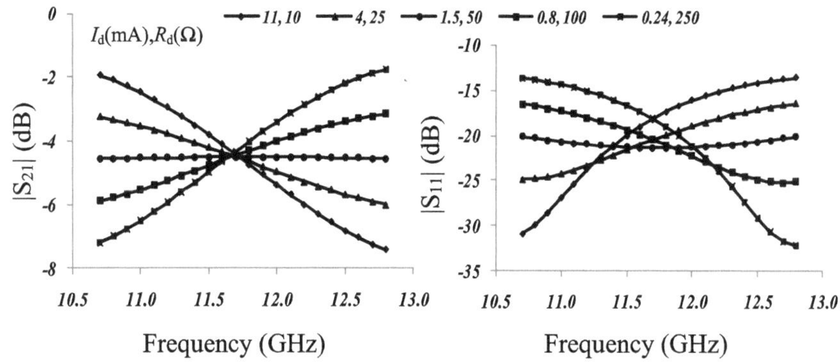

Fig. 12.17 Frequency response of the equalizer circuit of Fig. 10.13 for $n = 7$ and $R_a = 40\ \Omega$

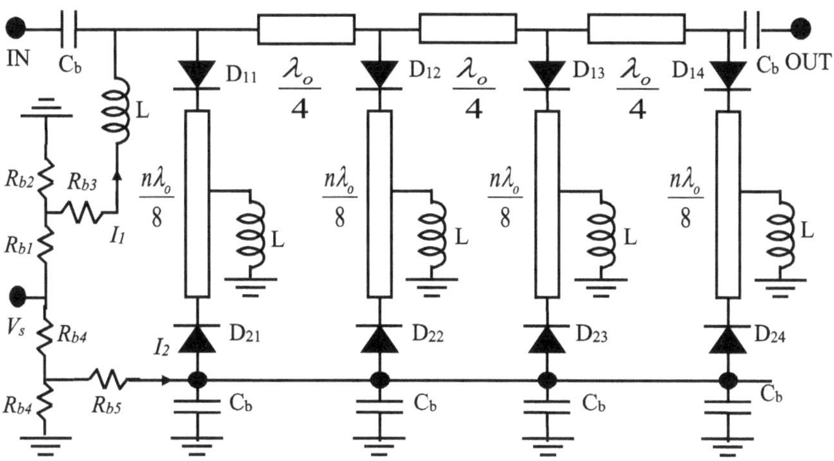

Fig. 12.18 Equalizer circuit with p-i-n diodes as variable resistor to achieve adjustable gain slope over frequency

the resistors R_a and R_b should be independently adjustable. An equalizer circuit diagram with resistors in each shunt network synthesized by p-i-n diodes is shown in Fig. 12.18. Here, four shunt networks are used with $\lambda_o/4$ separation to achieve more amounts of gain slopes with better ports return losses. Biases of the diodes are fed from a single DC voltage source V_s with resistive networks. Inductors (L) are used with all the shunt transmission lines to provide DC bias isolation in between the diodes (D_{11}–D_{14}) and diodes (D_{21}–D_{24}).

Frequency response ($|S_{21}|$) of the equalizer circuit, as shown in Fig. 12.18, is shown in Fig. 12.19 over the frequency range of 3–5 GHz, with $n = 3$ and $R_o = 50\ \Omega$. The different curves are correspond to different values of diode currents. Here, resistance R_a is for the diodes D_{11}, D_{12}, D_{13} and D_{14}. Whereas, resistance R_b is for the diodes D_{21}, D_{22}, D_{23} and D_{24}. For $R_b > 50\ \Omega$, $|S_{21}|$ increases with frequency. But

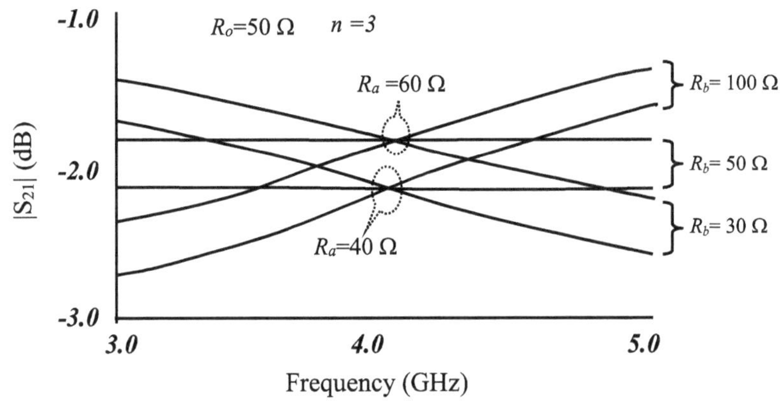

Fig. 12.19 Simulated $|S_{21}|$ as a function of frequency of the equalizer of Fig. 12.7 for different values of R_a and R_b

for $R_b < 50\ \Omega$, $|S_{21}|$ decreases. It can also be shown that for $n = 1$, the characteristic will be reversed. For a fixed value of R_b, with the increase of R_a, attenuation $|S_{21}|$ decreases without much affecting the slope of $|S_{21}|$ over the frequency range as shown in Fig. 12.19. Thus, by changing the value of resistor R_b of the equalizer, attenuation slope over the frequency range can be made positive, negative or zero. Whereas, attenuation at the centre frequency can be determined by changing the value of the resistor R_a.

To achieve better port matching, i.e. for better input and output return losses of the equalizer, more numbers of shunt networks can be used with the separation of $\lambda_o/4$ (λ_o corresponds to f_o). For example, the circuit shown in Fig. 12.18 consists of four shunt networks with separation of $\lambda_o/4$ corresponding to 4 GHz.

In Fig. 12.20, photograph of the equalizer circuit operating over the frequency range of 3–5 GHz is shown. 25 mil thick alumina substrate is used to realize the circuit using microstrip transmission line configuration. High-impedance $\lambda_o/4$ transmission lines are used to realize required inductors. Meander transmission lines are used for series and shunt transmission lines to realize the circuit on a 0.5 mm × 0.5 mm size of the substrate. Due to the negligible parasitic capacitance and inductance, beam lead p-i-n diodes are used to operate over the frequency range of 3–5 GHz.

Example 12.2 Compute S-parameter of a shunt network consists of a resistance R_a and open-circuit transmission line (Z_o) stub of length l as shown in Fig. 12.21. Also plot transmission coefficient and reflection coefficient of the network in dB over the frequency range of 1–20 GHz for $R_a = 30\ \Omega$ and $l = \lambda_o/8$, where λ_o is free space wavelength at 10 GHz.

Solution

The shunt network consists of a transmission line of length l and a resistor R_a as shown in Fig. 12.21. Using (7.80), the input impedance Z_{sn} of the shunt network can be written as

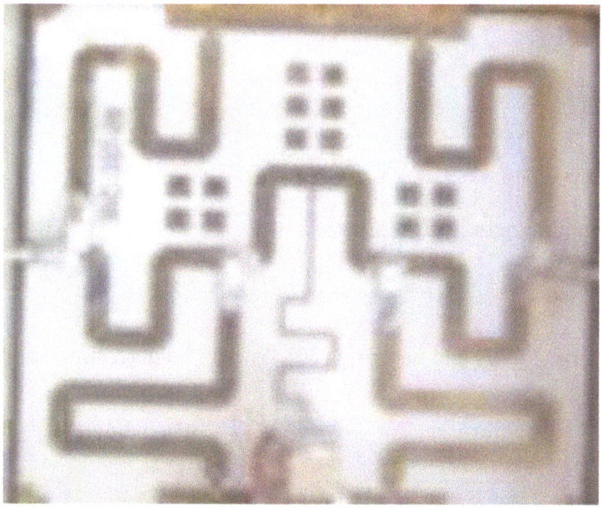

Fig. 12.20 Photograph of C-band equalizer with four shunt networks

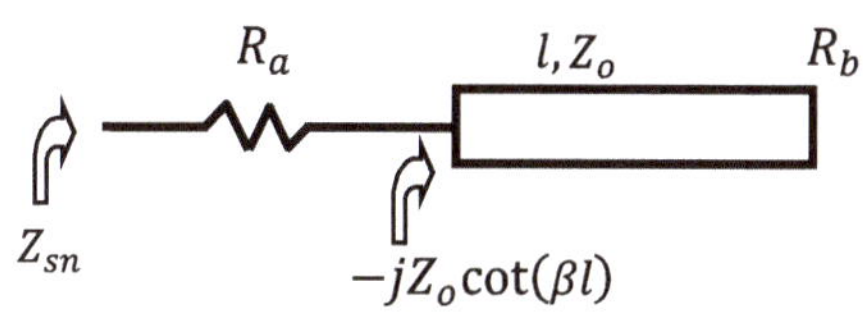

Fig. 12.21 The shunt network consists of a resistor and an open-circuit transmission line of Example 12.2

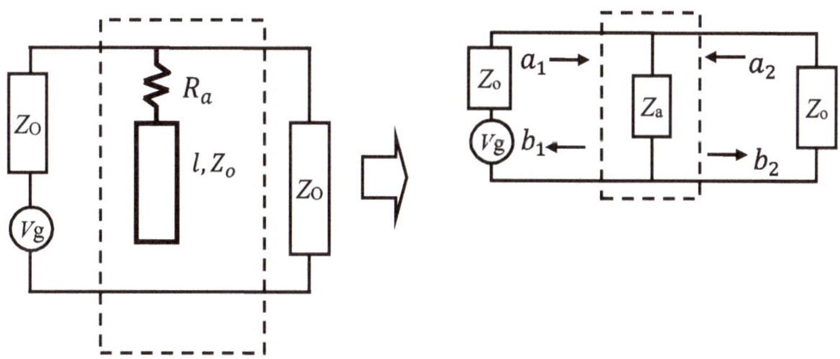

Fig. 12.22 The two-port network consists of the shunt network of Example 12.2

$$Z_{sn} = R_a - jZ_o\cot(\beta l)$$

Here, Z_o is the characteristic impedance of the shunt transmission line. The S-parameters of the shunt network can be determined from Fig. 12.22 using (7.134) as follows:

$$S_{11} = \frac{b_1}{a_1}, \text{ when } a_2 = 0$$
$$= \frac{V_1^-}{V_1^+}, \text{ when } V_2^+ = 0$$
$$= -Z_o/(Z_o + 2Z_{sn})$$

It is a symmetric network, therefore:

$$S_{11} = S_{22} = -Z_o/(Z_o + 2Z_{sn})$$

Similarly from (7.79) and using (7.136), S_{21} can be determined as

$$S_{21} = \frac{b_2}{a_1} = \frac{V_2^-}{\sqrt{Z_o}}\frac{\sqrt{Z_o}}{V_1^+} = \frac{V_2^-}{V_1^+}, \text{ when } V_2^+ = 0$$
$$= 2Z_{sn}/(Z_o + 2Z_{sn}) = S_{12}$$

Thus, in matrix form, S-parameter of the shunt network can be written as

$$\begin{bmatrix} S_{11} & S_{12} \\ S_{21} & S_{22} \end{bmatrix} = \begin{bmatrix} \frac{-Z_o}{Z_o+2Z_{sn}} & \frac{2Z_{sn}}{Z_o+2Z_{sn}} \\ \frac{2Z_{sn}}{Z_o+2Z_{sn}} & \frac{-Z_o}{Z_o+2Z_{sn}} \end{bmatrix}, \text{ where } Z_{sn} = R_a - jZ_o\cot(\beta l)$$

The transmission coefficient of the shunt network is given by

$$S_{21} = \frac{2Z_{sn}}{Z_o + 2Z_{sn}} = \frac{2R_a - j2Z_o\cot(\beta l)}{Z_o + 2R_a - j2Z_o\cot(\beta l)}$$

Transmission coefficient in dB can be written as

$$= 10 \times \log\left(\frac{4R_a^2 + 4Z_o^2\cot^2(\beta l)}{(Z_o + 2R_a)^2 + 4Z_o^2\cot^2(\beta l)}\right)$$

Putting $R_a = 30\ \Omega$, $Z_o = 50\ \Omega$, $l = \lambda_o/8$, the transmission coefficient can be written as

$$\text{Transmission coefficient} = 10 \times \log\left(\frac{3600 + 10{,}000\cot^2\left(\frac{\pi}{40} f_{\text{GHz}}\right)}{12{,}100 + 2500\cot^2\left(\frac{\pi}{40} f_{\text{GHz}}\right)}\right)\text{dB}$$
$$= 10 \times \log\left(\frac{36 + 100\cot^2\left(\frac{\pi}{40} f_{\text{GHz}}\right)}{121 + 25\cot^2\left(\frac{\pi}{40} f_{\text{GHz}}\right)}\right)\text{dB}$$

Similarly, the reflection coefficient in dB can be written as

$$S_{21} = \frac{-Z_o}{Z_o + 2Z_{sn}} = \frac{-Z_o}{(Z_o + 2R_a) - j2Z_o\cot(\beta l)}$$

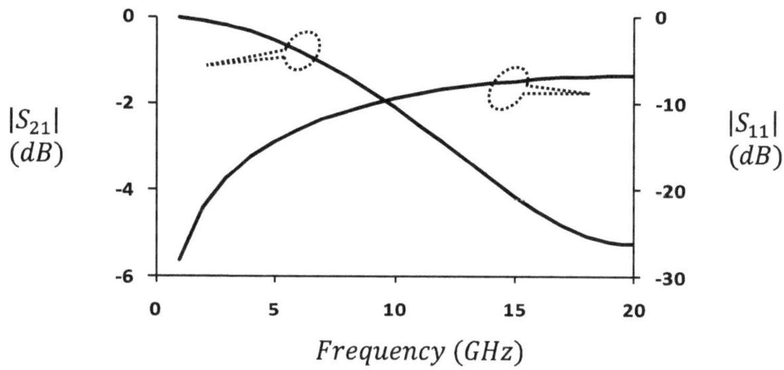

Fig. 12.23 Transmission and reflection coefficients of Example 12.2

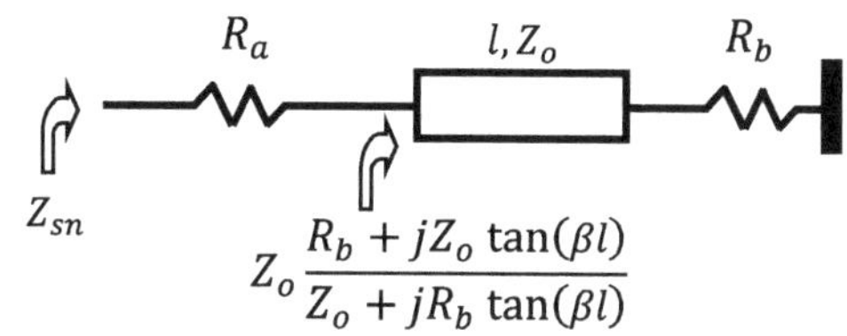

Fig. 12.24 The shunt network consists of a transmission line between two resistors of Example 12.3

Reflection coefficient in dB can be written as

$$= 10 \times \log\left(\frac{Z_o^2}{(Z_o + 2R_a)^2 + 4Z_o^2 \cot^2(\beta l)}\right)$$

$$\text{Reflection coefficient} = 10 \times \log\left(\frac{2500}{12{,}100 + 10{,}000 \cot^2\left(\frac{\pi}{40} f_{\text{GHz}}\right)}\right) \text{dB}$$

$$= 10 \times \log\left(\frac{25}{121 + 100 \cot^2\left(\frac{\pi}{40} f_{\text{GHz}}\right)}\right) \text{dB}$$

The variation of transmission and reflection coefficients over the frequency range of 1–20 GHz is plotted in Fig. 12.23 using the above expressions. This network provides increase of insertion loss with the increase of frequency.

Example 12.3 Compute *S*-parameter of a shunt network consists of transmission line (Z_o) of length l in between two resistances R_a and R_b as shown in Fig. 12.24. Also plot transmission coefficient of the network in dB over the frequency range of 5–15 GHz for $l = \lambda_o/4$, where λ_o is free space wavelength at 10 GHz, $R_a = 20\ \Omega$, and R_b as the parameter of values 20 Ω, 30 Ω, 50 Ω, 70 Ω and 100 Ω.

Solution

The shunt network consists of a transmission line of length l in between the resistors R_a and R_b is shown in Fig. 7.2.4. Using (7.16), the input impedance Z_{sn} of the shunt network can be written as

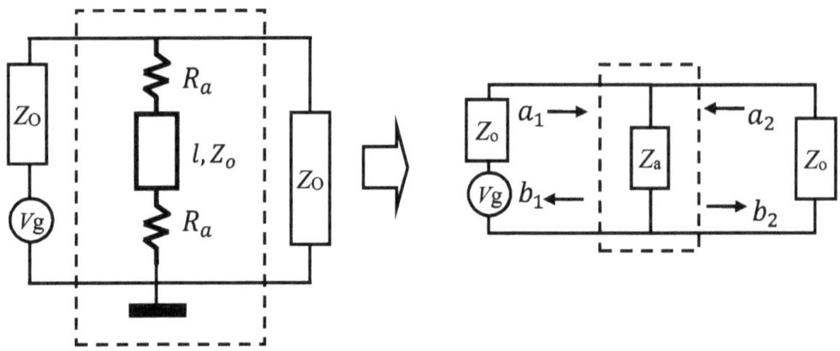

Fig. 12.25 The two port network consists of the shunt network of Example 12.3

$$Z_{sn} = R_a + Z_o \frac{R_b + jZ_o \tan(\beta l)}{Z_o + jR_b \tan(\beta l)}$$

Here, Z_o is the characteristic impedance of the shunt transmission line. The two-port network consists of the shunt network is shown in Fig. 12.25. The S-parameters of the two port network can be determined using (7.134) as follows:

$$\begin{aligned} S_{11} &= \frac{b_1}{a_1}, \text{ when } a_2 = 0 \\ &= \frac{V_1^-}{V_1^+}, \text{ when } V_2^+ = 0 \\ &= -Z_o/(Z_o + 2Z_{sn}) \end{aligned}$$

It is a symmetric network, therefore:

$$S_{11} = S_{22} = -Z_o/(Z_o + 2Z_{sn})$$

Similarly from (7.79) and using (7.136), S_{21} can be determined as

$$\begin{aligned} S_{21} &= \frac{b_2}{a_1} = \frac{V_2^-}{\sqrt{Z_o}} \frac{\sqrt{Z_o}}{V_1^+} = \frac{V_2^-}{V_1^+}, \text{ when } V_2^+ = 0 \\ &= 2Z_{sn}/(Z_o + 2Z_{sn}) = S_{12} \end{aligned}$$

Thus, in matrix form, S-parameter of the shunt network can be written as

$$\begin{bmatrix} S_{11} & S_{12} \\ S_{21} & S_{22} \end{bmatrix} = \begin{bmatrix} \frac{-Z_o}{Z_o+2Z_{sn}} & \frac{2Z_{sn}}{Z_o+2Z_{sn}} \\ \frac{2Z_{sn}}{Z_o+2Z_{sn}} & \frac{-Z_o}{Z_o+2Z_{sn}} \end{bmatrix}, \text{ where } Z_{sn} = R_a + Z_o \frac{R_b + jZ_o \tan(\beta l)}{Z_o + jR_b \tan(\beta l)}$$

The transmission coefficient of the shunt network is given by:

$$S_{21} = \frac{2Z_{sn}}{Z_o + 2Z_{sn}} = \frac{2Z_o(R_a + R_b) + j2\left(R_a R_b + Z_o^2\right)\tan(\beta l)}{Z_o(Z_o + 2R_a + 2R_b) + j\left(R_b(Z_o + 2R_a) + 2Z_o^2\right)\tan(\beta l)}$$

Transmission coefficient in dB can be written as

$$= 10 \times \log\left(\frac{4Z_o^2(R_a + R_b)^2 + 4\left(R_a R_b + Z_o^2\right)^2 \tan^2(\beta l)}{Z_o^2(Z_o + 2R_a + 2R_b)^2 + \left(R_b(Z_o + 2R_a) + 2Z_o^2\right)^2 \tan^2(\beta l)}\right)$$

Similarly, the reflection coefficient can be written as

$$S_{21} = \frac{-Z_o}{Z_o + 2Z_{sn}} = \frac{-Z_o^2 - jZ_o R_b \tan(\beta l)}{Z_o(Z_o + 2R_a + 2R_b) + j\left(R_b(Z_o + 2R_a) + 2Z_o^2\right)\tan(\beta l)}$$

Reflection coefficient in dB can be written as

$$= 10 \times \log\left(\frac{Z_o^4 + (Z_o R_b)^2 \tan^2(\beta l)}{Z_o^2(Z_o + 2R_a + 2R_b)^2 + \left(R_b(Z_o + 2R_a) + 2Z_o^2\right)^2 \tan^2(\beta l)}\right)$$

Putting $Z_o = 50\ \Omega$, $R_a = 20\ \Omega$, $\beta l = (\pi f_{\text{GHz}}/20)$, the transmission and reflection coefficients in terms of parameter R_b can be written as

$$|S_{21}|_{\text{dB}} = 10 \times \log\left(\frac{4(20 + R_b)^2 + 4(0.4R_b + 50)^2 \tan^2\left(\frac{\pi}{20} f_{\text{GHz}}\right)}{(90 + 2R_b)^2 + (1.8R_b + 100)^2 \tan^2\left(\frac{\pi}{20} f_{\text{GHz}}\right)}\right) \text{dB}$$

$$|S_{11}|_{\text{dB}} = 10 \times \log\left(\frac{2500 + R_b^2 \tan^2\left(\frac{\pi}{20} f_{\text{GHz}}\right)}{(90 + 2R_b)^2 + (1.8R_b + 100)^2 \tan^2\left(\frac{\pi}{20} f_{\text{GHz}}\right)}\right) \text{dB}$$

Plot of the transmission and reflection coefficients of the network for different values of R_b are shown in Figs. 12.26 and 12.27.

Example 12.4 Compute S-parameter of a shunt network consists of a resistance R_a and inductance L_a connected in series as shown in Fig. 12.28. Also plot transmission and reflection coefficients of the network in dB over the frequency range of 1–20 GHz for $R_a = 40\ \Omega$ and $L_a = 0.5$ nH.

Solution

The shunt network consists of a resistance R_a and capacitance L_a connected in series as shown in Fig. 12.28. The input impedance Z_{sn} of the shunt network can be written as

$$Z_{sn} = R_a + j2\pi f L_a$$

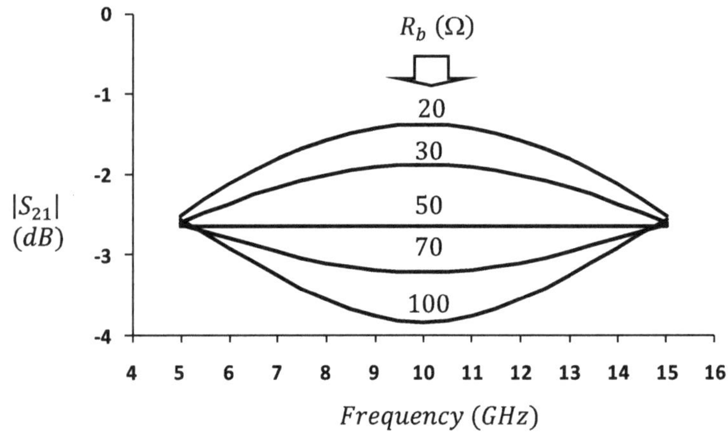

Fig. 12.26 Transmission coefficients of Example 12.3

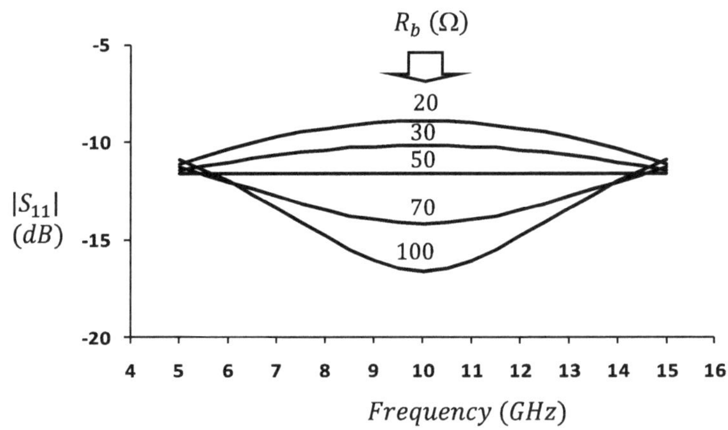

Fig. 12.27 Reflection coefficients of Example 12.3

Fig. 12.28 The shunt network of Example 12.4

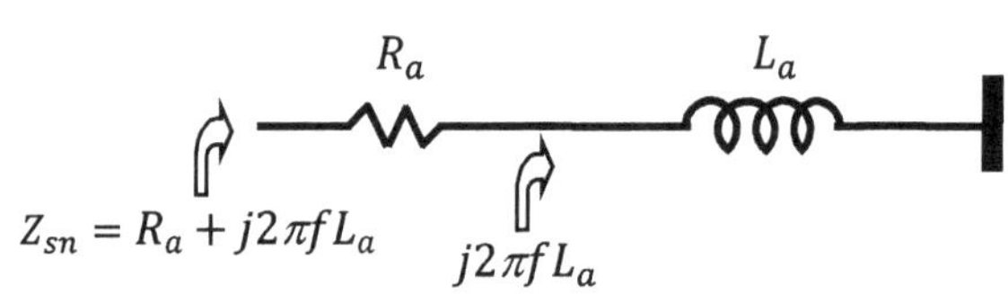

From (7.163d), matrix form S-parameters of the shunt network can be written as

$$\begin{bmatrix} S_{11} & S_{12} \\ S_{21} & S_{22} \end{bmatrix} = \begin{bmatrix} \frac{-Z_o}{Z_o+2Z_{sn}} & \frac{2Z_{sn}}{Z_o+2Z_{sn}} \\ \frac{2Z_{sn}}{Z_o+2Z_{sn}} & \frac{-Z_o}{Z_o+2Z_{sn}} \end{bmatrix}, \text{ where } Z_{sn} = R_a + j2\pi f L_a$$

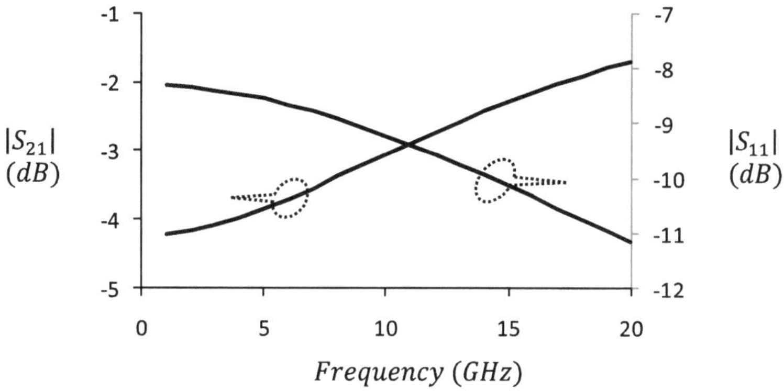

Fig. 12.29 Transmission and reflection coefficients of Example 12.4

Putting $Z_{sn} = R_a + j2\pi f L_a$, the S-parameters matrix of the network can be written as

$$\begin{bmatrix} S_{11} & S_{12} \\ S_{21} & S_{22} \end{bmatrix} = \begin{bmatrix} \frac{-Z_o}{(Z_o+2R_a)+j4\pi f L_a} & \frac{2R_a+j4\pi f L_a}{(Z_o+2R_a)+j4\pi f L_a} \\ \frac{2R_a+j4\pi f L_a}{(Z_o+2R_a)+j4\pi f L_a} & \frac{-Z_o}{(Z_o+2R_a)+j4\pi f L_a} \end{bmatrix}$$

Therefore, the transmission coefficient in dB can be written as

$$|S_{21}|_{\mathrm{dB}} = 10 \times \log\left(\frac{4R_a^2 + (4\pi f L_a)^2}{(Z_o + 2R_a)^2 + (4\pi f L_a)^2}\right)$$

Putting $Z_o = 50\ \Omega$, $R_a = 40\ \Omega$ and $L_a = 0.5$ nH:

$$|S_{21}|_{\mathrm{dB}} - 10 \times \log\left(\frac{1600 + (\pi f_{\mathrm{GHZ}})^2}{4225 + (\pi f_{\mathrm{GHZ}})^2}\right)$$

Similarly, the reflection coefficient in dB can be written as

$$|S_{11}|_{\mathrm{dB}} = 10 \times \log\left(\frac{Z_o^2}{(Z_o + 2R_a)^2 + (4\pi f L_a)^2}\right)$$

Putting $Z_o = 50\ \Omega$, $R_a = 40\ \Omega$ and $L_a = 0.5$ nH:

$$|S_{11}|_{\mathrm{dB}} = 10 \times \log\left(\frac{625}{4225 + (\pi f_{\mathrm{GHZ}})^2}\right)$$

Transmission and reflection coefficients of the shunt network of Fig. 12.28 are shown in Fig. 12.29 over the frequency range of 1–20 GHz.

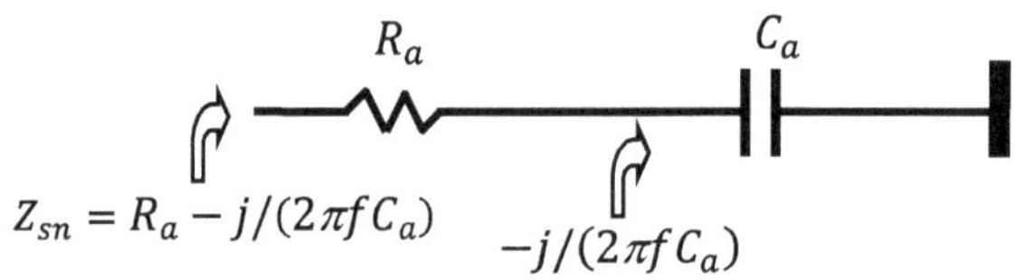

Fig. 12.30 The shunt network of Example 12.5

Example 12.5 Compute S-parameter of a shunt network consists of a resistance R_a and capacitance C_a connected in series as shown in Fig. 12.30. Also plot transmission and reflection coefficients of the network in dB over the frequency range of 5–15 GHz for $R_a = 50\ \Omega$ and $C_a = 0.25$ pF.

Solution
The shunt network consists of a resistance R_a and capacitance C_a connected in series as shown in Fig. 12.29. The input impedance Z_{sn} of the shunt network can be written as

$$Z_{sn} = R_a - j/(2\pi f C_a)$$

From (7.163d), S-parameters matrix of the shunt network can be written as

$$\begin{bmatrix} S_{11} & S_{12} \\ S_{21} & S_{22} \end{bmatrix} = \begin{bmatrix} \frac{-Z_o}{Z_o+2Z_{sn}} & \frac{2Z_{sn}}{Z_o+2Z_{sn}} \\ \frac{2Z_{sn}}{Z_o+2Z_{sn}} & \frac{-Z_o}{Z_o+2Z_{sn}} \end{bmatrix}, \text{ where } Z_{sn} = R_a - j/(2\pi f C_a)$$

Putting $Z_{sn} = R_a - j/(2\pi f C_a)$, the S-parameters matrix of the network can be written as

$$\begin{bmatrix} S_{11} & S_{12} \\ S_{21} & S_{22} \end{bmatrix} = \begin{bmatrix} \frac{\pi f Z_o C_a}{(Z_o+2R_a)\pi f C_a - j} & \frac{2\pi f C_a R_a - j}{(Z_o+2R_a)\pi f C_a - j} \\ \frac{2\pi f C_a R_a - j}{(Z_o+2R_a)\pi f C_a - j} & \frac{\pi f Z_o C_a}{(Z_o+2R_a)\pi f C_a - j} \end{bmatrix}$$

Therefore, the transmission coefficient in dB can be written as:

$$|S_{21}|_{\text{dB}} = 10 \times \log\left(\frac{(2\pi f C_a R_a)^2 + 1}{((Z_o + 2R_a)\pi f C_a)^2 + 1}\right)$$

Putting $Z_o = 50\ \Omega$, $R_a = 50\ \Omega$ and $C_a = 0.25$ pF:

$$|S_{21}|_{\text{dB}} = 10 \times \log\left(\frac{(0.025\pi f_{\text{GHZ}})^2 + 1}{(0.0375\pi f_{\text{GHZ}})^2 + 1}\right)$$

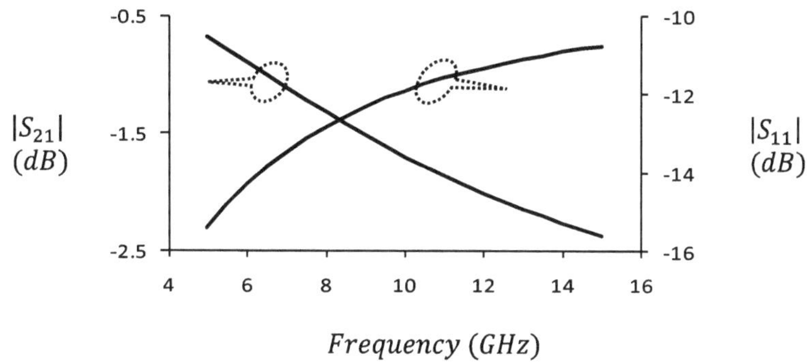

Fig. 12.31 Transmission and reflection coefficients of Example 12.5

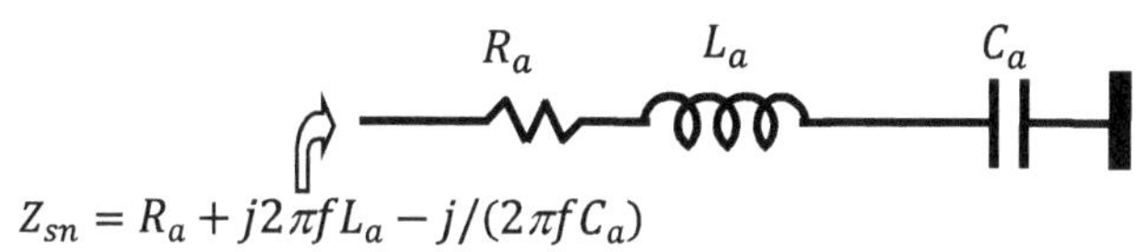

Fig. 12.32 The shunt network of Example 12.6

Similarly, the reflection coefficient in dB can be written as

$$|S_{11}|_{\text{dB}} = 10 \times \log\left(\frac{(\pi f Z_o C_a)^2}{((Z_o + 2R_a)\pi f C_a)^2 + 1}\right)$$

Putting $Z_o = 50\ \Omega$, $R_a = 50\ \Omega$ and $C_a = 0.25$ pF:

$$|S_{11}|_{\text{dB}} = 10 \times \log\left(\frac{(0.0125\pi f)^2}{(0.0375\pi f_{\text{GHZ}})^2 + 1}\right)$$

Transmission and reflection coefficients of the shunt network of Fig. 12.30 are shown in Fig. 12.31 over the frequency range of 5–15 GHz.

Example 12.6 Compute S-parameter of a shunt network consists of a resistance R_a, capacitance C_a and inductance L_a connected in series as shown in Fig. 12.32. Also plot transmission and reflection coefficients of the network in dB over the frequency range of 5–15 GHz for $R_a = 50\ \Omega$, $L_a = 1.25$ nH and $C_a = 0.25$ pF.

Solution

The shunt network consists of a resistance R_a, inductance L_a and capacitance C_a connected in series as shown in Fig. 12.32. The input impedance Z_{sn} of the shunt network can be written as

$$Z_{sn} = R_a + j2\pi f L_a - j/(2\pi f C_a)$$

From (7.163d), S-parameters of the shunt network can be written as:

$$\begin{bmatrix} S_{11} & S_{12} \\ S_{21} & S_{22} \end{bmatrix} = \begin{bmatrix} \frac{-Z_o}{Z_o+2Z_{sn}} & \frac{2Z_{sn}}{Z_o+2Z_{sn}} \\ \frac{2Z_{sn}}{Z_o+2Z_{sn}} & \frac{-Z_o}{Z_o+2Z_{sn}} \end{bmatrix}, \text{ here } Z_{sn} = R_a + j2\pi f L_a - j/(2\pi f C_a)$$

Putting $Z_{sn} = R_a + j2\pi f L_a - j/(2\pi f C_a)$, the S-parameters matrix of the network can be written as

$$\begin{bmatrix} S_{11} & S_{12} \\ S_{21} & S_{22} \end{bmatrix} = \begin{bmatrix} \frac{-Z_o\pi f C_a}{(Z_o+2R_a)\pi f C_a + j(4\pi^2 f^2 C_a L_a - 1)} & \frac{2\pi f C_a R_a + j(4\pi^2 f^2 C_a L_a - 1)}{(Z_o+2R_a)\pi f C_a + j(4\pi^2 f^2 C_a L_a - 1)} \\ \frac{2\pi f C_a R_a + j(4\pi^2 f^2 C_a L_a - 1)}{(Z_o+2R_a)\pi f C_a + j(4\pi^2 f^2 C_a L_a - 1)} & \frac{-Z_o f C_a}{(Z_o+2R_a)\pi f C_a + j(4\pi^2 f^2 C_a L_a - 1)} \end{bmatrix}$$

Therefore, the transmission coefficient in dB can be written as

$$|S_{21}|_{\text{dB}} = 10 \times \log\left(\frac{(2\pi f C_a R_a)^2 + \left(4\pi^2 f^2 C_a L_a - 1\right)^2}{((Z_o + 2R_a)\pi f C_a)^2 + \left(4\pi^2 f^2 C_a L_a - 1\right)^2}\right)$$

Putting $Z_o = 50\ \Omega$, $R_a = 50\ \Omega$, $L_a = 1.25$ nH and $C_a = 0.25$ pF:

$$|S_{21}|_{\text{dB}} = 10 \times \log\left(\frac{(0.025\pi f_{\text{GHZ}})^2 + \left(0.00125\pi^2 f_{\text{GHZ}}^2 - 1\right)^2}{(0.0375\pi f_{\text{GHZ}})^2 + \left(0.00125\pi^2 f_{\text{GHZ}}^2 - 1\right)^2}\right)$$

Similarly, the reflection coefficient in dB can be written as

$$|S_{11}|_{\text{dB}} = 10 \times \log\left(\frac{(Z_o\pi f C_a)^2}{((Z_o + 2R_a)\pi f C_a)^2 + \left(4\pi^2 f^2 C_a L_a - 1\right)^2}\right)$$

Putting $Z_o = 50\ \Omega$, $R_a = 50\ \Omega$, $L_a = 1.25$ nH and $C_a = 0.25$ pF:

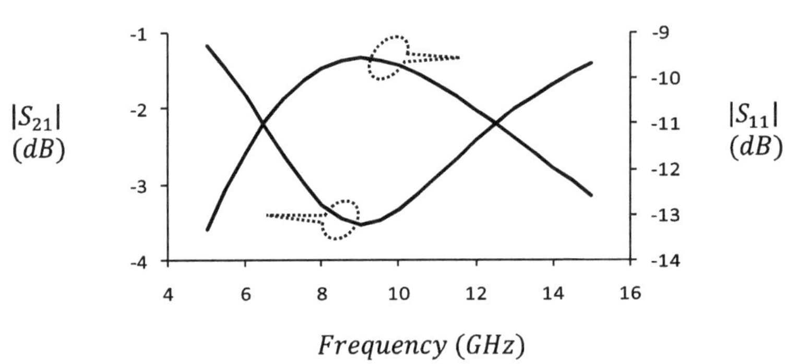

Fig. 12.33 Transmission and reflection coefficients of Example 12.6

$$|S_{11}|_{\mathrm{dB}} = 10 \times \log\left(\frac{(0.0125\pi f)^2}{(0.0375\pi f_{\mathrm{GHZ}})^2 + \left(0.00125\pi^2 f_{\mathrm{GHZ}}^2 - 1\right)^2}\right)$$

Transmission and reflection coefficients of the shunt network of Fig. 12.32 are shown in Fig. 12.33 over the frequency range of 5–15 GHz.

References

1. Morgan M, Newton T, Hayward B, Boyd T (2007) Non-reflective transmission-line filters for gain slope equalization. In: IEEE MTT-S international microwave symposium digest, June 2007, pp 545–548
2. Ma Q, Ma M (2008) Broadband amplifier gain slope equalization filter. In: Proceedings of progress in electromagnetic research symposium, Hangzhou, China, Mar 2008, pp 21–25
3. Madjar A, Gerfer EE (1991) A novel 2–4 GHz multi pass band tunable and gain controlled miniature active equalizer/filter. In: IEEE MTT-S international microwave symposium digest, July 1991, pp 451–454
4. Vassilev V, Angelov I, Belitsky V (1999) Design and performance of a 3.4 to 4.6 GHz active equalizer with controlled gain slope. Appl Microw Wireless 11(12):51–53
5. Vassilev V, Angelov I, Kovtonyuk S, Belitsky V (2000) Low-noise band selective amplifier and active equalizer with controlled gain-slope for 3.4–4.6 GHz. In: Proceedings of GHz'00 symposium of digest, Goteborg, Sweden, Mar 13–14, 2000, pp 177–180
6. Fejzuli A, Kaarsberg R, Roldan N (2006) Broadband amplifier gain slope equalization with a single passive component. High Freq Electron 5(6):22–26
7. Bera SC, Jain VK, Singh RV, Garg VK (2009) Circuit for compensating gain variation over operating frequency and/or temperature range. India Patent, IN284294
8. Bera SC (2011) Amplitude tilt active equalizer for frequency and temperature compensation. IEEE Microwave Wirel Compon Lett 21(7):344–346

Chapter 13
Microwave Detectors

Abstract This chapter presents design and analysis of various detectors for microwave power level measurements. Basic concepts of average, RMS, peak and envelope power are discussed, and methods of their measurements are presented. Use of Schottky barrier diodes and backward diodes as detectors; with their advantages and disadvantages are also presented.

13.1 Introduction

Signal power level is an important performance parameter of any microwave communication system. Output and input power levels are critical parameters for design of any circuit, system or component. Power level in a microwave system and subsystem is so important for its proper function that each stage in the system must receive predetermined proper signal power level from the previous stage and also it should provide proper output power level which will be the input power level of the next stage. At lower frequencies, signal power is calculated from measured voltage and current. Direct measurement of power is more convenient and accurate starting from the microwave frequency range to optical spectrum. Thus, as the frequency increases, power is measured by which voltage and current are calculated. Microwave power detectors are used to detect/measure the power level.

Schottky barrier and tunnel (backward) diodes are two popularly used devices which are used as microwave power level detector. Depending upon the signal structure and applications average, envelope, peak or pulse powers are measured. Selection of detector circuit also depends upon the bandwidth and dynamic power range of the signal.

The average or root-mean-square (RMS) power level is important in case where the total power is important. In systems such as radar, where basic performance mainly depends on the pulse power, it is important to measure pulse power of the microwave signal. The instantaneous envelope output power is used for RF system protection against excessive RF power overdrive. This is also used for power amplifier linearization and realization of enhanced efficiency power amplifier by envelope tracking. In

S. C. Bera, *Microwave Active Devices and Circuits for Communication*, Lecture Notes in Electrical Engineering 533, https://doi.org/10.1007/978-981-13-3004-9_13

order to adjust automatic gain control circuits in an RF chain, received signal strength of a communication system is measured to maintain the desired output signal fed to a transmitter. It protects the sensitive power amplifiers from excessive power overdrive. It also ensures stringent requirements for the transmitted output power from a radio transmitter meeting the specified communications standards and legal regulations. For microwave and mmwave signal amplitude measurements, Schottky barrier detector diodes are used for its faster response compared to p–n junction diodes [1, 4]. Faster response of Schottky barrier diodes is due to the absence of minority carrier storage delay time in Schottky junction, i.e. due to its operation with majority carrier only. Therefore, Schottky barrier diode can switch from its conduction to non-conduction condition almost instantaneously with the reversal of driving signals polarity. Thus, the Schottky diode's impedance can change almost instantaneously with the change of driving signals polarity that leads to its very high rectification efficiency extending to microwave and mmwave frequency range.

13.2 Microwave Power

Before designing a microwave detector, it is important to know the significance of instantaneous, average, envelope and pulse power. Suppose RF voltage $v(t)$ across a load and current $i(t)$ through the load at a time t are given by

$$v(t) = v_m \cos(\omega t) \tag{13.1a}$$

and

$$i(t) = i_m \cos(\omega t + \theta) \tag{13.1b}$$

Here, θ is the phase difference in between the voltage and current. The instantaneous power across the load is given by

$$p(t) = v(t) \times i(t) = v_m \cos(\omega t) \times i_m \cos(\omega t + \theta) \tag{13.2a}$$

This expression gives the power at any instant of time. Figure 13.1 shows the instantaneous voltage, current and instantaneous power. From (13.2a), the instantaneous power can also be written as

or

$$p(t) = \frac{1}{2} v_m i_m \cos(\theta) + \frac{1}{2} v_m i_m \cos(2\omega t + \theta) \tag{13.2b}$$

The first term is the DC component, i.e. independent of time. The second term is the time-dependent sinusoidal component of the instantaneous power with double the frequency of voltage and current waveform.

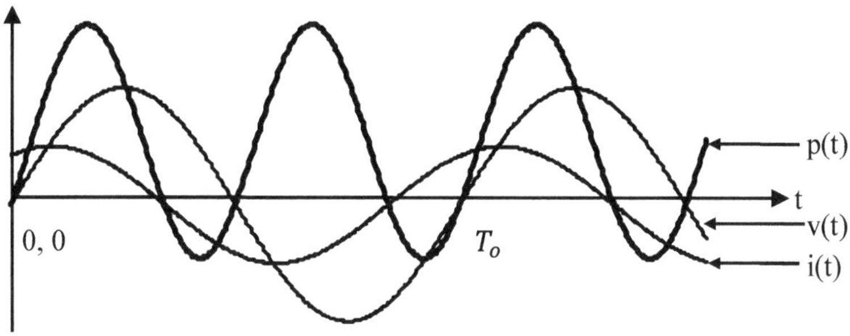

Fig. 13.1 Instantaneous RF voltage, current and power

Average RF/microwave power is the area under the power versus time plot. Therefore, it is determined by integrating and dividing the length of integration time. The length of the integration time should be an integer number of periods. Thus, the average RF power is given by

$$p_{av} = \frac{1}{nT_o} \int_0^{nT_o} p(t)dt \tag{13.3a}$$

Putting the value of $p(t)$ from (13.2b) to (13.3a), the average power can be written as

$$p_{av} = \frac{1}{2} v_m i_m \cos(\theta) \frac{1}{nT_o} \int_0^{nT_o} dt + \frac{1}{2} v_m i_m \frac{1}{nT_o} \int_0^{nT_o} \cos(2\omega t + \theta)dt \tag{13.3a}$$

$$= \frac{1}{2} v_m i_m \cos(\theta) + 0 \tag{13.3b}$$

$$= \frac{v_m}{\sqrt{2}} \frac{i_m}{\sqrt{2}} \cos(\theta) \tag{13.3c}$$

Therefore, the first term of (13.2b) is the average RF power. In case of resistive load, the phase difference in between the voltage $v(t)$ and the current $i(t)$ is zero, i.e. $\theta = 0$. Thus, for resistive load, the average RF power is given by

$$p_{av} = \frac{v_m}{\sqrt{2}} \frac{i_m}{\sqrt{2}} \tag{13.3d}$$

$$= v_{\text{rms}} \times i_{\text{rms}} \text{ in case of sinusoidal signal} \tag{13.3e}$$

Instantaneous RF power and its two components, i.e. average power and time-dependent sinusoidal power, are shown in Fig. 13.2. Ideally, for measurement of average RF power, it should be integrated over integer numbers of RF cycles. In practice, integration time is taken over many RF periods to avoid the restriction of integration over exact integer multiple cycles to avoid any significant error. For a continuous wave signal (i.e. sinusoidal waveform), the lowest and highest frequency

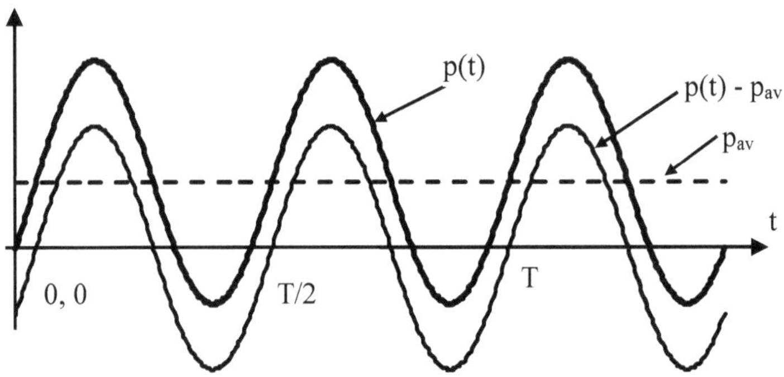

Fig. 13.2 Instantaneous RF power with its two components

are same; however, for a multicarrier and amplitude-modulated wave, it is required to average the power over many periods. Thus, in case of amplitude-modulated or multicarrier signal to measure the average power, it is required to take average over many periods of the lowest frequency involved.

In communication systems, it is very much convenient to express RF power level in dB with respect to a known reference power level. The unit dBm is used when the power level expressed with respect to 1 mW power and unit dBW is used for power level when it is expressed with respect to 1 W power level.

$$\text{Power(in dBm)} = 10 \times \log\left(\frac{\text{absolute power}}{1\ \text{mW}}\right) \tag{13.4a}$$

$$\text{Power(in dBW)} = 10 \times \log\left(\frac{\text{absolute power}}{1\ \text{W}}\right) \tag{13.4b}$$

The use of dB scale leads to easy calculation of gain/attenuation for a system just by algebraic addition among its output and input power level when both are expressed in dB. Overall gain/attenuation of a system consists of several number of cascaded matched networks is just the algebraic sum of individual network's gain. In this case, just addition of power gain in dB for each network replaces the multiplication of numeric gain.

Example 13.1 Convert 100 mW RF power in dBm and dBW.

Solution
Using (13.4a),

$$\begin{aligned} 100\ \text{mW Power(in dBm)} &= 10 \times \log\left(\frac{100\ \text{mW}}{1\ \text{mW}}\right) \\ &= 10 \times \log\left(10^2\right)\text{dBm} \\ &= 20\ \text{dBm} \end{aligned}$$

Using (13.4b),

$$\begin{aligned}100\ \text{mW Power(in dBW)} &= 10 \times \log\left(\frac{100\ \text{mW}}{1\ \text{W}}\right)\text{dBW}\\ &= 10 \times \log\left(\frac{10^2}{10^3}\right)\text{dBW}\\ &= -10\,\text{dBW}\end{aligned}$$

Example 13.2 Convert 30 dBm and 20 dBW RF power to mW and W respectively.

Solution

From (13.4a), absolute power can be expressed as

$$\text{Absolute power} = 10^{\frac{\text{Power in dBm}}{10}}\ \text{mW}$$

Thus,

$$\begin{aligned}30\ \text{dBm power} &= 10^{\frac{30}{10}}\ \text{mW}\\ &= 10^3\ \text{mW}\\ &= 1000\ \text{mW}\end{aligned}$$

Similarly, from (13.4b), absolute power can be expressed as

$$\text{Absolute power} = 10^{\frac{\text{Power in dBW}}{10}}\ \text{W}$$

Thus,

$$\begin{aligned}20\,\text{dBW power} &= 10^{\frac{20}{10}}\,\text{mW}\\ &= 10^2\text{W}\\ &= 100\ \text{W}\end{aligned}$$

Example 13.3 An attenuator of 15 dB attenuation is fed with 100 W input. Find the output power of the attenuator in dBW and Watt.

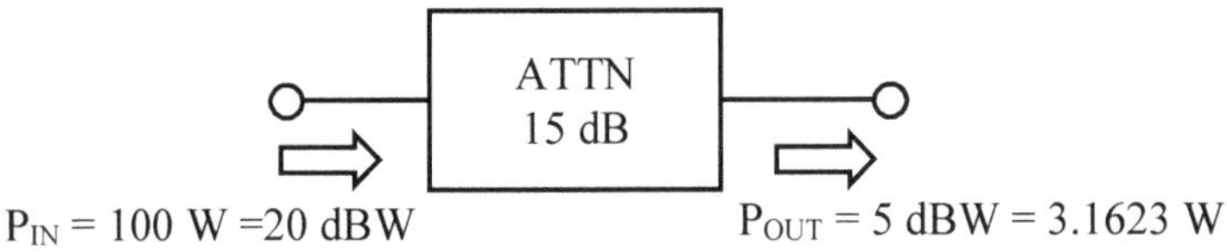

Fig. 13.3 Power levels of an attenuator

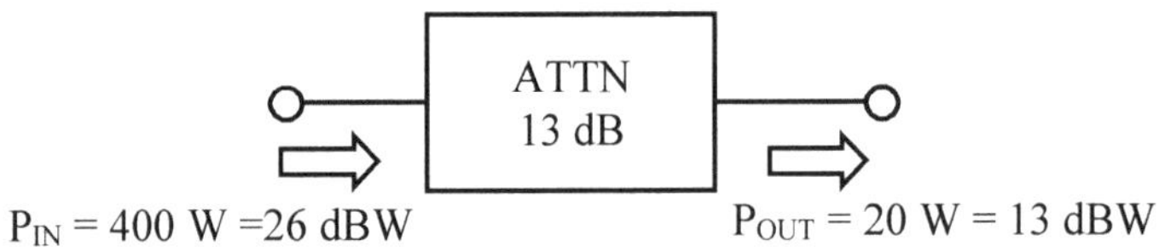

Fig. 13.4 Power levels of an attenuator

Solution

$$100\text{ W Power(in dBW)} = 10 \times \log\left(\frac{100\text{ W}}{1\text{ W}}\right)\text{dBW} = 20\,\text{dBW}$$

The output of the attenuator will be (Fig. 13.3)

$$\begin{aligned} P_{\text{OUT}}(\text{dBW}) &= P_{\text{IN}}(\text{dBW}) - \text{ATTN(dB)} \\ &= (20 - 15)\,\text{dBW} \\ &= 5\,\text{dBW} \\ P_{\text{OUT}}(\text{W}) &= 10^{P_{\text{OUT}}(\text{dBW})/10} \\ &= 10^{0.5}\,\text{dBW} \\ &= 3.1623\,\text{W} \end{aligned}$$

Example 13.4 The input power given to an attenuator is 400 W. The power output of the attenuator is 20 W. Calculate the attenuation value of the attenuator in dB.

Solution
Input power of the attenuator is 400 W (Fig. 13.4).

$$400\text{ W Power(in dBW)} = 10 \times \log\left(\frac{400\text{ W}}{1\text{ W}}\right)\text{dBW} = 26\,\text{dBW}$$

The output power of the attenuator is 20 W.

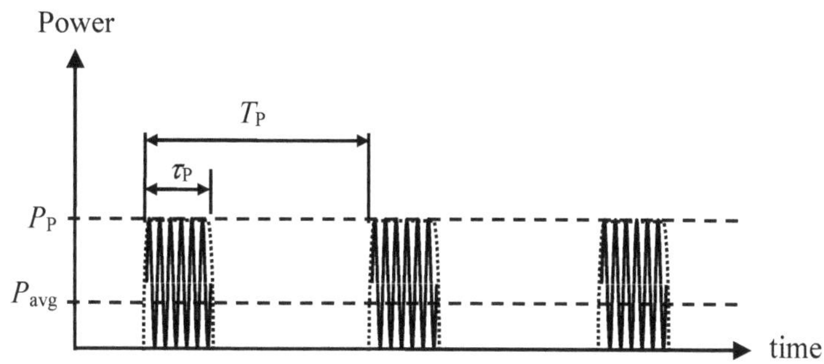

Fig. 13.5 Pulsed RF power

$$20\text{ W Power(in dBW)} = 10 \times \log\left(\frac{20\text{ W}}{1\text{ W}}\right)\text{dBW} = 13\,\text{dBW}$$

The attenuation of the attenuator in dB is given by

$$\begin{aligned}\text{Attenuation (dB)} &= P_{\text{IN}}(\text{dBW}) - P_{\text{OUT}}(\text{dBW}) \\ &= (26 - 13)\text{ dB} \\ &= 13\text{ dB}\end{aligned}$$

Though the average power is very popular for communication systems; however, the basic performance of radar, navigation and TDMA wireless communication systems depend primarily on the peak power radiation. The time constants of peak power instruments and sensors should be in the sub-microsecond region to measure the 'pulse power' and 'peak envelope power'. The pulsed RF power waveform is shown in Fig. 13.5. The power is averaged over the time duration of pulse width, τ_P, for pulse power. Thus, the pulse power P_P can be represented as

$$P_P = \frac{i}{\tau_p} \int_{t=0}^{\tau_p} v(t) \times i(t)\mathrm{d}t \tag{13.5a}$$

The pulse power is average power over the pulse duration. Thus, it averages out any overshoot or ringing within this duration. The pulse power also can be expressed in terms of its duty cycle as

$$P_P = \frac{P_{\text{avg}}}{\text{duty cycle}} \tag{13.5b}$$

Here, P_{avg} is the RF power averaged over integer numbers of full pulse cycles. The duty cycle is the pulse width times the repetition frequency. From Fig. 13.5, the duty cycle of the pulsed power can be written as

$$\text{Duty Cycle} = \frac{\tau_P}{T_P} \tag{13.6}$$

Therefore, the pulse power can be expressed as

$$P_P = \frac{P_{\text{avg}}}{\text{duty cycle}} = P_{\text{avg}} \times \frac{T_P}{\tau_P} \tag{13.7}$$

Using this definition of pulse power, calculation of pulse power can be done by the measurement of the duty cycle and the average power. This technique avoids the use of more complicated and expensive direct pulse power measurement scheme.

Example 13.5 The pulse power of an RF pulse power source is 1 KW. The duty cycle is 10^{-2}. Find its average power.

Solution
The average RF power can be expressed in terms of its pulse power as (13.7)

$$\begin{aligned} P_{\text{avg}} &= \text{pulse power} \times \text{duty cycle} \\ &= 1\ \text{kW} \times 10^{-2} \\ &= 10\ \text{W} \end{aligned}$$

Example 13.6 A microwave pulse has an average power of 200 W and duration is 5 μs. The time interval between pulses is 2 ms. Determine the value of pulse power.

Solution
Pulse power is given by (13.7)

$$\begin{aligned} P_P &= \frac{P_{\text{avg}}}{\text{duty cycle}} = P_{\text{avg}} \times \frac{T_P}{\tau_P} \\ &= 200\ \text{W} \times \frac{2\ \text{ms}}{5\ \mu\text{s}} \\ &= 80\ \text{kW} \end{aligned}$$

13.3 Diode Detectors

Point contact diodes were the first high-frequency diodes though they were extremely fragile and did not have repeatable performance [1]. For microwave frequency power measurement, low-barrier Schottky contact technology is used to construct diodes with metal–semiconductor junctions [2, 4]. These diodes are very rugged and provide very consistent performance. These diodes are able to detect and measure very low power up to very high frequencies. Metal–semiconductor diode also known as Schottky barrier diode is very popular as detector diode for its operation with majority carriers only. With the change in polarity of the driving signal, the impedance of a Schottky diode changes almost instantaneously. This property makes the Schottky diode suitable as rectifier not only at microwave frequency range but also used in mmwave and terahertz frequency range. By suitable construction, it is possible to achieve very low cut-in voltage of the Schottky barrier diodes. This makes the Schottky barrier diode as very sensitive detector to measure low level of microwave power compared to the p–n junction diodes.

Under forward-bias condition of junction voltage V_{dj}, the diode current is given by

$$I_d = I_o\left[\exp\left(\frac{qV_{dj}}{\eta kT}\right) - 1\right] \tag{13.8}$$

Here, the ideality factor $\eta \cong 1.0$–1.2, because of the negligible recombination in the depletion region, and I_o is the reverse saturation current. Due to the absence of charge storage delay effect of the Schottky diodes, its DC transfer characteristic (13.8) is also valid for AC transfer characteristic up to very high frequency.

Suppose, RF signal is applied in a finite biased Schottky diode. The diode voltage can be represented by

$$V_{dj} = V_o + V_s\cos(\omega t) \tag{13.9}$$

Combining (13.8) and (13.9), the diode current can be written as

$$I_d = I_o\left[\exp\left(\frac{V_o + V_s\cos(\omega t)}{V_t}\right) - 1\right] \tag{13.10}$$

$$= I_o\left[\exp\left(\frac{V_o}{V_t}\right)\exp\left(\frac{V_s\cos(\omega t)}{V_t}\right) - 1\right] \tag{13.11}$$

where $V_t = (\eta kT/q)$. Expanding the exponential term containing AC part by series expansion, the diode current can be written as

$$I_d = I_o\left[\exp\left(\frac{V_o}{V_t}\right) - 1\right] + I_o\exp\left(\frac{V_o}{V_t}\right) \times\left[\frac{V_s}{V_t}\cos(\omega t) + \frac{V_s^2}{2V_t^2}\cos^2(\omega t) + \frac{V_s^3}{6V_t^3}\cos^3(\omega t) + \frac{V_s^4}{24V_t^4}\cos^4(\omega t) + \cdots\right] \tag{13.12}$$

This can be written as

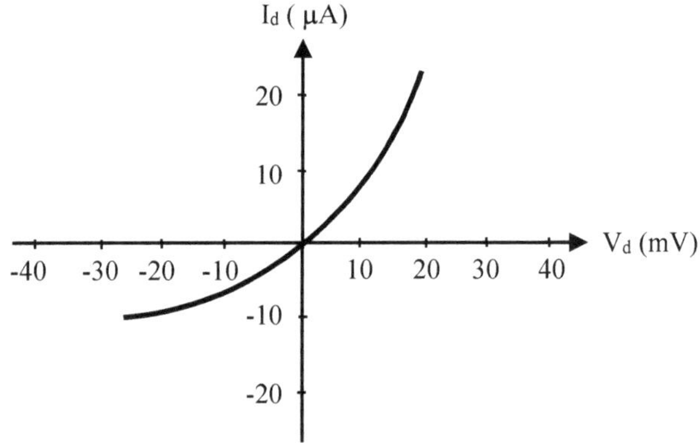

Fig. 13.6 Square-law characteristic region of Schottky barrier diode

$$
\begin{aligned}
I_d = {} & I_o\left[\exp\left(\frac{V_o}{V_t}\right) - 1\right] \Rightarrow \text{this term is DC current due to the bias voltage} \\
& + I_o \exp\left(\frac{V_o}{V_t}\right)\left[\frac{V_s}{V_t}\cos(\omega t)\right] \Rightarrow \text{is due to the fundamental AC component} \\
& + I_o \exp\left(\frac{V_o}{V_t}\right)\left[\frac{V_s^2}{4V_t^2} + \frac{V_s^2}{4V_t^2}\cos(2\omega t)\right] \Rightarrow \text{rectified dc and second harmonic} \\
& + I_o \exp\left(\frac{V_o}{V_t}\right)\left[\frac{V_s^3}{8V_t^2}\cos(\omega t) + \frac{V_s^3}{24V_t^2}\cos(3\omega t)\right] \Rightarrow \text{fundamental and third harmonic} \\
& + I_o \exp\left(\frac{V_o}{V_t}\right)\left[\frac{V_s^4}{64V_t^4} + \frac{V_s^4}{48V_t^4}\cos(2\omega t) + \frac{V_s^4}{192V_t^4}\cos(4\omega t)\right] \Rightarrow \text{dc + second and fourth} \\
& + \cdots
\end{aligned}
\tag{13.13}
$$

This expression is written under the assumption that the exponential I–V relationship is also valid under RF signal operation. This assumption is true if $V_s \ll V_t$. Otherwise, exponential I–V characteristic of the Schottky diode will not remail valid due to the significant rectification of the RF signal. This expression shows that second and all other even-order nonlinearity produces DC components due to the rectification effect of Schottky diodes. The property of generating DC term related to RF signal by the Schottky diode is used as power detector.

The region of I–V characteristic around the zero voltage across a low-barrier Schottky diode is shown in Fig. 13.6. Over this region, the I–V characteristic follows the square law, i.e. output current of the diode is proportional to the square of the input signal voltage. For a typical Schottky diode, the range of RF power from noise level to about −20 dBm considered as small signal level over which the diode's I–V characteristic obeys the square law. It is to be noted that for 50 Ω system, about 22.4 mV (RMS) corresponds to −20 dBm (10 μW) RF power. Figure 13.7 shows the input signal voltage and output diode current waveform over the operating square low region.

From (13.13), the DC components of the diode current can be written as

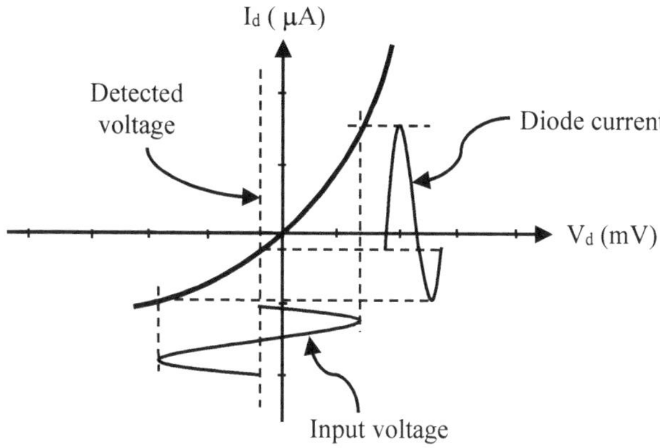

Fig. 13.7 Working of diode detector over square-law region

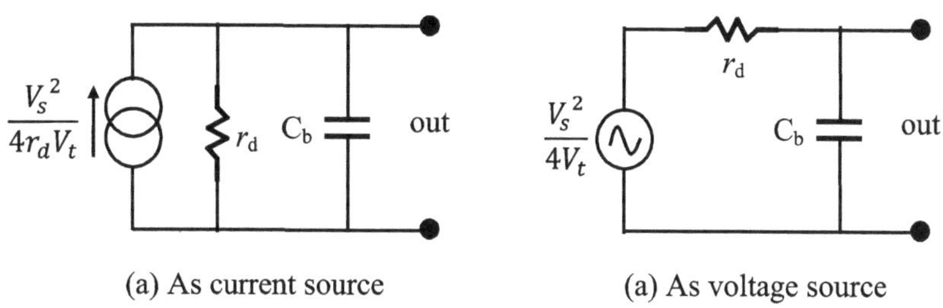

Fig. 13.8 Equivalent circuits of diode detector with output capacitor as filter

$$I_{dc} = I_O\left[\exp\left(\frac{V_O}{V_t}\right) - 1\right] + I_O\exp\left(\frac{V_O}{V_t}\right)\left(\frac{V_S^2}{4V_t^2}\right) + I_O\exp\left(\frac{V_O}{V_t}\right)\left(\frac{V_S^4}{64V_t^4}\right) + \cdots \tag{13.14}$$

Here, the first term is the DC bias current, second term is the rectified DC component proportional to the square of the RF input voltage generated due to the second-order nonlinearity and third term is the rectified DC component generated due to the fourth-order nonlinearity, and so on.

At small-signal condition, i.e. for $V_s \ll V_t$, up to the second-order term can be considered as significant. In this region, the diode current is proportional to the square of the RF input voltage. So the diode is said to be operating in the square-law region. In this condition, the DC components of the current can be written as

$$I_{dc} = I_O\left[\exp\left(\frac{V_O}{V_t}\right) - 1\right] + I_O\exp\left(\frac{V_O}{V_t}\right) \times \left(\frac{V_S^2}{4V_t^2}\right) \tag{13.15}$$

The presence of bias-dependent $\exp(V_O/V_t)$ in the rectified DC voltage increases the sensitivity of the detector diode. The DC current can be rewritten as

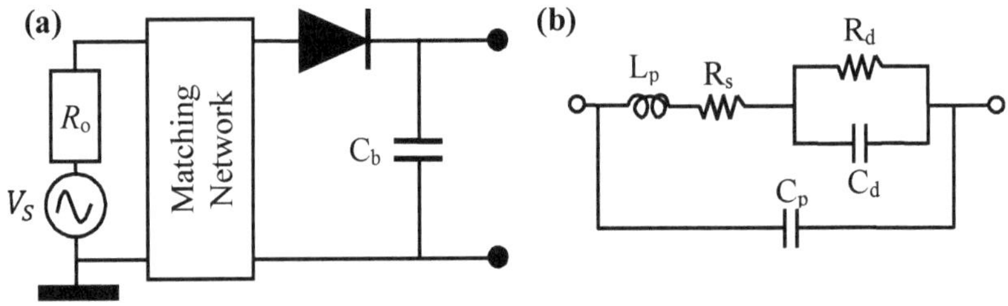

Fig. 13.9 Detector circuit diagram with matching network and the diode equivalent circuit

$$I_{\mathrm{dc}} = I_o\left[\exp\left(\frac{V_o}{V_t}\right) - 1\right] + \frac{V_s^2}{4r_d V_t} \tag{13.16}$$

where r_d is the RF resistance of the diode and is given by

$$r_d = 1/\frac{\partial I_d}{\partial V_{dj}} = \frac{V_t}{I_o \exp(V_o/V_t)} \tag{13.17}$$

Therefore, the detector diode with output capacitor to filter out the RF components can be considered as a current source driving a shunt resistor. The shunt resistor is equal to the diode resistance r_d. The basic equivalent circuits of the diode detector are shown in Fig. 13.8a, b considering the detector diode as current and voltage source respectively. Here, the fixed DC bias current of the diode is not shown for simplicity. The bypass capacitor C_b acts as low-pass filter which filters out the RF signals passing through the diode.

At higher power level of signal, the diode cannot behave as square low device due to the significant fourth-order and higher order component of V_s. The device then rectifies the signal in a quasi-square-law I–V region which is called the transition region. Above that range, the exponential I–V relationship will no longer valid and practically the detector relationship moves into the linear detection region where detected output voltage/current will be proportional to the input RF voltage/current.

To transfer maximum power from source to the detector diode, the RF resistance of the diode must be matched to the generator source resistance. The detector diode will detect RF voltage as long as the RF voltage falls across the diode. Practically, the RF resistance of diode is typically much larger than 50 Ω. To avoid reflection of RF power to the source, a separate matching network is used to set the detector's input termination impedance matched to the source resistance. Circuit diagram of a diode detector circuit with a matching network is shown in Fig. 13.9a, and the complete equivalent circuit diagram of a Schottky diode is also shown in Fig. 13.9b. Matching networks can be either resistive or reactive. A shunt resistor R_m, as shown in Fig. 13.10, is used as resistive matching network to achieve broadband matching with low reflection and flat frequency response. As discussed in Chap. 3, the RF resistance of Schottky diodes is a function of its operating temperature. Thus, the sensitivity of the diode detector and its reflection coefficient are also strong functions

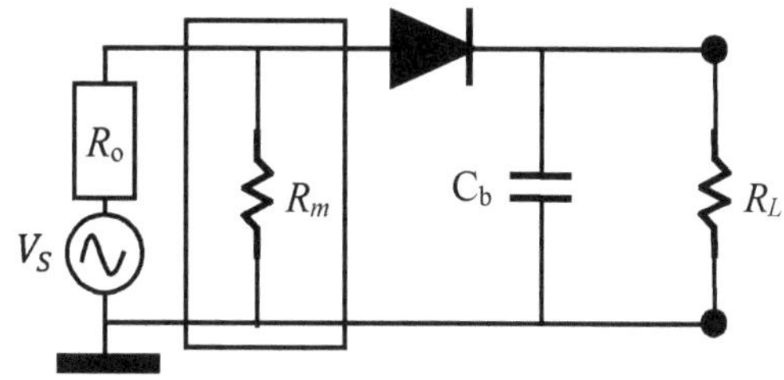

Fig. 13.10 Resistive matching network of the detector

of temperature. The RF impedance of the diodes should be very high compared to the matching resistor R_m to achieve its less temperature dependence. In that case, the matching resistor R_m acts as the termination of the generator due to very high RF impedance of the diodes. If RF resistance of the diode becomes too large, there would be poor power conversion from RF signal to detected DC voltage. The inverse dependency of diodes RF resistance on reverse saturation current leads to decrease of detector sensitivity with the increase of diode impedance. A compromise between sensitivity to RF signals and temperature dependency of the performance can be achieved by suitable construction of the diode using suitable materials to achieve low potential barrier across the junction. A low-barrier Schottky barrier diode can meet this requirement.

In case of reactively matching network, detector works over only moderate bandwidth. It also has drawback of poor matching compared to resistive matching and the performance of the circuit is also influenced by the change of diode characteristic. However, the advantage of using reactive matching network is that it provides better sensitivity to the RF signal. Sometimes combination of resistive and reactive components is used for the matching network to obtain better sensitivity to RF signal with lesser influence by the change of diode characteristic.

Tunnel and backward diodes have inherent I–V characteristic which leads to higher rectification current sensitivity [4]. This enables the tunnelling effect diodes to rectify small signals of amplitude which is smaller than the forward-bias cut-in voltage. The current is due to the tunnelling of the carriers, and it is fast in the absence of minority carriers charge storage effect which leads to broadband performance capability. Tunnelling current is insensitive to the change of temperature. Therefore, the performance of the backward and tunnel diodes is also not sensitive to its operating temperature. Therefore, tunnel and backward diode detectors have excellent temperature stability, very fast pulse response time and broadband frequency flatness. Open-circuit voltage sensitivity and high-power burnout are lesser than silicon-based Schottky detectors, but the tunnel detector's relatively low video impedance with no DC bias requirement enables DC and AC coupling with video and log video post-amplifiers.

Build-in detectors are required for embedded on-chip measurement of internal signal levels in analog and mixed-signal circuits. In this case mostly, RMS power detectors are more useful than peak or envelope detectors, because RMS detection provides signal levels independent of signal shape, i.e. modulation, number of carriers, etc. To realize on-chip detectors, it is required to develop detectors using the same technology in which full circuit/systems are made such as Si [2–4], GaAs [5] and GaN-based different types of FETs [6–8]. GaN-based Schottky barrier diodes offer superior performance in terms of high-power and high-frequency operation over a wide range of operating temperature due to their higher bandgap and higher breakdown electric field [6–8]. The GaN-based Schottky barrier diodes can operate as microwave detector with non-zero DC biasing due to its 0-V turn-on voltage, thus easy to implement in MMIC technology without any bulky passive components. High reverse breakdown voltage of the GaN-based diodes provides very wide dynamic range which is desired as RF power detector.

13.4 RMS Power (Average) Detector

One of the major applications of diode detectors is to measure the true RMS power which is nothing but the average signal power. A true RMS detector should be designed to measure the actual RMS power of the input signal, ideally independent of the modulation complexity or the modulation scheme. Generally, a RMS detector consists of Schottky barrier diodes operating over its square-law region and an integrator acting as low-pass filter. The RMS output signal is directly proportional to the time-averaged of the square of the input signal voltage. Under the small-signal detector operation, the output of the detector is determined by the slope of the I–V characteristic of the diode, and thus the detector output current is proportional to the square of the input voltage and thus proportional to the input power.

From the power series expansion of the diode characteristic, it is already discussed that the rectified output voltage is a function of the square of the input signal voltage. This relationship is valid up to a power level of about -20 dBm for practically used Schottky diode detectors. That means the rectified output voltage is proportional to the RF signal power regardless of the signal content. That means, under the small-signal diode operation region, as long as square-law detection is valid, the detected output is proportional to the input power which is independent of the signal content, i.e. complexity and modulation scheme.

As the power level increases above about -20 dBm, the higher order terms of the power series will be comparable with the second-order term and the rectification process becomes more and more linear. Thus, under large signal condition, the output voltage becomes proportional to the input voltage. For complex signals of power level corresponding to large signal condition, the output is then dependent upon the phase relationships among the various components of the input signal. Thus, the measured output will not be the true RMS power representation of the complex input signal. In case of very small power levels, i.e. below -50 dBm, the diode detector input is about 700 μV (RMS) which may correspond to detector DC output voltage of about 700 nV. This low signal level requires low-noise amplifier and chopper circuit to measure reliably in the presence of noise.

Schottky diode-based sophisticated detector circuit is capable to operate as true RMS power detector over the power range of about −70 to −20 dBm. Beyond −20 dBm, power level correction factor may be used to enhance the dynamic range. Though the use of corrections for the CW signals is applicable, however, for other signals which are modulated or multicarrier, the use of correction is not reliable. In this case with the expense of detection sensitivity, in place of single diode stacks (multiple numbers) of detector diodes, it can be used to extend the square-law operation extended to higher power levels. In practice, multiple stacks of a different number of detector diodes are used. For low-level power measurement, stacks of lesser number of diodes and for higher power measurement stacks of more number of diodes are used. Electronic RF switches are used to select the RF paths based on the RF power level to operate the diode over its square-law region over lowest to extended power levels.

Thermal-based detectors are also used for microwave power detection [9]. Using resistive components, it first converts AC power to thermal power, and then using a temperature sensor, e.g. thermistor, it measures the change of temperature. The change of temperature is proportional to the thermal power or the dissipated AC power, which gives the measured RMS power level. These kinds of detectors have the advantages of good accuracy, wide bandwidth and independent of signal shapes, i.e. independent of number of carriers and modulations. However, these types of detectors are not suitable to implement in chip level.

13.5 Envelope and Peak Power Detector

For multicarrier signals and complex digital-modulated single-carrier signals, such as wideband code division multiple access (WCDMA), instantaneous power measurement is required. This enables to measure the peak power or peak-to-average power ratio (PAPR). Envelope power can be measured using a diode detector in its square-law range.

The maximum value of the envelope power is the peak envelope power. In order to measure envelope power accurately for a fast-changing envelope power, the detector should be fast enough to track the envelope power. To provide this capability for the measurement of peak envelope power, the detector circuit should have sufficient video bandwidth. However, with the increase of the video bandwidth, there will be a reduction in dynamic range of the detector.

13.6 Applications

Microwave diode detectors are used for high accuracy RF signal measurement and control applications [10, 11]. The envelope power detectors can also be used to create fast, excessive RF power protection, power amplifier linearization and efficiency-enhancing envelope tracking power amplifier realizations [12]. Diode detector circuits are the building block of many microwave instruments like power meter, spectrum analyzer, network analyzer, etc. One of the main differences among different power measuring instruments is the frequency selectivity. In case of frequency selective measurement, the RF power within a predetermined bandwidth is measured. This is generally done by spectrum analyzer instrument. Generally, microwave power meters are not frequency selective. Power meter measures the average power over the full frequency range of the detector including present noise and any generated harmonics. However, spectrum analyzer measures power over a selected frequency band. Frequency selectivity in spectrum analyzer makes it enable to measure much lower power level compared to general power meters.

References

1. Bose JC (1904) Detector for electrical disturbances. US Patent 755840
2. Wu Y (1996) Design of Schottky barrier detector diode for silicon MMIC's. Int Rep, NISRC, Belfast, Northern Ireland
3. Wu Y, Armstrong BM, Gamble HS, Hu Z, Chen Q, Yang S, Fusco VF, Stewart JAC (1998) Microwave PtSi–Si Schottky-barrier-detector diode fabrication using an implanted active layer on high-resistivity silicon substrate. IEEE Trans Microw Theor Tech 46(5):641–646
4. Park SY, Yu R, Chung SY, Berger PR, Thompson PE, Fay P (2007) Sensitivity of Si-based zero-bias backward diodes for microwave detection. Electron Lett 43(5):53–54
5. De Bo W, Ping LX, Liu T (2012) A novel thermoelectric and capacitive power sensor with improved dynamic range based on GaAs MMIC technology. IEEE Electron Device Lett 33(2):269–271
6. Zhang AP, Johnson JW, Luo B, Ren F, Pearton SJ, Park SS, Park YJ, Chyi JI (2001) Vertical and lateral GaN rectifiers on freestanding GaN substrates. Appl Phys Lett 79:1555–1557
7. Zhou Y, Wang D, Ahyi C, Che C-T, Williams J, Park M, Williams NM, Hanser A (2006) High breakdown voltage Schottky rectifier fabricated on bulk n-GaN substrate. Solid State Electron 50(11/12):1744–1747
8. Zhou Q, Wong K-Y, Chen W, Chen KJ (2010) Wide-dynamic-range zero-bias microwave detector using AlGaN/GaN heterojunction field-effect diode. IEEE Microwave Wirel Compon Lett 20(5):277–279
9. Yi Z, Liao X (2012) Packaging test-fixture research for 10–12 GHz frequency detector based on thermoelectric microwave power sensor. Electron Lett 48(2):103
10. Bera SC, Singh RV (2004) A temperature-compensated closed loop overdrive level controller for microwave solid-state power amplifiers. Microw J 47(4):114–122
11. Bera SC, Singh RV, Garg VK (2006) Design and temperature compensation of a Ku-Band channel amplifier with ALC for a satellite transponder. Microw J 49(4):68–82
12. Bera SC, Singh RV, Garg VK (2008) Diode-based predistortion linearizer for power amplifiers. Electron Lett 44(2):125–126

Chapter 14
Microwave Solid-State Amplifiers

Abstract This chapter presents design and analysis of low-noise, high-gain and high power microwave solid-state amplifiers. Stability analysis and design of matching networks for all the types of amplifiers using Smith chart is presented. The dynamic range of amplifiers and its dependence on noise figure and nonlinearities of the amplifiers are discussed. Using conditional and unconditional stable devices, design examples for transducer power gain, operating power gain and available power gain are also presented with various examples. Design and analysis of linear and nonlinear Class-A, Class-B, Class-C and Class-F high power amplifiers are discussed in detail.

14.1 Introduction

A microwave amplifier is a circuit that takes a signal of low or intermediate level over a range of frequency band and significantly boosts its power level to a predetermined level. An amplifier is one of the most common elements in any communication system to boost the signal level wherever it is required. In addition to amplification of the desired signal, a practical amplifier-added noise is generated within it and also distorts the amplified signal. Based on the signal power level to be amplified, there are three types of microwave amplifiers: low-noise amplifier, high-gain amplifier and high power amplifier. Microwave communication systems include receiver and transmitter. Most of the elements of a transmitter and receiver are amplifiers operating at different power levels. Low-noise amplifiers (LNA) are used as the front end of a receiver where the strength of the desired signal is very low. High-gain amplifiers are used in the intermediate stages of a transmitter–receiver section to boost the signal strength to a desired level to drive other circuitry, e.g., high power amplifiers of a transmitter. The High Power Amplifier (HPA) is used as the final power amplifying stage of a transmitter. The design of matching networks and biasing criterions are different for LNA, high gain and high power amplifiers. However, there are common requirements for all types of the amplifiers such as bandwidth, gain and gain flatness over frequency, stability, efficiency, reliability, etc. In most of the cases, design

S. C. Bera, *Microwave Active Devices and Circuits for Communication*, Lecture Notes in Electrical Engineering 533, https://doi.org/10.1007/978-981-13-3004-9_14

trade-offs are required to achieve any one parameter compared to the other to meet the specific requirements of a system.

Based on the devices used for the realization of an amplifier, there are two general classes of microwave solid-state amplifiers (SSAs): diode based and transistor based. Due to the simplicity of fabrication and lower cost, two terminals diode-based SSAs were most popular. Varactor-diode-based parametric amplification, transferred electron device, i.e., Gunn diode's negative resistance based amplification and avalanche transit time diode (IMPATT, TRAPATT, BARITT) based pulsed power amplifications were used for various applications. With the advancement of microwave transistors fabrication technologies, today solid-state amplification is dominated by the use of three terminal devices, transistors. Among the microwave transistors, all the low-noise amplifiers for communication systems are dominated by psuedomorphic and metamorphic high electron mobility transistors (pHEMTs and mHEMTs). High gain with moderate output power amplifiers are also dominated by MESFET and HFETs. With the advancement of high bandgap semiconductor materials such as GaN on SiC, single-ended high power amplification beyond 100 W is achievable using solid-state technology.

At low frequencies, the design of an amplifier is trivial which involves designing a proper DC bias circuit for maximum power output. However, at microwave frequency range, in addition to the design of proper DC bias circuit, it involves the design of proper input and output matching networks based on the requirements of noise figure, gain and output power over the desired band of frequency. A typical block diagram of a transistor-based single-stage microwave amplifier is shown in Fig. 14.1. It consists of bias networks and input and output matching networks. The bias network provides proper terminal voltages/currents to the device without affecting the in-band RF performance of the amplifier. It also provides out-off band stability to the amplifier. Input and output matching networks provide proper impedance matching of the device for achieving the required RF performance of the amplifier. Based on the performance requirements, i.e. low noise figure, high gain or high power, the bias voltages/currents as well as impedances offered by the matching networks

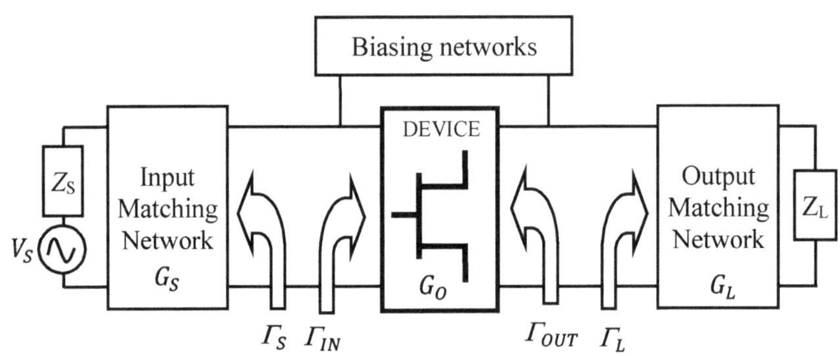

Fig. 14.1 Microwave amplifier with its associated networks

are different. Designing a stable amplifier is the prime requirement for any type of amplifier irrespective of its other performance requirements.

For designing an amplifier which will operate in its linear operating region, it is convenient to use S-parameter model of the device. Front-end amplifiers of a receiver such as LNA and other intermediate amplifiers such as IF amplifier used in the satellite receiver and channel amplifiers operate in their linear operating region. These amplifiers can be designed using S-parameter model of the transistors. However, the final stage of a transmitter section, in general, operates in its nonlinear operating region to achieve high output power with higher DC to RF efficiency. Thus, for designing the power amplifiers, it is required to use the nonlinear model of the device. Design methodology for different types of amplifiers will be discussed in their respective sections.

14.2 Types of Microwave Amplifiers

Design philosophy of a microwave amplifier depends on the performance requirements and its operating condition. Design of an amplifier operating in small signal condition is different from the design of an amplifier operating in large signal condition. It is very important to understand the difference between the terms large signal and high power, similarly the difference between small signal and low power. In case of FET-based amplifiers, large signal refers to a power level in which the RF signal level (voltage) across the gate-to-source layer is a significant fraction of the applied DC gate bias voltage. High power refers to an arbitrary power level which is independent of any amplifier, whereas large signal refers to a particular power level related to the particular amplifier. For example, achieving one watt output power from an amplifier of maximum output power capability of one watt is considered as the large signal operation of the amplifier. Whereas, the same one watt output power taken from an amplifier of maximum output power capability of ten watts is considered as small signal operation. Even an amplifier of maximum output power capability of few mW operates in large signal condition when it is operated with output power level of the few mW. The operating condition of an amplifier is determined by the signal level handled by the amplifier compared to its maximum power capability. The output power capability of an amplifier is determined by the applied bias voltage and current to the device which is limited by the absolute maximum rating of applied bias voltage (V_{DS}) and current (I_{DSS}) or power dissipation of the device (in terms of channel temperature).

An amplifier operating in large signal operating condition is considered operating in nonlinear region because of the nonlinear relationship between the input power and output power level. Thus, gain and other performance parameters become a function of input power level and the device performance cannot be modelled by a single set of S-parameter values. There is a technique of using large signal S-parameters for designing amplifiers operating in large signal operating conditions. However, it

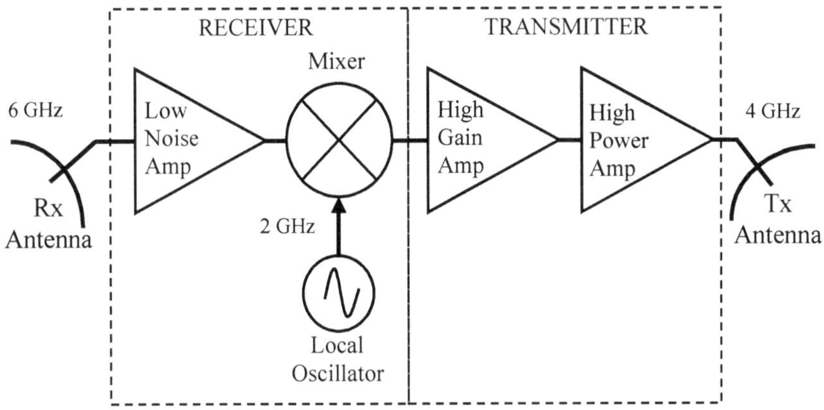

Fig. 14.2 Schematic diagram of a simple transceiver

is more convenient to use large signal circuit model or load-pull measurement and design methods for an amplifier operating in large signal condition.

Block diagram of a simple transceiver (transmitter + receiver) for a communication system is shown in Fig. 14.2. In a communication system, basically, three types of amplifiers: low-noise, high-gain and high power amplifiers are used. Low-noise amplifier is used at the front end of a receiver to amplify very weak signal received by the receive antenna. The high-gain amplifier further amplifies the signal to a moderate power level required to drive the final high power amplifier for signal transmission. The low-noise amplifiers are used where the signal levels are very small. Thus, low-noise amplifiers should have the capability to amplify very weak signal without much addition of its own noise. Generally, LNAs and high gain amplifiers are always operated in small signal condition. Whereas, high power amplifiers need to operate in large signal condition to provide high output power with high DC to RF efficiency.

Input and output matching networks as well as DC bias conditions of a transistor are different which depends on the operating condition of the amplifier. As long as the transistors operate in the linear region, i.e. operate in small signal condition, the output is a linear reproduction of the input. However, when the transistors operate in large signal condition, output no longer changes linearly with the input. This nonlinear behaviour is the main cause of signal distortion in microwave circuits. Other effects of nonlinearity are gain reduction and phase change in an amplifier resulting in the modulation transfer from one signal to another and generation of various frequencies related to frequencies of the input signals.

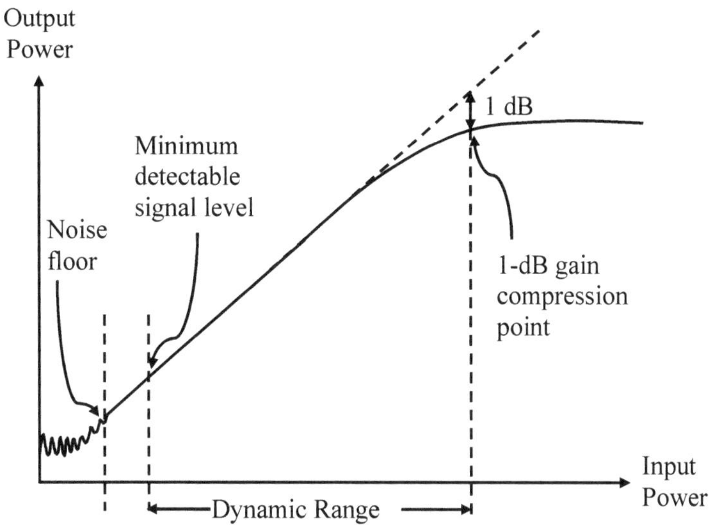

Fig. 14.3 Typical input–output characteristic of amplifier

14.2.1 Dynamic Range of Amplifier

An amplifier amplifies signal only over a certain range of its input power level maintaining acceptable linearity. Beyond this region, the signal output level does not follow the acceptable linearity with the input and thus the amplified signal becomes distorted beyond an acceptable limit. Amplifier also adds noise to the amplified signal during amplification. Therefore, noise dominates if the level of the input signal becomes below a certain level. The minimum input signal level above the noise level that can be detected is considered as the lower limit of the dynamic range.

Dynamic range of an amplifier is considered as the range of power level over which an amplifier amplifies the signal maintaining specified linearity. The lower level of the dynamic range is limited by the contribution of amplifiers noise , whereas the higher level is limited by the allowable nonlinearity of the amplifier. A typical input–output characteristic of an amplifier is shown in Fig. 14.3. Above the minimum detectable power level, amplifier amplifies the signal linearly, i.e. 1 dB increase of input power level results in 1 dB increase of output power level. This linear relationship continues until the RF output power reaches to a certain value, and beyond that power level, the rate of increase of output power level decreases due to the reduction of gain of the amplifier. Over this nonlinear characteristic, the output signal v_o of an amplifier can be represented by the power series function of its input signal v_i as

$$v_o = a_0 + a_1 v_i + a_2 v_i^2 + a_3 v_i^3 + \cdots \tag{14.1}$$

Here, $a_0, a_1, a_2, a_3, \ldots$ are constants having amplitude and phase term. Suppose, a sinusoidal signal $v_i = A\cos\omega t$ is applied to input of the amplifier, and considering the nonlinearities up to cubic term, the output signal can be written as

$$v_o = a_0 + a_1 A\cos\omega t + a_2 A^2\cos^2\omega t + a_3 A^3\cos^3\omega t \tag{14.2}$$

$$= a_0 + a_1 A\cos\omega t + \frac{a_2 A^2}{2}(1+\cos 2\omega t) + \frac{a_3 A^3}{4}(3\cos\omega t - \cos 3\omega t)$$

$$= \left(a_0 + \frac{a_2 A^2}{2}\right) + \left(Aa_1 + \frac{3a_3 A^3}{4}\right)\cos\omega t + \frac{a_2 A^2}{2}\cos 2\omega t + \frac{a_3 A^3}{4}\cos 3\omega t \tag{14.3}$$

The output signal level corresponding to the fundamental frequency component is given by

$$Aa_1\left(1 + \frac{3a_3 A^2}{4a_1}\right)$$

Therefore, gain of the amplifier corresponding to the fundamental frequency is a function of the input signal level and can be written as

$$G(A) = a_1\left(1 + \frac{3a_3 A^2}{4a_1}\right) \tag{14.4}$$

For $a_3/a_1 < 0$, gain of the amplifier decreases with the increase of input signal level. With further increase of input power level, output of the amplifier saturates. The reduction of gain and saturation of the output power of an amplifier is due to the limitation of bias voltage and current to the amplifier which is determined by the maximum absolute rating of applied voltage and current handling capability of the device used for the amplifier. Generally, the upper limit of dynamic range is considered as the power level that causes 1 dB gain reduction with respect to the linear gain of the amplifier.

The noise is generated within a solid-state device itself as well as comes from an external source. To detect a signal reliably, the signal strength should be much higher (generally considered as 3 dB) than the noise floor. The noise source in solid-state devices may be categorized into three types: thermal or Johnson noise, shot noise and flicker noise.

The thermal noise is generated due to the random fluctuations of signal produced by thermal agitation of the bound charges in the device. The RMS value of open-circuit thermal noise voltage V_n with resistance R over a frequency band of B is given by

$$V_n = \sqrt{4kTBR} \tag{14.5}$$

Thus, a noisy resistor can be modelled as a Thevenin circuit consisting of a noiseless resistance R together with a Thevenin voltage source given by (14.5). Thus, the available noise power at the terminal of any resistor is simply given by

$$P_n = \frac{V_n^2}{4R} = kTB \tag{14.6}$$

The spectrum of Johnson noise is white because the power spectral density is independent of frequency.

Shot noise is due to the fluctuation in kinetic energy of the current carrying discrete electrons in a device. The energy fluctuation modulates the arrival rate of electrons at a node where the current is measured. The shot noise is also considered as constant power spectral density. The Flicker noise is called $1/f$ noise or pink noise. The power spectral density of pink noise increases with the decrease of frequency, and at low enough frequencies it dominates Johnson (thermal) and shot noise. Though the pink noise is largely responsible for the "phase noise" in oscillators, it is negligible in case of FET-based microwave amplifiers.

The noise generated in an amplifier is quantitatively expressed by noise factor (NF) and noise factor expressed in dB called noise figure ($\mathrm{NF_{dB}}$). Thus, the noise figure can be written as

$$\text{Noise Figure } (\mathrm{NF_{dB}}) = 10 \times \log(\text{Noise Factor})$$

The noise factor is defined as

$$\begin{aligned} \mathrm{NF} &= \frac{\text{Total available output noise power}}{\text{Available output noise power due to the input source}} \\ &= \frac{P_{no}}{G_A \times P_{ni}} \end{aligned} \tag{14.7a}$$

where P_{no} is the total available output noise power, G_A is the available gain of the amplifier and P_{ni} is the available noise power of the source.

Thus, the total available output noise power can be written as

$$P_{no} = G_A \times P_{ni} \times \mathrm{NF} \tag{14.7b}$$

Replacing the available gain by the ratio of signal power available from the network P_{AVN} and signal power available from the source P_{AVS}, the noise factor can be written as

$$\mathrm{NF} = \frac{P_{\mathrm{AVS}}/P_{ni}}{P_{\mathrm{AVN}}/P_{no}} \tag{14.7c}$$

Thus,

$$\text{NF} = \frac{\text{Available input signal-to-noise power ratio}}{\text{Available output signal-to-noise power ratio}} \tag{14.8}$$

Therefore, the noise factor is the signal-to-noise ratio (SNR) of the amplifier at its input divided by the SNR at its output.

Noise figure of a system can also be represented by an effective input noise temperature T_e. It is the thermodynamic temperature of a resistance which is placed at the input of the system, and establishes the same available noise power at the output of the element as the actual element assuming the system is noise free. Suppose the system is driven by a reference source of noise temperature T_o, then the total available noise power at the output of the system will be

$$P_{no} = (k\,T_eB + k\,T_oB)G_A \tag{14.9a}$$

And, the available noise power at the input will be

$$P_{ni} = k\,T_oB \tag{14.9b}$$

Thus, noise factor of the system can be represented by

$$\text{NF} = \frac{(k\,T_eB + k\,T_oB)G_A}{k\,T_oBG_A} = \left(1 + \frac{T_e}{T_o}\right)$$

Therefore, noise temperature T_e of the system can be expressed as

$$T_e = (\text{NF} - 1)T_o \tag{14.9c}$$

Example 14.1 Calculate output noise power over 4.0 kHz bandwidth of an amplifier. The available gain and noise figure of the amplifier are 50, 3 dB, respectively.

Solution

From (14.7b), the output noise power is given by

$$P_{no} = P_{ni} \times G_A \times \text{NF}$$

Suppose, the amplifier is driven by a signal source consisting of a Thevenin resistance *R*, which is held at the standard room temperature of 300 K. In this condition, the available output noise power density is given by

$$\begin{aligned} kT_o \times G_A \times \text{NF} &= 10\log(k) + 10\log(T_o) + 10\log(G_A) + 10\log(\text{NF}) \\ &= \left[10\log(1.38 \times 10^{-23}) + 10\log(300) + 50 + 3\right]\text{dBW/Hz} \\ &= [-228.6 + 24.77 + 53]\,\text{dBW/Hz} \\ &\approx [-204 + 53]\,\text{dBW/Hz} = -151\ \text{dBW/Hz} \\ &= -121\ \text{dBm/Hz} \end{aligned}$$

Therefore, the total output noise power over the 4.0 kHz bandwidth is as follows:

$$
\begin{aligned}
&= -121 + 10\log(4.0 \times 10^3)\,\text{dBm} \\
&= -85\ \text{dBm}
\end{aligned}
$$

An input signal of an amplifier is detectable only if its amplified output signal power is above the noise power level by certain amount (considered as 3 dB). Suppose, the minimum detectable signal (MDS) level is X-dB above the noise power level. Thus, the minimum detectable output power level ($P_{o\text{MDS}}$) can be written as

$$
\begin{aligned}
P_{o\text{MDS}}(\text{in dBm}) &= P_{no}(\text{in dBm}) + X(\text{in dB}) \\
&= 10\log(kT_o) + 10\log(B) + 10\log(G_A) + 10\log(\text{NF}) + X(\text{in dB})
\end{aligned}
$$

Therefore, the minimum detectable input signal level $P_{i\text{MDS}}$ can be expressed as

$$
\begin{aligned}
P_{i\text{MDS}}(\text{in dBm}) &= 10\log(kT_o) + 10\log(B) + 10\log(\text{NF}) + X(\text{in dB}) \\
&= -174\ \text{dBm} + 10\log(B) + 10\log(\text{NF}) + X(\text{in dB}) \qquad (14.10)
\end{aligned}
$$

This is the lower limit of input power level for the dynamic range of an amplifier of noise factor NF and bandwidth B in Hz. Generally, for estimation of dynamic range, X-dB is considered as 3-dB.

In general, input power level corresponding to 1-dB gain compression point is considered as the higher limit of the input dynamic range. The input power level $P_{i1\text{dB}}$ corresponding to 1-dB gain compression point can be written as

$$
P_{i1\text{dB}}(\text{in dBm}) = P_{o1\text{dB}}(\text{in dBm}) - G_{1\text{dB}}(\text{in dB}) \qquad (14.11)
$$

Here, $P_{i1\text{dB}}$, $P_{o1\text{dB}}$ and $G_{1\text{dB}}$ are the input power level, output power level and gain of the amplifier corresponding to 1-dB gain compression point, respectively. $G_{1\text{dB}}$ is nothing but the 1-dB less gain compared to the amplifier's linear gain G_L.

Therefore, the dynamic range (DR) of an amplifier can be written as

$$
\text{DR}(\text{in dB}) = P_{i1\text{dB}}(\text{in dBm}) - P_{i\text{MDS}}(\text{in dBm}) \qquad (14.12a)
$$

Using (14.10), (14.11) and considering $X(\text{in dB}) = 3$ dB, the dynamic range can be written as

$$
\begin{aligned}
\text{DR}(\text{in dB}) &= P_{o1\text{dB}}(\text{in dBm}) - [G_L(\text{in dB}) - 1] \\
&\quad - \left[10\log(kT_o) + 10\log(B) + 10\log(\text{NF}) + 3\right] \qquad (14.12b)
\end{aligned}
$$

Example 14.2 Calculate the dynamic range of an amplifier of output power 1-W at 1-dB gain compression point. The available linear gain and noise figure of the amplifier is 44 and 5 dB, respectively. The bandwidth of the amplifier is 40 MHz.

Solution

The higher limit on input power level of the dynamic range:

$$\begin{aligned} P_{i1\text{dB}}(\text{in dBm}) &= P_{o1\text{dB}}(\text{in dBm}) - [G_L(\text{in dB}) - 1] \\ &= 10\log(10^3) - [44 - 1] \\ &= 30 - 43 = -13 \end{aligned}$$

The lower limit of the dynamic range:

$$\begin{aligned} P_{i\text{MDS}}(\text{in Bm}) &= \left[10\log(kT_o) + 10\log(B) + 10\log(\text{NF}) + 3\right] \\ &= -174 + 10\log(40 \times 10^6) + 5 + 3 \\ &= -174 + 76 + 5 + 3 = -90 \end{aligned}$$

Therefore,

$$\text{DR}(\text{in dB}) = P_{i1\text{dB}}(\text{in dBm}) - P_{i\text{MDS}}(\text{in dBm}) = -13 - (-90) = 77$$

Therefore, dynamic range of the amplifier is 77 dB.

14.2.2 Spurious-Free Dynamic Range

Another term spurious-free dynamic range (SFDR) is also used to specify the useful operating region of an amplifier and other microwave communication circuits. The spurious-free dynamic range measures the available dynamic range of a signal, which is above the spurious as well as the noise floor. A signal with the spurious products is shown in Fig. 14.4. In an amplifier spurious level increases with the increase of output power level. Thus, in a communication system, the largest output power from an amplifier is limited by the allowable spurious power level. Therefore, the higher limit of the SFDR is determined by the spurious acting as noise as shown in Fig. 14.4.

In case of intermodulation distortion-limited applications, the largest output signal level of an amplifier is determined by the signal level for which the intermodulation spurious level determines the minimum detectable signal as shown in Fig. 14.5. The third-order spurious signal level is proportional to the cubic term of the input signal level. Therefore, for each dB increase of input power level, there will be a 3-dB increase of third-order frequency power level as shown in Fig. 14.5. Generally, spurious-free dynamic range corresponding to the third-order intermodulation limited case is smaller than the DR range considering 1-dB gain compression point as the higher power level.

The spurious-free dynamic range (SFDR) for third-order intermodulation limited system can be computed considering the input–output relationship of an amplifier which is governed by the following nonlinear equation up to cubic term:

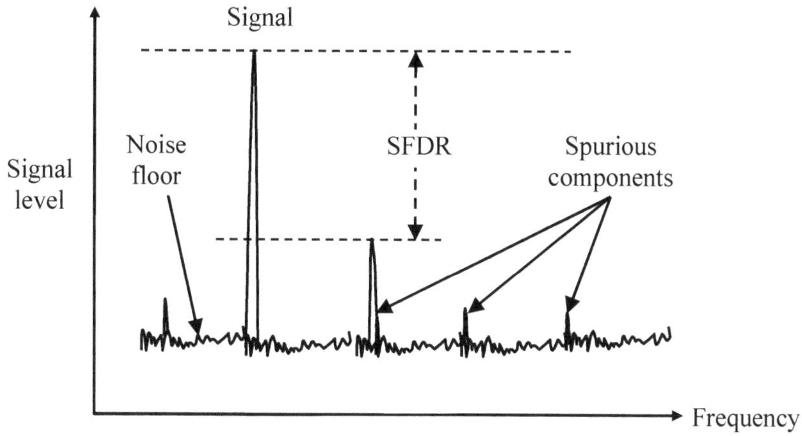

Fig. 14.4 Spurious-free dynamic range (SFDR)

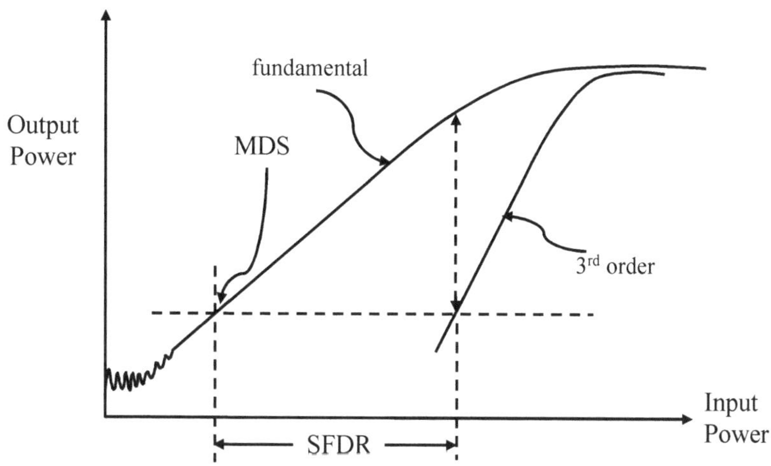

Fig. 14.5 Typical input–output characteristic of amplifier

$$v_o = a_0 + a_1 v_i + a_2 v_i^2 + a_3 v_i^3$$

Here, a_0, a_1, a_2, a_3 are the constants. Suppose the following signal with two different frequencies $\omega_1 = 2\pi f_1$ and $\omega_2 = 2\pi f_2$ are applied to the amplifier.

$$v_i = A\cos\omega_1 t + B\cos\omega_2 t$$

Here, A and B are amplitudes of the signal. The output is given by

$$v_o = a_0 + a_1(A\cos\omega_1 t + B\cos\omega_2 t) + a_2(A\cos\omega_1 t + B\cos\omega_2 t)^2$$

$$+ a_3(A\cos\omega_1 t + B\cos\omega_2 t)^3$$

or,

$$\begin{aligned}v_o = a_0 &+ a_1(A\cos\omega_1 t + B\cos\omega_2 t)\\ &+ a_2\left(A^2\cos^2\omega_1 t + B^2\cos^2\omega_2 t + 2AB\cos\omega_1 t\cos\omega_2 t\right)\\ &+ a_3\left(A^3\cos^3\omega_1 t + B^3\cos^3\omega_2 t\right.\\ &\left.+3A^2B\cos^2\omega_1 t\cos_2 t + 3AB^2\cos\omega_1 t\cos^2\omega_2 t\right)\end{aligned}$$

or,

$$\begin{aligned}v_o = a_0 &+ a_1(A\cos\omega_1 t + B\cos\omega_2 t)\\ &+ \frac{a_2A^2}{2}(1+\cos 2\omega_1 t) + \frac{a_2B^2}{2}(1+\cos 2\omega_2 t)\\ &+ a_2AB[\cos(\omega_1+\omega_2)t + \cos(\omega_1-\omega_2)t]\\ &+ \frac{a_3A^3}{4}(3\cos\omega_1 t - \cos 3\omega_1 t) + \frac{a_3B^3}{4}(3\cos\omega_2 t - \cos 3\omega_2 t)\\ &+ \frac{3a_3A^2B}{2}(1+\cos 2\omega_1 t)\cos\omega_2 t\\ &+ \frac{3a_3AB^2}{2}(1+\cos 2\omega_2 t)\cos\omega_1 t\end{aligned}$$

or,

$$\begin{aligned}v_o = a_0 &+ \frac{a_2A^2}{2} + \frac{a_2B^2}{2}\\ &+ \left(a_1A + \frac{3a_3A^3}{4} + \frac{3a_3AB^2}{2}\right)\cos\omega_1 t + \left(a_1B + \frac{3a_3B^3}{4} + \frac{3a_3A^2B}{2}\right)\cos\omega_2 t\\ &+ \frac{a_2A^2}{2}\cos 2\omega_1 t + \frac{a_2B^2}{2}\cos 2\omega_2 t\\ &+ a_2AB[\cos(\omega_1+\omega_2)t + \cos(\omega_1-\omega_2)t]\\ &- \frac{a_3AB}{4}\cos 3\omega_1 t + \frac{a_3B^3}{4}\cos 3\omega_2 t\\ &+ \frac{3a_3AB}{4}[A\cos(2\omega_1+\omega_2)t + B\cos(2\omega_2+\omega_1)t]\\ &+ \frac{3a_3AB}{4}[A\cos(2\omega_1-\omega_2)t + B\cos(2\omega_2-\omega_1)t]\end{aligned} \quad (14.13)$$

From (14.13), it is clear that the nonlinearity of the amplifier generates harmonic components $2\omega_1, 2\omega_2, 3\omega_1, 3\omega_2$ as well as intermodulation components $(\omega_1+\omega_2)$, $(\omega_1-\omega_2)$, $(2\omega_1+\omega_2)$, $(2\omega_2+\omega_1)$, $(2\omega_1-\omega_2)$, $(2\omega_2+\omega_1)$. Among all these, third-order intermodulation components $(2\omega_1-\omega_2)$ and $(2\omega_2+\omega_1)$ are most concerned due to their higher amplitudes compared to other intermodulation products and its existence within the pass band of the communication systems which cannot be rejected by using any filter. In case of equal level of the two input carriers, i.e., $A = B$, the level of each third-order intermodulation components is derived from (14.13) as follows:

$$\text{3rd order intermodulation level} = \frac{3a_3A^3}{4} \tag{14.14}$$

This shows that the levels of the third-order intermodulation products (IM3) are proportional to the cubic power of the input signal level. There will be 3-dB increase of absolute power level of IM3 products for each dB increase of input signal level. This implies that there will be 2-dBc increase of IM3 level for 1-dB increase of input power level.

Input–output power relationships of fundamental and third-order harmonics are shown in Fig. 14.5. From the slope relationship, it can be written that

$$\left(\text{slope of } \mathrm{IM}_3 \text{ power}\right) = 3 \times (\text{slope of fundamental power})$$

From Fig. 14.6

$$\frac{\left(P_{oIP3} - P_{o2f_1-f_2}\right)}{\Delta P_{\text{in}}} = 3\frac{\left(P_{oIP3} - P_{of_1}\right)}{\Delta P_{\text{in}}}$$

Here, P_{oIP3} is the third-order intercept output power level, $P_{o2f_1-f_2}$ is the minimum output detectable power level and P_{of_1} is the fundamental component output power level corresponding to $P_{o2f_1-f_2}$ as shown is Fig. 14.6.

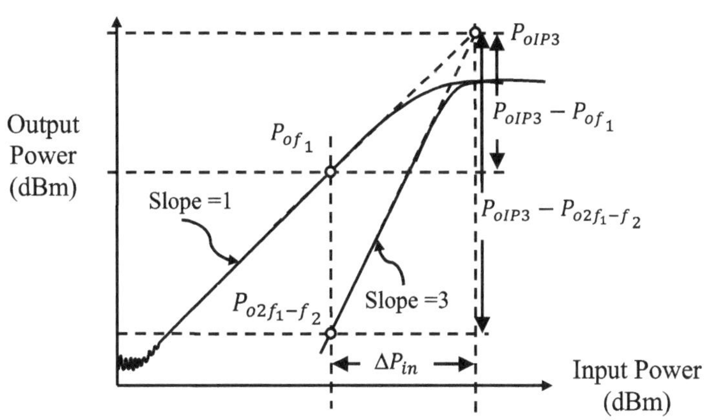

Fig. 14.6 Typical input–output characteristic of amplifier

$$\text{or,}\quad P_{of_1} - P_{o2f_1-f_2} = \frac{2}{3}\left(P_{oIP3} - P_{o2f_1-f_2}\right) \tag{14.15a}$$

$$\text{or,}\quad P_{of_1} - P_{o2f_1-f_2} = 2\left(P_{oIP3} - P_{of_1}\right) \tag{14.15b}$$

Therefore, for third-order intermodulation limited system, putting $P_{o\text{MDS}} = P_{o2f_1-f_2}$ in (14.15a), the SFDR can be written as

$$\text{SFDR} = P_{of_1} - P_{o2f_1-f_2} = \frac{2}{3}(P_{oIP3} - P_{o\text{MDS}}) \tag{14.16}$$

Putting the value of $P_{o\text{MDS}}$ from (14.10), the SFDR can be written as

$$\text{SFDR} = \frac{2}{3}(P_{oIP3} - 10\log(kT_o) - 10\log B - 10\log G_A - 10\log \text{NF} - X) \tag{14.17}$$

A communication transponder consists of several amplifiers as shown in Fig. 14.2. The front-end amplifier receives and amplifies very weak signal. Thus, the front-end amplifier is designed with minimum noise figure to achieve minimum possible detectable signal level. Intermediate amplifiers in a communication system operate over a moderate power level. Thus, these amplifiers are designed with maximum achievable power gain with moderate noise figure and output power capability. Final amplifier in a communication system is the high power amplifier to increase higher level of the dynamic range with the target of highest DC to RF efficiency.

Example 14.3 Calculate SFDR of an amplifier with output third-order intercept point 40 dBm. The available linear gain and noise figure of the amplifier is 44 and 5 dB, respectively. The bandwidth of the amplifier is 40 MHz.

Solution
Here,

$$P_{oIP3} = 40\ \text{dBm}$$
$$10\log(kT_o) = -174\ \text{dBm}$$
$$10\log B = 76\ \text{dBHz}$$
$$10\log G_A = 44\ \text{dB}$$
$$10\log \text{NF} = 5\ \text{dB}$$

Considering $X = 3$ dB

$$\text{SFDR} = \frac{2}{3}(P_{oIP3} - 10\log(kT_o) - 10\log B - 10\log G_A - 10\log \text{NF} - 3)$$
$$= \frac{2}{3}(40 + 174 - 76 - 44 - 5 - 3)\,\text{dB} = 57.3\ \text{dB}$$

14.3 Stability of Microwave Amplifier

While designing an amplifier, one important requirement is to ensure the stability of the amplifier, i.e., to get rid of any oscillations. Stability in a microwave circuit can be determined by its S-parameters. For a two-port network described by S-parameters, the input reflection coefficient Γ_{IN} and the output reflection coefficient Γ_{OUT} with the arbitrary load of reflection coefficient Γ_L and arbitrary source of reflection coefficient Γ_S can be written as (from 7.177a, 7.177b):

$$\Gamma_{\text{IN}} = S_{11} + \frac{\Gamma_L S_{12} S_{21}}{1 - \Gamma_L S_{22}} \tag{14.18a}$$

$$\Gamma_{\text{OUT}} = S_{22} + \frac{\Gamma_S S_{12} S_{21}}{1 - \Gamma_S S_{11}} \tag{14.18b}$$

A two-port network will oscillate if either or both of the input and the output ports have negative resistance. That means the network will oscillate if $|\Gamma_{\text{IN}}| > 1$ or $|\Gamma_{\text{OUT}}| > 1$ or both greater are than one. Though any passive network has the magnitude of the reflection coefficients less than one but for any active device with positive gain there is a fair possibility of having the magnitude of reflection coefficients greater than one.

There are two types of stability for a network, unconditional stability and conditional stability or potential unstability. A network is unconditional stable if the real part of the input and output impedances of the network are positive for all passive load and source impedances. Whereas, a network is said to be conditionally stable or potentially unstable if the real part of the input or output impedances of the network is negative for at least one passive load or source impedances. Also, the stability should be checked not only over the operating frequency range but also for all other frequencies.

The condition for unconditional stability of a network for any arbitrary passive load ($|\Gamma_L|<1$) and passive source ($|\Gamma_S|<1$) impedances is given by

$$|\Gamma_{\text{IN}}| < 1 \tag{14.19a}$$

and

$$|\Gamma_{\text{OUT}}| < 1 \tag{14.19b}$$

From (14.19a) and (14.19b) the condition for unconditional stability for a two-port network can be derived as [1–3]

$$K = \frac{1 - |S_{11}|^2 - |S_{22}|^2 + |\Delta|^2}{2|S_{12} S_{21}|} > 1 \tag{14.20a}$$

and,

$$|\Delta| < 1 \tag{14.20b}$$

where

$$|\Delta| = |S_{11}S_{22} - S_{12}S_{21}| \tag{14.20c}$$

This is known as the Rollett's condition for stability of a two-port network. These two conditions are the necessary and sufficient conditions for knowing the stability of a two-port network. These set of conditions are valid for any passive source and load impedances, i.e. source and load reflection coefficients of magnitude less than one. However, it should be remembered that these conditions are not necessarily sufficient for a source and load that have reflection gain, i.e. reflection coefficients greater than one.

Alternatively, condition for unconditional stability that involves only one condition of single parameter is the geometrically derived criterion as [4]

$$\mu_1 = \frac{1 - |S_{22}|^2}{|S_{11} - S_{22}^*\Delta| + |S_{12}S_{21}|} > 1 \tag{14.21a}$$

or

$$\mu_2 = \frac{1 - |S_{11}|^2}{|S_{22} - S_{11}^*\Delta| + |S_{12}S_{21}|} > 1 \tag{14.21b}$$

Either μ_1 or μ_2 alone greater than unity will ensure unconditional stability of a two-port network for any arbitrary passive source and load impedances. These stability factors measure the distance from the centre of the Smith chart to the nearest unstable regions.

For an unilateral network, $S_{12} = 0$ and thus from (14.18a) and (14.18b), the condition for unconditional stability of a unilateral network will be $|S_{11}| < 0$ and $|S_{22}| < 0$. Practically, there is no such transistor having $S_{12} = 0$, though it is very small. Thus, $|S_{11}| < 0$ and $|S_{22}| < 0$ are not sufficient conditions for unconditional stability of any practical device.

Practically, most microwave transistors used for amplifiers are potentially unstable due to the existence of internal feedback. The existence of nonzero-valued gate-to-drain capacitance C_{gd} in the FET and collector-to-base capacitance C_{cb} in a BJT provides the internal feedback which make the device potentially unstable.

A typical *S*-parameter file of a pHEMT device operating in small signal condition is shown in Table 14.1 over the frequency range of 0.5–18 GHz for the device DC bias of $V_{DS} = 2V, I_D = 15$ mA. We can check the stability of the device at different frequencies.

Example 14.4 Check the stability at frequency 6.0 GHz of the device for which *S*-parameters are given in Table 14.1.

Solution

From Table 14.1, the *S*-parameters of the device at 6.0 GHz are

Table 14.1 A typical S-parameter of a pHEMT device under bias condition: $V_{DS} = 2$ V, $I_D = 15$ mA, reference impedance: $Z_o = 50\ \Omega$

Freq. (GHz)	S_{11} (MAG)	$\angle S_{11}$ (DEG)	S_{21} (MAG)	$\angle S_{21}$ (DEG)	S_{12} (MAG)	$\angle S_{12}$ (DEG)	S_{22} (MAG)	$\angle S_{22}$ (DEG)
0.5	0.962	−13	6.112	166	0.0111	76	0.539	−15
1.0	0.937	−22	5.956	159	0.0211	69	0.525	−19
1.5	0.913	−33	5.810	150	0.0302	64	0.511	−24
2.0	0.889	−41	5.690	142	0.0394	58	0.498	−30
2.5	0.86	−51	5.522	133	0.0484	53	0.484	−36
3.0	0.834	−61	5.386	124	0.0567	48	0.469	−43
3.5	0.810	−71	5.236	116	0.0637	43	0.456	−49
4.0	0.784	−80	5.067	107	0.0702	38	0.440	−55
4.5	0.761	−90	4.911	99	0.0760	33	0.423	−61
5.0	0.740	−99	4.752	91	0.0809	28	0.410	−67
5.5	0.720	−107	4.586	84	0.0851	24	0.397	−73
6.0	0.701	−116	4.420	76	0.0889	19	0.385	−79
6.5	0.682	−124	4.260	69	0.0918	15	0.373	−84
7.0	0.663	−131	4.107	62	0.0941	11	0.362	−89
7.5	0.644	−139	3.974	55	0.0962	7	0.351	−93
8.0	0.627	−148	3.852	49	0.0980	3	0.343	−98
8.5	0.611	−157	3.747	42	0.0995	−1	0.333	−102
9.0	0.595	−165	3.659	35	0.1008	−5	0.323	−107
9.5	0.581	−173	3.571	29	0.1022	−9	0.313	−112
10.0	0.567	178	3.497	22	0.1039	−13	0.303	−116
10.5	0.556	170	3.430	16	0.1049	−17	0.293	−121
11.0	0.546	163	3.368	9	0.1064	−21	0.284	−127
11.5	0.537	155	3.317	3	0.1078	−26	0.274	−131
12.0	0.528	149	3.265	−4	0.1093	−30	0.265	−135
12.5	0.52	142	3.216	−10	0.1105	−35	0.255	−139
13.0	0.513	135	3.169	−17	0.1116	−39	0.246	−143
13.5	0.506	128	3.120	−24	0.1126	−44	0.235	−146
14.0	0.498	121	3.080	−30	0.1137	−49	0.225	−150
14.5	0.492	113	3.044	−37	0.1151	−54	0.215	−155
15.0	0.489	106	3.014	−44	0.1160	−59	0.207	−159
15.5	0.484	98	2.990	−51	0.1171	−65	0.200	−163
16.0	0.485	91	2.967	−58	0.1185	−71	0.193	−167
16.5	0.485	83	2.945	−65	0.1197	−77	0.187	−171
17.0	0.485	75	2.947	−71	0.1206	−82	0.182	−175
17.5	0.487	69	2.961	−77	0.1215	−87	0.177	−178
18.0	0.490	64	2.979	−81	0.1230	−90	0.174	179

$$S_{11} = 0.701\angle -116^\circ$$
$$S_{12} = 0.0889\angle 19^\circ$$
$$S_{21} = 4.420\angle 76^\circ$$
$$S_{22} = 0.385\angle -79^\circ$$

To calculate $|\Delta|$:

$$\begin{aligned}\Delta &= S_{11}S_{22} - S_{12}S_{21} \\ &= 0.701\angle -116^\circ \times 0.385\angle 79^\circ - 0.0889\angle 19^\circ \times 4.420\angle 76^\circ \\ &= 0.393\angle -125.2^\circ\end{aligned}$$

Thus, $|\Delta| = 0.393 < 1$

To calculate Rollett's stability factor K:

$$\begin{aligned}K &= \frac{1 - |S_{11}|^2 - |S_{22}|^2 + |\Delta|^2}{2|S_{12}S_{21}|} \\ &= \frac{1 - (0.701)^2 - (0.385)^2 + (0.393)^2}{2(0.0889 \times 4.420)} \\ &= 0.655\end{aligned}$$

Thus, $K = 0.655 < 1$

From (14.20a) and (14.20b), it is clear that the device at the given bias condition is potentially unstable or conditionally stable at 6.0 GHz. This can also be verified by computing either μ_1 or μ_2 as given below.

To compute μ_1:

$$\begin{aligned}\mu_1 &= \frac{1 - |S_{22}|^2}{\left|S_{11} - S_{22}^*\Delta\right| + |S_{12}S_{21}|} \\ &= \frac{1 - (0.385)^2}{|0.701\angle -116^\circ - 0.385\angle 79^\circ \times 0.393\angle -125.2^\circ| + |0.0889 \times 4.420|} \\ &= 0.806 < 1\end{aligned}$$

$\mu_1 < 1$ confirms the potential instability of the device at 6.0 GHz.

To compute μ_2:

$$\begin{aligned}\mu_2 &= \frac{1 - |S_{11}|^2}{\left|S_{22} - S_{11}^*\Delta\right| + |S_{12}S_{21}|} \\ &= \frac{1 - (0.701)^2}{|0.385\angle -79^\circ - 0.701\angle 116^\circ \times 0.393\angle -125.2^\circ| + |0.0889 \times 4.420|} \\ &= 0.651 < 1\end{aligned}$$

$\mu_2 < 1$ also confirms the potential instability of the device at 6.0 GHz.

Example 14.5 Check the stability at frequency 14.0 GHz of the same device using S-parameters given in Table 14.1.

Solution

From Table 14.1, the S-parameters of the device at 14.0 GHz are

$$\begin{aligned} S_{11} &= 0.498\angle 121^\circ \\ S_{12} &= 0.1137\angle -49^\circ \\ S_{21} &= 3.080\angle -30^\circ \\ S_{22} &= 0.225\angle -150^\circ \end{aligned}$$

To calculate $|\Delta|$:

$$\begin{aligned} \Delta &= S_{11}S_{22} - S_{12}S_{21} \\ &= 0.498\angle 121^\circ \times 0.225\angle -150^\circ - 0.1137\angle -49^\circ \times 3.080\angle -30^\circ \\ &= 0.291\angle 83.9^\circ \end{aligned}$$

Thus, $|\Delta| = 0.393 < 1$

To calculate Rollett's stability factor K:

$$\begin{aligned} K &= \frac{1 - |S_{11}|^2 - |S_{22}|^2 + |\Delta|^2}{2|S_{12}S_{21}|} \\ &= \frac{1 - (0.498)^2 - (0.225)^2 + (0.291)^2}{2(0.1137 \times 3.080)} \\ &= 1.122 \end{aligned}$$

Thus, $K = 1.122 > 1$.

From (14.21a) and (14.21b), it is clear that the device at the given bias condition is unconditionally stable at 14.0 GHz. This can also be verified by computing μ_1 and μ_2 as given below.

To compute μ_1 at 14.0 GHz:

$$\begin{aligned} \mu_1 &= \frac{1 - |S_{22}|^2}{\left|S_{11} - S_{22}^*\Delta\right| + |S_{12}S_{21}|} \\ &= \frac{1 - (0.225)^2}{|0.498\angle 121^\circ - 0.225\angle 150^\circ \times 0.291\angle 83.9^\circ| + |0.1137 \times 3.080|} \\ &= 1.082 > 1 \end{aligned}$$

$\mu_1 > 1$, confirms that the device is unconditionally stable at 14.0 GHz.

To compute μ_2 at 14.0 GHz:

$$\mu_2 = \frac{1 - |S_{11}|^2}{\left|S_{22} - S_{11}^*\Delta\right| + |S_2 S_{21}|}$$

$$= \frac{1 - (0.498)^2}{|0.225\angle - 150^\circ - 0.498\angle - 121^\circ \times 0.393\angle - 125.2^\circ| + |0.0889 \times 4.420|}$$

$$= 1.137 > 1$$

$\mu_2 > 1$ also confirms that the device is unconditionally stable at 14.0 GHz.

These two examples show that, though the device is potentially unstable at 6.0 GHz, the same device at the same bias condition is unconditionally stable at 14 GHz. Practically, all the microwave devices at lower operating frequency are potentially unstable and the same device is unconditionally stable at its higher operating frequency range. Figure 14.7 shows the plot of different stability factors of the device from 0.5 to 18 GHz based on its *S*-parameters written in the Table 14.1. It shows that the device is potentially unstable over the frequency range below 11.5 GHz and the same device is unconditionally stable over the frequency range above 11.5 GHz.

The stability condition is frequency dependent, as it is shown in the above discussions for certain matching networks it is possible for an amplifier to be stable at its design frequency and still be unstable at other frequencies. Thus, care must be taken to design the matching networks to ensure stability not only over the operating frequency range but also outside the operating frequencies. A stable amplifier can be designed using potentially unstable or conditional stable device only for certain range of passive source and load impedances, i.e. for the amplifier both the conditions $|\Gamma_{IN}| < 1$ and $|\Gamma_{OUT}| < 1$ can be satisfied.

To address the stability problems, graphical analysis is also used to determine the regions in the Smith chart where the values of load impedances (Γ_L) and source impedances (Γ_S) produce stable amplifier, i.e. $|\Gamma_{IN}| < 1$ and $|\Gamma_{OUT}| < 1$. Based on

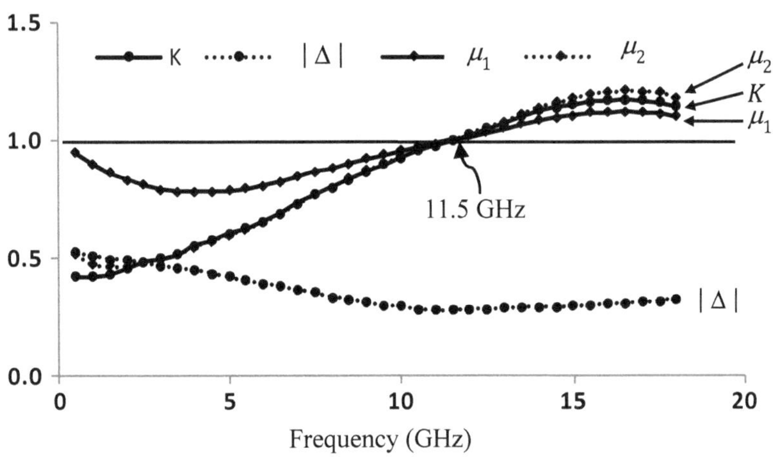

Fig. 14.7 Plot of stability factors of the pHEMT device over frequency

these two conditions, i.e. $|\Gamma_{\text{IN}}| < 1$ and $|\Gamma_{\text{OUT}}| < 1$, there are two stability circles, source stability circle and load stability circle. The source or input stability circle is defined as the loci of source reflection coefficient (Γ_S) for which $|\Gamma_{\text{OUT}}| = 1$. And the load or output stability circle is defined as the loci of load reflection coefficient (Γ_L) for which $|\Gamma_{\text{IN}}| = 1$. The input and output stability circles define the boundary between stable and unstable regions for Γ_S and Γ_L, respectively.

To determine the input stability circle, the magnitude of the output reflection coefficient will be equal to one for the source reflection coefficient. Thus, it can be written as

$$|\Gamma_{\text{OUT}}| = \left| S_{22} + \frac{\Gamma_S S_{12} S_{21}}{1 - \Gamma_S S_{11}} \right| = 1 \tag{14.22a}$$

or

$$|S_{22}(1 - \Gamma_S S_{11}) + \Gamma_S S_{12} S_{21}| = |1 - \Gamma_S S_{11}|$$

or

$$|S_{22} - \Delta \Gamma_S| = |1 - \Gamma_S S_{11}|$$

Squaring both the sides, it can be written as

$$|S_{22}|^2 + |\Delta|^2 |\Gamma_S|^2 - \Delta \Gamma_S S_{22}^* + \Delta^* \Gamma_S^* S_{22} = 1 + |S_{11}|^2 |\Gamma_S|^2 - \left(S_{11}^* \Gamma_S + \Gamma_S S_{11}\right)$$

or

$$\left(|S_{11}|^2 - |\Delta|^2\right) \Gamma_S \Gamma_S^* - \left(S_{11} - \Delta S_{22}^*\right) \Gamma_S - \left(S_{11}^* - \Delta^* S_{22}\right) \Gamma_S^* = |S_{22}|^2 - 1$$

or,

$$\Gamma_S \Gamma_S^* - \frac{\left(S_{11} - \Delta S_{22}^*\right) \Gamma_S + \left(S_{11}^* - \Delta^* S_{22}\right) \Gamma_S^*}{\left(|S_{11}|^2 - |\Delta|^2\right)} = \frac{|S_{22}|^2 - 1}{\left(|S_{11}|^2 - |\Delta|^2\right)}$$

now adding $\frac{|S_{11} - \Delta S_{22}^*|^2}{\left(|S_{11}|^2 - |\Delta|^2\right)^2}$ to both the sides:

$$\left| \Gamma_S - \frac{\left(S_{11} - \Delta S_{22}^*\right)^*}{|S_{11}|^2 - |\Delta|^2} \right|^2 = \frac{|S_{22}|^2 - 1}{|S_{11}|^2 - |\Delta|^2} + \frac{|S_{11} - \Delta S_{22}^*|^2}{\left(|S_{11}|^2 - |\Delta|^2\right)^2}$$

or

$$\left| \Gamma_S - \frac{\left(S_{11} - \Delta S_{22}^*\right)^*}{\left(|S_{11}|^2 - |\Delta|^2\right)} \right| = \left| \frac{S_{12} S_{21}}{\left(|S_{11}|^2 - |\Delta|^2\right)} \right| \tag{14.22b}$$

This is an equation of circle in the Γ_S plane with centre c_S and radius r_S as follows:

$$c_S = \frac{\left(S_{11} - \Delta\, S_{22}^*\right)^*}{|S_{11}|^2 - |\Delta|^2} \tag{14.23a}$$

$$r_S = \left|\frac{S_{12}\, S_{21}}{|S_{11}|^2 - |\Delta|^2}\right| \tag{14.23b}$$

In a similar way, centre and radius of the output stability circle can be derived. To determine the output stability circle, the magnitude of the input reflection coefficient will be equal to one for the load reflection coefficient. Thus, it can be written as

$$|\Gamma_{\text{IN}}| = \left|S_{11} + \frac{\Gamma_L\, S_{12}\, S_{21}}{1 - \Gamma_L\, S_{22}}\right| = 1 \tag{14.24a}$$

or

$$|S_{11}(1 - \Gamma_L\, S_{22}) + \Gamma_L\, S_{12}\, S_{21}| = |1 - \Gamma_L\, S_{22}|$$

or,

$$|S_{11} - \Delta\, \Gamma_L| = |1 - \Gamma_L\, S_{22}|$$

Squaring both the sides, it can be written as

$$\begin{aligned} &|S_{11}|^2 + |\Delta|^2 |\Gamma_L|^2 - \Delta\, \Gamma_L\, S_{11}^* + \Delta^*\, \Gamma_L^*\, S_{11} \\ &= 1 + |S_{22}|^2 |\Gamma_L|^2 - \left(S_{22}^*\, \Gamma_L^* + \Gamma_L\, S_{22}\right) \end{aligned}$$

or

$$\left(|S_{22}|^2 - |\Delta|^2\right)\Gamma_L\, \Gamma_L^* - \left(S_{22} - \Delta\, S_{11}^*\right)\Gamma_L - \left(S_{22}^* - \Delta^*\, S_{11}\right)\Gamma_L^* = |S_{11}|^2 - 1$$

or

$$\Gamma_L\, \Gamma_L^* - \frac{\left(S_{22} - \Delta\, S_{11}^*\right)\Gamma_L + \left(S_{22}^* - \Delta^*\, S_{11}\right)\Gamma_L^*}{\left(|S_{22}|^2 - |\Delta|^2\right)} = \frac{|S_{11}|^2 - 1}{\left(|S_{22}|^2 - |\Delta|^2\right)}$$

now adding $\frac{|S_{22} - \Delta\, S_{11}^*|^2}{\left(|S_{22}|^2 - |\Delta|^2\right)^2}$ to both the sides:

$$\left|\Gamma_L - \frac{\left(S_{22} - \Delta\, S_{11}^*\right)^*}{|S_{22}|^2 - |\Delta|^2}\right|^2 = \frac{|S_{11}|^2 - 1}{|S_{22}|^2 - |\Delta|^2} + \frac{|S_{22} - \Delta\, S_{11}^*|^2}{\left(|S_{22}|^2 - |\Delta|^2\right)^2}$$

or

$$\left| \Gamma_L - \frac{\left(S_{22} - \Delta\, S_{11}^*\right)^*}{\left(|S_{22}|^2 - |\Delta|^2\right)} \right| = \left| \frac{S_{12}\, S_{21}}{\left(|S_{22}|^2 - |\Delta|^2\right)} \right| \tag{14.24b}$$

This is an equation of circle in the Γ_L plane with centre c_L and radius r_L as follows:

$$c_L = \frac{\left(S_{22} - \Delta\, S_{11}^*\right)^*}{|S_{22}|^2 - |\Delta|^2} \tag{14.25a}$$

$$r_L = \left| \frac{S_{12}\, S_{21}}{|S_{22}|^2 - |\Delta|^2} \right| \tag{14.25b}$$

Using Eqs. (14.23a), (14.23b) and (14.25a), (14.25b), input and output stability circles can be drawn and stable and unstable regions in the Γ_S and Γ_L plane can be determined.

For a conditionally stable device, input and output stability circles are shown in Fig. 14.8a, b, respectively. In the Γ_S plane, on one side of the boundary of the input stability circle, $|\Gamma_{\mathrm{OUT}}| < 1$ and on the other side of the boundary, $|\Gamma_{\mathrm{OUT}}| > 1$. At the origin of the Smith chart, $\Gamma_S = 0$ thus from (14.18b) the output reflection coefficient, $\Gamma_{\mathrm{OUT}} = S_{22}$. Therefore, if $|S_{22}| < 1$ then $|\Gamma_{\mathrm{OUT}}| < 1$, i.e. the origin of the Smith chart represents a stable operating point. On the other hand, if $|S_{22}| > 1$ then $|\Gamma_{\mathrm{OUT}}| > 1$, i.e. the origin of the Smith chart represents an unstable operating point. This is illustrated in Fig. 14.8a.

Similarly, in the Γ_L plane, on one side of the input stability circle boundary $|\Gamma_{\mathrm{IN}}| < 1$ and on the other side of the boundary $|\Gamma_{\mathrm{IN}}| > 1$. At the origin of the Smith chart, $\Gamma_L = 0$, and thus from (14.18a) the input reflection coefficient $\Gamma_{\mathrm{IN}} = S_{11}$. Therefore, if $|S_{11}| < 1$ then $|\Gamma_{\mathrm{IN}}| < 1$, i.e. the origin of the Smith chart represents a stable operating point. On the other hand, if $|S_{11}| > 1$ then $|\Gamma_{\mathrm{IN}}| > 1$, i.e. the origin of the Smith chart represents an unstable operating point. This is illustrated in Fig. 14.8(b).

For an unconditionally stable device, any passive load and source impedance produce stable amplifier. For an unconditionally stable device, $|S_{11}| > 1$ or $|S_{22}| > 1$ is not possible. Otherwise, the source and load impedance corresponding to $\Gamma_S = 0$ or $\Gamma_L = 0$ will produce $|\Gamma_{\mathrm{OUT}}| > 1$ or $|\Gamma_{\mathrm{IN}}| > 1$, which is contradictory to the condition for unconditional stable device.

Therefore, for $|S_{11}| < 1$ and $|S_{22}| < 1$, the input and the output stability circles should fall completely outside the Smith chart or to completely enclose the Smith chart as shown in Fig. 14.9. Therefore, the conditions for unconditional stability for all passive source and load impedances can be written as

$$||c_S| - r_S| > 1 \quad \text{for} \quad |S_{22}| < 1 \tag{14.26a}$$

$$||c_L| - r_L| > 1 \quad \text{for} \quad |S_{11}| < 1 \tag{14.26b}$$

Example 14.6 Compute the stability circles at 6.0 GHz of the device for S-parameters which are given in Table 14.1.

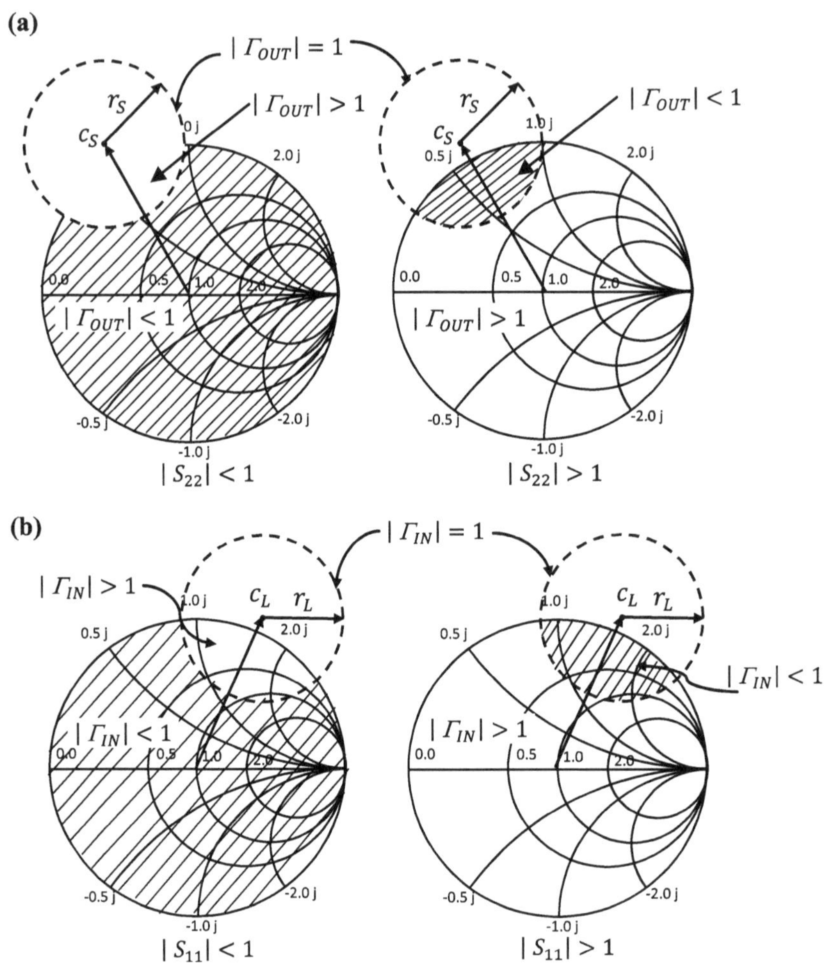

Fig. 14.8 Stability circles for conditionally stable device, **a** Γ_S plane and **b** Γ_L plane

Solution

For input stability circle at 6.0 GHz:

$$
\begin{aligned}
c_S &= \frac{\left(S_{11} - \Delta S_{22}^*\right)^*}{|S_{11}|^2 - |\Delta|^2} \\
&= \frac{(0.701\angle - 116^\circ - 0.393\angle - 125.2^\circ \times 0.385\angle 79^\circ)^*}{(0.701)^2 - |0.393|^2} \\
&= \frac{(0.701\angle - 116^\circ - 0.151\angle - 46.2^\circ)^*}{0.337} \\
&= \frac{0.664\angle 128.4^\circ}{0.337} = 1.973\angle 128.4^\circ
\end{aligned}
$$

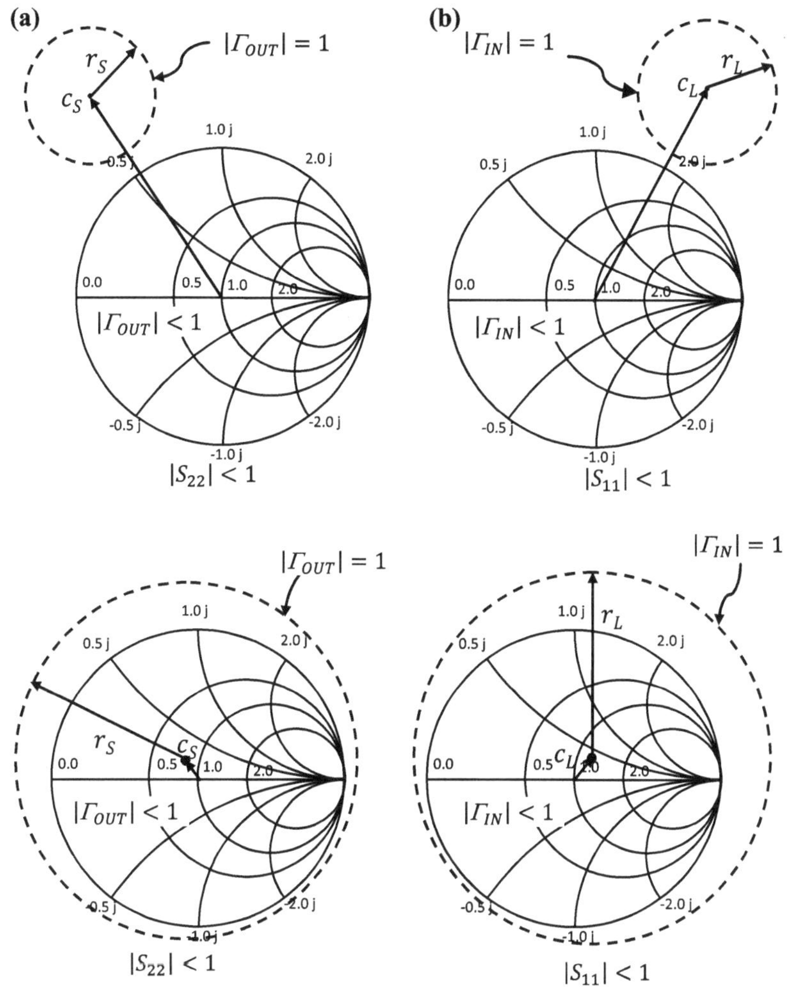

Fig. 14.9 Stability circles for unconditionally stable devices, **a** Γ_S plane and **b** Γ_L plane

$$r_S = \left| \frac{S_{12}\,S_{21}}{|S_{11}|^2 - |\Delta|^2} \right|$$
$$= \frac{0.0889 \times 4.420}{(0.701)^2 - |0.393|^2}$$
$$= \frac{0.3929}{0.337} = 1.167$$

Therefore, $||c_S| - r_S| = 1.973 - 1.167 = 0.806$ (= μ_1 (!!) Example 14.4).

The input stability circle and stable region on the Smith chart is shown in Fig. 14.10a.

For output stability circle at 6.0 GHz:

$$c_L = \frac{\left(S_{22} - \Delta S_{11}^*\right)^*}{|S_{22}|^2 - |\Delta|^2}$$
$$= \frac{(0.385\angle - 79° - 0.393\angle - 125.2° \times 0.701\angle 116°)^*}{(0.385)^2 - |0.393|^2}$$
$$= -60.063\angle 120.7°$$
$$r_L = \left| \frac{S_{12}\,S_{21}}{|S_{22}|^2 - |\Delta|^2} \right|$$
$$= \frac{0.0889 \times 4.420}{(0.385)^2 - |0.393|^2}$$
$$= \frac{0.3929}{0.00622} = 60.713$$

Therefore, $||c_L| - r_L| = 0.651$ (= μ_2 (!!) Example 14.4).

The output stability circle and the stable region on the Smith chart are shown in Fig. 14.10b.

Example 14.7 Compute the stability circles at 14.0 GHz of the device for *S*-parameters which are given in Table 14.1.

Solution

For input stability circle at 14.0 GHz:

$$c_S = \frac{\left(S_{11} - \Delta\, S_{22}^*\right)^*}{|S_{11}|^2 - |\Delta|^2}$$

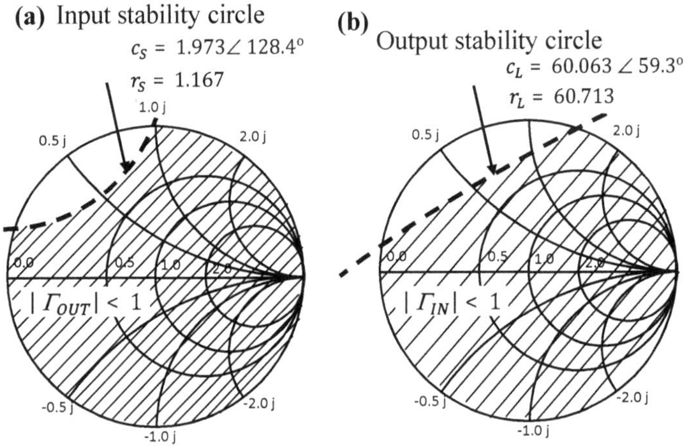

Fig. 14.10 Stable and unstable regions of the device

$$= \frac{(0.498\angle 121° - 0.291\angle 83.9° \times 0.225\angle 150°)^*}{(0.498)^2 - |0.291|^2}$$

$$= \frac{0.527\angle - 114.42°}{0.163} = 3.227\angle - 114.42°$$

$$r_S = \left|\frac{S_{12}\,S_{21}}{|S_{11}|^2 - |\Delta|^2}\right|$$

$$= \frac{0.1137 \times 3.080}{(0.498)^2 - |0.291|^2}$$

$$= \frac{0.3502}{0.163} = 2.145$$

$||c_S|-r_S| = 3.227 - 2.145 = 1.082\,(= \mu_1$ (!!) Example 14.5).

The input stability circle is shown in Fig. 14.11a. It shows that for the unconditionally stable device, the input stability circle is outside the Smith chart.

For output stability circle at 14.0 GHz:

$$c_L = \frac{\left(S_{22} - \Delta S_{11}^*\right)^*}{|S_{22}|^2 - |\Delta|^2}$$

$$= \frac{(0.225\angle - 150° - 0.291\angle 83.9° \times 0.498\angle - 121°)^*}{(0.225)^2 - |0.291|^2}$$

$$= -9.126\angle 175.4°$$

$$r_L = \left|\frac{S_{12}\,S_{21}}{|S_{22}|^2 - |\Delta|^2}\right|$$

$$= \left|\frac{0.1137 \times 3.080}{(0.225)^2 - |0.291|^2}\right|$$

$$= \frac{0.350}{0.034} = 10.263$$

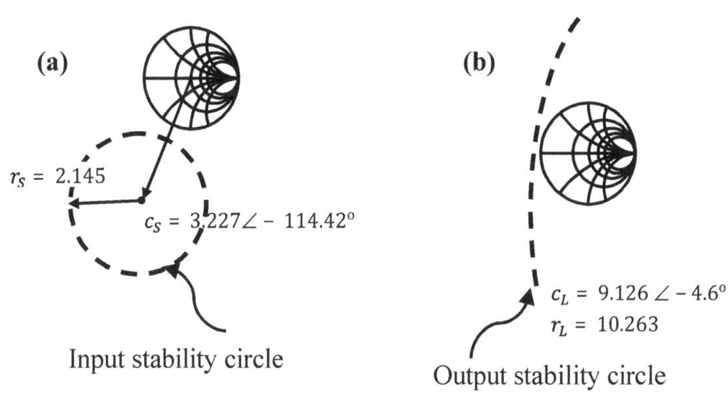

Fig. 14.11 Stability circles for unconditionally stable devices

Therefore, $||c_L| - r_L| = 1.137$ (= μ_2 (!!) Example 14.5).

The output stability circle is shown in Fig. 14.11(b). It shows that for the unconditionally stable device, the output stability circle encloses the Smith chart.

14.4 Single-Stage Amplifier Design

Knowing the stability of a device, input and output matching networks are designed to ensure unconditional stability of the amplifier over its entire frequency range where the device has gain. Stability of an amplifier using an unconditionally stable device is not an issue as long as the passive matching networks are used and there is no positive feedback. To design an amplifier using an unconditional stable device with maximum gain, complex conjugate matching at its input and output is done. However, to realize a stable amplifier care must be taken to avoid any unwanted positive feedback due to microwave radiation from the output network that couples to its input network.

There are many ways to design a stable amplifier using a potentially unstable device. As already discussed, use of lossless appropriate input and output matching networks that ensure the source and load impedances outside the unstable region in the Smith chart makes the amplifier stable. Other ways to design stable amplifier are the use of resistive loading at the device input and/or output, and use of negative feedback.

Transducer power gain G_T is defined as the ratio of power delivered to the load (P_L) and power available (P_{AVS}) from the source. This is already derived in Chap. 7 and is written here as follows:

$$G_T = \frac{P_L}{P_{\text{AVS}}} = \frac{1 - |\Gamma_S|^2}{|1 - \Gamma_S\,\Gamma_{\text{IN}}|^2}|S_{21}|^2\frac{1 - |\Gamma_L|^2}{|1 - S_{22}\,\Gamma_L|^2} \tag{14.27}$$

Thus, the transducer power gain depends on both the source and the load reflection coefficients. With reference to Fig. 14.1 and (14.27), the transducer power gain G_T of an amplifier can be expressed as separate gain factors as

$$G_T = G_S \times G_o \times G_L \tag{14.28}$$

where G_S, the first term, is gain factor due to the input (source) matching network, G_o, the middle term, is the gain factor for the transistor itself, and G_L, the last term, is the gain factor for the output (load) matching network. For a given transistor with particular bias condition, while input of the device is fed by a source of reference impedance (R_o) and output is terminated with the reference load impedance (R_o), then the gain of the device is fixed and is given by, $G_T = G_o = |S_{21}|^2$. Thus, the overall gain of the amplifier is controlled by the gains G_S and G_L which are determined by the input and output matching networks. Maximum gain may be realized when the matching networks provide a conjugate match to the input and output of the device.

However, the conjugate matching at both the terminals are possible only when the device is unconditionally stable at the operating frequency band. If the device is not unconditionally stable, then the conjugate matching will not be permissible at both the device terminals. In this case, input and output matching networks must be selected to ensure load and source impedances within the respective stable regions of the Smith chart.

14.4.1 Amplifiers Using Unconditionally Stable Device

A transistor over its unconditionally stable frequency range provides stable gain for any passive source and load impedances. Therefore, depending on the performance requirements, the input and output matching networks can be designed without any concern of stability of the amplifier. For example, to design a low-noise amplifier, the input matching network should provide input termination to the device corresponding to its minimum noise figure requirement and output can be conjugate matched for achieving maximum possible gain. In case of high-gain amplifier, both the input and output ports of the device can be conjugately matched. Therefore, to achieve maximum gain, simultaneous conjugate matching is done for an unconditional stable transistor. The conditions for simultaneous conjugate match at input and output are

$$\Gamma_{\mathrm{IN}} = \Gamma_S^*, \quad \text{and } \Gamma_{\mathrm{OUT}} = \Gamma_L^* \tag{14.29}$$

Therefore, from (14.18a) and (14.18b):

$$\left(S_{11} - \Gamma_S^*\right)(1 - S_{22}\,\Gamma_L) + S_{12}\,S_{21}\,\Gamma_L = 0 \tag{14.30a}$$

$$\left(S_{22} - \Gamma_L^*\right)(1 - S_{11}\,\Gamma_S) + S_{12}\,S_{21}\,\Gamma_S = 0 \tag{14.30b}$$

Or,

$$\Gamma_L = \frac{\Gamma_S^* - S_{11}}{\Gamma_S^* S_{22} - \Delta} \tag{14.31a}$$

$$\Gamma_S = \frac{\Gamma_L^* - S_{22}}{\Gamma_L^* S_{11} - \Delta} \tag{14.31b}$$

Computing Γ_L^* using (14.31a) and putting it into (14.30b), the source reflection coefficient $\Gamma_S = \Gamma_{SM}$, corresponding to the simultaneous conjugate matching of both the ports can be written as

$$\left(S_{11} - \Delta S_{22}^*\right)\Gamma_{SM}^2 - \left(1 + |S_{11}|^2 - |S_{22}|^2 - |\Delta|^2\right)\Gamma_{SM} + \left(S_{11}^* - S_{22}\,\Delta^*\right) = 0$$

or,

$$C_1\Gamma_{SM}^2 - B_1\Gamma_{SM} + C_1^* = 0 \quad (14.32) \tag{14.32}$$

where

$$B_1 = 1 + |S_{11}|^2 - |S_{22}|^2 - |\Delta|^2 \quad \mathit{and} \quad C_1 = S_{11} - \Delta S_{22}^*$$

Similarly, computing Γ_S^* using (14.31b) and putting it into (14.30a), the load reflection coefficient $\Gamma_L = \Gamma_{LM}$, corresponding to the simultaneous conjugate matching of both the ports can be written as

$$\left(S_{22} - \Delta S_{11}^*\right)\Gamma_{LM}^2 - \left(1 + |S_{22}|^2 - |S_{11}|^2 - |\Delta|^2\right)\Gamma_{LM} + \left(S_{22}^* - S_{11}\Delta^*\right) = 0$$

or,

$$C_2\Gamma_{LM}^2 - B_2\Gamma_{LM} + C_2^* = 0 \quad (14.33) \tag{14.33}$$

where

$$B_2 = 1 + |S_{22}|^2 - |S_{11}|^2 - |\Delta|^2 \quad \text{and} \quad C_2 = S_{22} - \Delta\, S_{11}^*$$

From (14.32) and (14.33), values of source reflection coefficient Γ_{SM} and load reflection coefficients Γ_{LM} corresponding to simultaneous conjugate matching at input and output can be written as

$$\Gamma_{SM} = \frac{B_1}{2C_1} \pm \frac{|C_1|}{C_1}\sqrt{\left(\frac{B_1}{2|C_1|}\right)^2 - 1} \tag{14.34a}$$

$$\Gamma_{LM} = \frac{B_2}{2C_2} \pm \frac{|C_2|}{C_2}\sqrt{\left(\frac{B_2}{2|C_2|}\right)^2 - 1} \tag{14.34b}$$

For passive matching networks at both the input and output, $|\Gamma_{SM}| < 1$ and $|\Gamma_{LM}| < 1$. The plus (+) or minus (−) sign in the expressions should be selected accordingly.

Putting the value of Γ_{IN} from (14.18a) in (14.27), the transducer power gain can be written as

$$G_T = \frac{\left(1 - |\Gamma_S|^2\right)|S_{21}|^2\left(1 - |\Gamma_L|^2\right)}{|(1 - S_{11}\Gamma_S)(1 - S_{22}\Gamma_L) - S_{12}S_{21}\,\Gamma_S\Gamma_L|^2} \tag{14.35}$$

Under simultaneous conjugate math conditions ($\Gamma_S = \Gamma_{SM}$ and $\Gamma_L = \Gamma_{LM}$), the maximum transducer gain can be expressed as

$$G_{T\max} = \frac{\left(1 - |\Gamma_{SM}|^2\right)|S_{21}|^2\left(1 - |\Gamma_{LM}|^2\right)}{|(1 - S_{11}\Gamma_{SM})(1 - S_{22}\Gamma_{LM}) - S_{12}S_{21}\,\Gamma_{SM}\Gamma_{LM}|^2} \tag{14.36}$$

$$= \frac{1 - |\Gamma_{SM}|^2}{\left|1 - |\Gamma_{SM}|^2\right|^2} |S_{21}|^2 \frac{1 - |\Gamma_{LM}|^2}{|1 - S_{22}\Gamma_{LM}|^2}$$

$$= \frac{1}{1 - |\Gamma_{SM}|^2} |S_{21}|^2 \frac{1 - |\Gamma_{LM}|^2}{|1 - S_{22}\Gamma_{LM}|^2} = G_{P\max} \tag{14.37}$$

Comparing with (7.182), this expression (14.37) is nothing but the maximum operating power gain $G_{P\max}$. This can also be expressed as

$$G_{T\max} = \frac{1 - |\Gamma_{SM}|^2}{|1 - S_{11}\Gamma_{SM}|^2} |S_{21}|^2 \frac{1}{\left(1 - |\Gamma_{LM}|^2\right)} = G_{A\max} \tag{14.38}$$

Comparing with (7.183), this is the expression of maximum available gain $G_{A\max}$. Therefore, a two-port network under conjugate match conditions at its both the ports provides maximum available gain. The maximum available gain for an unconditional stable device can also be expressed in terms of its stability factor as (to be proved in Sect. 14.5.3):

$$G_{T\max} = \frac{|S_{21}|}{|S_{12}|}\left(k - \sqrt{k^2 - 1}\right) \tag{14.39}$$

This can be expressed in dB as

$$G_{A\max}(\mathrm{dB}) = 10\log\left(\frac{|S_{21}|}{|S_{12}|}\right) + 10\log\left(k - \sqrt{k^2 - 1}\right) \tag{14.40}$$

For unilateral transistor, $S_{12} = 0$ which leads to

$$\Gamma_{SM} = S_{11}^* \quad \text{and} \quad \Gamma_{LM} = S_{22}^*$$

Therefore, from (14.38) the maximum available gain for a unilateral device can be expressed as

$$(G_{A\max})_{\mathrm{UNILATERAL}} = \frac{1}{1 - |S_{11}|^2} |S_{21}|^2 \frac{1}{1 - |S_{22}|^2} \tag{14.41}$$

Example 14.8 Calculate maximum available gain at 14.0 GHz of the device of S-parameters as in Table 14.1.

Solution

From (14.39), the maximum available gain:

$$G_{T\max} = \frac{|S_{21}|}{|S_{12}|}\left(k - \sqrt{k^2 - 1}\right)$$

From Table 14.1, the S-parameters of the device at 14.0 GHz are

$$\begin{aligned} S_{11} &= 0.498\angle 121^\circ \\ S_{12} &= 0.1137\angle -49^\circ \\ S_{21} &= 3.080\angle -30^\circ \\ S_{22} &= 0.225\angle -150^\circ \end{aligned}$$

Therefore,

$$\begin{aligned} \Delta &= S_{11}S_{22} - S_{12}S_{21} \\ &= 0.498\angle 121^\circ \times 0.225\angle -150^\circ - 0.1137\angle -49^\circ \times 3.080\angle -30^\circ \\ &= 0.291\angle 83.9^\circ \end{aligned}$$

Thus, stability factor:

$$\begin{aligned} K &= \frac{1 - |S_{11}|^2 - |S_{22}|^2 + |\Delta|^2}{2|S_{12}S_{21}|} \\ &= \frac{1 - (0.498)^2 - (0.225)^2 + (0.291)^2}{2(0.1137 \times 3.080)} \\ &= 1.122 \end{aligned}$$

$$\begin{aligned} G_{A\,\max}(\mathrm{dB}) &= 10\log\left(\frac{3.080}{0.1137}\right) + 10\log\left(1.122 - \sqrt{1.122^2 - 1}\right) \\ &= 14.43\ \mathrm{dB} - 1.922\ \mathrm{dB} = 12.5\ \mathrm{dB} \end{aligned}$$

14.4.2 Amplifier Using Conditionally Stable Device

Conjugate matching is not allowed for an amplifier at a frequency where the device is potentially unstable. Either input or output or both ports of the device should be terminated with an impedance which is not corresponding to conjugate match. The load and source impedances should be within the stable region of the Smith chart. Maximum stable gain from a device can be achieved when the stability factor $k = 1$. Thus, the maximum stable gain MSG can be expressed as

$$\mathrm{MSG} = \frac{|S_{21}|}{|S_{12}|} \tag{14.42a}$$

This can be expressed in dB as:

$$\mathrm{MSG(dB)} = 10\log\left(\frac{|S_{21}|}{|S_{12}|}\right) \tag{14.42b}$$

Example 14.9 Compute maximum stable gain at 6.0 GHz of the device (Ref. Table 14.1).

Solution

From Table 14.1, the S-parameters of the device at 6.0 GHz are

$$S_{12} = 0.0889\angle 19^\circ$$
$$S_{21} = 4.420\angle 76^\circ$$

From (14.42b), the maximum stable gain at 6 GHz:

$$\text{MSG(dB)} = 10\log\left(\frac{4.420}{0.0889}\right)$$
$$= 17.32 \text{ dB}$$

Figure 14.12 shows the plot of gain over frequency for a pHEMT device whose S-parameters are written in Table 14.1. The maximum stable gain (MSG) of the device is shown over the frequency range where it is conditionally stable and maximum available gain (MAG) is written over the unconditionally stable frequency range. The plot shows an important characteristic of decrease of gain with increase of operating frequency. This creates a challenge for an amplifier designer to design an amplifier with a flat gain response over a broad range of frequency bands.

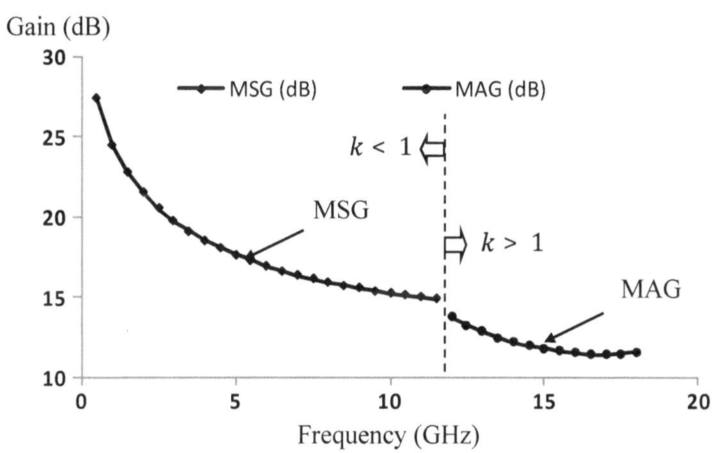

Fig. 14.12 Plot of maximum stable gain (MSG) and maximum available gain (MAG) of the pHEMT device (S-parameter Table 14.1)

14.5 Amplifier with Specific Gain

Under both input and output conjugate matched condition, a transistor over its unconditional stable frequency region provides maximum gain. Similarly, over its conditional stable frequency range, the transistor provides maximum gain for the input and output matching condition for which stability factor of the amplifier becomes one. In many cases, it is desirable to design an amplifier for less than the maximum obtainable gain to improve other performance such as operating bandwidth, minimization of the effect of device parameters variations on amplifier's performance, noise figure, output power capability, stability, etc. Therefore, in many cases, it is required to design the input and/or output matching networks which purposely introduce impedance mismatches to reduce the overall gain. For designing a low-noise amplifier, it is required to match the input terminal of a transistor other than corresponding to maximum gain, i.e. it is required to intentionally introduce input mismatch to achieve the best noise figure. Similarly, conditions for achieving highest output power and highest DC to RF efficiency from a transistor are different from its output matching network corresponding to achieving maximum gain. Therefore, in case of power amplifier design, it is required to intentionally introduce some mismatch at device output. The design for achieving specific gain can be done by plotting constant gain circles on a Smith chart, by representing loci of the specific fixed values of gain for the input and output sections.

14.5.1 Amplifier with Specific Transducer Power Gain

The transducer power gain G_T is given by (14.27) and (14.28):

$$G_T = \frac{1 - |\Gamma_S|^2}{|1 - \Gamma_S\,\Gamma_{\mathrm{IN}}|^2} \times |S_{21}|^2 \times \frac{1 - |\Gamma_L|^2}{|1 - S_{22}\,\Gamma_L|^2} = G_S \times G_o \times G_L$$

where G_S is the source gain factor and is given by

$$G_S = \frac{1 - |\Gamma_S|^2}{|1 - \Gamma_S\,\Gamma_{\mathrm{IN}}|^2} \tag{14.43a}$$

G_L is the load gain factor and is given by

$$G_L = \frac{1 - |\Gamma_L|^2}{|1 - S_{22}\,\Gamma_L|^2} \tag{14.43b}$$

and $G_o = |S_{21}|^2$ is the device gain factor. It is the gain of the device when source and load impedances are the reference impedance (R_o).

The amplifier will provide maximum gain when the source gain and load gain factors become maximum. Therefore, the maximum obtainable gain will be

$$G_{T_{\max}} = G_{S_{\max}} \times G_o \times G_{L_{\max}} \tag{14.44}$$

The transducer power gain G_T depends upon both the source and load impedances. Thus, to achieve a specific gain, which is less than the maximum achievable gain, the source gain factor (G_S) and load gain factor (G_L) need to be adjusted lower than their maximum values $G_{S_{\max}}$ and $G_{L_{\max}}$, respectively. Practically all the transistors are bilateral, i.e. $S_{12} \neq 0$. In this case, the input and output reflection coefficient, Γ_{IN} and Γ_{OUT} depend on the load and source impedances, respectively. Therefore, for a specific transducer power gain using a bilateral transistor, it is difficult to determine the source and load impedances. To simplify the procedure here we consider the device as unilateral device, i.e. $S_{12} = 0$.

For a unilateral transistor, $\Gamma_{\text{IN}} = S_{11}$ and $\Gamma_{\text{OUT}} = S_{22}$. Therefore, transducer power gain in unilateral case G_{TU} can be written as (14.35):

$$G_{TU} = \frac{1-|\Gamma_S|^2}{|1-S_{11}\Gamma_S|^2} \times |S_{21}|^2 \times \frac{1-|\Gamma_L|^2}{|1-S_{22}\Gamma_L|^2} = G_{SU} \times G_o \times G_{LU} \tag{14.45}$$

Thus, for unconditional stable unilateral device, the maximum transducer power gain $G_{TU_{\max}}$ will be achieved when $\Gamma_S = S_{11}^*$ and $\Gamma_L = S_{22}^*$ and is given by

$$G_{TU_{\max}} = \frac{1}{1-|S_{11}|^2} \times |S_{21}|^2 \times \frac{1}{1-|S_{22}|^2} = G_{SU_{\max}} \times G_o \times G_{LU_{\max}}$$

The normalized gain factors g_{SU} and g_{LU} can be written as

$$g_{SU} = \frac{G_{SU}}{G_{SU_{\max}}} = \frac{1-|\Gamma_S|^2}{|1-S_{11}\Gamma_S|^2} \times \left(1-|S_{11}|^2\right) \tag{14.46a}$$

$$g_{LU} = \frac{G_{LU}}{G_{LU_{\max}}} = \frac{1-|\Gamma_L|^2}{|1-S_{22}\Gamma_L|^2} \times \left(1-|S_{22}|^2\right) \tag{14.46b}$$

Expanding (14.46a):

$$g_{SU}|1-S_{11}\Gamma_S|^2 = \left(1-|\Gamma_S|^2\right) \times \left(1-|S_{11}|^2\right)$$

or

$$\left(g_{SU}|S_{11}|^2 + 1 - |S_{11}|^2\right)|\Gamma_S|^2 - g_{SU}\left(S_{11}\Gamma_S + S_{11}^*\Gamma_S^*\right) = 1 - |S_{11}|^2 - g_{SU}$$

or

$$\Gamma_S\Gamma_S^* - \frac{g_{SU}\left(S_{11}\Gamma_S + S_{11}^*\Gamma_S^*\right)}{1-(1-g_{SU})|S_{11}|^2} = \frac{1-|S_{11}|^2 - g_{SU}}{1-(1-g_{SU})|S_{11}|^2}$$

or

$$\left| \Gamma_S - \frac{g_{SU} S_{11}^*}{1-(1-g_{SU})|S_{11}|^2} \right|^2 = \left| \frac{\sqrt{1-g_{SU}}\left(1-|S_{11}|^2\right)}{1-(1-g_{SU})|S_{11}|^2} \right|^2 \tag{14.47}$$

This is the form of $|\Gamma_S - C_S|^2 = |R_S|^2$, which is equation of a circle.

Therefore, the expression (14.47) represents circle, called constant source gain circle, in the Γ_S plane for a fixed value of source gain factor g_{SU}. The centre C_S and radius R_S of the constant gain circle in the Γ_S plane are given by

$$C_S = \frac{g_{SU}\ S_{11}^*}{1-(1-g_{SU})|S_{11}|^2} \tag{14.48a}$$

$$R_S = \frac{\sqrt{1-g_{SU}}\left(1-|S_{11}|^2\right)}{1-(1-g_{SU})|S_{11}|^2} \tag{14.48b}$$

In a similar way expanding (14.46b), it can be written as

$$\left| \Gamma_L - \frac{g_{LU}\ S_{22}^*}{1-(1-g_{LU})|S_{22}|^2} \right|^2 = \left| \frac{\sqrt{1-g_{LU}}\left(1-|S_{22}|^2\right)}{1-(1-g_{LU})|S_{22}|^2} \right|^2 \tag{14.49}$$

The expression (14.49) also represents circle, called constant load gain circle, in the Γ_L plane for a fixed value of load gain factor g_{LU}. The centre C_L and radius R_L of the constant gain circle in the Γ_L plane are given by

$$C_L = \frac{g_{LU}\ S_{22}^*}{1-(1-g_{LU})|S_{22}|^2} \tag{14.50a}$$

$$R_L = \frac{\sqrt{1-g_{LU}}\left(1-|S_{22}|^2\right)}{1-(1-g_{LU})|S_{22}|^2} \tag{14.50b}$$

From (14.48a) and (14.50a), it is clear that the centres of each family of circles corresponding to different values of g_{SU}, and g_{LU} lie along straight lines given by the angle of S_{11}^* and S_{22}^*, respectively. Corresponding to the maximum source gain factor, $g_{SU} = 1$, the radius of the constant input gain circle $R_S = 0$ and centre of the circle become $C_S = S_{11}^*$. Similarly, for the maximum- load gain factor, $g_{LU} = 1$ the radius of the constant output gain circle $R_L = 0$ and centre of the circle become $C_L = S_{22}^*$. It can also be shown that the circles correspond to 0 dB source and load gain factors will pass through the centre of the Smith chart.

Example 14.10 Computation of constant transducer power gain circles and plotting them in Smith charts of a unilateral device with following S-parameters:

$$S_{11} = 0.498\angle 121^\circ$$

$$S_{12} = 0$$

$$S_{21} = 3.080\angle - 30^\circ$$
$$S_{22} = 0.225\angle - 150^\circ$$

Solution

$$G_{SU_{\max}} = \frac{1}{1 - |S_{11}|^2} = \frac{1}{1 - 0.498^2} = 1.33 = 1.238 \text{ dB}$$
$$G_o = |S_{21}|^2 = 3.080^2 = 9.486 = 9.771 \text{ dB}$$
$$G_{LU_{\max}} = \frac{1}{1 - |S_{22}|^2} = \frac{1}{1 - 0.225^2} = 1.053 = 0.226 \text{ dB}$$

Thus,

$$\begin{aligned} G_{TU_{\max}} &= G_{SU_{\max}} \times G_o \times G_{LU_{\max}} \\ &= 1.33 \times 9.486 \times 1.953 = 13.288 \\ &= (1.238 + 9.771 + 0.226) \text{ dB} \\ &= 11.234 \text{ dB} \end{aligned}$$

Thus, the maximum obtainable gain from the device is 11.234 dB when both input and outputs are conjugate matched. To achieve specific gain which is other than this maximum gain, it is required to introduce mismatches at the input and/or output of the device. We can draw constant input and output gain circles for different values of input and output gain factors.

Tables 14.2 and 14.3 show the centre and radius of circles that correspond to different values of source and load gain factors, respectively. Constant source gain circles are shown in Fig. 14.13 on a Γ_S plane corresponding to specific source gain of 1.238, 1.0, 0.5, 0.229, 0.0, −0.5 and −1.0 dB using the parameters shown in Table 14.2. The 0 dB gain circle passes through the centre of the Smith chart. The centre of all the circles lies on a straight line (dotted line) passing through the centre of the Smith chart. For a specific source gain, the source reflection coefficient may be selected from values lying on the corresponding circle. In the same way, constant load gain circles corresponding to 0.226, 0, −0.5, −1.0 and −2.0 dB are drawn in the Γ_L plane as shown in Fig. 14.14.

To obtain a specific transducer gain from the unilateral device, the source and load reflection coefficients should be selected from the specific values of constant gain circles. For example, to achieve 10 dB transducer gain, $G_{SU} = 0.229$ dB and $G_{LU} = 0$ dB can be selected. So that the transducer gain will be

$$G_{TU} = (0.229 + 9.771 + 0) \text{ dB} = 10.0 \text{ dB}$$

To achieve this, the source reflection coefficient should be chosen any value on the 0.229 dB source gain factor circle and load reflection coefficient will be zero.

Table 14.2 Parameters for specific transducer power gain circles corresponding to specific source gain factors of a unilateral device : $S_{11} = 0.498\angle 121°$, $S_{12} = 0$, $S_{21} = 3.080\angle -30°$, $S_{22} = 0.225\angle -150°$

G_{SU} (dB)	G_{SU} (ratio)	$g_{SU} = \left[\frac{G_{SU}}{G_{SU\max}}\right]$	$C_S = \left[\frac{g_{SU}S_{11}^*}{1-(1-g_{SU})\|S_{11}\|^2}\right]$	$R_S = \left[\frac{\sqrt{1-g_{SU}}(1-\|S_{11}\|^2)}{1-(1-g_{SU})\|S_{11}\|^2}\right]$
1.238	1.33	1.0	$\frac{1 \times 0.498\angle -121°}{1-(1-1)0.498^2} = 0.498\angle -121°$	$\frac{\sqrt{1-1}(1-0.498^2)}{1-(1-1)0.498^2} = 0$
1.0	1.259	0.947	$\frac{0.947 \times 0.498\angle -121°}{1-(1-0.947)0.498^2} = 0.478\angle -121°$	$\frac{\sqrt{1-0.947}(1-0.498^2)}{1-(1-0.947)0.498^2} = 0.176$
0.5	1.122	0.844	$\frac{0.844 \times 0.498\angle -121°}{1-(1-0.844)0.498^2} = 0.437\angle -121°$	$\frac{\sqrt{1-0.844}(1-0.498^2)}{1-(1-0.844)0.498^2} = 0.309$
0.229	1.054	0.793	$\frac{0.793 \times 0.498\angle -121°}{1-(1-0.793)0.498^2} = 0.416\angle -121°$	$\frac{\sqrt{1-0.793}(1-0.498^2)}{1-(1-0.793)0.498^2} = 0.361$
0	1.0	0.752	$\frac{0.752 \times 0.498\angle -121°}{1-(1-0.752)0.498^2} = 0.399\angle -121°$	$\frac{\sqrt{1-0.752}(1-0.498^2)}{1-(1-0.752)0.498^2} = 0.399$
−0.5	0.891	0.670	$\frac{0.670 \times 0.498\angle -121°}{1-(1-0.670)0.498^2} = 0.363\angle -121°$	$\frac{\sqrt{1-0.670}(1-0.498^2)}{1-(1-0.670)0.498^2} = 0.470$
−1.0	0.794	0.597	$\frac{1 \times 0.597\angle -121°}{1-(1-597)0.498^2} = 0.330\angle -121°$	$\frac{\sqrt{1-597}(1-0.498^2)}{1-(1-597)0.498^2} = 0.530$

However, for achieving broadband gain performance, it is preferable to select the source and load reflection coefficients on the corresponding gain circles near the centre of the Smith chart.

14.5.2 *Amplifier with Specific Available Power Gain*

Transducer power gain depends on both load and source matching networks. Due to the inter-dependency of the load and source impedances in case of bilateral transistors ($S_{21} \neq 0$), it is practically difficult to design an amplifier with specific gain. It becomes more complicated in case of a conditionally stable device. The available

Table 14.3 Parameters for specific transducer power gain circles corresponding to specific load gain factors of a unilateral device: $S_{11} = 0.498\angle 121°$, $S_{12} = 0$, $S_{21} = 3.080\angle -30°$, $S_{22} = 0.225\angle -150°$

G_{LU} (dB)	G_{LU} (ratio)	$g_{LU} = \left[\frac{G_{LU}}{G_{LU\max}}\right]$	$C_L = \left[\frac{g_{LU} S_{22}^*}{1-(1-g_{LU})\lvert S_{22}\rvert^2}\right]$	$R_L = \left[\frac{\sqrt{1-g_{LU}}(1-\lvert S_{22}\rvert^2)}{1-(1-g_{LU})\lvert S_{22}\rvert^2}\right]$
0.226	1.053	1.0	$\frac{1 \times 0.225\angle 150°}{1-(1-1)0.225^2} = 0.225\angle 150°$	$\frac{\sqrt{1-1}(1-0.225^2)}{1-(1-1)0.225^2} = 0$
0	1.0	0.949	$\frac{0.949 \times 0.225\angle 150°}{1-(1-949)0.225^2} = 0.214\angle 150°$	$\frac{\sqrt{1-0.949}(1-0.225^2)}{1-(1-949)0.225^2} = 0.214$
−0.5	0.891	0.846	$\frac{0.846 \times 0.225\angle 150°}{1-(1-0.846)0.225^2} = 0.192\angle 150°$	$\frac{\sqrt{1-0.846}(1-0.225^2)}{1-(1-0.846)0.225^2} = 0.375$
−1.0	0.794	0.754	$\frac{0.754 \times 0.225\angle 150°}{1-(1-0.754)0.225^2} = 0.172\angle 150°$	$\frac{\sqrt{1-0.754}(1-0.225^2)}{1-(1-0.754)0.225^2} = 0.477$
−2.0	0.631	0.599	$\frac{0.599 \times 0.225\angle 150°}{1-(1-0.599)0.225^2} = 0.138\angle 150°$	$\frac{\sqrt{1-0.599}(1-0.225^2)}{1-(1-0.599)0.225^2} = 0.614$

power gain is independent of the load impedance. Thus, it is convenient to use available power gain for designing a low-noise amplifier where specific source impedance is required for achieving a minimum noise figure.

Available power gain G_A of a network is defined as the ratio of power available from the network P_{AVN} and power available from the source P_{AVS}. Thus for a two-port network, from (7.183), the available power gain can be rewritten as

$$G_A = \frac{P_{\text{AVN}}}{P_{\text{AVS}}} = \frac{1-|\Gamma_S|^2}{|1-S_{11}\Gamma_S|^2}|S_{21}|^2\frac{1}{\left(1-|\Gamma_{\text{OUT}}|^2\right)} \tag{14.51a}$$

The available power gain depends on the input matching condition, Γ_S because actual power transferred to the transistor is not necessarily the same as available power from the source which happens only for conjugate match at the input. However, G_A is independent of the input reflection coefficient Γ_{IN} and load impedance Γ_L. The available power gain G_A can be written as

$$G_A = g_A|S_{21}|^2 \tag{14.51b}$$

where

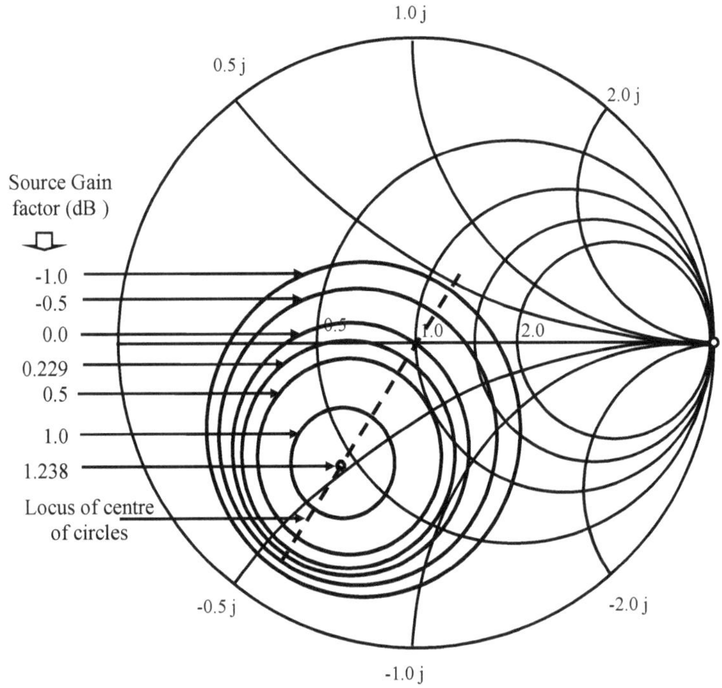

Fig. 14.13 Circles on Γ_S plane corresponding to constant source gain factors

$$\begin{aligned} g_A &= \frac{1-|\Gamma_S|^2}{|1-S_{11}\Gamma_S|^2\left(1-|\Gamma_{\mathrm{OUT}}|^2\right)} \\ &= \frac{1-|\Gamma_S|^2}{|1-S_{11}\Gamma_S|^2\left(1-\left|\frac{S_{22}-\Delta\,\Gamma_S}{1-S_{11}\Gamma_S}\right|^2\right)} \\ &= \frac{1-|\Gamma_S|^2}{|1-S_{11}\Gamma_S|^2|S_{22}-\Delta\,\Gamma_S|^2} \\ &= \frac{1-|\Gamma_S|^2}{1-|S_{22}|^2+|\Gamma_S|^2\left(|S_{11}|^2-|\Delta|^2\right)-2\,Re(\Gamma_S\,C_1)} \end{aligned} \tag{14.52}$$

Then (14.52) can be expressed as

$$\begin{aligned} &\left|\Gamma_S-\frac{g_A\,C_1^*}{1+g_A\left(|S_{11}|^2-|\Delta|^2\right)}\right|^2 \\ &\quad=\left|\frac{\left(1-2k\,g_A|S_{12}\,S_{21}|+g_A^2|S_{12}\,S_{21}|^2\right)^{\frac{1}{2}}}{1+g_A\left(|S_{11}|^2-|\Delta|^2\right)}\right|^2 \end{aligned} \tag{14.53}$$

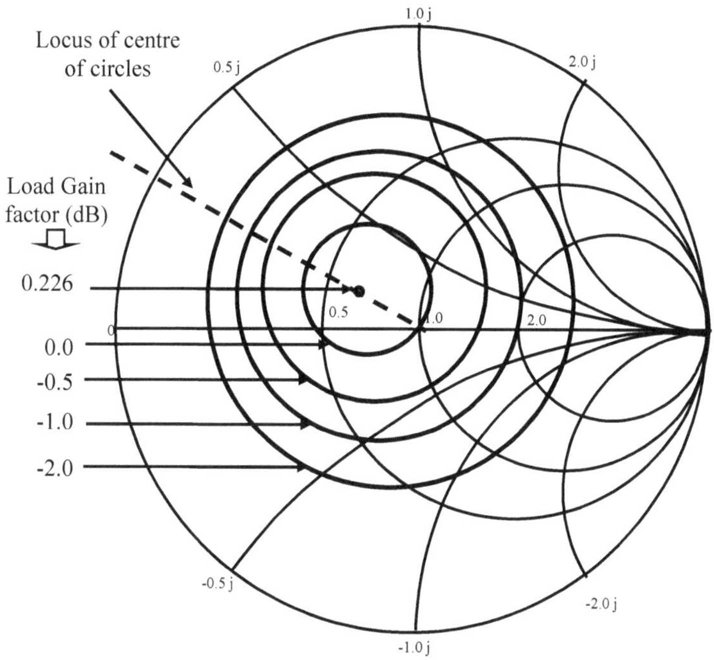

Fig. 14.14 Circles on Γ_L plane corresponding to constant load gain factors

The expression (14.53) represents circle in the Γ_S plane for a fixed value of normalized available gain g_A. The centre C_A and radius R_A of the constant gain circle in the Γ_S plane are given by

$$C_A = \frac{g_A\, C_1^*}{1 + g_A\left(|S_{11}|^2 - |\Delta|^2\right)} \tag{14.54a}$$

$$R_A = \frac{\left(1 - 2kg_A|S_{12}\, S_{21}| + g_A^2|S_{12}\, S_{21}|^2\right)^{\frac{1}{2}}}{\left|1 + g_A\left(|S_{11}|^2 - |\Delta|^2\right)\right|} \tag{14.54b}$$

The centres of the constant g_A circles are always on the line drawn between C_1^* and the origin of the Γ_S plane.

In case of unconditionally stable device, R_A becomes zero when g_A reaches to its maximum value.

In case of potentially unstable device, when g_A is infinite, the constant gain circle become input stability circle and its centre and radius are given by

$$C_A = \frac{C_1^*}{|S_{11}|^2 - |\Delta|^2} \tag{14.55a}$$

$$R_A = \frac{|S_{12}\, S_{21}|}{|S_{11}|^2 - |\Delta|^2} \tag{14.55b}$$

Example 14.11 Computation of constant available power gain circles of a bilateral device with S-parameters: $S_{11} = 0.498\angle 121°$, $S_{12} = 0.1137\angle -49°$, $S_{21} = 3.080\angle - 30°$, $S_{22} = 0.225\angle - 150°$.

Solution

$$\begin{aligned}\Delta &= 0.498\angle 121° \times 0.225\angle - 150° - 0.1137\angle - 49° \times 3.080\angle - 30° \\ &= 0.291\angle 83.9°, \text{ Thus, } |\Delta| = 0.291 < 1\end{aligned}$$

The stability factor:

$$K = \frac{1 - (0.498)^2 - (0.225)^2 + (0.291)^2}{2(0.1137 \times 3.080)} = 1.122 > 1$$

Thus, the device is unconditional stable. The maximum available gain:

$$\begin{aligned}G_{A\,\max}(\text{dB}) &= 10\log\left(\frac{3.080}{0.1137}\right) + 10\log\left(1.122 - \sqrt{1.122^2 - 1}\right) \\ &= 14.328 \text{ dB} - 2.127 \text{ dB} = 12.2 \text{ dB}\end{aligned}$$

$$\begin{aligned}C_1 &= S_{11} - \Delta S_{22}^* \\ &= 0.498\angle 121° - (0.291\angle 83.9°) \times (0.225\angle 150°) \\ &= 0.498\angle 121° - 0.291\angle 83.9° \times 0.225\angle 150° = 0.527\angle 114.4° \\ C_1^* &= 0.527\angle - 114.4°\end{aligned}$$

And,

$$\begin{aligned}|S_{12}S_{21}| &= |0.1137\angle - 49° \times 3.080\angle - 30°| = 0.350 \\ |S_{11}|^2 - |\Delta|^2 &= 0.498^2 - 0.291^2 = 0.163\end{aligned}$$

Using the values of these parameters, centre and radius of constant available power gain circles given by (14.54a) and (14.54b) corresponding to 12.2, 10.0, 8.0 and 6.0 dB are calculated as shown in Table 14.4. It shows that for maximum available gain, the radius of the constant gain circle becomes zero and angle of the centre of all the circles are same as the angle of C_1^*.

The constant available gain circles of the device are plotted on Smith chart shown in Fig. 14.15. The centres of all the circles are always on the line drawn between C_1^* and origin of the Γ_S plane as shown by a dashed line. The radii of constant gain circles are getting smaller with the increase of gain.

To achieve constant available gain, the input matching should provide any source reflection coefficient Γ_S which lie on that constant available gain circle.

Table 14.4 Parameters for available power gain circles of a bilateral device: $S_{11} = 0.498\angle 121°$, $S_{12} = 0.1137\angle -49°$, $S_{21} = 3.080\angle -30°$, $S_{22} = 0.225\angle -150°$

G_A (dB)	$g_A = \left[\frac{G_A}{\lvert S_{21}\rvert^2}\right]$	$C_A = \left[\frac{g_A C_1^*}{1+g_A\left(\lvert S_{11}\rvert^2-\lvert\Delta\rvert^2\right)}\right]$	$R_A = \left[\frac{\left(1-2kg_A\lvert S_{12}S_{21}\rvert+g_A^2\lvert S_{12}S_{21}\rvert^2\right)^{\frac{1}{2}}}{\lvert 1+g_A\left(\lvert S_{11}\rvert^2-\lvert\Delta\rvert^2\right)\rvert}\right]$
12.2	$\frac{16.596}{3.080^2} = 1.749$	$\frac{0.922\angle -114.4°}{1.286} = 0.717\angle -114.4°$	$\frac{0}{1.286} = 0$
10.0	$\frac{10}{3.080^2} = 1.054$	$\frac{0.555\angle -114.4°}{1.172} = 0.474\angle -114.4°$	$\frac{0.555}{1.172} = 0.473$
8.0	$\frac{6.310}{3.080^2} = 0.665$	$\frac{0.350\angle -114.4°}{1.109} = 0.316\angle -114.4°$	$\frac{0.729}{1.109} = 0.658$
6.0	$\frac{3.981}{3.080^2} = 0.420$	$\frac{0.221\angle -114.4°}{1.069} = 0.207\angle -114.4°$	$\frac{0.832}{1.069} = 0.778$

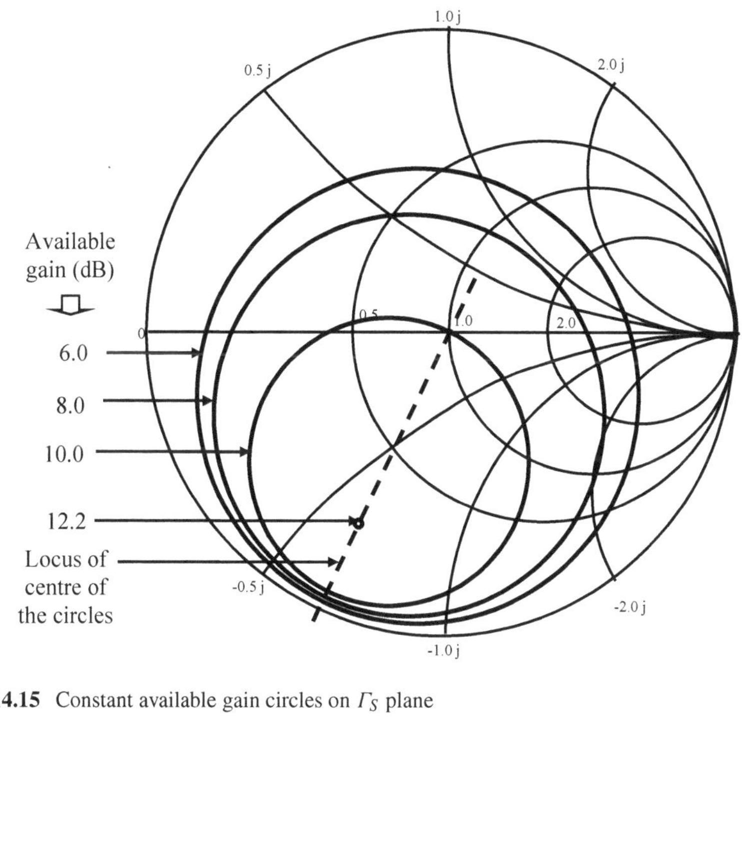

Fig. 14.15 Constant available gain circles on Γ_S plane

Example 14.12 Compute available gain circles of the device at frequency 6.0 GHz, where S-parameters are given by $S_{11} = 0.701\angle-116°$, $S_{12} = 0.0889\angle 19°$, $S_{21} = 4.420\angle 76°$, $S_{22} = 0.385\angle-79°$

Solution

$$\begin{aligned}\Delta &= 0.701\angle-116° \times 0.385\angle-79° - 0.0889\angle 19° \times 4.420\angle 76° \\ &= 0.393\angle-125.2°\end{aligned}$$

Thus, $|\Delta| = 0.393 < 1$

$$K = \frac{1-(0.701)^2-(0.385)^2+(0.393)^2}{2(0.0889\times 4.420)} = 0.655 < 1$$

Thus the device is conditionally stable. It is required to draw the input stability circle for the device. The source and load reflection coefficients should be in the stable region on Γ_S and Γ_L plane, respectively.

The centre and radius of the input stability circle is given by

$$\begin{aligned}c_S &= \frac{(0.701\angle-116° - 0.393\angle-125.2° \times 0.385\angle 79°)^*}{(0.701)^2-|0.393|^2} \\ &= 1.973\angle 128.4° \\ r_S &= \frac{0.0889\times 4.420}{(0.701)^2-|0.393|^2} = 1.167\end{aligned}$$

The maximum available gain of the device:

$$\begin{aligned}G_{A\max}(\text{dB}) &= 10\log\left(\frac{4.420}{0.0889}\right) = 16.97\text{ dB} \\ C_1 &= S_{11} - \Delta S_{22}^* \\ &= 0.701\angle 116° - (0.393\angle-125.2°)\times(0.385\angle-79°) \\ &= 0.664\angle-128.4° \\ C_1^* &= 0.664\angle 128.4°\end{aligned}$$

And,

$$\begin{aligned}|S_{12}S_{21}| &= |0.0889\angle 19° \times 4.420\angle 76°| = 0.393 \\ |S_{11}|^2 - |\Delta|^2 &= 0.701^2 - 0.393^2 = 0.337\end{aligned}$$

Using the values of these parameters, centre and radius of the constant available power gain circles given by (14.54a) and (14.54b) corresponding to 16.0, 14.0, 12.0, 10.0 and 8.0 dB are tabulated in Table 14.5. It shows that angle of the centre of all the constant available gain circles are the same as angle of C_1^*.

Table 14.5 Parameters for available power gain circles of a bilateral device: $S_{11} = 0.701\angle -116°$, $S_{12} = 0.0889\angle 19°$, $S_{21} = 4.420\angle 76°$, $S_{22} = 0.385\angle -79°$

G_A(dB)	g_A	C_A	R_A
16.0	$\frac{39.811}{4.420^2} = 2.038$	$\frac{1.354\angle 128.4°}{1.686} = 0.803\angle 128.4°$	$\frac{0.770}{1.686} = 0.457$
14.0	$\frac{25.119}{4.420^2} = 1.286$	$\frac{0.854\angle 128.4°}{1.433} = 0.596\angle 128.4°$	$\frac{0.771}{1.433} = 0.538$
12.0	$\frac{15.849}{4.420^2} = 0.811$	$\frac{0.539\angle 128.4°}{1.273} = 0.423\angle 128.4°$	$\frac{0.827}{1.273} = 0.650$
10.0	$\frac{10.000}{4.420^2} = 0.512$	$\frac{0.340\angle 128.4°}{1.172} = 0.290\angle 128.4°$	$\frac{0.882}{1.172} = 0.752$
8.0	$\frac{6.310}{4.420^2} = 0.323$	$\frac{0.215\angle 128.4°}{1.109} = 0.193\angle 128.4°$	$\frac{0.922}{1.109} = 0.832$

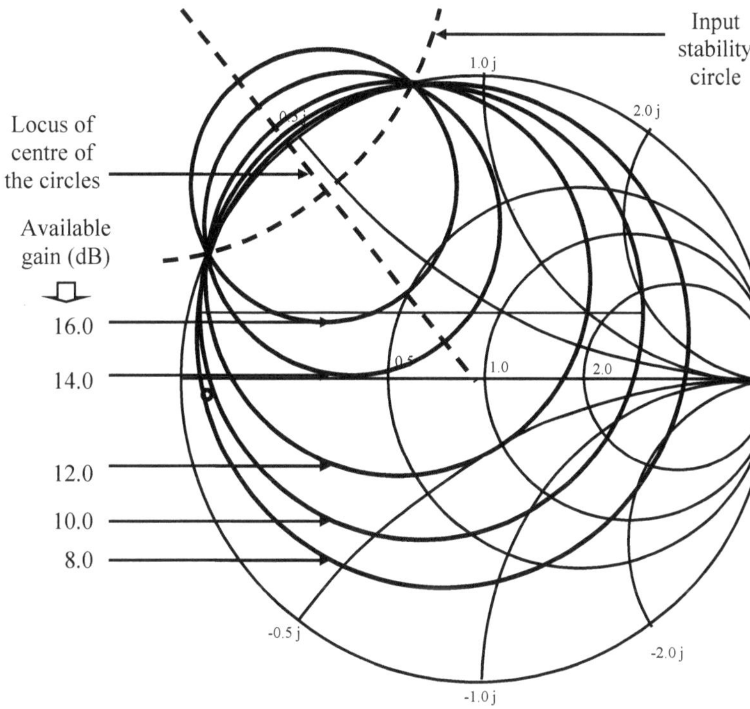

Fig. 14.16 Available gain circles on Γ_S plane for a potential unstable device

The constant available gain circles of the device are plotted on Smith chart as shown in Fig. 14.16. The centres of all the circles are always on the line drawn between C_1^* and origin of the Γ_S plane as shown by a dotted straight line. The radii of constant gain circles are getting larger with the decrease of available gain.

The input stability circle is also shown in the same Γ_S plane. The stability circle and all the constant gain circles intersect at two same points on the $|\Gamma_S| = 1$ circle of the Smith chart.

To achieve constant available gain, the input matching network should provide a source reflection coefficient Γ_S which lie on that corresponding constant available gain circle and the point should be within the stable region in the Γ_S plane.

14.5.3 Amplifier with Specific Operating Power Gain

Operating power gain G_P of a network is defined as the ratio of power delivered to the load, P_L and power input, P_{IN} to the network. Thus, for a two-port network, from (7.182) G_P can be rewritten as

$$G_P = \frac{P_L}{P_{\text{IN}}} = \frac{1}{\left(1 - |\Gamma_{\text{IN}}|^2\right)} |S_{21}|^2 \frac{1 - |\Gamma_L|^2}{|1 - S_{22}\Gamma_L|^2} \tag{14.56}$$

Thus, the operating power gain depends on the load impedance Γ_L; however, it is independent of the source impedance Γ_S and output impedance Γ_{OUT}. Therefore, for designing a high power amplifier, output of the device should be matched corresponding to achieving maximum output power or efficiency or linearity.

The operating power gain G_P can be written as

$$G_P = g_P |S_{21}|^2 \tag{14.57}$$

where

$$\begin{aligned} g_P &= \frac{1 - |\Gamma_L|^2}{(1 - |\Gamma_{\text{IN}}|^2)|1 - S_{22}\Gamma_L|^2} \\ &= \frac{1 - |\Gamma_L|^2}{\left(1 - \left|\frac{S_{11} - \Delta\, \Gamma_L}{1 - S_{22}\, \Gamma_L}\right|^2\right)|1 - S_{22}\, \Gamma_L|^2} \\ &= \frac{1 - |\Gamma_L|^2}{|1 - S_{22}\, \Gamma_L|^2 - |S_{11} - \Delta\, \Gamma_L|} \\ &= \frac{1 - |\Gamma_L|^2}{1 - |S_{11}|^2 + |\Gamma_L|^2\left(|S_{22}|^2 - |\Delta|^2\right) - 2\, Re(\Gamma_L\, C_2)} \end{aligned} \tag{14.58}$$

Thus (14.58) can be expressed as

$$\begin{aligned} &\left|\Gamma_L - \frac{g_P\, C_2^*}{1 + g_P\left(|S_{22}|^2 - |\Delta|^2\right)}\right|^2 \\ &= \left|\frac{\left(1 - 2k g_P |S_{12}\, S_{21}| + g_P^2 |S_{12}\, S_{21}|^2\right)^{\frac{1}{2}}}{1 + g_P\left(|S_{22}|^2 - |\Delta|^2\right)}\right|^2 \end{aligned} \tag{14.59}$$

The expression (14.59) represents circle in the Γ_L plane for a fixed value of normalized operating power gain g_P. The centre C_P and radius R_P of the constant gain circle in the Γ_L plane are given by

$$C_P = \frac{g_P\, C_2^*}{1 + g_P\left(|S_{22}|^2 - |\Delta|^2\right)} \tag{14.60a}$$

$$R_P = \frac{\left(1 - 2kg_P|S_{12}\, S_{21}| + g_P^2|S_{12}\, S_{21}|^2\right)^{\frac{1}{2}}}{\left|1 + g_P\left(|S_{22}|^2 - |\Delta|^2\right)\right|} \tag{14.60b}$$

The centres of the constant g_P circles are always on the line drawn between C_2^* and the origin of the Γ_L plane.

In case of unconditionally stable device, when output is matched for maximum gain, i.e. conjugate matched, the radius of the constant gain circle R_P becomes zero. Thus from (14.60b), the normalized maximum operating power gain $g_{P\max}$ can be expressed as

$$1 - 2kg_{P\max}|S_{12}\, S_{21}| + g_P^2|S_{12}\, S_{21}|^2 = 0 \tag{14.61}$$

or

$$g_{P\max} = \frac{k - \sqrt{k^2 - 1}}{|S_{12}\, S_{21}|} \tag{14.62}$$

Therefore, the maximum operating power gain $G_{P\max}$ becomes

$$G_{P\max} = \frac{|S_{21}|}{|S_{12}|}\left(k - \sqrt{k^2 - 1}\right) \tag{14.63}$$

This is the same result as mentioned in (14.39). In case of potentially unstable device, g_P becomes infinite. The constant operating power gain circles become output stability circles with centre and radius given by (14.60a and 14.60b):

$$C_P = \frac{C_2^*}{|S_{22}|^2 - |\Delta|^2} \tag{14.64a}$$

$$R_P = \frac{|S_{12}\, S_{21}|}{|S_{22}|^2 - |\Delta|^2} \tag{14.64b}$$

Example 14.13 Computation of constant operating power gain circles of a bilateral device with S-parameters: $S_{11} = 0.498\angle 121°$, $S_{12} = 0.1137\angle -49°$, $S_{21} = 3.080\angle -30°$, $S_{22} = 0.225\angle -150°$.

Solution

$$\Delta = 0.498\angle 121° \times 0.225\angle -150° - 0.1137\angle -49° \times 3.080\angle -30°$$

$$= 0.291\angle 83.9^\circ,\ \text{Thus,}\ |\Delta| = 0.291 < 1$$

The stability factor:

$$K = \frac{1 - (0.498)^2 - (0.225)^2 + (0.291)^2}{2(0.1137 \times 3.080)} = 1.122 > 1$$

Thus, the device is unconditional stable. The maximum operating power gain:

$$\begin{aligned}
G_{P\,\max}(\text{dB}) &= 10\log\left(\frac{3.080}{0.1137}\right) + 10\log\left(1.122 - \sqrt{1.122^2 - 1}\right) \\
&= 14.328\ \text{dB} - 2.127\ \text{dB} = 12.2\ \text{dB} \\
C_2 &= S_{22} - \Delta S_{11}^* \\
&= 0.225\angle 150^\circ - (0.291\angle 83.9^\circ) \times (0.498\angle - 121^\circ) \\
&= 0.311\angle - 175.4^\circ \\
C_2^* &= 0.311\angle + 175.4^\circ
\end{aligned}$$

And

$$\begin{aligned}
|S_{12}S_{21}| &= |0.1137\angle - 49^\circ \times 3.080\angle - 30^\circ| = 0.350 \\
|S_{11}|^2 - |\Delta|^2 &= 0.498^2 - 0.291^2 = 0.163
\end{aligned}$$

Using the values of these parameters, centre and radius of constant operating power gain circles given by (14.60a) and (14.60b) corresponding to 12.2, 10.0, 8.0, and 6.0 dB are calculated as tabulated in Table 14.6. It shows that for maximum operating power gain (12.2 dB), the radius of the constant gain circle becomes zero and angle of the centre of all the circles are the same as angle of C_2^*.

The constant operating power gain circles of the device are plotted on Smith chart as shown in Fig. 14.17. The centres of all the circles are always on the line drawn between C_2^* and origin of the Γ_L plane as shown by a dotted line. The radii of the constant gain circles are getting smaller with the increase of gain.

To achieve specific operating power gain, the output matching network should provide load reflection coefficients Γ_L which lie on that constant operating power gain circle.

Example 14.14 Computation of operating gain circles of the device at frequency 6.0 GHz, with S-parameter as $S_{11} = 0.701\angle - 116^\circ, S_{12} = 0.0889\angle 19^\circ, S_{21} = 4.420\angle 76^\circ, S_{22} = 0.385\angle - 79^\circ$

Solution

$$\begin{aligned}
\Delta &= 0.701\angle - 116^\circ \times 0.385\angle - 79^\circ - 0.0889\angle 19^\circ \times 4.420\angle 76^\circ \\
&= 0.393\angle - 125.2^\circ
\end{aligned}$$

Table 14.6 Parameters for operating power gain circles of a bilateral device: $S_{11} = 0.498\angle 121°, S_{12} = 0.1137\angle -49°, S_{21} = 3.080\angle -30°, S_{22} = 0.225\angle -150°$

G_P (dB)	$g_P = \left[\frac{G_P}{\lvert S_{21}\rvert^2}\right]$	$C_P = \left[\frac{g_P C_2^*}{1+g_P(\lvert S_{22}\rvert^2-\lvert\Delta\rvert^2)}\right]$	$R_P = \left[\frac{\left(1-2kg_P\lvert S_{12}S_{21}\rvert+g_p^2\lvert S_{12}S_{21}\rvert^2\right)^{\frac{1}{2}}}{\lvert 1+g_p(\lvert S_{11}\rvert^2-\lvert\Delta\rvert^2)\rvert}\right]$
12.2	$\frac{16.596}{3.080^2} = 1.749$	$\frac{0.545\angle 175.4°}{0.940} = 0.579\angle 175.4°$	$\frac{0}{0.940} = 0$
10.0	$\frac{10}{3.080^2} = 1.054$	$\frac{0.328\angle 175.4°}{0.964} = 0.341\angle 175.4°$	$\frac{0.555}{0.964} = 0.575$
8.0	$\frac{6.310}{3.080^2} = 0.665$	$\frac{0.207\angle 175.4°}{0.977} = 0.212\angle 175.4°$	$\frac{0.729}{0.977} = 0.746$
6.0	$\frac{3.981}{3.080^2} = 0.420$	$\frac{0.131\angle 175.4°}{0.986} = 0.133\angle 175.4°$	$\frac{0.832}{0.986} = 0.844$

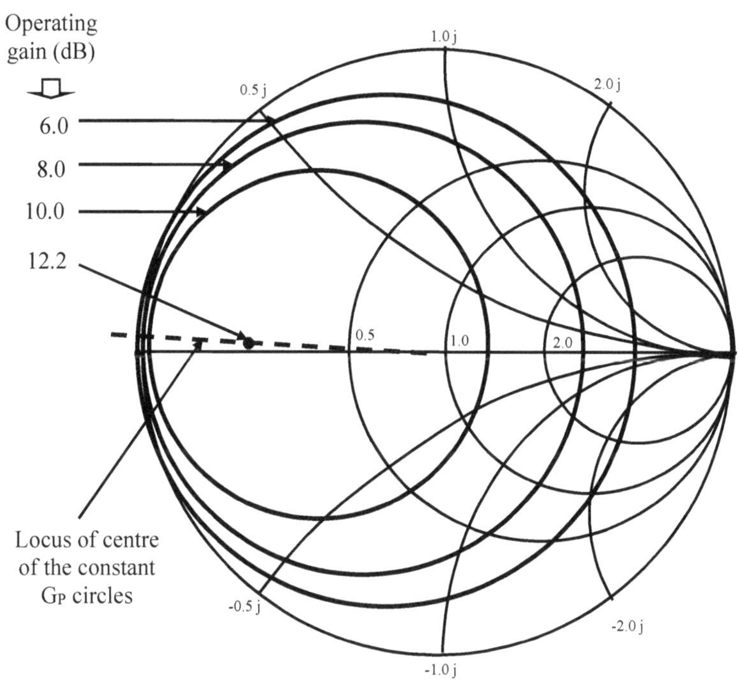

Fig. 14.17 Constant operating power gain circles on Γ_L plane

Thus, $|\Delta| = 0.393 < 1$

$$K = \frac{1 - (0.701)^2 - (0.385)^2 + (0.393)^2}{2(0.0889 \times 4.420)} = 0.655 < 1$$

Thus the device is conditionally stable. It is required to draw the output stability circle for the device. The source and load reflection coefficients should be within the stable region on Γ_S and Γ_L plane, respectively.

The centre and radius of the output stability circle is given by

$$c_L = \frac{(0.385\angle - 79° - 0.393\angle - 125.2° \times 0.701\angle 116°)^*}{(0.385)^2 - |0.393|^2}$$
$$= 60.063\angle 59.3°$$
$$r_L = \frac{0.0889 \times 4.420}{(0.385)^2 - |0.393|^2} = 60.713$$

The maximum stable gain MSG of the device (14.42b):

$$\text{MSG(dB)} = 10\log\left(\frac{4.420}{0.0889}\right) = 16.97 \text{ dB}$$
$$C_2 = S_{22} - \Delta S_{11}^*$$
$$= 0.385\angle - 79° - (0.393\angle - 125.2°) \times (0.701\angle 116°)$$
$$= 0.389\angle - 120.7°$$

Therefore,

$$C_2^* = 0.389\angle 120.7°$$

And

$$|S_{12}S_{21}| = |0.0889\angle 19° \times 4.420\angle 76°| = 0.393$$
$$|S_{11}|^2 - |\Delta|^2 = 0.701^2 - 0.393^2 = 0.337$$

Using the values of these parameters, centre and radius of constant operating power gain circles given by (14.64a) and (14.64b) corresponding to 16.0, 14.0, 12.0, 10.0 and 8.0 dB are tabulated in Table 14.7. It shows that angle of the centre of all the constant operating power gain circles are the same as angle of C_2^* i.e. 120.7°.

The constant operating power gain circles of the device are plotted on Smith chart as shown in Fig. 14.18. The centres of all the circles are always on the line drawn between C_2^* and origin of the Γ_L plane of the Smith chart as shown by a dotted straight line. The radii of the constant gain circles are getting higher with the decrease of gain.

Table 14.7 Parameters for operating power gain circles of a bilateral device: $S_{11} = 0.498\angle 121°$, $S_{12} = 0.1137\angle -49°$, $S_{21} = 3.080\angle -30°$, $S_{22} = 0.225\angle -150°$

G_P(dB)	g_P	C_P	R_P
16.0	$\frac{39.811}{4.420^2} = 2.038$	$\frac{0.792\angle 120.7°}{0.987} = 0.803\angle 120.7°$	$\frac{0.769}{0.987} = 0.779$
14.0	$\frac{25.119}{4.420^2} = 1.286$	$\frac{0.500\angle 120.7°}{0.992} = 0.504\angle 120.7°$	$\frac{0.770}{0.992} = 0.777$
12.0	$\frac{15.849}{4.420^2} = 0.811$	$\frac{0.315\angle 120.7°}{0.995} = 0.317\angle 120.7°$	$\frac{0.827}{0.995} = 0.831$
10.0	$\frac{10.000}{4.420^2} = 0.512$	$\frac{0.199\angle 120.7°}{0.997} = 0.200\angle 120.7°$	$\frac{0.881}{0.997} = 0.884$
8.0	$\frac{6.310}{4.420^2} = 0.323$	$\frac{0.126\angle 120.7°}{0.998} = 0.126\angle 120.7°$	$\frac{0.922}{0.998} = 0.924$

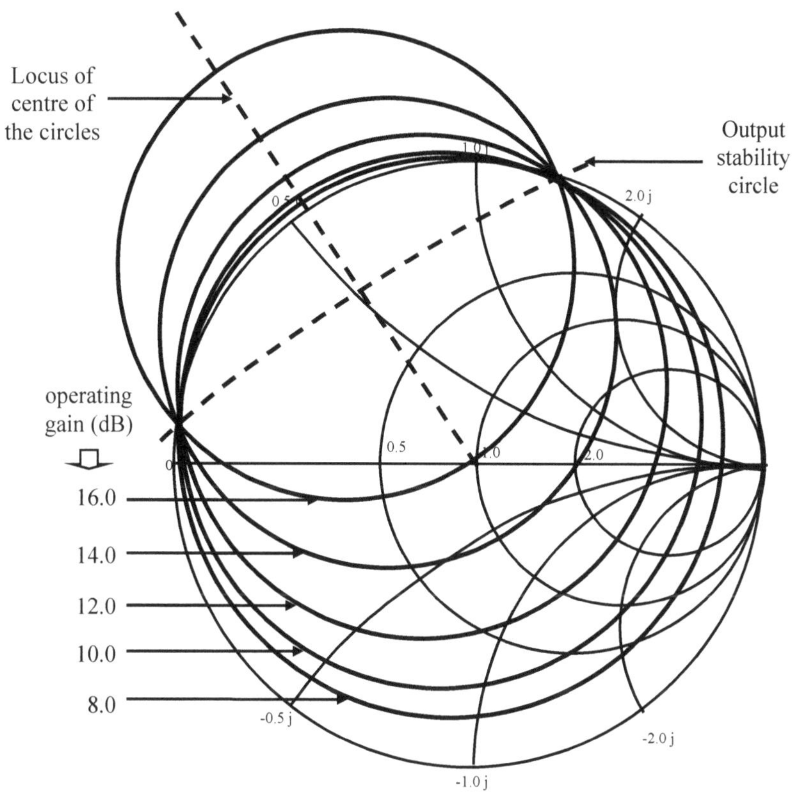

Fig. 14.18 Operating gain circles on Γ_L plane for a potential unstable device

The output stability circle is also shown in the same Γ_L plane. The output stability circle and all the constant operating power gain circles intersect at same two points on the $|\Gamma_L| = 1$ circle of Smith chart.

To achieve constant operating power gain, the output matching network should provide any load reflection coefficient Γ_L which lie on that corresponding constant operating power gain circle and the point should be within the stable region in the Γ_L plane.

14.6 Small Signal Amplifiers

Small signal amplifiers operate in their linear region. S-parameter model is valid to model the devices in this operating region. The following sections will present the design of low-noise and high-gain amplifiers based on S-parameter models which operate in their linear region.

14.6.1 Low Noise Amplifier (LNA) Design

Front end of any receiver handles most weak signal in a communication system. Therefore, the receiver needs a front-end amplifier with as low a noise figure as possible. Practically, input matching circuits are different to obtain minimum noise figure and maximum gain for an amplifier [5–9]. Some sort of compromise must be made with the gain of the amplifiers to obtain the lowest noise figure. This can be done by using constant available gain circles and circles of constant noise figure to select a usable trade-off between noise figure and gain.

For microwave amplifiers, the noise factor can be written as [5]

$$\mathrm{NF} = \mathrm{NF}_{\min} + \frac{R_N}{G_{\mathrm{Source}}} |Y_{\mathrm{Source}} - Y_{\mathrm{OPT}}|^2 \tag{14.65}$$

Here, Y_{Source} is the source admittance presented to the transistor, Y_{OPT} is the optimum source admittance that results in minimum noise figure $\mathrm{NF}_{\min}$. R_N is the equivalent noise resistance of the transistor and G_{Source} is the real part of the source admittance, i.e. $Y_{\mathrm{Source}} = G_{\mathrm{Source}} + jB_{\mathrm{Source}}$. The equivalent noise resistance, R_N, determines the sensitivity or dependency of the noise factor, NF, with the deviation of source admittance, Y_{Source} from its optimum value, Y_{OPT}, corresponding to the minimum noise figure, $\mathrm{NF}_{\min}$. The parameters $\mathrm{NF}_{\min}$, Y_{OPT} and R_N are characteristics of a particular transistor and known as noise parameters of the device. Noise parameters of a pHEMT device are shown in Table 14.8. The S-parameters of the same device are shown in Table 14.1.

The admittance parameters Y_{Source} and Y_{OPT} of (13.65) can be replaced by the corresponding reflection coefficients Γ_S and Γ_{OPT} as:

$$Y_{\mathrm{Source}} = \frac{1}{Z_o} \frac{1 - \Gamma_S}{1 + \Gamma_S} \tag{14.66a}$$

Table 14.8 Typical noise parameter of the pHEMT device, bias condition: $V_{DS} = 2$ V, $I_D = 15$ mA, reference impedance: $Z_o = 50\ \Omega$

Freq.	NF_{min}	Γ_{OPT}		R_N
(GHz)	(dB)	(MAG)	(DEG)	(Ω)
1.0	0.29	0.756	14	15.60
2.0	0.30	0.690	28	14.65
3.0	0.34	0.643	43	13.56
4.0	0.38	0.606	58	12.10
5.0	0.41	0.578	73	10.53
6.0	0.46	0.553	87	8.86
7.0	0.50	0.534	102	7.16
8.0	0.55	0.518	116	5.62
9.0	0.60	0.505	131	4.29
10.0	0.64	0.495	145	3.23
11.0	0.69	0.486	159	2.53
12.0	0.73	0.476	173	2.22
13.0	0.78	0.467	−173	2.37
14.0	0.84	0.455	−160	2.96
15.0	0.88	0.443	−146	4.01
16.0	0.93	0.428	−132	5.47
17.0	0.99	0.412	−118	7.26
18.0	1.05	0.394	−103	9.61

$$Y_{OPT} = \frac{1}{Z_o}\frac{1 - \Gamma_{OPT}}{1 + \Gamma_{OPT}} \tag{14.66b}$$

Using (14.66a) and (14.66b), expression (14.65) of noise figure can be written as

$$NF = NF_{min} + \frac{4R_N}{G_{Source}Z_o^2}\frac{|\Gamma_S - \Gamma_{OPT}|^2}{|1 - \Gamma_S|^2|1 - \Gamma_{OPT}|^2} \tag{14.67}$$

Expressing G_{Source}, the real part of Y_{Source} in terms of the reflection coefficient and substituting in (14.67):

$$NF = NF_{min} + \frac{4R_N}{Z_o}\frac{|\Gamma_S - \Gamma_{OPT}|^2}{\left(1 - |\Gamma_S|^2\right)|1 + \Gamma_{OPT}|^2} \tag{14.68}$$

Defining the noise figure parameter N, as

$$N = \frac{|\Gamma_S - \Gamma_{OPT}|^2}{1 - |\Gamma_S|^2} = \frac{Z_o(NF - NF_{min})}{4R_N}|1 + \Gamma_{OPT}|^2 \tag{14.69}$$

This is a constant, for a given noise figure and set of noise parameters NF_{min} for Γ_{OPT}. Thus, (14.69) can be written as

$$\Gamma_S\Gamma_S^* + \Gamma_{\text{OPT}}\Gamma_{\text{OPT}}^* - \Gamma_S\Gamma_{\text{OPT}}^* - \Gamma_S^*\Gamma_{\text{OPT}} = N\left(1 - \Gamma_S\Gamma_S^*\right) \tag{14.70}$$

or,

$$(1+N)\Gamma_S\Gamma_S^* + \Gamma_{\text{OPT}}\Gamma_{\text{OPT}}^* - \Gamma_S\Gamma_{\text{OPT}}^* - \Gamma_S^*\Gamma_{\text{OPT}} = N$$

or,

$$\Gamma_S\Gamma_S^* + \frac{\Gamma_{\text{OPT}}\Gamma_{\text{OPT}}^*}{(N+1)^2} - \frac{\Gamma_S^*\Gamma_{\text{OPT}}}{N+1} - \frac{\Gamma_S\Gamma_{\text{OPT}}^*}{N+1} = \frac{N}{N+1} + \frac{\Gamma_{\text{OPT}}\Gamma_{\text{OPT}}^*}{(N+1)^2} - \Gamma_{\text{OPT}}\Gamma_{\text{OPT}}^*$$

or,

$$\left|\Gamma_S - \frac{\Gamma_{\text{OPT}}}{N+1}\right|^2 = \left|\frac{\sqrt{N\left(N+1-|\Gamma_{\text{OPT}}|^2\right)}}{N+1}\right|^2 \tag{14.71}$$

This is equation of a circle of constant noise figures. The centre C_F and radius R_F are given by

$$C_F = \frac{\Gamma_{\text{OPT}}}{N+1} \tag{14.72a}$$

$$R_F = \frac{\sqrt{N\left(N+1-|\Gamma_{\text{OPT}}|^2\right)}}{N+1} \tag{14.72b}$$

For special case of minimum noise figure, $\text{NF} = \text{NF}_{\min}$ and $\Gamma_S = \Gamma_{\text{OPT}}$, thus, $N = 0$. In this condition, the centre and radius of the noise circle become $C_F = \Gamma_{\text{OPT}}$ and $R_F = 0$ respectively. Thus, for minimum noise figure, the circle becomes a point.

Example 14.15 Noise parameters of a device in 50 Ω system are as follows: $\text{NF}_{\min} = 0.84$ dB, $\Gamma_{\text{OPT}} = 0.455\angle-160°$, $R_N = 3.0\ \Omega$.

Compute the noise figure of the amplifier in case of the following source reflection coefficients: (i) $\Gamma_S = \Gamma_{\text{OPT}}$ (ii) $\Gamma_S = 0$ (iii) $\Gamma_S = 0.718\angle-114°$.

Solution

$$\begin{aligned}\text{NF}_{\min} &= 0.84\ \text{dB} = 1.213\\ \Gamma_{\text{OPT}} &= 0.455\angle-160° = -0.428 - j0.156\\ R_N &= 3.0\ \Omega\end{aligned}$$

The noise factor is given by (14.68):

$$\text{NF} = \text{NF}_{\min} + \frac{4R_N}{Z_o}\frac{|\Gamma_S - \Gamma_{\text{OPT}}|^2}{\left(1-|\Gamma_S|^2\right)|1+\Gamma_{\text{OPT}}|^2}$$

(i) In case of $\Gamma_S = \Gamma_{\text{OPT}}$, the noise factor

$$\text{NF} = \text{NF}_{\min} + \frac{4R_N}{Z_o}\frac{|\Gamma_{\text{OPT}} - \Gamma_{\text{OPT}}|^2}{\left(1 - |\Gamma_{\text{OPT}}|^2\right)|1 + \Gamma_{\text{OPT}}|^2} = \text{NF}_{\min} = 0.84\ \text{dB}$$

(ii) In case of $\Gamma_S = 0$, the noise factor

$$\begin{aligned}\text{NF} &= 1.213 + \frac{4 \times 3.0}{50}\frac{|0 - 0.455\angle - 160°|^2}{\left(1 - |0|^2\right)|1 + 0.455\angle - 160°|^2} \\ &= 1.213 + \frac{12}{50} \times \frac{0.207}{0.352} = 1.354 \\ &= 1.32\ \text{dB}\end{aligned}$$

(iii) In case of $\Gamma_S = 0.718\angle - 114°$, the noise factor

$$\begin{aligned}\text{NF} &= 1.213 + \frac{4 \times 3.0}{50}\frac{|0.718\angle - 114° - 0.455\angle - 160°|^2}{\left(1 - |0.718\angle - 114°|^2\right)|1 + 0.455\angle - 160°|^2} \\ &= 1.213 + \frac{12}{50} \times \frac{0.269}{0.170} = 1.592 \\ &= 2.02\ \text{dB}\end{aligned}$$

Example 14.16 Noise parameters of a device in 50 Ω system are as follows: $\text{NF}_{\min} = 0.84$ dB, $\Gamma_{\text{OPT}} = 0.455\angle - 160°$, $R_N = 9.0\ \Omega$.

Compute the noise figure of the amplifier in case of following source reflection coefficients: (i) $\Gamma_S = \Gamma_{\text{OPT}}$ (ii) $\Gamma_S = 0$ (iii) $\Gamma_S = 0.718\angle - 114°$.

Solution

$$\begin{aligned}\text{NF}_{\min} &= 0.84\ \text{dB} = 1.213 \\ \Gamma_{\text{OPT}} &= 0.455\angle - 160° = -0.428 - j0.156 \\ R_N &= 9.0\ \Omega\end{aligned}$$

The noise factor is given by (14.68):

$$\text{NF} = \text{NF}_{\min} + \frac{4R_N}{Z_o}\frac{|\Gamma_S - \Gamma_{\text{OPT}}|^2}{\left(1 - |\Gamma_S|^2\right)|1 + \Gamma_{\text{OPT}}|^2}$$

(i) In case of $\Gamma_S = \Gamma_{\text{OPT}}$, the noise factor

$$\text{NF} = \text{NF}_{\min} + \frac{4R_N}{Z_o}\frac{|\Gamma_{\text{OPT}} - \Gamma_{\text{OPT}}|^2}{\left(1 - |\Gamma_{\text{OPT}}|^2\right)|1 + \Gamma_{\text{OPT}}|^2} = \text{NF}_{\min} = 0.84\ \text{dB}$$

(ii) In case of $\Gamma_S = 0$, the noise factor

$$\mathrm{NF} = 1.213 + \frac{4 \times 9.0}{50} \frac{|0 - 0.455\angle - 160°|^2}{(1 - |0|^2)|1 + 0.455\angle - 160°|^2}$$
$$= 1.213 + \frac{36}{50} \times \frac{0.207}{0.352} = 1.637$$
$$= 2.14 \text{ dB}$$

(iii) In case of $\Gamma_S = 0.718\angle - 114°$, the noise factor

$$\mathrm{NF} = 1.213 + \frac{4 \times 3.0}{50} \frac{|0.718\angle - 114° - 0.455\angle - 160°|^2}{(1 - |0.718\angle - 114°|^2)|1 + 0.455\angle - 160°|^2}$$
$$= 1.213 + \frac{36}{50} \times \frac{0.269}{0.170} = 2.349$$
$$= 3.71 \text{ dB}$$

Examples 14.15 and 14.16 show the effects of equivalent noise resistance R_N on the noise figure when the source matching deviates from its optimum value. Though R_N does not affect the minimum noise figure corresponding to Γ_{OPT}, it affects the noise figure when the source impedance is other than the value of Γ_{OPT}. Therefore, it is always desirable to have the value of R_N as low as possible.

Example 14.17 Compute and draw the noise figure circles for NF = $\mathrm{NF_{min}}$, 1.0, 1.5 and 2.0 dB at 14 GHz of the device with S-parameters and noise parameters tabulated in Tables 14.1 and 14.8 respectively.

Solution

The noise parameters of the device at 14 GHz are (from Table 14.8):

$$\mathrm{NF_{min}} = 0.84 \text{ dB} = 1.213$$
$$\Gamma_{\mathrm{OPT}} = 0.455\angle - 160° = -0.428 - j0.156$$
$$R_N = 2.96\ \Omega$$

$$N = \frac{Z_o(\mathrm{NF} - \mathrm{NF_{min}})}{4R_N}|1 + \Gamma_{\mathrm{OPT}}|^2 = \frac{50(\mathrm{NF} - 1.213)}{4 \times 2.96}|1 - 0.428 - j0.156|^2$$

or,

$$N = 8.708 \times (\mathrm{NF} - 1.213)$$

Therefore, from (14.72a) and (14.72b):

$$C_F = \frac{0.455\angle - 160°}{8.708 \times (\mathrm{NF} - 1.213) + 1}$$

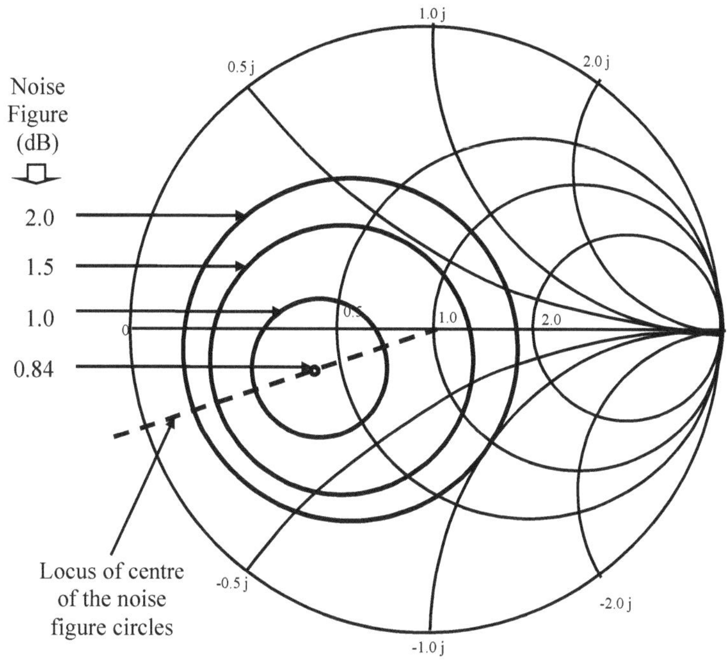

Fig. 14.19 Noise figure circles on Γ_S plane

$$
\begin{aligned}
R_F &= \frac{\sqrt{N\left(N+1-|\Gamma_{\mathrm{OPT}}|^2\right)}}{N+1} \\
&= \frac{\sqrt{8.708 \times (\mathrm{NF}-1.213)(8.708 \times (\mathrm{NF}-1.213)+0.793)}}{8.708 \times (\mathrm{NF}-1.213)+1}
\end{aligned}
$$

Putting the values of NF in the above expressions of center and radius of the noise figure circles:

$$
\begin{aligned}
&\text{for, NF} = \mathrm{NF_{min}}, && N = 0, && R_F = 0, && C_F = 0.455\angle -160^\circ \\
&\text{for, NF} = 1\ \text{dB}, && N = 0.068, && R_F = 0.226, && C_F = 0.426\angle -160^\circ \\
&\text{for, NF} = 1.5\ \text{dB}, && N = 0.296, && R_F = 0.438, && C_F = 0.351\angle -160^\circ \\
&\text{for, NF} = 2.0\ \text{dB}, && N = 0.552, && R_F = 0.555, && C_F = 0.293\angle -160^\circ
\end{aligned}
$$

Using these data, the constant noise figure circles of the device are plotted on Smith chart as shown in Fig. 14.19. It shows that angle of the centre of all the constant noise figure circles from centre of Smith chart are the same as the angle of Γ_{OPT}, i.e. $-160°$.

Any source reflection coefficient on a particular constant noise figure circle will provide the same noise figure. And any source reflection coefficient inside a constant noise figure circle will provide better noise figure. In general, it is not possible to achieve maximum gain and minimum noise figure simultaneously from a device, the constant noise figure circles and constant available gain circles are used to select a source reflection coefficient for particular noise figure and available gain.

Example 14.18 Determine the source and load reflection coefficients and also the source and load impedances to design a LNA at 14 GHz, using a pHEMT having S-parameters and noise parameters given as follows:

$$S_{11} = 0.498\angle 121^\circ, S_{12} = 0.1137\angle -49^\circ$$
$$S_{21} = 3.080\angle -30^\circ, S_{22} = 0.225\angle -150^\circ$$
$$\text{NF}_{\min} = 0.84 \text{ dB}, \Gamma_{\text{OPT}} = 0.455\angle -160^\circ, R_N = 2.96\ \Omega$$

Solution
From the S-parameters,

$$\Delta = 0.291\angle 83.9^\circ, \text{ and } K = 1.122$$

Here $|\Delta| < 1$ and $K > 1$, therefore, the device is unconditional stable. The maximum available gain:

$$G_{Amax}(\text{dB}) = 10\log\left(\frac{3.080}{0.1137}\right) + 10\log\left(1.122 - \sqrt{1.122^2 - 1}\right) = 12.2 \text{ dB}$$

Constant associated gain circles for constant gain of 12.2, 11.6, 10.65, 10, 8 and 6 dB are shown in Fig. 14.20 on Γ_S plane (Refer Example 14.11). Constant noise figure circles for noise figure of 0.84, 1.0, 1.32, 1.5 and 2.0 dB are also drawn in the same Γ_S plane (Refer Example 14.16). Here, the source reflection coefficient for minimum noise figures, $\Gamma_{\text{OPT}} = 0.455\angle -160^\circ$ (marked as A), is different from the source reflection coefficient for maximum available gain, i.e. $0.717\angle -114.4^\circ$ (marked as B). Therefore, it is not possible to design the amplifier with simultaneous minimum noise figure and maximum gain. From the figure it is clear that, if the amplifier is designed with source reflection coefficient of $\Gamma_{\text{OPT}} = 0.455\angle -160^\circ$, noise figure of the amplifier will be minimum, i.e. 0.84 dB, however the available gain of the amplifier will be 10.65 dB. On the other hand, if the amplifier is designed with source reflection coefficient of $0.717\angle -114.4^\circ$, available gain of the amplifier will be maximum, i.e. 12.2 dB; however noise figure of the amplifier will be 2.0 dB. For a source reflection coefficient within the shaded region, as shown in Fig. 14.20, the noise figure of the amplifier will be within 0.84–1 dB and available gain of the amplifier will be within 10.65–11.6 dB.

Depending on the requirement of noise figure and gain of the amplifier, the source reflection coefficient can be determined. Since the device is unconditionally stable at this frequency, the output can be complex conjugate matched for attaining maximum possible gain.

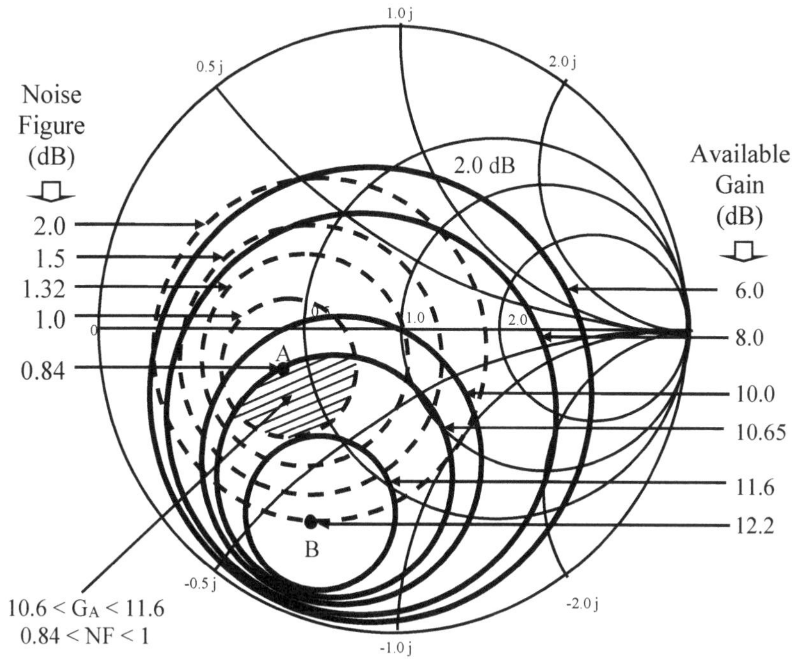

Fig. 14.20 Constant available gain and noise figure circles on Γ_S plane

Designing the amplifier with minimum noise figure, the source reflection coefficient should be Γ_{OPT}. Thus, noise figure will be 0.84 dB. And gain of the amplifier will be 10.65 dB.

$$\Gamma_S = \Gamma_{OPT} = 0.455\angle - 160°$$

Thus, the source impedance Z_S will be given by

$$Z_S = Z_o \frac{1 - \Gamma_S}{1 + \Gamma_S} = 50 \times \frac{1 - 0.455\angle - 160°}{1 + 0.455\angle - 160°}$$
$$= 50 \times \frac{0.572 - j0.156}{1.428 + j0.156} = 19.227 - j7.546$$

The output reflection coefficient will be

$$\Gamma_{OUT} = S_{22} + \frac{\Gamma_S S_{12} S_{21}}{1 - \Gamma_S S_{11}}$$
$$= 0.225\angle - 150° + \frac{0.455\angle - 160° \times 0.1137\angle - 49° \times 3.080\angle - 30°}{1 - 0.455\angle - 160° \times 0.498\angle 121°}$$
$$= (-0.195 - j0.113) + \frac{-0.082 + j0.137}{0.824 + j0.143}$$

Table 14.9 Noise figure and gain for different matching conditions

Γ_S (mag, ang)	Z_S (Ω)	NF (dB)	$\Gamma_L = \Gamma^*_{\text{out}}$ (mag, ang)	Z_L (Ω)	Gain (GHz)
$0.455\angle-160°$	$19.2-j7.55$	0.84	$0.272\angle-166°$	$28.9-j4.07$	10.65
$0.718\angle-114°$	$11.5-j31.0$	2.0	$0.580\angle175°$	$13.3+j1.84$	12.2
0	$50-j0$	1.32	$0.225\angle150°$	$32.9+j7.81$	10.0

$$= (-0.195 - j0.113) + (-0.069 + j0.178)$$
$$= -0.264 + j0.065$$

For complex conjugate match of output port:

$$\Gamma_L = \Gamma^*_{\text{OUT}}$$
$$= -0.264 - j0.065$$
$$= 0.272\angle-166°$$

Thus, the load impedance Z_L will be given be

$$Z_L = Z_o \frac{1-\Gamma_L}{1+\Gamma_L} = 50 \times \frac{1-0.272\angle-166°}{1+0.272\angle-166°}$$
$$= 50 \times \frac{1.264 + j0.065}{0.736 - j0.065} = 28.922 - j4.071$$

The source and load reflection coefficients for minimum noise figure as well as other two cases are shown in Table 14.9.

One of the other two cases is the complex conjugate match for both the ports, to achieve a maximum available gain of 12.2 dB. Since both the ports are complex conjugate matched at this frequency, there will be no reflection of power from both the ports, i.e. VSWR of both the ports will be one at this frequency. However, in this case, the noise figure of the amplifier becomes 2.0 dB.

Just to show the effect of matching network, another case is shown where the source impedance seen by the device is just 50 Ω. In Fig. 14.20, it is shown that the noise figure and gain of the amplifier become 1.32 and 10 dB, respectively. In this case, the load reflection coefficient$\Gamma_L = \Gamma^*_{\text{OUT}} = S^*_{22} = 0.225\angle150°$.

Example 14.19 Synthesize input and output matching networks using series and shunt transmission lines for the amplifier of Example 14.18, where the desired source impedance $Z_{S\text{desired}}$ and load impedance $Z_{L\text{desired}}$ are

$$Z_{S\text{desired}} = (11.5 - 31j)\ \Omega$$
$$Z_{L\text{desired}} = (13.3 + 1.84j)\ \Omega$$

Solution

Synthesis of input matching network:

The desired normalized source impedance is

$$z_{S\text{desired}} = (0.23 - 0.62j)$$

Reflection coefficient corresponding to this desired normalized source impedance is

$$\Gamma_{S\text{desired}} = 0.718\angle - 114°$$

The desired source reflection coefficient is plotted in Fig. 14.21a on a Smith chart and marked as P. The function of the matching network is to transfer the reference source impedance say 50 Ω (marked as A in the Smith chart) to the desired normalized source impedance $z_{S\text{desired}}$.

Though there are many combinations of transmission lines to synthesize a matching network, here the considered input matching network using series and shunt transmission lines is shown in Fig. 14.21a. The lengths and characteristic impedances (here it is chosen as 50 Ω) of the lines are to be properly selected to transfer the impedance from point A to point P. A series transmission line transfers impedance along a constant VSWR circle. The constant VSWR circle is drawn passing through the point P. The shunt stub must transfer the 50 Ω impedance to a point on the constant VSWR circle. Here, a short-circuited shunt stub is chosen to transfer 50 Ω to a point B on the constant VSWR circle along the constant conductance circle. The electrical length of the short-circuit stub is 25.9° (0.072λ) as shown. The function of the series transmission line is to transfer the point B to the impedance point P moving along the constant VSWR circle as shown in the diagram. The electrical length of the series transmission line is 125° (0.347λ).

Synthesis of output matching network:

The desired normalized load impedance is

$$z_{L\text{desired}} = (0.266 + 0.0368j)$$

Reflection coefficient corresponding to this desired normalized load impedance is

$$\Gamma_{L\text{desired}} = 0.58\angle + 175°$$

The desired load reflection coefficient is plotted in Fig. 14.21b on a Smith chart and marked as P. The function of the desired matching network is to transfer the reference load impedance say 50 Ω (marked as A in the Smith chart) to the desired load impedance $z_{L\text{desired}}$.

The output matching network using series and shunt transmission lines is shown in Fig. 14.21b. The lengths and characteristic impedances (here it is chosen as 50 Ω) of the lines are to be properly selected to transfer the impedance from point A to point

Fig. 14.21 **a** Synthesis of input matching network, **b** synthesis of output matching network, **c** amplifier circuit with input and output matching networks

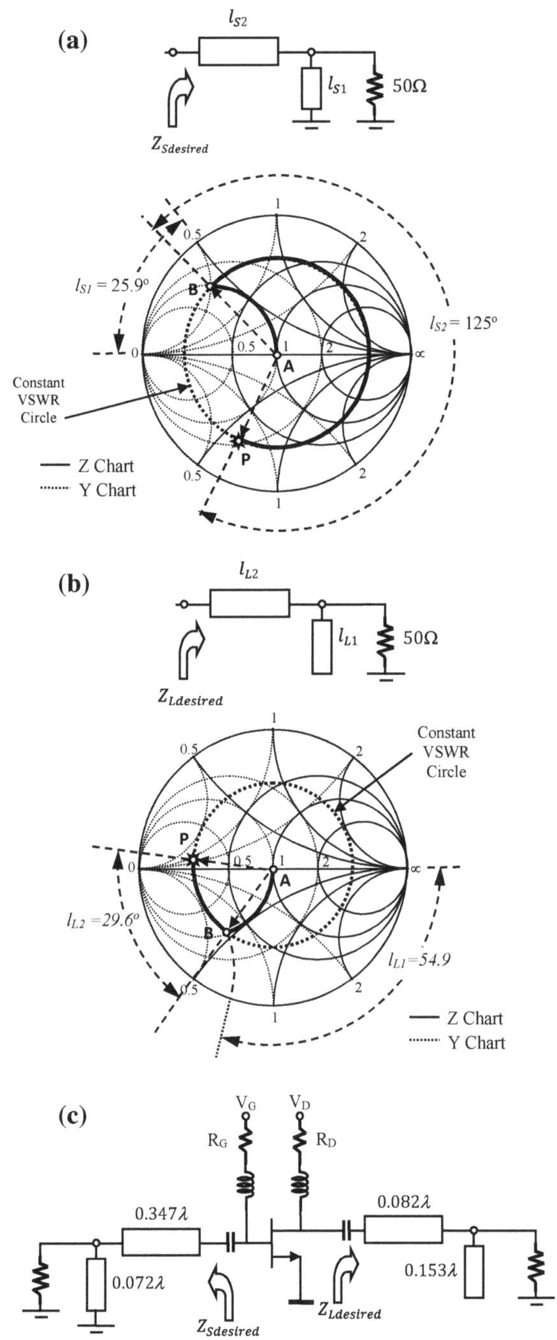

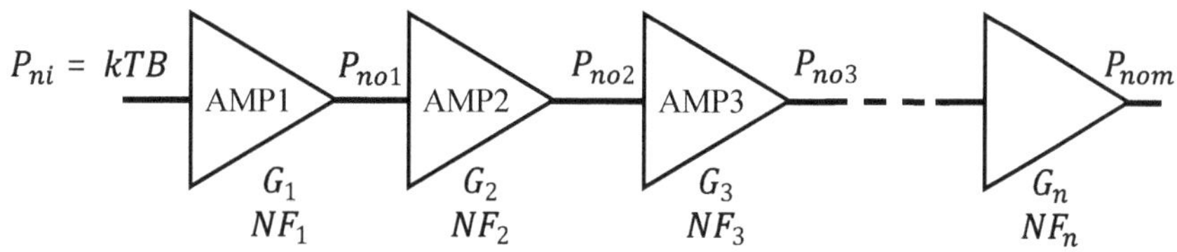

Fig. 14.22 Schematic diagram of a multistage amplifier

P. A constant VSWR circle is drawn passing through the point P. Thus, the shunt stub must transfer the 50 Ω impedance to a point on this constant VSWR circle. An open-circuit stub is chosen to transfer the 50 Ω to a point B on the constant VSWR circle along the constant conductance circle. The electrical length of the open-circuit stub is 54.9° (0.153λ) as shown. The function of the series transmission line is to transfer the point B to the desired impedance point P moving along the constant VSWR circle as shown in the diagram. The electrical length of the series transmission line is 29.6° (0.082λ). The input and output matching networks along with the bias networks of the device are shown in Fig. 14.21c.

14.6.2 High-Gain Amplifier Design

In the previous section, it is shown that the matching requirements for minimum noise figure and maximum gain are different. Thus, practically high-gain amplifiers have comparatively higher noise figure than that of an LNA using the same device. Amplifiers are designed with highest possible gain compromising its noise figure performance for applications where poor NF of the amplifier does not affect the overall system performance. Design philosophy of all high-gain amplifiers is same irrespective of their operating frequency and realization technology [10–13]. A communication receiver consists of multistage amplifier to achieve required gain as shown in Fig. 14.22. Suppose, gain and noise factors of the first, second and third amplifiers are G_1 and NF_1, G_2 and NF_2 and G_3 and NF_3, respectively, and bandwidth of each amplifiers are the same, i.e., B. Suppose the system is driven by a reference source of noise temperature T_o. Then the available noise power at the input of the first amplifier is

$$P_{ni} = kT_oB$$

Here, k is the Boltzmann constant. Thus, from (14.7) the noise power output, P_{no1} of the first amplifier can be written as

$$P_{no1} = (P_{ni}) \times (G_1\mathrm{NF}_1) = (kT_oB) \times (G_1\mathrm{NF}_1) \tag{14.73a}$$

$$= kT_oBG_1 + (\text{NF}_1 - 1)kT_oBG_1 \tag{14.73b}$$

The first term is the noise power due to the presence of input noise power P_{ni} and the second term is the noise power which is added by the first amplifier. In a similar way, the total noise power at the output of the second amplifier can be calculated considering P_{no1} as the noise input to the second amplifier. Thus, the noise power output from the second amplifier will be

$$P_{no2} = (kT_oBG_1\text{NF}_1)G_2 + (\text{NF}_2 - 1)(kT_oB)G_2 \tag{14.74}$$

The noise power P_{no2} is the input noise power for the third amplifier. Thus, the total noise power at the output of the third amplifier will be

$$\begin{aligned} P_{no3} &= [(kT_oBG_1\text{NF}_1)G_2 + (\text{NF}_2 - 1)(kT_oB)G_2]G_3 + (\text{NF}_3 - 1)(kT_oB)G_3 \\ &= kT_oBG_1G_2G_3\text{NF}_1 + (\text{NF}_2 - 1)kT_oBG_2G_3 + (\text{NF}_3 - 1)kT_oBG_3 \end{aligned} \tag{14.75}$$

Suppose, NF_{123} is the overall noise figure of the three stage amplifier. The overall gain of the three-stage amplifier is $G_1 \times G_2 \times G_3$. Thus, the noise power output of the third stage amplifier will be

$$P_{no3} = kT_oBG_1\, G_2\, G_3\, \text{NF}_{123} \tag{14.76}$$

From (14.75) and (14.76), the overall noise figure can be expressed as

$$\text{NF}_{123} = \text{NF}_1 + \frac{(\text{NF}_2 - 1)}{G_1} + \frac{(\text{NF}_3 - 1)}{G_1\, G_2}$$

In this way, the analysis can be extended for multistage amplifiers and the overall noise figure can be expressed as

$$\text{NF}_{123\ldots m} = \text{NF}_1 + \frac{(\text{NF}_2 - 1)}{G_1} + \frac{(\text{NF}_3 - 1)}{G_1G_2} + \cdots + \frac{(\text{NF}_m - 1)}{G_1G_2G_3\ldots G_{m-1}} \tag{14.77}$$

This is known as Friis's formula for overall noise figure for a cascaded multistage network. This formula shows that the overall noise factor will be dominated by the noise factor of the first stage if the gain of the first stage is sufficiently high. Contribution in the overall noise factor by other stages will be reduced by the overall gain of the preceding stages.

Example 14.20 Determine the overall noise figure of a cascaded three-stage amplifier. The individual noise figure and gain of the three stages are as follows:

$$\begin{aligned} \text{NF}_1 &= 0.84\text{ dB}, G_1 = 10.65\text{ dB} \\ \text{NF}_2 &= 2.0\text{ dB}, G_2 = 12.2\text{ dB} \\ \text{NF}_3 &= 2.0\text{ dB}, G_1 = 12.2\text{ dB} \end{aligned}$$

Solution
Noise and gain in factors of the individual amplifiers are

$$\begin{aligned} NF_1 &= 1.21,\ G_1 = 11.61 \\ NF_2 &= 1.58,\ G_2 = 16.60 \\ NF_3 &= 1.58,\ G_1 = 16.60 \end{aligned}$$

The overall noise figure:

$$\begin{aligned} NF_{123} &= 1.21 + \frac{(1.58 - 1)}{11.61} + \frac{(1.58 - 1)}{11.6 \times 16.6} \\ &= 1.21 \text{ (contribution from first stage)} \\ &\quad + 0.0504 \text{ (contribution from second stage)} \\ &\quad + 0.003 \text{ (contribution from third stage)} \\ &= 1.267 \end{aligned}$$

Therefore the overall noise figure of the cascaded three-stage amplifier in dB is

$$NF_{123}(\text{in dB}) = 10\log(1.267)\text{ dB} = 1.027\text{ dB}$$

The above example shows how the contribution in overall noise figure of the following stages diminishes. In communication transponders, high-gain amplifiers are used after a low-noise amplifier of sufficiently high gain so that poor noise figure of the following stages does not affect the overall noise performance.

Example 14.21 Noise figure and gain of two amplifiers AMP1 and AMP2 are as follows:

$$\begin{aligned} NF_1 &= 2\text{ dB}, \quad G_1 = 6\text{ dB} \\ NF_2 &= 2.2\text{ dB}, G_2 = 15\text{ dB} \end{aligned}$$

Determine which combination of these two amplifiers will provide better cascaded noise figure.

Solution
Noise and gain in factors of the individual amplifiers are

$$\begin{aligned} NF_1 &= 1.5849, \quad G_1 = 3.98 \\ NF_2 &= 1.6596, \quad G_2 = 31.62 \end{aligned}$$

Two possible combinations are shown in Fig. 14.23a, b. The cascaded noise factor for the first combination (AMP1 as input amplifier):

$$NF_{12} = 1.5849 + \frac{(1.6596 - 1)}{3.98}$$

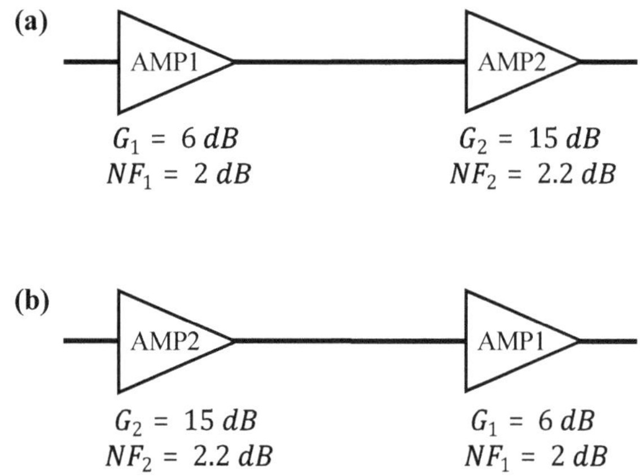

Fig. 14.23 Cascaded two-stage amplifiers

$$\begin{aligned} &= 1.5849 + 0.1657 \\ &= 1.7506 \\ &= 2.432\ \text{dB} \end{aligned}$$

The cascaded noise factor for the second combination (AMP2 as input amplifier):

$$\begin{aligned} \text{NF}_{21} &= 1.6596 + \frac{(1.5849 - 1)}{31.62} \\ &= 1.6596 + 0.0185 \\ &= 1.6781 \\ &= 2.248\ \text{dB} \end{aligned}$$

The second combination, i.e. the amplifier of higher noise figure as input stage of the cascaded two-stage amplifier provides better overall noise figure. This is due to the lower gain of the AMP1. Thus, to achieve overall lowest noise figure from a multistage amplifier, not only noise figure of the first stage, its gain is also important. The gain of the first stage amplifier should not be too low.

For high-gain amplifiers, it is always targeted for a complex conjugate match for both the ports. However, this is only possible at the frequency where the device is unconditional stable. For the frequency where the device is conditionally stable, conjugate matching is not possible. With the help of input and output stability circles, the source and load reflection coefficients are selected to ensure the design of amplifier with stable gain.

Example 14.22 Determine source and load reflection coefficient for a high gain amplifier of gain 16 dB at frequency 4.0 GHz, where S-parameters of the device are $S_{11} = 0.784\angle - 80°$, $S_{12} = 0.0702\angle 38°$, $S_{21} = 5.067\angle 107°$, $S_{22} = 0.440\angle - 55°$.

Also estimate the noise figure of the amplifier for the following noise parameters: $NF_{min} = 0.38$ dB, $\Gamma_{OPT} = 0.606\angle 58°$, $R_N = 12.1\ \Omega$.

Solution

For the device:

$$\begin{aligned}\Delta &= 0.784\angle -80° \times 0.440\angle -55° - 0.0702\angle 38° \times 5.067\angle 107° \\ &= 0.450\angle -83.95°\end{aligned}$$

$$K = \frac{1 - (0.784)^2 - (0.440)^2 + (0.450)^2}{2(0.0702 \times 5.067)} = 0.555$$

The device is conditionally stable because $|\Delta| = 0.450 < 1$ and $K = 0.555 < 1$. Therefore, it is required to draw the stability circles for the device. The source and load reflection coefficients should lie within the stable region on Γ_S and Γ_L plane, respectively.

The centre and the radius of the input stability circle are given by

$$c_S = \frac{\left(S_{11} - \Delta\, S_{22}^*\right)^*}{|S_{11}|^2 - |\Delta|^2}$$

$$\begin{aligned}c_S &= \frac{(0.784\angle -80° - 0.450\angle -83.95° \times 0.440\angle 55°)^*}{(0.784)^2 - |0.450|^2} \\ &= \frac{(0.784\angle -80° - 0.198\angle -28.95°)^*}{0.412} \\ &= 1.645\angle 93.16°\end{aligned}$$

$$\begin{aligned}r_S &= \left|\frac{S_{12}\, S_{21}}{|S_{11}|^2 - |\Delta|^2}\right| \\ &= \frac{0.0702 \times 5.067}{(0.784)^2 - |0.450|^2} \\ &= \frac{0.356}{0.412} = 0.864\end{aligned}$$

The centre and the radius of the output stability circle are given by

$$\begin{aligned}c_L &= \frac{\left(S_{22} - \Delta\, S_{11}^*\right)^*}{|S_{22}|^2 - |\Delta|^2} \\ &= \frac{(0.440\angle -55° - 0.450\angle -83.95° \times 0.784\angle 80°)^*}{(0.440)^2 - |0.450|^2} \\ &= -37.67\angle 106.6°\end{aligned}$$

$$\begin{aligned}r_L &= \left|\frac{S_{12}\, S_{21}}{|S_{22}|^2 - |\Delta|^2}\right| \\ &= \frac{0.0702 \times 5.067}{(0.440)^2 - |0.450|^2}\end{aligned}$$

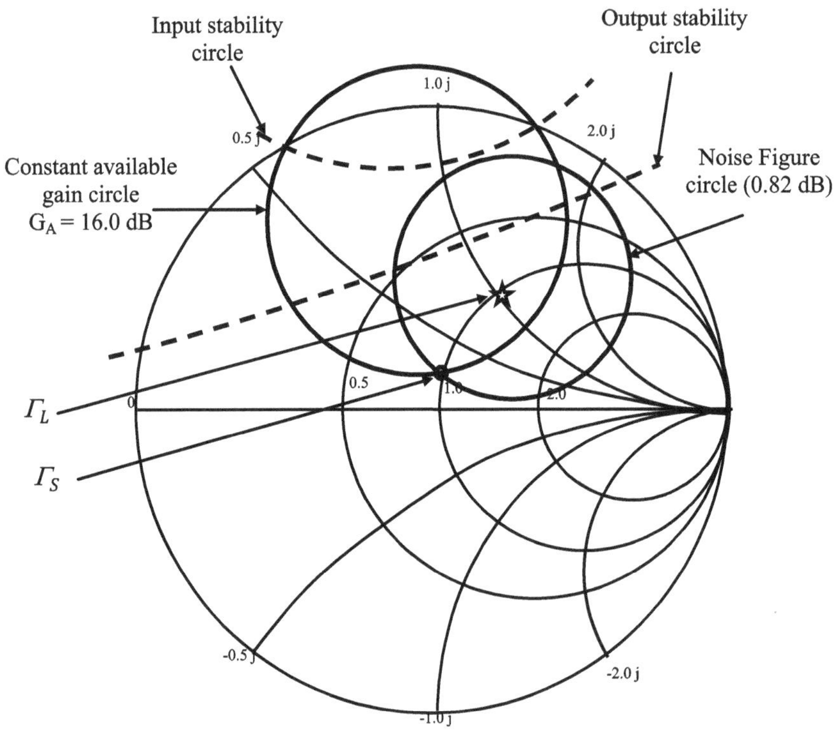

Fig. 14.24 Source and load reflection coefficients for the conditional stable device using constant available gain circle

$$= \frac{0.356}{0.0093} = 38.2$$

The input and output stability circles are shown in Fig. 14.24. Since S_{11} and S_{22} are both less than 1, the region containing the centre of the Smith chart are the stable regions.

The maximum stable gain of the device:

$$\text{MSG} = 10\log\left(\frac{S_{21}}{S_{12}}\right) = 10\log\left(\frac{5.067}{0.0702}\right) = 18.54 \text{ dB}$$

To design a stable amplifier using a conditional stable transistor, it is required to mismatch either input or output or both the ports of the device. There are two options to select the source and load reflection coefficients. The targeted design gain of the amplifier is 16 dB. In one option, mismatch can be introduced in the input section and complex conjugate match can be provided to the output port. In other option, mismatch can be introduced in the output terminal and input port can be complex conjugate matched. For the former case, constant available gain circles approach and for the latter case constant operating gain circle approach should be used.

Constant Available Gain Approach:

In this approach, source reflection coefficient will be selected and the load reflection coefficient will be selected for complex conjugate matching of the output port. Thus, amplifier will provide better return loss at the output port compared to the input port. A constant available gain circle is drawn for $G_A = 16$ dB as shown in Fig. 14.24. Source reflection coefficient Γ_S on any point on the circle will provide the available gain of 16 dB. To ensure better stability, it is advisable to select the Γ_S as far as possible from the input stability circle. Here, $\Gamma_S = 0.141\angle 83.88^\circ$ is selected on the constant available gain circle as shown. Thus, the source impedance Z_S is given by

$$Z_S = Z_o \frac{1 - \Gamma_S}{1 + \Gamma_S} = 50 \times \frac{1 - 0.141\angle 83.88^\circ}{1 + 0.141\angle 83.88^\circ} = 49.51 + j14.14$$

Thus, output reflection coefficient:

$$\begin{aligned}\Gamma_{\text{OUT}} &= S_{22} + \frac{\Gamma_S S_{12} S_{21}}{1 - \Gamma_S S_{11}} \\ &= 0.440\angle -55^\circ + \frac{0.141\angle 83.88^\circ \times 0.0702\angle 38^\circ \times 5.067\angle 107^\circ}{1 - 0.141\angle 83.88^\circ \times 0.784\angle -80^\circ} \\ &= 0.457\angle -61.82\end{aligned}$$

For complex conjugate match of output port:

$$\Gamma_L = \Gamma_{\text{OUT}}^* = 0.457\angle 61.82^\circ$$

The load reflection coefficient $\Gamma_L = 0.457\angle 61.82^\circ$ is shown in Fig. 14.24. Since it is within the stable region on the Γ_L plane, the complex conjugate match for the output port is feasible. The load impedance Z_L is given by

$$Z_L = Z_o \frac{1 - \Gamma_L}{1 + \Gamma_L} = 50 \times \frac{1 - 0.457\angle 61.82}{1 + 0.457\angle 61.82} = 50.89 + j51.82$$

Here, noise figure circle is also drawn. It is shown that for the selected source reflection coefficient, the noise figure is 0.82 dB which is more than the minimum noise figure of 0.38 dB.

Constant Operating Gain Approach:

In this approach, load reflection coefficient is selected at first then the source reflection coefficient is selected for complex conjugate matching of the input port. Thus, amplifier provides better return loss at the input port compared to the output port. A constant operating gain circle is drawn for $G_P = 16$ dB as shown in Fig. 14.25. The load reflection coefficient Γ_L on any point on the circle will provide the operating

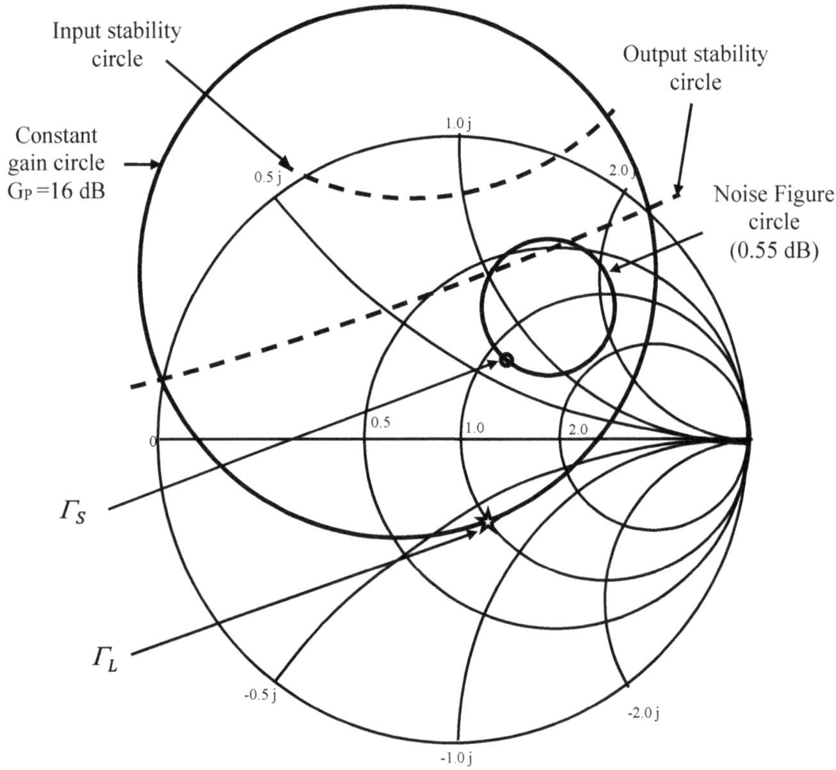

Fig. 14.25 Source and load reflection coefficients for conditional stable device using constant operating gain approach

gain of 16 dB. To ensure better stability, it is preferable to select the Γ_L as far as possible from the output stability circle. Here, $\Gamma_L = 0.288\angle - 72.82$ is selected on the constant operating gain circle as shown. Thus, the load impedance Z_L will be given by

$$Z_L = Z_o \frac{1 - \Gamma_L}{1 + \Gamma_L} = 50 \times \frac{1 - 0.288\angle - 72.82^\circ}{1 + 0.288\angle - 72.82^\circ} = 50.24 - j30.122$$

Thus the input reflection coefficient will be

$$\begin{aligned}\Gamma_{\text{IN}} &= S_{11} + \frac{\Gamma_L S_{12} S_{21}}{1 - \Gamma_L S_{11}} \\ &= 0.784\angle - 80^\circ + \frac{0.288\angle - 72.82^\circ \times 0.0702\angle 38^\circ \times 5.067\angle 107^\circ}{1 - 0.288\angle - 72.82^\circ \times 0.440\angle - 55^\circ} \\ &= 0.324\angle - 57.61^\circ\end{aligned}$$

For complex conjugate match of input port:

$$\Gamma_S = \Gamma_{\mathrm{IN}}^* = 0.324\angle 57.61^\circ$$

The source reflection coefficient $\Gamma_S = 0.324\angle 57.61^\circ$ is shown in Fig. 14.25. Since it is within the stable region on the Γ_S plane, the complex conjugate match for the input port is feasible. The source impedance Z_S is given by

$$Z_S = Z_o \frac{1 - \Gamma_S}{1 + \Gamma_S} = 50 \times \frac{1 - 0.324\angle 57.61^\circ}{1 + 0.324\angle 57.61^\circ} = 59.05 + j36.1$$

Here, noise figure circle passing through the selected source reflection coefficient is also drawn. It is shown that for the selected source reflection coefficient, the noise figure is 0.55 dB which is more than the minimum noise figure of 0.38 dB.

Example 14.23 Synthesize input and output matching networks using series and shunt transmission lines for the amplifier where the desired source impedance $Z_{S\text{desired}}$ and load impedance $Z_{L\text{desired}}$ are

$$Z_{S\text{desired}} = (59.05 + j36.1)\ \Omega$$
$$Z_{L\text{desired}} = (50.24 - j30.122)\ \Omega$$

Solution

Synthesis of input matching network:

The desired normalized source impedance corresponding to the $Z_{S\text{desired}}$ is

$$z_{S\text{desired}} = (1.181 + 0.722j)$$

The source reflection coefficient corresponding to this source impedance is

$$\Gamma_{S\text{desired}} = 0.324\angle 57.61^\circ$$

The desired source reflection coefficient is plotted in Fig. 14.26a on a Smith chart and marked as P. The function of the matching network is to transfer the reference source impedance, say 50 Ω (marked as A in the Smith chart) to the desired source reflection coefficient $\Gamma_{S\text{desired}}$ (marked as P in the Smith chart).

Series transmission line transfers impedance along a constant VSWR circle. The constant VSWR circle is drawn in the Smith chart passing through the point P. Thus, the shunt stub must transfer the 50 Ω impedance to a point on the constant VSWR circle. Here, a short-circuited shunt stub is chosen to transfer 50 Ω to a point B on the constant VSWR circle along the constant conductance circle. The electrical length of the short-circuit stub is 55.59° (0.154λ) as shown. Now, the function of the series transmission line is to transfer the point B to the desired impedance point P moving along the constant VSWR circle as shown in the diagram. The electrical length of the series transmission line is 25.65° (0.0713λ).

Synthesis of output matching network:
The desired normalized load reflection coefficient corresponding to $Z_{L\text{desired}}$ is

$$z_{L\text{desired}} = (1.005 - 0.6024j)$$

The load reflection coefficient corresponding to this load impedance is

$$\Gamma_{L\text{desired}} = 0.288\angle - 72.82$$

The load reflection coefficient is plotted in Fig. 14.26b on a Smith chart and marked as P. The function of the desired network is to transfer the load impedance say 50 Ω (marked as A in the Smith chart) to the desired load impedance $z_{L\text{desired}}$.

The output matching network using series and shunt transmission lines is shown in Fig. 14.26b. A constant VSWR circle is drawn passing through the point P. Thus, the shunt stub must transfer the 50 Ω impedance to a point on this constant VSWR circle. Here, a short-circuited stub is chosen to transfer the 50 to a point B on the constant VSWR circle along the constant conductance circle. The electrical length of the short-circuit stub is 58.98° (0.164λ) as shown. Now, the function of the series transmission line is to transfer the point B to the desired impedance point P moving along the constant VSWR circle as shown in the diagram. The electrical length of the series transmission line is 89.78° (0.249λ).

The input and output matching networks along with the gate and drain bias networks of a device are shown in Fig. 14.26c.

14.7 Large Signal Amplifiers

Power amplifiers are designed to deliver maximum power output for a given device with selected DC bias condition. To achieve optimum output power, the output of the device needs to be matched other than the complex conjugate matching. Thus, a device cannot be matched for maximum power output with maximum gain. In case of a power amplifier, the target is to get maximum output power with the sacrifice of some amount of gain to maximize efficiency. However, when the signal gain drops below a certain value the required extra RF drive power will often cancel out any efficiency advantage. Power amplifiers are designed to operate under the large signal condition to achieve high power. A typical DC I–V characteristic of a MESFET/HEMT is shown in Fig. 14.27. It is shown how the transconductance gain, g_m, and drain-to-source conductance g_{ds} varies with the gate-to-source voltage. The negative value of g_{ds} at higher V_{gs} and V_{ds} is mainly due to the heating effect of the device at a high current level. Other than these resistive equivalent circuit components, reactive components such as gate-to-source capacitance C_{gs} and gate–to-drain capacitance C_{gd} are also have nonlinear characteristic. Under small amplitude of sinusoidal signal, the output voltage and current waveforms also remain sinusoidal. However, under large signal operation, the output voltage and current of the device

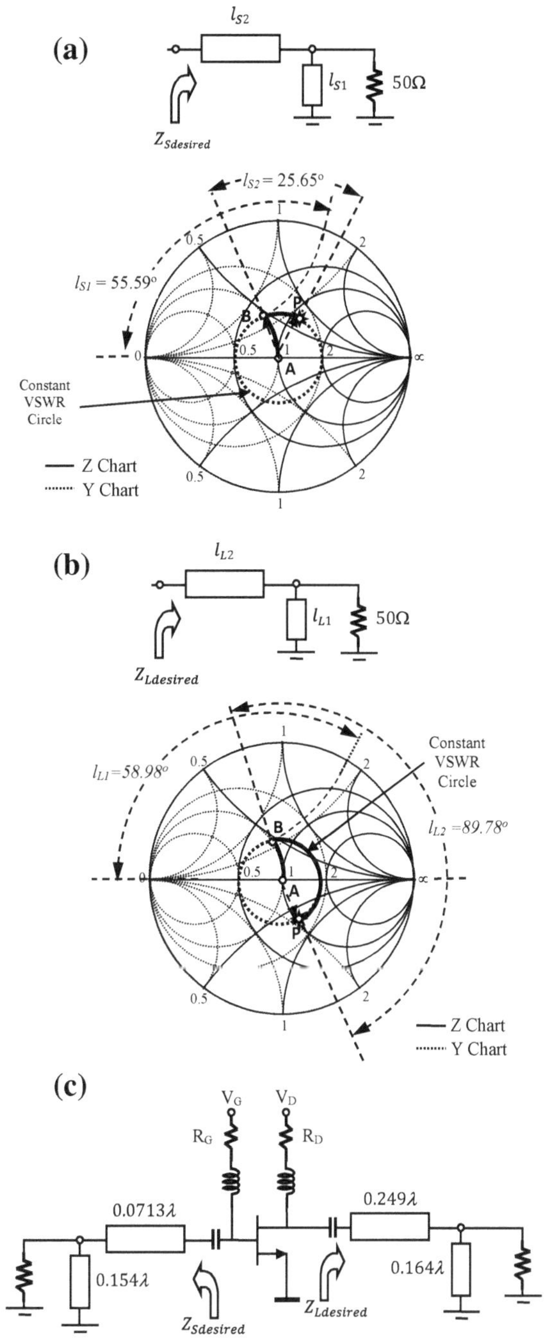

Fig. 14.26 **a** Synthesis of input matching network, **b** synthesis of output matching network, **c** amplifier circuit with input and output matching networks

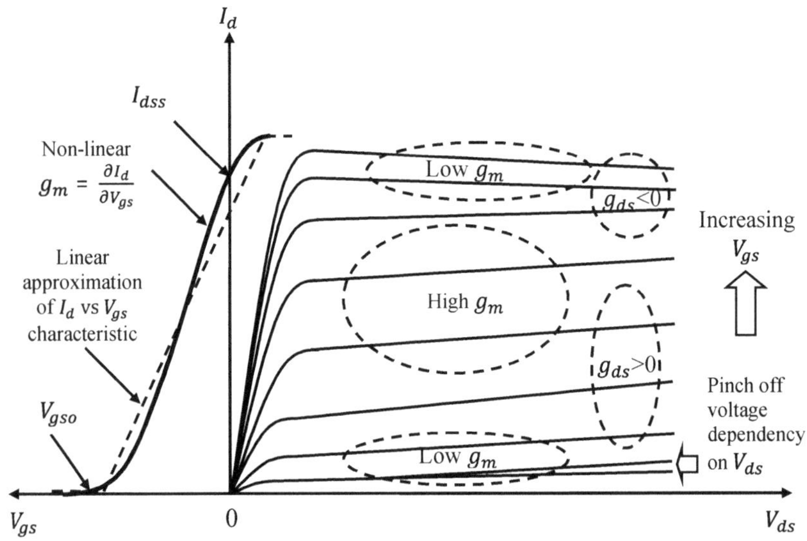

Fig. 14.27 A typical DC I–V characteristics of power MESFET/HEMT

swings over a wide range of I–V characteristics of the device. Traditionally, a load line is used to determine output signal waveforms under large signal operating condition. In a realistic scenario, the load curve is an ellipse due to the reactive elements at the output of the device. The reactive elements may be due to the equivalent reactive elements of the device output or part of the load impedance. The load curve remains elliptic as long as the signal amplitude is small. The inverse slope of the major axis of the elliptic curve is the load resistance and the minor axis is related to the reactive part of the load impedance. The dynamic load line under large signal condition is clipped at both the ends and the shape deviates from elliptic shape as shown in Fig. 14.28. However, considering the reactive elements within the device output as part of the output matching network of the amplifier, the load curve can be considered as a straight line as shown in the figure as a dotted line.

At large signal operating condition, the output voltage and current of the device swing over a wide range of I–V characteristics of the device. This results in distorted output voltage and current waveforms as shown in Fig. 14.29. This is the reason for the power amplifiers even in Class-A operation cannot be considered as a linear amplifier. The output current waveform is not only non-sinusoidal it is also unsymmetrical which contributes to DC current thus changes the device operating point.

A power amplifier delivers the highest output power only when the output voltage as well as the current swings become maximum as shown in Fig. 14.29. The maximum voltage and current swings are determined by the bias voltage, current and selection of load line of an amplifier. It is also very important that under large signal operating condition, the drain current I_d and gate-to-drain voltage V_{ds} remain within the allowable maximum ratings of the device. The drain voltage of a

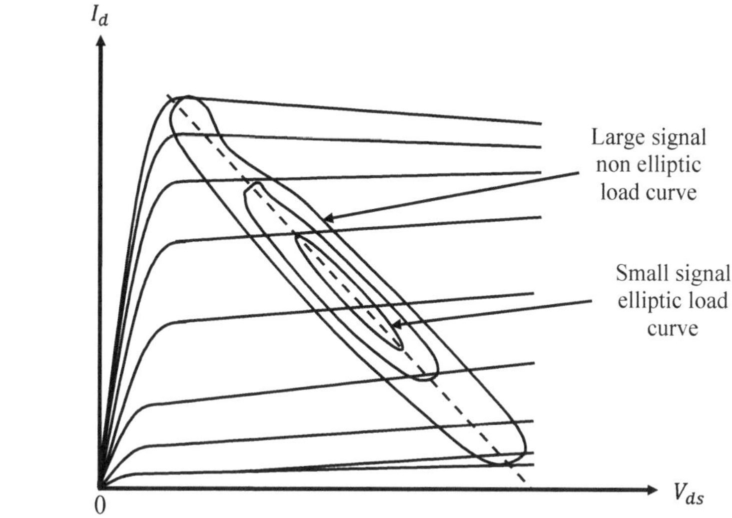

Fig. 14.28 A typical dynamic load line of power amplifiers

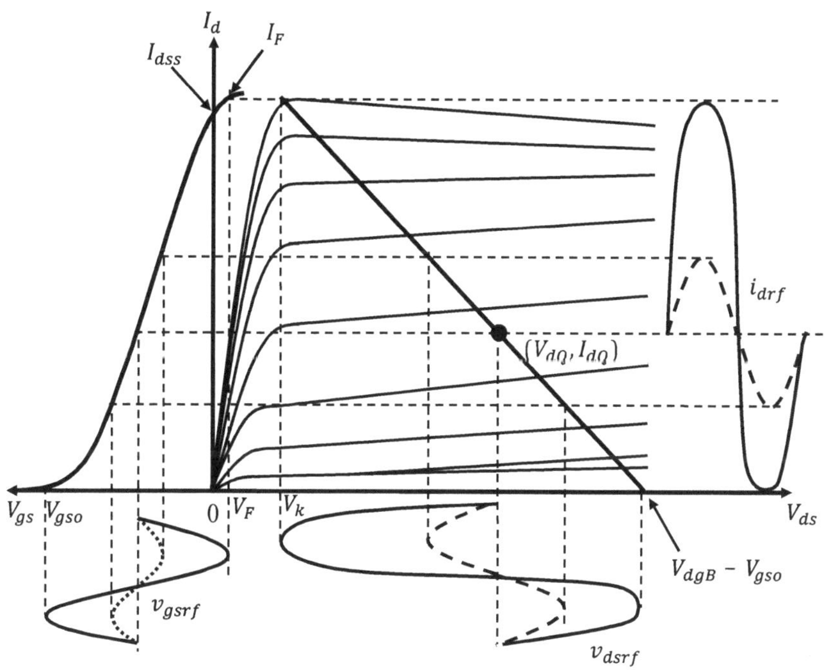

Fig. 14.29 Signal amplification MESFET/HEMT device

MESFET/HEMT device is mainly limited by its gate-to-drain breakdown voltage V_{dgB}. The maximum drain current I_F flows when the channel of the device is fully open which occurs at small forward gate bias voltage V_F as shown in Fig. 14.29. In case of many high power devices, the maximum allowable drain current may be limited by the maximum allowable temperature of the device due to thermal dissipation and to keep the channel temperature of the device within allowable limit. The lower limit of the voltage swing is limited by the knee voltage V_k as shown in Fig. 14.29. To achieve maximum output RF power for the Class-A amplifier, the load line, i.e. output impedance of the FET device should be chosen so that the drain voltage becomes maximum when the gate voltage is at its pinch-off point V_{gso} and the drain voltage becomes minimum when the drain current is at its maximum allowable value.

14.7.1 Linear Power Amplifier

To derive the output power and other performance parameters of a Class-A power amplifier, consider an amplifier circuit using an ideal FET of I–V characteristic with infinite input impedance and output conductance as shown in Fig. 14.30. Also consider that the inductor and the DC blocking capacitor provide infinite and zero impedances, respectively at the operating RF frequency. Thus, to achieve maximum RF voltage and current swing the drain bias voltage and drain current of the device will be

$$\begin{aligned} V_{dQ} = V_S &= \frac{V_{dgB} - V_{gso} - V_K}{2} + V_K \\ &= \frac{V_{dgB} - V_{gso} + V_K}{2} \end{aligned} \tag{14.78a}$$

and

$$I_{dQ} = \frac{I_F}{2} \tag{14.78b}$$

At this condition the gate bias voltage will be

$$\begin{aligned} V_{gs} &= -\frac{V_{gso} + V_F}{2} + V_F \\ &= \frac{-V_{gso} + V_F}{2} \end{aligned} \tag{14.78c}$$

Thus, the DC power delivered by the supply to the device is given by

$$\begin{aligned} P_{\text{DC}} &= V_{dQ} \times I_{dQ} \\ &= \frac{V_{dgB} - V_{gso} + V_K}{2} \times \frac{I_F}{2} \end{aligned} \tag{14.79a}$$

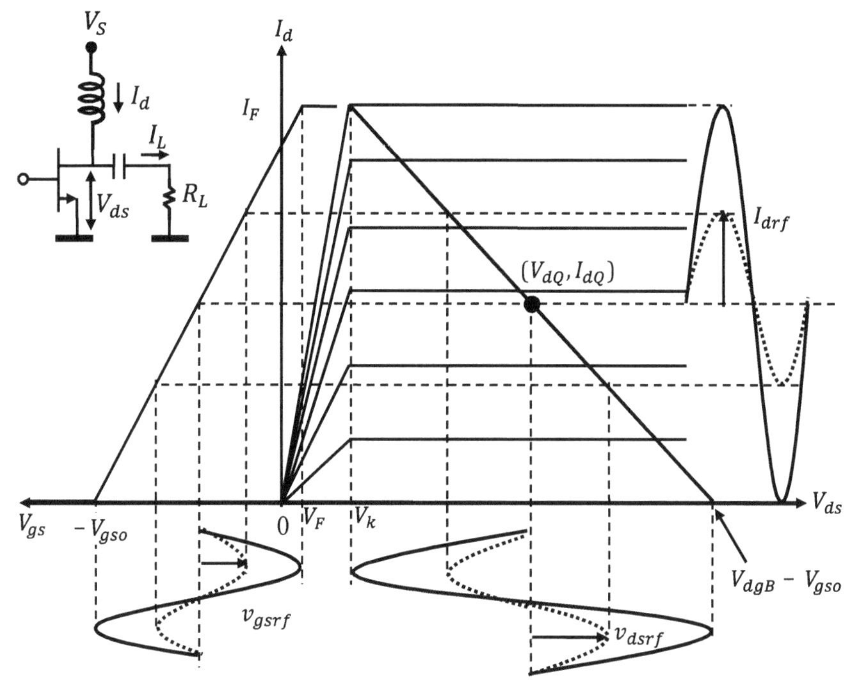

Fig. 14.30 Signal amplification through an ideal linear amplifier

This is independent of the RF power level. Thus, without any RF input power, the amplifier will consume this amount of DC power which is one major drawback of Class-A amplifiers.

The optimum load impedance $R_{L\mathrm{opt}}$ for which the voltage and current swings are maximum is given by

$$R_{L\mathrm{opt}} = \frac{V_{dgB} - V_{gso} - V_K}{I_F} \tag{14.79b}$$

Suppose, i_{drf} and v_{dsrf} are the peak of the drain current and drain-to-source voltage of the device. The output RF power can be expressed as

$$P_{rf} = \frac{v_{dsrf}}{\sqrt{2}} \times \frac{i_{drf}}{\sqrt{2}} \tag{14.80a}$$

The maximum voltage and current swings of a Class-A amplifier is shown in Figs. 14.30 and 14.31.

The maximum amplitude of the output voltage:

$$v_{dsrf}(\max) = \frac{V_{dgB} - V_{gso} - V_K}{2} \tag{14.80b}$$

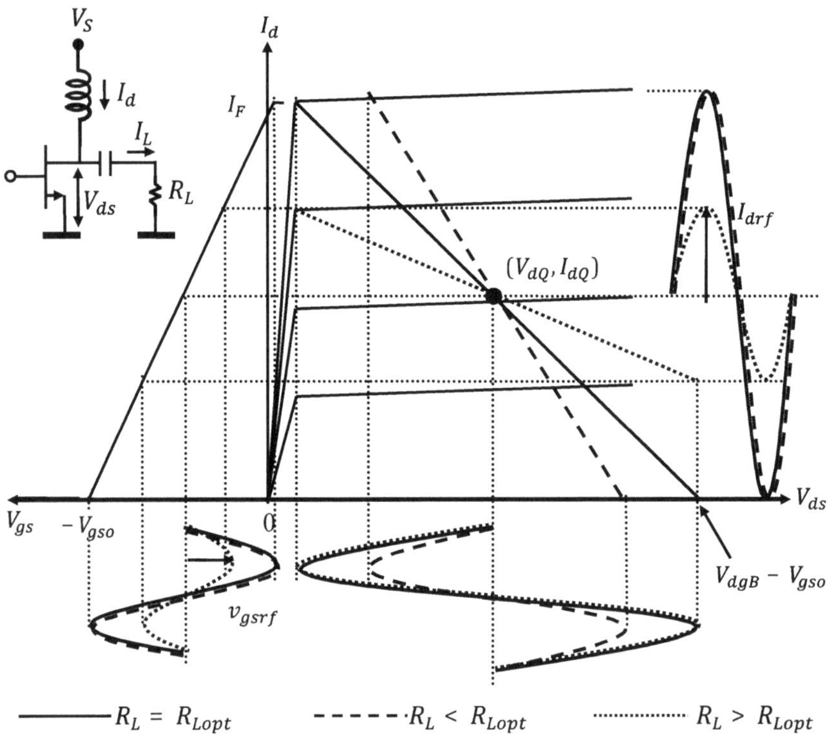

Fig. 14.31 Input output waveforms under different load impedances

The maximum amplitude of the output current:

$$i_{drf}(\max) = \frac{I_F}{2} \tag{14.80c}$$

Thus, the maximum RF output power of Class-A amplifier is given by

$$\begin{aligned} P_{rf}(\max) &= \frac{v_{dsrf}(\max)}{\sqrt{2}} \times \frac{i_{drf}(\max)}{\sqrt{2}} \\ &= \frac{I_F\left(V_{dgB} - V_{gso} - V_K\right)}{8} \end{aligned} \tag{14.81}$$

This is the maximum linear output power from a FET based Class-A amplifier under the assumption of idealized I–V characteristic of the device.

To account the effect of gain of an amplifier, power added efficiency is used in place of simple efficiency, which is the ratio of output RF power over the DC power consumption. The power added efficiency η_{add} of an amplifier is defined as

$$\eta_{\text{add}} = \frac{\text{Output RF power} - \text{Input RF power}}{\text{DC power consumption}}$$

or

$$\eta_{\text{add}} = \left(1 - \frac{1}{G}\right)\frac{\text{Output RF power}}{\text{DC power consumption}} \quad (14.82)$$

Here, G is the ratio of output RF power to the input RF power, i.e. RF gain of the amplifier. With the increase of output RF power, the efficiency of Class-A amplifier increases. From (14.79a) and (14.80a) the power added efficiency can be written as

$$\eta_{\text{add}} = \left(1 - \frac{1}{G}\right)\frac{2 \times v_{dsrf} \times i_{drf}}{I_F\left(V_{dgB} - V_{gso} + V_K\right)} \quad (14.83)$$

The power added efficiency is maximum at the maximum output RF power condition. Putting the maximum values of v_{dsrf} and i_{drf} from (14.80b) and (14.80c), the maximum power added efficiency $\eta_{\text{add}}(\max)$ can be expressed as

$$\eta_{\text{add}}(\max) = \frac{1}{2} \times \left(1 - \frac{1}{G}\right) \times \frac{\left(V_{dgB} - V_{gso} - V_K\right)}{\left(V_{dgB} - V_{gso} + V_K\right)} \quad (14.84)$$

Therefore, classical maximum power added efficiency of 50% from a Class-A amplifier is only possible in the limiting case of infinite (practically very high) gain ($G = \infty$) and zero Knee voltage ($V_K = 0$).

Example 14.24 Calculate the maximum output power and power added efficiency from a GaAs FET power amplifier operating in Class-A with optimum load impedance for achieving maximum output RF power. Where $V_K = 1.5$ V, $I_F = 1A$, $V_{dgB} = 24$ V, $V_{gso} = 2.5$ V and $G = 13$ dB. Also, calculate the optimum output load impedance of the device for achieving maximum voltage and current swings.

Solution

Here, $G = 13\text{ dB} = 10^{13/10} = 20$

The maximum RF output power from (14.81):

$$\begin{aligned} P_{rf}(\max) &= \frac{I_F\left(V_{dgB} - V_{gso} - V_K\right)}{8} \\ &= \frac{1 \times (24 - 2.5 - 1.5)}{8}\text{ W} \\ &= 2.5\text{ W} \end{aligned}$$

The DC power consumption from (14.79a):

$$P_{\text{DC}} = \frac{I_F \times \left(V_{dgB} - V_{gso} + V_K\right)}{4}$$

$$= \frac{1 \times (24 - 2.5 + 1.5)}{4}$$
$$= 5.75 \text{ W}$$

Therefore, the maximum power added efficiency:

$$\eta_{\text{add}}(\text{max}) = \left(1 - \frac{1}{20}\right)\frac{2.5}{5.75}$$
$$= 41.3\%$$

The optimum output load impedance is given by (14.79b)

$$R_{L\text{opt}} = \frac{V_{dgB} - V_{gso} - V_K}{I_F}$$
$$= \frac{24 - 2.5 - 1.5}{1} = 20\ \Omega$$

The voltage and current swings are maximum when the output load impedance is optimum to provide maximum RF output power. When the output load impedance is lesser than the optimum load impedance (i.e. $R_L < R_{L\text{opt}}$) then though the current swing becomes maximum, the voltage swing becomes lesser than the maximum possible voltage swing as shown in Fig. 14.31. On the other hand when the output of the device is terminated with load impedance which is greater than the optimum load impedance (i.e. $R_L > R_{L\text{opt}}$) then to keep the voltage swing within the allowable limit, the current swing should be less than the values obtained in optimum load condition as shown in Fig. 14.31.

Example 14.25 Calculate the maximum output power and gain from a GaAs FET power amplifier operating in Class-A when (a) the amplifier is matched for maximum RF output power, (b) the amplifier is matched for maximum RF gain. The equivalent circuit of the device is shown in Fig. 14.32. Where, $1/2\pi f c_{gs} \gg R_s$, $r_{ds} = 100\ \Omega$, other parameters are $I_F = 1$ A, $V_{dgB} = 20$ V, $V_{gso} = 4$ V, $V_K \ll V_{dgB}$, $V_F \ll V_{gso}$, $g_m = 0.25$ mho.

Solution

(a) To match the device for maximum RF output power, the output impedance to be selected for maximum drain-to-source voltage and drain current swings:

$$v_{dsrf}(\text{max}) = \frac{V_{dgB} - V_{gso}}{2} = \frac{20 - 4}{2} \text{ V} = 8 \text{ V}$$
$$i_{drf}(\text{max}) = \frac{I_F}{2} = \frac{1}{2} \text{ A}$$

The optimum output impedance is

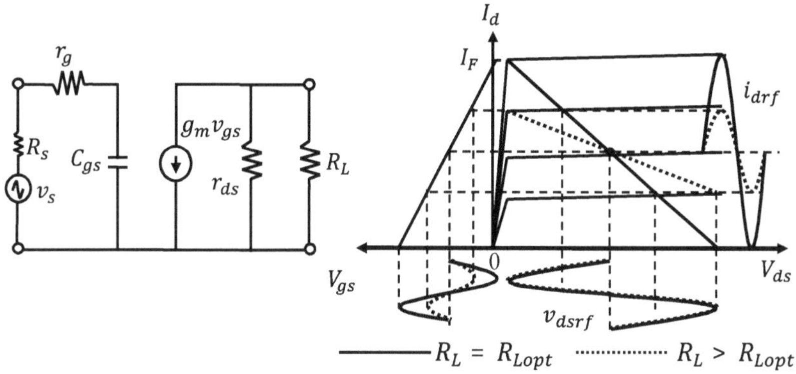

Fig. 14.32 Signal amplification through an ideal linear amplifier

$$
\begin{aligned}
R_{o\mathrm{OPT}} &= \frac{V_{dgB} - V_{gso} - V_K}{I_F} \\
&= \frac{20 - 4}{1}\,\Omega = 16\,\Omega
\end{aligned}
$$

The optimum load impedance is

$$
\begin{aligned}
R_{L\mathrm{OPT}} &= \frac{r_{ds} \times R_{o\mathrm{OPT}}}{r_{ds} - R_{o\mathrm{OPT}}} \\
&= \frac{100 \times 16}{100 - 16} = 19.05\,\Omega
\end{aligned}
$$

Therefore, the maximum RF current through the load resistance:

$$
\begin{aligned}
i_{Lrf}(\max) &= \frac{i_{drf}(\max) \times r_{ds}}{r_{ds} + R_{L\mathrm{opt}}} = \frac{100}{2 \times 119.05} = 0.42\ \mathrm{A} \\
P_{rf}(\max) &= \frac{v_{dsrf}(\max)}{\sqrt{2}} \times \frac{i_{Lrf}(\max)}{\sqrt{2}} \\
&= \frac{8}{\sqrt{2}} \times \frac{0.42}{\sqrt{2}}\ \mathrm{W} = 1.68\ \mathrm{W}
\end{aligned}
$$

With reference to Fig. 14.32, the available RF power from the source is

$$
P_{\mathrm{avs}} = \frac{v_s^2}{8R_s}
$$

Since, $1/2\pi f c_{gs} \gg R_s$, it can be considered that voltage across the capacitor c_{gs} will be v_s. Thus, the drain current i_{drf} can be written as

$$i_{drf} = g_m v_s$$

The RF load current can be written as

$$i_{Lrf} = g_m v_s \frac{r_{ds}}{r_{ds} + R_{LOPT}}$$

The output RF power can be written as

$$P_{rf} = \left(\frac{i_{Lrf}}{\sqrt{2}}\right)^2 \times R_{LOPT}$$
$$= \left(\frac{g_m v_s}{\sqrt{2}} \frac{r_{ds}}{r_{ds} + R_{LOPT}}\right)^2 \times R_{LOPT}$$

The gain of the amplifier is given by

$$G = \frac{P_{rf}}{P_{avs}} = 4R_{LOPT} R_s \left(g_m \frac{r_{ds}}{r_{ds} + R_{LOPT}}\right)^2$$
$$= 4 \times 19.05 \times 50 \left(0.25 \frac{100}{100 + 19.05}\right)^2$$
$$= 168.6 = 22.25 \text{ dB}$$

(b) To match the device for maximum gain, both the output and input should be conjugate matched. Thus, the load impedance R_L will be equal to the output impedance r_{ds}, i.e. 100 Ω.

$$v_{dsrf}(\max) = \frac{V_{dgB} - V_{gso}}{2} = \frac{20 - 4}{2} \text{ V} = 8 \text{ V}$$
$$i_{drf}(\max) = \frac{v_{dsrf}(\max)}{R_L \| r_{ds}} = \frac{8}{50} \text{ A}$$

Therefore, the maximum RF current through the load resistance:

$$i_{Lrf}(\max) = \frac{i_{drf}(\max) \times r_{ds}}{r_{ds} + R_L} = \frac{8 \times 100}{50 \times 200} \text{ A}$$
$$P_{rf}(\max) = \frac{v_{dsrf}(\max)}{\sqrt{2}} \times \frac{i_{Lrf}(\max)}{\sqrt{2}}$$
$$= \frac{8}{\sqrt{2}} \times \frac{8}{100\sqrt{2}} \text{ W} = 0.32 \text{ W}$$

Which is $10 \log(1.68/0.32) = 7.2$ dB less power compared to the output power obtained by optimum output impedance matching.
Thus, gain of the amplifier will be given by

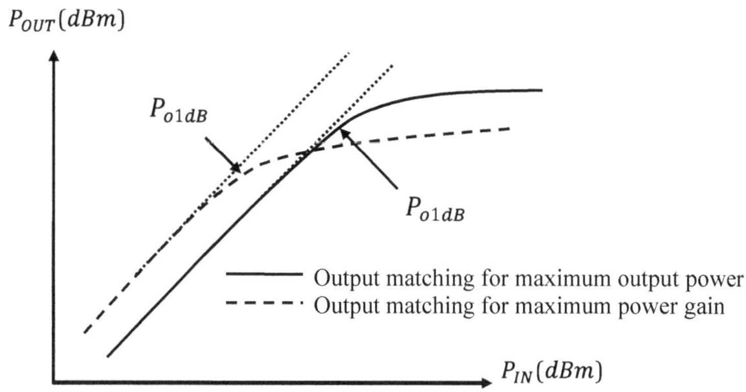

Fig. 14.33 Transfer characteristic for two different output matching conditions

$$\begin{aligned} G &= 4R_L R_s \left(g_m \frac{r_{ds}}{r_{ds} + R_L} \right)^2 \\ &= 4 \times 100 \times 50 \left(0.25 \times \frac{100}{100 + 100} \right)^2 \\ &= 312.5 = 24.9 \text{ dB} \end{aligned}$$

This gain is about 2.65 dB more than the gain when the device is matched for maximum output RF power.

Input–output transfer characteristic of two amplifiers one with output matching corresponding to maximum output power and another with output matching corresponding to maximum gain is shown in Fig. 14.33.

The linear analysis of the amplifier operating in large signal condition is done considering that the linear characteristic of the devices holds good over its entire operating regions. Equivalent circuit elements of practical transistors not only depend on the DC bias level, these are also variations with the instantaneous signal levels. The practical dependence of equivalent circuit parameters on the signal level produces distortion in the output signal waveform before starting of cut-off and clipping behaviour of the transistor.

The Class-A power amplifier operating up to 1–2 dB gain compression point may be due to the nonlinearities of the circuit elements and the output signal can be expressed by a power series of the input signal up to the third-order terms as

$$v_o = a_0 + a_1 v_i + a_2 v_i^2 + a_3 v_i^3$$

Suppose, $v_i = A \cos \omega t$, then the output voltage can be written as

$$v_o = a_0 + a_1 A \cos \omega t + a_2 A^2 \cos^2 \omega t + a_3 A^3 \cos^3 \omega t$$

$$= a_0 + a_1 A \cos \omega t + \frac{a_2 A^2}{2}(1 + \cos 2\omega t) + \frac{a_3 A^3}{4}(3 \cos \omega t - \cos 3\omega t)$$

$$= \left(a_0 + \frac{a_2 A^2}{2}\right) + \left(Aa_1 + \frac{3a_3 A^3}{4}\right) \cos \omega t + \frac{a_2 A^2}{2} \cos 2\omega t + \frac{a_3 A^3}{4} \cos 3\omega t$$

Thus, the output signal level corresponding to fundamental frequency is given by

$$\left(Aa_1 + \frac{3a_3 A^3}{4}\right) = Aa_1 \left(1 + \frac{3a_3 A^2}{4a_1}\right)$$

Here, the second term is contribution from the third-order distortion of the output signal. This term causes gain compression of the amplifier with the increase of input signal level. Suppose, the gain of the amplifier at 1-dB gain compression point ($G_{1\text{dB}}$) is corresponding to input signal level $A_{1\text{dB}}$, then we can write

$$20 \log(a_1 A_{1\text{dB}}) = 20 \log\left(a_1 A_{1\text{dB}} \left(1 + \frac{3a_3 A_{1\text{dB}}^2}{4a_1}\right)\right) + 1$$

or

$$20 \log(a_1 A_{1\text{dB}}) = 20 \log(a_1 A_{1\text{dB}}) + 20 \log\left(1 + \frac{3a_3 A_{1\text{dB}}^2}{4a_1}\right) + 1$$

or

$$20 \log\left(1 + \frac{3a_3 A_{1\text{dB}}^2}{4a_1}\right) = -1$$

or

$$1 + \frac{3a_3 A_{1\text{dB}}^2}{4a_1} = 10^{\frac{-1}{20}} = 0.89$$

or

$$\frac{3a_3 A_{1\text{dB}}^2}{4a_1} = 0.89 - 1 = -0.11$$

or

$$|A_{1\text{dB}}| = \sqrt{\frac{4a_1}{3a_3}} \times \sqrt{0.11}$$

or

$$|A_{1\text{dB}}|(\text{in dB}) = 20\log\left(\sqrt{\frac{4a_1}{3a_3}}\right) - 9.63 \tag{14.85}$$

The third-order output intercept point is defined as the output power level at which the third-order intermodulation power level becomes equal to the power level corresponding to the fundamental frequency. From (14.14), the condition for third-order intercept point can be written as

$$A_{IP3}a_1 = \frac{3a_3A_{IP3}^3}{4}$$

or

$$A_{IP3} = \sqrt{\frac{4a_1}{3a_3}} \tag{14.86}$$

Therefore, input power in dB corresponding to third-order intercept point P_{iIP3} is given by

$$P_{iIP3} = 20\log\left(\sqrt{\frac{4a_1}{3a_3}}\right) \tag{14.87}$$

Now, the output power at 1-dB gain compression point $P_{o1\text{dB}}$ can be expressed as

$$P_{o1\text{dB}} = A_{1\text{dB}} + G_{1\text{dB}} = A_{1\text{dB}} + G_{L\text{dB}} - 1 \tag{14.88}$$

Here, $G_{L\text{dB}}$ is the linear gain of the amplifier. Combining (14.85), (14.87) and (14.88), the $P_{o1\text{dB}}$ can be expressed as

$$P_{o1\text{dB}} = P_{iIP3} + G_{L\text{dB}} - 10.63 \tag{14.89}$$

or

$$P_{o1\text{dB}} = P_{oIP3} - 10.63 \tag{14.90}$$

This is a very useful expression and frequently used by manufactures and engineers to define the output power capability of power devices and amplifiers.

Example 14.26 Calculate the output power at 1-dB gain compression point of an amplifier having third order output intercept point 1-W. Also, calculate the third order intermodulation when the amplifier operates at its 1-dB gain compression point and 10 dB input back-off compared to its $P_{1\text{dB}}$ (consider the amplifier characteristic can be expressed by a power series up to third order).

Solution

Output power at 1-dB gain compression point is given by

$$P_{o1\text{dB}} = P_{oIP3} - 10.63$$

Here, $P_{oIP3} = 1\ \text{W} = 30\ \text{dBm}$.

Thus, $P_{o1\text{dB}} = (30 - 10.63) = 19.37\ \text{dBm}$.

From the definition, at the third-order intercept point the power of fundamental frequency and third-order intermodulation is the same and each dB decrease/increase of input power level corresponding to 1-dB decrease/increase of fundamental output power and 3 dB decrease/increase of third-order intermodulation output power level.

Since $P_{o1\text{dB}}$ is 10.63 dB less than the third-order intercept point, the power level of the third-order intermodulation product at $P_{o1\text{dB}}$ will be $(30 - 3 \times 10.63)\ \text{dBm} = -1.89\ \text{dBm}$, which is $(19.37 + 1.89)\ \text{dBc} = 21.26\ \text{dBc}$ less compared to fundamental power level.

Thus, $\text{IM}_3(\text{at } P_{o1\text{dB}}) = -21.26\ \text{dBc}$.

Similarly the third order intermodulation product at 10-dB input back-off compared to 1-dB gain compression point will be

$$\text{IM}_3(\text{at } P_{o1\text{dB}} - 10\ \text{dB}) = (-21.26 - 2 \times 10)\ \text{dBc} = -41.26\ \text{dBc}.$$

It should be kept in mind that this calculation is valid up to the power level over which the transfer characteristic is expressed by a power series up to third-order term. For Class-A power amplifiers, operating regions up to 1–2 dB gain compression point are considered to be expressed by a power series of the input signal up to the third-order terms.

14.7.2 Load-Pull Characterization Technique

Load-pull technique was developed to practically tune the output termination to determine the required output impedance for maximum output power from a device. This technique is not only used for output tuning, but also used to characterize the device as well as to determine source and load impedances for targeted output power, gain, noise figure, efficiency, etc. The load-pull characterization involves tuning the input and output impedances using tuners, recording performance and impedances and then plotting contours of constant performances such as noise figure, output power, efficiency, etc., on the impedance plane. To determine optimum load impedance of microwave power transistor corresponding to maximum output power, the load-pull technique is used. In case of power amplifiers, this technique is used for large signal characterization and determination of load as well as source impedances of microwave power transistors to map contours of constant power and efficiency on a Smith chart. Schematic diagram of a typical load-pull test setup is shown in Fig. 14.34. Here, output RF power is delivered to the load

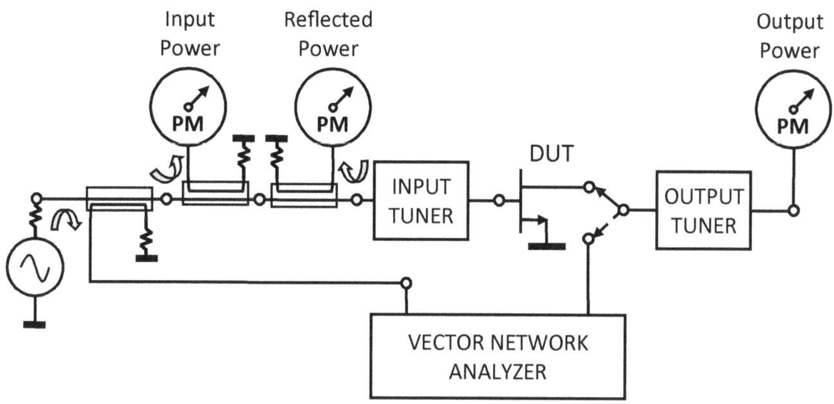

Fig. 14.34 Load-pull setup for power device characterization

and input power to the amplifier and power reflected from the amplifier is measured using RF power meter. In load-pull measurement technique for obtaining maximum output power/efficiency, the load impedance is modified/adjusted using a stub tuner until the output RF power/efficiency becomes maximum. A vector network analyzer (VNA) is used to measure the real and imaginary parts of the tuned load impedance (complex load) with the help of RF switch. The gain of the amplifier is maximized by adjusting the input tuner. For a bidirectional device, it is required to adjust both the tuner iteratively to maximize output power/efficiency and gain of the amplifier due to the dependence of output impedance on its input source impedance.

The requirement of load impedances for achieving maximum output power and maximum efficiency is different. Constant output power contours are the loci of load impedances for constant output power. The constant output power contours are circles when the amplifier operates at small signal conditions as discussed and computed using S-parameters of the device earlier in this chapter. With the increase of power level, the constant output power contours deviate from circles and look like potato shapes as shown in Fig. 14.35. The output impedance contours consist of several curves on a Smith chart representing constant output power and become a point for the maximum output power. In practical case, constant efficiency contours are also derived by experiment and drawn on the same Smith chart with the constant power contours. Depending on applications, proper load impedance is selected for targeted output power and efficiency.

14.7.3 *Amplifiers with Reduced Conduction Angle*

One of the techniques to improve power amplifiers efficiency is to operate the device for reduced conduction angle. Power amplifiers are categorized into Class-A, Class-AB, Class-B and Class-C based on the conduction angle. Efficient RF/microwave

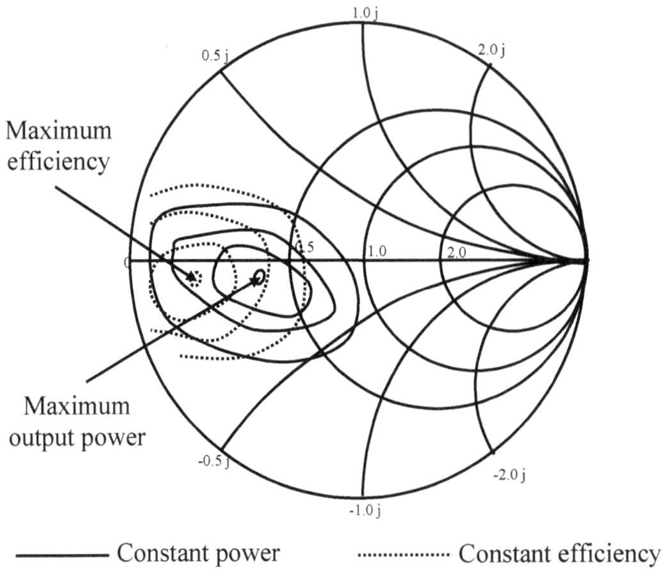

Fig. 14.35 Constant output power contours

Table 14.10 Amplifier classes

Class of amplifier	Bias current	Conduction angle
Class-A	$I_F/2$	2π
Class-AB	$I_F/2$–0	2π–π
Class-B	0	π
Class-C	0	π–0

amplifiers are designed by biasing the transistor to a low quiescent current, thus allowing the device for conduction over lesser angle as shown in Fig. 14.36. Beyond Class-A operation, the device is biased with a quiescent point more towards cut-off.

The conduction angle and operating bias conditions for different classes of amplifiers are shown in Table 14.10. It is clear from Fig. 14.36, that larger amplitude of RF drive is required for the power amplifiers beyond Class-A operation to achieve maximum current swing. Therefore, the gain of the amplifier reduces when the device operates beyond Class-A condition.

The current waveform with reduced conduction angle is shown in Fig. 14.37. The clipped drain current waveform i_{drf} can be expressed as Fourier series:

$$i_{drf}(t) = a_0 + \sum_{n=1}^{\infty} a_n \cos n\omega t + \sum_{n=1}^{\infty} b_n \sin n\omega t \tag{14.91}$$

Here, ω is the angular frequency of the RF signal, a_0, a_n and b_n are constants and can be determined by

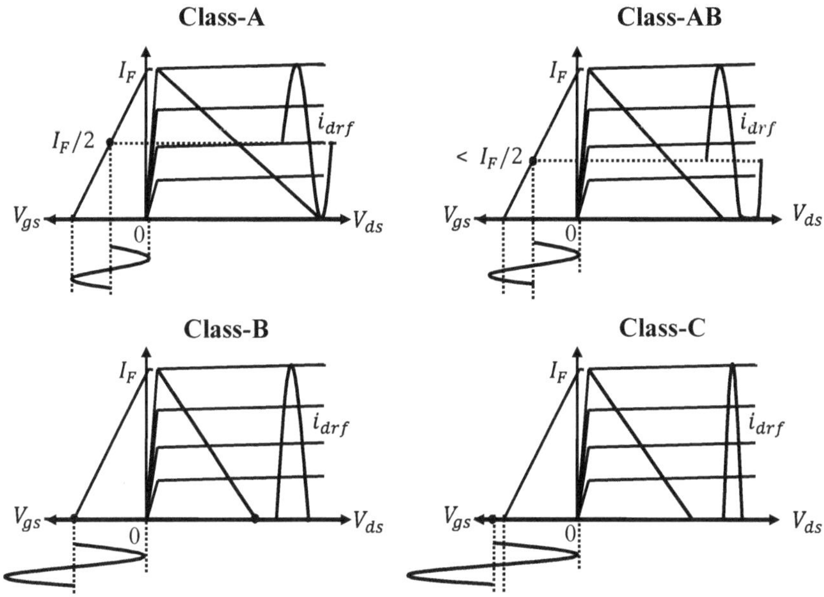

Fig. 14.36 Device current for different classes of amplifiers

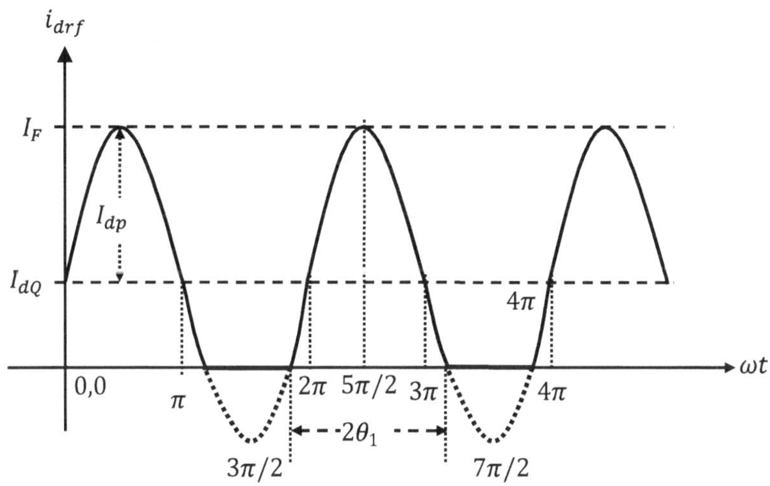

Fig. 14.37 Current waveform with reduced conduction angle ($2\theta_1$)

$$a_0 = \frac{1}{2\pi}\int_0^{2\pi} i_{drf}(\theta)\mathrm{d}\theta \tag{14.92a}$$

$$a_n = \frac{1}{\pi}\int_0^{2\pi} i_{drf}(\theta)\cos n\theta\mathrm{d}\theta \tag{14.92b}$$

$$b_n = \frac{1}{\pi}\int_0^{2\pi} i_{drf}(\theta)\sin n\theta\mathrm{d}\theta \tag{14.92c}$$

Here, $\omega t = \theta$. Suppose, conduction angle of the drain current waveform is $2\theta_1$ as shown in Fig. 14.37. The coefficients can be determined by performing the integration over any 2π radians without loss of any generality. Here, for ease of mathematical derivation the integration will be performed over the period of $3\pi/2$ to $7\pi/2$ where the function can be expressed as

$$\begin{aligned} i_{drf}(\theta) &= 0, && \text{over } 3\pi/2 \le \theta \le 5\pi/2 - \theta_1 \\ &= I_{dQ} + I_{dp}\sin\theta && \text{over } 5\pi/2 - \theta_1 \le \theta \le 5\pi/2 + \theta_1 \\ &= 0, && \text{over } 5\pi/2 - \theta_1 \le \theta \le 7\pi/2 \end{aligned} \tag{14.93}$$

From Fig. 14.37, I_{dp} can be expressed as

$$I_F = I_{dQ} + I_{dp} \tag{14.94a}$$

Putting $i_{drf} = 0$ for $\theta = 5\pi/2 - \theta_1$ in (14.93):

$$I_{dQ} + I_{dP}\sin(5\pi/2 - \theta_1) = 0 \tag{14.94b}$$

Therefore,

$$I_{dp} = \frac{I_F}{1 - \cos\theta_1} \quad \text{and,} \quad I_{dQ} = \frac{-I_F\cos\theta_1}{1 - \cos\theta_1} \tag{14.95}$$

The coefficients can be determined as follows:

$$\begin{aligned} a_0 &= \frac{1}{2\pi}\int_{3\pi/2}^{7\pi/2} i_{drf}(\theta)\mathrm{d}\theta \\ &= \frac{1}{2\pi}\int_{5\pi/2-\theta_1}^{5\pi/2+\theta_1} \left(I_{dQ} + I_{dp}\sin\theta\right)\mathrm{d}\theta \\ &= \frac{1}{2\pi}\left[I_{dQ}\theta - I_{dp}\cos\theta\right]_{5\pi/2-\theta_1}^{5\pi/2+\theta_1} \end{aligned}$$

$$= \frac{1}{2\pi}\left[2I_{dQ}\theta_1 - 2I_{dp}\sin\theta\right]$$

Putting the values of I_{dp} and I_{dQ} from (14.95)

$$a_0 = \frac{I_F}{\pi}\frac{(\sin\theta_1 - \theta_1\cos\theta_1)}{(1-\cos\theta_1)} \tag{14.96}$$

This gives the DC component of the current waveform.

$$a_n = \frac{1}{\pi}\int\limits_0^{2\pi} i_{drf}(\theta)\cos n\theta \mathrm{d}\theta = 0$$

$$= \frac{1}{\pi}\int\limits_{5\pi/2-\theta_1}^{5\pi/2+\theta_1}\left(I_{dQ} + I_{dp}\sin\theta\right)\cos n\theta \mathrm{d}\theta = 0$$

$$= 0$$

And,

$$b_n = \frac{1}{\pi}\int\limits_{5\pi/2-\theta_1}^{5\pi/2+\theta_1}\left(I_{dQ} + I_{dp}\sin\theta\right)\sin n\theta \mathrm{d}\theta$$

Therefore, the fundamental component of the current waveform can be expressed as

$$b_1 = \frac{1}{\pi}\int\limits_{5\pi/2-\theta_1}^{5\pi/2+\theta_1}\left[I_{dQ}\sin\theta + \frac{I_{dp}}{2}(1-\cos 2\theta)\right]\mathrm{d}\theta$$

$$= \frac{1}{\pi}\left[-I_{dQ}\cos\theta + \frac{I_{dp}}{2}\theta - \frac{I_{dp}}{4}\sin 2\theta\right]_{5\pi/2-\theta_1}^{5\pi/2+\theta_1}$$

$$= \frac{1}{\pi}\left[2I_{dQ}\sin\theta_1 + I_{dp}\theta_1 + \frac{I_{dp}}{2}\sin 2\theta_1\right]$$

Putting the values of I_{dp} and I_{dQ} from (14.95)

$$b_1 = \frac{I_F}{2\pi}\frac{(2\theta_1 - \sin 2\theta_1)}{(1-\cos\theta_1)} \tag{14.97}$$

The maximum output power and efficiency of the amplifier for different conduction angles can be calculated using the DC component a_0 and amplitude of the fundamental component b_1 of the drain current waveform. To compute these parameters, it will be assumed that the harmonic of the current waveforms are perfectly short and the current swing is linear and up to the maximum value I_F. It will also be assumed that FET is used as the amplifying device of zero Knee voltage and the load resistance R_L will remain constant with the following value:

$$R_L = \frac{2V_{DC}}{I_F} \tag{14.98}$$

Here, V_{DC} is the drain supply voltage of the FET device.

For Class-A amplifier configuration, there will be no clipping of the current waveform, thus conduction angle is $2\theta_1 = 2\pi$. Thus, from (14.96) and (14.97) the DC current (I_{DC}) and amplitude of the fundamental component (I_{d1}) of the drain current waveform can be written as

$$I_{DC} = a_0 = \frac{I_F}{\pi} \frac{(\sin\pi - \pi\cos\pi)}{(1 - \cos\pi)} = \frac{I_F}{2} \tag{14.99a}$$

$$I_{d1} = b_1 = \frac{I_F}{2\pi} \frac{(2\pi - \sin 2\pi)}{(1 - \cos\pi)} = \frac{I_F}{2} \tag{14.99b}$$

Therefore, the maximum RF output power:

$$P_{Arf\,\max} = \left(\frac{I_F}{2\sqrt{2}}\right)^2 \frac{2V_{DC}}{I_F} = \frac{I_F V_{DC}}{4} \tag{14.99c}$$

The maximum efficiency $\eta_{\max}$ will be

$$\eta_{A\,\max} = \frac{P_{rf\,\max}}{V_{DC} I_{DC}} = \frac{1}{2} = 50\% \tag{14.99d}$$

Maximum RF output power and efficiency for conduction angle of 2π to zero which covers Class-A to Class-C amplifiers are derived based on (14.96), (14.97) and (14.98) and are shown in Fig. 14.38. Here, output RF power is shown in dB with respect to maximum RF output power corresponding to Class-A condition. It shows that efficiency of the amplifier increases while configuring the amplifier from Class-A towards Class-C amplifier. However, RF output power drastically reduces beyond Class-B amplifier configuration

For Class-B amplifier configuration, the conduction angle $2\theta_1 = \pi$. Thus, from (14.96) and (14.97), the DC current (I_{DC}) and amplitude of the fundamental component (I_{d1}) of the drain current waveform can be written as

$$I_{DC} = a_0 = \frac{I_F}{\pi} \frac{\left(\sin(\frac{\pi}{2}) - \frac{\pi}{2}\cos(\frac{\pi}{2})\right)}{\left(1 - \cos(\frac{\pi}{2})\right)} = \frac{I_F}{\pi} \tag{14.100a}$$

$$I_{d1} = b_1 = \frac{I_F}{2\pi} \frac{(\pi - \sin\pi)}{(1 - \cos(\pi/2))} = \frac{I_F}{2} \tag{14.100b}$$

Therefore, the maximum RF output power:

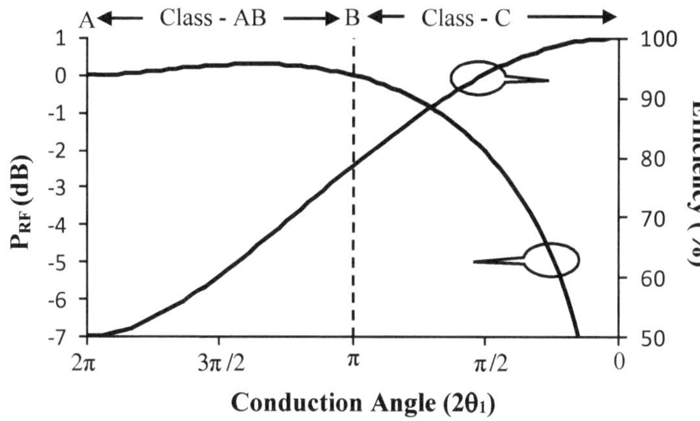

Fig. 14.38 RF power and efficiency of different classes of amplifier

$$P_{Brf\,\max} = \left(\frac{I_F}{2\sqrt{2}}\right)^2 \frac{2V_{\mathrm{DC}}}{I_F} = \frac{I_F\, V_{\mathrm{DC}}}{4} \tag{14.100c}$$

The maximum output RF power of a Class-B amplifier is same as Class-A amplifier. The maximum efficiency of Class-B amplifier $\eta_{B\,\max}$ is given by

$$\eta_{B\,\max} = \frac{P_{Brf\,\max}}{V_{\mathrm{DC}}\, I_{\mathrm{DC}}} = \frac{\pi}{4} = 78.5\%$$

The amplifier operating in Class-B condition delivers the same output power as of Class A condition at their maximum current and voltage swings. However, the DC current is reduced to I_F/π, which is reduced by a factor of $\pi/2$ compared to DC current in Class A amplifier.

The conduction angle of the device operating in Class-C condition is further reduced which thus increases efficiency. However, with the reduction of conduction angle, RF output power decreases substantially. For the reduced conduction angle, the current waveform becomes pulse like which contain high level of harmonic components. Thus, in addition to the generation of high level of harmonic components, the drastic reduction of RF output power makes Class-C amplifier not favourable for RF/microwave communication applications.

14.7.3.1 Back-off Operation of Class-A and Class-B Amplifiers

Depending on the applications, power amplifiers are also operated under different output power conditions. For example, power amplifiers are operated under power back-off condition under multicarrier as well as the applications where higher modulation codes are used for communication. Thus, it is important to know the output

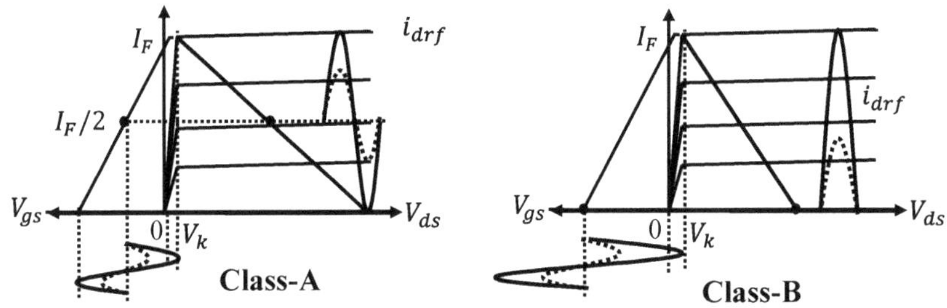

Fig. 14.39 Signal waveforms of Class-A and Class-B amplifier

RF power and implication on the efficiency of Class-A and Class-B amplifiers when they are operated under power back-off condition. Input voltage and output current swings for Class-A and Class-B amplifiers at the maximum input and back-off swing conditions are shown in Fig. 14.39.

Suppose FET device of gate-to-drain breakdown voltage V_{dgB} and Knee voltage V_k is used for the power amplifier. The gate-to-source pinch-off voltage of the FET is V_{gsO}. Bias conditions for Class-A amplifiers to achieve maximum drain voltage and current swings under maximum RF operating condition:

$$
\begin{aligned}
V_{gs} &= -\left(V_{gsO} - V_F\right)/2 \\
I_{dQ} &= I_F/2 \\
V_{dQ} &= \left(V_{dgB} - V_{gsO} + V_K\right)/2
\end{aligned}
$$

The load resistance:

$$R_L = \left(V_{dgB} - V_{gsO} - V_K\right)/I_F$$

The RF output power for Class-A amplifier is given by

$$P_{Arf} = \left(\frac{i_{drf}}{\sqrt{2}}\right)^2 R_L = \frac{\left(i_{drf}\right)^2\left(V_{dgB} - V_{gsO} - V_K\right)}{2I_F} \tag{14.101a}$$

The DC power:

$$P_{\rm dc} = V_{dQ} \times I_{dQ} = I_F \times \frac{\left(V_{dgB} - V_{gsO} + V_K\right)}{4} \tag{14.101b}$$

Thus, the power added efficiency:

$$\eta_{\rm add} = 2\left(1 - \frac{1}{G}\right)\left(\frac{i_{drf}}{I_F}\right)^2 \frac{\left(V_{dgB} - V_{gsO} - V_K\right)}{\left(V_{dgB} - V_{gsO} + V_K\right)} \tag{14.101c}$$

Considering, $V_{dgB} \gg V_{gsO}$ and $V_{dgB} \gg V_K$, the power added efficiency of the Class-A amplifier of very high gain can be approximated as

$$\eta_{\text{add}} = 2\left(\frac{i_{drf}}{I_F}\right)^2 \tag{14.101d}$$

Thus, the efficiency of Class-A amplifier increases with the increase of RF current swing and it becomes the maximum of value 50% when the RF current swing becomes $I_F/2$.

Bias conditions for Class-B amplifiers to achieve maximum drain voltage and current swings under maximum RF operating condition is

$$V_{gs} = -(V_{gsO})$$
$$I_{dQ} = 0$$

and

$$V_{dQ} = (V_{dgB} - 2V_{gsO} + V_K - V_F)/2$$

Thus load resistance will be

$$R_L = (V_{dgB} - 2V_{gsO} - V_K - V_F)/I_F$$

For Class-B amplifier, the current waveform is half sinusoidal. Suppose, peak value of the half sinusoidal waveform is i_{drf}. From (14.100b), the amplitude of the fundamental component of the half sinusoidal waveform of peak value i_{drf} is $i_{drf}/2$. Thus, the RF output power P_{Brf} of Class-B amplifier:

$$P_{Brf} = \left(\frac{i_{drf}}{2\sqrt{2}}\right)^2 R_L = \frac{(i_{drf})^2 (V_{dgB} - 2V_{gsO} - V_K - V_F)}{8I_F} \tag{14.102a}$$

From (14.100a), the DC component I_{dc} of the half sinusoidal waveform of peak value i_{drf} will be

$$I_{\text{DC}} = \frac{i_{drf}}{\pi}$$

Thus, DC power will be

$$P_{\text{dc}} = I_{\text{DC}} \times V_{dQ} = \frac{i_{drf}}{\pi} \frac{(V_{dgB} - 2V_{gsO} + V_K - V_F)}{2} \tag{14.103a}$$

The power added efficiency:

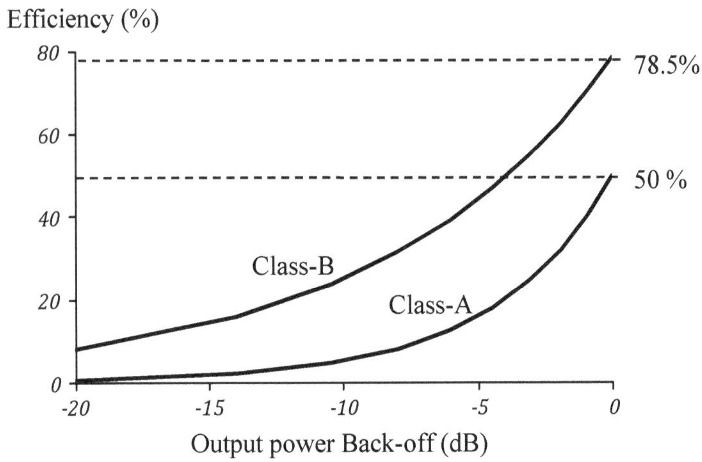

Fig. 14.40 Efficiency of Class-A and Class-B power amplifier with output power back-off

$$\eta_{\text{add}} = \left(1 - \frac{1}{G}\right)\frac{\pi}{4}\left(\frac{i_{drf}}{I_F}\right)\frac{\left(V_{dgB} - 2V_{gsO} - V_K - V_F\right)}{\left(V_{dgB} - 2V_{gsO} + V_K - V_F\right)} \tag{14.103b}$$

Considering $V_{dgB} \gg V_{gsO}$, $V_{dgB} \gg V_K$ and $V_{dgB} \gg V_F$ the power added efficiency of Class-B amplifier with vary high gain can be approximated as

$$\eta_{\text{add}} = \frac{\pi}{4}\left(\frac{i_{drf}}{I_F}\right) \tag{14.103c}$$

Thus, efficiency of Class-B amplifier also increases with the increase of RF current swing and it reaches a maximum value of 78.54% when the RF current swing becomes equal to value of I_F.

From (14.101d) and (14.103c), it is clear that power-added efficiency is linearly proportional to i_{drf} for Class-B amplifier whereas proportional to the square of i_{drf} for Class-A amplifier. Figure 14.40 shows the efficiency variation of Class-A and Class-B amplifiers with output power back-off. Output power back-off is defined as the decrease of RF output power compared to the maximum RF output power.

Thus, for Class-A amplifier the RF output power back-off (OBO) in dB can be expressed from (14.101a) as

$$\text{OBO}_{\text{Class-A}} = 10 \times \log_{10}\left(\frac{i_{drf}}{i_{drf\,\max}}\right)^2 = 20 \times \log_{10}\left(\frac{2 \times i_{drf}}{I_F}\right)$$

Similarly, for Class-B amplifier the RF output power back-off in dB can be expressed as

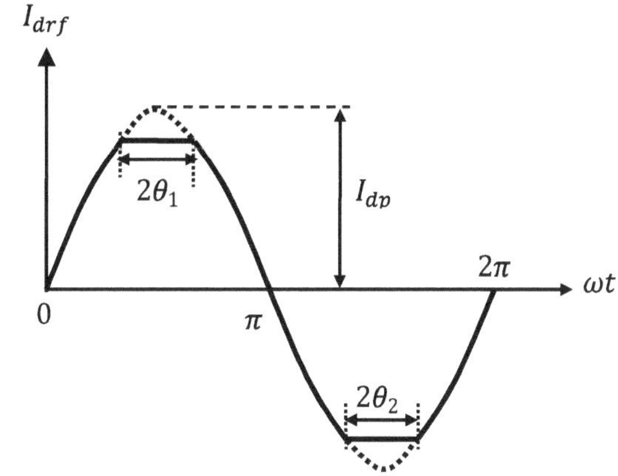

Fig. 14.41 Drain current waveform of overdriven FET power amplifier

$$\mathrm{OBO}_{\text{Class-B}} = 10 \times \log_{10}\left(\frac{i_{drf}}{i_{drf\,\max}}\right)^2 = 20 \times \log_{10}\left(\frac{i_{drf}}{I_F}\right)$$

Figure 14.40 shows that the efficiency of Class-A amplifier decreases sharply compared to Class-B amplifier with output power back-off. This is one of the drawbacks of Class-A amplifier compared to other classes of amplifier.

14.7.4 Nonlinear Power Amplifiers

Practically, it is required to maintain linearity of an amplifier for communication systems. To maintain the linearity under the assumption of an ideal characteristic of the device, it is required to restrict the input signal level of a power amplifier to avoid clipping of the output voltage and current waveforms. However, there are certain applications in communication systems where linearity can be traded for achieving better efficiency of the power amplifiers. Therefore, Class-A and Class-B amplifiers are also operated under higher input signal level which drives the amplifiers towards saturation.

Under overdriven condition, both upper and lower portion of the drain current waveform may clip due to the limitation of current from the device. A typical clipped RF output current waveform is shown in Fig. 14.41. This periodic output current waveform can be expressed with Fourier series as

$$i_{drf}(t) = a_0 + \sum_{n=1}^{\infty} a_n \cos n\omega t + \sum_{n=1}^{\infty} b_n \sin n\omega t \tag{14.104}$$

Here, ω is the angular frequency of the RF signal a_0, a_n and b_n are constants and can be determined by

$$a_0 = \frac{1}{2\pi}\int_0^{2\pi} i_{drf}(\theta)\mathrm{d}\theta \tag{14.105a}$$

$$a_n = \frac{1}{\pi}\int_0^{2\pi} i_{drf}(\theta)\cos n\theta\mathrm{d}\theta \tag{14.105b}$$

$$b_n = \frac{1}{\pi}\int_0^{2\pi} i_{drf}(\theta)\sin n\theta\mathrm{d}\theta \tag{14.105c}$$

As shown in the figure, the clippings of the current waveform are over $2\theta_1$ and $2\theta_2$. Therefore, the drain current waveform can be expressed as

$$\begin{aligned} i_{drf}(\theta) &= I_{dQ} + I_{dp}\sin\theta, && \text{over } 0 \le \theta \le \pi/2 - \theta_1 \\ &= I_F, && \text{over } \pi/2 - \theta_1 \le \theta \le \pi/2 + \theta_1 \\ &= I_{dQ} + I_{dp}\sin\theta, && \text{over } \pi/2 + \theta_1 \le \theta \le 3\pi/2 - \theta_2 \\ &= 0, && \text{over } 3\pi/2 - \theta_2 \le \theta \le 3\pi/2 + \theta_2 \\ &= I_{dQ} + I_{dp}\sin\theta, && \text{over } 3\pi/2 + \theta_2 \le \theta \le 2\pi \end{aligned} \tag{14.106}$$

The constants can be determined as follows:

$$\begin{aligned} a_0 &= \frac{1}{2\pi}\int_0^{2\pi} i_{drf}(\theta)\mathrm{d}\theta \\ &= \frac{1}{2\pi}\left[\int_0^{\frac{\pi}{2}-\theta_1}\left(I_{dQ} + I_{dp}\sin\theta\right)\mathrm{d}\theta + \int_{\frac{\pi}{2}-\theta_1}^{\frac{\pi}{2}+\theta_1} I_F\mathrm{d}\theta \right. \\ &\quad \left. + \int_{\frac{\pi}{2}+\theta_1}^{\frac{3\pi}{2}-\theta_2}\left(I_{dQ} + I_{dp}\sin\theta\right)\mathrm{d}\theta + \int_{\frac{3\pi}{2}+\theta_2}^{2\pi}\left(I_{dQ} + I_{dp}\sin\theta\right)\mathrm{d}\theta\right] \end{aligned}$$

or

$$a_0 = \frac{\pi - \theta_1 - \theta_2}{\pi} I_Q + \frac{\sin\theta_2 - \sin\theta_1}{\pi} I_{dp} + \frac{\theta_1}{\pi} I_F \tag{14.107}$$

Similarly,

$$a_1 = \frac{1}{\pi}\int_0^{2\pi} i_{drf}\cos\theta\mathrm{d}\theta$$

$$= \frac{1}{\pi}\left[\int_{0}^{\frac{\pi}{2}-\theta_1} \left(I_{dQ} + I_{dp}\sin\theta\right)\cos\theta \mathrm{d}\theta + \int_{\frac{\pi}{2}-\theta_1}^{\frac{\pi}{2}+\theta_1} I_F \cos\theta \mathrm{d}\theta\right.$$

$$\left. + \int_{\frac{\pi}{2}+\theta_1}^{\frac{3\pi}{2}-\theta_2} \left(I_{dQ} + I_{dp}\sin\theta\right)\cos\theta \mathrm{d}\theta + \int_{\frac{3\pi}{2}+\theta_2}^{2\pi} \left(I_{dQ} + I_{dp}\sin\theta\right)\cos\theta \mathrm{d}\theta\right]$$

or

$$a_1 = 0 \tag{14.108}$$

$$b_1 = \frac{1}{\pi}\int_{0}^{2\pi} i_{drf}(\theta)\sin\theta \mathrm{d}\theta$$

$$= \frac{1}{\pi}\left[\int_{0}^{\frac{\pi}{2}-\theta_1} \left(I_{dQ} + I_{dp}\sin\theta\right)\sin\theta \mathrm{d}\theta + \int_{\frac{\pi}{2}-\theta_1}^{\frac{\pi}{2}+\theta_1} I_F \sin\theta \mathrm{d}\theta\right.$$

$$\left. + \int_{\frac{\pi}{2}+\theta_1}^{\frac{3\pi}{2}-\theta_2} \left(I_{dQ} + I_{dp}\sin\theta\right)\sin\theta \mathrm{d}\theta + \int_{\frac{3\pi}{2}+\theta_2}^{2\pi} \left(I_{dQ} + I_{dp}\sin\theta\right)\sin\theta \mathrm{d}\theta\right]$$

or,

$$b_1 = \frac{1}{\pi}\left(\pi - \theta_1 - \theta_2 - \frac{1}{2}(\sin 2\theta_1 + \sin 2\theta_2)\right)I_{dp} + \frac{2\sin\theta_2 - 2\sin\theta_1}{\pi}I_Q + \frac{2\sin\theta_1}{\pi}I_F \tag{14.109a}$$

Using these expressions, output power and efficiency of Class-A and Class-B amplifiers when operated under overdriven condition can be derived.

14.7.4.1 Overdriven Class-A Power Amplifiers

Class-A amplifiers are operated over the RF power level in which output current of the device conducts over full RF cycle. However, when the Class-A amplifier operates in RF overdriven condition, output current waveform of the device clips at extreme levels. Input and output signal current waveforms of a Class-A amplifier when it is operated under such large RF signal level condition, is shown in Fig. 14.42. Here, it is considered that the device is biased to its exactly midpoint of the current and

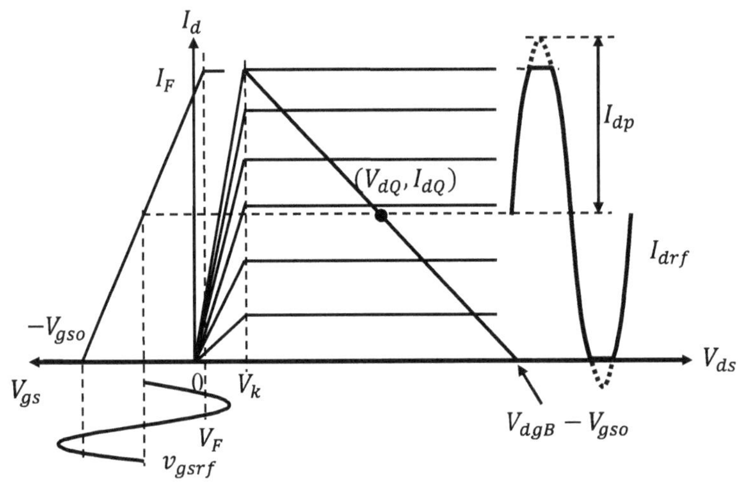

Fig. 14.42 Current waveform of overdriven power amplifier

voltage capacity which ensures symmetric clipping of the output current waveform, i.e.

$$\theta_1 = \theta_2 = \alpha \text{ (say)}$$
$$I_Q = I_F/2$$

Putting this condition to (14.107), the DC current can be expressed as

$$I_{DC} = a_0 = \frac{\pi - 2\alpha}{\pi}\frac{I_F}{2} + \frac{\alpha}{\pi}I_F$$
$$= \frac{I_F}{2}$$

Putting the values of $\theta_1 = \theta_2 = \alpha$ and $I_Q = I_F/2$ to (14.109a), the fundamental component of RF output current can be written as
or

$$i_{drf} = b_1 = \frac{1}{\pi}\left(\pi - 2\alpha - \frac{1}{2}(\sin 2\alpha + \sin 2\alpha)\right)I_{dp} + \frac{2\sin\alpha}{\pi}I_F. \quad (14.109b)$$

Putting $i_{drf}(\pi/2 - \theta_1) = I_F$ in (14.106):

$$I_{dp} = \frac{I_F - I_Q}{\cos\alpha} = \frac{I_F}{2\cos\alpha} \quad (14.110)$$

Putting the value of I_{dp} in (14.109b):

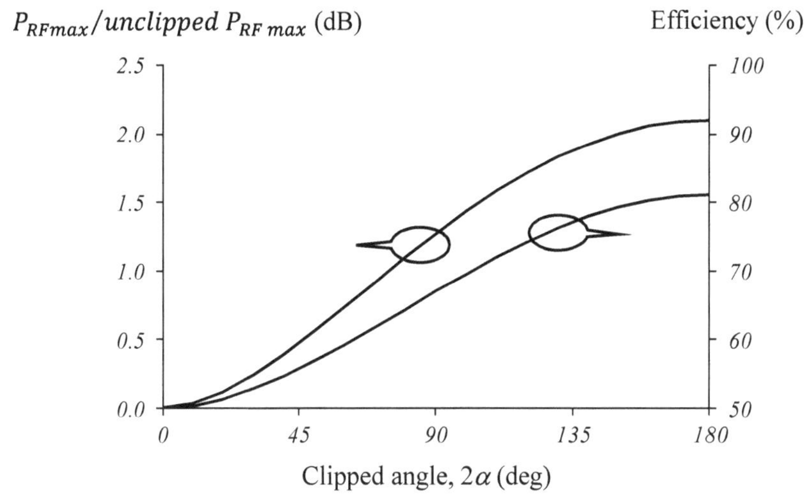

Fig. 14.43 RF output power and efficiency for clipped sinusoidal output (overdriven Class-A amplifier)

$$i_{drf} = \frac{I_F}{\pi}\left(\frac{\pi - 2\alpha + \sin 2\alpha}{2\cos\alpha}\right) \tag{14.111}$$

Therefore, the output RF power and efficiency can be written as

$$\begin{aligned} P_{rf} &= \left(\frac{i_{drf}}{\sqrt{2}}\right)^2 R_L \\ &= \frac{I_F}{2\pi^2}\left(\frac{\pi - 2\alpha + \sin 2\alpha}{2\cos\alpha}\right)^2 \left(V_{dgB} - V_{gsO} - V_K\right) \end{aligned} \tag{14.112}$$

$$\eta = \frac{2}{\pi^2}\left(\frac{\pi - 2\alpha + \sin 2\alpha}{2\cos\alpha}\right)^2 \frac{\left(V_{dgB} - V_{gsO} - V_K\right)}{\left(V_{dgB} - V_{gsO} + V_K\right)} \tag{14.113}$$

RF output power and efficiency of overdriven Class-A amplifier is shown in Fig. 14.43. The saturated RF output power, i.e. power corresponding to α approaching 90°, is about 2-dB more than the maximum RF output power compared to unclipped Class-A operation. Under this condition, the efficiency will be about 81%.

It is very important and interesting, that the output power as well as efficiency of an overdriven Class-A amplifier improves significantly. However, to achieve square wave output waveform, injection of a very high level of input is required which not only decreases the gain of the amplifier but also degrades device reliability due to high gate current for excessive forward gate voltage. One practical way for achieving high power and efficiency from the Class-A amplifier is the use of square wave input signal in place of the sinusoidal signal waveform. Practically, this is done to some

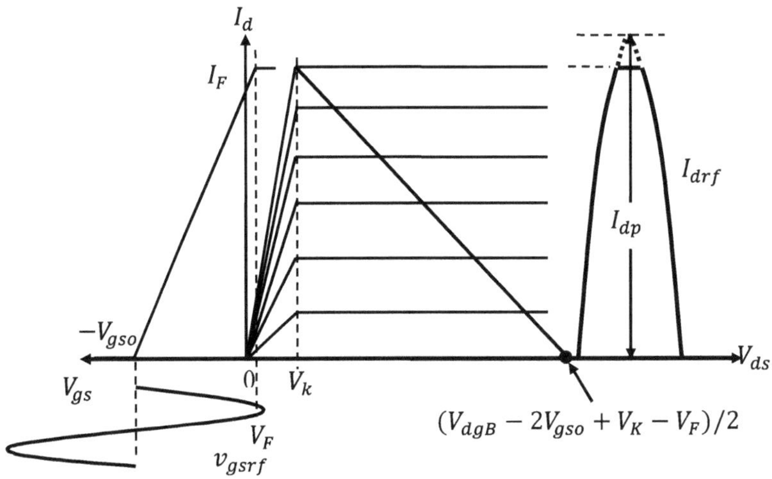

Fig. 14.44 Current waveform of overdriven Class-B power amplifier

extent by clipping the signal waveform in the preceding amplifier stage before feeding it to the high power amplifier.

14.7.4.2 Overdriven Class-B Power Amplifiers

Input and output signal waveforms of a class-B amplifier when operating under higher signal level condition which drives the amplifier to clip the output RF current is shown in Fig. 14.44. With reference to Fig. 14.41, for current waveform of Fig. 14.44:

$$\theta_2 = \pi/2, \quad I_Q = 0$$

and

$$\theta_1 = \alpha \text{ (say)}$$

Putting $i_{drf}(\pi/2 - \alpha) = I_F$ in (14.106):

$$I_{dp} = \frac{I_F}{\cos\alpha} \tag{14.114}$$

Putting this condition to (14.114), the DC current can be expressed as

$$I_{\text{DC}} = a_0 = \frac{I_F}{\pi}\left(\frac{\alpha\cos\alpha - \sin\alpha + 1}{\cos\alpha}\right) \tag{14.115}$$

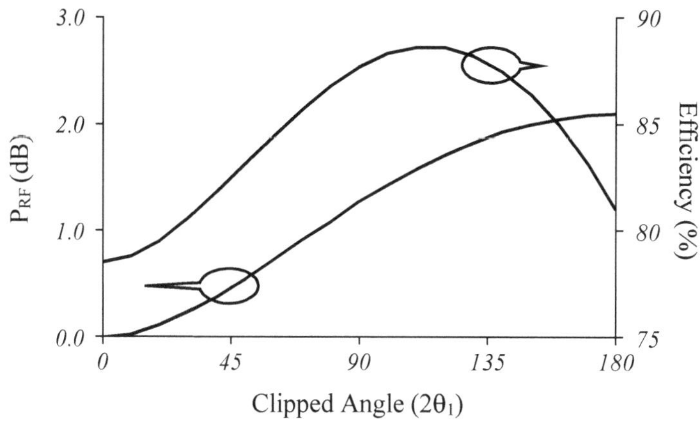

Fig. 14.45 RF power and efficiency of overdriven Class-B power amplifier

From (14.109a), the fundamental component of RF output current can be written as

$$i_{drf} = b_1 = \frac{I_F}{\pi}\frac{(\pi - 2\alpha + \sin 2\alpha)}{2\cos\alpha} \tag{14.116}$$

Therefore, output RF power P_{rf} and efficiency η can be written as

$$\begin{aligned} P_{rf} &= \left(\frac{i_{drf}}{\sqrt{2}}\right)^2 R_L \\ &= \frac{I_F}{2\pi^2}\left(\frac{\pi - 2\alpha + \sin 2\alpha}{2\cos\alpha}\right)^2 \left(V_{dgB} - 2V_{gsO} - V_K - V_F\right) \end{aligned} \tag{14.117}$$

$$\eta = \frac{1}{4\pi\cos\alpha}\frac{(\pi - 2\alpha + \sin 2\alpha)^2}{(\alpha\cos\alpha - \sin\alpha + 1)} \tag{14.118}$$

RF output power and efficiency of overdriven Class-B amplifier is shown in Fig. 14.45. The saturated RF output power is about 2 dB more than the maximum RF output power. At this condition, the efficiency is about 81%. However, it is clear from the plot that the required amount of waveform clipping is not same for the maximum output power and maximum efficiency i.e. efficiency does not go on increasing as in case of overdriven Class-A amplifier.

14.7.5 Class-F Power Amplifier

In the case of amplifiers operating in an ideal Class-B operation mode with the load tuned at the fundamental frequency, the maximum achievable efficiency is 78.5%. At

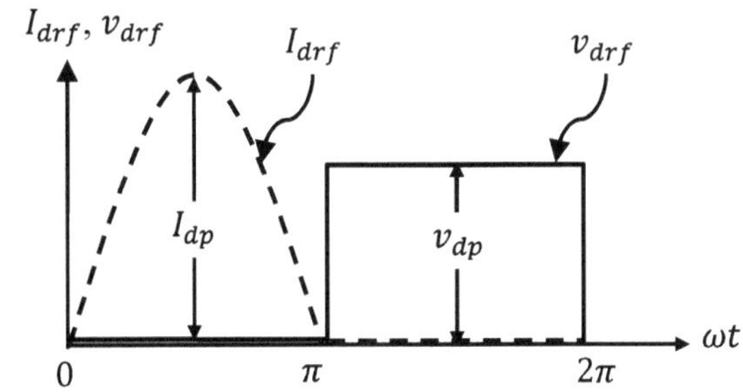

Fig. 14.46 Device output voltage and current waveforms for Class-F amplifier

this condition, the device output voltage waveform is sinusoidal and current waveform is half-sinusoidal. In case of Class-F amplifier, efficiency can be maximized up to the theoretical limit of 100%, by shaping the device voltage and current waveforms by putting additional resonators which are tuned to the corresponding harmonic components [14–29].

A Class-F power amplifier provides improved efficiency with higher RF power output compared to Class-A and Class-B configuration by proper termination of harmonics at the device output which shapes its voltage and current waveforms. Purpose of loading for harmonics is to shape the device output voltage in a way to avoid overlap or to minimize the overlap with its current, thus greatly reducing the power dissipation within the device, and hence increasing the efficiency. Here, amplitude of voltage and current harmonics are appropriately added to the respective amplitude of fundamental components to keep resultant voltage and current swings within the allowable limit of the device. This ensures to operate the device within its alloable voltage and current limit and without driving the amplifier into hard compression. In general, either voltage or current waveform (e.g. voltage) approximates a square wave, while the other (e.g. current) approximates a half sine wave with minimum overlap over the time by providing proper termination to the harmonics at the output of the device. The ideal voltage and current waveforms of a device operating in Class-F configuration is shown in Fig. 14.46. Zero overlap between voltage and current of the device signifies that the power dissipation within the device is zero. Thus, efficiency of the power amplifier operating in Class-F condition is 100%. The voltage waveform v_{drf} of the device can be expressed using Fourier series as follows:

$$v_{drf}(t) = a_0 + \sum_{n=1}^{\infty} a_n \cos n\omega t + \sum_{n=1}^{\infty} b_n \sin n\omega t \tag{14.119}$$

Here, ω is the angular frequency of the RF signal a_0, a_n and b_n are constants and can be determined by

$$a_0 = \frac{1}{2\pi}\int_0^{2\pi} v_{drf}(\theta)\mathrm{d}\theta$$

$$a_n = \frac{1}{\pi}\int_0^{2\pi} v_{drf}(\theta)\cos n\theta\mathrm{d}\theta$$

$$b_n = \frac{1}{\pi}\int_0^{2\pi} v_{drf}(\theta)\sin n\theta\mathrm{d}\theta$$

Here, the drain voltage waveform can be expressed as

$$\begin{aligned} v_{drf}(\theta) &= 0, \quad \text{over } 0 \le \theta \le \pi \\ &= v_{dp}, \text{ over } \pi \le \theta \le 2\pi \end{aligned}$$

Therefore, constants can be determined as follows:

$$\begin{aligned} a_0 &= \frac{1}{2\pi}\int_0^{2\pi} v_{drf}(\theta)\mathrm{d}\theta \\ &= \frac{1}{2\pi}\int_\pi^{2\pi} v_{dp}\mathrm{d}\theta \end{aligned}$$

or

$$a_0 = \frac{v_{dp}}{2}$$

Similarly,

$$\begin{aligned} a_n &= \frac{1}{\pi}\int_\pi^{2\pi} v_{dp}\cos n\theta\mathrm{d}\theta \\ &= \frac{v_{dp}}{n\pi}\left[v_{dp}\sin n\theta\right]_\pi^{2\pi} = 0 \text{ for all values of } n \end{aligned}$$

Similarly,

$$\begin{aligned} b_n &= \frac{1}{\pi}\int_\pi^{2\pi} v_{dp}\sin n\theta\, d\theta \\ &= \frac{v_{dp}}{n\pi}[-\cos n\theta]_\pi^{2\pi} \end{aligned}$$

or

$$b_n = -\tfrac{2v_{dp}}{n\pi} \text{ for all odd values of } n$$
$$= 0 \quad \text{for all even values of } n$$

Therefore, the output voltage waveform can be expressed as

$$v_{drf}(t) = \frac{v_{dp}}{2} - \frac{2v_{dp}}{\pi}\sin\omega t - \sum_{n=1,3,5,\ldots}^{\infty} \frac{2v_{dp}}{n\pi}\sin n\omega t \tag{14.120}$$

The first term of (14.120) is the DC component, second term is the fundamental component and others are all odd harmonic components. Therefore, the normalized amplitude of fundamental and harmonic components with respect to the DC components can be expressed as

$$v_n = \frac{4}{n\pi} \tag{14.121}$$

The half sinusoidal output current waveform is given by

$$i_{drf}(\theta) = I_{dp}\sin\theta, \text{ over } 0 \le \theta \le \pi$$
$$= 0, \qquad \text{over } \pi \le \theta \le 2\pi$$

Thus, the current waveform can be expressed in Fourier series as

$$i_{drf}(t) = \frac{I_{dp}}{\pi} + \frac{I_{dp}}{2}\sin\omega t - \frac{2I_{dp}}{\pi}\sum_{n=2,4,6,\ldots}^{\infty} \frac{\cos n\omega t}{n^2-1} \tag{14.122}$$

Here, the first and second terms are the DC and fundamental components, respectively. Other terms are all the even-order harmonic components.

Current and voltage waveforms have never simultaneous nonzero values, and only the fundamental frequency component is present in both voltage and current waveforms and thus power is generated only in fundamental frequency. Whereas the current contains only even harmonics, the voltage contains only odd harmonics, therefore there will be no generation of harmonic power.

Output matching network and bias condition of a Class-F power amplifier should be such that it supports square wave voltage waveform and half sinusoidal output current wave with no overlapping in the time axis. A generalized Class-F power amplifier using FET and its drain voltage v_{drf} and drain current i_{drf} waveforms are shown in Fig. 14.47 when the device is biased as for the case of Class-B amplifier configuration. From the above analysis, it is clear that the drain voltage and current waveform of the fundamental frequency are shifted in phase by 180 from each other. Also, the drain voltage waveform consists of all odd harmonics (no even harmonics) to build its square wave shape whereas the drain current waveform adds all even har-

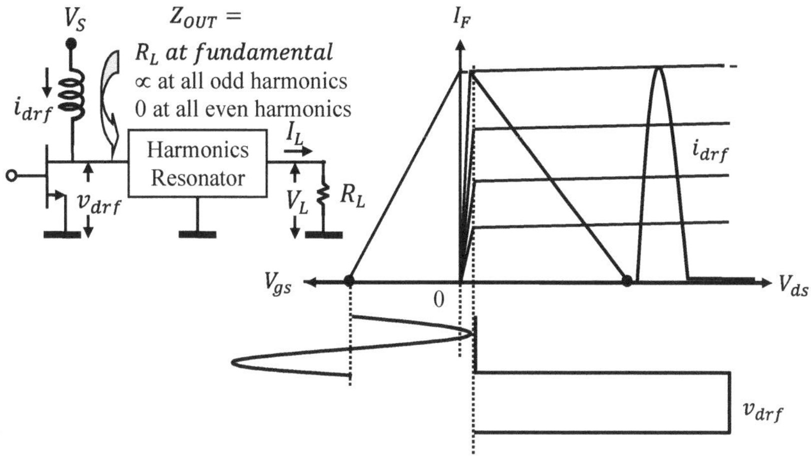

Fig. 14.47 Dain voltage and current waveform of Class-F amplifier

monics (no odd harmonics) to build its half sine wave shape. Thus, there will be no power dissipation within the device and also there will be no power generation at the harmonic frequencies since there is either no voltage or no current present at a given harmonic. To achieve these voltage and current waveforms, output impedance must provide either zero or infinite impedances at the harmonic frequencies. In case of an ideal Class-F power amplifier with square shape voltage waveform and half sinusoidal current waveform, the output impedance should provide an infinite impedance for all odd harmonics and zero impedance to all even harmonics and optimum load impedance corresponding to maximum allowable voltage and current swings at the fundamental frequency. Under this condition, the ratio of voltage amplitude of fundamental frequency to the DC voltage is given from (14.121) as

$$v_1 = \frac{4}{\pi} = 1.273$$

Since the current amplitude of the fundamental frequency is the same as in case of Class-A amplifier, the maximum output power of the ideal Class-F power amplifier is 1.273 times the power corresponding to Class-A amplifier configuration, which is about 1-dB higher. The efficiency of the Class-F power amplifier can be calculated in the following way:

Suppose the amplifier is biased as in Class-B. Therefore, from (14.120) and (14.122) the DC power P_{DC} is given by

$$P_{\mathrm{DC}} = V_{\mathrm{DC}} \times I_{\mathrm{DC}} = \frac{v_{dp}}{2} \times \frac{I_{dp}}{\pi}$$

Whereas, the maximum RF power $P_{\mathrm{RF\,max}}$ from (14.120) and (14.122) is given by

$$P_{\text{RF max}} = \frac{2v_{dp}}{\pi\sqrt{2}} \times \frac{I_{dp}}{2\sqrt{2}} = \frac{v_{dp}}{2} \times \frac{I_{dp}}{\pi}$$

Thus, maximum efficiency in Class-F amplifier:

$$\eta_{\max} = P_{\text{RF max}}/P_{\text{DC}} = 100\%$$

Here, the device considered is an ideal device with zero Knee voltage. This highly efficient operation of Class-F power amplifier can be obtained only when resonant circuit tuned to an infinite number of odd and even harmonics of the fundamental frequency is added into the load network.

Another version of Class-F power amplifier mode of operation is inverse Class-F mode. In case of inverse Class-F mode of operation, the device output RF voltage waveform is half sinusoidal and the current waveform is square shaped. Design of a Class-F and inverse Class-F amplifier involves the design of matching network at the fundamental frequency, and harmonic tuned load networks.

In conventional Class-F or inverse Class-F power amplifiers, the fundamental and harmonic load-network impedances are optimized by short-circuit termination and open-circuit peaking corresponding to finite number of harmonics to control the voltage and current waveforms at the device output to obtain maximum efficiency and output power. Practically, few harmonics such as up to third- or fifth-order harmonics are considered to reduce the complexity for improvement of efficiency and RF output power. To achieve maximum possible output power and efficiency considering proper termination of up to a finite number of harmonics, the relative amplitude should be chosen such that the amplitude of the fundamental component becomes maximum keeping the composite voltage and current waveform within the capability of the device. In this way, without overstressing the device, higher RF output power and efficiency can be achieved from Class-F power amplifier with a finite number of harmonic tuning. Figure 14.48 shows the dependence of RF output power and efficiency of Class-F power amplifier on the amplitude of third harmonic voltage component when up to third harmonic is considered for proper termination. It shows that in absence of any harmonics in voltage waveform, the efficiency is nothing but the classical value of 78.5% corresponding to the Class-B amplifier. With the increase of the amplitude of the third harmonic component, the efficiency increases up to a value of about 90.7%. In this condition, the RF output power is about 0.63 dB more compared to maximum power available from Class-A configuration. With further increase of amplitude of the third harmonic component, the efficiency as well as RF output power decreases.

Table 14.11 shows the normalized amplitudes of fundamental and harmonic components for different cases when up to 3rd, 5th, 7th, 9th, 11th, 13th, 15th and infinite number of harmonics are considered. These values are determined numerically by optimizing the amplitude of harmonic components to maximize the signal amplitude of fundamental components keeping the normalize waveform envelope within the allowable limit of unity.

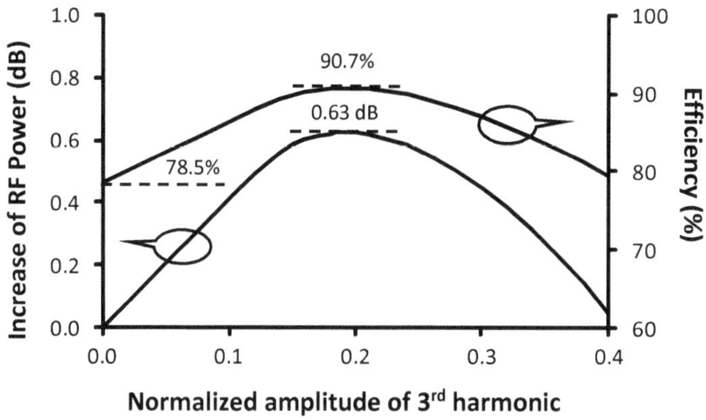

Fig. 14.48 RF output power and efficiency dependence on amplitude of third harmonic component

Table 14.11 Normalized amplitude of fundamental and odd harmonics

Harm	v_1	v_3	v_5	v_7	v_9	v_{11}	v_{13}	v_{15}
1	1	–	–	–	–	–	–	–
3	1.155	0.194	–	–	–	–	–	–
5	1.207	0.283	0.075	–	–	–	–	–
7	1.229	0.336	0.132	0.046	–	–	–	–
9	1.240	0.364	0.170	0.081	0.035	–	–	–
11	1.245	0.381	0.191	0.105	0.055	0.026	–	–
13	1.250	0.390	0.208	0.125	0.076	0.045	0.026	–
15	1.252	0.398	0.216	0.133	0.084	0.053	0.032	0.018
Infinite	$4/\pi$	$4/3\pi$	$4/5\pi$	$4/7\pi$	$4/9\pi$	$4/11\pi$	$4/13\pi$	$4/115\pi$

Figure 14.49 shows the device voltage waveforms corresponding to fundamental frequency as well as composite waveform when up to different harmonic components are considered for waveform shaping to maximize RF output power and efficiency. The normalized amplitude of the fundamental frequency component becomes 1.155, 1.207, 1.229, 1.240, 1.245, 1.250 and 1.252 when the harmonics are considered up to 3rd, 5th, 7th, 9th, 11th, 13th and 15th, respectively. Thus, with the increase of the considered number of harmonics, the amplitude of the fundamental frequency component increases though the maximum value of the composite waveform is within the allowable limit of unity. It is also to be noted that, with the increase of a number of considered harmonics, the composite voltage waveform becomes more and more flatten, i.e. similar to a square waveform.

The improvement in efficiency and RF output power level with the increase of number of harmonics considered are shown in Figs. 14.50 and 14.51, respectively. From the plots, it is evident that theoretical efficiency of about 95% and increase of RF output power level of about 0.8 dB is possible when only up to the 5th harmonic components are considered. Thus, practically without going into the realization of

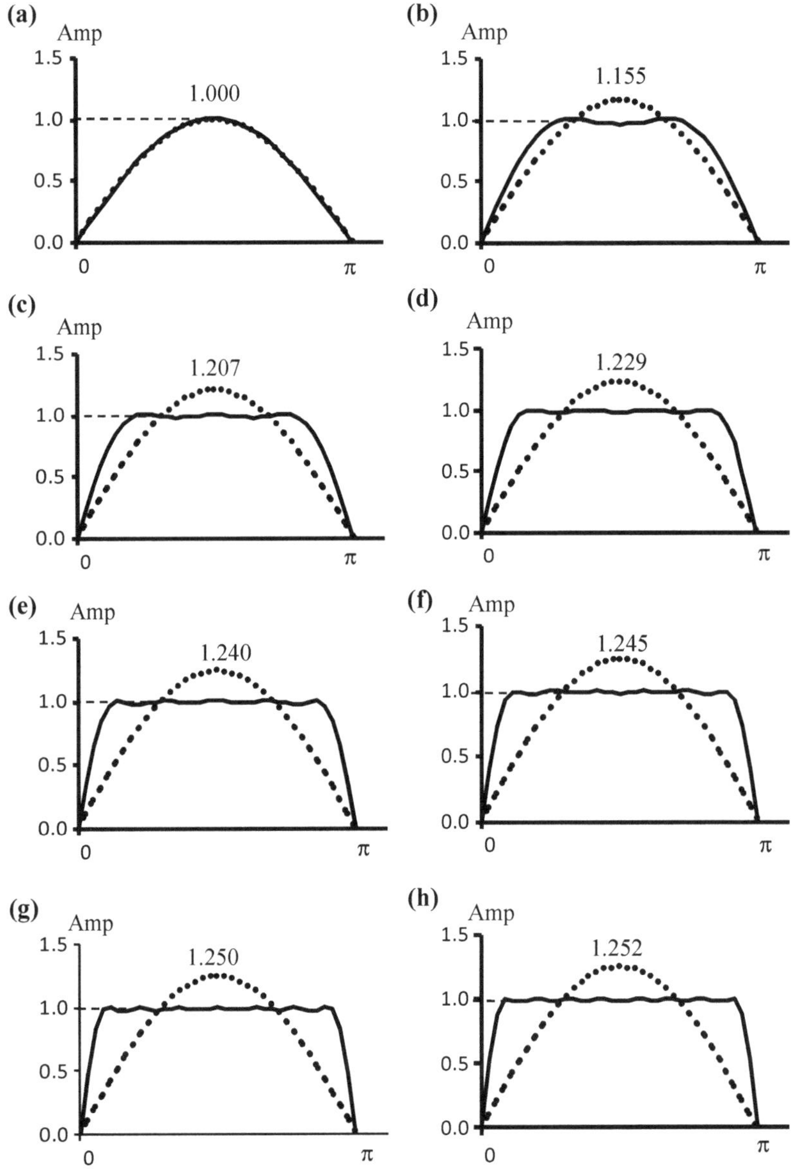

Fig. 14.49 Device voltage waveform (composite and fundamental component) considering, **a** no harmonic component tuned, harmonic components tuned up to **b** 3rd, **c** 5th, **d** 7th and **e** 9th, **f** 11th, **g** 13th. **h** 15th

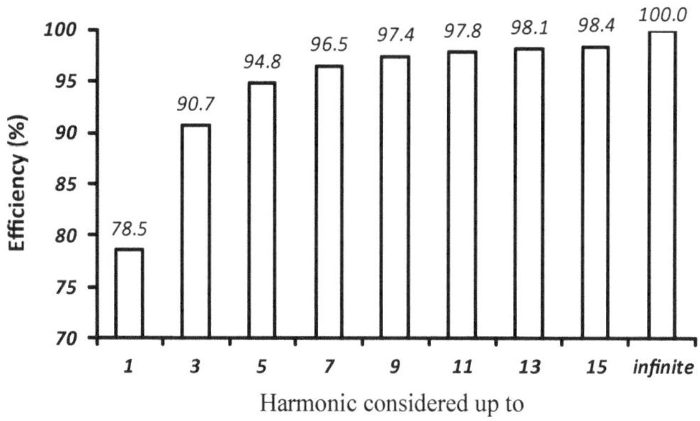

Fig. 14.50 Efficiency of Class-F power amplifier

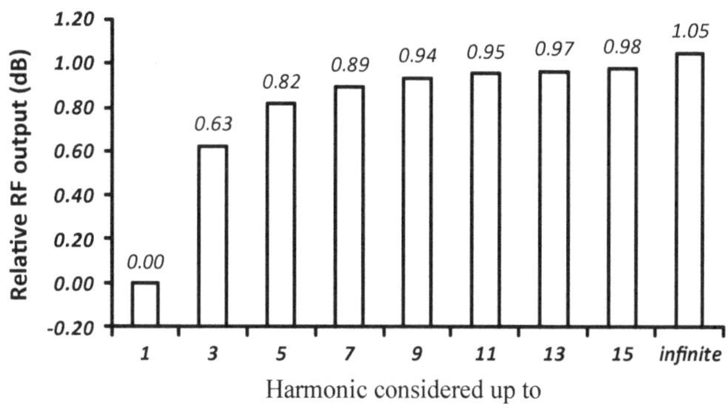

Fig. 14.51 RF output power of Class-F power amplifier

very complex output networks, higher efficiency of the order of 90% with at least 0.5 dB more output power compared to Class-A is expected from Class-F microwave power amplifiers.

14.8 FETs Output Power Capability

The maximum linear RF output power from a Class-A FET amplifier is given by (14.81):

$$P_{rf}(\max) = \frac{I_F\left(V_{dgB} - V_{gsO} - V_K\right)}{8}$$

Neglecting V_{gsO} and V_K compared to the gate-to-drain breakdown voltage $\left(V_{dgB}\right)$ the maximum RF output power is given by

$$P_{rf}(\max) = \frac{I_F \times V_{dgB}}{8} \tag{14.123}$$

Assuming uniform channel doping of density N_D, the maximum drain current (I_F) can be written using (6.3) as

$$I_F = q N_D d v_{n\max} W_g$$

Here, d is channel depth, W_g is gate width and $v_{n\max}$ is the maximum velocity of the channel carriers. The breakdown voltage is given by

$$V_{dgB} = \frac{\varepsilon E_{dgB}^2}{2qN_D}$$

Here, E_{dgB} is the breakdown field and ε is the dielectric constant of the FET channel material. The maximum RF output power (14.123) can be expressed as

$$P_{rf}(\max) = \frac{d\varepsilon W_g}{16} v_{nmax} E_{dgB}^2 \tag{14.124}$$

Suppose effective gate length of the FET is l_{geff}. The maximum frequency of operation $f_{\max}$ can be expressed as

$$f_{\max} = \frac{v_{nmax}}{(2\pi\, l_{\text{geff}})} \tag{14.125}$$

Combining (14.124) and (14.125):

$$\frac{P_{rf}(\max)}{W_g} \times f_{\max} = \frac{\varepsilon}{32\pi} \frac{v_{nmax}^2 E_{dgB}^2}{l_{\text{geff}}/a} \tag{14.126a}$$

The term, $(P_{rf}(\max)/W_g)$ is the RF power density of a FET. Suppose, $p_{rf}(\max) = P_{rf}(\max)/W_g$. Then

$$p_{rf}(\max) \times f_{\max} = \frac{\varepsilon}{32\pi} \frac{v_{nmax}^2 E_{dgB}^2}{l_{\text{geff}}/a} \tag{14.126b}$$

Here, (l_{geff}/a) is the aspect ratio of FET and practically is about 10–5. Considering the factor (l_{geff}/a) is a constant, the product of maximum RF power density and maximum frequency of operation depends only on the material properties $(v_{nmax} \times E_{dgB})^2$. Thus, the product (power × frequency) is a constant for a particular material technology and is figure of merit for power FET technology. Power amplifiers based on GaN technology provides better (power × frequency) product due to its greater gate-to-drain breakdown voltage and higher saturated electron velocity compared to others [30–32].

Table 14.12 Properties of GaN and GaAs materials

Material	Dielectric constant (ε_r)	Breakdown field (V/cm)	Electron saturation velocity (cm/s)
GaAs	13.1	0.65×10^6	1.3×10^7
GaN	9.0	3.5×10^6	2.5×10^7

Example 14.27 Compare the power frequency product between GaAs and GaN FETs technologies.

Solution

Material properties of GaAs and GaN required for computing the power frequency products of power FET technologies are shown in Table 14.12.

Putting parameter values in (14.126b), the ratio of power frequency products of GaN and GaAs is

$$\frac{\left[p_{rf}(\max) \times f_{\max}\right]_{\mathrm{GaN}}}{\left[p_{rf}(\max) \times f_{\max}\right]_{\mathrm{GaAs}}} = \frac{9.0}{13.1}\left[\frac{2.5 \times 3.5}{1.3 \times 0.65}\right]^2 = 73.7$$

This shows that power density of GaN HEMTs are about 74 times that of RF GaAs FETs with both the devices having same maximum frequency of operation.

Example 14.28 Two GaAs FETs are designed with different gate lengths, one for $f_{\max} = 1$ GHz and another for $f_{\max} = 10$ GHz maintaining their aspect ratio as 8. Calculate RF power density of both the FETs.

Solution

The power density of a FET is given from (14.126b) as follows:

$$\text{Power density} = p_{rf}(\max) = \frac{\varepsilon}{32\pi} \frac{v_{nmax}^2 E_{dgB}^2}{l_{\mathrm{geff}}/a} \frac{1}{f_{\max}}$$

From Table 14.12 for GaAs FETs:
Dielectric constant, $\varepsilon = 13.1 \times \varepsilon_o = 13.1 \times 8.854 \times 10^{-12}$ F/m

$$v_{nmax} = 1.3 \times 10^7 \text{ cm/s} = 1.3 \times 10^5 \text{ m/s}$$

$$E_{dGB} = 0.65 \times 10^6 \text{ V/cm} = 0.65 \times 10^8 \text{ V/cm}$$

Aspect Ratio $= l_{\mathrm{geff}}/a = 8$ (as given)

$$\text{Power density} = \frac{13.1 \times 8.854 \times 10^{-12}}{32\pi} \frac{\left(1.3 \times 10^5 \times 0.65 \times 10^8\right)^2}{10} \frac{1}{f_{\max}(\mathrm{Hz})} \text{ W/m}$$

$$= 10.3 \frac{1}{f_{\max}(\text{in GHz})} \text{ W/mm}$$

Thus, for GaAs FET with $f_{\max} = 1$ GHz power density is 10.3 W/mm and for GaAs FET with $f_{\max} = 10$ GHz, power density is 1.03 W/mm.

Example 14.29 Two GaN FETs are designed with different gate lengths, one for $f_{\max} = 10$ GHz and another for $f_{\max} = 100$ GHz maintaining their aspect ratio as 8. Calculate RF power density of both the FETs.

Solution

The power density of a FET is given by (14.126b):

$$\text{Power density} = p_{rf}(\max) = \frac{\varepsilon}{32\pi}\frac{v_{nmax}^2 E_{dgB}^2}{l_{\text{geff}}/a}\frac{1}{f_{\max}}$$

From Table 14.12, for GaN FETs:
Dielectric constant, $\varepsilon = 9 \times \varepsilon_o = 9 \times 8.854 \times 10^{-12}$ F/m

$$v_{nmax} = 2.5 \times 10^7 \text{ cm/s} = 2.5 \times 10^5 \text{ m/s}$$
$$E_{dgB} = 3.5 \times 10^6 \text{ V/cm} = 3.5 \times 10^8 \text{ V/cm}$$

Aspect ratio $= l_{\text{geff}}/a = 8$ (as given)

$$\text{Power density} = \frac{9 \times 8.854 \times 10^{-12}}{32\pi}\frac{\left(2.5 \times 10^5 \times 3.5 \times 10^8\right)^2}{10}\frac{1}{f_{\max}(Hz)} \text{ W/m}$$
$$= 758.6\frac{1}{f_{\max}(\text{in GHz})} \text{ W/mm}$$

Thus, for GaN FET with $f_{\max} = 10$ GHz, power density is 75.86 W/mm and for GaN FET with $f_{\max} = 100$ GHz power density is 7.58 W/mm.

The RF output power per unit gate width of a FET which is designed for a particular maximum frequency $f_{\max}$ is practically determined by its material technology. For a particular $f_{\max}$, with the increase of device gate width, output current increases, thus increases the output RF power.

A practical power amplifier designed to operate at about 4 GHz has $f_{\max} =$ 20 GHz. Thus, from Example 14.27, power density of the GaAs power FET intended to be operated at about 4 GHz is around 0.5 W/mm. Considering the gate-drain breakdown voltage = 20 V, maximum current can be determined from (14.123) as $I_F = 8 \times P_{rf}(\max)/V_{dgB} = 8 \times 0.5/20 = 200$ mA/mm. Neglecting V_{gsO} and V_K compared to V_{dgB}, the load resistance for maximum output power in Class-A operation is, $R_L = V_{dgB}/I_F = (20/200)\,\text{k}\Omega = 100\,\Omega$.

Thus, a GaAs power FET designed for $f_{\max} = 20$ GHz with gate width of 1.0 mm can provide RF output power of about maximum 0.5 W when it is operated in linear Class-A operation with a load resistance of 100 Ω. A typical power device cell can have about 1.5 mm gate width, thus the maximum RF output power from a power device cell will be about 0.75 W with the required load resistance of 67 Ω. Practically, several such cells are connected in parallel to increase the effective gate width as shown in Fig. 14.52a. Suppose 10 such cells are connected in parallel to design power FET with 15 mm of effective gate width. The achieved maximum RF power is about 7.5 Watts, and the load impedance is about 10 times less, i.e. about 6.7 Ω. Thus, with the increase of FET gate width to increase RF output power capability, the required value of output load resistance decreases. So the conversion ratio to transfer 50 Ω to the required load resistance will be larger, making matching more difficult over a wide band of frequency. Output powers of the individual chips are combined using chips with pre-matched ports before parallel combination, or without prior port impedance matching. To avoid impedance level problems, in general, chip-level FETs are individually matched and transferred to the reference impedance (Z_o = 50 Ω) before combining power. In wide bandwidth applications, it is preferable to use lesser gate width FETs and using of proper power combining technique, as shown in Fig. 14.52b, instead of just paralleling the FETs to achieve more RF output power. In the case of GaN HEMTs, higher RF output power density allows much smaller size devices with the same output power. Therefore, for the same output power capability, output impedance of GaN device is much higher compared to GaAs devices. Higher output impedance of GaN power devices enables design of broadband output matching network for realization of broadband power amplifiers.

14.8.1 Microwave Power Combing Techniques

There are many applications where higher RF power is required compared to what is available from a single transistor. To meet the higher power requirement it is required to combine the output power of several transistors. One technique to get higher output power is to connect more and more number of transistors in parallel as shown in Fig. 14.52a. This leads to lower terminal impedances mainly at the output

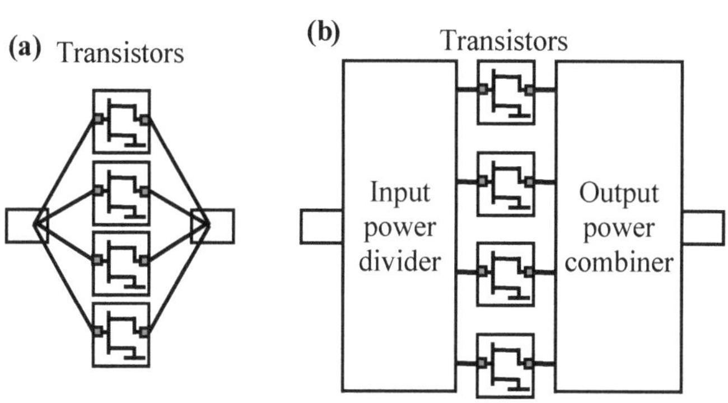

Fig. 14.52 Microwave power combining techniques, **a** parallel connection of transistors, **b** true power combining with proper impedance transformation

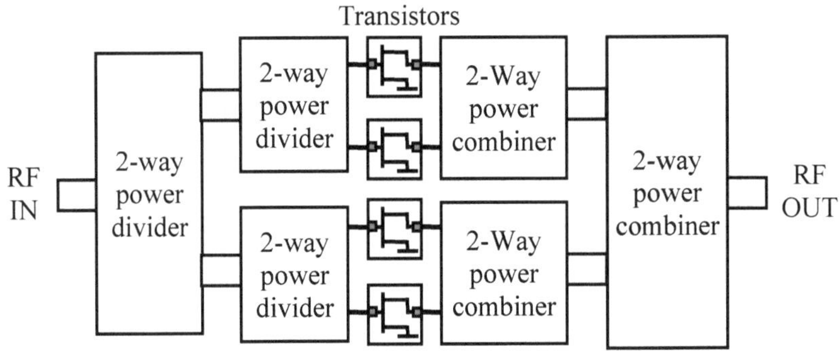

Fig. 14.53 Schematic block diagram of corporate-type power combining technique for combining four transistors

of the combined amplifier which makes the impedance matching more difficult over a broadband of frequency. This problem can be addressed by partially matching individual transistors to transfer terminal impedances to higher values before combining by parallel connection. Another problem is the phase difference among the signals combining at its output due to the path differences, which leads to reduced gain and reduced combined output power. This problem becomes more severe with the increase of operating frequency and numbers of transistors to be connected in parallel. The phase imbalance problem can be reduced by placing the transistors close to each other. However, this leads to the problem of heat dissipation and increase the channel temperature of the device which reduces device reliability. To avoid the problems of lowering port-impedances and phase imbalance in the simple parallel combining of transistors, true power combining techniques using input power divider and output power combiner are used as shown in Fig. 14.52b.

There are various types of power combining techniques such as serial power combining, N-way power combining and corporate power combining. Among all these techniques, corporate power combining also called tree method power combining is mostly used. Schematic block diagram of a corporate-type power combining for four transistor chips is shown in Fig. 14.53. In this case, the number of devices changes with 2^n, where n is an integer, thus it is also called binary power combining. In the ideal case, where all the power chips are identical and perfectly matched and there is no loss of the power divider and combiner networks, combined output power will be 2^n times of one chip but the gain will be equal to the gain of one device.

14.8.2 In-Phase Power Combiners

The Wilkinson splitter is the most widely used two-way in-phase power divider/combiner used for microwave power combining. Schematic diagram of Wilkinson power splitter is shown in Fig. 14.54. The power splitter consists of two

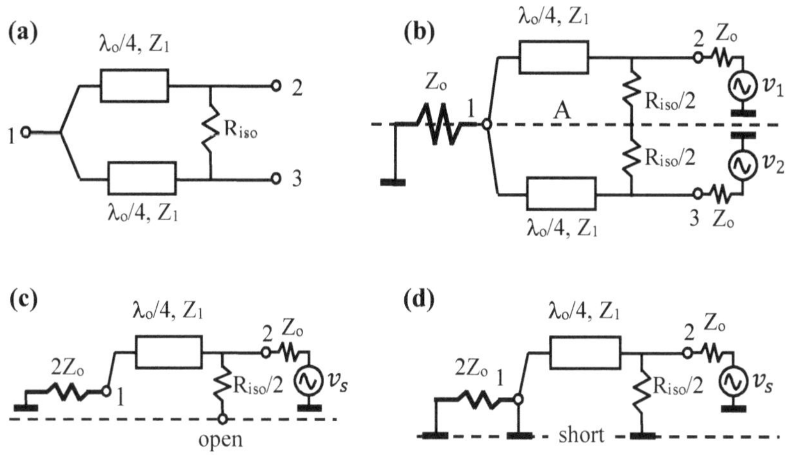

Fig. 14.54 Wilkinson power divider, **a** schematic diagram, **b** shown as symmetric structure, **c** equivalent circuit under even-mode signal ($v_1 = v_2 = v_s$), **d** equivalent circuit for odd-mode signal ($v_1 = -v_2 = v_s$)

branch lines of length $\lambda_o/4$ and a resistor in between the branches to provide isolation between the output ports. The Wilkinson splitter has symmetric structure and thus ideally it provides zero amplitude and phase difference between two outputs irrespective of operating frequency. The characteristic impedance, Z_1 of the lines and value of the isolation resistor R_{iso} can be determined by even- and odd-mode analysis as follows:

With reference to Fig. 14.54b, under even mode excitation of the output ports, i.e. when $v_1 = v_2 = v_s$, electrical open can be considered along the line of symmetry as shown by the dotted line "A". Under the even mode excitation, the equivalent circuit can be considered as shown in Fig. 14.54c. For impedance matching at the input port, we can write:

$$2Z_o = Z_1^2/Z_o$$

Thus,

$$Z_1 = \sqrt{2}Z_o \tag{14.127a}$$

Similarly, under odd-mode excitation among the output ports, i.e. when $v_1 = -v_2 = v_s$, electrical short can be considered along the line of symmetry as shown. Thus, under the odd-mode excitation, the equivalent circuit will be as shown in Fig. 14.54d. For impedance matching at the output port, we can write:

$$R_{\text{iso}}/2 = Z_o$$

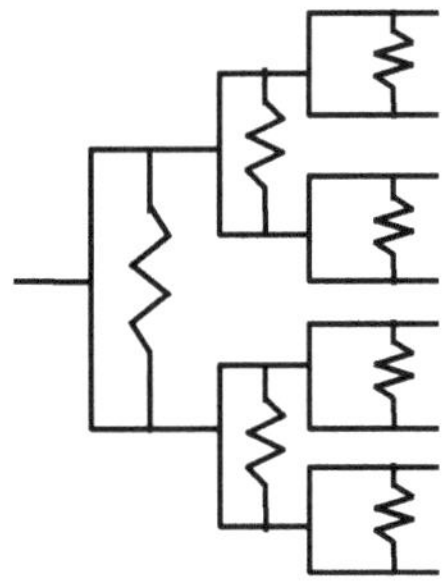

Fig. 14.55 Schematic block diagram of corporate-type power combining technique for combining 8 transistors

Thus,

$$R_{\text{iso}} = 2Z_o \tag{14.127b}$$

Four and other 2^n ways corporate-type power combiner/divider can be realized by just connecting more and more 2-way splitters in cascade configuration as shown in Fig. 14.55.

Schematic diagram of a special type 4-way power combiner/divider suitable for high power combing with low loss is shown in Fig. 14.56 [33, 34]. Using even- and odd-mode analysis, branch impedances and values of isolation resistors can be determined as follows:

Under the even-mode signal excitation, i.e. for $v_1 = v_2 = v_3 = v_4 = v_s$, all the isolation resistors (R_{iso1} and R_{iso2}) of Fig. 14.56a can be considered as open. Thus, for input impedance matching, we can write:

$$Z_o = \frac{1}{2} \times \frac{Z_1^2}{Z_2^2/2Z_o}$$

This leads to

$$Z_1 = Z_2 \tag{14.128a}$$

This implies that the required characteristic impedances of the branches are independent of the reference impedance Z_o. Therefore, in case of planner circuit, wider (low impedance) transmission lines can be used instead of narrow transmission line of impedance $\sqrt{2}Z_o$ to minimize conductor loss of the power combiner.

For even–odd-mode signal excitation with $v_1 = v_2 = -v_3 = -v_4 = v_s$, the line of symmetry "*A*" can be considered as short but the line of symmetry "*B*" can be considered as open. In this condition, the equivalent circuit can be represented as shown in Fig. 14.56 b. From this

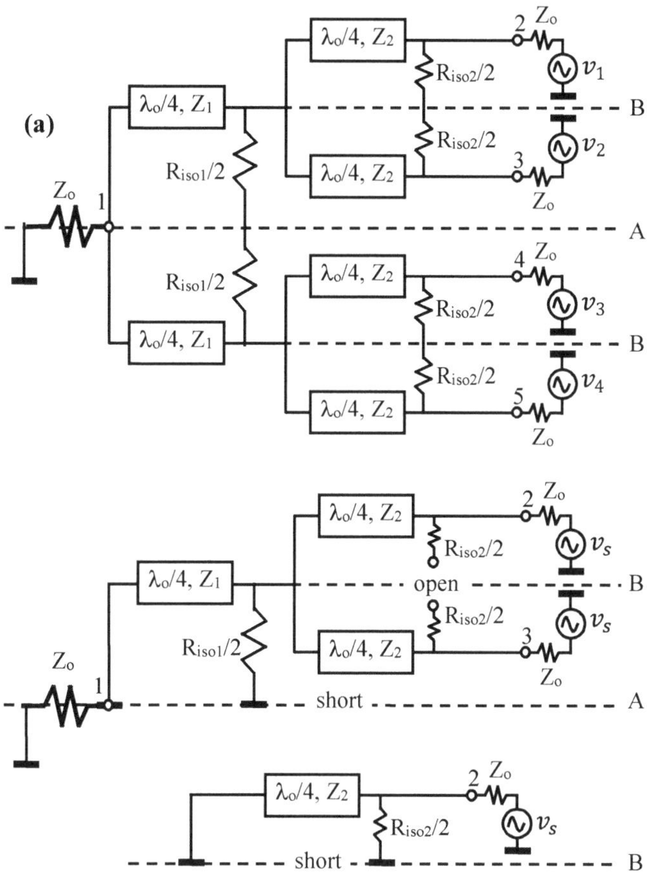

Fig. 14.56 **a** 4-way power divider, **b** equivalent circuit under even mode signal ($v_1 = v_2 = v_3 = v_4 v_s$), **d c** equivalent circuit for odd-mode signal ($v_1 = -v_2 = v_s$)

$$\frac{R_{\text{iso1}}}{2} = \frac{1}{2} \times \frac{Z_2^2}{Z_o}$$

This leads to

$$R_{\text{iso1}} = Z_2^2 / Z_o \tag{14.128b}$$

For other combinations of even–odd-mode signal excitation with $v_1 = -v_2 = v_3 = -v_4 = v_s$, the line of symmetry "*B*" can be considered as short. At this condition, the equivalent circuit can be represented as shown in Fig. 14.56c. From this

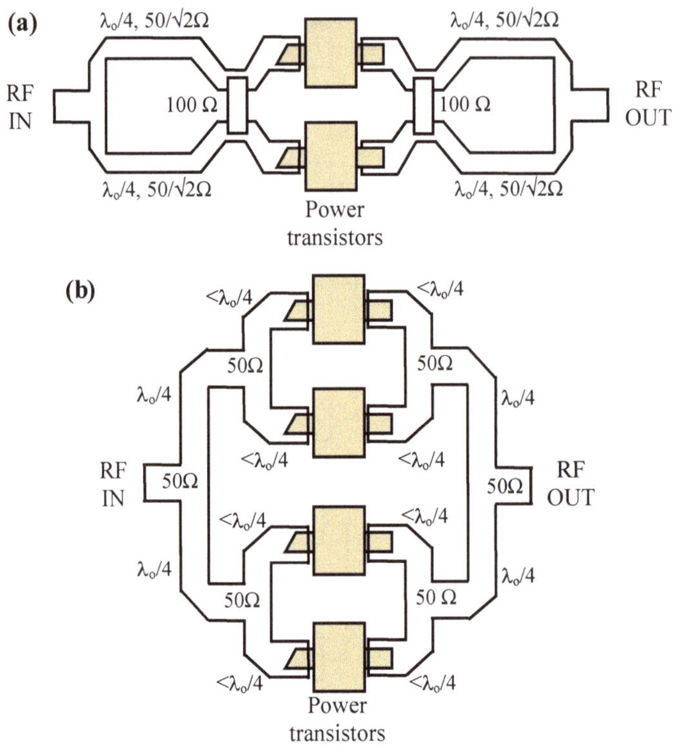

Fig. 14.57 Schematic diagram of combining, **a** two power transistors, **b** four power transistors using 4-way splitters with low impedance (wider transmission line) output combiner used

$$\frac{R_{\text{iso2}}}{2} = Z_o$$

This leads to

$$R_{\text{iso2}} = 2Z_o \tag{14.128c}$$

Thus, the design parameters for this special 4-way power splitter are $Z_1 = Z_2, R_{\text{iso1}} = Z_2^2/Z_o, R_{\text{iso2}} = 2Z_o$.

For a special case of $Z_1 = Z_2 = Z_o = 50\ \Omega$, $R_{\text{iso1}} = Z_o = 50\ \Omega$ and $R_{\text{iso1}} = 2Z_o = 100\ \Omega$.

Schematic diagram of a power amplifier combining two transistors using two-way Wilkinson power splitters is shown in Fig. 14.57a. Though the Wilkinson power splitters are excellent in terms of amplitude and phase balance of the two paths, this combiner suffers from the requirement of a very good port VSWRs of the transistors and higher power handling requirement of the output isolation resistor.

Any amplitude and/or phase imbalance between the two transistors lead to power dissipation through the isolation resistor. Thus, high power RF resistance is required to realize the power module using Wilkinson power splitter. In practical applications, particularly for chip level combining matched transistors are used without using isolation resistors as shown in Fig. 14.57b. Here, for example, the special 4-way power splitter is shown without using isolation resistors. Here impedances of all the branch line impedances are kept the same as reference impedance 50 Ω with shorter length of the two branches as shown to minimize the splitter losses and to realize the amplifier module very compact. Loss of the input power divider and output power combiner reduces the gain of the combined power module, whereas only the loss of the output power combiner is responsible to reduce the output power of the power module. Therefore, loss of the output power combiner is the most critical factor to realize highly efficient power module. Another in-phase combiner is the rat-race power combiner. Its frequency bandwidth is very narrow thus it is not widely used as Wilkinson power splitter.

14.8.3 Balanced Power Combining

In case of in-phase power combining, VSWR of all the transistors should be very good. The VSWR of a combined power amplifier module is never better than the VSWR of the transistors. In general, the output port of power amplifiers are matched for maximum output power, which is different from conjugate matching thus, output VSWR is poor. To make a power amplifier module having better VSWR of both the ports using devices with poor port VSWRs, balanced power combining technique is used. To realize balanced power amplifier, quadrature power splitters are used. Schematic diagram of a balanced power amplifier with signal levels using 3-dB 90° power splitters is shown in Fig. 14.58. Two amplifiers (AMP_a and AMP_b) having transmission coefficients $S_{21a} = k_1\angle\theta_1$ and $S_{21b} = k_2\angle\theta_2$ and input reflection coefficients $S_{11a} = \rho_1\angle\phi_1$ and $S_{11b} = \rho_2\angle\phi_2$, respectively, are used. When port-1 of the power amplifier module is excited by signal v_s through the reference resistance R_o and all the other terminals are terminated with the same reference resistance R_o, the signal reflected to the input port is given by

$$v_{r1} = \frac{v_s}{4}(\rho_1(0° + \phi_1) + \rho_2(-180° + \phi_2)) \tag{14.129a}$$

$$= \frac{v_s}{4}(\rho_1\angle\phi_1 - \rho_2\angle\phi_2) \tag{14.129b}$$

Therefore, the input reflection coefficient of the combined amplifier is given by

$$\Gamma_{r1} = \frac{2v_{r1}}{v_s} = \frac{(\rho_1\angle\phi_1 - \rho_2\angle\phi_2)}{2} \tag{14.129c}$$

Thus, in terms of S-parameters, the input reflection coefficient can be written as

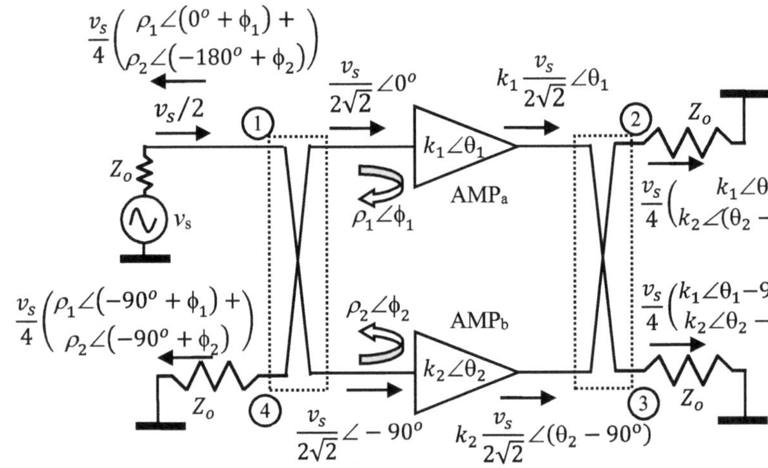

Fig. 14.58 Balanced power amplifier configuration with signal levels

$$S_{11}(\mathrm{AMP}) = \Gamma_{r1} = \frac{1}{2}(S_{11a} - S_{11b}) \tag{14.129d}$$

For the batch-matched amplifiers AMP$_a$ and AMP$_b$, only with $\rho_1 = \rho_2$, and $\phi_1 = \phi_2$, i.e. with $S_{11a} = S_{11b}$, the input refection coefficient of the balanced amplifier, $\Gamma_{r1} = 0$, i.e. $S_{11}(\mathrm{AMP}) = 0$. Thus, the balanced amplifiers provide ideally no reflection of RF power to its input port even if individual amplifiers have poor return losses. In a similar way, it can be shown that the output reflection coefficient of balanced amplifiers is ideally zero for matched pair of amplifiers.

The signal level at the isolated port of the input power divider (port-4) is given by

$$v_{\mathrm{iso1}} = \frac{v_s}{4}(\rho_1\angle(-90^\circ + \phi_1) + \rho_2\angle(-90^\circ + \phi_2)) \tag{14.130a}$$

$$= \frac{-jv_s}{4}(S_{11a} + S_{11b}) \tag{14.130b}$$

For the batch-matched amplifiers with $\rho_1 = \rho_2 = \rho$, and $\phi_1 = \phi_2 = \phi$, i.e. with $S_{11a} = S_{11b}$ the signal level at the isolated port is

$$v_{\mathrm{iso1}} = \frac{-jv_s}{2}(\rho\angle\phi) = \frac{-jv_s}{2}S_{11a} = \frac{-jv_s}{2}S_{11b} \tag{14.131}$$

The reflected power will be dissipated through the isolation resistor without going to the input port.

The signal level at port-3 of the combined amplifier is given by

$$v_{t4} = \frac{v_s}{4}(k_1\angle\theta_1 - 90^\circ + k_2\angle\theta_2 - 90^\circ) \tag{14.132a}$$

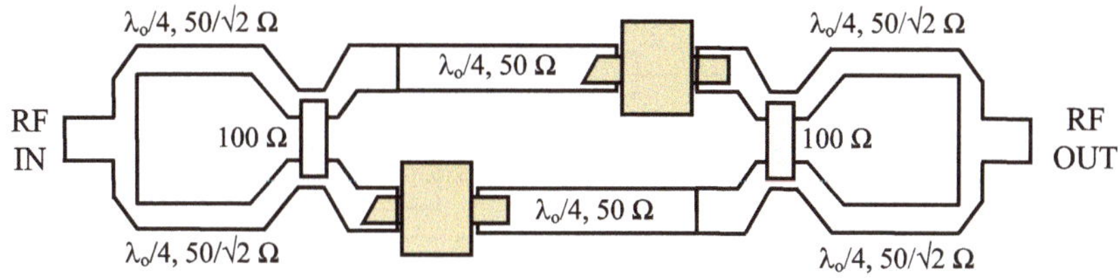

Fig. 14.59 Schematic diagram of balanced power amplifier using Wilkinson power splitter

Therefore, transmission coefficient from port-1 to port-3, i.e. gain of the balanced amplifier is given by

$$g = \frac{v_{t4}}{\frac{v_s}{2}} = \frac{1}{2}(k_1\angle(\theta_1 - 90°) + k_2\angle(\theta_2 - 90°)) \tag{14.132b}$$

$$= \frac{-j}{2}(S_{21a} + S_{21b}) \tag{14.132c}$$

For the matched pair amplifiers with $k_1 = k_2 = k$ and $\theta_1 = \theta_2 = \theta$, the gain can be written as

$$g = k = S_{21a} = S_{21b} \tag{14.132d}$$

Thus, gain of combined amplifiers with same gain coefficients is the same as the gain of the individual amplifiers.

The signal level at the port-2 of the combined amplifier is

$$v_{\text{iso2}} = \frac{v_s}{4}(k_1\angle\theta_1 + k_2\angle(\theta_2 - 180°)) \tag{14.133a}$$

$$= \frac{v_s}{4}(S_{21a} - S_{21b}) \tag{14.133b}$$

For the matched batch amplifiers with $k_1 = k_2 = k$ and $\theta_1 = \theta_2 = \theta$, the signal level at the port-2 is

$$v_{\text{iso2}} = 0 \tag{14.133c}$$

Thus, for batch matched amplifiers there will be no power at the output isolated port.

There are various types of quadrature power splitters. The Wilkinson power splitter with $\lambda_o/4$ more length of one arm than the length of other arm works as quadrature-phase power splitter. Schematic drawing of a balanced amplifier using Wilkinson power splitter is shown in Fig. 14.59.

Among the different types of quadrature couplers, branch line coupler as shown in Fig. 14.60a is mostly used for high power applications. Suppose the lengths of all the branch lines are quarter wavelength, i.e. $\lambda_o/4$ corresponding to the operating

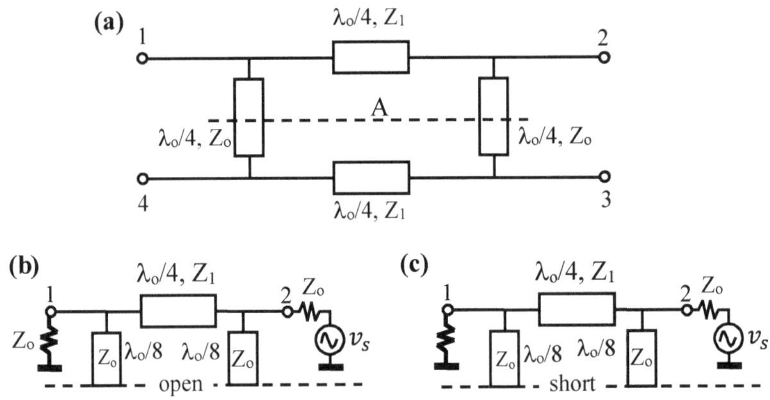

Fig. 14.60 **a** Branch line power divider, **b** equivalent circuit under even-mode signal ($v_1 = v_2 = v_s$), **c** equivalent circuit for odd-mode signal ($v_1 = -v_2 = v_s$)

frequency. The characteristic impedance of the shunt lines are considered as Z_o. Using even- and odd-mode analysis, branch impedances of the coupler can be determined as follows:

Under even-mode signal excitation at port-2 and port-3, along the line of symmetry, "A" can be considered as open and the equivalent circuit can be considered as shown in Fig. 14.60b. Similarly, under odd-mode signal excitation, the equivalent circuit can be considered as shown in Fig. 14.60c. With reference to Fig. 14.60b, using (7.80) the input impedance of the open circuited $\lambda_o/8$ length transmission line is given by

$$Z_{\text{in}(\lambda_o/8)} = -jZ_o$$

Thus, for input impedance matching, we can write

$$Z_o = -jZ_o \| \frac{Z_1^2}{-jZ_o \| Z_o}$$

This leads to

$$Z_o^2 = 2Z_1^2$$

or,

$$Z_1 = Z_o/\sqrt{2} \tag{14.134a}$$

Similarly, with reference to Fig. 14.60c, using (7.79) the input impedance of the short-circuit $\lambda_o/8$ length transmission line is given by

Fig. 14.61 Photograph of a branch line 3 dB 90° power divider/combiner

$$Z_{\text{in}(\lambda_o/8)} = +jZ_o$$

Thus, for input impedance matching we can write

$$Z_o = +jZ_o \| \frac{Z_1^2}{+jZ_o \| Z_o}$$

This also leads to

$$Z_o^2 = 2Z_1^2 \tag{14.134b}$$

or

$$Z_1 = Z_o/\sqrt{2} \tag{14.134b}$$

Therefore, characteristic impedances of the series quarter wave length transmission liners are $Z_o/\sqrt{2}$.

Photograph of a branch line coupler using microstrip transmission line on a 20 mil-thick alumina substrate is shown in Fig. 14.61. Here, meander transmission lines are used to realize the coupler over a very small area at lower microwave frequency band. Schematic diagram of a balanced amplifier using branch line couplers is shown in Fig. 14.62. Branch line couplers are widely used for realization of high power amplifiers due to the feasibility of using wider planner transmission line and possibility for mounting high-power loads to serve as isolation resistors.

There are also various types of quadrature power splitters such as broadside and edge-coupled transmission line couplers. Among them, Lange couplers are widely used due to its very wide frequency band performance. The balanced amplifiers are not only used for achieving high RF output power, these configurations are also used for low-power amplifiers to realize very broadband amplifiers.

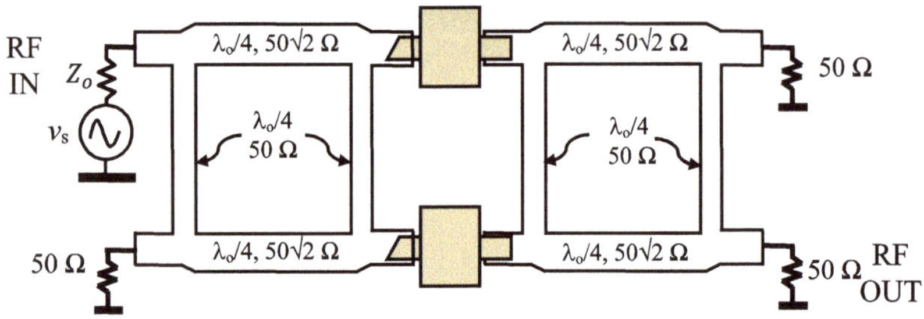

Fig. 14.62 Schematic diagram of balanced power amplifier using branch line couplers

Example 14.30 Amplifier-A is having 6 dB gain and 3.8 dB NF. Amplifier-B is having 20 dB gain and 4 dB NF. What should be the sequence of the cascaded two-stage amplifier for lowest NF. What will be the value of the lowest NF for this arrangement?

Solution

Gain of the amplifier-A and amplifier-B are

$$G_A = 6 \text{ dB} = 4 \text{ (in factor)}$$
$$G_B = 20 \text{ dB} = 100 \text{ (in factor)}$$

Noise figure of amplifier-A and amplifier-B are

$$\text{NF}_A = 3.8 \text{ dB} = 2.4 \text{ (in factor)}$$
$$\text{NF}_B = 4 \text{ dB} = 2.5 \text{ (in factor)}$$

Case-1: Amplifier-A as first amplifier: Overall noise figure NF_{AB}

$$\begin{aligned}\text{NF}_{AB} &= \text{NF}_A + (\text{NF}_B - 1)/G_A \\ &= 2.4 + (2.5 - 1)/4 = 2.775 \\ &= 4.43 \text{ dB}\end{aligned}$$

Case-2: Amplifier-B as first amplifier: Overall noise Figure NF_{BA}

$$\begin{aligned}\text{NF}_{BA} &= \text{NF}_B + (\text{NF}_A - 1)/G_B \\ &= 2.5 + (2.4 - 1)/100 = 2.524 \\ &= 4.024 \text{ dB}\end{aligned}$$

Thus, cascaded lowest noise figure is 4.024 dB where amplifier-B is used as first amplifier in cascaded two-stage amplifier.

Example 14.31 For a Class-A FET power amplifier with 10 V drain supply and 2 A drain current bias providing RF load current of 1 A amplitude. What is the DC to RF efficiency for a load resistance of 5 Ω?

Solution
DC power consumption of the Class-A power amplifier:

$$P_{\text{DC}} = 10\ \text{V} \times 2\ \text{A} = 20\ \text{W}$$

RF power delivered to the load:

$$P_{\text{RF}} = \left(\sqrt{1/2} \times 1\right)^2 \times 5 = 2.5\ \text{W}$$

Thus, efficiency:

$$\eta = \frac{2.5}{20} = 12.5\%$$

Example 14.32 For a Class-B FET power amplifier with 15 V drain supply providing 12 V peak signal to 15 Ω load. What is DC to RF efficiency of the amplifier?

Solution
In Class-B operation, the efficiency:

$$\begin{aligned}\eta &= \frac{\pi}{4} \times \frac{V_{drf}}{V_{DD}} \\ &= \frac{\pi}{4} \times \frac{12}{15} = 62.8\ \%\end{aligned}$$

Example 14.33 A high-gain MESFET packaged device of case-to-channel thermal resistance 4.5 °C/Watt provides 8 W RF output taking 20 W DC power. What will be the channel temperature of the device if the case temperature of the device is maintained at 55 °C?

Solution
Case-to-channel thermal resistance:

$$R_{\text{th}} = 4.5\ ^\circ\text{C/W}$$

Dissipated power within the device:

$$P_{\text{dissi}} = (20 - 8)\ \text{W} = 12\ \text{W}$$

Increase of temperature:

$$\Delta T = 12\ \text{W} \times 4.5\ ^\circ\text{C/W} = 54\ ^\circ\text{C}$$

Thus, channel temperature:

$$T_{\text{ch}} = (55 + 54)\ ^\circ\text{C} = 109\ ^\circ\text{C}$$

14.9 Temperature Compensation of Microwave Amplifiers

Linear gain, saturated output power and efficiency of GaAs- and GaN-based FETs depend strongly on its temperature [35–39]. At fixed drain and current and voltage bias, transconductance as well as unity current gain cut-off frequency decreases with the increase of temperature. One of the most important parameters determining the gain of a microwave amplifier is unity current gain cut-off frequency, which decreases steeply, irrespective of bias condition, with the increase of temperature. Therefore, at a fixed voltage and current bias, the performance parameters such as gain, output power and efficiency of the amplifiers decrease with the increase of temperature. Practically, the linear gain of these amplifiers decreases approximately by the amount 0.015 dB/°C with the increase of device temperature.

There are many techniques to achieve temperature compensated RF performance of the FET-based amplifiers. Active-controlled voltage/current bias is used to vary the drain current and/or voltage to compensate gain and output power of the amplifiers. However, care must be taken to avoid any instability due to the change of bias with the temperature. Another technique is to use temperature-dependent resistance as feedback element to compensate linear gain variation over temperature. This type of compensation is more effective and suitable for MMIC based amplifier modules [36, 37]. In most of the practical applications, where amplifiers are operated in linear region, a separate voltage variable amplifier/attenuator with temperature-dependent bias is used to compensate linear gain of the overall system.

References

1. Rollett JM (1962) Stability and power gain invariants of linear two-ports. IRE Trans Circuit Theory 29–32
2. Owens PJ, Woods D (1970) Reappraisal of the unconditional stability criteria for active 2-port networks in terms of S parameters. Electron Lett 6:315
3. Woods D (1976) Reappraisal of the unconditional stability criteria for active 2-port networks in terms of S-parameters. IEEE Trans Circuits Syst 23(2):73–81
4. Jackson RW (2006) Rollett proviso in the stability of linear microwave circuits—a tutorial. IEEE Trans Microw Theory Tech 54(3):993–1000
5. Rothe H, Dahlke W (1956) Theory of Noisy Fourports. Proc Inst Radio Eng 44:811–815
6. Haus H, Atkinson W, Branch G, Davenport W, Fonger W, Harris W, Harrison S, McLeod W, Stodola E, Talpey T (1960) Representation of noise in linear twoports. Proc IRE 48:69–74

7. Fukui H (1966) Available power gain, noise figure, and noise measure of two-ports and their graphical representations. IEEE Trans Circuit Theory CT-13:137–142
8. Tucker R (1975) Low-noise design of microwave transistor amplifiers. IEEE Trans Microw Theory Tech 23(8):697–700
9. Hillbrand H, Russer P (1976) An efficient method for computer aided noise analysis of linear amplifier networks. IEEE Trans Circuits Syst 23(4):235–238
10. Garg VK, Singh RV, Jain VK, Bera SC (1999) New amplifier design eliminates chip capacitors. IETE Tech Rev 16(6):197–201
11. Deal W, Mei X, Radisic V, Liu P, Uyeda J, Barsky M, Gaier T, Fung A, Lai R (2007) Demonstration of a S-MMIC LNA with 16-dB gain at 340 GHz. In: 2007 IEEE compound semiconductor integrated circuits symposium, Oct 2007, pp 1–4
12. Samoska L, Gaier T, Fung A, Mei XB, Yoshida W, Lee J, Uyeda J, Liu PH, Deal WR, Radisic V, Lai R (2008) Submillimeter-wave InP MMIC amplifiers from 300–345 GHz. IEEE Microw Wirel Compon Lett 18(1):61–63
13. Bera SC, Shah LB, Raval DU, Pandey S, Kumar V, Singh S, Das DK (2012) Design and development of V&W band amplifier modules. In: Proceedings of international conference on microwaves, antenna, propagation & remote sensing, Jodhpur, India, Dec 2012
14. Raab FH (1997) Class-F power amplifiers with maximally flat waveforms. Proc IEEE Conf Microw Theory Technol 45:2007–2012
15. Pukala D, Raab LF (2001) Maximum efficiency and output of Class-F power amplifiers. IEEE Trans Microw Theory Tech 49(6):1162–1166
16. Kuroda K, Ishikawa R, Honjo K (2008) High efficiency GaN-HEMT Class-F amplifier operating at 5.7 GHz. In: Proceedings of 38th European microwave conference, Amsterdam, Netherland, Oct 2008, pp 440–443
17. Ando A, Takayama Y, Yoshida T, Ishikawa R, Honjo K (2008) A high-efficiency class-F GaN HEMT power amplifier with a diode pre-distortion linearizer. In: Proceedings of Asia-Pacific microwave conference, Macau, China, Dec 2008, pp 1–4
18. Khan FN, Mohammadi FA, Yagoub MCE (2008) A GaNHEMT Class-F amplifier for UMTS/WCDMA applications. In: Proceedings of IEEE international RF and microwave conference, Kuala Lumpur, Malaysia, Dec 2008, pp 478–482
19. Yoo C-U, Kim M-S, Yang Y (2009) High-efficiency Class-F amplifier design using defected ground structure. In: Proceedings of international conference on new trends in information and service science, Beijing, China, July 2009, pp 700–703
20. Wu D, Korolkiewicz E, Lu Q, Liu L (2010) To design and model a Class-F amplifier and investigate the effect of losses on the efficiency of DC to AC power conversion. In: Proceedings of 7th international symposium on communication systems networks and digital signal processing, New Castle upon Tyne, UK, July 2010, pp 119–122
21. Kuroda K, Ishikawa R, Honjo K (2010) Parasitic compensation design technique for a C-Band GaN HEMT Class-F amplifier. IEEE Trans Microw Theory Tech 58(11):2741–2750
22. Kim J, Jo G, Oh J, Kim Y, Lee K, Jong J (2011) Modeling and design methodology of high-efficiency Class-F and Class-F^{-1} power amplifiers. IEEE Trans Microw Theory Tech 59(1):153–165
23. Grebennikov A (2011) Load network design technique for Class F and inverse Class F power amplifiers. High Freq Electron 10(5):58–76
24. Tuffy N, Guan L, Zhu A, Brazil TJ (2012) A simplified broadband design methodology for linearized high-efficiency continuous class-F power amplifiers. IEEE Trans Microw Theory Tech 60(6):1952–1963
25. Carrubba V et al (2012) The continuous inverse class-F mode with resistive second-harmonic impedance. IEEE Trans Microw Theory Tech 60(6):1928–1936
26. Kenle C, Dimitrios P (2012) Design of broadband highly efficient harmonic-tuned power amplifier using in-band continuous Class-F1/F mode transferring. IEEE Trans Microw Theory Tech 60(12):4107–4116
27. Lu Z, Chen W (2013) Resistive second-harmonic impedance continuous class-F power amplifier with over one octave bandwidth for cognitive radios. IEEE J Emerg Sel Top Circuits Syst 3(4):489–497

28. Arnous MT et al (2014) Highly efficient and wideband harmonically tuned GaN-HEMT power amplifier. In: Proceedings of 20th International Conference on Microwave, Radar, and Wireless Communication (MIKON), June 2014, pp 1–4
29. Yefymovych A, Krizhanovski V, Giofre R, Colantonio P (2014) Load network design technique for Class-F power amplifier. In: Proceedings of 20th international conference on microwaves, radar and wireless communications, Gdansk, Poland, June 2014, pp 1–3
30. Kawano A et al (2005) High-efficiency and wide-band single-ended 200 W GaN HEMT power amplifier for 2.1 GHz W-CDMA base station application. In: Asia-Pacific Microwave Conference
31. Kimball DF et al (2006) High-efficiency envelope-tracking W-CDMA base-station amplifier using GaN HFETs. IEEE Trans Microw Theory Tech 54(11):3848–3856
32. Wu YF, Wood SM, Smith RP, Sheppard S, Allen ST, Parikh P, Milligan J (2006) An internally-matched GaN HEMT amplifier with 550-watt peak power at 3.5 GHz. In: International electron devices meeting, Dec 2006, pp 1–3
33. Bera SC, Singh RV, Gagr VK, Arora NS, Nair SS (2005) PBG effect on a modified Wilkinson power divider. In: Proceedings of Indian conference on microwave and remote sensing-2005, pp 303–307
34. Bera SC, Singh RV, Garg VK (2008) Modified Wilkinson power divider with harmonic suppression characteristic. Microw J (2008)
35. Anholt RE, Swirhun SE (1992) Experimental investigation of the temperature dependence of GaAs FET equivalent circuits. IEEE Trans Electron Devices 39(9)
36. Yamauchi K et al (2001) X-band MMIC power amplifier with an on-chip temperature compensation circuit. IEEE Trans Microw Theory Tech 49(12):2501–2506
37. Huang CW, Chang SJ, Wu W, Wu CL, Chang CS (2005) A Ku-Band four-stage temperature compensated pHEMT MMIC power amplifier. Microw Opt Technol Lett 44(5):480–485
38. Torregrosa-Penalva G, Asensio-Lopez A, Ortega-Gonzalez FJ, Bianco-del-Campo A (2003) PAE improvement and compensation of small-signal gain drift due to temperature on power amplifiers through active biasing. Microw Opt Technol Lett 38(5):389–392
39. Bera SC (2006) Remarks on 'PAE improvement and compensation of small-signal gain drift due to temperature on power amplifiers through active biasing'. Microw Opt Technol Lett 48(4):827–828

Chapter 15
Microwave Limiters

Abstract This chapter presents design, analysis and characteristics of various microwave limiter circuits. Both open- and closed-loop-type limiters based on p-i-n diodes, Schottky barrier diodes and transistors are discussed. Complete design, analysis and application of a closed-loop-limiter system with temperature compensation mechanism is presented for microwave systems.

15.1 Introduction

To protect sensitive devices from unintentional or intentional exposure to high level of RF power, limiters are used in receivers, solid-state power amplifiers, detectors, etc. Basically, there are two types of limiters; open-loop limiter and closed-loop or feedback-type limiter. In case of open-loop limiter, the same device detects the RF power level and acts to limit the power level. Here, limiting device works due to its self-bias effect, whereas a closed-loop limiter circuit uses separate detector circuit to detect the RF power level and the detected signal is applied to other devices/circuits to limit the power level in closed-loop configuration. Diodes being two terminal devices are very stable in operation and are commonly used as limiting devices. Schottky and p-i-n diodes are the most suitable limiting device for microwave applications [1–11]. Transistor-based microwave amplifiers are also used as limiter under suitable bias condition. The amplifiers are used as limiter by operating them in gain compression/output power saturation region.

Limiter may be considered as an attenuator which provides attenuation beyond a predefined RF power level. The predefined power level is called threshold power level. Beyond the threshold power level, amount of attenuation will be governed by self-biasing of the limiting device which is determined by the RF power level. Ideally, the amount of self-biasing will be such that there will be 1-dB attenuation for 1-dB increase of power level. Therefore, beyond the threshold power level, ideally, there will be no change of output power with the increase of input power level. However, below the predetermined threshold power level, output power changes linearly with the input power level. That is, output power level changes by 1-dB with change of

S. C. Bera, *Microwave Active Devices and Circuits for Communication*, Lecture Notes in Electrical Engineering 533, https://doi.org/10.1007/978-981-13-3004-9_15

1-dB input power level. In case of zero-bias limiter circuit, the threshold power level is determined by the construction of the device itself. To achieve adjustable threshold power level, external control bias may be applied to the limiting device.

The absence of minority carrier storage effect in Schottky barrier diodes makes them efficient to rectify microwave and mmwave signal. Therefore, Schottky diodes s are superior than p-i-n diodes as microwave limiters and suitable for realization in MMIC technology. Though Si- and GaAs-based Schottky diode limiters are suitable for low power level applications, SiC-based Schottky diodes can be used as high power limiter [8–10]. To increase the overall limiting power range, multistage limiters with different threshold power levels are also used. Small-signal microwave amplifiers using MESFETs with appropriate values of gate and drain bias resistors have capability for overdrive protection. Thus, small-signal amplifiers are also used as limiter in microwave systems to protect other devices and circuits in the system.

15.2 Limiter Characteristics

Sensitive microwave/RF devices such as BJTs, FETs and HEMTs are used in receivers, channel amplifiers and solid-state power amplifiers as well as in many measurement instruments. These devices are very sensitive and susceptible to damage from input signals having amplitudes of more than the safe limit of these devices. The safe limits are of the order of few mW for low-power devices which are used in receivers, channel amplifiers and small-signal stages of solid-state power amplifiers, whereas the safe limit of power level may be of the order of Watts for high-power devices which are used in the medium- and high-power stages of power amplifiers. The unwanted higher power level may come from signals which are within the operating frequency band or may be out of the normal operating frequency band of the device/circuit. Practically, a limiter should work over a very wide band of frequency extending beyond the system's operating frequency bands in both the sides to protect sensitive devices used in communication subsystems and other instruments.

A limiter should provide very low insertion loss over the operating frequency band when the signal levels are within the system's operating dynamic range. This requirement is more critical when it is used at the front end of a receiver to avoid its impact on system noise figure. It should provide very high attenuation to the incoming signals over any frequency band which exceeds the predefined threshold level of the respective devices and units. The limiter should be very fast to provide protection within about nanoseconds of arrival of the high-level signal. Another important requirement of a limiter circuit is that it must survive exposure to high-power signal level providing sufficient attenuation and it must return to its low insertion loss state within nanoseconds after the disappearance of high-level signal power.

Important performance parameters of a limiter are operating frequency range, insertion loss, flat and spike leakage and limiting power range. Operating frequency range defines the range over which the limiter must protect the succeeding devices or systems. In general, the operating frequency range of a limiter is broader than the

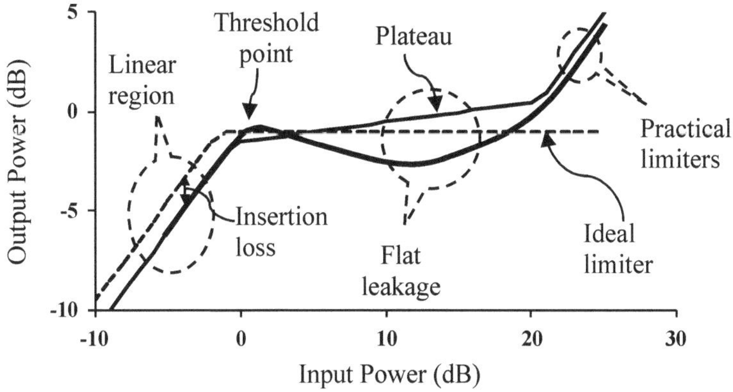

Fig. 15.1 Typical characteristic of limiter

operating frequency range of the circuits or subsystems required to be protected from high level of RF power levels. Insertion loss of a limiter is defined as the insertion loss over its linear region, i.e. small-signal operating region over its entire dynamic range as shown in Fig. 15.1.

When high input power is applied to a limiter, most of the power is absorbed or reflected by the limiter and only remaining small portion of the power can pass through it. The small portion of the applied power which passes through the limiter is called flat leakage or only leakage power. Generally, the leakage power is specified as an absolute level, i.e. in dBm or Watts. The input limiting power range is defined from the threshold power level to the maximum power level up to which the limiter provides attenuation to keep its output power level within specified limit. Ideally, over this region the output power of the limiter remains constant. However, in most of the limiter, over the input dynamic range, the output increases gradually with its input as shown in Fig. 15.1 as plateau. The region between the initiation of limiting and the region of maximum isolation is termed the plateau region. For input signals above the maximum input power level, the output will begin to increase again at a greater slope. The output power variation over the input limiting power range may also initially decrease and then increase with greater slope as shown in the figure. This type of limiter characteristic is observed for fast response devices due to their efficient rectification of RF signal thereby providing more attenuation over the initial limiting power range compared to the higher level of signal power. Another limitation is the presence of spike leakage as shown in Fig. 15.2 [4–6]. For a sudden increase of input power level, limiter takes some time to settle the output at its leakage power level. This time is called recovery time. Spike leakage is a transient phenomenon and sometimes is a risk factor to sensitive electronic devices. This becomes critical for some sensitive devices such as microwave FETs and HEMTs when comparatively slow limiting diodes such as wide intrinsic region p-i-n diodes are used as limiters. In these cases, fast response devices such as Schottky barrier diodes as limiting devices are preferable in place of p-i-n diodes.

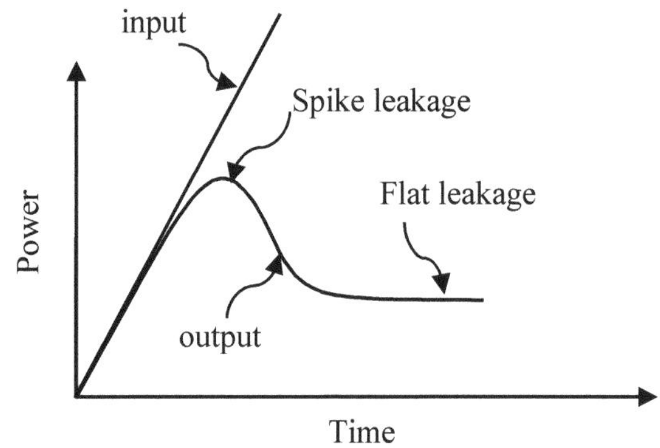

Fig. 15.2 Limiter characteristic

Ideally, both ports of a limiter should have good impedance matching over its entire input power range. To achieve this, diodes are needed to be mounted in a balanced configuration as it is used in case of diode-based attenuator circuits. Another aspect of a limiter is to precisely protect sensitive devices such as pHEMT, HFET, etc. To protect such sensitive devices from overdrive power situation, provision of threshold power level adjustment is required. This adjustability is required to protect the system without affecting the system's normal performance over its entire operating dynamic range.

15.3 P-I-N Diode Limiters

Nonlinear devices are required to achieve limiter function for protecting RF and microwave devices and circuits from high-level signals while maintaining very low loss, below the limiting threshold power level. In the RF and lower microwave frequency range, p-i-n diodes are popularly used as limiting power device. Generally p-i-n diodes have slow response time. In the microwave frequency range, special types of p-i-n diodes with relatively short carrier lifetime, called limiter diodes, are constructed with lower device width to achieve comparatively fast response time. Intrinsic layer thickness of the limiter diodes is between 2 to 7 μm. These diodes with relatively short carrier lifetime, efficiently rectify RF power when it exceeds a specified level and provide required attenuation by self-biasing.

Generally, p-i-n diodes are used to realize limiter due to their capability of handling high power levels and providing reasonable leakage level. Shunt-connected p-i-n diode limiters are very popular because of their very high impedance at OFF condition (low signal power level) and very low impedance when forward biased under high-level signal power. In the presence of small-level signals, the shunt diode does not conduct, and therefore represents high impedance and thus provides very low loss to the RF signal. When RF signal of high power level is applied, p-i-n diode goes into the forward-bias condition and provides attenuation. A schematic circuit of a

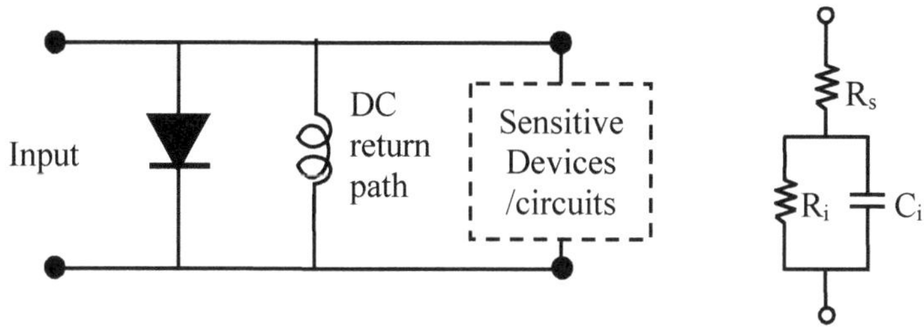

Fig. 15.3 Limiter circuit diagram of single shunt-connected p-i-n diode and its equivalent circuit

shunt-connected p-i-n diode limiter is shown in Fig. 15.3. The RF choke shown in the figure provides a DC return path for the self-bias current.

The diode is shunt mounted across the transmission line, and is provided with a DC bias return by an RF choke, as shown in Fig. 15.3. For RF input signal level which is below the predetermined threshold level, the diode acts as an ordinary unbiased p-i-n diode. Thus, the diode acts as parallel combination of high-value resistor with low-value capacitor and provides very low loss to the RF signal. However, when the incident signal level exceeds the threshold power level, the diode's I-layer is flooded with carriers during the positive half cycle of the input RF signal. Due to the large carrier lifetime, most of these carriers remain in the I-region through the negative half cycle. Thus, rectified DC current begins to flow through the diode and the diode biases itself to a low value of RF resistance. The diode's RF resistance falls to a very low value under the influence of this self-generated bias current, called the self-biasing effect. Reflecting the large signal back to its source, the limiter circuit then acts as a reflective switch, and thus limits the output RF power. Additionally, some part of the signal is also absorbed by the diode resistance. When the large amplitude input signal has disappeared, recombination of the carriers in the diode's I-region leads to decrease or diminishing of the circulating bias current. Due to the absence of self-bias current, the RF resistance of the p-i-n diode becomes very high, allowing small signals to pass without attenuation.

The S-parameter matrix of shunt-connected diode limiter is the same as derived in case of shunt-connected attenuator and is given by (9.5)

$$\begin{bmatrix} S_{11} & S_{12} \\ S_{21} & S_{22} \end{bmatrix} = \begin{bmatrix} \frac{-Z_o}{2Z_T+Z_o} & \frac{2Z_T}{2Z_T+Z_o} \\ \frac{2Z_T}{2Z_T+Z_o} & \frac{-Z_o}{2Z_T+Z_o} \end{bmatrix} \tag{15.1}$$

where Z_T is the equivalent impedance of the p-i-n diode and Z_o is the system reference impedance. Therefore, fraction of power that is reflected from the diode to source is given by

$$p_r = |S_{11}|^2 = \left| \frac{-Z_o}{2Z_T + Z_o} \right|^2 \tag{15.2}$$

And the fractional power transmitted to the output is given by

$$p_t = |S_{21}|^2 = \left|\frac{2Z_T}{2Z_T + Z_o}\right|^2 \tag{15.3}$$

Fractional power absorbed, i.e. dissipated within the diode impedance is given by

$$p_d = 1 - \left(|S_{11}|^2 + |S_{21}|^2\right) = 1 - (p_r + p_t) \tag{15.4}$$

Therefore, in the limiter circuit, some of the incident power is reflected back to the source, some is passed to the load, i.e. at the output load and some is absorbed and dissipated in the diode resistor as heat. The part of the incident power dissipated in the shunt device is a function of its impedance Z_T and is governed by (15.4).

From the equivalent circuit of the p-i-n diode as shown in Fig. 15.3, its impedance is given by

$$Z_T = (R_i||C_i) + R_s \tag{15.5}$$

Neglecting the parasitic components, i.e. series resistor R_s and shunt capacitor C_j, the p-i-n diode impedance can be considered as pure resistance. Different fractional power components are shown in Fig. 15.4. From the figure, it is clear that at a diode resistance value of 25 Ω in a 50 Ω system, the fractional dissipated power reaches a maximum of 50% of the incident power. This 25 Ω diode resistance corresponds to the limiter attenuation ($|S_{21}|$) of 6 dB and return loss ($|S_{11}|$) of 6 dB. Thus, a slow limiter which turns ON with some delay, may pass through this maximum RF power dissipation of 3-dB point after sufficient exposure of high RF power level. Accordingly, power rating of the limiting diode needs to be selected.

Typical characteristics of a shunt-connected p-i-n diode limiter circuit are shown in Fig. 15.5. All the plots are at steady-state condition and power levels are in dB scale with reference to the threshold input and output power levels. It may be noted that over the limiting region, the loss of the limiter increases and the return loss decreases. For the p-i-n diode limiter, the output power increases again with the increase of input signal levels above the maximum input power limiting level. The increase of output power beyond the maximum limiting point is due to the inability to provide further attenuation by the limiter circuit with the increase of RF power level.

Figure 15.6 shows different power components of the limiter circuit at steady-state condition over the input RF power range. It shows that the returned power as well as absorbed power increases with the increase of input power to the limiter. It is clear that though the fractional dissipated power becomes maximum at the 6 dB attenuation point; however, in steady-state condition, the power dissipated in the diode increases with the increase of input power level.

From the above characteristic, it is also clear that the return loss of the single shunt-connected p-i-n diode limiter decreases with the decrease of diode resistance, that is, with the increase of input RF power level. This is undesirable for any RF/microwave

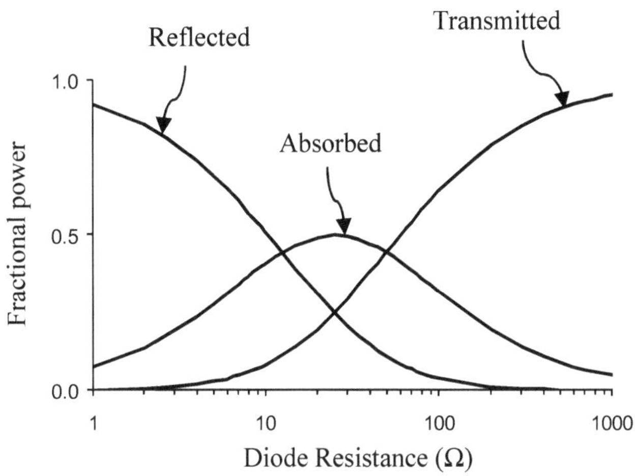

Fig. 15.4 Fractional power limiter with variation of diode resistance

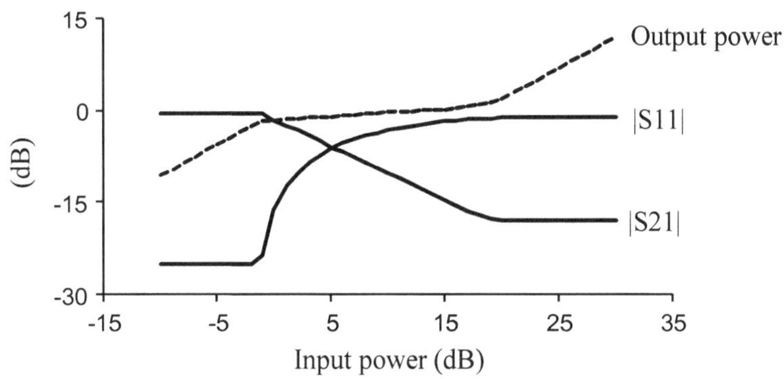

Fig. 15.5 Characteristic of shunt-connected p-i-n diode limiter with input power variation

circuits which affect the matching of cascaded circuits and systems with it. To achieve better input and output return losses, more number of p-i-n diodes are used with separation of $\lambda/4$ transmission line length which cancel the reflected power. Schematic circuit diagram of a limiter with two shunt-connected p-i-n diodes is shown in Fig. 15.7.

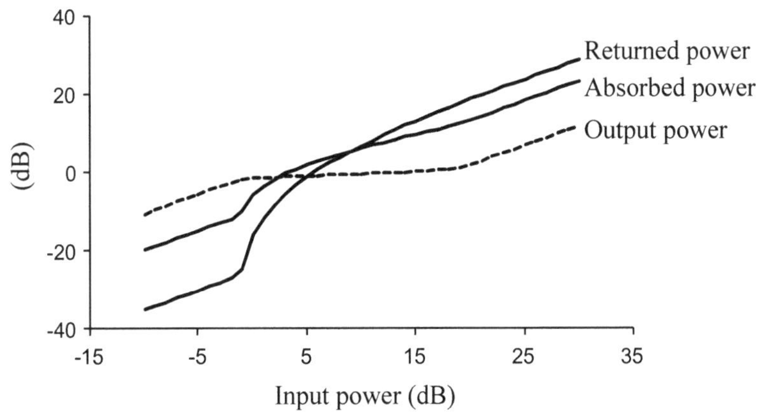

Fig. 15.6 Characteristic of shunt-connected p-i-n diode limiter with input power variation

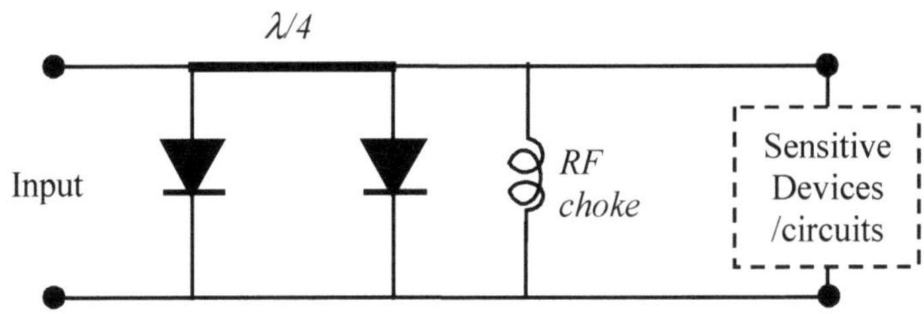

Fig. 15.7 Limiter circuit diagram with two shunt-connected p-i-n diodes

Example 15.1 Calculate maximum possible dissipated power within a p-i-n diode in a shunt-connected p-i-n diode limiter circuit operating at frequency of 4 GHz. Consider, the parasitic series resistance and capacitance of the diode are 5 Ω and 0.01 pF, respectively.

Solution

Figure 15.4 shows that maximum fraction of input power is dissipated through the diode when the diode resistance becomes 25 Ω. At the operating frequency of 4 GHz, the capacitive reactance is

$$\frac{1}{(j2\pi \times 4 \times 0.00001)} = -j39789\ \Omega$$

Thus, reactive part of the diode can be ignored compared to its resistive part. During its limiting action, when the diode passes to its net RF resistance of 25 Ω i.e. including its parasitic resistance (5 Ω), it will experience maximum power dissipation.

Fraction of input power that is reflected from the diode to the source is given by (15.2)

$$p_r = |S_{11}|^2 = \left|\frac{-Z_o}{2Z_T + Z_o}\right|^2$$
$$= \left|\frac{-50}{50+50}\right|^2 = 0.25$$

Therefore, reflected power = 10 W × 0.25 = 2.5 W
And the fractional power transmitted to the output is given by (15.3)

$$p_t = |S_{21}|^2 = \left|\frac{2Z_T}{2Z_T + Z_o}\right|^2$$
$$= \left|\frac{50}{50+50}\right|^2 = 0.25$$

Therefore, transmitted power = 10 W × 0.25 = 2.5 W
Thus, the maximum power dissipated within the p-i-n diode

= input power − (reflected power + transmitted power)
= 10 W − (2.5 W + 2.5 W)
= 5 W

15.4 Schottky Diode Limiters

At lower frequencies, p-i-n diodes are better choice for limiter circuits. Because of the absence of minority carrier storage effect, Schottky barrier diodes can rectify microwave and mmwave signals more effectively. Therefore, for better performance and suitability of realization in MMIC technology, Schottky diodes are preferable at microwave and mmwave frequency range as limiter [12].

Though GaAs-based Schottky diode limiters are suitable for low-power operation, SiC-based Schottky barrier diodes are capable to operate over higher power level. Practically, multistage limiters of different threshold power levels using Schottky and p-i-n diodes are used for high-power-limiting applications.

Under small-signal operating condition, RF resistance of Schottky barrier diodes can be written as

$$R_d(I_d) = 1/\frac{\partial I_d}{\partial V_d} = \frac{\eta k T}{q}\frac{1}{(I_d - I_o)} \tag{15.6}$$

where I_o is reverse saturation current, I_d is forward current, η is ideality factor, k is Boltzmann constant and q is electron charge.

The expression (15.6) of Schottky diodes RF resistance is valid only for small-signal levels. Operation of limiter in its limiting power region cannot be considered

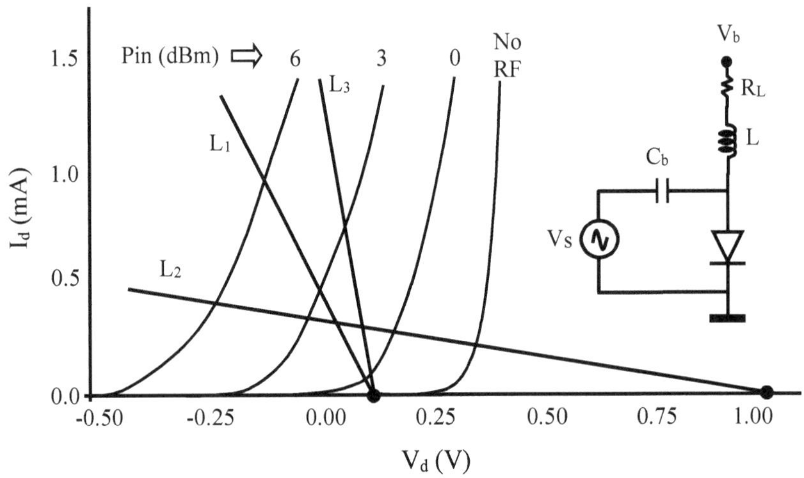

Fig. 15.8 I–V characteristic of Schottky diode in presence of RF signal

as small-signal operation. I–V characteristic of Schottky diodes deviate from the exponential relationship at large signal operating condition, as shown in Fig. 15.8. Typical I–V characteristic of a Schottky diode in presence of different RF power levels at microwave frequency is shown in Fig. 15.8. It is clear that with the increase of RF power level, the I–V curves deviate more and more from their exponential characteristics. In presence of RF signal, the deviation from the exponential characteristic is due to the self-biasing effect caused by rectification of RF signal. The load line L_1 is for low supply voltage and low-bias resistance, whereas the load line L_2 is for high supply voltage and high-bias resistance. From the plot, it is clear that RF resistance of the diode decreases with the increase of RF power level for the load line L_1, whereas for the load line L_2, RF resistance of the diode increases with the increase of RF power level. For the shunt-connected diode-based limiter circuits as shown in Figs. 15.3 and 15.7, RF resistance of the diode should decrease with the increase of RF power level. Thus, for these shunt diode limiter circuits, the diode should be biased through a very low series resistance with zero or very low voltage bias, i.e. bias condition similar to load line L_1.

To achieve good impedance matching of both the ports of a limiter over its entire input power range, diodes with balanced configuration are used as shown in Fig. 15.9. Here, for the limiter circuit, a 3-dB 90° coupler is used. The output port of the limiter circuit is the isolated port of the coupler. Two Schottky barrier diodes are connected to coupled and direct port of the hybrid. Other remaining port is used as input of the limiter. The control voltage V_c is applied through a very low value of bias resistor R_b (about 0–10 Ω) connected to the bias point of the RF circuit using high-impedance short-circuited $\lambda/4$ transmission lines.

The Schottky diode's RF equivalent impedance, as shown in Fig. 15.9, is a combination of RF resistance (R_d), parasitic capacitance (C_d) and parasitic series resistance

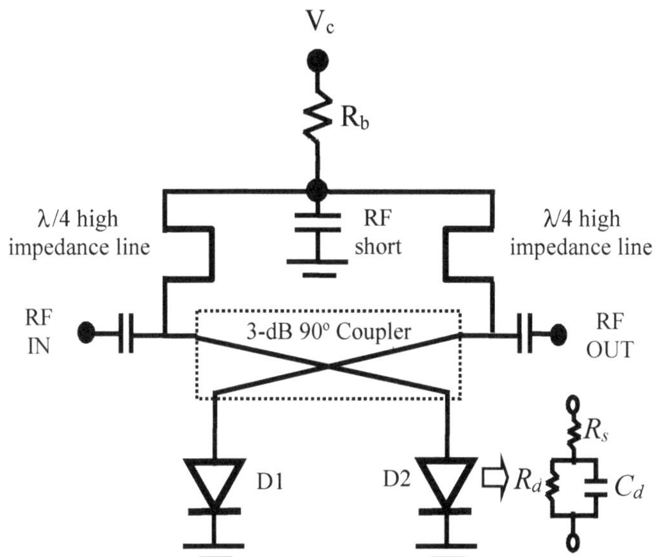

Fig. 15.9 Limiter circuit diagram based on balanced diode configuration

R_s. RF resistance (R_d) of the diode is determined by bias current and the RF power level at the diodes. Depending on its value, some portion of the RF power will be reflected back from the diode and rest of the power will be dissipated within the diode.

At the output port of the circuit, the reflected power from both the diodes will be combined. Considering same characteristic for both the diodes (batch-matched diodes), there will be no reflected power at the input port of the circuit. Thus, both the ports of the limiter circuit will remain in matched condition at any value of RF impedance of the diodes. Thus, there will be no reflected power to the input port.

A limiter circuit is realized on a 25-mil-thick alumina ($\varepsilon_r = 9.9$) substrate. Photograph of the realized circuit is shown in Fig. 15.10. For the 3-dB power divider, Lange coupler is used at the centre frequency of 11.5 GHz. Beam-lead Schottky barrier diodes are used for its low parasitic capacitances. Accordingly the circuit layout is made. Beam-lead Schottky diode of medium-barrier height is used for the circuit for its lower values of the parasitic resistor (<5 Ω) and capacitor (<0.05 pF).

Adjustable threshold power level is required for protecting very sensitive devices like HBTs and pHEMTs without affecting their performance over nominal operating range. To achieve this, adjustable bias voltage to the diodes is required. Behaviour of the limiter circuit for different supply voltages is shown in Fig. 15.11. The attenuation of the limiter increases with the increase of input power levels when the supply voltage is lower than certain value (here, it is 0.4 V). For higher supply voltages, the attenuation of the limiter decreases with the increase of RF power levels. The attenuation of the limiter should increase with the increase of RF power level for

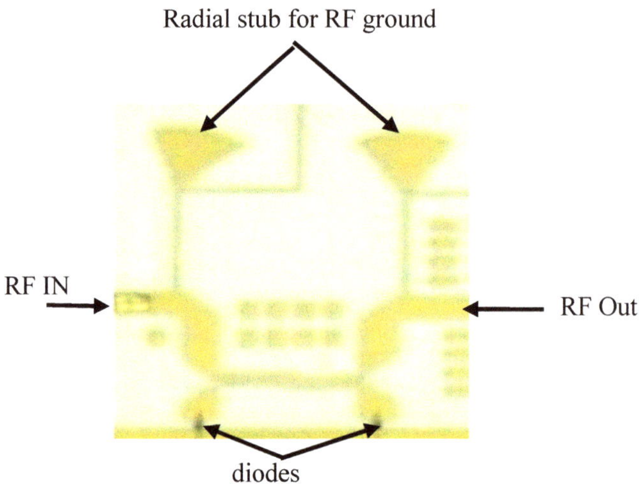

Fig. 15.10 Photograph of the limiter based on beam-lead diodes

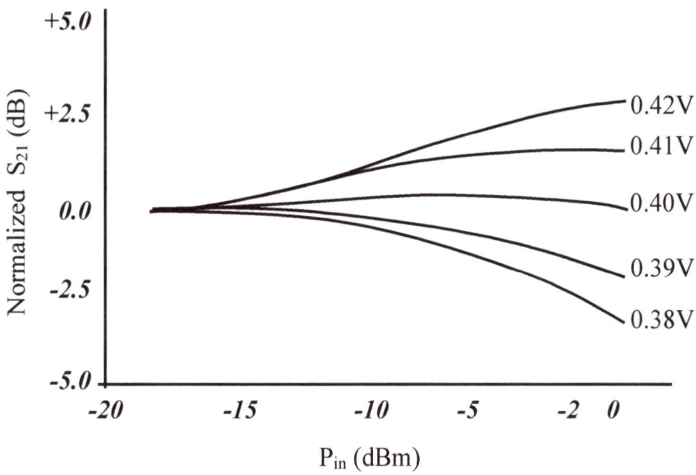

Fig. 15.11 Supply voltage dependency of limiter characteristics

limiting function. Thus, the adjustable supply voltage of the limiter diodes should be of lower value.

Figure 15.12 shows I–O characteristic (output power variation with input power) of the limiter for different values of bias resistor (R_b) at zero control voltage (V_c). Over the small-signal condition, the insertion loss of the limiter is nearly 0.9 dB for all the bias resistors. Over the limiting region of the limiter, the slope of the output power with the change of input power level depends on the value of bias resistor R_b. For higher values of bias resistors (e.g. 15 and 20 Ω), I–O characteristics

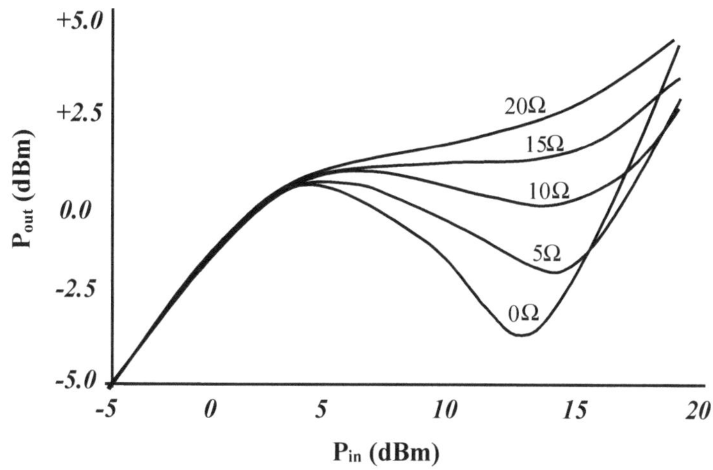

Fig. 15.12 Bias resistor dependency of Schottky diode limiter characteristic

have positive slope, whereas for zero and low values of bias resistor, initially the output power decreases then increases with the increase of input power level. This behaviour is due to the more self-biasing effect of the Schottky diodes at lower values of bias resistor compared to higher resistor values. From the plots, it is clear that the rate of change of attenuation is a function of input RF power level and bias resistor (R_b). From the bias load lines L_1 and L_3 of Fig. 15.8 this can be explained. With increase of RF power level, RF resistance of the diodes drop sharply in case of L_3, corresponding to lower value of bias resistor, than in case of L_1, corresponding to higher value of bias resistor. From the plot, it is also clear that there is an optimum bias resistor to keep output power nearly constant over the limiting power level. In this case, it is about 10 Ω.

For different values of control voltages (V_c), the input–output characteristic of the limiter at bias resistor of 10 Ω is shown in Fig. 15.13. The plot shows that by changing the control voltage, V_c, from −0.2 to +0.2 V, the threshold power level changes maintaining its linearity below the threshold power level. However, for higher positive control voltage, as shown in the figure for $V_c = +0.3$ V, the I–O characteristic shows nonlinearity and also increase of insertion loss. The nonlinearity below the threshold power level, i.e. above the normal operating power region is undesirable for any communication system.

With the control voltage from $V_c = -0.2$ to +0.2 V, the adjusted threshold power levels are from −10 to +5 dBm. The limiting power range is about 15 dB for each control voltage. It is also shown that beyond the range of limiting power level, output power level increases with the increase of input power. This is due to the decrease of diode impedance below 50 Ω.

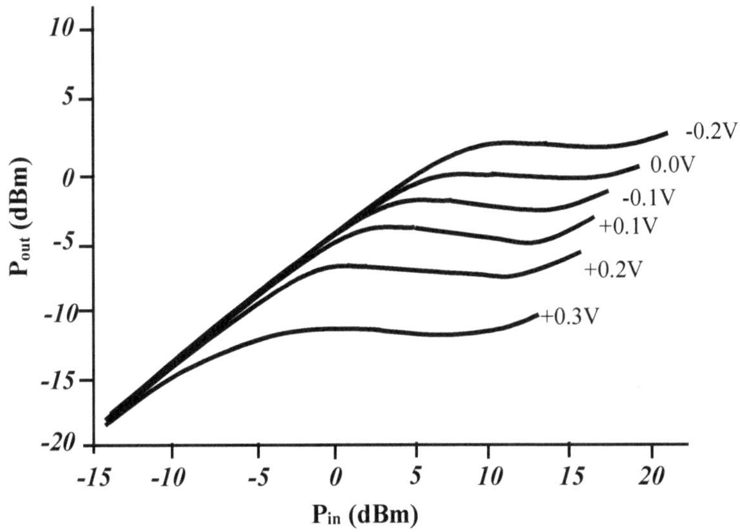

Fig. 15.13 Threshold power level adjustment by bias voltage

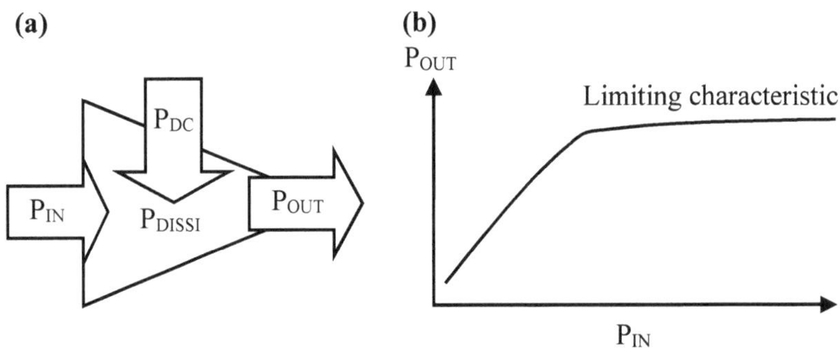

Fig. 15.14 Output power saturation of microwave amplifier

15.5 Amplifier-Based Limiter

Microwave amplifiers maintain linear characteristic over their small-signal operating region. However, with the increase in input power level beyond linear operating region, gain of the amplifiers starts to decrease and ultimately output power saturates. Suppose P_{IN}, P_{OUT}, P_{DC} and P_{DISSI} are the input RF power, output RF power, DC power and power dissipated within the amplifier, respectively, as shown in Fig. 15.14a. The relationship among these powers can be written as

$$P_{DC} + P_{IN} = P_{OUT} + P_{DISSI} \tag{15.7}$$

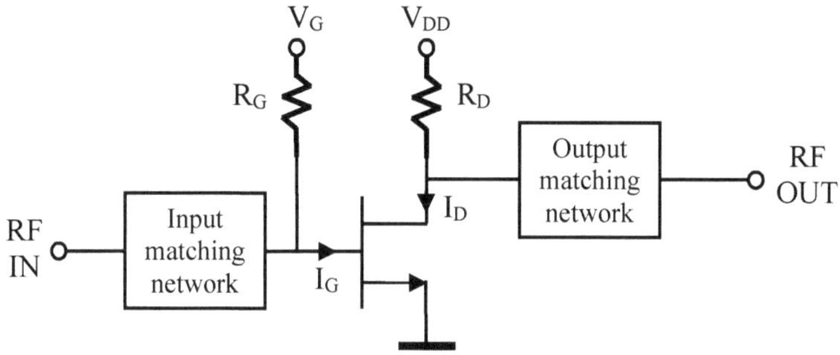

Fig. 15.15 FET amplifier with high value gate and drain resistors

or

$$P_{\text{DISSI}} = P_{\text{DC}} - P_{\text{IN}}(G - 1) \tag{15.8}$$

Here, G is the gain of amplifier. Suppose, DC power (P_{DC}) supplied to an amplifier is fixed and the input RF power level to the amplifier goes on increasing. The equation (15.8) suggest that beyond certain input power (P_{IN}) level, the power dissipated (P_{DISSI}) within the amplifier becomes negative if we assume gain of the amplifier (G) remains constant. However, dissipated power cannot be negative (implies generation of power), thus the gain of the amplifier must decrease with the increase of input RF power level beyond certain limit. Thus, due to the limited DC power supplied to an amplifier, which is governed by capability of the amplifying device, gain of an amplifier decreases and acts as limiter as shown in Fig. 15.14b.

Under high RF power overdrive condition, excessive gate current may flow due to the rectification effect at the gate-to-channel Schottky barrier junction of GaAs FETs which reduces device life. GaAs FET-based amplifiers are designed with high value of gate resistor (R_G) to limit the gate current (I_G) under high power level overdrive condition and used as limiter circuit as shown in Fig. 15.15. In some cases, moderate value of drain resistor (R_D) is also used to control the drain current (I_D) as well as voltage across the drain-to-source terminal under overdrive condition. GaAs FETs are used for low power amplifiers operating at microwave frequency bands to develop receivers, solid-state power amplifiers (SSPAs) and other subsystems. In addition to function as amplifier in its nominal operating power level, the low power amplifying stages can also function as limiter and protect the following high-power-sensitive devices during overdrive.

15.6 Closed-Loop Limiter

Very precise restriction of RF power level is required for highly sensitive microwave circuits and systems such as pHEMT-based receiver and power amplifiers to protect them from RF overdrive without affecting their RF performance over nominal operating power level. In this case, open-loop limiters based on p-i-n diode or Schottky diode cannot serve the purpose for their soft-limiting characteristics. A hard limiter, i.e. a limiter with very low input power difference from its highest linear operating RF power level to threshold power level is required to protect such sensitive subsystems. To set the threshold power level very precisely, the hard limiter should also have a provision of adjustable threshold power level. Schematic circuit diagram of such a closed-loop limiter configuration is shown in Fig. 15.16. An RF power sampler is used to sample the RF power level for detection. The detected voltage is amplified and used as control signal to control the attenuation of a voltage-controlled attenuator for controlling the signal level (Fig. 15.17).

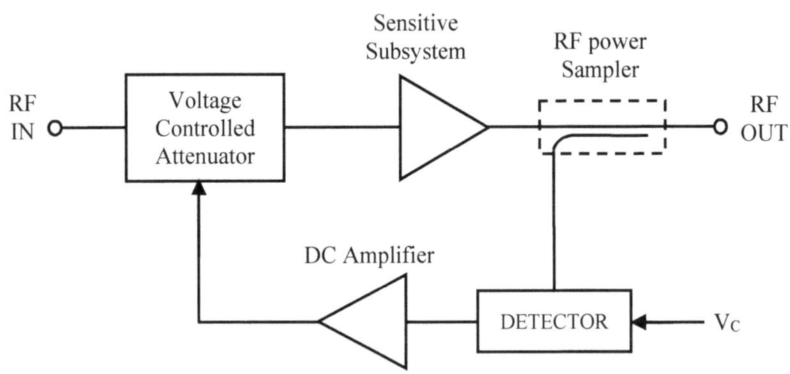

Fig. 15.16 Schematic diagram of closed-loop overdrive level control

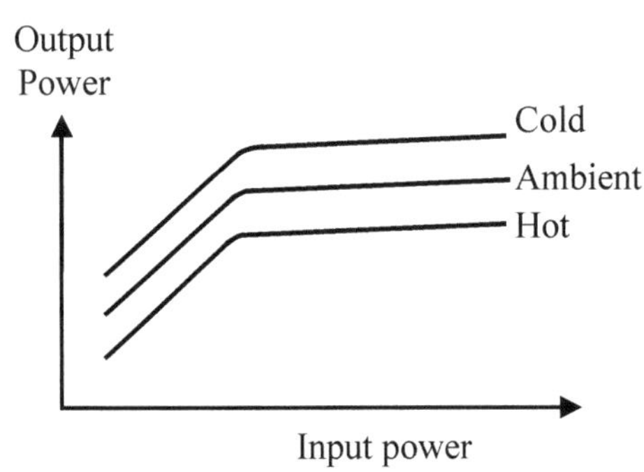

Fig. 15.17 Temperature behaviour of OLC system

The closed-loop overdrive level control scheme is capable of protecting all the sensitive subsystems which are connected after the power level control point [13, 14]. In Fig. 15.6, all the devices connected after the voltage-controlled attenuator circuit is protected under overdrive condition. Thus, any level of high power output can be controlled using this scheme just by changing the coupling factor of the coupler. Mainly, Schottky barrier diode is used as RF power detector at microwave frequency range. To adjust the limiting power level precisely, i.e. to adjust the threshold power level of the limiter, a control voltage (V_c) is applied to the detector diode. A DC amplifier is used to amplify the detected voltage and fed to a voltage-controlled attenuator. The voltage-controlled attenuator may be realized using p-i-n diodes or FETs. The limiting range of the limiter is determined by the attenuation range of the voltage-controlled attenuator. Single- or multistage analog voltage-variable attenuators are used depending on the required range of overdrive level protection.

15.6.1 Temperature Behaviour of OLC System

Attenuation of voltage-variable attenuators based on p-i-n diodes or FETs changes with change of the device's temperature. Gain of the sensitive subsystems used within a closed-loop system is also temperature dependent. Therefore, input power level of the detector changes with the change of temperature. Moreover, detected DC voltage not only depends on the RF power level but it is also very sensitive to the temperature of the detector diode itself. The gain requirement of the DC amplifier is very high, in case of closed-loop hard limiter. Thus, a small change of the detector diode voltage significantly changes the control voltage to the voltage-controlled attenuator. Typical I–O characteristic of an OLC system with the change of operating temperature is shown in Fig. 15.7.

The change of input power level due to the attenuation variation of the voltage-controlled attenuator, gain variation of the sensitive subsystem and DC amplifier are inherently compensated by the inherent property of the closed-loop negative feedback system. The negative feedback system automatically changes the output of the detector to keep its input power level constant by controlling the attenuation of the voltage-controlled attenuator. Thus, it is important to achieve constant/unchanged I–O characteristic of the detector circuit in terms of its detected voltage versus its input power level over the operating temperature range. The unchanged I–O characteristic of the detector over the operating temperature range, automatically adjusts the attenuation of the voltage-controlled attenuator by detecting change of the detector's input power level by any means and generating the required control voltage.

15.6.2 Temperature Compensation of OLC System

To compensate the variation of limiting power level over the change of operating temperature, a temperature-controlled bias voltage (V_c) is applied to the detector diode as shown in Fig. 15.18. The detector diode is biased using a temperature-dependent control voltage V_c, and a bias resistor R_c, through a bias network that consists of an RF decoupling network which in turn consists of a high-impedance $\lambda_o/4$ transmission line and a capacitor C_b.

The detected voltage is filtered by a low-pass filter (LPF) and is fed to one input of the differential DC amplifier. The control voltage V_c is connected to the other input of the differential amplifier. The output voltage of the differential amplifier controls the voltage-controlled attenuator (VCA). The threshold power level of the differential amplifier is determined by the diode characteristic and controlled voltage V_c. When the RF power level exceeds the threshold power level, the differential amplifier provides controlled bias voltage to the VCA. This controlled bias voltage controls the attenuation of the attenuator to limit the RF power to a predetermined limiting power level. To compensate the threshold power variation with the operating temperature, a temperature-dependent voltage drop across a p–n junction diode D_1 and associated resistive networks and supply voltages are used to generate the temperature-dependent control voltage V_c. Here, the diode D_1 is used as temperature sensor. The op-amp-based DC amplifier is used to achieve the required sensitivity of the controlled voltage over the operating temperature range. Both positive and

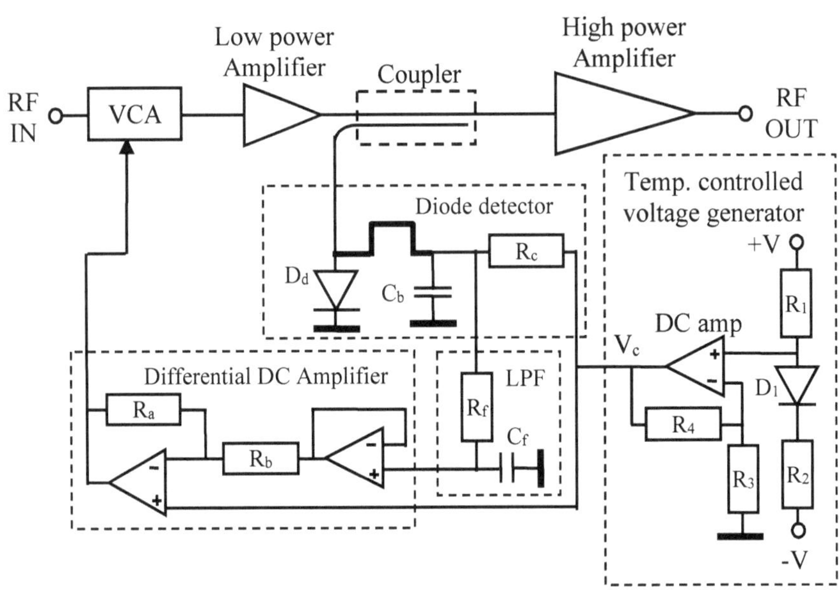

Fig. 15.18 Temperature-compensated OLC system for high power amplifiers

negative supply voltage sources are used to bias the p–n junction diode to get the control voltage with both polarities in case it is required to compensate the limiting power level over temperature.

15.7 Applications

Limiters are used to protect sensitive devices used in receivers, solid-state power amplifiers and other microwave subsystems from intentional or unintentional exposure to high power levels. Such a scheme is shown in Fig. 15.19. Here, a limiter is used before a predistortion linearizer cascaded with a travelling wave tube amplifier (TWTA). In this case, the limiter circuit not only protects the TWTA from RF overdrive condition but it also improves the linearity of the system beyond and near saturation of the TWTA by preventing the output power drop beyond its saturation.

Limiters protect the subsystems from severe EMI conditions. Limiters are also used in narrow sub-band channels to avoid communication disruption of other sub-

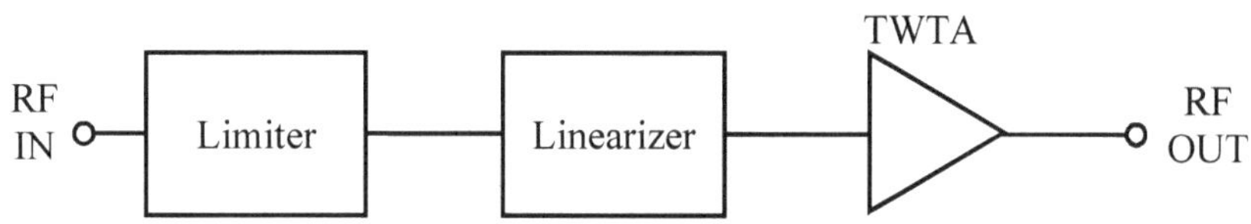

Fig. 15.19 Schematic diagram of an application of limiter to protect linearized TWTA

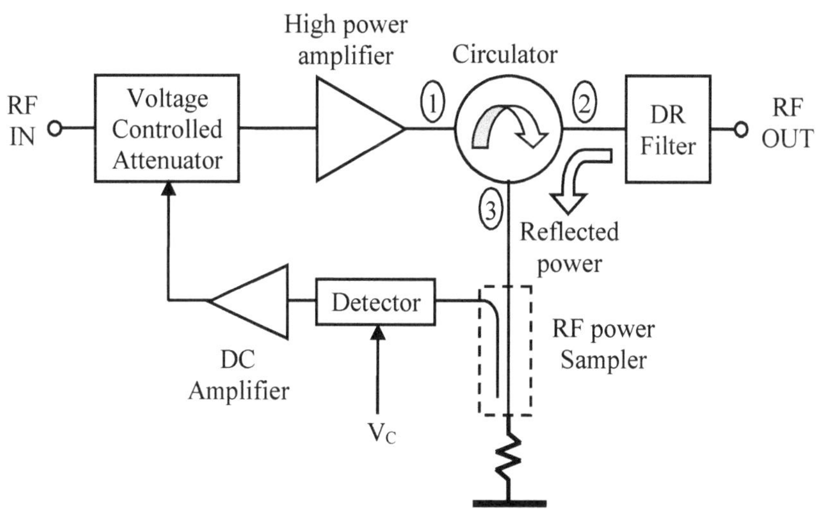

Fig. 15.20 Application of closed-loop OLC to protect DR filter from high power dissipation under OOB carrier

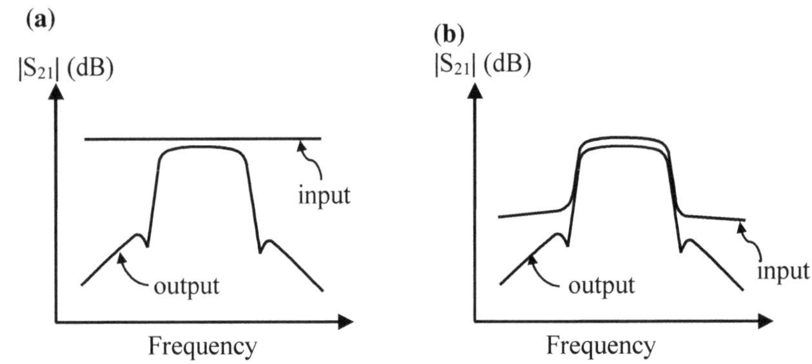

Fig. 15.21 Input and output of the filter **a** without OLC and **b** with OLC

bands within the same channel. In some applications, both Schottky and p-i-n diode based limiters are used to provide better protection than a p-i-n diode-only limiter due to its lower limiting threshold [15]. However, it has more insertion loss compared to either only p-i-n diode or only Schottky diode-based limiters.

15.7.1 OLC for Out-of-Band Carrier Mitigation in Filters

Power handling capability of dielectric-resonator-based filter is limited by power dissipation when it is exposed to out-of-band (OOB) carriers. It becomes more severe in case of narrowband high-power filters when realized based on dielectric resonator (DR). Dielectric-resonator-based high-power filters are preferred option for satellite transponders due to its compact size and less mass along with stable RF performance over temperature. However, average power handling is limited due to the excessive power dissipation at OOB frequency, particularly in the frequency range where the group delay is maximum. A closed-loop limiter system is shown in Fig. 15.20 to mitigate the OOB carrier heating problem. The DR filter is used in a satellite transponder as output filter and connected at the output of high power amplifier such as TWTA. Poor return loss of the DR filter over its OOB frequency region is used to detect and limit the RF power level. A circulator is used between the TWTA and the filter to get the reflected power from the DR filter at port-3. The out-of-band frequency return loss of the filter is poorer than its in-band frequency return loss. Thus, suitably setting the threshold limiting power level by proving control bias to the detector, the presence of OOB carrier beyond certain level can be detected. The higher in-band return loss of the filter compared to its out-of band return loss ensures very less in-band power return compared to OOB carrier power. Thus, in absence of power corresponds to OOB frequency, closed-loop system will not provide any attenuation to the signal. A typical response of filter with and without closed-loop limiter circuit is shown in Fig. 15.21. It shows that power level at the input of the filter for out-of-band frequency range becomes lower than the in-band frequency power level due to the OLC function.

References

1. Leenov D (1965) The silicon PIN diode as a microwave radar protector at megawatt levels. IEEE Trans Electron Dev 12:53–61
2. Smith DG, Heston DD, Heston J, Heimer B, Decker K (1979) Designing high-power limiter circuit with GaAs PIN diodes. In: IEEE MTT-S international microwave symposium digest, pp 329–331, Apr 1979
3. Alpha Application Note 80300 (1985) Characteristics of semiconductor limiter diodes. Specifications, application notes, packages, Alpha Industries Inc., Semiconductor Division, pp 3–45
4. Tan RJ, Ward AL, Garver RV, Brisker H (1988) PIN diode limiter spike leakage, recovery time and damage. In: IEEE international microwave symposium digest, vol 1, pp 275–278, May 1988
5. Tan RJ, Ward AL, Kaul R (1989) Transient response of PIN limiter diodes. In: IEEE international microwave symposium digest, vol 3, pp 1303–1306, June 1989
6. Ward AL, Tan RJ, Kaul R (1994) Spike leakage of thin Si PIN limiters. IEEE Trans Microw Theory Tech 42(10):1879–1885
7. Smith DG, Heston DD, Allen DL (1999) Designing high-power limiter circuits with GaAs PIN diodes. In: IEEE international microwave symposium digest, vol 1, pp 329–332, June 1999
8. Erikson J, Rorsman N, Zirath H (2001) Microwave silicon carbide Schottky diodes. Electron Lett 37:250–252
9. Erikson J, Rorsman N, Zirath H (2003) 4H-silicon carbide Schottky barrier diodes for microwave applications. IEEE Trans Microw Theory Tech 51:796–804
10. Sudow M, Andersson K, Billstrom N, Grahn J, Hjelmgren H, Nilsson J, Nilsson PA, Stahl J, Zirath H, Rorsman N (2006) An SiC MESFET-based MMIC process. IEEE Trans Microw Theory Tech 54:4072–4078
11. Coaker BM, Dowthwaite (2007) Planar limiters and receiver protectors. In: Proceedings of the 4th European radar conference, pp 405–408, Oct 2007
12. Bera SC, Basak K, Jain VK, Singh RV, Garg VK (2010) Schottky diode-based microwave limiter with adjustable threshold power level. Microw Optical Technol Lett 52(7):1671–1673, July 2010
13. Miyazaki IK, Ohata T, Ono T, Kamikokura A, Hayashi R, Seino K, Hirose H, Kurebayashi H (1998) L-band 50 W SSPA incorporating overdrive limitation circuit and power saving circuit for MTSAT. In: 17th international communications satellite systems conference and exhibit, Yokohama, Japan, 23–27 Feb 1998
14. Bera SC, Singh RV (2004) A temperature-compensated closed loop overdrive level controller for microwave solid-state power amplifiers. Microw J 47(4):114–122
15. Lim CL (2012) Reduce losses in RF Schottky-PIN limiters circuits. Electron Des

Chapter 16
Microwave Linearizers

Abstract This chapter presents design and analysis of different types of linearizers for microwave power amplifiers. Working principles of feedback-, feedforward- and predistortion-type linearizers are discussed. In addition to series and shunt diode-based predistortion linearizers, complete design details of a diode-based predistortion linearizer in vector modulator configuration with temperature-compensated performance suitable for space use is presented. Different circuit topologies for digital linearizer and analog broadband linearizers using equalizers are also presented.

16.1 Introduction

Linearity of amplifiers plays a major role in design of communication systems. Throughout transmit and receive chain linear amplification of modulated signals is required to accurately decode the signals to extract information. Any amplitude and/or phase distortions of the modulated signal reduce the ability to decode these signals properly. Distortion cannot be dissociated from a signal because it modifies the signal. Distortion also cannot be detected when the signal source is shut down. Amplifier's nonlinearity becomes more severe when more bandwidth efficient higher order modulations such as QPSK, OQPSK, and DQPSK schemes are used. Even higher order modulation codes such as 8APSK, 16APSK, 16QAM, etc. exhibit greatly improved bandwidth efficiency. However, these are more susceptible to amplifier's nonlinearity due to their large non-constant envelope. Signal levels in receiving sections are very low, thus amplifiers in the receiving section operate in linear region. However, problem arises when linear amplification is required in the transmitting section where power amplifiers are used. To operate the power amplifiers in linear region it is required to operate them in power back-off condition which reduces the amplifier's DC to RF efficiency as discussed in Chap. 14. Most of the DC power is consumed in the transmitting power amplifiers. Thus, it is very important to operate the transmitting power amplifiers near to their saturation point to get highest possible efficiency. In other way, it is required to use more expensive higher power amplifier and operate it at back-off condition to achieve required transmitted

S. C. Bera, *Microwave Active Devices and Circuits for Communication*, Lecture Notes in Electrical Engineering 533, https://doi.org/10.1007/978-981-13-3004-9_16

power with required linearity. In addition to the requirement of higher DC power and higher overall system cost, another drawback of the system is the thermal management and reliability of the final power device. With the reduction of efficiency, power dissipation within the amplifying device will be more which demands more efficient cooling system to operate the device within its allowable temperature limit otherwise the device will operate at higher temperature which will lead to reduction of its reliability.

The efficiency of a power amplifier can be improved by operating a Class-A amplifier towards hard saturation, i.e. clipping the signal waveform. This is done by biasing the device at lower quiescent current level such as in case of Class-AB, B and C configuration. When the devices are biased at lower quiescent current, the amplifiers are operated in reduced conduction angle. Class-A amplifiers operating towards hard saturation and the amplifiers with reduced conduction angle that are operating in their nonlinear region generates distortion on modulated signals. Therefore, some linearization technique is required which improves the overall linearity of the system even though the amplifiers in the transmitter section operate in their nonlinear region for achieving higher DC to RF efficiency [1–15].

16.2 Types of Distortion and Amplifier Nonlinearity

Signal distortion is linear or nonlinear transformation of the signal. Linear distortion is the change in signal form due to the variation of gain and time delay over the frequency range of interest. Linear distortion causes different gains and time delays for different frequencies present in the signal which causes change in signal form. Linear distortion in a communication system mainly arises in dynamic circuits such as filters and can result in severe signal spectrum shaping in case of analog transmission chain and inter-symbol interference in digital transmission chain. To minimize the linear distortions, amplitude tilt equalizer and group delay equalizers are used in the transmission chain.

Nonlinear distortion produces modifications of gain and signal phase with the change of signal strength which results in the generation of new frequency components that were not present in the original signal. The generated new components behave as random noise to the desired signal. In addition to the original components, nonlinear distortion produces many other spectral components which is known as spectral regrowth. Thus, linear distortion only modifies the gain and phase of each spectral component, whereas nonlinear distortions modify the spectrum qualitatively by generating new spectral components which act as random noise to the original signal. The spectral regrowth happens not only within the usable bandwidth of the signal but it also falls outside the allocated bandwidth which may violate the regulations about spectrum transmission outside the allocated bandwidth which defines the maximum allowable adjacent channel interference levels. Another effect of the nonlinearity is the shifting of bias point, i.e. quiescent point of the operating device from its original bias point at no RF condition. This effect modifies the DC power

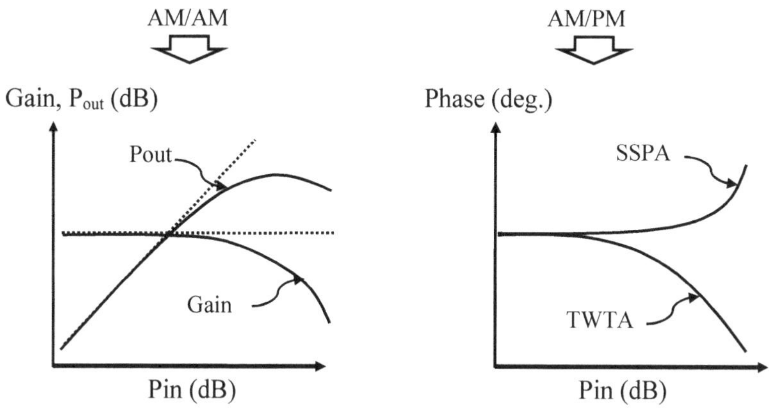

Fig. 16.1 Amplitude and phase characteristic of power amplifier

consumption depending on the signal power strength. Any linear amplifier is not perfectly linear; there are always nonlinear effects which can be modelled as RF signal level dependent equivalent circuit elements. Also, it can be remembered that any amplifier can be considered as linear amplifier up to certain signal level, but the same amplifier will have nonlinear behavior when the signal level crosses certain limit.

Nonlinearities of power amplifiers are characterized by output power or gain against the variation of input power level, called AM/AM conversion, and output phase variation against the input power variation called AM/PM conversion. First is the input power dependent output power or gain and second is the input power dependent output signal phase. Typical amplitude and phase characteristics of solid-state power amplifier (SSPA) and travelling wave tube amplifier (TWTA) are shown in Fig. 16.1. The characteristic of TWTAs is more nonlinear compared to the characteristic of SSPAs. Practically, gain compression (decrease of gain) of TWTA is about 6 dB at its maximum output power, whereas it is about 2 dB (recommended maximum operating gain compression) in case of solid-state power amplifiers. Phase decrease of TWTAs is approximately 45°, whereas it is approximately 20° for SSPAs. Another important point may be noted that, in general, with the increase in input power level, the phase of the output signal decreases for TWTAs, whereas it increases in case of SSPA as shown in Fig. 16.1.

A very common solution to minimize distortion due to power amplifier nonlinearity is backing off their operating region away from the saturation point. The output power is much smaller than its saturated output power when the power amplifier operates in the linear region. This solution not only increases the cost of the system and reduces the dynamic range of an amplifier but also drastically reduces its DC to RF efficiency which is undesirable. Power efficiency is a critical consideration in case of portable- and space-based systems where batteries and solar cells are used. More heat dissipation may also reduce the reliability of the device used in the system.

More elegant solution is the use of linearization techniques to improve overall linearity of the system and operating the power amplifier nearer to its saturation to achieve better DC to RF efficiency. Use of linearizer reduces in-band distortion and as well as adjacent channel interference by reducing the overall nonlinearity of the system. Thus, it is desirable to design a moderately linear PA and employ some linearization technique for further improvement of linearity. The linearization system maximizes the spectral efficiency operating the power amplifier in its near-saturated region. Thus, linearization allows a high power amplifier (HPA), for a given level of distortion, to operate at a higher power level with greater efficiency. At microwave frequencies, the use of a linearizer has an even greater advantage since available HPA power level is more limited and cost is greater than at lower frequencies.

16.3 Types of Linearizers

Different types of linearization techniques are used for microwave power amplifiers in communication systems. These techniques are broadly classified into three groups: (1) feedback, (2) feedforward and (3) predistortion. The feedback and predistortion linearization techniques can be implemented at intermediate frequency (IF), RF or at baseband level, whereas the feedforward linearization technique is realized directly at RF level. Depending on frequency and bandwidth, different linearizers can be realized in digital or in analog domain. Some linearization techniques are combination of two or more different techniques. For example, adaptive baseband predistortion is the combination of both the feedback and predistortion techniques.

Among all the linearization techniques, predistortion technique is the most commonly used technique where a nonlinear circuit module is inserted between input signal and power amplifier to improve overall linearity of the system. Following sections will discuss the different linearization techniques.

16.3.1 Feedback Linearizers

To carry more traffic over a given spectrum, it is required to linearize high power amplifiers used in a transmitter section. This leads to increased use of spectrally efficient modulation schemes. Feedback linearization technique is a widespread and successful application to meet this requirement. A generic feedback scheme for linearization of power amplifiers is shown in Fig. 16.2. Here, input signal v_i is compared with feedback signal to generate error signal. The error signal passes through the loop compensation network of gain $G(s)$ and the power amplifier of gain G_A. Here, the distortion generated within the amplifier is modelled as signal v_d. The output of the amplifier is sampled and feedback for compensation is given through compensation network. The feedback factor including the coupling network

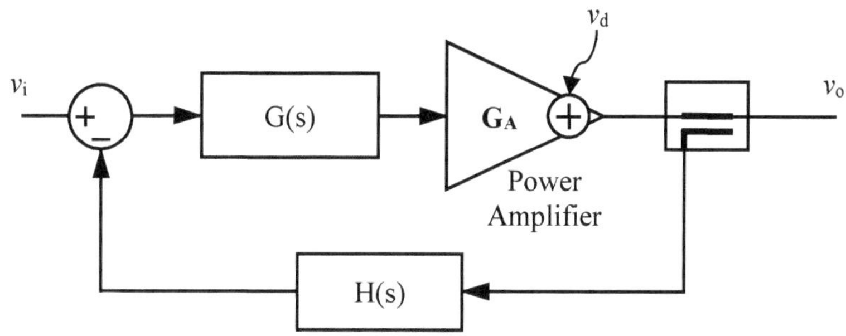

Fig. 16.2 A generic feedback scheme for linearization of power amplifier

is considered as $H(s)$. Due to the feedback action, the error will be minimized by forcing the output v_o to track the input signal.

Here, the output signal can be written as

$$v_o = \frac{G_A \times G}{1 + G_A \times G \times H} v_i + \frac{1}{1 + G_A \times G \times H} v_d \tag{16.1a}$$

Here, G_AGH is the loop gain. In case of large loop gain, where $G_AGH \gg 1$, the output signal can be written as

$$v_o = \frac{1}{H} \times v_i + \frac{1}{G_AGH} \times v_d \tag{16.1b}$$

In case of large loop gain, the second term of the output signal corresponds to the distortion and hence is minimized. The amount of reduction of the distortion signal is by the factor of $1/(1 + G_AGH)$. Increasing the loop gain, the effect of the distortion can be made very small. However, making the loop gain too large can result in the instability of the feedback loop. Another advantage of the feedback linearization technique is that it tracks the input signal (v_i) more accurately by the feedback signal regardless of the forward gain. It is clear from (16.1b) that the overall gain for the input signal is primarily determined by the feedback factor $1/H$. Therefore, realizing H using only stable passive components such as resistors will stabilize the overall gain of the feedback system significantly.

16.3.2 Feedforward Linearizers

Feedforward linearization technique is realized directly at RF level. In the scheme, at first the distortion generated in the main power amplifier is extracted by comparing the output of the power amplifier with its delayed input. Then, this extracted distortion is amplified using an auxiliary power amplifier and destructively combined with the

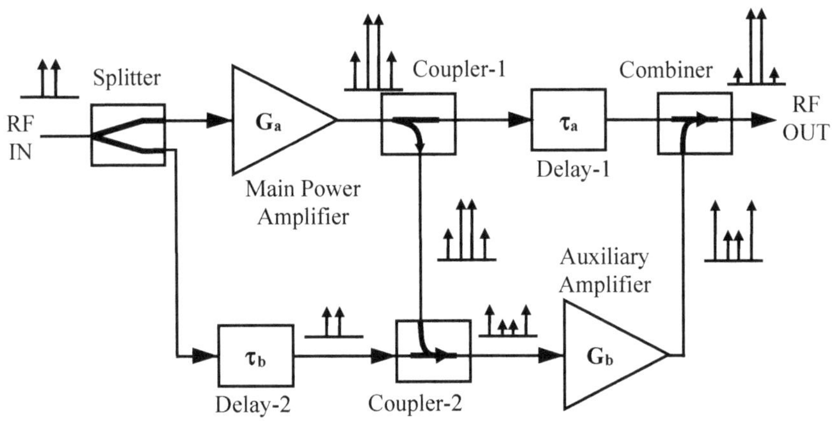

Fig. 16.3 A generic feedforward scheme for linearization of power amplifier

delayed output of the main power amplifier using a power combiner. Thus, ideally, the output of the system will be distortion free.

Schematic diagram of a typical feedforward scheme for linearization of microwave power amplifiers is shown in Fig. 16.3. The working principle of the scheme is demonstrated in the diagram for a two-tone input signal. Two time delay circuits, one each in the main amplifier path and another in the auxiliary amplifier path are used to match delay for achieving proper extraction and subtraction of the distortion. Any deviation from the required amplitude and phase matching will degrade the extraction of distortion component and subsequently distortion cancellation after combination at the output of the system. Amplitude and phase matching becomes critical due to the drift of amplifier characteristics with temperature.

Main drawback of the feedforward linearization technique is the requirement of auxiliary amplifier with output power comparable to the main amplifier to overcome the loss through the output coupler/combiner. The efficiency of the feedforward system is reduced by the power consumption of the auxiliary amplifier which must be operated in its linear region. Another reason for reduction of efficiency of the system is the loss of the output coupler (coupler-1), combiner and delay circuit (delay-1) which are connected at the output of the main power amplifier. However, feedforward linearization technique provides broad bandwidth performance with reasonable improvement in linearity. Also, this technique has the advantage of inherent stability.

16.3.3 Predistortion Linearizers

Among the various types of linearization techniques, predistortion (PD) linearization is the most popular for satellite communication. The popularity of predistortion linearizers is due to its low power consumption, simplicity and ability to linearize over wide bandwidth of power amplifiers operating at near saturation condition. Predistortion linearizers provide relatively wide-band performance and are able to function as stand-alone unit, which makes it preferable for microwave systems. Predistortion linearizer creates inverse nonlinearity of the transmitting amplifiers such as of TWTAs or SSPAs in order to compensate for the distortion. Working principle of the predistortion linearization scheme is shown in Fig. 16.4. A predistortion linearizer creates amplitude and phase distortions of the signal opposite to the distortion of the HPA which is to be linearized. At first, the signal to be amplified by an amplifier is passed through the linearizer to introduce opposite distortion to the signal. When the signal distorted by the linearizer passes through the amplifier, it negates the distortion and produces a near-linear transfer characteristic as shown in Fig. 16.4. Amplitude and phase nonlinearities are generally considered as unwanted. In case of predistorter, it is required to generate the amplitude and phase nonlinearity intentionally with the same amount in opposite direction with that of the power amplifier. Thus, in this technique, a nonlinear transfer function is generated which is complimentary to the amplitude and phase nonlinearities of the power amplifier to be linearized. This is used to predistort the incoming signal such that when the distorted signal passes through the nonlinear amplifier, it is further restored in the opposite way to get its original waveform back. This is done by increasing the gain of the linearizer by the same amount as the amplifier gain decreases with the increase of input power level. Likewise, the phase shift introduced by the linearizer is equal and opposite to that of the amplifier and results in the phase improvement as shown in Fig. 16.4. In general, there is no feedback in the predistortion linearization technique, thus it is an open-loop system. Therefore, the distortion characteristics (amplitude and phase variation with the power level) of the power amplifier to be cancelled must be known in advance.

16.4 Implementation of Predistortion Linearizers

Predistortion linearizers can be implemented at RF as well as intermediate frequency (IF) level. The advantage of IF predistortion is that the same design can be used for different carrier frequencies (RF) by simply varying the local oscillator (LO) frequencies of up- and downconverters. RF predistortion is done immediately before the RF power amplifier. However, in case of IF predistortion, there will be upconverter and amplifying stages in between the RF power amplifier and the predistortion linearizer. Thus, characteristics of the IF predistortion linearizer needs to be designed by considering the nonlinearity as well as gain variation of the in-between subsystems.

A predistorter should have its amplitude and phase characteristics opposite to that of the power amplifier. Therefore, the device for designing the predistortion linearizer must be biased in such a way that it should operate in its nonlinear region at the operating power level. Practically, Schottky barrier diodes, varactor diodes and transistors (FETs and BJTs) are used to generate nonlinear characteristics for designing predistortion linearizers [1–15]. These devices are biased in such a way

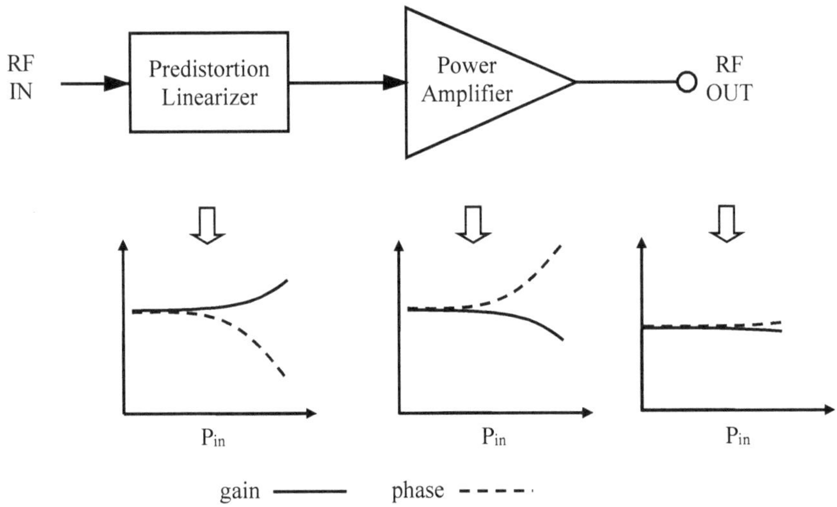

Fig. 16.4 Predistortion scheme for linearization of power amplifier

that some of its parameters vary with the change of its input RF power level. In case of Schottky barrier diodes, RF resistance variation with its input RF power level is used to generate proper amplitude and phase nonlinearities, whereas capacitance variation of a varactor diode with its input RF power level is used for designing a varactor diode-based linearizer. Gain compression and phase shift of a low power transistor-based amplifier can also be used to linearize performance of a high power RF amplifier. However, amplifier in a transistor-based linearizer needs to operate at very high level of compression to achieve required amplitude and phase nonlinearities that may be a concern for high-reliability operation over an extended lifetime. Among all types of predistortion linearizers, diode-based linearizers are most suitable for its compactness, low power consumption and high-reliability operation. For microwave as well as mmwave frequency range linearizers, Schottky diode is most suitable for its high-frequency operation to generate amplitude and phase distortions due to its self-biasing effect.

16.4.1 Schottky Diode Predistortion Linearizers

Diode-based predistortion linearizers are most suitable for many applications including for space use due to its compactness, less power consumption and higher reliability [8–15]. Schottky diodes are preferable for designing microwave/mmwave linearizers for its superior detection capability over these frequency ranges. Dependency of RF resistance of the Schottky diodes on RF power level is used to generate amplitude and phase nonlinearities required for predistortion linearizers.

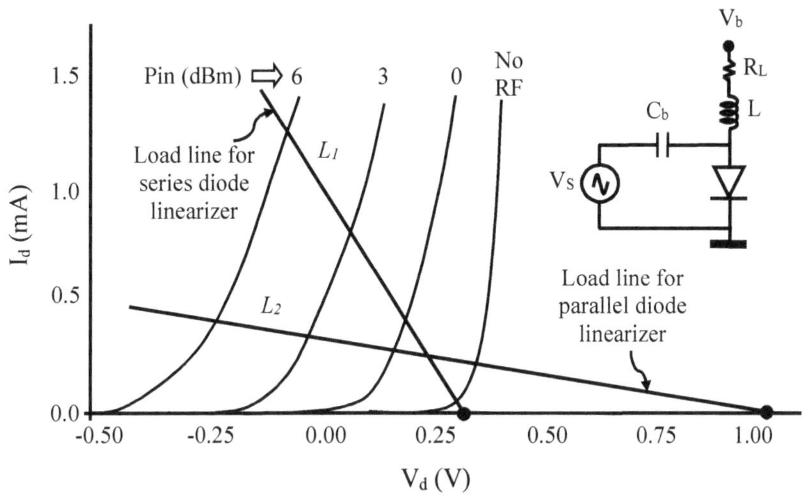

Fig. 16.5 *I–V* characteristic of Schottky diodes in presence of RF power

A typical *I–V* characteristic of Schottky diodes in presence of different RF power levels including schematic circuit diagram is shown in Fig. 16.5. The exponential *I–V* characteristic deviates more and more from its exponential nature with the increase of RF signal level [11–15]. RF resistance of Schottky diodes is given by inverse slope of the *I–V* curve at particular operating condition and is given by

$$R_d(I_d) = 1/\frac{\partial I_d}{\partial V_d} \tag{16.2}$$

Under large signal RF operation, DC voltage and current of the diode are not only the function of RF signal level but are also dependent on bias load line as shown in Fig. 16.5. The two types of bias load lines, L_1 and L_2, as shown in the diagram correspond to nearly constant-voltage-bias and constant-current-bias conditions, respectively. It is clear from the figure that for the bias load line L_1, RF resistance of the diode decreases with the increase of RF signal level, whereas in case of the bias load line L_2, RF resistance of the diode increases with the increase of RF signal level. Thus, RF resistance of the diode may increase or decrease with the increase of RF power level depending on the bias load line. Based on this property of the RF resistance of Schottky diodes, two types (1) series diode and (2) shunt diode predistortion linearizer circuits are designed.

Among all types of predistortion linearizers, series and shunt diode linearizers are simple and least expensive to implement. However, both types of linearizers offer modest improvement in linearity suitable for solid-state power amplifiers. Thus, these types of linearizers are good candidates for mobile applications such as for small handheld radios or cell phones due to their simplicity in circuitry.

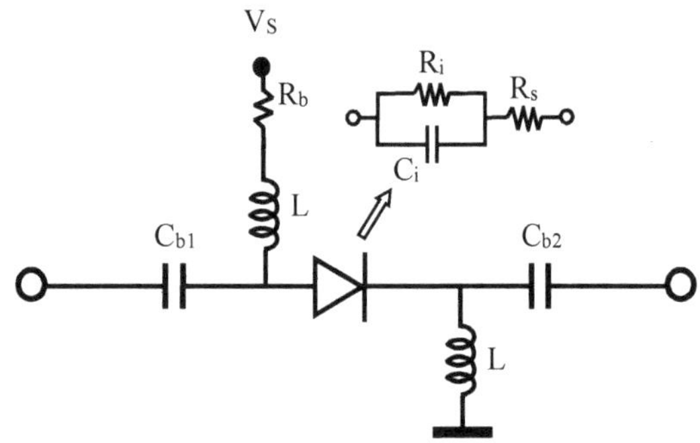

Fig. 16.6 Schematic of a series diode linearizer circuit

16.4.1.1 Series Diode Predistortion Linearizer

Schematic circuit diagram of a series diode linearizer is shown in Fig. 16.6. Here, diode is biased through a very small bias resistance R_b to operate the diode at nearly constant-voltage-bias condition to achieve decrease of RF resistance of the diode with the increase of RF power level. Two RF chokes are used to provide DC current path without affecting the RF performance. Two capacitors C_{b1} and C_{b2} are used for DC blocking. The Schottky diode can be represented by its equivalent circuit as shown in the diagram. Here, diode equivalent capacitance C_i and the series resistance R_s can be considered as constant over the operating RF power level, whereas RF resistance R_i of the diode varies with the RF power level. The equivalent impedance Z_d of the diode can be written as

$$Z_d = R_s + R_i||C_i \tag{16.3}$$

Neglecting the effect of bias networks and blocking capacitors on the RF performance, the S-parameters of the series diode linearizer can be written as from (7.162c)

$$\begin{bmatrix} S_{11} & S_{12} \\ S_{21} & S_{22} \end{bmatrix} = \begin{bmatrix} \frac{Z_d}{2Z_o+Z_d} & \frac{2Z_o}{2Z_o+Z_d} \\ \frac{2Z_o}{2Z_o+Z_d} & \frac{Z_d}{2Z_o+Z_d} \end{bmatrix} \tag{16.4}$$

Therefore, the amplitude and phase of the series diode linearizer can be written as

$$|S_{21}| = \frac{\sqrt{(2Z_o)^2 + (4\pi f Z_o C_i R_i)^2}}{\sqrt{(R_i + R_s + 2Z_o)^2 + [2\pi(R_s + 2Z_o) f C_i R_i]^2}} \tag{16.5a}$$

$$\angle S_{21} = \tan^{-1}\frac{4\pi\ f Z_o C_i R_i}{2Z_o} - \tan^{-1}\frac{2\pi(R_s + 2Z_o) f C_i R_i}{R_i + R_s + 2Z_o} \tag{16.5b}$$

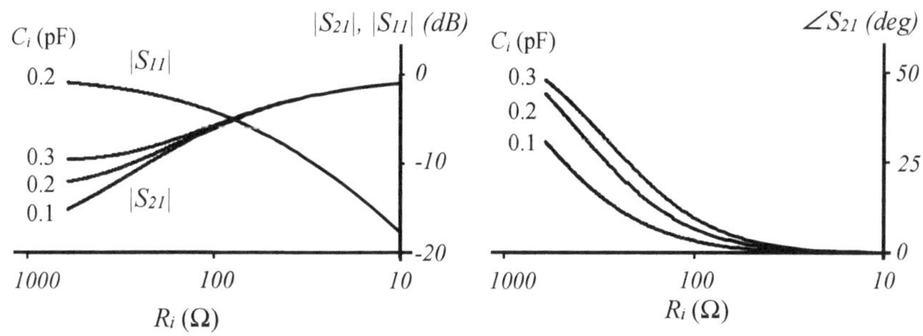

Fig. 16.7 Amplitude and phase characteristics of response of series diode linearizer

The amplitude and phase variation of the series diode predistortion linearizer with the variation of diode RF resistance are plotted in Fig. 16.7 for operating frequency of 2 GHz using (16.5a) and (16.5b), respectively, for different values of capacitance of the diode. It shows that the amplitude of the linearizer increases with the increase of diode resistance as required to linearize amplitude characteristic of a power amplifier. The phase of the linearizer decreases with the decrease of diode resistance which is required to linearize phase characteristic of a power amplifier, particularly for SSPAs. Here, dependency of shape as well as amount of amplitude and phase shift on the capacitance value C_i is also shown. It suggests that different combinations of amplitude and phase characteristics can be achieved to linearize different power amplifiers by selecting diodes of different capacitances. Practically, an external capacitor can be used in parallel to the diode to achieve adjustable amplitude and phase characteristics of the linearizers.

The series diode linearizer, biased nearly constant-voltage-bias condition, generates amplitude expansion and phase decrease characteristic with the increase of RF power level. Though the series diode linearizer is very simple with less power consumption, it has drawback of poor port return losses as shown in Fig. 16.7. Practically, it is desirable to independently adjust the phase and amplitude characteristic of a linearizer to compensate the amplifier's amplitude and phase nonlinearity of different combinations. However, the series diode linearizers are not capable to provide this flexibility.

16.4.1.2 Shunt Diode Predistortion Linearizer

Schematic circuit diagram of a shunt diode linearizer is shown in Fig. 16.8 [9, 10]. Here, diode is biased through a high-value bias resistance R_b to operate the diode at nearly constant-current-bias condition to achieve increase of RF resistance of the Schottky diode with the increase of RF power level. RF choke L is used to provide the bias to the diode without affecting RF performance.

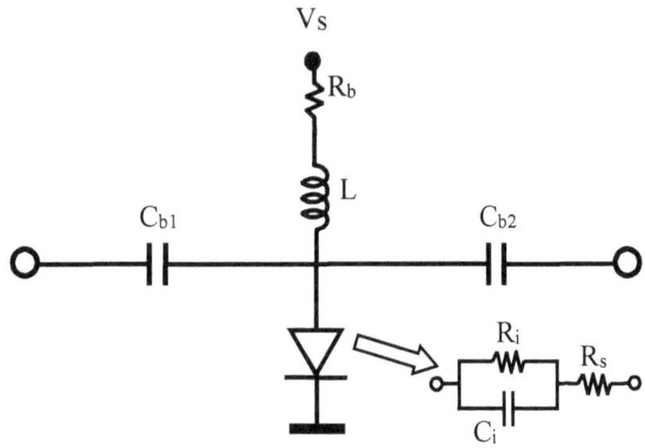

Fig. 16.8 Schematic circuit diagram of a shunt diode predistortion linearizer

Neglecting the effect of bias networks and blocking capacitors on the RF performance, the S-parameters of the shunt diode linearizer can be written as from (7.163d)

$$\begin{bmatrix} S_{11} & S_{12} \\ S_{21} & S_{22} \end{bmatrix} = \begin{bmatrix} \frac{Z_d||Z_o - Z_o}{Z_d||Z_o + Z_o} & \frac{2Z_d||Z_o}{Z_d||Z_o + Z_o} \\ \frac{2Z_d||Z_o}{Z_d||Z_o + Z_o} & \frac{Z_d||Z_o - Z_o}{Z_d||Z_o + Z_o} \end{bmatrix} \tag{16.6a}$$

$$= \begin{bmatrix} \frac{-Z_o}{2Z_d + Z_o} & \frac{2Z_d}{2Z_d + Z_o} \\ \frac{2Z_d}{2Z_d + Z_o} & \frac{-Z_o}{2Z_d + Z_o} \end{bmatrix} \tag{16.6b}$$

Therefore, the amplitude and phase of the shunt diode linearizer can be written as

$$|S_{21}| = \frac{\sqrt{(2R_i + 2R_s)^2 + (4\pi\, fC_i R_i R_s)^2}}{\sqrt{(2R_i + 2R_s + Z_o)^2 + [2\pi(2R_s + Z_o) f C_i R_i]^2}} \tag{16.7a}$$

$$\angle S_{21} = \tan^{-1}\frac{4\pi\, fC_i R_i R_s}{2(R_i + R_s)} - \tan^{-1}\frac{2\pi(2R_s + Z_o) f C_i R_i}{2(R_i + R_s) + Z_o} \tag{16.7b}$$

With the increase of RF resistance of the diode, the amplitude and phase variation of the shunt diode predistortion linearizer is shown in Fig. 16.9 based on (16.7a) and (16.7b) considering the frequency of operation 2 GHz. It shows that the amplitude as well as phase of the linearizer increases with the increase of RF resistance of the diode, i.e. with the increase of RF power level. It also shows that amount of phase expansion increases with the increase of capacitance of the diode.

The shunt diode predistortion linearizers are used in case of amplifiers where gain as well as phase expansions are required such as travelling wave tube amplifiers (TWTAs). These are not suitable for SSPAs where decrease of phase with the increase

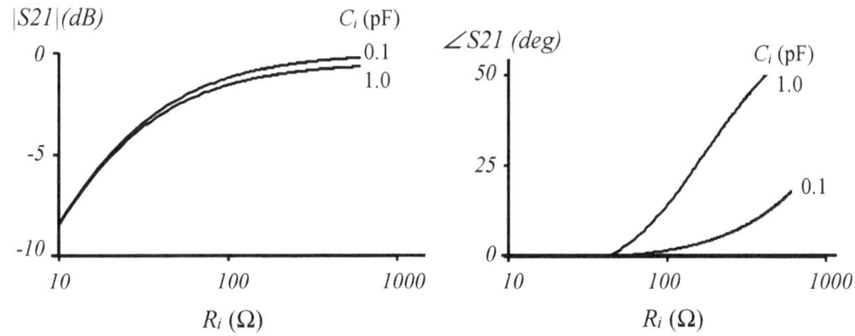

Fig. 16.9 Amplitude and phase characteristics of shunt diode linearizer

of RF power level is required. Moreover, shunt diode linearizers suffer for their poor input and output return losses and provide no flexibility for achieving various combinations of amplitude and phase characteristics.

16.4.2 Linearizer with Vector Modulator Configuration

The series and shunt diode linearizers are the simplest and least expensive to implement. However, both types of linearizers offer only modest improvement in linearity. These types of linearizers are good candidates for mobile applications such as small handheld radios or cell phones due to simplicity in circuitry. For applications, where efficiency and linearity are very critical like TWTAs in space communication, very accurate amplitude and phase compensations are required. To achieve this, flexible linearizer is needed to linearize different power amplifiers encountered in practice with the capability of generating adjustable amplitude and phase characteristics. Linearizer with vector modulator configuration can provide various combinations of adjustable amplitude and phase nonlinearities which are required for practical power amplifier linearization. This configuration is also having very good input and output return losses over the entire operating RF power level which is required for cascading the linearizer with power amplifier at its output and other circuit at its input.

Schematic block level diagram of a predistortion linearizer with vector modulator configuration is shown in Fig. 16.10 [11–16]. The input RF signal to the predistortion linearizer at first is divided into two parts and passes through two arms consists of different circuits. One of the arms consists of phase shifter and attenuator/amplifier operating in their linear region and the other arm consists of an attenuator or amplifier operating in nonlinear region to generate distortion. The signals coming from both the arms are combined by an RF combiner to get final output. Conceptual working principle of the predistortion linearizer in vector modulator configuration is shown in Fig. 16.11. Here, g_l is gain of the linear arm; g_{nl} and g'_{nl} are the gains of the nonlinear arm at small signal and large signal conditions respectively. Whereas, G_L

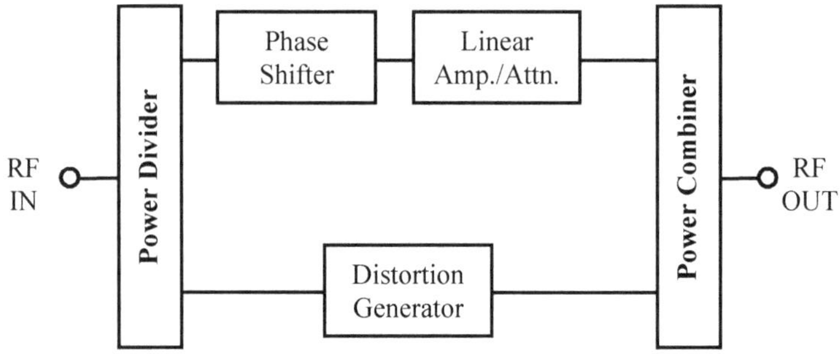

Fig. 16.10 Block schematic of predistortion linearizer with vector modulator configuration

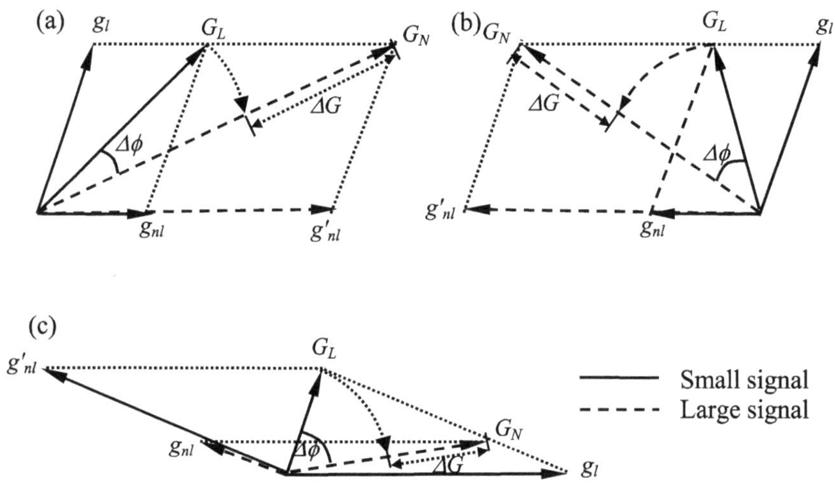

Fig. 16.11 Conceptual working principle of predistortion linearizers based on vector modulator configuration using phasor diagram

and G_N are the resultant gain of the linearizer at small-signal and large-signal conditions respectively. With the increase of RF power level, the gain/loss of the nonlinear circuit may increase or decrease depending on its characteristic. For example, generally gain of transistor based amplifiers decreases with the increase of RF power level, whereas loss of Schottky diode based circuit may increase or decrease with increase of RF power level depending on its bias condition. In Fig. 16.11a, it is shown that the expansion of gai (ΔG) and decrease of phase ($\Delta\phi$) are achieved with the increase of gain of the nonlinear arm with the increase of RF power. Figure 16.11b shows the expansion of gain (ΔG) and increase of phase ($\Delta\phi$) of the output signal due to the increase of gain of the nonlinear arm with the increase of RF power level, whereas Fig. 16.11c shows the expansion of gain and decrease of phase of the output signal due to the gain compression of the nonlinear circuit with the increase of RF power level.

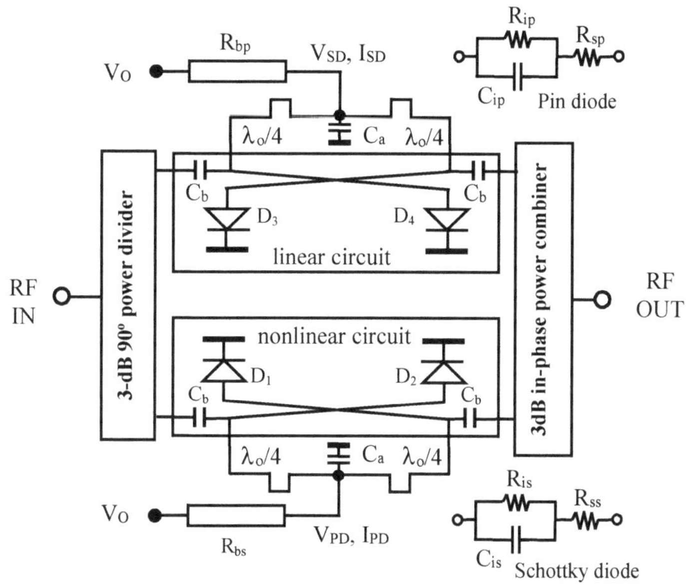

Fig. 16.12 Circuit schematic of diode-based predistortion linearizers in vector modulator configuration

16.4.2.1 Diode Linearizer with Vector Modulator Configuration

Schematic RF circuit diagram of a linearizer based on Schottky and p-i-n diodes in vector modulator configuration is shown in Fig. 16.12. The input RF signal to the linearizer is divided into two equal and quadrature phase parts using a 3-dB 90° power divider. The one part passes through the linear circuit and another through nonlinear circuit and then combined using an in-phase power combiner. The nonlinear arm consists of a 3-dB 90° hybrid coupler in which two Schottky barrier diodes (D_1 and D_2) are connected to its direct and coupled ports, whereas the linear arm consists of another 3-dB 90° hybrid coupler in which two p-i-n diodes (D_3 and D_4) are connected to its direct and coupled ports. Isolated ports of both the couplers act as output of the nonlinear and linear arms. Considering both the diode pairs are matched, inherent property of the 3-dB 90° hybrid structure, always ensures the perfectly matched input and output ports of both the arms for all values of the diode resistance under any bias condition and RF power level. A simple equivalent circuit of p-i-n and Schottky barrier diodes consists of a parallel combination of RF resistance R_i and capacitance C_i with a series-connected parasitic resistance R_s. Therefore, amplitude ρ and phase ϕ of the linear and nonlinear attenuator circuits with 3-dB 90° coupler and two numbers of matched diodes can be expressed as

$$\rho = \left[\frac{(R_i + R_s - Z_o)^2 + (2\pi f C_i R_i (Z_o - R_s))^2}{(R_i + R_s + Z_o)^2 + (2\pi f C_i R_i (Z_o + R_s))^2} \right]^{1/2} \tag{16.8a}$$

$$\phi = -\tan^{-1}\left[\frac{4\pi f C_i R_i^2 Z_o}{(R_i + R_s + Z_o)(R_i + R_s - Z_o) - (2\pi f C_i R_i)^2\left(Z_o^2 - R_s^2\right)}\right] \tag{16.8b}$$

where $\rho = \rho_{sd}$ and $\phi = \phi_{sd}$ for $R_i = R_{is}$ and $C_i = C_{is}$ in case of the nonlinear attenuator with Schottky diodes.

And, $\rho = \rho_{pd}$ and $\phi = \phi_{pd}$ for $R_i = R_{ip}$ and $C_i = C_{ip}$ in case of linear attenuator with p-i-n diodes.

The amplitude $|S_{21}|$ and insertion phase $\angle S_{21}$ of the linearizer is given by

$$|S_{21}|_{LZ} = \left[\left(\frac{1}{2}\rho_{sd}\right)^2 + \left(\frac{1}{2}\rho_{pd}\right)^2 + \frac{1}{2}\rho_{sd}\rho_{pd}\cos\theta\right]^{1/2} \tag{16.9a}$$

$$\angle S_{21_{LZ}} = \tan^{-1}\left[\frac{\rho_{pd}\sin\theta}{\rho_{sd} + \rho_{pd}\cos\theta}\right] \tag{16.9b}$$

Here, θ is the phase angle in between the two signals combined at the output port of the linearizer and is given by

$$\theta = \phi_{sd} - \phi_{pd} - 90 \tag{16.10}$$

Figure 16.13 shows the plot of amplitude and phase of the linearizer with the variation of Schottky diode's resistance (R_{sd}) for different settings of the p-i-n diode's RF resistance (R_{pd}) taking the typical values of C_{pd} = 0.01 pF and C_{sd} = 0.1 pF at the frequency of f = 11.5 GHz. The plot shows that amplitude and phase expansion can be achieved when Schottky diode's resistance increases beyond 50 Ω with the increase of RF power level. Thus, Schottky diodes of the nonlinear arm used to be biased through a high resistance with small-signal current corresponding to its RF resistance more than or equal to 50 Ω when p-i-n diodes are biased corresponding to their RF resistance more than 50 Ω. It is also clear that depending upon the setting of p-i-n diode's resistance and initial biasing of the Schottky diodes, different combinations of amplitude and phase can be obtained. It can be shown that with different combinations of DC biasing of Schottky and p-i-n diodes, various possible combinations of amplitude and phase variations with respect to the RF power level can be obtained. In this regard, the linearizer with vector modulator configuration is very much versatile in nature.

Photograph of a Ku-band linearizer is shown in Fig. 16.14. Here, beam-lead medium-barrier Schottky diodes with a 3-dB 90° Lange coupler is used for nonlinear attenuator and beam-lead p-i-n diodes with another 3-dB 90° Lange coupler is used for linear attenuator of the linearizer. Input power divider is also 3-dB 90° Lange coupler and output power combiner is a Wilkinson power combiner. The entire circuit is realized on a single 0.75 in. × 0. 5 in. alumina substrate of thickness 25 mm. For different DC bias conditions, measured amplitude and phase characteristics of the linearizer are shown in Fig. 16.15. It shows that by the selection of only the DC bias currents of the Schottky diodes and p-i-n diodes used in the nonlinear arm and linear

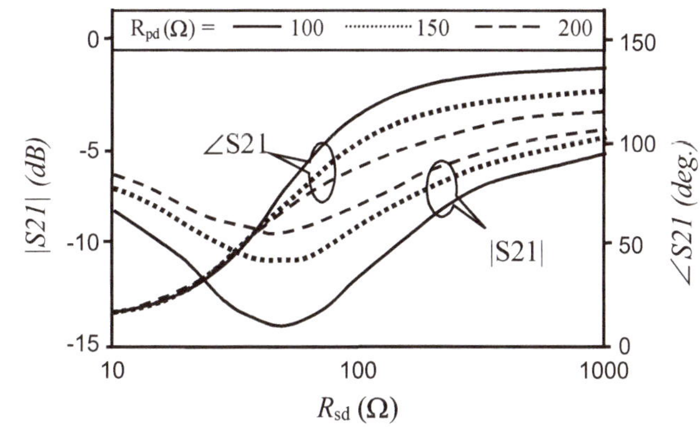

Fig. 16.13 Amplitude and phase characteristics of linearizer of Fig. 16.12

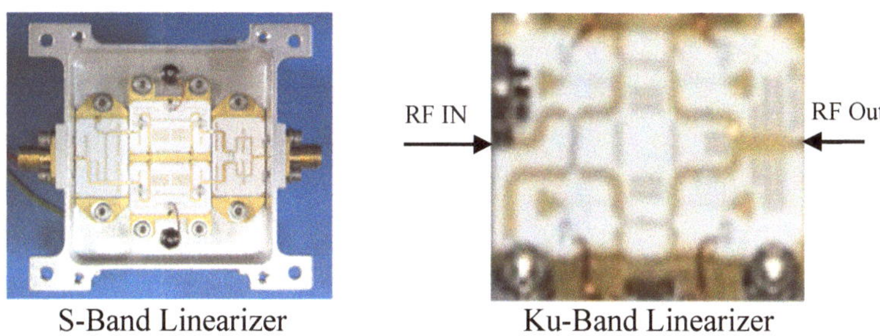

Fig. 16.14 Photograph of diode-based linearizer in vector modulator configuration

arm, respectively, various combinations of amplitude and phase characteristics over the RF power level can be obtained. Thus, the diode-based linearizer with its vector modulator configuration is very versatile to generate required I–O characteristic to linearize different power amplifiers with different nonlinear characteristics.

16.4.3 Temperature Behaviour of Diode Linearizer

Among the various types of linearizers, the vector modulator configuration based predistortion linearizers using diodes are most suitable for generating various combinations of amplitude and phase characteristics with the RF power level. Thus, this configuration is very useful for linearization of different power amplifiers with various nonlinear I–O characteristics.

Temperature-dependent forward-biased *I–V* characteristic of Schottky barrier and p-i-n diodes can be written as

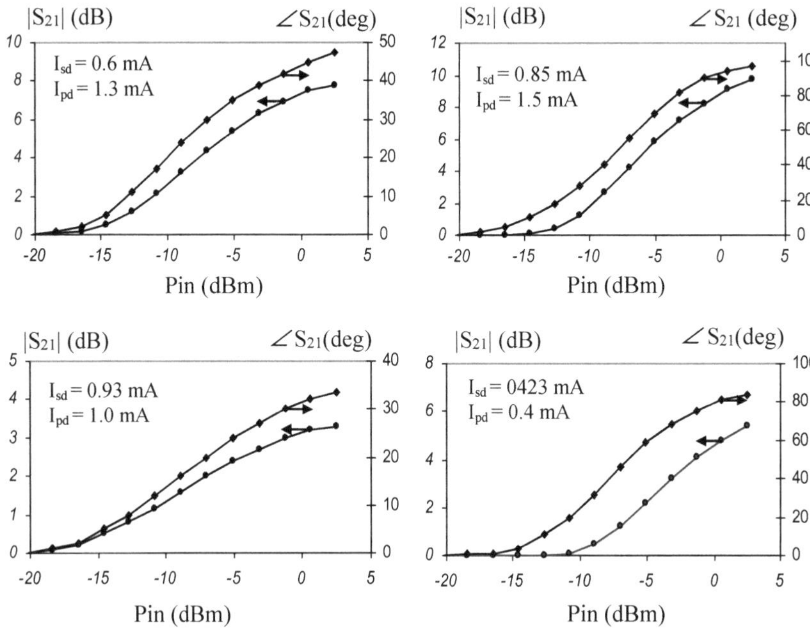

Fig. 16.15 Various combinations of measured amplitude and phase of the linearizer of Fig. 16.14 for different bias conditions

$$I_d(V_d(T), T) = I_o \exp\left(\frac{qV_d}{\eta kT}\right) = aT^{\nu} \exp^{\left(\frac{q(V_d - E_b)}{\eta kT}\right)} \tag{16.11}$$

Here, $I_o = aT^{\nu} \exp(-qE_b/\eta kT)$, the reverse-bias saturation current of the diode which is negligible compared to its forward current I_d. k is Boltzmann constant, q is the electron charge, and a is a temperature independent constant. E_b is the bandgap potential of the semiconductor in volts for p-i-n diodes and barrier potential ϕ_b in case of Schottky barrier diode. η is the ideality factor of the diode.

The temperature- and bias current-dependent RF resistance of p-i-n diodes is given by

$$R_d(I_d(T), T) = z\,\frac{T^{n-m}}{I_d^p} \tag{16.12a}$$

Here, z, m, p, ν, n all are temperature-invariant constants. For p-i-n diodes, the approximate values of the parameter are $m = 1, p = 1, \nu = 3, n = 2$.

The small-signal RF resistance of Schottky diodes is determined by

$$\frac{1}{\partial I_d/\partial V} = \frac{\eta kT}{qI_d} \tag{16.12b}$$

Thus, (16.12b) is the same as the expression of p-i-n diode's RF resistance given by (16.12a) with $z = \eta k/q$, $m = 1$, $p = 1$, $\nu = 2$, $n = 2$. From (16.12a), it is clear that under constant-current-bias condition (constant I_d), with the increase of operating temperature, the RF resistance of both the diodes increases linearly. Thus, amplitude and phase characteristics of the linearizer change with the operating temperature of the diodes [16].

To achieve temperature-insensitive amplitude and phase characteristics, RF resistance of the diodes should be temperature independent, i.e. $\partial R_d/\partial T = 0$. Differentiating (16.12a) with respect to T and putting the temperature independency condition

$$\frac{\partial I_d}{\partial T} = \frac{(n-m)I_d}{Tp} \tag{16.13}$$

Suppose, the diodes are biased by a voltage source V_O through a resistor R_b. The current of the diode can be written as

$$I_d = (V_O - V_d)/R_b \tag{16.14}$$

Differentiating (16.14) with respect to T and combining with (16.13), we can write the condition as

$$\partial V_d/\partial T = -(n-m)(V_O - V_d)/Tp \tag{16.15}$$

Combining (16.11) with (16.12), then differentiating it, we can write (16.16).

$$\frac{\partial V_d}{\partial T} = -\frac{E_b - V_d}{T} - \frac{\nu p + m - n}{p} . \frac{\eta k}{q} \tag{16.16}$$

Combining (16.15) and (16.16), we can write

$$V_O = \left(\frac{p}{n-m}\right)E_b + \left(\frac{n-m-p}{n-m}\right)V_b + \left(\frac{\nu p - n + m}{n-m}\right)\frac{\eta k T}{q} \tag{16.17a}$$

Thus, for p-i-n diodes, putting $m = 1$, $p = 1$, $\nu = 3$, $n = 2$, we can write

$$V_O = E_b + \frac{2\eta k T}{q} \tag{16.17b}$$

And for Schottky diodes, putting $m = 1$, $p = 1$, $\nu = 2$, $n = 2$, we can write

$$V_O = \phi_b + \frac{\eta k T}{q} \tag{16.17c}$$

This (given by 16.17b and 16.17c) is the optimum voltage (open-circuit voltage) required to achieve temperature-insensitive RF resistance of p-i-n diode and small-signal Rf resistance of Schottky barrier diode. The bias resistor R_b is determined by

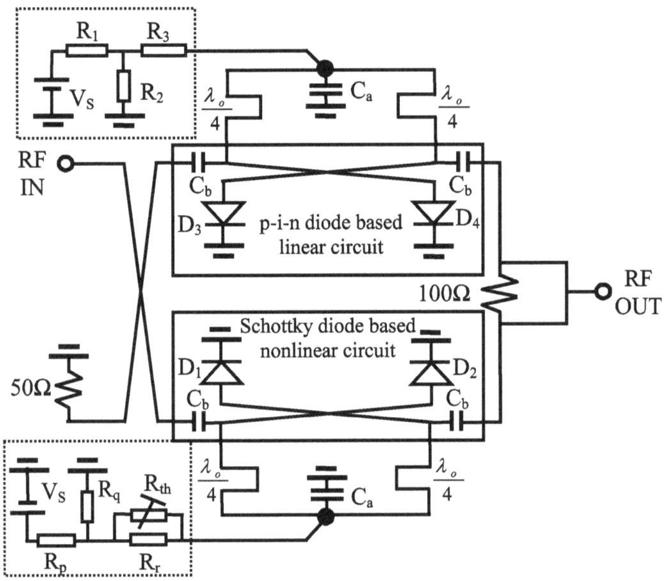

Fig. 16.16 Schematic circuit diagram of the diode linearizer with temperature compensation networks

(16.14), where diode current (I_d) is at ambient temperature. The p-i-n diodes operate in linear region over the full operating RF power level. Thus, it is valid for p-i-n diodes, whereas Schottky barrier diodes operate in nonlinear region, thus rectify the signal, and thus the diode voltage and current changes with the RF power level. Practically, the required optimum open-circuit voltage is different from (16.17c), and increases with the increase of RF power level. Therefore, bias resistor for the p-i-n diodes is determined by (16.14), whereas temperature-dependent bias resistor is used for Schottky barrier diodes for achieving temperature-insensitive RF resistance. Figure 16.16 shows the schematic circuit diagram of the linearizer with bias networks for temperature-dependent voltage/current to the Schottky diodes. Voltage source V_s (+5 V) in combination with resistors R_1 and R_2 are used to get $V_O = V_s R_2/(R_1 + R_2)$ and $R_b = R_1||R_2 + R_3$ for p-i-n diodes. Resistors R_p and R_q are used to set optimum voltage $V_O = V_s R_q/(R_p + R_q)$ for Schottky diodes. Resistor R_r in parallel combination with a negative temperature coefficient thermistor R_{th} is used to get temperature-dependent bias resistor for Schottky diodes.

16.4.4 Broadband Linearizers

Predistortion linearizers are cascaded with a power amplifier such as solid-state power amplifiers (SSPAs) and travelling wave tube amplifiers (TWTAs) to improve nonlinear performance of a communication system. The predistortion linearizer generates inverse amplitude and phase nonlinearity with respect to the power amplifier's amplitude and phase nonlinearity. Diode-based linearizers use Schottky barrier diodes as nonlinear attenuator and p-i-n diodes as linear attenuator to generate the amplitude and phase nonlinearity. Radio frequency (RF) impedance of the diodes determined by RF resistance (R_i) in combination with a parallel equivalent capacitor (C_i) determines the nonlinear amplitude and phase characteristic of the linearizer. Therefore, gain expansion and phase expansion of the diode-based linearizers vary with the change of operating frequency. The frequency dependencies of the linearizer's performances are not only due to the frequency dependency of diodes impedance. It is also due to the frequency-dependent characteristics of other used circuit elements such as attenuators and the input and output couplers of the distortion generation bridge. Moreover, nonlinear responses of the power amplifiers to be linearized also vary over the operating frequency range that needs compensation. Schematic circuit of a broadband diode linearizer using amplitude tilt equalizers is shown in Fig. 16.17. The amplitude tilt equalizers provide RF amplitude slope over operating frequency range. The equalizer circuit may be based on active or passive components. The advantage of active equalizer is its ability to generate slope adjustability by adjusting its bias current/voltage. In case of active equalizer, depending on bias voltage/current it can generate positive, negative or zero amplitude slope. Due to the amplitude slope of the RF signal, linearizer can provide different types of amplitude, amplitude expansion and phase expansion over the broad operating frequency range required for linearization of broadband amplifiers. Here, equalizers are used outside the distortion generation bridge, one at the input and another at the output of the bridge. The equalizers can also be put within the distortion generation bridge as shown in Fig. 16.18, to develop broadband linearizers. The equalizer used within the distortion generator bridge may be placed before the nonlinear circuit as shown in Fig. 16.18a or after the nonlinear circuit as shown in Fig. 16.18b.

In case of linear attenuator with p-i-n diodes, suitable optimum open-circuit bias voltage can serve the purpose of gain compensation over the operating temperature range. Temperature-controlled biasing circuitry is made by exploiting the property of diode's RF resistance over temperature. It is known that RF resistance of the diode may increase or decrease or remain zero depending on the supply voltage or open-circuit voltage of the bias circuit of the diodes. Therefore, by providing a bias circuit with optimum open-circuit voltage and current, any temperature coefficient of the attenuators and equalizers can be obtained for achieving temperature-invariant amplitude, amplitude expansion and phase expansion of the linearizer. Here, another linear attenuator at the output of the distortion generation unit is used to compensate the overall gain/loss variation of the linearizer over the operating temperature range.

The component values of the bias networks correspond to temperature-compensated performance and can be determined experimentally. RF circuit of the diode-based linearizers to be temperature compensated is located inside a chamber in which the temperature can be maintained at a required level in the range varying from lowest (say −20 °C) to highest (say +80 °C). The amplitude, amplitude expansion and phase expansion of the linearizer are monitored using a vector network

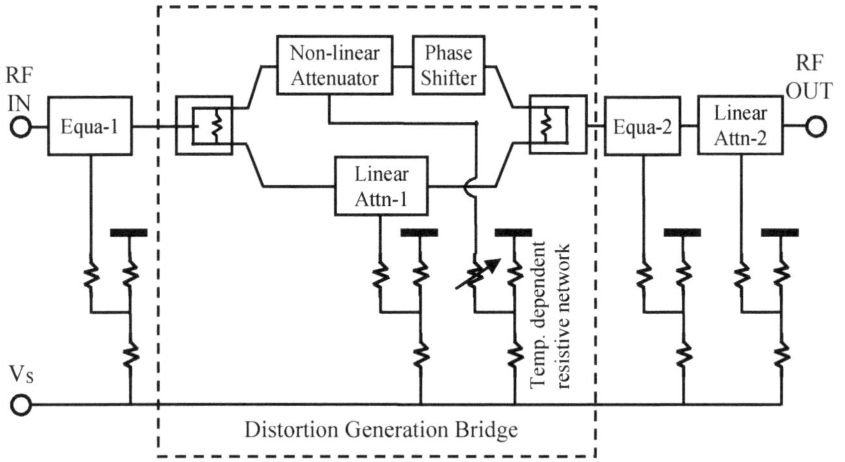

Fig. 16.17 Schematic block diagram of a broadband diode linearizer with temperature compensation networks

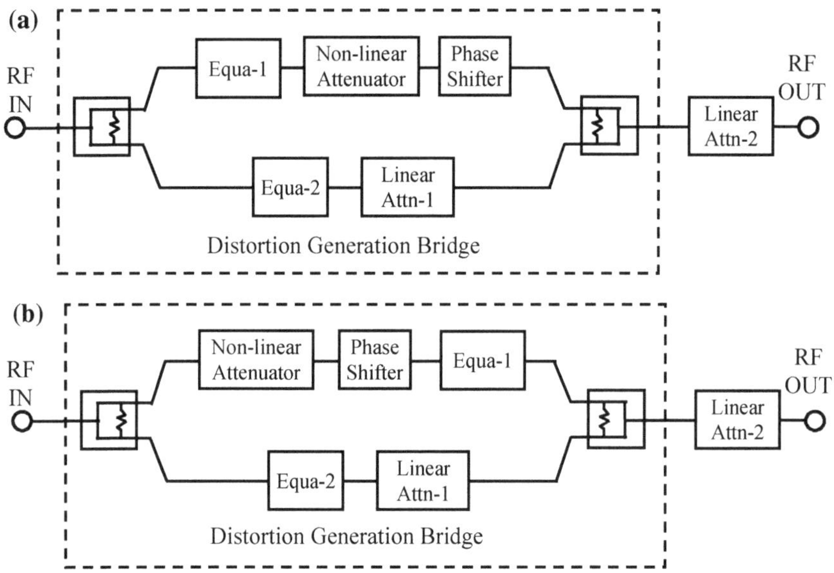

Fig. 16.18 Schematic block diagram of a broadband diode linearizer with equalizer within the distortion generation bridge, equalizer **a** at input **b** at output of the nonlinear attenuator

analyzer (VNA). The voltage and current supply to the nonlinear attenuator, linear attenuators and equalizers at different operating temperatures can be obtained. The recorded voltage and current data of the linear attenuator are plotted as shown in Fig. 16.19. From the best-fitted straight load line data, the open-circuit voltage and equivalent open-circuit resistance can be obtained as shown in the plot. Similarly,

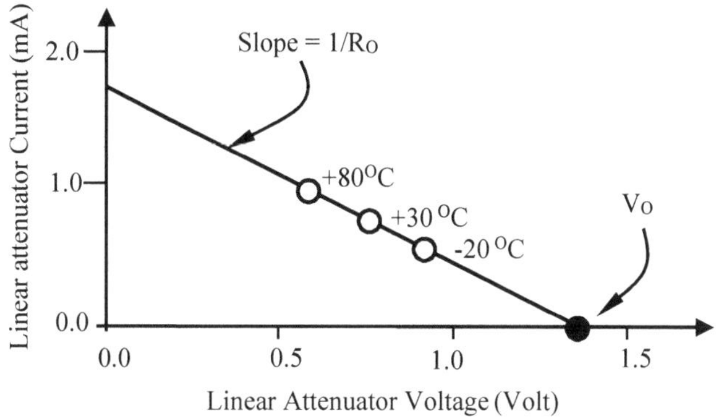

Fig. 16.19 Experimental determination of optimum bias resistance and open-circuit voltage for the linear attenuator

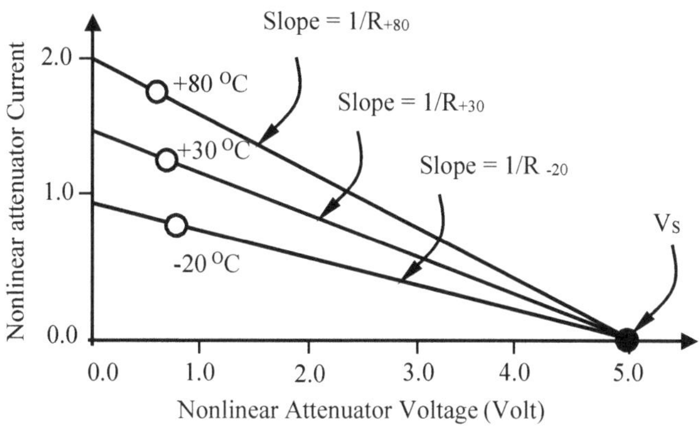

Fig. 16.20 Experimental determination of optimum bias resistance and open-circuit voltage for the nonlinear attenuator

recorded voltage and current data for the nonlinear attenuator are plotted as shown in Fig. 16.20. Three load lines are drawn corresponding to three different temperatures. The value of the temperature-dependent resistive network can be obtained to achieve the resistances R_{+80}, R_{+30} and R_{-20} at the temperatures of +80, +30, −20 °C, respectively.

16.5 Digital Predistortion Linearizers

Using analog circuit technique, it is very difficult to generate exact opposite amplitude and phase nonlinearities to that of a power amplifier to be linearized. Digital predistortion linearizers with advanced digital processing techniques are capable

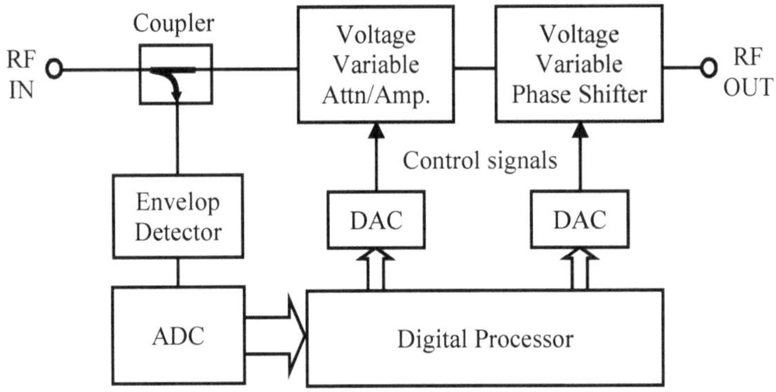

Fig. 16.21 Schematic block diagram of a digital predistortion linearizer variable amplifier and phase shifter

of generating nearly ideal nonlinear amplitude and phase characteristics required to linearize different types of power amplifiers.

Schematic diagram of a digital predistortion linearizer is shown in Fig. 16.21. Here, input RF signal is sampled and information about its envelope is extracted using an envelope detector circuit. The detected signal is then digitized using an analog-to-digital converter (ADC). The digitized signal is then processed using a high-speed digital processor and converted to analog control signals using digital-to-analog converters (DACs) for controlling a voltage-variable attenuator/amplifier and a phase shifter. The variable attenuator/amplifier and the phase shifter are connected in the signal path which controls the amplitude and phase of the signal to generate proper amplitude and phase nonlinearities to linearize the power amplifier.

Schematic diagram of another digital predistortion linearizer is shown in Fig. 16.22. The scheme is similar to as shown in Fig. 16.21. Here, the distortion generator circuit is implemented using two voltage-variable amplifier/attenuators, (I-VVA) and (Q-VVA), in vector modulator (I–Q) configuration instead of cascaded amplifier and phase shifter. By controlling the amplitude of I-VVA and Q-VVA signals, both amplitude and phase of the signal can be controlled according to the envelope of the input RF signal.

The circuit elements used in the RF signal path, i.e. VVA, phase shifter, power divider and combiner, power coupler and detectors of the digital linearizers need to work at the operating RF frequency over its bandwidth. However, the digital processing unit, ADCs and DACs shall process envelope of the modulated signal. Thus, bandwidth of the RF signal will be limited by the processing speed of ADC, DAC, and signal processing units.

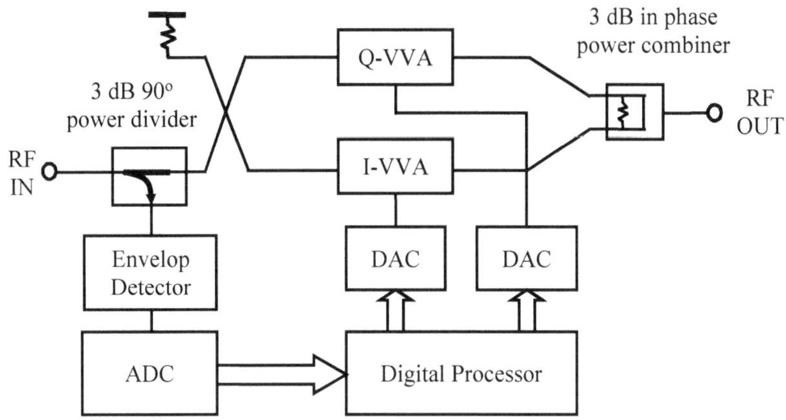

Fig. 16.22 Schematic block diagram of a digital linearizer using variable amplifiers/attenuators in I & Q signals

References

1. Inada R, Ogawa H, Kitazume S, Desantis P (1986) A compact 4-GHz linearizer for space use. IEEE Trans Microw Theory Tech 34(12):1327–1332
2. Imai N, Nojima T, Murase T (1989) Novel linearizer using balanced circulators and its application to multilevel digital radio systems. IEEE Trans Microw Theory Tech 37(8):1237–1243
3. Yamauchi K, Mori K, Nakayama M, Itoh Y, Mitsui Y, Ishida O (1996) A novel series diode linearizer for mobile radio power amplifiers. In: IEEE MTT-S international microwave symposium digest, pp 831–834
4. Zhang W-M, Yuen C (1998) A broadband linearizer for Ka-band satellite communication. In: IEEE MTT-S international microwave symposium digest, pp 1203–1206
5. Haskins C, Winslow T, Raman S (2000) FET diode linearizer optimization for amplifier predistortion in digital radios. IEEE MGWL 10(1):21–23
6. Yi J, Yang Y, Park M, Kang W, Kim B (2000) Analog predistortion linearizer for high-power RF amplifiers. IEEE Trans Microw Theory Tech 48(12):2709 2713
7. Yang Y, Woo YY, Kim B (2002) New predistortion linearizer using low-frequency even-order intermodulation components. IEEE Trans Microw Theory Tech 50(2):446–452
8. Katz A, Sudarsanam R, Aubert C (1985) A reflective diode linearizer for spacecraft applications. In: IEEE MTT-S international microwave symposium digest, pp 661–664, June 1985
9. Yamauchi K, Mori K, Nakayama M, Mitsui Y, Takagi T (1997) A microwave miniaturized linearizer using a parallel diode. IEEE MTT-S digest, pp 1199–1202, June 1997
10. Yamauchi K, Mori K, Nakayama M, Mitsui Y, Takagi T (1997) A microwave miniaturized linearizer using a parallel diode with a bias feed resistance. IEEE Trans Microw Theory Tech 45(12):2431–2435
11. Bera SC, Bhardhwaj PS, Singh RV, Garg VK (2003) A diode linearizer for microwave power amplifiers. Microw J 46(11):102–113
12. Bera SC, Singh RV, Garg VK (2004) A compact Ku-band linearizer for space application. In: Proceedings of Asia Pacific microwave conference, pp 37–38
13. Bera SC, Singh RV, Garg VK (2008) Diode-based predistortion lineariser for power amplifiers. IEE Electron Lett 44(2):125–126

14. Garg VK, Bera SC (2009) Predistortion linearizer with novel temperature compensation technique for satellite transponder. In: 60th international science congress, Korea
15. Bera SC, Jain A, Raval DU, Shah LB (2010) L-band predistortion linearizer for satellite transponder. In: Proceedings of international conference on microwaves, antenna, propagation & remote sensing, Jodhpur, India
16. Bera SC, Kumar V, Singh S, Das DK (2013) Temperature behavior and compensation of diode-based predistortion linearizer. IEEE Microwave Wirel Compon Lett 23(4):211–213

Chapter 17
Microwave Frequency Multipliers

Abstract This chapter presents the working principle, design and analysis of various microwave frequency multipliers based on diodes and transistors. Different frequency multiplier circuits based on Varactor diode, Schottky barrier diode, step recovery diodes (SRDs) and FETs are presented. The chapter also presents diode and transistor-based balanced configuration of frequency multiplier circuits.

17.1 Introduction

A frequency multiplier generates an output signal whose frequency is a multiple of its input frequency. The basic realization principle for frequency multiplication is the generation of multiple higher frequencies by distorting (amplitude or phase) a sinusoidal signal and then filtering out the desired higher frequency component [1, 2]. The distortion of a sinusoidal waveform is carried out using non-linear characteristic of devices such as diodes, transistors, etc. Various techniques exist for the realization of frequency multipliers. At radio frequencies, these techniques generally use a non-linear device to generate the desired multiplication of standard/reference frequency. The non-linear elements such as varactor diodes, Schottky barrier diodes (SBDs), step recovery diodes (SRDs) and bipolar junction transistors (BJTs) or field effect transistors (FETs) are used as non-linear devices [1–14].

Frequency multipliers are used to realize RF source, e.g. local oscillator (LO) of higher frequency from low-frequency standard. With the use of higher frequency for wireless communications, to take advantage of available spectrum, the required LO frequencies are also increasing. The performances of microwave solid-state oscillators degrade with the increase of frequency. To overcome this problem, highly stable oscillators operating at a lower frequency are used, followed by frequency multipliers with low conversion loss for realization of microwave frequency generator. In case of multiplier-based local oscillator, a crystal oscillator followed by multipliers, filters and amplifiers is used to generate high frequency source with required output power. Other approaches for realizing local oscillator at higher frequencies

S. C. Bera, *Microwave Active Devices and Circuits for Communication*, Lecture Notes in Electrical Engineering 533, https://doi.org/10.1007/978-981-13-3004-9_17

are the use of transistor-based oscillator stabilized by a dielectric resonator or use of phase-locked loop (PLL) using crystal source, frequency divider, phase detector and voltage-controlled oscillator. Among all these approaches, multiplier-based local oscillators are simple in circuitry and lower in cost.

17.2 Principle of Multiplier Operation

A pure sinusoidal waveform is distorted without affecting its zero crossing point in the time axis, i.e. maintaining its periodicity when it passes through a non-linear (amplitude and/or phase) system. Thus, the distorted output signal consists of only the fundamental and other components of frequencies exactly integer multiple of the fundamental frequency, called harmonics. In this process, any component of frequencies other than the exact multiple of the fundamental frequency cannot be generated maintaining the zero crossing point unaltered. Therefore, any non-sinusoidal periodic waveform (distorted sine wave) contains energy at fundamental and harmonics of the fundamental frequency. The design of multiplier is to design a non-linear circuit with filtering that produces a waveform with significant signal strength at the desired harmonic. Conversion gain G_c is used to characterize microwave frequency multiplier's conversion efficiency and is defined as the ratio of the output power delivered to the load to the available power from the input source. The goal of the circuit design is to maximize the conversion gain for a given device and input/output frequencies.

Generalized block diagram with associated input and output spectrum of a frequency multiplier is shown in Fig. 17.1. Here, devices such as diodes and transistors are biased to operate in strongly non-linear region to distort the input sinusoidal waveform of frequency f_o for the generation of higher order harmonics in which one desired higher order harmonic passes through the bandpass filter, suppressing other frequency components [1, 2]. The input and output matching networks provide required terminations at each harmonic frequency to get output sinusoidal waveform of frequency nf_o with maximum possible amplitude, i.e. with maximum conversion gain G_c. The use of LPF and BPF provide high impedance to all the unwanted harmonic components. However, it is desirable to allow the currents of the undesired harmonics to flow so that the inter-modulation products of those harmonics will contribute to the desired output harmonic. Therefore, to maximize conversion efficiency it is desirable to short the currents of the undesired components. Another performance requirement is the suppression of fundamental as well as harmonic components other than the wanted frequency component, i.e. nf_o at the output of the multiplier. The worst case suppression S, as shown in Fig. 17.1, should be as high as possible. The importance of the suppression obviously depends on the particular application. In general, the input matching network should provide proper matching to the fundamental frequency so that most of the input power reaches the non-linear device and the output matching network should provide proper matching to the desired multiplied frequency so that most of the output power transferred to the load. Except the

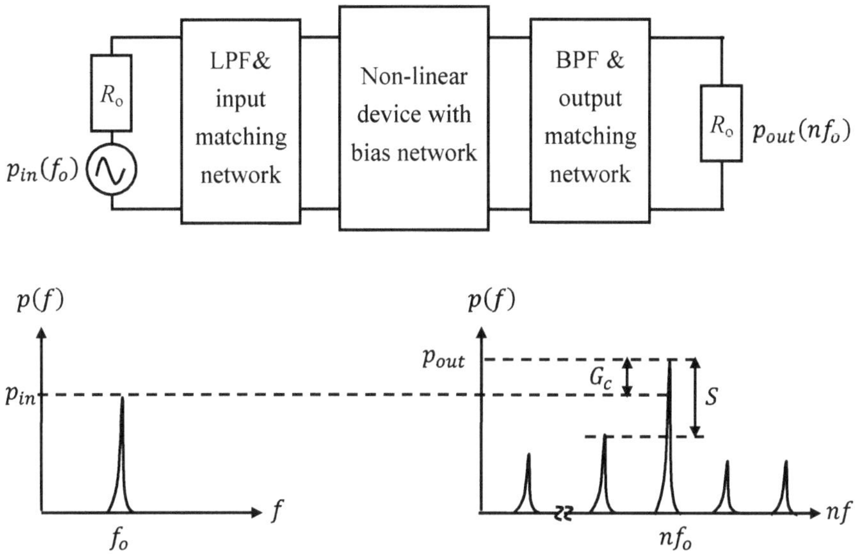

Fig. 17.1 Generalized block schematic of a frequency multiplier

fundamental and desires multiplied frequencies, all other frequencies are referred to as idle frequencies. The matching networks should provide proper termination to the idle frequencies to transfer maximum output power to the load corresponds to desired multiplied frequency and suppress the undesired frequency components. In principle, reactive loads at idle frequencies are desirable so that they do not dissipate active power.

Any periodic non-sinusoidal wave can be expressed by Fourier series as

$$v_o(\omega t) = a_0 + \sum_{n=1}^{\infty} a_n \cos n\omega t + \sum_{n=1}^{\infty} b_n \sin n\omega t \tag{17.1}$$

Here, ω is the angular frequency of the RF signal a_0, a_n and b_n are constants. Here, a_0 represents the average DC value of the waveform, and it is not important for multiplier application, thus may be ignored. The other constants can be determined by performing the following integrations over any 2π radians:

$$a_n = \frac{1}{\pi} \int_0^{2\pi} v_o(\theta) \cos n\theta \, d\theta \tag{17.2a}$$

$$b_n = \frac{1}{\pi} \int_0^{2\pi} v_o(\theta) \sin n\theta \, d\theta \tag{17.2b}$$

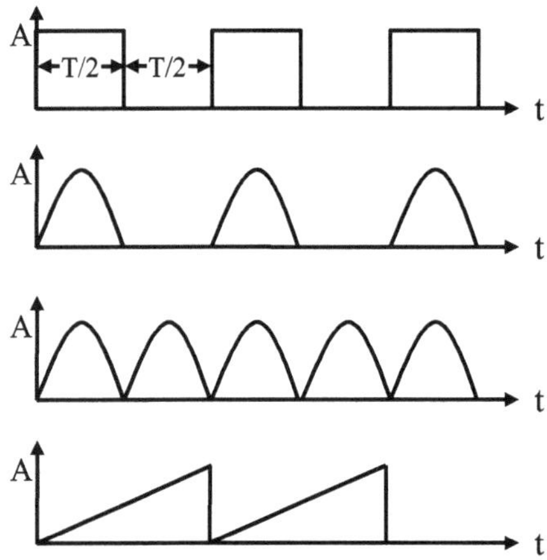

$$a_n = 0, \qquad b_n = \frac{2A}{n\pi}, n = 1, 3, 5, \ldots$$

$$a_n = \frac{2A}{(n^2-1)\pi}, n = 2, 4, 6, \ldots, \; b_n = 0$$

$$a_n = \frac{4A}{(n^2-1)\pi}, n = 2, 4, 6, \ldots, b_n = 0$$

$$a_n = 0, \qquad b_n = -\frac{A}{n\pi}, n = 1, 2, 3, \ldots$$

Fig. 17.2 Harmonic components for various waveforms

Here, $\omega t = \theta$. These integrals can be easily calculated and some may be computed by inspection. As example, a very narrow pulse has about the same amplitude for all the harmonics, i.e. comblike spectrum. Similarly, any waveform of 'half-wave symmetry' like symmetrical clipped sinusoidal waveform has the only odd harmonic components. Some other waveforms with their harmonic components are shown in Fig. 17.2. It is to be noted that for waveforms with sharp edges (e.g. square wave) have higher harmonics with larger amplitudes due to $1/n$ dependency of amplitudes compared to other waveforms (e.g. rectified sine wave) of $1/n^2$ dependency. It is also to be noted that some waveforms contain only even or only odd harmonic components, which reduces the requirement of filtering out the undesired harmonic to get desired frequency component. The harmonic components not only depend upon the shape of the waveform, but also depend on the duty cycle. Figure 17.3 shows the waveform of different pulse and their spectral components. This type of half and full wave current waveforms are generated when a sinusoidal signal applied to diode-based circuits. It shows that with the decrease in width of the waveform, level of the higher harmonic components increases. Suppose, a square waveform have a duty cycle T_1/T_2, as shown in Fig. 17.4. The coefficients a_n and b_n of the waveform can be derived as

$$a_n = \frac{1}{\pi} \int_0^{\theta_1} A \cos n\theta \, \mathrm{d}\theta \tag{17.3a}$$

Here, $\theta = \omega t$ and $\theta_1 = \omega T_1$. Therefore,

$$a_n = \frac{A}{n\pi} \sin(n\theta_1) \tag{17.3b}$$

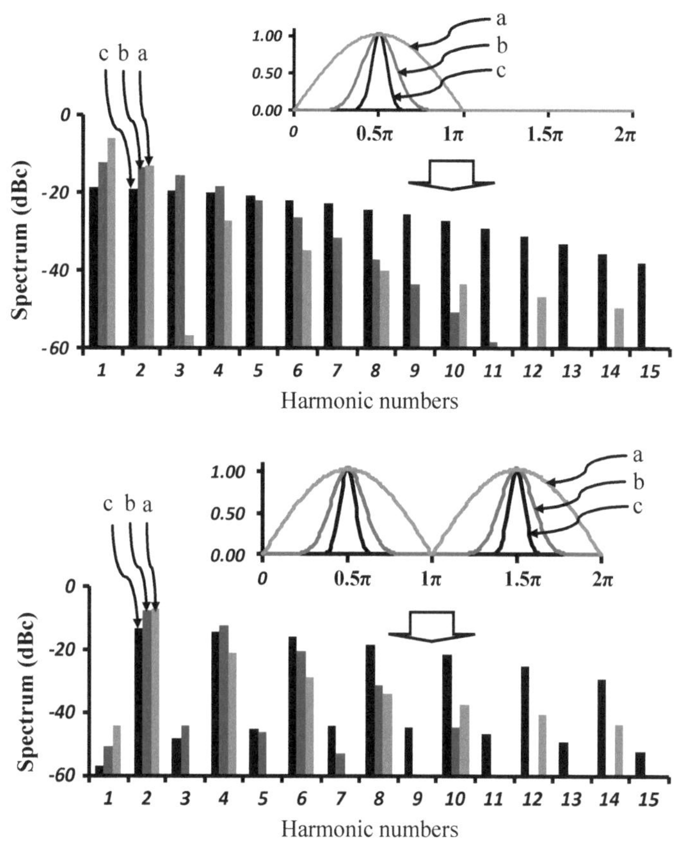

Fig. 17.3 Waveform and spectrum of different waveforms

Similarly,

$$b_n = \frac{1}{\pi}\int_0^{\theta_1} A \sin n\theta \, d\theta \tag{17.4a}$$

$$= \frac{A}{n\pi}[1 - \cos(n\theta_1)] \tag{17.4b}$$

Thus, the resultant amplitude of the nth harmonic C_n can be written as

$$C_n = \sqrt{a_n^2 + b_n^2} \tag{17.5}$$

Figure 17.4 shows the normalized harmonic amplitudes of square waves with different duty cycles corresponds to $\theta_1 = 0$ to π. The plot is very useful to decide duty

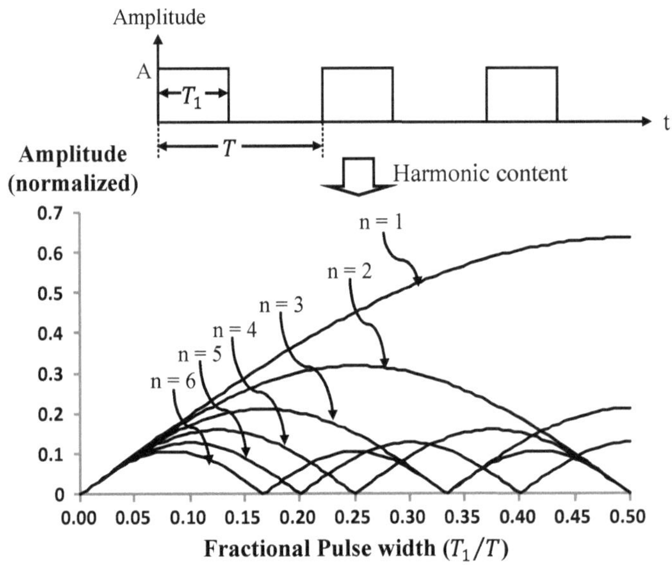

Fig. 17.4 Harmonic amplitudes of square waves with different duty cycles

cycle of a square waveform for particular multiplier. It shows that square wave of very small duty cycle contain all harmonics of nearly equal amplitudes, i.e. comblike spectrum whereas, square wave of 50% duty cycle has only the odd harmonic components. It also suggests that to design a frequency doubler using a square-shaped pulse, though most second harmonic energy will be generated for duty cycle of 25%; it is preferable to choose duty cycle of 33% for suppressing the third harmonic without any extra filter.

Other aspects of a frequency multiplier are the DC power consumption, bandwidth of operation, and reliability of the system. Bandwidth is important to provide variable output frequency with the use of variable fundamental frequency oscillator at input of the multiplier. Reliability is important since the device operates in highly non-linear region. Precaution in terms of bias condition as well as input RF level should be taken to prevent operation of the device goes beyond its safe operating limit.

In case of ideal multiplier considering noise less multiplication and no phase noise generation of its own, the process of frequency multiplication inevitably increase the phase noise. The reason for the increase of phase noise is that a frequency multiplier is in fact also a phase multiplier. Thus, it multiplies the frequency as well as phase deviations of the input signal. All frequency multipliers will increase the phase noise by the same factor 'n' that they multiply, in dB this is $20\log(n)$. However, generation of microwave frequency signal by multiplying a very stable low-frequency reference signal can have better phase noise than producing it directly in the microwave frequency range.

17.3 Diode Multiplier

Diode-based frequency multipliers are very popular in microwave frequency range for their simple structure and higher reliability. Uses of diode-based multipliers extend to millimetre and submillimetre wave systems [3–9]. Diode-based multipliers are the only convenient technique of realizing local oscillators for frequencies above 200 GHz. DC power consumption of diode multipliers is very less or zero; obviously, it does not provide any conversion gain. Therefore, cascading an amplifier with the diode multiplier is required to achieve the required RF power level of the multiplied frequency output. To obtain higher order frequency multiplication, it is required to cascade several multipliers which will increase complexity as well as conversion efficiency. At microwave and millimetre wave frequency regions, mainly varactor, Schottky and step recovery diodes are used as non-linear device for multiplier. In case of Schottky diode, strong non-linearity of the conduction current is the mechanism for frequency multiplication, called resistive multiplication. Mainly non-linearity of capacitance is the mechanism for frequency multiplication of varactor and step recovery diodes, known as reactive multiplication.

Ideally, resistive multipliers have infinite bandwidth. However, the presence of associated junction capacitance and reactive parasitic elements of the diodes limit the high-frequency operation. Moreover, input and output matching networks further limit the bandwidth. In addition to that, power loss within the resistive non-linear component increases the conversion loss. The conversion efficiency of a non-linear resistance multiplier is proportional to $1/n^2$, where n is the harmonic number. Thus, resistive multipliers are not suitable in case of applications where conversion loss is critical. Resistive multipliers are efficiently used in balanced configurations. Singly balanced or doubly balanced configurations have intrinsic property of fundamental and odd harmonic frequency rejection. This reduces the need for filtering out the odd harmonic frequency components.

Ideally, reactive multipliers have zero conversion losses. This requires proper reactive loading at all frequencies other than the input (fundamental) frequency and the output (multiplied) frequency. However, strong frequency dependency of the reactive impedances limits the operating bandwidth of the multiplier, i.e. reactive multipliers have narrowband performance.

17.3.1 Schottky Diode Frequency Multiplier

Schottky diodes for multiplier can be used as non-linear resistor, i.e. varistor or non-linear capacitor, i.e. varactor. There are tradeoffs in choosing a resistive vs. reactive mode of operation. Schottky diodes are used as non-linear resistor for multiplier design to provide high-frequency multiplication with very high bandwidth, with the penalty of higher conversion loss, whereas frequency multipliers using Schottky diode as a varactor have high conversion efficiency with the penalty of narrow

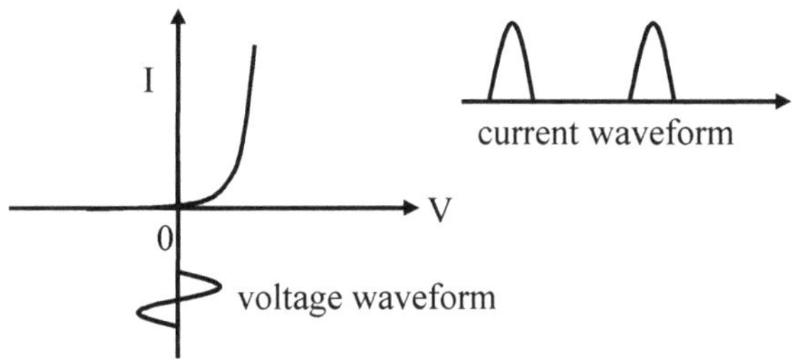

Fig. 17.5 Input signal and diode current waveform

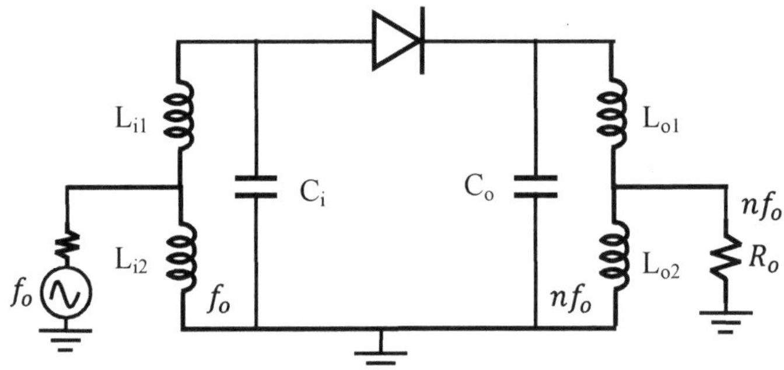

Fig. 17.6 A simple schematic circuit diagram of a frequency multiplier

bandwidth and need high Q circuit, thus very sensitive to small change of structural dimension.

Resistive frequency multipliers use non-linearity of *I*-*V* characteristic of a Schottky barrier diode to distort a sinusoidal waveform for generating harmonic components as shown in Fig. 17.5. Level of the harmonic components increases with the increase in amount of distortion. However, resistive frequency multipliers have higher conversion loss; thus, the level of maximum attainable power level of the harmonic components is less. A simplified schematic circuit diagram of a resistive Schottky diode frequency multiplier is shown in Fig. 17.6. Here, input inductors (L_{i1}, L_{i2}) and capacitor (C_i) match the diode impedance corresponding to frequency f_o to the source impedance for transferring maximum fundamental signal to the diode. Whereas, output inductors ($L_{o1,} L_{o2}$) and capacitor (C_o) match the diode impedance corresponds to desired *n*th harmonic to the load impedance for transferring maximum harmonic power to the load. Input inductor–capacitor combination also resonates at f_o and output inductors–capacitor combination resonates at nf_o. Thus, the resonators provide short circuits to the unwanted harmonics and decouple the input from the output. The resonators also put the diode in parallel with the input at the fundamental frequency and in parallel with the output at the desired harmonic.

A Schottky diode frequency doubler using microstrip distributed components suitable for microwave frequency range is shown in Fig. 17.7. Here, the input side of the diode is terminated to ground through a $\lambda_o/4$ short-circuited stub. It will provide

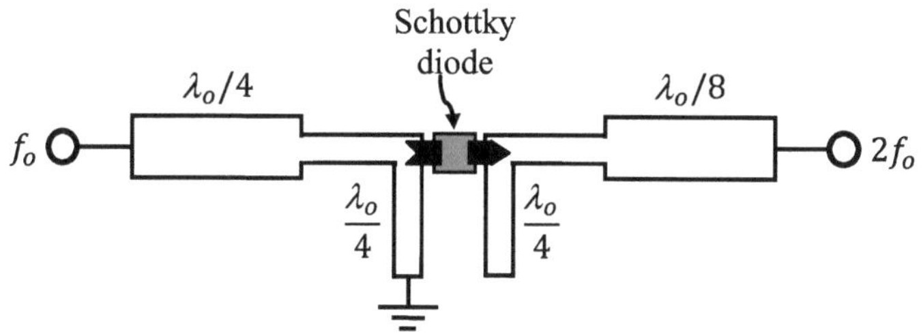

Fig. 17.7 Schottky diode frequency doubler realization using microstrip line distributed circuit elements

high impedance (ideally open) circuit at the input side of the diode thus it will allow the fundamental frequency to reach the diode. The short-circuited $\lambda_o/4$ transmission line will act as short-circuited half-wavelength transmission line corresponding to second harmonic, thus it will prevent the output power generated in the diode travelling towards the input. Similarly, the use of $\lambda_o/4$ open-circuited transmission line at the output side of the diode creates an RF short at the output for the fundamental frequency. Thus, it causes the input signal reaching the output of the device reflected back to the diode.

High impedance transmission line elements at the input and output of the diode are used as inductor to resonate the diode junction capacitance. Finally, $\lambda_o/4$, i.e. quarter wavelength impedance transformer is used at the input of the circuit to transform 50-Ω source impedance to the required diode impedance for transferring maximum signal to the diode. Similarly, $\lambda_o/8$ transformer which is quarter wavelength transmission line at $2f_o$ transform the 50 Ω load impedance to the diode output impedance, to transfer maximum second harmonic power to the load.

17.3.2 Varactor Diode Frequency Multiplier

Non-linear capacitive reactance of varactor diode distorts sinusoidal signal to generate its harmonics. A typical non-linear capacitance–voltage characteristic of varactor diodes is shown in Fig. 17.8. Considering zero series resistance, reactive multipliers have high conversion efficiency, when matching networks are designed using only reactive (loss less) elements. Here, frequency multiplication is based on the non-linear reactance, thus the multipliers have tuned input and output matching circuits and often tuned idler circuits are used for achieving higher conversion gain. Due to the reactive nature of the matching networks, it exhibits narrow bandwidth and high sensitivity to operating conditions, like bias voltage/current and component values. Practically, varactor diode-based frequency multipliers generate very little phase as well as amplitude noise. It only contributes to thermal noise due to the series resistance of the varactor and loss resistances associated with the matching networks.

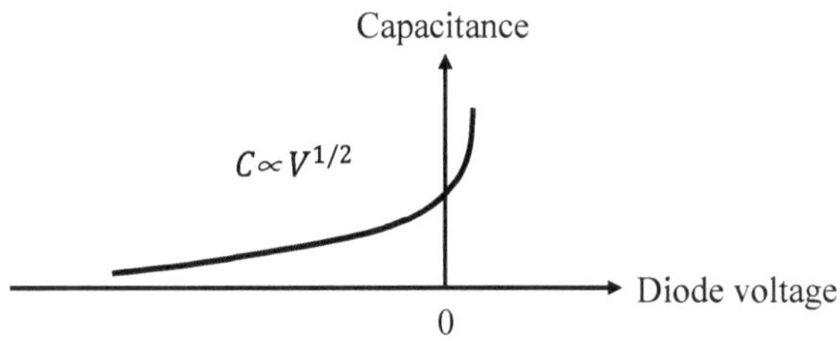

Fig. 17.8 Typical non-linear capacitance characteristic of a varactor diode

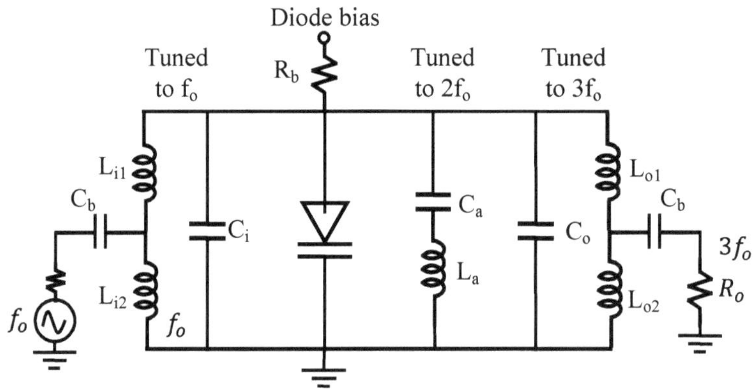

Fig. 17.9 A simple schematic circuit diagram of a frequency tripler using lumped components

To generate desired higher order harmonic components with sufficient level, it is required to provide short circuit to other harmonic components. The short circuit networks are known as idlers. Idlers are usually realized using a series resonating network using lumped components or using shunt short or open transmission line of proper length for a particular harmonic component. Without idlers, varactor multiplier does not generate harmonic components efficiently beyond the second order. An idler corresponds to second harmonic allowed second harmonic current through the diode which mixed with the fundamental and can generate third order harmonic. In the same way, providing idlers at second and third harmonics it is possible to generate fourth harmonic efficiently.

Metal–semiconductor junction diode, i.e. Schottky diode and p–n junction diode both are used as varactor diode for frequency multiplier. However, Schottky barrier varactor diode-based multiplier can generate frequencies to several hundred of GHz. Figure 17.9 shows a simple schematic circuit diagram of a multiplier using varactor diode as the non-linear device. The bias resistor R_b is used to bias the diode, which also changes the DC voltage across the diode with the change of RF power level, thus increases non-linearity. This is an example of tripler ($\times 3$) circuit. The series resonator circuit consists of C_a and L_a is used to provide short to the second harmonic. The input network consists of L_{i1}, L_{i2} and C_i is for matching the source impedance to the input

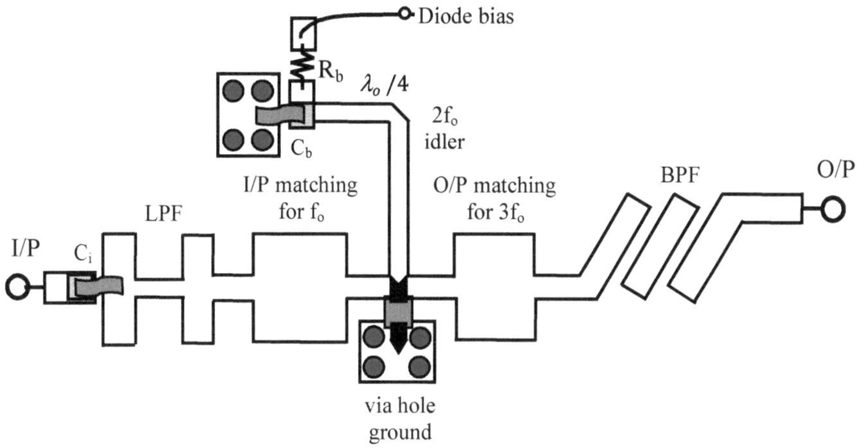

Fig. 17.10 Varactor diode frequency tripler realization using microstrip line distributed circuit elements

of the device corresponds to fundamental frequency f_o. It will also provide rejection to the third as well as other higher order harmonics at the input side. Whereas, the output network consists of L_{o1}, L_{o2} and C_o is for matching the load impedance to the output of the device corresponding to third harmonic component, $3f_o$. It will also provide rejection to fundamental as well as other higher order harmonics at the output.

Figure 17.10 shows a tripler circuit layout using varactor diode and microstrip distributed elements. The diode is grounded using via holes. The bias network is designed with $\lambda_o/4$ line with short circuit using capacitor C_b. At the device end it provides very high impedance (ideally open) for fundamental frequency, f_o but very low impedance (ideally short) for second harmonic, $2f_o$. Thus, the bias network will provide required short circuit to the second harmonic to generate higher order harmonics in addition to provide the bias voltage/current to the diode. The input and output matching networks are designed using $\lambda_o/4$ microstrip impedance transformer for f_o and $3f_o$, respectively. The required low-pass filter for input and bandpass filter for output are also designed using microstrip circuit elements as shown.

17.3.3 SRD Frequency Multiplier

Step recovery diodes are based on p-i-n configuration and behave as an ideal non-linear capacitor thus used for realization of higher order multipliers. Step recovery diodes are used as multipliers for all orders of multiplication, starting from 2 through 20 for both narrow and wide bandwidths. Ideally, it acts as a two-state capacitor with zero switching time between the states. Under forward charge storage condition, it

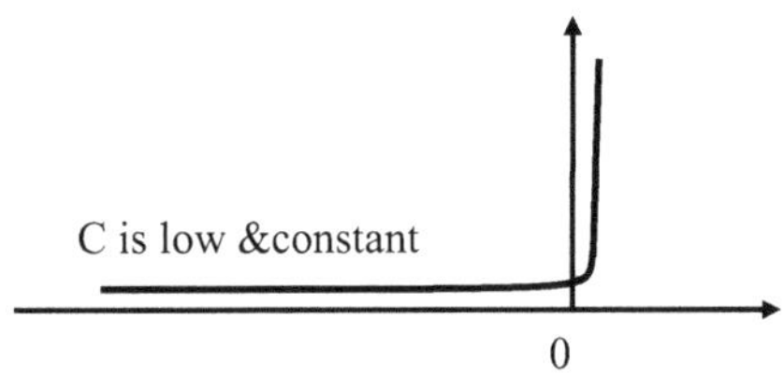

Fig. 17.11 Typical non-linear capacitance characteristic of SRD

acts as a large (ideally infinite) capacitance, thus behave as low impedance whereas under reverse bias condition it acts as a small capacitance (the depletion capacitance) thus behave as high impedance as shown in Fig. 17.11. The change from its low impedance state to high impedance state takes place very quickly, in a time interval called the diode transition time (T_T). Thus, when an RF signal applied to the diode, on the one half-cycle, it stores charge and appears as low impedance and on the second half-cycle, the diode conducts until the stored charge is removed and then switches off very rapidly at a speed governed by the transition time, T_T. This change produces a narrow pulse of voltage from the applied RF signal. This is equivalent to a number of frequencies which are multiples of the applied input frequency, i.e. harmonics of the fundamental frequency. In general, it is desirable that the minority carrier lifetime (τ) be greater than 10 times the period of the input frequency, while the transition time T_T should be less than the period of the output frequency. Thus, the highest order harmonic is limited by the narrowness of the pulse that is determined by the transition time, T_T, of the SRD.

SRD multipliers are in the category of switching reactance multiplier thus it does not have efficiency limitation ($1/n^2$) as in the case of resistive multipliers. In case of high-order frequency multiplication, the SRD multipliers provide much higher efficiency even compared to varactor multiplier.

A multiplier using step recovery diode consists of impulse generator, input and output matching networks, and input and output filters as shown in Fig. 17.12. The impulse generator converts the energy in each input cycle into a narrow large amplitude voltage pulse as shown in Fig. 17.13. It consists of a step recovery diode (SRD), a drive inductor L_3 and capacitor C_3. The capacitor C_3 is used to provide short at the required output frequency. Thus, the RF output signal cannot get back into the input section of the frequency multiplier circuit. Another function of the capacitor C_3 is to act as part of the input matching network. Here, bias resistor R_B is used for self-biasing the diode. Thus, it will provide reverse bias to the diode due to the rectification of the RF signal.

The input matching circuit provides a match to the source impedance at the fundamental frequency to the input of the impulse generator circuit. The output matching circuit is basically a resonator which converts the generated impulse into a damped ringing waveform at the desired output frequency. When the output of the multiplier is terminated directly to the load impedance R_o, it works as a broadband multiplier and its output waveform looks like a comb spectrum. But when the output of the mul-

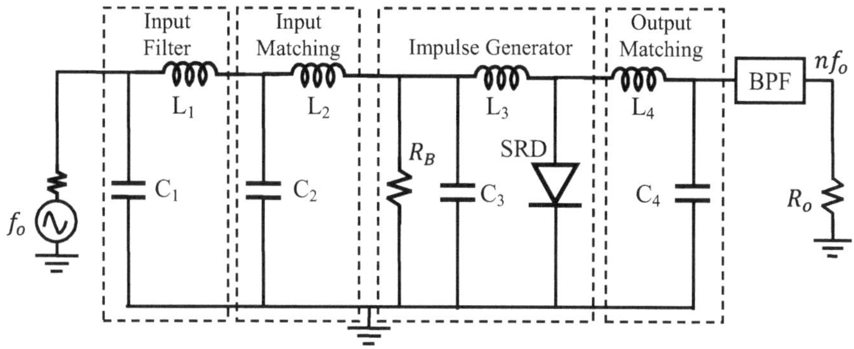

Fig. 17.12 Schematic of a series diode linearizer circuit

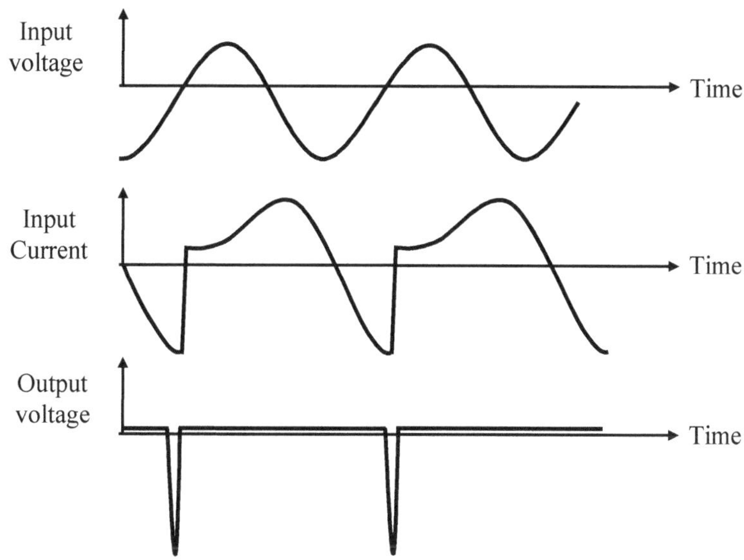

Fig. 17.13 Typical impulse response of SRD-based impulse generator

tiplier is terminated in a resonant network, i.e. a bandpass filter, it works as a narrow band multiplier, and the output level will be the same energy but will be concentrated around the resonant frequency, nf_o. The output filter reflects the unwanted harmonics back to the SRD where they mix to increase the power level at the desired harmonic frequency. The BPF ultimately pass the desired frequency to the load impedance rejecting the undesired frequency components. A typical spectrum at the output of the impulse generator circuit and at the output of a 15th-order frequency multiplier circuit is shown in Fig. 17.14.

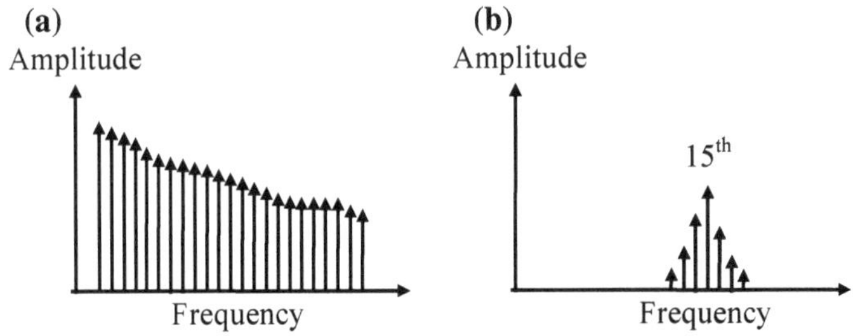

Fig. 17.14 Typical spectrum at the output of **a** impulse generator and **b** BPF

17.4 Transistor Multiplier

Frequency multipliers using transistors have better conversion efficiency compared to the diode-based frequency multipliers [10–14]. The diode multipliers always have conversion loss but FET and BJT multipliers provide conversion gain with good DC to RF efficiency. Transistor-based circuits generate harmonics of an input signal by distorting it when the level of the input signal and bias conditions are such that it drives the transistor in its non-linear region.

Origin of non-linearities of BJT and FETs are usually the non-linearities in the output current characteristics with the input voltage. Practically, output current clipping is the main contributor for generating harmonics of the input signal. This non-linearity is due to the non-linear characteristic of common emitter current gain and transconductance gain in case of BJTs and FETs, respectively. Figure 17.15 shows a typical output current characteristic and transconductance gain of FETs. Other than the non-linearities of the output current characteristic, non-linear input and output capacitances as well as feedback capacitance also contribute to the generation of harmonic components. For example, non-linearities of gate–source, drain–source and gate–drain capacitances of FETs under large signal operating condition contribute to the generation of harmonic components. Practically, the presence of these non-linearities enhances the harmonics level where the main contributor is the output current clipping caused by proper biasing and input signal level.

A simplified model of transistors can be considered to derive the harmonic contents of a transistor-based multiplier. It can be considered that the harmonics are generated due to only clipping of the output current waveform. Thus, the output current waveform can be described as piecewise linear model as shown in Fig. 17.16 as dotted line in case of FETs. The corresponding transconductance gain is constant with finite positive value over $-V_{gso}$ to V_F and is zero beyond these values as shown in the figure. This simple piecewise linear model can be used assuming the load line does not reach the ohmic and breakdown regions. Thus, it can be assumed that the operating voltage is always higher than the knee voltage and smaller than the breakdown voltage by proper drain voltage biasing and by suitable choice of the output load impedance.

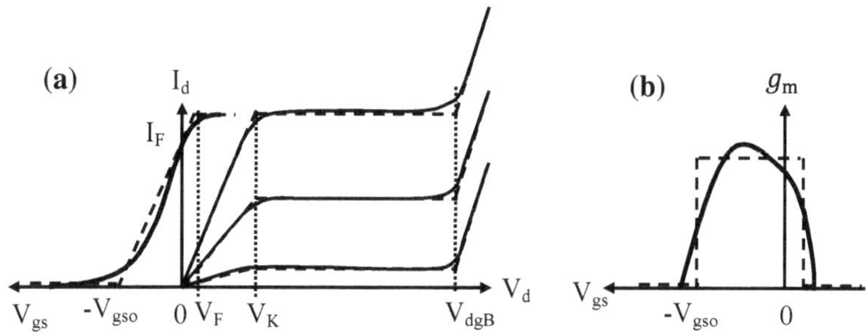

Fig. 17.15 Typical **a** drain current and **b** transconductance gain of FETs and their piecewise linear approximate plots

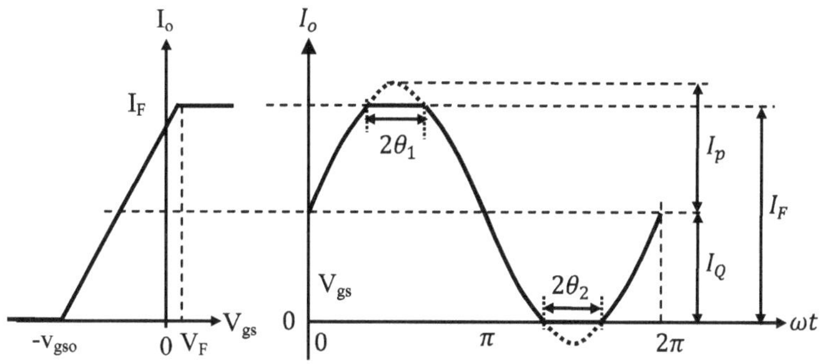

Fig. 17.16 Output current waveform with asymmetric clipping

Output current waveform may clip only in one side by the pinch-off of the channel in case of FETs and by cut-off region of the BJT. In this case, transistors are biased for Class A–C regions. One-sided clipping may also happen by gate–channel junction forward conduction in case of FETs and operating BJTs at saturation region biasing in Class-A. The disadvantage of this operating condition for frequency multiplier is the high DC power dissipation and possibility of device damage for excessive gate junction current for FETs. Both sided symmetric and asymmetric clipping may also happen depending on the biasing and input voltage/current swings.

A generalized output current waveform of a transistor-based multiplier with both side clipping is shown in Fig. 17.16. Here, I_Q is the quiescent bias current of the transistor, I_F is the maximum peak current and I_p is the amplitude of corresponding sinusoidal output current waveform. The output current waveform is clipped with amount of $2\theta_1$ and $2\theta_2$ as shown in the diagram. Thus, the sinusoidal waveform can be expressed as

$$i_o(\theta) = I_Q + I_p \sin \omega t \tag{17.6}$$

Here, ω is the angular frequency of the input RF signal and $\omega t = \theta$. The clipping of the output current waveform can be expressed as

$$i_o\left(\frac{\pi}{2} - \theta_1\right) = I_F \quad \text{and} \quad i_o\left(\frac{3\pi}{2} - \theta_2\right) = 0 \tag{17.7}$$

From (17.6) and (17.7), I_Q and I_p can be expressed in terms of I_F as

$$I_Q = \frac{\cos\theta_2}{\cos\theta_1 + \cos\theta_2} I_F \quad \text{and} \quad I_p = \frac{1}{\cos\theta_1 + \cos\theta_2} I_F \tag{17.8a}$$

The bias point of the transistor can be derived from (17.8a). For FET multiplier, gate bias voltage can be derived from:

$$V_{gs} = -V_{gso} + \frac{I_Q}{I_F}\left(V_{gso} + V_F\right) \tag{17.8b}$$

$$= -V_{gso} + \frac{\cos\theta_2}{\cos\theta_1 + \cos\theta_2}\left(V_{gso} + V_F\right) \tag{17.8c}$$

This periodic output current waveform can be expressed with Fourier series as

$$i_o(t) = a_0 + \sum_{n=1}^{\infty} a_n \cos n\omega t + \sum_{n=1}^{\infty} b_n \sin n\omega t \tag{17.9}$$

Here, ω is the angular frequency of the input RF signal; a_0, a_n and b_n are constants and can be determined by

$$a_0 = \frac{1}{2\pi}\int_0^{2\pi} i_o(\theta)\mathrm{d}\theta \tag{17.10a}$$

$$a_n = \frac{1}{\pi}\int_0^{2\pi} i_o(\theta)\cos n\theta \,\mathrm{d}\theta \tag{17.10b}$$

$$b_n = \frac{1}{\pi}\int_0^{2\pi} i_o(\theta)\sin n\theta \,\mathrm{d}\theta \tag{17.10c}$$

Here, $\theta = \omega t$. As shown in the figure, the clippings of the current waveform are over $2\theta_1$ and $2\theta_2$. Here, a_0 is DC component of the output current waveform which depends on the initial biasing of the transistor and clipping of the waveform. The constants a_n and b_n will tell us about the amplitude of the fundamental and harmonics at the output of the transistor. The resultant output amplitude I_{on} corresponds to nth harmonic can be expressed as

$$I_{on} = \sqrt{a_n^2 + b_n^2} \tag{17.11}$$

The output current waveform can be expressed as

$$\begin{aligned} i_o(\theta) &= I_Q + I_p \sin\theta, \quad \text{over}\, 0 \le \theta \le \pi/2 - \theta_1 \\ &= I_F, \quad \text{over}\, \pi/2 - \theta_1 \le \theta \le \pi/2 + \theta_1 \\ &= I_Q + I_p \sin\theta, \quad \text{over}\, \pi/2 + \theta_1 \le \theta \le 3\pi/2 - \theta_2 \\ &= 0, \quad \text{over}\, 3\pi/2 - \theta_2 \le \theta \le 3\pi/2 + \theta_2 \\ &= I_Q + I_p \sin\theta, \quad \text{over}\, 3\pi/2 + \theta_2 \le \theta \le 2\pi \end{aligned} \tag{17.12}$$

Therefore, the DC component of the waveform can be determined as follows:

$$\begin{aligned} a_0 &= \frac{1}{2\pi} \int_0^{2\pi} i_o(\theta) \mathrm{d}\theta \\ &= \frac{1}{2\pi} \left[\int_0^{\frac{\pi}{2}-\theta_1} (I_Q + I_p \sin\theta) \mathrm{d}\theta + \int_{\frac{\pi}{2}-\theta_1}^{\frac{\pi}{2}+\theta_1} I_F \,\mathrm{d}\theta + \int_{\frac{\pi}{2}+\theta_1}^{\frac{3\pi}{2}-\theta_2} (I_Q + I_p \sin\theta) \mathrm{d}\theta \right. \\ &\quad \left. + \int_{\frac{3\pi}{2}+\theta_2}^{2\pi} (I_Q + I_p \sin\theta) \mathrm{d}\theta \right] \end{aligned}$$

or,

$$a_0 = \frac{\pi - \theta_1 - \theta_2}{\pi} I_Q + \frac{\sin\theta_2 - \sin\theta_1}{\pi} I_p + \frac{\theta_1}{\pi} I_F \tag{17.13}$$

Similarly, coefficients a_n can be determined as

$$\begin{aligned} a_n &= \frac{1}{\pi} \int_0^{2\pi} i_o(\theta) \cos n\theta \,\mathrm{d}\theta \\ &= \frac{1}{\pi} \left[\int_0^{\frac{\pi}{2}-\theta_1} \left(I_Q + I_p \sin\theta\right) \cos n\theta \,\mathrm{d}\theta + \int_{\frac{\pi}{2}-\theta_1}^{\frac{\pi}{2}+\theta_1} I_F \cos n\theta \,\mathrm{d}\theta \right. \\ &\quad + \int_{\frac{\pi}{2}+\theta_1}^{\frac{3\pi}{2}-\theta_2} \left(I_Q + I_p \sin\theta\right) \cos n\theta \,\mathrm{d}\theta \\ &\quad \left. + \int_{\frac{3\pi}{2}+\theta_2}^{2\pi} \left(I_Q + I_p \sin\theta\right) \cos n\theta \,\mathrm{d}\theta \right] \end{aligned} \tag{17.14}$$

or,

$$\begin{aligned}
a_n = {} & \frac{I_Q}{n\pi}\left[\sin n\left(\frac{\pi}{2}-\theta_1\right)+\sin n\left(\frac{3\pi}{2}-\theta_2\right)-\sin n\left(\frac{\pi}{2}+\theta_1\right)-\sin n\left(\frac{3\pi}{2}+\theta_2\right)\right] \\
& + \frac{I_p}{2\pi(n-1)}\left[\cos\left\{(n-1)\left(\frac{\pi}{2}-\theta_1\right)\right\}+\cos\left\{(n-1)\left(\frac{3\pi}{2}-\theta_2\right)\right\}\right. \\
& - \cos\left\{(n-1)\left(\frac{\pi}{2}+\theta_1\right)\right\}+\cos(n-1)(2\pi)-1 \\
& \left. - \cos\left\{(n-1)\left(\frac{3\pi}{2}+\theta_2\right)\right\}\right] \\
& + \frac{I_p}{2\pi(n+1)}\left[\cos\left\{(n+1)\left(\frac{\pi}{2}+\theta_1\right)\right\}-\cos\left\{(n+1)\left(\frac{3\pi}{2}-\theta_2\right)\right\}\right. \\
& - \cos\left\{(n+1)\left(\frac{\pi}{2}-\theta_1\right)\right\}-\cos(n+1)(2\pi)+1 \\
& \left. + \cos\left\{(n+1)\left(\frac{3\pi}{2}+\theta_2\right)\right\}\right] \\
& + \frac{I_F}{n\pi}\left[\sin n\left(\frac{\pi}{2}+\theta_1\right)-\sin n\left(\frac{\pi}{2}-\theta_1\right)\right]
\end{aligned} \tag{17.15}$$

Similarly, the coefficients b_n can be determined as

$$\begin{aligned}
b_n &= \frac{1}{\pi}\int_0^{2\pi} i_o(\theta)\sin n\theta\,\mathrm{d}\theta \\
&= \frac{1}{\pi}\left[\int_0^{\frac{\pi}{2}-\theta_1}\left(I_Q+I_p\sin\theta\right)\sin n\theta\,\mathrm{d}\theta+\int_{\frac{\pi}{2}-\theta_1}^{\frac{\pi}{2}+\theta_1} I_F\sin n\theta\,\mathrm{d}\theta\right. \\
&\quad + \int_{\frac{\pi}{2}+\theta_1}^{\frac{3\pi}{2}-\theta_2}\left(I_Q+I_p\sin\theta\right)\sin n\theta\,\mathrm{d}\theta \\
&\quad \left. + \int_{\frac{3\pi}{2}+\theta_2}^{2\pi}\left(I_Q+I_p\sin\theta\right)\sin n\theta\,\mathrm{d}\theta\right]
\end{aligned} \tag{17.16}$$

or,

$$b_n = \frac{I_Q}{n\pi}\left[\cos n\left(\frac{\pi}{2}+\theta_1\right) - \cos n\left(\frac{\pi}{2}-\theta_1\right) + \cos n\left(\frac{3\pi}{2}+\theta_2\right) - \cos n\left(\frac{3\pi}{2}-\theta_2\right)\right]$$
$$+ \frac{I_p}{2\pi(n-1)}\left[\sin\left\{(n-1)\left(\frac{\pi}{2}-\theta_1\right)\right\} + \sin\left\{(n-1)\left(\frac{3\pi}{2}-\theta_2\right)\right\}\right.$$
$$\left. - \sin\left\{(n-1)\left(\frac{\pi}{2}+\theta_1\right)\right\} + \sin(n-1)(2\pi) - \sin\left\{(n-1)\left(\frac{3\pi}{2}+\theta_2\right)\right\}\right]$$
$$+ \frac{I_p}{2\pi(n+1)}\left[\sin\left\{(n+1)\left(\frac{\pi}{2}+\theta_1\right)\right\} - \sin\left\{(n+1)\left(\frac{\pi}{2}-\theta_1\right)\right\}\right.$$
$$\left. - \sin\left\{(n+1)\left(\frac{3\pi}{2}-\theta_2\right)\right\} - \sin(n+1)(2\pi) + \sin\left\{(n+1)\left(\frac{3\pi}{2}+\theta_2\right)\right\}\right]$$
$$+ \frac{I_F}{n\pi}\left[\cos n\left(\frac{\pi}{2}-\theta_1\right) - \cos n\left(\frac{\pi}{2}+\theta_1\right)\right] \tag{17.17}$$

In case of one-sided clipping by the pinch-off of the channel, amplitude of fundamental and up to fifth harmonics as well as DC current components are shown in Fig. 17.17 with the variation of the conduction angle using (17.13), (17.15) and (17.17). Thus, here $\theta_1 = 0$ and the conduction angle of the current waveform are $2\pi - 2\theta_2$. It shows that, for the best conversion gain, Class-C bias must be selected. For example, for optimum doubler operation, an operating angle of approximately 0.66π must be selected that yields a second harmonic current amplitude about 0.275 times of the maximum output current, I_F. For a frequency tripler, an operating angle of approximately 0.45π must be selected for a third harmonic current amplitude equal to about 0.185 times of the maximum output current. Similarly, for fourth- and fifth-order harmonics amplitude will be maximum of values 0.138 and 0.110 times of the maximum output current for the conduction angle of 0.34π and 0.26π, respectively.

The output load impedance must be such that the voltage swing corresponding to the required harmonic component becomes maximum without crossing the voltage

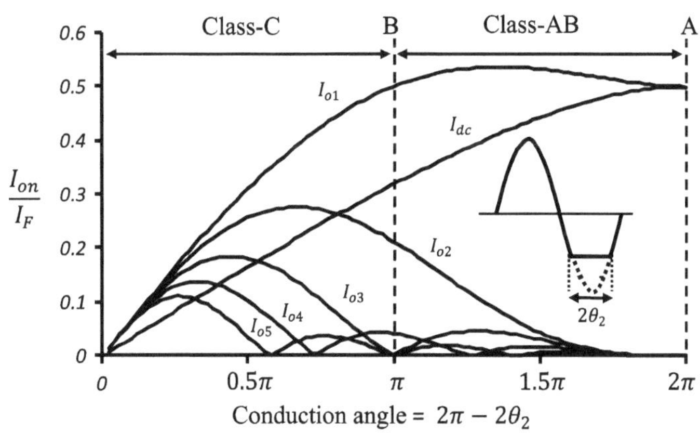

Fig. 17.17 Normalized amplitudes of harmonics with current conduction angle ($2\pi - 2\theta_2$), for one-sided clipping ($\theta_1 = 0$)

breakdown limit. All output harmonics must be shorted except the desired output harmonic. Thus, the optimum output load is the resistance that maximizes the voltage swing within the hard non-linear limits imposed by breakdown and knee voltage for FETs. Thus, In case of FET multiplier, the load impedance of nth harmonic is given by

$$R_{Ln} = \frac{V_{dsn}}{I_{dsn}} = \frac{V_{dgB} + 2V_{gs} - V_F - V_k}{2I_{dsn}} \tag{17.18}$$

Here, V_{gs} is given by (17.8c). Symmetric clipping of the output current waveform, one side by pinch-off and other side by saturation of the FET channel, happen when FET is biased for Class-A operation. Thus, here $\theta_1 = \theta_2$ and the conduction angle of the current waveform is $2\pi - 2\theta_2$. Using (17.13), (17.15) and (17.17), the normalized amplitude of the harmonic components for the symmetric clipped waveform are shown in Fig. 17.18. There is no existence of even order harmonic components due to the symmetric nature of the current waveform. However, relatively larger maximum current amplitude of all the odd order harmonics are obtained compared to the one side clipped current waveform. For example, maximum amplitudes of the third and fifth harmonic current components are 0.212 and 0.127 times, respectively, of the maximum current, I_F. Thus, more output power, i.e. higher conversion gain can be obtained for the odd harmonic components due to the symmetric current clipping compared to the one side clipped waveform but at the expense of a more critical reliability due to the forward gate current conduction and higher dissipated power. The direct generation of harmonics of higher order can be obtained by biasing the device at different conduction angles which are determined by the gate biasing of the transistors. For even order harmonic generation the device needs to be biased in Class-AB or C, in order to get rectified sinusoidal output current rich in even order harmonics. For odd order harmonics generation, the optimum bias condition is to generate an output current waveform with both sides clipping. Thus, Class-A biasing about centre of the output current is required and the input voltage/current should have high enough amplitude to clip the output current waveform in both sides. The magnitude of harmonic components decreases with the increase of its order thus it is required to increase the load impedance with the increase of harmonic order to get output voltage of higher amplitude.

By varying the bias voltage and the amplitude of the input voltage, all possible combinations of upper and lower current clippings can be obtained. The normalized amplitude of different harmonics in this general case of non-symmetric clipping can be obtained using Eqs. (17.13), (17.15) and (17.17). The results are plotted as contour plots of constant second harmonic current amplitude, as shown in Fig. 17.19. The plot is symmetric with respect to the bisector $2\theta_1 + 2\theta_2 = 2\pi$. The points on the bisector represent square waves with duty cycle from 0 ($2\theta_1 = 0, 2\theta_2 = 2\pi$) to 1 ($2\theta_1 = 2\pi, 2\theta_2 = 0$). The point at middle on this line ($2\theta_1 = 2\theta_2 = \pi$) corresponds to symmetric square wave; thus, there is no second harmonic component. The plot is anti-symmetric with respect to the bisector line $2\theta_1 = 2\theta_2$, which represents symmetric clipping of the output current waveform from a Class-A bias condition;

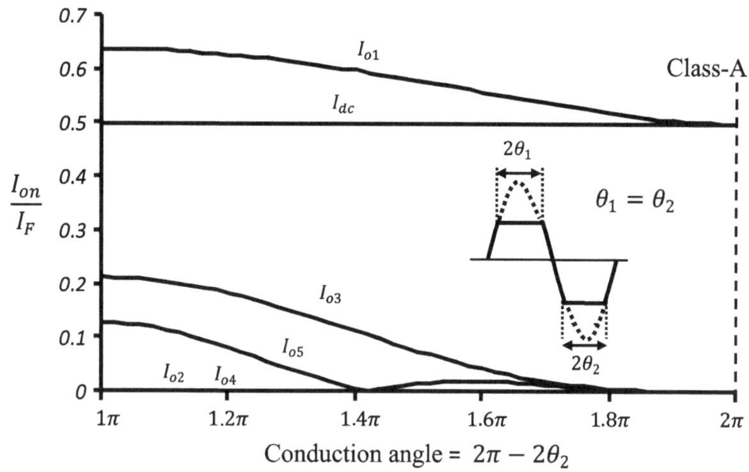

Fig. 17.18 Normalized amplitudes of harmonics with current conduction angle ($2\pi - 2\theta_2$) for symmetric clipping ($\theta_1 = \theta_2$)

thus, there is no second harmonic component along this bisector. It is clear from the constant second harmonic contour plot that the normalized maximum current component is about 0.318 which is higher than the one side clipped current waveform of value 0.275. From the contour, the maximum normalized amplitude (0.318) of the second harmonic component corresponds to asymmetric clipping around a square waveform with duty cycle equal to 1/4 (or symmetrically 3/4).

The constant amplitudes of the third harmonic current component for different values of clipping angles are shown in Fig. 17.20. In case of third harmonic component also, the plot is symmetric with respect to the bisector $2\theta_1 + 2\theta_2 = 2\pi$. The points on the bisector represent square waves with duty cycle from 0 ($2\theta_1 = 0, 2\theta_2 = 2\pi$) to 1 ($2\theta_1 = 2\pi, 2\theta_2 = 0$). The point at middle on this line ($2\theta_1 = 2\theta_2 = \pi$) corresponds to symmetric square wave; the amplitude of the third harmonic component becomes highest of normalized value 0.212. It is interesting to see that another two combinations of clippings are corresponded to the maximum third harmonic waveform exists. The maximum normalized amplitude (0.212) of the third harmonic component corresponds to asymmetric clipping around a square waveform with duty cycle equal to 1/6 (or symmetrically 5/6). The plot is asymmetric with respect to the bisector line $2\theta_1 = 2\theta_2$, which represents symmetric clipping of the output current waveform from a Class-A bias condition.

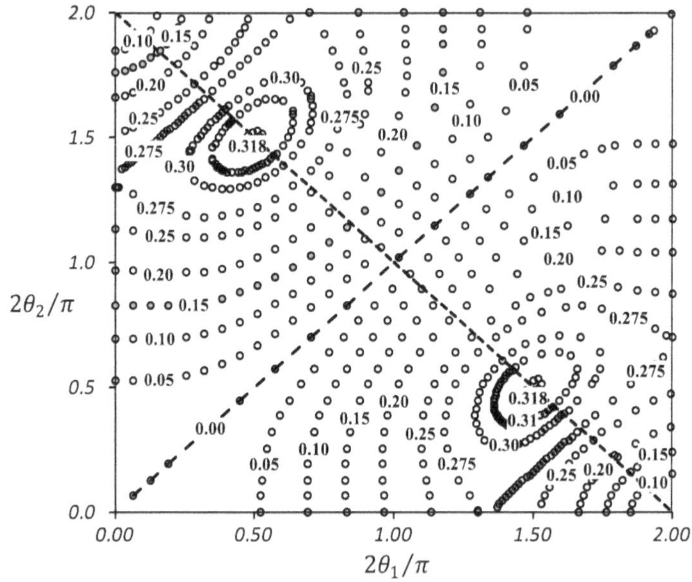

Fig. 17.19 Constant amplitude contour of second harmonic over clipping angles

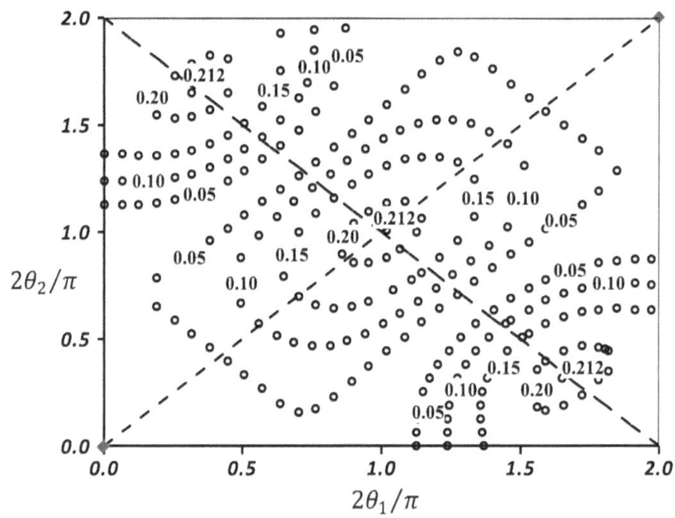

Fig. 17.20 Constant amplitude contours of third harmonic over clipping angles

17.5 Realization of FET Multipliers

A generalized schematic circuit diagram of a FET multiplier is shown in Fig. 17.21. Ideally, the input and output matching networks provide the required impedance to each harmonic component independently. The transistor is biased at sufficiently negative bias to operate it in Class-AB, B or C mode depending on the requirement of output current clipping as discussed in the previous section. To achieve higher conversion gain, in most of the cases the requirement of gate voltage will be smaller than the pinch-off voltage, V_{gso}, and the FET will remain OFF during most of the signal cycle. Thus, in most of the cases, the duty cycle will be smaller than 50%. The drain bias is applied depending on its breakdown voltage and required output voltage swing. The duty cycle of the output current waveform is adjusted by adjusting the gate bias voltage. The input matching network provides complex conjugate match at the input port of the device corresponds to fundamental frequency for maximum power transfer into the device if permissible from the stability point of view. Similarly, the output port is also complex conjugate matched at the desired harmonic frequency to transfer maximum power to the load for the desired harmonic. The input matching circuit also consists of resonators, also called idler circuit, which provides short circuit to the desired output as well as other harmonic frequencies. The output matching circuit also consists of resonators which provide fundamental and other harmonic frequency components except for the desired frequency. It is desirable to realize the idler circuits at input and output ports using reactive elements for minimizing power dissipation, i.e. for maximizing conversion gain. Also, reactive terminations reflect back the signal into the active device; this contribution can be exploited for increasing the conversion gain by constructive interference.

Microstrip circuit layout of a microwave frequency doubler using FET is shown in Fig. 17.22. Here, negative gate bias is fed to the device through a short-circuited $\lambda_o/4$ transmission line. On the other hand, drain bias is given to the device through a

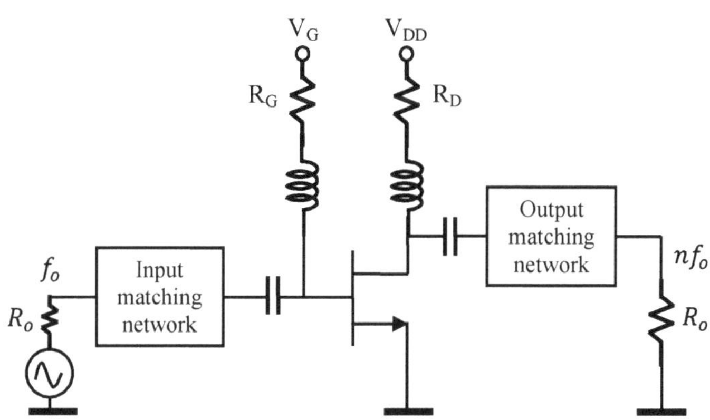

Fig. 17.21 Generalized circuit diagram of FET multiplier

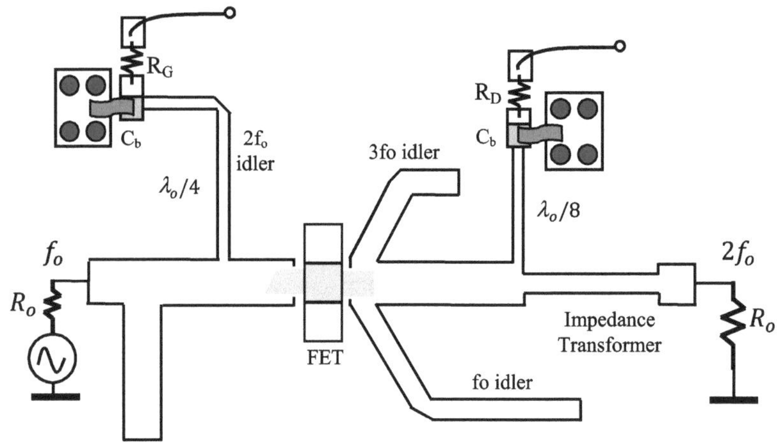

Fig. 17.22 Circuit layout of an FET doubler

short-circuited $\lambda_o/8$ high impedance transmission line which is quarter wavelength long corresponds to the desired second harmonic frequency. The gate bias network also provides short to the second harmonic frequency at the input of the device. The input of the device is complex conjugate matched to the source impedance by using series and stunt matching network. Two idler circuits are connected at the output of the device correspond to undesired fundamental and third harmonic components using open stub transmission lines. A high impedance transmission line is used as impedance transformer to match the output port for the desired second harmonic component.

Practical way to design a frequency multiplier without having its non-linear model is the use of non-linear harmonic load pull test method. In this case, the active device is inserted into a circuit where the input port is tuned for the fundamental frequency and the output port is tuned for the maximum possible harmonic components for achieving the maximum amplitude of the desired harmonic component. After that, the active device is removed and both the input and output matching networks are measured using a vector network analyzer to determine the desired impedances. Then the desired impedances can be synthesized using linear circuit analysis method.

17.6 Balanced Frequency Multipliers

Balanced frequency multipliers are preferred configuration at microwave and millimetre wave frequency for simplifying idler circuits and providing higher efficiency, i.e. more conversion gain at the desired frequency. Another property of the balanced frequency multipliers is the inherent rejection of certain unwanted harmonic frequency components. Balanced frequency-multiplier configuration can be realized with diodes as well as transistors.

17.6.1 Balanced Frequency Multipliers Using Diodes

A simple schematic circuit diagram of a diode-based balanced frequency multiplier is shown in Fig. 17.23. Here, two diodes with antiparallel connection are series connected to the transmission line. A low-pass filter (LPF) at the input side of the diodes is used to allow the fundamental frequency component to the diodes, and to prevent leaking the produced third order harmonic into the input side (i.e. source). The third harmonic produced by the diodes will be passed through the bandpass filter to reach the output load. The bandpass filter (BPF) will reject the fundamental as well as other unwanted harmonic components and prevents to reach the output load resistance.

The non-linear diode current can be written as

$$i_d = a_0 + a_1 v_d + a_2 v_d^2 + a_3 v_d^3 + a_4 v_d^4 + a_5 v_d^5 + \cdots \tag{17.19}$$

Here, v_d is the voltage across the diode from anode to cathode and i_d is the diode current also with the direction from anode to cathode. Therefore, the currents through the diode as shown in Fig. 17.23 can be expressed as

$$i_{d1} = a_0 + a_1 v_{d1} + a_2 v_{d1}^2 + a_3 v_{d1}^3 + a_4 v_{d1}^4 + a_5 v_{d1}^5 + \cdots \tag{17.20a}$$

$$i_{d2} = a_0 + a_1 v_{d2} + a_2 v_{d2}^2 + a_3 v_{d2}^3 + a_4 v_{d2}^4 + a_5 v_{d2}^5 + \cdots \tag{17.20b}$$

Here, the magnitude of the diode voltages are same but of opposite direction. Thus, $v_{d2} = -v_{d1} = v_d$, (say).

Therefore, the load current can be written as

$$i_o = i_{d1} - i_{d2} = 2a_1 v_d + 2a_3 v_d^3 + 2a_3 v_d^5 + \cdots \tag{17.21}$$

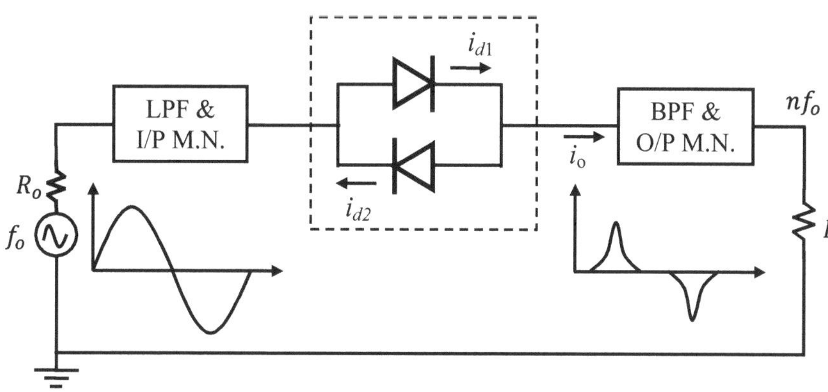

Fig. 17.23 Simple schematic circuit of a diode balanced frequency multiplier

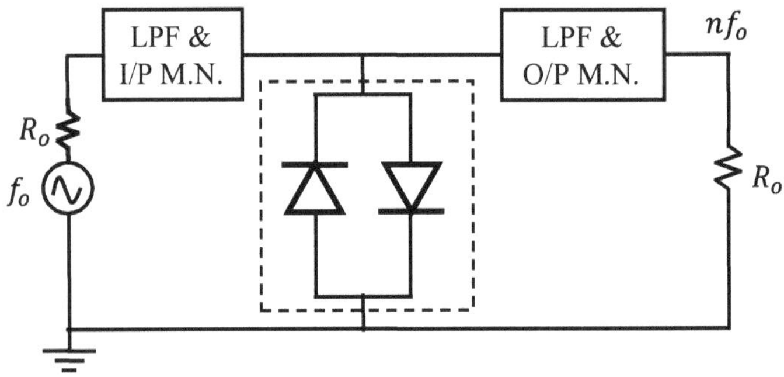

Fig. 17.24 Schematic circuit of a shunt connected antiparallel diode balanced frequency multiplier

This shows that the output current of the antiparallel diode contains only odd harmonics. The even harmonic components are not generated due to the balanced configuration of the multiplier. This is also clear from the output current waveform which has half-wave symmetry, thus contains only odd harmonic components. The bandpass filter at the output allows only the required odd harmonic component and will reject all other harmonic components. The output matching network transfer the output impedance of the diodes to the load resistance corresponds to the desired harmonic component.

Another scheme to design a balanced multiplier is shown in Fig. 17.24. Antiparallel diodes are shunt connected with the transmission line. From the same argument, it can be shown that this configuration also will generate only odd harmonic components. The input low-pass filter (LPF) allows the fundamental frequency signal to the diodes. It also rejects the higher order harmonics. The input matching network matches the source impedance to the input impedance of the diodes for maximum signal transfer to the diodes. The bandpass filter (BPF) only passes the desired harmonic and rejects the other harmonics.

17.6.2 Balanced Frequency Multiplier Using Transistors

Balanced frequency multipliers using transistors generally provide broadband performance due to the improved input and output impedance matching over the broad frequency range. The performance of balanced multipliers also becomes very stable since the devices are terminated with 50 Ω over the very wide frequency range. Balanced frequency multipliers provide 3-dB more harmonic power than an equivalent single-device multiplier circuit.

Schematic circuit diagram of a balanced frequency multiplier using 3-dB 90° coupler is shown in Fig. 17.25. Here, resistors R_g, R_s and drain supply voltage V_D used for self-biasing the transistors. Values of the resistors are chosen to operate the individual FETs in Class-B or C mode. The input coupler divides the signal

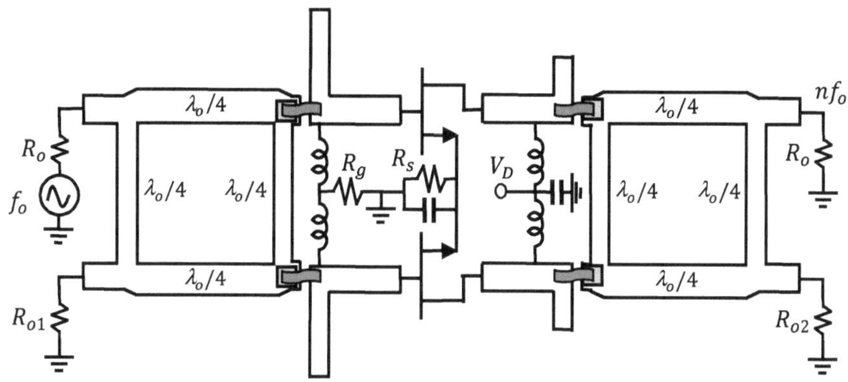

Fig. 17.25 A transistor (FET) based balanced multiplier

into two equal amplitudes with 90° phase shift before applied to the transistors. The output coupler will introduce another 90° phase shift in between the two signals before combining at the output port of load impedance R_o. Therefore, fundamental frequency component and all the odd harmonics generated by the transistors will have 180° phase shift in between the signals coming from the two paths at the output load and cancelled each other. These odd harmonic powers will reach to the isolation resistor R_{o2} and dissipated. In this configuration, only the even order harmonics coming from the two paths will reach the load termination with relative 0° phase shift and added in power. Using proper bandpass filter at the output, required even order harmonic frequency component is selected. Balanced multiplier circuit also can be realized using a 180° rat-race hybrid coupler at the input to drive the transistors in anti-phase and combining the outputs from both the transistors by using an in-phase combiner or simply by parallel connection.

References

1. Gilmore R (1987) Concepts in the design of frequency multipliers. Microw J 129–139
2. Thomas DG, Branner GR (1996) Optimization of active microwave frequency multiplier performance utilizing harmonic terminating impedances. IEEE Trans Microw Theory Tech 44(12):2617–2624
3. Cronin NJ, Law VJ (1985) Planar millimeter-wave diode mixer. IEEE Trans Microw Theory Tech 33(9):827–830
4. Benson K, Frerking MA (1985) Theoretical efficiency for triplers using real varister diodes at submillimeter wavelengths. In: IEEE MTT-S international microwave symposium digest, pp 315–318
5. East J, Kollberg E, Frerldng M (1993) Performance limitations of varactor multipliers. In: Presented at the fourth international conference on space terahertz technology, Los Angles, California, March 1993

6. Louhi J, Raisanen A (1994) On the modeling and optimization of varactor frequency multipliers at submillimeter wavelengths. In: Presented at the fifth international conference on space terahertz technology, Ann Arbor, Michigan, May 1994
7. Erickson NR, Tuovinen J, Rizzi BJ, Crowe TW (1994) A balanced doubler using a planar diode array for 270 GHz. In: 5th International symposium on space terahertz technology, pp 409–413
8. East J (1996) High frequency limitations of diode frequency multipliers. In: Seventh international symposium on space terahertz technology, Charlottesville, pp 116–124, March 1996
9. Zhao M, Fan Y, Wu D, Zhan J (2008) The investigation of W-band microstrip integrated high order frequency multiplier based on the nonlinear model of avalanche diode. Prog Electromagnetics Res 85:439–453
10. Gopinath A, Rankin JB (1982) Single-gate MESFET frequency doublers. IEEE Trans Microw Theory Tech MTT-30(6):869–875
11. Rauscher C (1983) High-frequency doubler operation of GaAs field-effect transistors. IEEE Trans Microw Theory Tech MTT-31(6):462–473
12. Borg M, Branner GR (1991) Novel MIC bipolar frequency doublers having high gain wide bandwidth and good spectral performance. IEEE Trans Microw Theory Tech 39(12):1936–1946
13. Chen SW et al (1993) A high performance 94-GHz MMIC doubler. IEEE Microw Guided Wave Lett 3(6):167–169
14. Augusto JS, Rosario MJ, Vaz JC, Freire JC (1993) Optimal design of MESFET frequency multipliers. In: Proceedings of the 23rd European microwave conference, Madrid, pp 402–405, September 1993

Chapter 18
Microwave Frequency Mixers

Abstract This chapter presents the working principle, design and analysis of various types of microwave frequency mixers based on diodes and transistors for communication systems. Noise performance analysis and importance of image reject filters in the design of mixers are discussed. Design details and analysis of single-ended, single-balanced, double-balanced, image reject and subharmonic mixers are also presented along with examples.

18.1 Introduction

Mixers are used in transmitters and receivers to translate signals in frequency domain. It mixes two input signals to provide an output signal of frequency either sum and/or difference frequency of the input signals. It is fundamentally a multiplier having three ports: a radio frequency (RF) port, a local oscillator (LO) port and an intermediate frequency (IF) port. Practically, the LO signal is CW, whereas the RF and IF signals are modulated carriers. In the transmitter section, mixers are used for modulation and upconversion to ease wireless transmission. In the receiver section, mixers are used for downconversion and demodulation. In principle, any nonlinear device can be used to design a mixer circuit. Mainly Schottky barrier diodes, GaAs FETs and BJTs are used as nonlinear device for realization of microwave mixers [1–15]. It is also possible to use CMOS technology to realize microwave and millimetre wave mixers. However, for high-performance applications at higher microwave and mmwave frequency range, Schottky diode mixers are preferable [1, 2].

18.2 Working Principle of Frequency Mixers

Frequency mixers produce output signal of frequency equal to the sum or difference of the frequencies of the two input signals, as shown in Fig. 18.1. Specifically, in case of downconversion, the frequency (f_{OUT}) of the output signal, called intermediate

S. C. Bera, *Microwave Active Devices and Circuits for Communication*, Lecture Notes in Electrical Engineering 533, https://doi.org/10.1007/978-981-13-3004-9_18

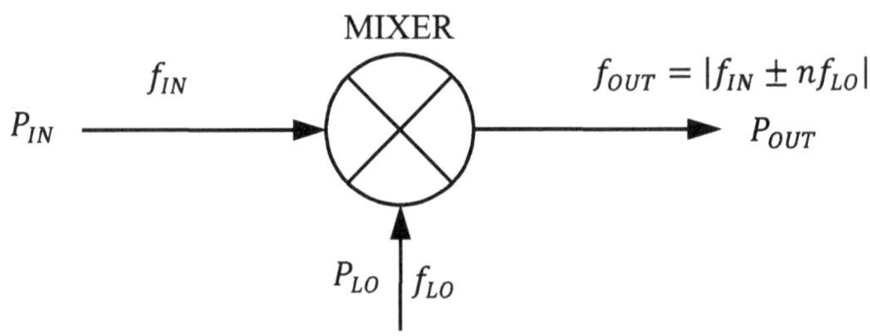

Fig. 18.1 Conceptual block diagram of the frequency mixer

frequency (f_{IF}), is the difference of frequencies of the two input signals namely input frequency (f_{IN}), called RF frequency (f_{RF}), and nth harmonic of local oscillator frequency (nf_{LO}) and is given by

$$f_{\mathrm{OUT}} = f_{\mathrm{IF}} = |nf_{\mathrm{LO}} - f_{\mathrm{RF}}| \tag{18.1}$$

Whereas in case of upconversion, frequency of the output signal (f_{OUT}), called RF frequency (f_{RF}), is the sum of frequencies of the input signal (f_{IN}) called IF frequency (f_{IF}) and nth harmonic of local oscillator frequency (nf_{LO}) and is given by

$$f_{\mathrm{OUT}} = f_{\mathrm{RF}} = nf_{\mathrm{LO}} + f_{\mathrm{IF}} \tag{18.2}$$

Mixers of desired output with fundamental frequency component of the local oscillator, i.e. for $n = 1$, are known as fundamental mode mixers. Whereas, mixers having desired output with harmonics of the local oscillator, i.e. for $n = 2$ or 3 or 4 . . ., are known as subharmonic mixers (SHM). In case of the subharmonic mixing, the LO frequency is internally multiplied, thus producing mixing components from the RF frequency and an integer multiple of the LO frequency. Practically, it is easier to design a lower frequency LO with the required higher output power than the design of higher frequency LO with same output power. In SHM, local oscillator of lower frequency is required. The reduction of LO frequency can simplify the LO design, and this makes the subharmonic mixing technique very attractive.

An important requirement for mixers is linearity in its RF performance. Thus, to maintain the input–output linear relationship for the RF signal, the device of the mixer needs to operate in nonlinear condition to produce frequency translation. For the mixer to operate in the linear mode, LO power should be sufficient to drive the device by a suitable sinusoidal signal in its saturated condition to generate nonlinearity and the RF signal level should be small enough compared to the LO level to maintain linearity.

A mixer consists of at least one device forced to operate in its nonlinear region to generate the frequency translation. The nonlinearity of a device can be expressed by the following nonlinear input–output relationship:

$$v_o = a_0 + a_1 v_i + a_2 v_i^2 + a_3 v_i^3 + \ldots \tag{18.3}$$

Here, a_0, a_1, a_2, a_3 are the constants. Suppose the input signal (v_i) with two different frequencies $\omega_{\mathrm{IN}} = 2\pi f_{\mathrm{IN}}$ and $\omega_{\mathrm{LO}} = 2\pi f_{\mathrm{LO}}$ is applied to the device. Thus,

$$v_i = v_{\mathrm{IN}} \cos \omega_{\mathrm{IN}} t + v_{\mathrm{LO}} \cos \omega_{\mathrm{LO}} t$$

Here, v_{IN} and v_{LO} are amplitude of the signals. Considering up to third-order terms, the output will be given by

$$\begin{aligned}
v_o &= a_0 + a_1(v_{\mathrm{IN}} \cos \omega_{\mathrm{IN}} t + v_{\mathrm{LO}} \cos \omega_{\mathrm{LO}} t) + a_2(v_{\mathrm{IN}} \cos \omega_{\mathrm{IN}} t + v_{\mathrm{LO}} \cos \omega_{\mathrm{LO}} t)^2 \\
&\quad + a_3(v_{\mathrm{IN}} \cos \omega_{\mathrm{IN}} t + v_{\mathrm{LO}} \cos \omega_{\mathrm{LO}} t)^3 \\
&= \left(a_0 + \frac{a_2 v_{\mathrm{IN}}^2}{2} + \frac{a_2 v_{\mathrm{LO}}^2}{2}\right) \\
&\quad + \left(a_1 v_{\mathrm{IN}} + \frac{3a_3 v_{\mathrm{IN}}^3}{4} + \frac{3a_3 v_{\mathrm{IN}} v_{\mathrm{LO}}^2}{2}\right) \cos \omega_{\mathrm{IN}} t \\
&\quad + \left(a_1 v_{\mathrm{LO}} + \frac{3a_3 v_{\mathrm{LO}}^3}{4} + \frac{3a_3 v_{\mathrm{IN}}^2 v_{\mathrm{LO}}}{2}\right) \cos \omega_{\mathrm{LO}} t \\
&\quad + \frac{a_2 v_{\mathrm{IN}}^2}{2} \cos 2\omega_{\mathrm{IN}} t + \frac{a_2 v_{\mathrm{LO}}^2}{2} \cos 2\omega_{\mathrm{LO}} t \\
&\quad + a_2 v_{\mathrm{IN}} v_{\mathrm{LO}} \cos(\omega_{\mathrm{IN}} + \omega_{\mathrm{LO}}) t \Rightarrow \text{desired for upconversion} \\
&\quad + a_2 v_{\mathrm{IN}} v_{\mathrm{LO}} \cos(\omega_{\mathrm{IN}} - \omega_{\mathrm{LO}}) t \Rightarrow \text{desired for downconversion} \\
&\quad - \frac{a_3 v_{\mathrm{IN}} v_{\mathrm{LO}}}{4} \cos \omega 3_{\mathrm{IN}} t + \frac{a_3 v_{\mathrm{LO}}^3}{4} \cos 3\omega_{\mathrm{LO}} t \\
&\quad + \frac{3a_3 v_{\mathrm{IN}}^2 v_{\mathrm{LO}}}{4} \cos(2\omega_{\mathrm{IN}} + \omega_{\mathrm{LO}}) t \\
&\quad + \frac{3a_3 v_{\mathrm{IN}} v_{\mathrm{LO}}^2}{4} \cos(2\omega_{\mathrm{LO}} + \omega_{\mathrm{IN}}) t \Rightarrow \text{desired } 2 \times \text{subharmonic mixing} \\
&\quad + \frac{3a_3 v_{\mathrm{IN}}^2 v_{\mathrm{LO}}}{4} \cos(2\omega_{\mathrm{IN}} - \omega_{\mathrm{LO}}) t \\
&\quad + \frac{3a_3 v_{\mathrm{IN}} v_{\mathrm{LO}}^2}{4} \cos(2\omega_{\mathrm{LO}} - \omega_{\mathrm{IN}}) t \Rightarrow \text{desired } 2 \times \text{subharmonic mixing}
\end{aligned} \tag{18.4}$$

From (18.4), it is clear that a large number of spectral components are produced by the device nonlinearity. Output frequencies are shown in Fig. 18.2 up to higher order mixing considering higher LO signal level compared to the RF signal level. For example, the squared term in (18.3) produces the sum and difference frequencies, ($\omega_{\mathrm{IN}} + \omega_{\mathrm{LO}}$) and ($\omega_{\mathrm{IN}} - \omega_{\mathrm{LO}}$), respectively, which are desired for fundamental frequency mixing as well as undesired double frequency terms, $2\omega_{\mathrm{IN}}$ and $2\omega_{\mathrm{LO}}$. The cubic term in (18.3) produces frequency components ($2\omega_{\mathrm{LO}} \pm \omega_{\mathrm{IN}}$), which are desired for $2\times$ subharmonic mixing in addition to the undesired frequency compo-

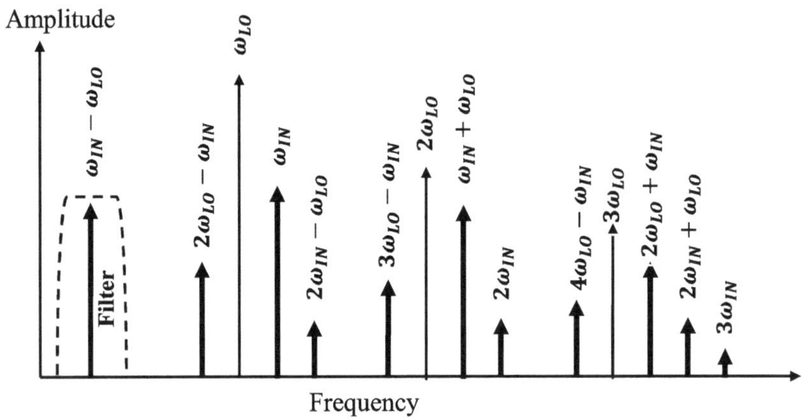

Fig. 18.2 Output spectrum of mixer

nents $3\omega_{IN}$, $3\omega_{LO}$, $(2\omega_{IN}-\omega_{LO})$ and $(2\omega_{IN}+\omega_{LO})$. Therefore, square term of the nonlinearity produces fundamental mixing whereas higher order terms can generate subharmonic mixing. It is also clear that in addition to the desired frequency component, other undesired frequency components are also generated. In the linear region operation corresponding to the RF signal level, the components corresponding to the harmonic of RF frequency, e.g. $2\omega_{IN}$, $3\omega_{IN}$, $(2\omega_{IN}\pm\omega_{LO})$, etc., can be neglected due to their negligible level compared to the desired component. Other undesired components need to be eliminated by proper filtering and circuit topology. For example, in case of fundamental frequency mixing, a bandpass filter may be used to get only $(\omega_{IN}-\omega_{LO})$ as shown in Fig. 18.2. Most practical mixers use filters to reduce the levels of the unwanted spurious outputs. Another commonly used technique, for the reduction of some unwanted spurious, is the use of balanced mixer topology.

18.2.1 Image Frequency in Mixer

For the design of frequency mixer, special care shall be taken for the frequency band which is the image of the desired RF band with respect to the LO frequency [13]. The image frequency is twice the IF frequency away from the RF and will be converted directly to the same IF frequency as the RF as shown in Fig. 18.3. Practically, it is required to translate/detect a particular band of frequency. Thus, noise and unwanted signals present at this frequency band can severely degrade the system performance due to the interference with the desired signal. One way to get rid of from the image frequency is the use of a preselector filter preceding the mixer input to reject the image frequency band allowing the required frequency to the mixer input. Another way is the use of mixer topology with image reject property.

In the case of downconverting fundamental mixers, IF frequency is given by

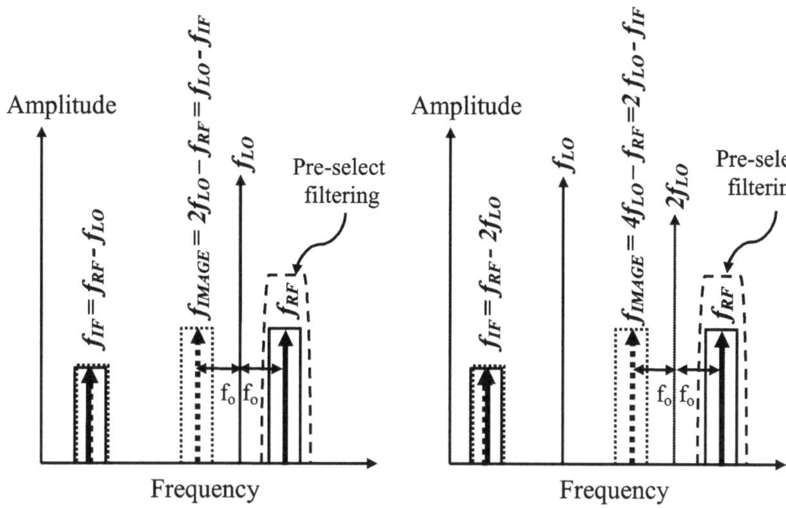

Fig. 18.3 Image spectrum for fundamental and 2× subharmonic mixing

$$f_{\mathrm{IF}} = f_{\mathrm{RF}} - f_{\mathrm{LO}} \tag{18.5a}$$

Thus the image frequency is given by

$$f_{\mathrm{IMAGE}} = f_{\mathrm{RF}} - 2f_{\mathrm{IF}} \tag{18.5b}$$

$$f_{\mathrm{IMAGE}} = |2f_{\mathrm{LO}} - f_{\mathrm{RF}}| \tag{18.5c}$$

Similarly for 2× subharmonic mixer, the image frequency will be given by

$$f_{\mathrm{IMAGE}} = |4f_{\mathrm{LO}} - f_{\mathrm{RF}}| \tag{18.5d}$$

Practically, desired RF frequency is a band of frequency around the centre frequency. Correspondingly, the image frequency will also be a band of frequency. The passband of the preselect filter should have proper passband performance without introducing any distortion in the desired RF frequency band and very high rejection over the entire image frequency band.

Example 18.1 The receive frequency band for mobile satellite communication system is 6450–6485 MHz. For processing the spectrum, it is translated to IF band of 50–85 MHz using a local oscillator of frequency 6400 MHz. Find out the image frequency band.

Solution

The image frequency band can be determined from (18.5b) as

$$
\begin{aligned}
f_{\mathrm{IMAGE}} &= f_{\mathrm{RF}} - 2f_{\mathrm{IF}} \\
&= (6450 \text{ to } 6485)\,\mathrm{MHz} - 2 \times (50 \text{ to } 85)\,\mathrm{MHz} \\
&= (6350 \text{ to } 6315)\,\mathrm{MHz}
\end{aligned}
$$

18.2.2 Conversion Loss of Frequency Mixer

The conversion gain of mixers is an important parameter. It is defined as the ratio of output power (P_{OUT}) of the desired output frequency to the input power (P_{IN}) of the input frequency as shown in Fig. 18.1. Basic scheme of an idealistic mixer is shown in Fig. 18.4. At the RF (input) and LO sides, a balun is necessary in order to convert unbalanced inputs into balanced signals. Conversely, no adapter is needed at the IF output, which is already unbalanced. In low-frequency mixers, the baluns may be implemented with small transformers. However, at higher frequencies, microstrip networks may be the preferred option for realization of baluns. The basic assumptions are that the LO signal $v_{\mathrm{LO}}(t)$ is large enough so that the device (e.g. Schottky diodes or FETs) acts as an ideal commutator under its control, and the input RF signal $v_{\mathrm{IN}}(t)$ is very small compared to it. Thus, a mixer is almost always operated with the LO input saturated at the nominal operating condition, whereas the input (RF) power is usually well below the saturation level to get linear RF performance. During the positive half-cycle of $v_{\mathrm{LO}}(t)$, one switch will be ON and other switch will remain OFF. During the other half-cycle, the roles are interchanged. Therefore, the output signal $v_{\mathrm{OUT}}(t)$ switches between $+v_{\mathrm{IN}}(t)$ and $-v_{\mathrm{IN}}(t)$ depending on the sign of $v_{\mathrm{LO}}(t)$. This is equivalent to multiplying $v_{\mathrm{IN}}(t)$ by a square wave of amplitude ± 1 that takes the sign of $v_{\mathrm{LO}}(t)$. Considering matched RF and IF ports and resistively termination of all the intermodulation (IM) products and no parasitic resistive or reactive losses, the output voltage can be represented mathematically by multiplying the input signal voltage with the Fourier series for the square wave function as

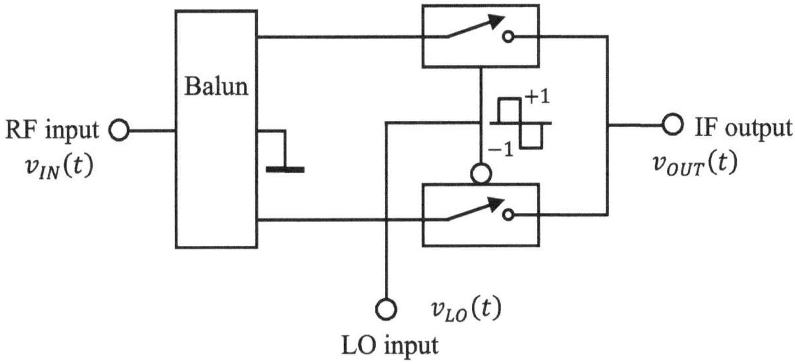

Fig. 18.4 Idealistic mixer configuration

$$v_{OUT}(t) = v_{IN}\sin(\omega_{IN}t)\left[\sum_{n=1,3,5,\ldots}\frac{4}{\pi n}\sin(n\omega_{LO}t)\right] \tag{18.6}$$

The terms within the square bracket are the Fourier series representation of square wave function (from Fig. 17.2, putting $A = 2$). Thus, inside the mixer, the local oscillator signal can be considered as the sum of the fundamental and all its odd harmonic components with amplitudes $(4/n\pi)$, where $n = 1, 3, 5, \ldots\infty$. Using trigonometry identity,

$$\begin{aligned} v_{OUT}(t) = &\frac{2v_{IN}}{\pi}[\sin(\omega_{LO}-\omega_{IN})t - \sin(\omega_{LO}+\omega_{IN})t] \\ &+\frac{2v_{IN}}{3\pi}[\sin(3\omega_{LO}-\omega_{IN})t - \sin(3\omega_{LO}+\omega_{IN})t] \\ &+\frac{2v_{IN}}{5\pi}[\sin(5\omega_{LO}-\omega_{IN})t - \sin(5\omega_{LO}+\omega_{IN})t] \\ &+\ldots \end{aligned} \tag{18.7}$$

Output of the mixer consists of two sidebands corresponding to the fundamental and all odd harmonics of the LO frequency. The signal amplitudes of each sideband of all the products are

$$\frac{2v_{IN}}{n\pi}, \quad n = 1, 3, 5, \ldots, \infty$$

Therefore, output signal power of each sideband is given by

$$P_{OUTn} = \frac{1}{R_o}\frac{4v_{IN}^2}{n^2\pi^2}, \quad n = 1, 3, 5, \ldots, \infty \tag{18.8}$$

Thus, the total output power is the sum of both the sidebands of all the harmonics power:

$$P_{OUT}(\text{Total}) = 2\times\sum_{n=1,3,\ldots\infty}\frac{1}{R_o}\frac{4v_{IN}^2}{n^2\pi^2} = \frac{v_{IN}^2}{R_o} \tag{18.9}$$

This is nothing but equal to the input RF power. Thus, the total power is conserved. The conversion loss is the reciprocal of the conversion gain and can be written as

$$\text{Conversion Loss} = \frac{P_{IN}}{P_{OUT}} \tag{18.10}$$

Here P_{IN} is the power of the RF input, and P_{OUT} is the power of the selected frequency at the output. For the fundamental frequency mixer, one of the sideband will be selected. The conversion loss of the fundamental mixer corresponds to one sideband is

$$\text{Conversion Loss} = \frac{v_{\text{IN}}^2}{R_o} / \frac{1}{R_o} \frac{4v_{\text{IN}}^2}{\pi^2} = \frac{\pi^2}{4} \tag{18.11}$$

$$\text{Conversion Loss(in dB)} = 10 \log \frac{\pi^2}{4} \text{dB} = 3.92 \text{ dB} \tag{18.12}$$

This is the minimum loss of the single-sideband corresponds to output frequency of $(\omega_{\text{LO}} - \omega_{\text{IN}})$ or $(\omega_{\text{LO}} + \omega_{\text{IN}})$.

The above analysis shows that when matched loads are presented to each IM product, and the RF, IF, and image signals are matched terminated, then the theoretical minimum conversion loss of a fundamental sideband is 3.92 dB. The theoretical conversion loss is 3 dB when all the IM products are reactively terminated and the IF and RF ports are matched terminated and also the image signals are properly terminated. In this case, the input RF power is equally divided in between the image band and required IF band. Thus, it can be concluded that the theoretical limit for the conversion loss is 0 dB in case all the IM products are reactively terminated and the image signals are short or open circuited. However, practically conversion loss has three major components: (a) mismatch loss at RF and IF ports, (b) loss due to the finite value of diode spreading and junction resistance, and (c) loss due to the generation of sideband and IM products.

18.2.3 Noise Performance of Frequency Mixer

Noise figure of a mixer is also defined as the signal-to-noise power ratio (SNR) at the input to the SNR at the output. All practical mixers fold the RF spectrum around the local oscillator frequency [1–7]. Thus, it creates an output which contains the summation of the spectrum on both sides since $f_{\text{IF}} = |f_{\text{RF}} - f_{\text{LO}}|$. Thus, a mixer (down converter) converts both the desired and the image bands to the desired output frequency (IF) band. This implies folding the noise present at the image frequency on top of the desired output frequency as shown in Fig. 18.5. Therefore, the overall noise at the translated output frequency band is (a) the noise at desired input frequency band converted to desired output frequency band, (b) the noise at input image frequency band converted to desired output frequency band and (c) the noise added by the mixer circuit itself.

In case of heterodyne receivers, only one side frequency band is used thus, in this case, one of these sides acts as spurious, i.e. noise and the other side as desired signal. Therefore, image reject filtering (in practice a bandpass filter) as shown in Fig. 18.6 or image cancelling mixer circuit topology is used to remove one of these responses. However, in case of direct-conversion receivers, both the sidebands, i.e. above and below $f_{\text{IF}} = f_{\text{LO}}$ are converted and utilized for the intended signal. Based on the applications and circuitry, noise figures (NF) of the mixer can be categorized as (a) single-sideband noise figure without image rejection mechanism (b) single-sideband noise figure with image rejection mechanism and (c) double-sideband noise figure.

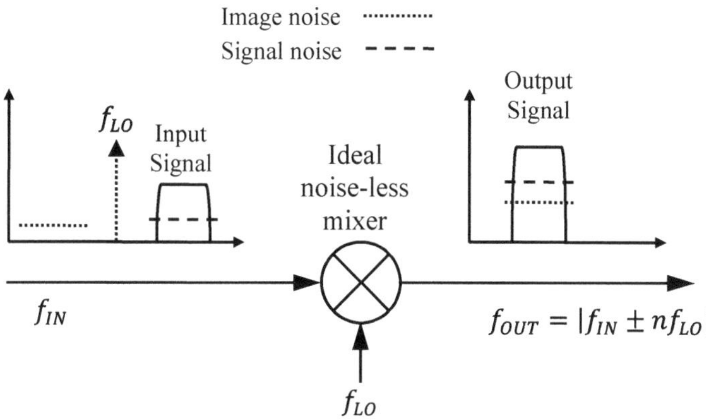

Fig. 18.5 Contribution of image frequency noise to IF band

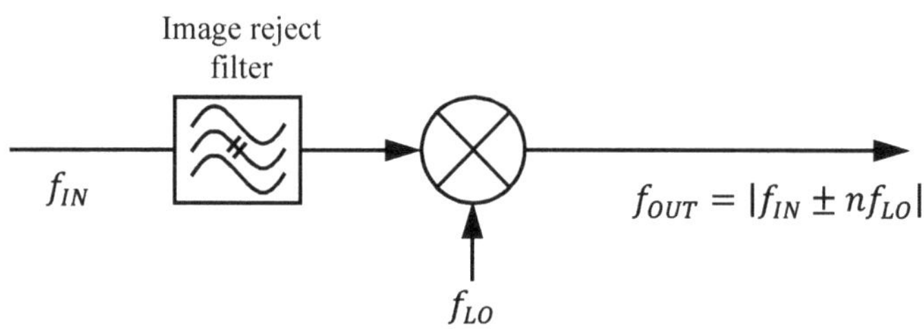

Fig. 18.6 Block diagram of mixer with image reject filter

In case of single-sideband NF without image frequency band rejection mechanism, it is considered that there is no signal at the image frequency. Thus, the image frequency band contributes to noise power. Suppose, over the desired frequency band, the input signal power and gain of the mixer are S_{IN} and G_{IN}. Then the output signal power over the desired frequency band is

$$S_{OUT} = S_{IN} \times G_{IN} \tag{18.13}$$

The total output noise power over the desired single-sideband is

$$N_{OUT} = N_{IN} \times G_{IN} + N_{IM} \times G_{IM} + N_{MIX_{IN}} \times G_{IN} + N_{MIX_{IM}} \times G_{IM} \tag{18.14}$$

Here, N_{IN} and N_{IM} are the input noise in the desired frequency band and image frequency band, respectively. $N_{MIX_{IN}}$, $N_{MIX_{IM}}$ are the noise power generated by the mixer at the desired and image frequency band, respectively, referred to the input of the mixer. And G_{IM} is the gain of the mixer at the image frequency band.

Therefore, ratio of the output signal power to the output noise power is

$$\frac{S_{OUT}}{N_{OUT}} = \frac{S_{IN} \times G_{IN}}{N_{IN} \times G_{IN} + N_{IM} \times G_{IM} + N_{MIX_{IN}} \times G_{IN} + N_{MIX_{IM}} \times G_{IM}} \tag{18.15}$$

Assuming $G_{IN} = G_{IM}, N_{MIX_{IN}} = N_{MIX_{IM}}, N_{IN} = N_{IM}$, the output signal-to-noise ratio can be written as

$$\frac{S_{OUT}}{N_{OUT}} = \frac{S_{IN}}{2 \times (N_{IN} + N_{MIX_{IN}})} \tag{18.16}$$

Therefore, the single-sideband noise factor without image frequency band rejection F_{SSB1} can be written as

$$F_{SSB1} = \frac{S_{IN}}{N_{IN}} / \frac{S_{OUT}}{N_{OUT}} = 2\left(1 + \frac{N_{MIX_{IN}}}{N_{IN}}\right) \tag{18.17}$$

In case of single-sideband NF, with image frequency band rejection mechanism (e.g. using image reject filter as in Fig. 18.6), it can be assumed that there is no signal power at the image frequency and no contribution of the image band noise at the output of the mixer. Then, the output signal power over the desired frequency band is

$$N_{OUT} = N_{IN} \times G_{IN} + N_{MIX_{IN}} \times G_{IN} + N_{MIX_{IM}} \times G_{IM} \tag{18.18}$$

Therefore, ratio of the output signal power to the output noise power is

$$\frac{S_{OUT}}{N_{OUT}} = \frac{S_{IN} \times G_{IN}}{N_{IN} \times G_{IN} + N_{MIX_{IN}} \times G_{IN} + N_{MIX_{IM}} \times G_{IM}} \tag{18.19}$$

Assuming $G_{IN} = G_{IM}, N_{MIX_{IN}} = N_{MIX_{IM}}$, the output signal-to-noise ratio can be written as

$$\frac{S_{OUT}}{N_{OUT}} = \frac{S_{IN}}{N_{IN} + 2 \times N_{MIX_{IN}}} \tag{18.20}$$

Therefore, the single-sideband noise factor with image frequency band rejection F_{SSB2} can be written as

$$F_{SSB2} = \frac{S_{IN}}{N_{IN}} / \frac{S_{OUT}}{N_{OUT}} = 1 + \frac{2 \times N_{MIX_{IN}}}{N_{IN}} \tag{18.21}$$

Therefore, in case of SSB signal, the use of image reject filter improves the noise figure by about 3-dB when contribution of noise by the mixer itself is less.

In case of double-sideband NF, it is considered that the image band contains both noise and an image signal which is identical to the desired input signal. Here, the image frequency band contributes to noise N_{IM} as well as signal power S_{IM}. Thus, the output signal power is

$$S_{\text{OUT}} = S_{\text{IN}} \times G_{\text{IN}} + S_{\text{IM}} \times G_{\text{IN}} \tag{18.22}$$

The total output noise power over the desired single-sideband is

$$N_{\text{OUT}} = N_{\text{IN}} \times G_{\text{IN}} + \times G_{\text{IM}} + N_{\text{MIX}_{\text{IN}}} \times G_{\text{IN}} + N_{\text{MIX}_{\text{IM}}} \times G_{\text{IM}} \tag{18.23}$$

Therefore, ratio of the output signal power to the output noise power is

$$\frac{S_{\text{OUT}}}{N_{\text{OUT}}} = \frac{S_{\text{IN}} \times G_{\text{IN}} + S_{\text{IM}} \times G_{\text{IN}}}{N_{\text{IN}} \times G_{\text{IN}} + N_{\text{IM}} \times G_{\text{IM}} + N_{\text{MIX}_{\text{IN}}} \times G_{\text{IN}} + N_{\text{MIX}_{\text{IM}}} \times G_{\text{IM}}} \tag{18.24}$$

Assuming $G_{\text{IN}} = G_{\text{IM}}, N_{\text{MIX}_{\text{IN}}} = N_{\text{MIX}_{\text{IM}}}, N_{\text{IN}} = N_{\text{IM}}$, the output signal-to-noise ratio can be written as

$$\frac{S_{\text{OUT}}}{N_{\text{OUT}}} = \frac{S_{\text{IN}}}{\left(N_{\text{IN}} + N_{\text{MIX}_{\text{IN}}}\right)} \tag{18.25}$$

Therefore, the double-sideband noise factor without image frequency band rejection F_{DSB} can be written as

$$F_{\text{DSB}} = \frac{S_{\text{IN}}}{N_{\text{IN}}} \Big/ \frac{S_{\text{OUT}}}{N_{\text{OUT}}} = 1 + \frac{N_{\text{MIX}_{\text{IN}}}}{N_{\text{IN}}} \tag{18.26}$$

It is clear from (18.17) and (18.26) that the single-sideband noise factor F_{SSB1} without any image frequency mechanism is exactly double of the double-sideband noise factor F_{DSB}, i.e. 3 dB higher. However, the difference of noise figure in between single-sideband noise factor with and without image frequency mechanism is not exactly 3-dB, actually the difference approached 3-dB as the mixer noise contribution of mixer itself is very low.

18.3 Single-Ended Mixers

Schottky barrier diodes are used as nonlinear device for the microwave as well as mmwave mixers because of its low forward threshold voltage and fast switching capability [1, 2]. Mostly unbiased diodes are used for mixers. However, sometimes small forward current bias is given to the diodes to achieve lower conversion loss. This is necessary when limited LO drive level is available. The diode is forward biased to have a quiescent operating point close to the region of maximum nonlinear characteristic, as shown in Fig. 18.7a, to generate maximum possible distortion by lower levels of the LO drive. The equivalent circuit of a Schottky diode is shown in Fig. 18.7b. Here, R_S is the parasitic resistance and R_d and C_d are the nonlinear junction resistance and capacitance of the Schottky diode, respectively. Under positive half-cycle of LO signal of sufficient level, the diode will be forward biased and

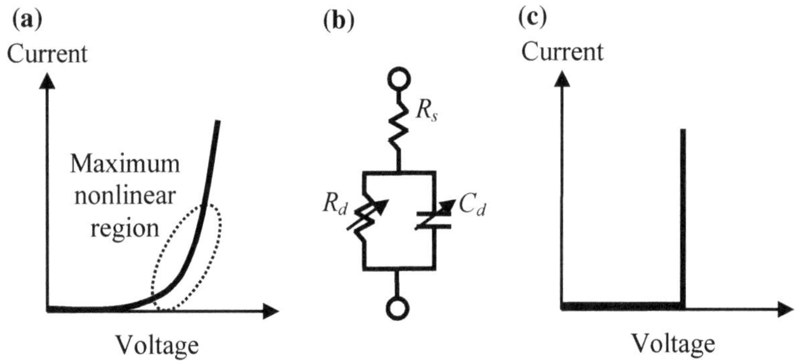

Fig. 18.7 *I-V* characteristic of (a) Schottky diode and (b) its equivalent circuit (c) idealistic commutator characteristic

thus the signal will experience very small impedance and during its other half-cycle diode resistance will be very high thus it will experience very high diode impedance. An idealized diode can be considered as ideal commutator which switches transconductance instantaneously as shown in Fig. 18.7c and yields the theoretically optimal diode mixer performance.

Single-ended mixers consist of single device and are unbalanced, thus, in general, there are no inherent isolations in between LO and RF ports. These mixers do not have any intrinsic property to eliminate any undesired frequency components generated by the mixing process. Diplexer may be used to combine LO and RF signal to provide the isolation before feeding to the mixer device. However, it is difficult to do this when the LO and RF frequencies are very close.

Basic block schematic of a single-ended mixer using single diode is shown in Fig. 18.8. Here, the LO, RF and IF ports are electrically same, and they are only separated by filters to prove port to port isolations. The required bandwidths of the filters must be sufficiently apart to provide high isolation. An LO signal of large amplitude compared to the RF signal level is combined at the anode of the diode after passing through respective bandpass filters. For idealistic operation, stronger LO signal needs to be applied so that the LO signal level has the capability to switch the transconductance of the diode in between high and low values. Level of the RF signal is low enough compared to the LO level to achieve linear RF performance. Practically no Schottky diode behaves as an ideal commutator. It has finite turn ON transition as shown in Fig. 18.7a. In addition to this, though the RF signal level is small compared to LO level, it will also modulate the diode transconductance to some extent, leading to nonlinearity in RF signals.

Suppose, RF and LO signal level v_{IN} and v_{LO}, with frequencies $\omega_{\mathrm{IN}} = 2\pi f_{\mathrm{IN}}$ and $\omega_{\mathrm{LO}} = 2\pi f_{\mathrm{LO}}$, respectively. Then the diode voltage

$$v_d = v_{\mathrm{IN}} \cos \omega_{\mathrm{IN}} t + v_{\mathrm{LO}} \cos \omega_{\mathrm{LO}} t$$

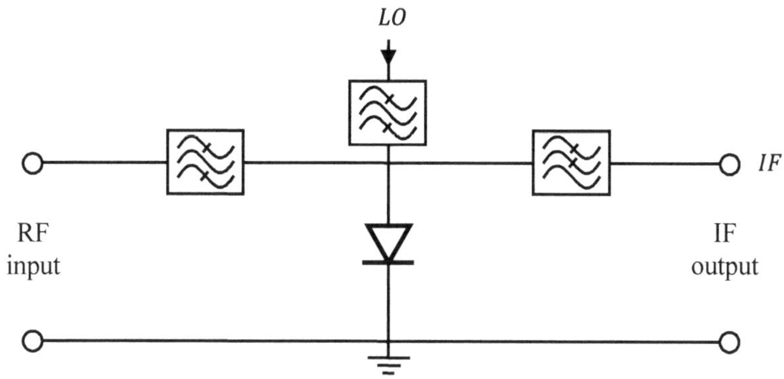

Fig. 18.8 Simple schematic circuit of a diode balanced frequency multiplier

The diode current can be written as

$$\begin{aligned} i_d &= I_o\left[\exp\left(\frac{v_d}{V_t}\right) - 1\right] \\ &= I_o\left[\frac{v_d}{V_t} + \frac{1}{2!}\left(\frac{v_d}{V_t}\right)^2 + \frac{1}{3!}\left(\frac{v_d}{V_t}\right)^3 + \frac{1}{4!}\left(\frac{v_d}{V_t}\right)^4 + \ldots\right] \end{aligned} \quad (18.27)$$

Here, I_o is the reverse saturation current and V_t is the diode thermal voltage. From (18.27), it is clear that exponential characteristic of the diode produces large number of spectral components due to the presence of square, cubic and other higher order terms. Thus, practical mixer circuit will produce all possible harmonic mixing components. Since, the single-ended mixer topology does not have any inherent spurious suppression capability, its output contain all harmonic components. Proper filtering at the output of the mixer is used to filter out the desired frequency band. This is the reason for unpopularity of single-ended mixer topology for modern mixers. However, in case of higher mmwave frequency design, single-ended mixer circuits are realized for their simplicity. Microstrip layout of a typical single-ended diode mixer is shown in Fig. 18.9.

Transistors are used for active as well as passive mixers [9, 14, 15]. Transistors are operated as switch for passive mixers. Drain–source resistance of a FET behaves as voltage-variable RF resistance. A simple schematic circuit diagram of a FET passive mixer is shown in Fig. 18.10. Here, the FET is operated at zero drain and source bias condition. LO signal acts as controlling the FET switch. The RF signals are applied at the drain terminal after passing through a bandpass filter. The IF signals are taken from the source terminal. Capacitor C_o at the drain terminal is used to provide a short circuit to the RF frequency. At IF port, an LPF is used to filter out the undesired frequency components which are generated due to the mixing of LO and RF components. Large enough LO signal is required to turn the FET OFF on the negative cycle in case of zero gate bias. Thus, it will drive the gate–source junction into forward bias on its

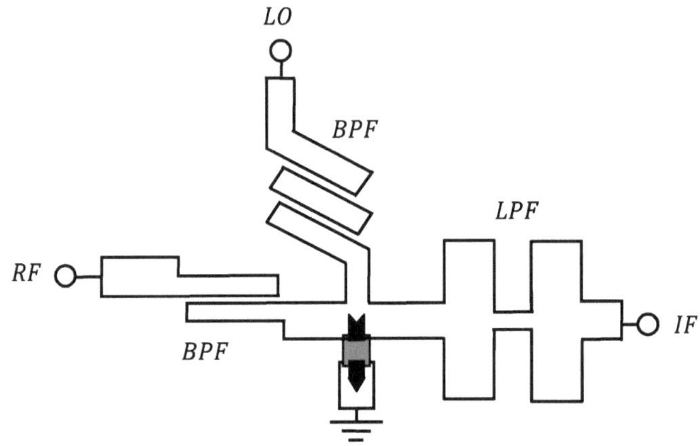

Fig. 18.9 Microstrip circuit layout of a simple single-ended diode mixer

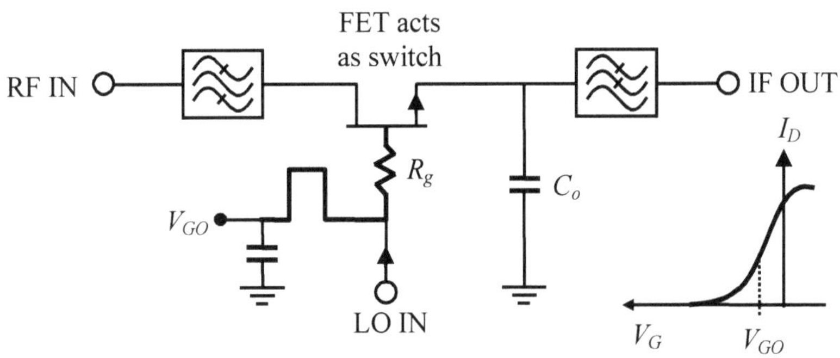

Fig. 18.10 Simple schematic circuit of passive mixer using FET as switch

positive half-cycle, which degrades the device reliability. To operate the FET as a switch by the LO signal without reliability problem, proper gate biasing is required. Therefore, it is necessary to set the gate bias voltage appropriately to operate the FET in switching mode keeping forward gate current within the allowable limit and without crossing the reverse gate breakdown voltage.

In case of active FET mixers, the nonlinear transconductance of the transistor is used for mixing the RF signals with LO signal. The LO signal is used to vary the transconductance of the transistor. Active FET mixers provide conversion gain rather than loss and also have lower noise figures than passive mixers. A simple circuit diagram of an active FET mixer is shown in Fig. 18.11. Gate and drain terminals of the FET are biased through $\lambda_{RF}/4$ and $\lambda_{IF}/4$ transmission lines, respectively. Output capacitor C_o is used to ensure the short circuit for RF and LO signal at the drain ter-

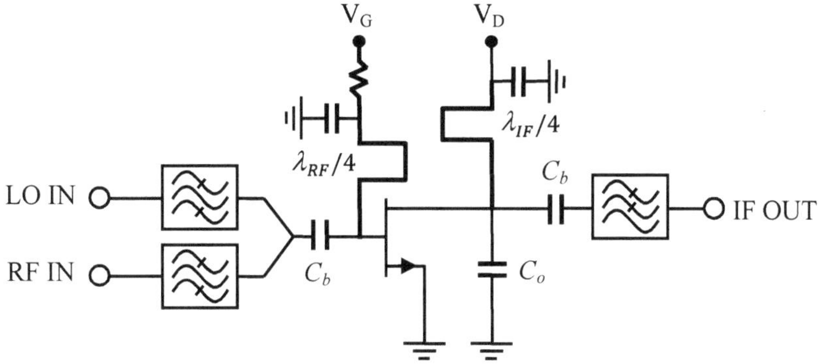

Fig. 18.11 A simple schematic circuit of a FET active mixer

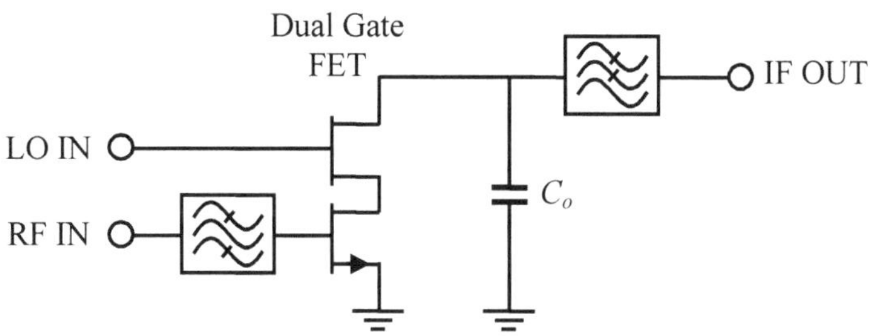

Fig. 18.12 Simple schematic circuit of a diode balanced frequency multiplier

minal of the FET to keep nearly constant drain bias with and without LO signal which maximizes time-varying transconductance for achieving maximum conversion gain. The difficulty of the circuit of Fig. 18.11 is the isolation in between RF and LO ports. Here, some form of diplexing of the RF and LO signal is required to separate the RF and LO inputs which are incident on the single gate terminal. To avoid this problem, dual-gate FET is used to realize mixer circuit as shown in Fig. 18.12. Dual-gate FET consists of two gate terminals to control its channel properties. Transconductance of the FET can be controlled independently by applying signal in both the gate terminals. Here, RF and LO signals are applied to the two separate gate terminals. Thus, the dual-gate FET structure provides isolation in between RF and LO for its inherent characteristics. Very compact active mixer circuit is designed using the dual-gate FET structure with the disadvantage of lower linearity compared to passive mixers.

18.4 Balanced Mixers

The mixer topologies so far discussed consist of a single device and are unbalanced. Thus, it is very difficult to achieve isolation in between LO and RF ports. It is not possible to design single-ended frequency mixers with isolation among RF, LO and IF ports, resulting in undesired signal interactions and one signal reaches to another port. Moreover, such circuitry does not have any intrinsic property to eliminate any undesired frequency components generated by the mixing process. Diplexer may be used to combine LO and RF signal to provide the isolation before feeding to the mixer device. However, it is difficult to do this when the LO and RF frequencies are very close. To provide isolations among LO, RF and IF ports and also to eliminate some unwanted mixing products from the output of a mixer, it is required to balance the signals. Balancing is out-phasing the signal to create reactive cancellation. The reactive cancellation is accomplished with a channelization technique using phase-correlating power dividers. The cancellation of undesired mixing products at the output of a mixer is determined by its balancing. At the RF and LO ports, a network called balun is necessary in order to convert the unbalanced inputs into the balanced signals. However, no such type of balun is required at the IF output, which we need unbalanced. Single-balanced and double-balanced mixers are popularly used which are classified depending on their configuration. Single-balanced mixers strongly attenuate either the input signal or the LO signal at the output port, but not both the signals simultaneously. Thus, less numbers of the above mixing products go to its output compared to single-ended mixers. A double-balanced mixer exhibits superior isolation among its LO, RF and IF ports. Here, only the sum and difference frequencies of the input and LO frequencies go to the output, while attenuating both the LO and RF signals, and provide significant rejection to three-quarters of the possible mixer products at the output. Thus, the balancing of the mixers minimizes the requirement of filtering out the undesired mixing products.

The balanced mixer uses more number of devices (diodes, transistors) thus, the requirement of LO power increases [8, 10, 11, 14, 15]. This increases the linearity of a balanced mixer due to the sharing of the power among the more number of devices working in the nonlinear region. A higher barrier height diode or more numbers of diodes in one arm may be used to improve linearity further. This enables higher drive level without overcoming the maximum limit of individual devices.

18.4.1 Single-Balanced Mixers

Single-balanced mixer uses two devices. It provides isolation in between one pair of signals. Here, either the LO signal or the RF signal is balanced. Any one of the signals added destructively at the IF port of the mixer and is rejected inherently. The level of rejection depends on the amplitude and phase balance of the balun,

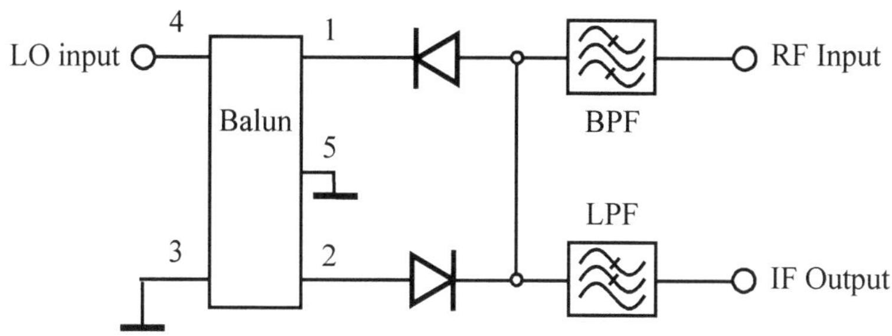

Fig. 18.13 Schematic circuit of a single-balanced diode mixer using balun

and the matching between the two devices. Depending on the configuration, single-balanced mixer rejects certain unwanted mixing products at the IF port.

Schematic circuit diagram of a single-balanced mixer shown in Fig. 18.13 consists of one balun two matched diodes. The balun converts the unbalanced LO signal to a balanced signal before feeding to the diodes. A balanced balun has characteristic to maintain phase angles of $\pm 90°$ at port-1, $\pm 90°$ at port-2 with respect to ground and $0°$ at port-5. Other important parameters of a balun are insertion loss and output-to-input impedance ratios. The single-balanced mixers have better LO-to-IF and LO-to-RF isolations due to the balanced characteristic of the balun and matched diode pair. Matched diode pair means the DC and RF characteristics of both the diodes are same. Due to the matched property of the diodes, a virtual ground is created between the diodes, which makes a null for the LO signal, thus providing isolation to the RF and IF ports. Thus, single balance mixers cancel the LO power which thus prevents LO leakage to the RF and IF ports. Thus, single-balanced mixers provide superior LO-to-IF and LO-to-RF isolation compared to single-ended mixers. The two filters BPF for RF port and LPF for IF port are used to provide isolations in between RF and IF ports. Thus, these filters are used not only to pass the required frequencies; they also provide rejections to unwanted mixing components.

Schematic circuit diagram of a single-balanced mixer using 180° hybrid is shown in Fig. 18.14. Microstrip layout diagram of a rat-race 180° hybrid is shown in the figure. Here, LO signal is applied at port-4 and RF signal is applied at port-3 of the hybrid. Thus, the LO is applied 180° out of phase and RF is applied in phase to the two diodes. A low-pass filter at the output passes only the IF frequency and rejects other frequency components, also it provides a short circuit to the LO and RF frequencies to isolate IF from LO and RF. An $\lambda_o/4$ open-circuit stub may also be used as shown to provide a short circuit to the RF signal at the IF port.

Suppose, the amplitudes and frequency of the LO and RF signals are A_L, A_R and ω_L, ω_R, respectively. Thus, the resultant voltages v_{d1} and v_{d2} at port-1 and port-2, respectively, can be expressed as (eliminating common phase shifts):

$$v_{d1} = \frac{A_L}{\sqrt{2}} \cos \omega_L t + \frac{A_R}{\sqrt{2}} \cos \omega_R t \tag{18.28a}$$

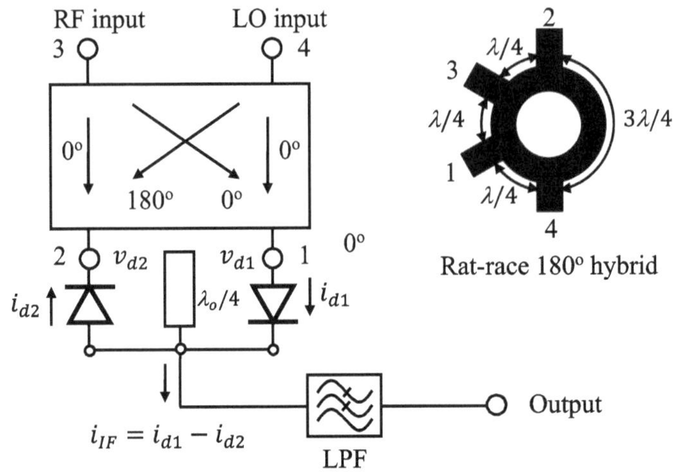

Fig. 18.14 Schematic circuit of a single-balanced diode mixer using 180° hybrid

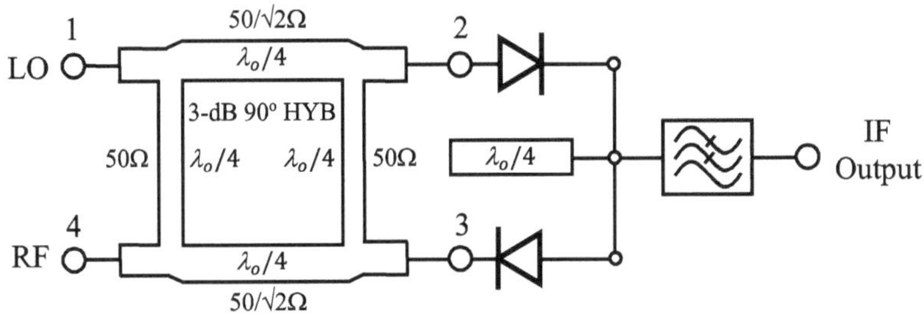

Fig. 18.15 Circuit schematic of a single-balanced mixer with 90° hybrid

$$v_{d2} = -\frac{A_L}{\sqrt{2}}\cos\omega_L t + \frac{A_R}{\sqrt{2}}\cos\omega_R t \tag{18.28b}$$

The diode *I-V* ($i_d - v_d$) characteristic from (18.27) can be expressed as

$$i_d = a_1 v_d + a_2 v_d^2 + a_3 v_d^3 + a_4 v_d^4 + \ldots \tag{18.28c}$$

Thus, the diode currents i_{d1} and i_{d2} can be written as

$$\begin{aligned} i_{d1} = i_d(v_{d1}) = a_1&\left(\frac{A_L}{\sqrt{2}}\cos\omega_L t + \frac{A_R}{\sqrt{2}}\cos\omega_R t\right) \\ &+ a_2\left(\frac{A_L}{\sqrt{2}}\cos\omega_L t + \frac{A_R}{\sqrt{2}}\cos\omega_R t\right)^2 \\ &+ a_3\left(\frac{A_L}{\sqrt{2}}\cos\omega_L t + \frac{A_R}{\sqrt{2}}\cos\omega_R t\right)^3 \end{aligned}$$

$$+\,a_4\left(\frac{A_L}{\sqrt{2}}\cos\omega_L t+\frac{A_R}{\sqrt{2}}\cos\omega_R t\right)^4+\cdots$$

Whereas

$$\begin{aligned}
i_{d2}=i_d(-v_{d2})=&-a_1\left(-\frac{A_L}{\sqrt{2}}\cos\omega_L t+\frac{A_R}{\sqrt{2}}\cos\omega_R t\right)\\
&+a_2\left(-\frac{A_L}{\sqrt{2}}\cos\omega_L t+\frac{A_R}{\sqrt{2}}\cos\omega_R t\right)^2\\
&-a_3\left(-\frac{A_L}{\sqrt{2}}\cos\omega_L t+\frac{A_R}{\sqrt{2}}\cos\omega_R t\right)^3\\
&+a_4\left(-\frac{A_L}{\sqrt{2}}\cos\omega_L t+\frac{A_R}{\sqrt{2}}\cos\omega_R t\right)^4+\cdots
\end{aligned}$$

Therefore, the IF current $i_{\mathrm{IF}}=i_{d1}-i_{d2}$ can be written as

$$\begin{aligned}
i_{\mathrm{IF}}=&\,a_1\left(\frac{A_L}{\sqrt{2}}\cos\omega_L t+\frac{A_R}{\sqrt{2}}\cos\omega_R t\right)+a_2\left(\frac{A_L}{\sqrt{2}}\cos\omega_L t+\frac{A_R}{\sqrt{2}}\cos\omega_R t\right)^2\\
&+a_3\left(\frac{A_L}{\sqrt{2}}\cos\omega_L t+\frac{A_R}{\sqrt{2}}\cos\omega_R t\right)^3+a_4\left(\frac{A_L}{\sqrt{2}}\cos\omega_L t+\frac{A_R}{\sqrt{2}}\cos\omega_R t\right)^4+\cdots\\
&+a_1\left(-\frac{A_L}{\sqrt{2}}\cos\omega_L t+\frac{A_R}{\sqrt{2}}\cos\omega_R t\right)-a_2\left(-\frac{A_L}{\sqrt{2}}\cos\omega_L t+\frac{A_R}{\sqrt{2}}\cos\omega_R t\right)^2\\
&+a_3\left(-\frac{A_L}{\sqrt{2}}\cos\omega_L t+\frac{A_R}{\sqrt{2}}\cos\omega_R t\right)^3-a_4\left(-\frac{A_L}{\sqrt{2}}\cos\omega_L t+\frac{A_R}{\sqrt{2}}\cos\omega_R t\right)^4+\cdots\\
=&\,\sqrt{2}a_1A_R\cos\omega_R t+2a_2A_LA_R\cos\omega_L t\cos\omega_R t\\
&+1.5\sqrt{2}a_3A_L^2A_R\cos^2\omega_L t\cos\omega_R t+0.2\sqrt{2}a_3A_R^3\cos^3\omega_R t\\
&+2a_4A_L^3A_R\cos^3\omega_L t\cos\omega_R t+2a_4A_LA_R^3\cos\omega_L t\cos^3\omega_R t\\
&+\cdots
\end{aligned}\qquad(18.28\text{d})$$

This shows that there is no fundamental component of LO frequency at the output. Thus, due to the balance configuration, there is no LO leakage to the IF port. Also, there are no even-order harmonic components of the RF frequency. Thus, the requirement of the rejection of undesired components at the diode output is lesser than the requirement in case of single-ended mixers. Moreover, considering very less RF level signal compared to the level of the LO, higher order RF components will have less contribution to the overall harmonic power level.

In the 3-dB 90° hybrid version single-balanced mixer, the RF and LO signals are applied to the mutually isolated two ports as shown in Fig. 18.15. Here, RF and LO signals are applied to port-1 and port-2 which are mutually isolated over its designed frequency band. However, due to the 90° phase difference of the hybrid, there will be no cancellation of the LO and RF as well as intermodulation components at the IF port. To provide a short circuit to the LO signal at the IF port, a $\lambda_o/4$ open-circuit

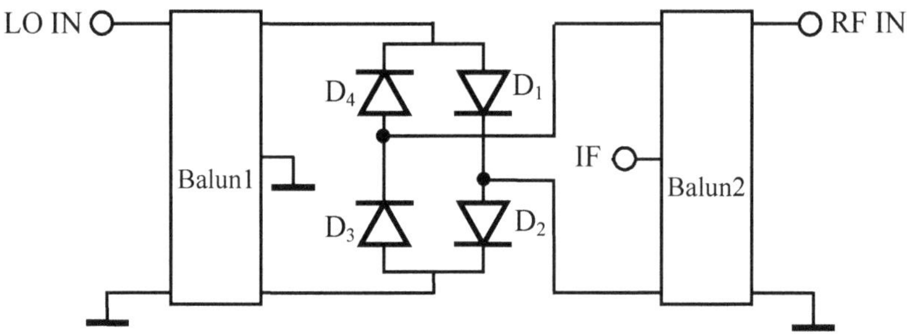

Fig. 18.16 Schematic circuit of a double-balanced diode mixer

stub is used. In case of RF frequency near to the LO frequency, this stub will also provide a low impedance to the RF signal. Thus, IF will also be isolated from the RF and LO signals. The low-pass filter at the IF port is used to pass only the desired IF frequency component to the output eliminating the generated undesired mixing components.

18.4.2 Double-Balanced Mixers

Single-balanced mixers can provide inherent isolation among either LO to IF or RF to IF but not both simultaneously. Double-balanced mixers can be considered as they consist of two single-balanced mixers with different configurations [8, 10, 11, 16–18]. Schematic configuration of a double-balanced mixer is shown in Fig. 18.16. The configuration consists of two baluns with four devices (here diodes). Here, LO port isolation from the RF and IF ports is achieved in the same way as it is achieved in single-balanced mixers due to the virtual ground in between the matched diode pairs. The IF signal is taken from the RF balun instead of from the LO balun to achieve better isolation in between LO and IF ports. The isolation between the RF and the IF ports depends on the balancing of the balun. Theoretically, double-balanced mixers generate only one-fourth of the possible intermixing products, which consist of only odd harmonics of LO and odd harmonics of RF frequencies. The other mixing products suppressed due to the balanced configurations. The degree of suppression depends on the amount of balance of the baluns, as well as matching among the diodes. However, practically, the suppression of even-even products will be more than the even-odd and odd-even products. Double-balanced mixers consist of four diodes, thus LO drive will be more than the single-ended as well as single-balanced mixers. Therefore, the upper limit of the RF level will be higher for the same amount of IM3 level, i.e. its 1-dB compression point will be higher than the other two configurations.

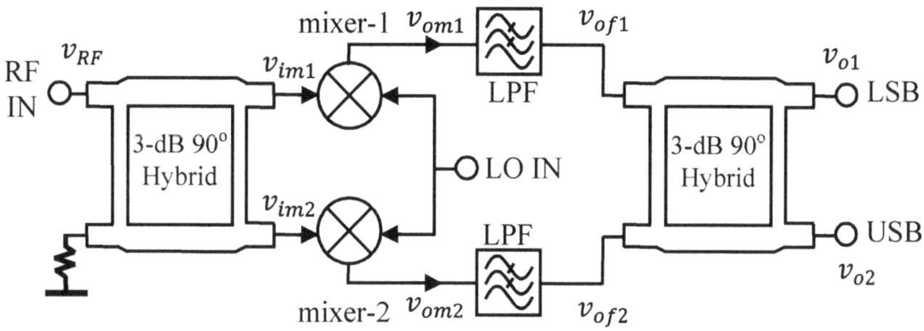

Fig. 18.17 Circuit schematic of an image rejection mixer

18.4.3 Image Rejection Mixers

Image rejection mixers (IRM) are used as a converter which isolated lower sideband frequency spectrum from upper sideband spectrum [13]. Schematic diagram of an IRM using two 3-dB 90° couplers is shown in Fig. 18.17. Suppose, the LO and IF frequencies are ω_{LO} and ω_{IF}, respectively. The two sideband RF frequencies can be represented as

$$v_{\mathrm{RF}} = v_L \cos(\omega_{\mathrm{LO}} - \omega_{\mathrm{IF}})t + v_U \cos(\omega_{\mathrm{LO}} + \omega_{\mathrm{IF}})t \tag{18.29}$$

Here, v_L and v_U are the amplitudes of lower and upper RF sidebands. Suppose the LO signal is $v_{\mathrm{LO}} \cos\ \omega_{\mathrm{LO}}t$.

The inputs of the mixer-1 and mixer-2 respectively can be written as

$$v_{im1} = \frac{v_L}{\sqrt{2}} \cos(\omega_{\mathrm{LO}} - \omega_{\mathrm{IF}})t + \frac{v_U}{\sqrt{2}} \cos(\omega_{\mathrm{LO}} + \omega_{\mathrm{IF}})t \tag{18.30a}$$

and

$$v_{im2} = \frac{v_L}{\sqrt{2}} \sin(\omega_{\mathrm{LO}} - \omega_{\mathrm{IF}})t + \frac{v_U}{\sqrt{2}} \sin(\omega_{\mathrm{LO}} + \omega_{\mathrm{IF}})t \tag{18.30b}$$

Considering only the second-order term of the mixers with multiplication factor k_m, the output of the mixer-1 can be written as

$$\begin{aligned} v_{om1} &= k_m\left(\frac{v_L}{\sqrt{2}} \cos(\omega_{\mathrm{LO}} - \omega_{\mathrm{IF}})t + \frac{v_U}{\sqrt{2}} \cos(\omega_{\mathrm{LO}} + \omega_{\mathrm{IF}})t\right) v_{\mathrm{LO}} \cos \omega_{\mathrm{LO}}t \\ &= \frac{k_m v_L v_{\mathrm{LO}}}{2\sqrt{2}} \cos(2\omega_{\mathrm{LO}} - \omega_{\mathrm{IF}})t + \frac{k_m v_U v_{\mathrm{LO}}}{2\sqrt{2}} \cos(2\omega_{\mathrm{LO}} + \omega_{\mathrm{IF}})t \\ &\quad + \frac{k_m v_L v_{\mathrm{LO}}}{2\sqrt{2}} \cos \omega_{\mathrm{IF}}t + \frac{k_m v_U v_{\mathrm{LO}}}{2\sqrt{2}} \cos \omega_{\mathrm{IF}}t \end{aligned}$$

The first two terms will be eliminated by the LPF connected at the output of the mixer. Thus, the LPF output will be

$$v_{of1} = \frac{k_m v_L v_{LO}}{2\sqrt{2}} \cos \omega_{IF} t + \frac{k_m v_U v_{LO}}{2\sqrt{2}} \cos \omega_{IF} t \tag{18.31a}$$

In the same way, the output of the second mixer after LPF will be

$$v_{of2} = -\frac{k_m v_L v_{LO}}{2\sqrt{2}} \sin \omega_{IF} t + \frac{k_m v_U v_{LO}}{2\sqrt{2}} \sin \omega_{IF} t \tag{18.31b}$$

Thus, one of the outputs of the output hybrid will be

$$\begin{aligned} v_{o1} &= \frac{k_m v_L v_{LO}}{4} \cos \omega_{IF} t + \frac{k_m v_L v_{LO}}{4} \cos \omega_{IF} t \\ &= \frac{k_m v_L v_{LO}}{2} \cos \omega_{IF} t \end{aligned} \tag{18.32a}$$

From (18.32a), it is clear that the output v_{o1} contains the IF component which corresponds to only lower sideband of the RF signal.

Another output of the second hybrid will be

$$\begin{aligned} v_{o2} &= \frac{k_m v_U v_{LO}}{4} \sin \omega_{IF} t + \frac{k_m v_U v_{LO}}{4} \sin \omega_{IF} t \\ &= \frac{k_m v_U v_{LO}}{2} \sin \omega_{IF} t \end{aligned} \tag{18.32b}$$

From (18.32b), it is clear that the output v_{o2} contains the IF component which corresponds to only upper sideband of the RF signal.

Thus, the image rejection mixer separates lower and upper sideband RF frequency components into its two output ports. Depending upon applications, the IF frequency band corresponding to either the lower sideband ($\omega_{LO} - \omega_{IF}$) or the upper sideband ($\omega_{LO} + \omega_{IF}$) can be selected for use.

18.5 Subharmonic Mixers

The desired output of a subharmonic mixer (SHM) is a mixing component of RF frequency and an integer multiple (2, 3, ...) of LO frequency. In SHM, the LO frequency is internally multiplied and mixes with the RF frequency to produce the desired output frequency [6, 7, 12, 19–21]. Thus, the requirement of frequency of the LO is lower. The reduction of LO frequency simplifies the LO design, particularly at high frequencies where it is difficult to design an LO with the required output power. Therefore, the subharmonic mixer is advantageous and is more attractive at higher microwave and mmwave frequency range. With reference to Fig. 18.1, the output frequency of SHM is given by

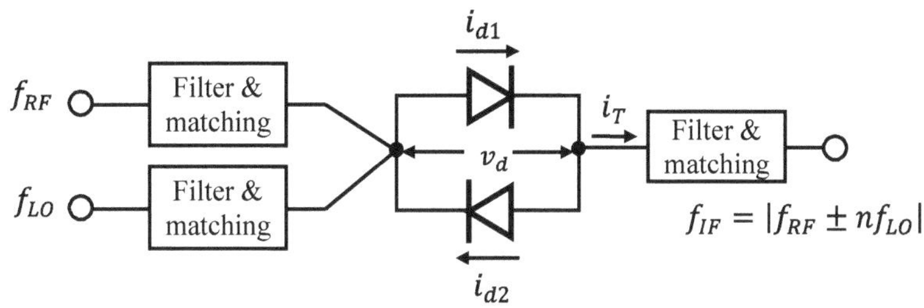

Fig. 18.18 Scheme of subharmonic mixer using an antiparallel diode pair

$$f_{\mathrm{OUT}} = |f_{\mathrm{IN}} \pm nf_{\mathrm{LO}}| \tag{18.33}$$

Here, n is the order of the subharmonic mixer. For 4× subharmonic mixer, $n = 4$, which requires LO with one-fourth the frequency that would be required with a fundamental frequency mixer.

As the subharmonic mixers are more advantageous at higher microwave and mmwave frequency range, mostly Schottky barrier diodes are used as nonlinear device to realize the circuits as it is difficult to realize transistor-based mixer circuits at such high frequencies. Schematic circuit of a subharmonic mixer using antiparallel diode pair is shown in Fig. 18.18. Here, sum of LO and RF signals is applied to the diodes and can be written as

$$v_d = v_{\mathrm{RF}} \cos \omega_{\mathrm{RF}} t + v_{\mathrm{LO}} \cos \omega_{\mathrm{LO}} t \tag{18.34a}$$

Considering the individual diode current represented by

$$i_d(v_d) = a_1 v_d + a_2 v_d^2 + a_3 v_d^3 + a_4 v_d^4 + \cdots \tag{18.34b}$$

The total current can be expressed as

$$\begin{aligned}
i_T &= i_{d1}(v_d) - i_{d2}(-v_d) \\
&= a_1(v_{\mathrm{RF}} \cos \omega_{\mathrm{RF}} t + v_{\mathrm{LO}} \cos \omega_{\mathrm{LO}} t) + a_1(v_{\mathrm{RF}} \cos \omega_{\mathrm{RF}} t + v_{\mathrm{LO}} \cos \omega_{\mathrm{LO}} t) \\
&\quad + a_2(v_{\mathrm{RF}} \cos \omega_{\mathrm{RF}} t + v_{\mathrm{LO}} \cos \omega_{\mathrm{LO}} t)^2 - a_2(v_{\mathrm{RF}} \cos \omega_{\mathrm{RF}} t + v_{\mathrm{LO}} \cos \omega_{\mathrm{LO}} t)^2 \\
&\quad + a_3(v_{\mathrm{RF}} \cos \omega_{\mathrm{RF}} t + v_{\mathrm{LO}} \cos \omega_{\mathrm{LO}} t)^3 + a_3(v_{\mathrm{RF}} \cos \omega_{\mathrm{RF}} t + v_{\mathrm{LO}} \cos \omega_{\mathrm{LO}} t)^3 \\
&\quad + a_4(v_{\mathrm{RF}} \cos \omega_{\mathrm{RF}} t + v_{\mathrm{LO}} \cos \omega_{\mathrm{LO}} t)^4 - a_4(v_{\mathrm{RF}} \cos \omega_{\mathrm{RF}} t + v_{\mathrm{LO}} \cos \omega_{\mathrm{LO}} t)^4 \\
&\quad + a_5(v_{\mathrm{RF}} \cos \omega_{\mathrm{RF}} t + v_{\mathrm{LO}} \cos \omega_{\mathrm{LO}} t)^5 + a_5(v_{\mathrm{RF}} \cos \omega_{\mathrm{RF}} t + v_{\mathrm{LO}} \cos \omega_{\mathrm{LO}} t)^5 \\
&\quad + \cdots \\
&= 2a_1(v_{\mathrm{RF}} \cos \omega_{\mathrm{RF}} t + v_{\mathrm{LO}} \cos \omega_{\mathrm{LO}} t) + 2a_3(v_{\mathrm{RF}} \cos \omega_{\mathrm{RF}} t + v_{\mathrm{LO}} \cos \omega_{\mathrm{LO}} t)^3 \\
&\quad + 2a_5(v_{\mathrm{RF}} \cos \omega_{\mathrm{RF}} t + v_{\mathrm{LO}} \cos \omega_{\mathrm{LO}} t)^5 + \cdots
\end{aligned} \tag{18.35}$$

This shows that there are no even-order products; it only generates odd-order frequency mixing components. Generally, the amplitude of LO signal is much higher than the amplitude of the RF signal. Therefore, using the condition $v_{LO} \gg v_{RF}$, the higher order terms of RF components can be neglected compared to the fundamental component of RF signal level and higher order LO signal levels. Thus, the total current can be written as

$$\begin{aligned} i_T &= 2a_1 v_{RF} \cos\omega_{RF}t + 2a_1 v_{LO} \cos\omega_{LO}t + 6a_3 v_{RF} v_{LO}^2 \cos\omega_{RF}t \cos^2\omega_{LO}t \\ &+ 2a_3 v_{LO}^3 \cos^3\omega_{LO}t + 20a_5 v_{RF}^2 v_{LO}^3 \cos^2\omega_{RF}t \cos^3\omega_{LO}t \\ &+ 10a_5 v_{RF} v_{LO}^4 \cos\omega_{RF}t \cos^4\omega_{LO}t + 2v_{LO}^5 \cos^5\omega_{LO}t + \cdots \end{aligned} \quad (18.36)$$

The output current contains $(2\omega_{LO} - \omega_{RF})$, $(2\omega_{LO} + \omega_{RF})$, $(4\omega_{LO} - \omega_{RF})$, $(4\omega_{LO} + \omega_{RF})$ and other mixing products including fundamental RF and LO frequency components. This shows that the antiparallel diode pair is capable to generate mixing products which are required for subharmonic mixer. Though the circuit structure inherently eliminates the even-order intermodulation products, the fundamental RF and LO signals will pass to the output. Therefore, an output filter is required to pass only the desired subharmonic mixing products eliminating other undesired mixing products.

Example 18.2 A system consists of a low-noise amplifier (LNA) cascaded with a mixer without any image rejection filter in between them. The gain of the LNA at its desired and image frequency bands are 30 and 32 dB, respectively. The noise figures of the LNA at its desired and image frequency bands are 2.0 and 3.0 dB, respectively. Calculate the noise figure of the system at its desired frequency band considering ideal noise-less mixer of 0 dB insertion loss.

Solution

The system with low-noise amplifier (LNA) and mixer without any image frequency band rejection filter is shown in Fig. 18.19.

This is the case of single-sideband signal. Thus, there is no signal at the image frequency band. Without image frequency band rejection mechanism, the noise of the image frequency band contributes to total output noise at the desired frequency band.

Suppose, over the desired frequency band the input signal power, gain and noise figure of the LNA are S_{IN}, G_{IN} and NF_{IN}, respectively. Also suppose, over the image frequency band, the gain and noise figure of the LNA are G_{IM} and NF_{IM}, respectively.

Then, the signal power at the output of LNA over the desired frequency band is

$$S_{OUT} = S_{IN} \times G_{IN} \quad (18.37)$$

The mixer is lossless. Thus, signal power at the output of the mixer is the same and can be written as

$$S_{OUT\_MIXER} = S_{IN} \times G_{IN} \quad (18.38)$$

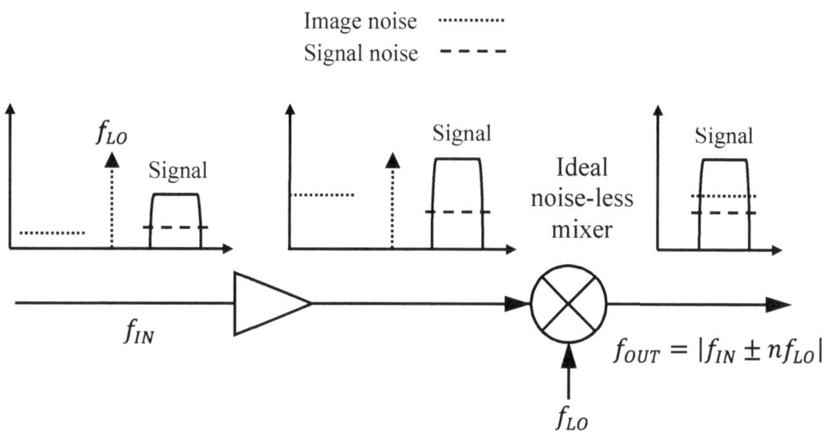

Fig. 18.19 Contribution of image frequency noise to IF band in the absence of image frequency band reject filter

Suppose, noise power input to the LNA at the desired and image frequency bands are same and is denoted as N_{IN}.

Thus, the noise power at the output of the LNA.

Over the desired frequency band is $N_{LNA\_OUT\_IN} = N_{\mathrm{IN}} \times G_{\mathrm{IN}} \times \mathrm{NF_{IN}}$

Over the image frequency band is $N_{LNA\_OUT\_IM} = N_{\mathrm{IN}} \times G_{\mathrm{IM}} \times \mathrm{NF_{IM}}$

Thus, the total output noise power over the desired single-sideband at the output of the mixer is

$$\begin{aligned} N_{\mathrm{OUT\_MIXER}} &= N_{\mathrm{LNA_{OUT_{IN}}}} + N_{\mathrm{LNA_{OUT_{IM}}}} \\ &= N_{\mathrm{IN}} \times G_{\mathrm{IN}} \times \mathrm{NF_{IN}} + N_{\mathrm{IN}} \times G_{\mathrm{IM}} \times \mathrm{NF_{IM}} \end{aligned} \tag{18.39}$$

Therefore, ratio of the output signal power to the output noise power is

$$\frac{S_{\mathrm{OUT\_MIXER}}}{N_{\mathrm{OUT\_MIXER}}} = \frac{S_{\mathrm{IN}} \times G_{\mathrm{IN}}}{N_{\mathrm{IN}} \times G_{\mathrm{IN}} \times \mathrm{NF_{IN}} + N_{\mathrm{IN}} \times G_{\mathrm{IM}} \times \mathrm{NF_{IM}}} \tag{18.40}$$

Thus, the single-sideband noise factor without image frequency band rejection can be written as

$$\mathrm{NF_{MIXER}} = \frac{S_{\mathrm{IN}}}{N_{\mathrm{IN}}} / \frac{S_{\mathrm{OUT\_MIXER}}}{N_{\mathrm{OUT\_MIXER}}} = \frac{G_{\mathrm{IN}} \times \mathrm{NF_{IN}} + G_{\mathrm{IM}} \times \mathrm{NF_{IM}}}{G_{\mathrm{IN}}} \tag{18.41}$$

$$= \mathrm{NF_{IN}}\left(1 + \frac{G_{\mathrm{IM}} \times \mathrm{NF_{IM}}}{G_{\mathrm{IN}} \times \mathrm{NF_{IN}}}\right) \tag{18.42}$$

Here,

$$\mathrm{NF_{IN}} = 2\ \mathrm{dB} = 1.585$$

$$\mathrm{NF_{IM}} = 3\ \mathrm{dB} = 1.995$$
$$G_{\mathrm{IN}} = 30\ \mathrm{dB} = 1000$$
$$G_{\mathrm{IM}} = 32\ \mathrm{dB} = 1585.893$$

Thus from (18.42),

$$\begin{aligned}\mathrm{NF_{MIXER}} &= 1.585\left(1 + \frac{1585.893 \times 1.995}{1000 \times 1.585}\right)\\ &= 4.747 = 6.764\ \mathrm{dB}\end{aligned}$$

Thus, the absence of image reject filter degrades the system noise figure by (6.764 − 2) dB = 4.764 dB!!

References

1. Mark R (1967) Barber, noise figure and conversion loss of the Schottky barrier mixer diode. IEEE Trans Microw Theor Tech 15(11):629–635
2. Gewartowski JW (1971) Noise figure for a mixer diode. IEEE Trans Microw Theor Tech 29(5):481
3. Held DN, Kerr AR (1978) Conversion loss and noise of microwave and millimeter-wave mixers: Part 1—theory. IEEE Trans Microw Theory Tech 26(2):49–55
4. Held DN, Kerr AR (1978) Conversion loss and noise of microwave and millimeterwave mixers: Part 2—experiment. IEEE Trans Microw Theor Tech 26(2):55–61
5. Kerr AR (1979) Shot-noise in resistive-diode mixer and the attenuator noise model. IEEE Trans Microw Theory Tech 27(2):135–140
6. Kerr AR (1979) Noise and loss in balanced and subharmonically pumped mixers: Part I—theory. IEEE Trans Microw Theor Tech 27(12):938–943
7. Kerr AR (1979) Noise and loss in balanced and subharmonically pumped mixers: Part II—application. IEEE Trans Microw Theor Tech 27(12):944–950
8. Ogawa HA, Masayoshi A, Kozo M (1980) K-band integrated double-balanced mixer. IEEE Trans Microw Theory Tech 28(3):180–185
9. Maas SA (1987) A GaAs MESFET mixer with very low intermodulation. IEEE Trans Microw Theor Tech 35(4):425
10. Basu S, Maas SA (1993) Design and performance of a planar star mixer. IEEE Trans Microw Theor Tech 41(11):2028–2030
11. Maas SA, Chang KW (1993) A broadband, planar, doubly balanced monolithic Ka-band diode mixer. IEEE Trans Microw Theor Tech 41:2330–2335
12. Madjar A (1997) A novel general approach for the optimum design of microwave and millimeter wave subharmonic mixers. IEEE Trans Microw Theor Tech 44(11):1997–2000
13. Marki FA (1977) Miniature image reject mixers and their use in low noise front ends in conjunction with GaAs FET amplifiers. In: 11th Asilomar Conference on Circuits, Systems and Computers, pp 159–162, Nov 1977
14. Fong KL, Meyer RG (1999) Monolithic RF active mixer design. IEEE Trans Circ Syst-II: Analog Digital Signal Process 46(3):231
15. Rahman NA, Majlis BY (2004) A GaAs PHEMT single-ended mixers for 28 GHz applications. In: The 4th annual seminar of national science fellowship
16. Kuo C, Kuo C, Kuo S, Maas HW (2008) Novel miniature and broadband millimeter-wave monolithic star mixers. IEEE Trans Microw Theor Tech 56(4):793–802

17. Lai Y-A, Chen C-N, Su C-C, Hung S-H, Wu C-L, Wang Y-H (2011) A compact double-balanced star mixer with novel dual 180° hybrid. Progr Electromag Res C 24:147–159
18. Guo J, Xu Z, Qian C, Dou W-B (2011) Design of a microstrip balanced mixer for satellite communication. Progr Electromagn Res 115:289–301
19. Kallfass I, Massler H, Leuther A (2007) A 210 GHz, Subharmonically pumped active FET mixer MMIC for radar imaging applications. In: 2007 IEEE CSIC symposium digest, pp 71–74, Oct 2007
20. Gunnarsson S (2008) Analysis and design of a novel X4 subharmonicaliy pumped resistive HEMT mixer. IEEE Trans Microw Theor Tech 56(4):809–816
21. Lee Y-C, Liu C-H, Hung S-H, Su C-C, Wang Y-H (2012) A 20–31 GHz high image rejection ratio subharmonic mixer. Progr Electromagn Res C 27:197–207

Chapter 19
Microwave Communication Systems

Abstract This chapter presents terrestrial mobile communication and satellite communication systems with emphasis on design and analysis of various subsystems used in receiver and transmitter sections. Homodyne and heterodyne receivers and different transmitter architectures for wireless mobile communication systems are described. Satellite communication systems defining important parameters like EIRP, G/T and SFD are also presented. Detailed design, analysis and realization of various microwave subsystems for satellite transponders like receivers, channel amplifiers, linearizers, TWTAs, SSPAs, microwave power module (MPM) and multiport power amplifier (MPA) are also described.

19.1 Introduction

Wireless communication technology is the fastest growing technology in the world. The successful development of wireless technologies has greatly improved people's ability to remain connected socially as well as for business. The phenomenal success of wireless mobile communications is reflected by the rapid pace of technology innovation. From the first-generation (1G) analog mobile communication system in 1980 to the 3G system which was first launched in 2001 through the second-generation (2G) mobile communication system in 1991, the wireless mobile network has transformed from a pure telephony system to a rich multimedia content transport system. The fourth-generation (4G) wireless system can support data rates of up to 1 Gbps for low mobility and up to 100 Mbps for high mobility. The target of a wireless communication system is to connect and achieve seamless communications between anybody, anything, wherever they are, whenever they need and by whatever electronic means they wish with affordable cost by consuming less power. Upcoming wireless communication systems are aiming to fulfil these requirements by configuring next-generation terrestrial wireless and satellite communication systems. As the lower frequency bands become congested, the communication systems need to shift towards higher and higher frequency bands to meet the capacity and data rate targets. Higher microwave and millimetre wave frequency bands are much less congested

S. C. Bera, *Microwave Active Devices and Circuits for Communication*, Lecture Notes in Electrical Engineering 533, https://doi.org/10.1007/978-981-13-3004-9_19

and can potentially provide multi-gigahertz spectrum. This is the reason why the satellite communication systems are already using higher microwave and mmwave frequency bands for high-throughput applications. It is also envisaged that the next-generation (5G) mobile communication system needs to use higher microwave and mmwave frequency bands to meet the requirement of spectrum [1]. This chapter will discuss terrestrial mobile communication and satellite communication systems with associated microwave circuits and subsystem.

19.2 Mobile Communication Systems

Mobile communication system is one of the greatest innovations in the history of technology. The rapid adoption of smartphones by a wide range of users and the explosive growth of data traffic of these devices have been phenomenal. A simplified block diagram of a generic transceiver for wireless communication system is shown in Fig. 19.1. It consists of transmit and receive RF section and a baseband processing section. The transmitter modulates the carrier signal with particular scheme and transmits through an antenna. The receiver recovers the information from the signal received by the antenna.

A bandpass signal $v(t)$ can be considered as a sinusoidal waveform of frequency f_c, with varying amplitude $A(t)$ and/or phase angle $\varphi(t)$ with time, which can be written as

$$v(t) = A(t)\cos\left[2\pi f_c t + \varphi(t)\right] \tag{19.1}$$

Suppose

$$v_i(t) = A(t)\cos\varphi(t) \tag{19.2a}$$

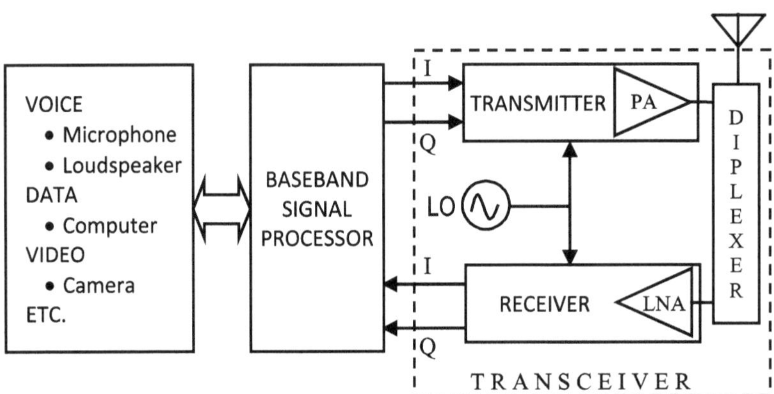

Fig. 19.1 Block diagram of generic wireless communication system

$$v_q(t) = A(t)\cos(\varphi(t) + 90^\circ) \tag{19.2b}$$

Then, (19.1) can be expressed as

$$v(t) = v_i(t)\cos(2\pi f_c t) + v_q(t)\cos(2\pi f_c t + 90^\circ) \tag{19.3}$$

This shows that a bandpass signal can be expressed completely by its in-phase and quadrature-phase components. The components $v_i(t)$ and $v_q(t)$, respectively, are the in-phase (I) and quadrature-phase (Q) components of the signal. In a generic transmitter, the baseband signal processor converts the information signal into an in-phase (I) and quadrature-phase (Q) component in the digital domain. The properties of the I–Q signals determine the type of modulation. Thus, the actual data symbols to be transmitted are coded to the I and Q signals in baseband circuitry. The I–Q signals are finally amplified by the power amplifier and transmitted through an antenna. In the generic receiver, the received signal from the antenna is amplified by a low-noise amplifier and finally the transmitted symbols are decoded by recovering the original I and Q signals from the received RF signal. The amplitude and phase information of the signal can be extracted from the I–Q signals as follows:

$$A(t) = \sqrt{v_i(t)^2 + v_q(t)^2}$$

$$\varphi(t) = \tan^{-1}\left(\frac{v_q(t)}{v_i(t)}\right)$$

19.2.1 Receiver Architecture for Mobile Communication

The basic function of a wireless communication receiver is to distinguish the desired signal from all other signals and to amplify the signal to a sufficient level for reliable detection by its baseband signal processing section. The simplest topology of its RF section is a bandpass filter tuned to the same frequency with the signal. The frequency response of a band select filter receiver in the presence of adjacent channels/interferences is shown in Fig. 19.2. The filter should be a tunable or there should be a bank of selectable bandpass filters with different centre frequencies to select one channel at a time among desired multiple channels. This scheme is suitable only if desired channels are separated sufficiently, and there should not be any interference signal near to the desired signals to design a practical filter to select only one channel in the presence of other channels. However, in practice, many narrow frequency bands spaced very closely to utilize the total allocated frequency bands with minimum wastage. Thus, it becomes very difficult to design a practical filter to select only the desired channel suppressing all other closely spaced channels and interferences. In some situations, the signal strength of the neighbouring channels becomes much higher than the selected channel due to the path loss and/or channel fading. In a practical scenario, a channel filter with very sharp cut-off response is

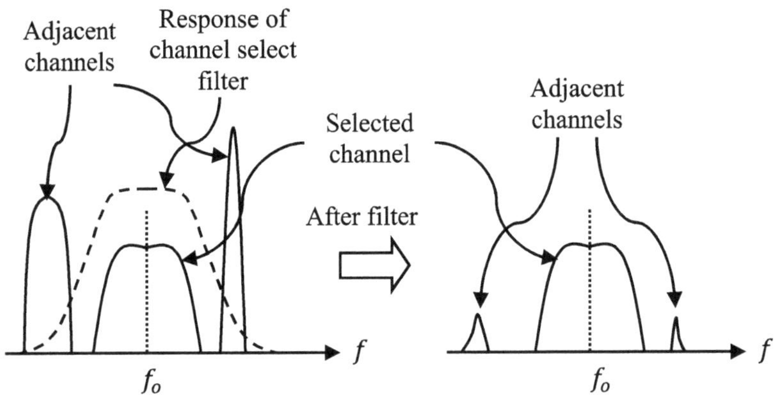

Fig. 19.2 Band-select filter receiver response

required to select a narrow bandwidth signal at higher carrier frequency. This needs a filter with very high-quality factor which is practically impossible at least in an integrated form. To overcome this problem practically, heterodyne as well as homodyne receiver architectures are used for wireless communication receivers.

In heterodyne receiver architecture, the requirement of an extremely narrowband channel select filter centred around a higher carrier frequency is eliminated by performing the channel selection at a lower frequency. Block diagram of simple heterodyne receiver architecture is shown in Fig. 19.3. In this case, the channel select filter operates at a lower frequency with unchanged pass bandwidth leading to practical realizable filter with achievable quality factor. Another advantage of this architecture is that the different channels can be selected using fixed channel select filter simply by changing the frequency of the local oscillator. The frequency translation is achieved by a process called heterodyning. In heterodyning, the original signal $v_i(t)$ at frequency f_{o1} is translated to a new intermediate frequency (IF) $f_{o1} - f_{\mathrm{LO}}$ by multiplying it in the time domain with another signal $v_{\mathrm{LO}}(t)$ at frequency f_{LO}. Cascading several mixing stages with an appropriate filter, the signal spectrum can be translated gradually to any other frequency band. Such receivers are called superheterodyne receivers. Though the heterodyne receiver architecture solves the problem of channel selection, it suffers from the requirement of image reject filter and associated insertion loss of it. There are some other receiver architectures to overcome the image rejection problem such as image reject mixer, direct conversion, low IF and wideband IF. Among them, image reject mixers are popularly used and already discussed in the previous chapter.

The direct-conversion receiver is also known as zero-IF receiver and homodyne receiver. Block diagram of zero-IF receiver architecture is shown in Fig. 19.4. Here, the centre frequency of the incoming signal and the LO signal are same, and thus the mixer translates the input spectrum directly around DC. The low-pass filters at outputs of the mixers attenuate all unwanted signals outside the desired frequency band and divide the signal into I and Q components for further processing by

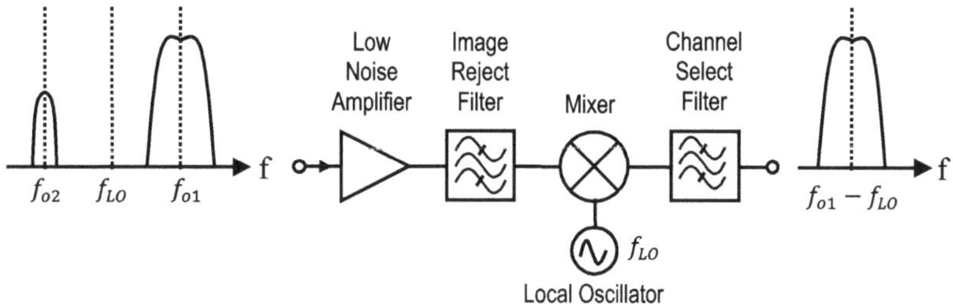

Fig. 19.3 Block diagram of heterodyne receiver

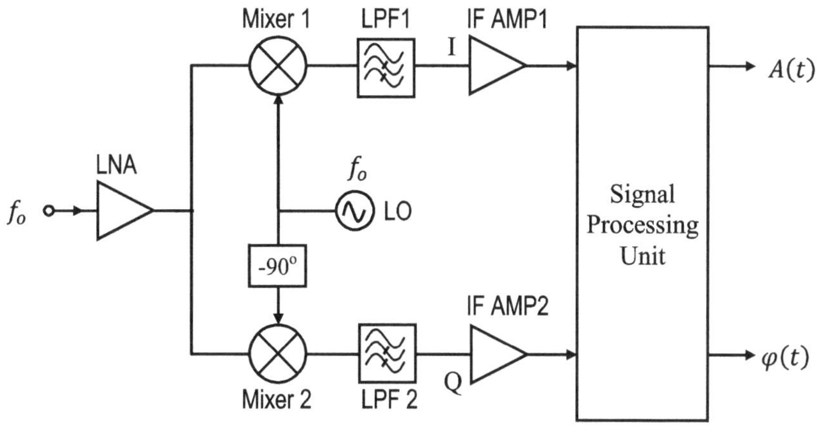

Fig. 19.4 Block diagram of a homodyne receiver

digital processor for information extraction. The homodyne receivers have several advantages over the conventional heterodyne architecture. These architectures do not require any highly selective image reject filters; this leads to savings in size. Here, simple low-pass filters are designed at lower frequencies as an integrated unit to select the desired channel. The homodyne architecture is attractive for the possibility of realizing integrating receiver using low-frequency amplifiers and digital-to-analog converter (DAC) followed by digital signal processing unit just after the low-pass filters. However, there are some practical issues to implement the homodyne receivers since the desired signal translated around DC. The architecture is highly sensitive to all sources of DC power, such as LO self-mixing, re-receiving of LO signal after reflection from other circuitry (antenna), DC offsets in the circuitry, down-mixing of signals due to even-order distortion in LNA and mixer, etc. This architecture also sufferers from $1/f$ noise of active devices working in the low-frequency regions. Thus, to achieve full benefit from the homodyne architecture, both the LNA and the mixers should be highly linear to avoid even-order distortion.

The low IF architecture receiver is the combination of good properties of the image reject receiver and the zero-IF architecture. In this architecture, the intermediate frequency (IF) is non-zero, but it is much lower than the conventional image reject receiver. The low IF frequency leads to the use of low-frequency/low-power circuit configuration which is more suitable for integration than the structures of the image reject receiver. It also helps to circumvent the DC offset and $1/f$ noise problems of the zero-IF architecture.

19.2.2 Transmitter Architecture for Mobile Communication

The function of a wireless communication transmitter is to modulate a carrier frequency by information signal with particular scheme and to transmit through an antenna. Contrary to the receiver, the transmitter circuitry processes signals of well-known strength and spectral contents. The challenges in transmitter design are mainly related to the spectral purity of the final power-amplified modulated signal entering the antenna, particularly in a system where many channels are closely spaced in frequency domain. Practically, the final power amplifier which amplifies the modulated signals generally operates in nonlinear region to achieve better DC-to-RF efficiency. Thus, the most severe source of spectral deterioration is due to nonlinearity of the power amplifier that generates spectral re-growth which disturbs in-band as well as adjacent channel signals. The power amplifier is generally designed using III–V group semiconductors for achieving high RF power level with better efficiency and it is integrated with linearizer to minimize the overall nonlinear effect.

Block diagram of a simple architecture direct-conversion transmitter is shown in Fig. 19.5. The direct-conversion transmitter consists of an I/Q modulator, bandpass filter and power amplifier. The simple architecture suffers severely from sensitivity to the imbalance in between the mixers and to the phase and amplitude errors between the LO signals. These error mechanisms will lead to increased LO leakage and non-optimal suppression of the image frequency. Another shortcoming of the direct-conversion transmitter is the sensitivity to LO pulling, where the strong output signal of the power amplifier disturbs the operation of local oscillator due to their same frequency. This effect becomes worse if power amplifier and local oscillator are integrated into the same chip. One way to minimize LO pulling is to offset the local oscillator from the output frequency. Two local oscillators with offset frequency may be used to produce final frequency using a mixer. In this case, it is possible to keep frequency of the LOs far away from the output frequency. Another way to reduce the LO pulling is the use of heterodyne transmitter architecture as shown in Fig. 19.6. Here, the I/Q modulation is performed at a lower LO frequency and the final frequency translation is done by a second mixer stage to translate the spectrum to output frequency.

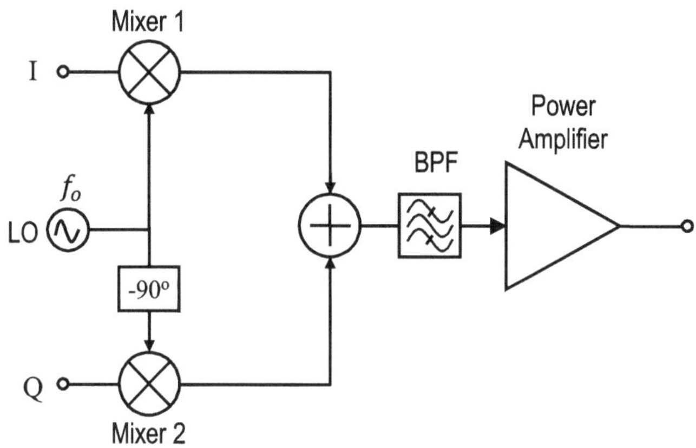

Fig. 19.5 Block diagram of a direct-conversion transmitter

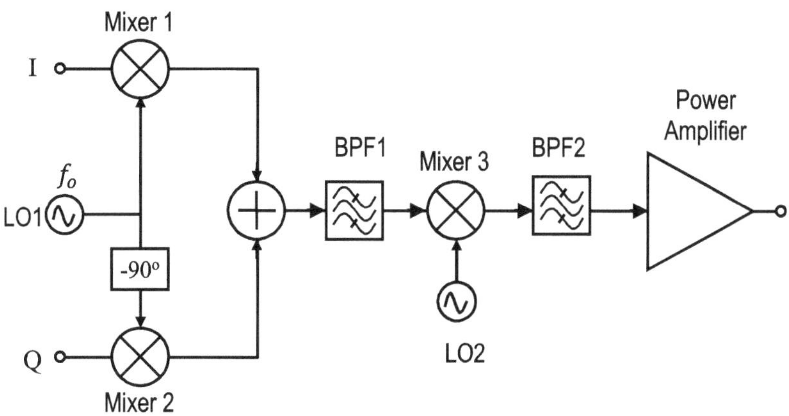

Fig. 19.6 Block diagram of a heterodyne transmitter

19.2.3 Transceiver for Mobile Communication

Though the functionalities of the transmitter and receiver for base station and for the mobile units are same, the practical realizations are different due to the different driving forces behind their development. The integration of the mobile transmitter and receiver units is extremely desirable to extend battery life, reducing weight and size which make the handheld mobile unit more desirable and attractive to the customer. This can be done by increasing the level of integration. Reliability of the integrated complete unit is also directly influenced by the integration of multiple functions into a single chip and the mass production of ICs.

The RF section of a transmitter usually consists of an oscillator, a modulator, an upconverter, filters and power amplifier as shown in Fig. 19.7. It also consists of a phase-locked oscillator or synthesizer. The signals to be transmitted modulate

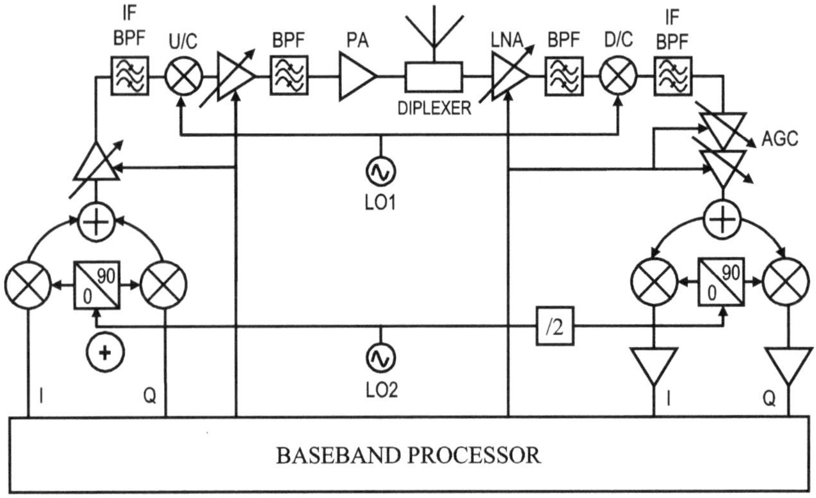

Fig. 19.7 Transceiver block diagram for mobile communication systems

the oscillator frequency through a suitable modulation scheme such as amplitude modulation, frequency modulation, phase modulation or other digital modulation schemes such as PSK, FSK, etc. In WCDMA system, QPSK modulation scheme is used as shown in the Fig. 19.7. Here, the data stream from baseband processor is split into odd and even streams and are separately modulated before combining into a QPSK signal. The signals are then amplified by amplifier and upconverted to a higher desired carrier frequency by an upconverter. The upconverted RF signals are amplified by a power amplifier and transmitted by the antenna. The function of the receiver is to extract the useful information from the received RF signals from antenna. Receiver section in an RF transceiver consists of a low-noise amplifier, downconverter, IF amplifier, bandpass filters and demodulator. The demodulated signal processed in the baseband processor for extraction of the information.

Enormous growth of data traffic volume is one of the main drivers behind the next-generation (5G) mobile communication systems [1]. Mobile wireless communication with ultra-wide bandwidth will be the key requirement for the next-generation mobile system to meet the ever-increasing needs for higher capacity. Presently, wireless communication demands tens of billions of Internet of things (IoT) and machine-to-machine (M2M) communications. The next-generation wireless communication system needs enormous spectrum to fulfil the requirements for smart city, smart home, smart grid, smart remotely controlled and self-driven vehicle, e-health which is available only at higher microwave and millimetre wave frequency bands. In addition to the mmwave frequency bands, the fundamental technologies that enable next-generation mobile communications include massive multiple-input-multiple-output (MIMO) system and small-cell configuration. The mmwave technologies enable to utilize the wideband spectrum; massive MIMO extends the range and increases spectral efficiency in these frequencies by employing a large number of antennas, and small-cell technologies provide the means to deploy a wide area mobile network with a large number of small cells, scalable backhaul and proper interference mitigation techniques. All these technologies together enable next-generation mobile communication systems with much larger capacity, much higher data rate and much denser deployment than 4G systems. It is envisaged that the next-generation mobile communication systems (5G) will work in higher microwave and mmwave frequency bands.

The main functions of the RF section of the transceiver chain are amplification of signal, conversion of frequencies and analog beamforming. The signal amplification is done by power amplifiers in transmit chain and by LNA in receive chain. The frequency conversions are done by the mixers using suitable local oscillators as shown in Fig. 19.8. At higher microwave and mmwave frequencies, patch antenna arrays are most viable option for implementing antenna arrays with large number of elements in a small size to achieve sufficient range. A simplified block schematic of a mobile transceiver with hybrid beamforming architecture using 64 patch elements is shown in Fig. 19.8. Maintaining most of the capability to dynamically form beams and adapt the MIMO processing schemes according to the channel conditions, it is required to minimize the number of RF chains for reduction of cost and complexity. Here, this is shown by using antenna sub-arrays, analog beamforming and digital MIMO processing. Analog beamforming can be implemented using RF, IF or LO beamforming technology. Here, it is shown in RF section. The 64 antenna elements are grouped into 16 numbers of 4×1 sub-arrays, and thus only 16 numbers of RF chains are required for 64 elements. Each RF signal is phase shifted and combined in a group of four before frequency conversion. This reduces the number of converter and MIMO streams. This scheme needs to handle only four MIMO streams with four converters (each in transmit and receive section) to drive the entire 64-element antenna array which reduces the complexity in the digital processing unit. For RF signal phase shifting, reflective-type phase shifter, loaded line phase shifter or switched delay phase shifters can be implemented at microwave and mmwave frequency range.

RF section of the mobile transceiver mainly consists of low-noise amplifier (LNA), power amplifier (PA), up- and downconverters, IF amplifiers, phase shifters, modulators, switches and filters as shown in Fig. 19.8. There are several viable semiconductor device technologies to realize RF transceiver circuits working at microwave and mmwave frequency bands. Gallium arsenide (GaAs) and indium phosphide (InP) and high-electron-mobility transistor (HEMT) and hetero-junction bipolar transistor (HBT) devices based on GaAs and SiGe can be used for solid-state power amplifiers to transmit few Watts of power. GaAs and InP p-HEMTs are among the best choices for RF Tx/Rx switches and LNAs. SiGe HBTs and RF CMOS have also been used for low-cost transceiver circuits in low-to-medium RF power amplifications. The mobile transceiver can be realized in multi-chip modules and surface mount packages using MMIC chips. Thus, the whole transceiver can be integrated on a single printed circuit board.

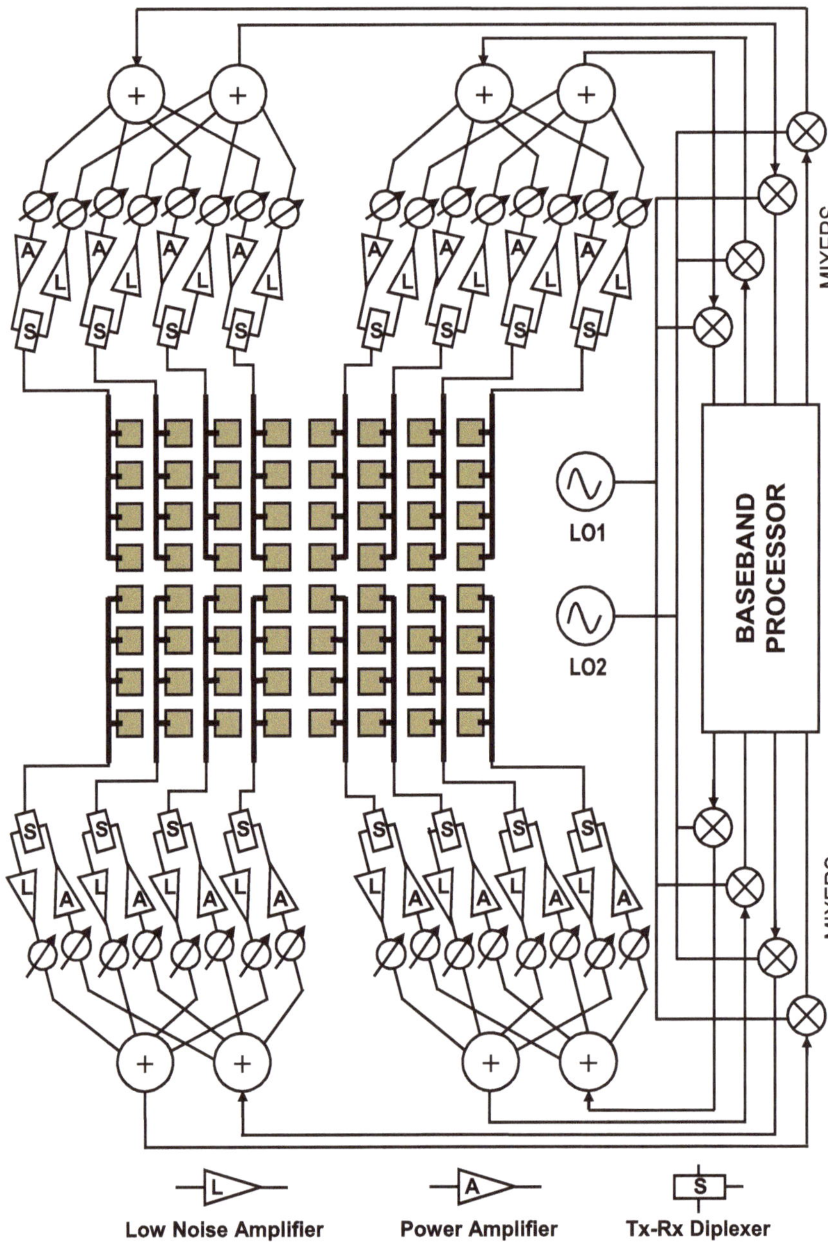

Fig. 19.8 Block diagram of mobile transceiver with MIMO system

Another critical technology required for the 5G mobile communication is the high-speed analog-to-digital converter (ADC). To support multi-Gbps data rate, it is required to use multiple channels of ADC with a sampling rate of few hundred MHz to few GHz.

19.3 Satellite Communication Systems

Microwave frequencies are used in satellite communication due to its higher penetration capability through the atmosphere, higher percentage bandwidth and higher directive coverage. A microwave communication system mainly consists of a receiver, a transmitter and an antenna system. Block diagram of a satellite communication system is shown in Fig. 19.9. Like any other communication systems, the space and the ground segments consist of a transmitter, a receiver and an antenna system. A simple transponder for satellite communication is a repeater system. It amplifies the signal received by a receive antenna with minimum addition of noise and then translates the signal frequency band to other bands. The frequency-translated signal further amplified to high power level by transmitter and then transmitted to ground by a transmit antenna. For example, Fig. 19.9 shows a $C \times S$-band transponder. It receives the signal in C-band and transmits it in S-band. Frequency translation provides high level of isolation to the sensitive receive sections from the high level of transmitted RF power. This ensures the stability of the transponder avoiding reinjection of transmitted output signal to the receive section by using filtering. A satellite transponder has three most important performance parameters: saturation flux density (SFD), receive antenna gain (G_R) to system temperature (T_R) ratio (G_R/T_R) and satellite-saturated effective isotropic radiated power (EIRP). Saturation flux density is the received power flux density at the satellite which saturates the transponder, i.e. to saturate final high power amplifier (HPA) used in the transmitter. Practically, the satellite SFD is not an independent parameter; its minimum value is limited by the ratio G_R/T_R. Suppose the equivalent noise temperature of a satellite receiver (including losses in between receiver and receive antenna output) is T_e, and the total equivalent system noise power P_n can be written as

$$P_n = k(T_e + T_o)B \tag{19.1a}$$

Here, B is the noise bandwidth of the system and T_o is the antenna noise temperature which is about 290 K and k is Boltzmann constant. Thus, the overall system noise temperature T_R is $(T_e + T_o)$. The received power P_R at the output of the satellite receive antenna should be more than the total noise power P_n. Considering the received power flux density at the satellite is PFD, the received power at the output of the receive antenna can be written as

$$P_R = \text{PFD} \times A_e$$

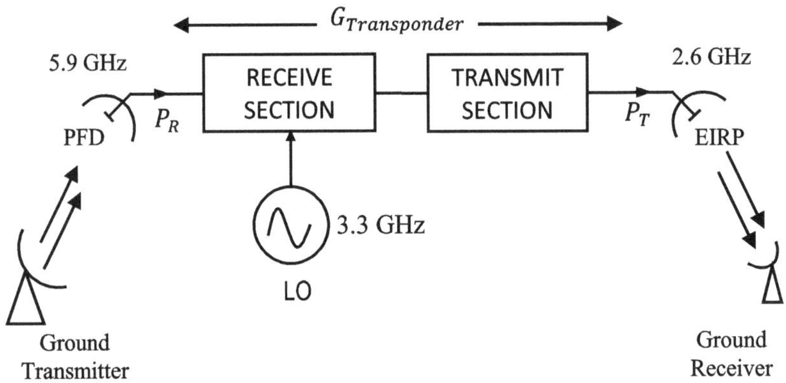

Fig. 19.9 A simplified block diagram of a satellite communication system

Here, A_e is the effective aperture of the receive antenna, and it is related to the gain of the antenna G_R by

$$G_R = 4\pi A_e/\lambda_R^2$$

Here, λ_R is the wavelength corresponding to the receive signal frequency. Thus, the received power is given by

$$P_R = \frac{\text{PFD} \times G_R \times \lambda_R^2}{4\pi} \tag{19.2}$$

For reliable communication, the received power must be more than the total noise power. The amount will be determined by the required signal-to-noise power ratio needed for particular data rate, used modulation code, etc. for a communication. Here, to derive the expression for minimum SFD level, we will impose the condition

$$P_R > P_n \tag{19.3}$$

Combining (19.1), (19.2) and (19.3),

$$\frac{\text{PFD} \times G_R \times \lambda_R^2}{4\pi} > kT_R B \tag{19.4}$$

or

$$\text{PFD} > \frac{4\pi}{\lambda_R^2} \times k \times B \times \frac{T_R}{G_R}$$

Expressing all the parameters in dB (19.4), it can be written as

$$[\text{PFD}]_{\text{dBW/m}^2} > \left[\frac{4\pi}{\lambda_R^2}\right]_{\text{dB/m}^2} + [k]_{\text{dBW/Hz-K}} + [B]_{\text{dBHz}} - \left[\frac{G_R}{T_R}\right]_{\text{dB/K}} \tag{19.5}$$

This shows that for a transponder operating at particular frequency and bandwidth, the minimum operating PFD limit is determined by its receive antenna gain-to-system noise temperature ratio, i.e. (G_R/T_R). Depending upon the requirement of uplink carrier-to-noise power ratio, which is determined by the communication data rate, modulation codes, etc.; the operating PFD limit is selected which is higher than the value given by (19.5).

Example 19.1 A satellite transponder operates at receive frequency 5.9 GHz of bandwidth 35 MHz. The gain of the receive antenna is 27 dBi and the receiver noise figure (including loss in between receive antenna and receiver) is 3.5 dB. Calculate the minimum received power flux density limit of the transponder.

Solution

Receive frequency = 5.9 GHz

Thus, $\lambda_R = (0.3/5.9)\,\text{m} = 0.05085$ m

Receiver noise figure

$$\text{NF} = 3.5\text{ dB} = 2.2387(\text{in factor})$$

Receiver noise temperature

$$T_e = (\text{NF} - 1)T_o = (2.2387 - 1)290\text{ K} = 359\text{ K}$$

Thus, the system noise temperature

$$T_R = T_e + T_o = (290 + 359)\text{ K} = 649\text{ K} = 28\text{ dBK}$$

Thus,

$$\left[\frac{G_R}{T_R}\right]_{\text{dB/K}} = (27 - 28)\,\text{dB/K} = -1\text{ dB/K}$$

$$\left[\frac{4\pi}{\lambda_R^2}\right]_{\text{dB}} = 10\log\left[\frac{4\pi}{\lambda_R^2}\right] = 36.87\text{ dB/m}^2$$

$$[k]_{\text{dBW}} = -228.6\text{ dBW/Hz} - \text{K}$$

$$[B]_{\text{dBHz}} = 10\log(35 \times 106)\,\text{dB} - \text{Hz} = 75.4\text{ dBHz}$$

$$[\text{PFD}]_{\text{dBW/m}^2} > \left[\frac{4\pi}{\lambda_R^2}\right]_{\text{dB/m}^2} + [k]_{\text{dBW/Hz-K}} + [B]_{\text{dBHz}} - \left[\frac{G_R}{T_R}\right]_{\text{dB/K}}$$

$$> 36.87 - 228.6 + 75.4 - \left[\frac{G_R}{T_R}\right]_{\text{dB/K}}$$

$$> -116 - \left[\frac{G_R}{T_R}\right]_{\text{dB/K}}$$
$$> -115 \text{ dBW/m}^2$$

Thus, the minimum PFD is -115 dBW/m^2.

The saturated power flux density (SFD) also should be higher than the value given by (19.5). Otherwise, the transponder will be saturated with the noise power itself.

Saturated effective isotropic radiated power (saturated EIRP) of a satellite transponder is determined by the final high power amplifiers output power capability (P_{sat}), output loss (L_{OUT}) and satellite's transmit antenna gain (G_T). The output loss is the RF loss of all the elements in between output of the power amplifier and input of the antenna. Therefore, saturated RF power at the input of the antenna ($P_{T\text{sat}}$) is given by

$$P_{T\text{sat}} = P_{\text{sat}} - L_{\text{OUT}} \tag{19.6a}$$

Thus, the saturated EIRP is given by

$$\text{EIRP}_{\text{sat}} = P_{T\text{sat}} \times G_T \tag{19.6b}$$

The overall gain of a transponder is determined by its saturated EIRP and SFD. From (19.2), under saturated flux density condition, the received power at the input of the receiver is given by

$$P_{R\text{sat}} = \frac{\text{SFD} \times G_R \times \lambda_R^2}{4\pi} \tag{19.7}$$

Therefore, the overall gain ($G_{\text{Transponder}}$) of the transponder is given by

$$G_{\text{Transponder}} = \frac{P_{T\text{sat}}}{P_{R\text{sat}}} = \frac{\text{EIRP}_{\text{sat}}}{G_T \times P_{R\text{sat}}} \tag{19.8}$$

$$\left[G_{\text{Transponder}}\right]_{\text{dB}} = [\text{EIRP}_{\text{sat}}]_{\text{dBW}} - [G_T]_{\text{dBi}} - [P_{R\text{sat}}]_{\text{dBW}} \tag{19.9}$$

Example 19.2 A satellite transponder operates at receive frequency 5.9 GHz and transmit frequency 2.6 GHz of bandwidth 35 MHz. The gain of receive and transmit antennae are 27 and 42 dBi, respectively. Calculate the gain of the transponder if SFD is -95 dBW/m^2 and saturated EIRP is 65 dBW.

Solution

Receive frequency = 5.9 GHz

Thus, $\lambda_R = (0.3/5.9)\,\text{m} = 0.05085$ m

Receive antenna gain = 27 dBi

From (19.6b), the saturated transmit power

$$P_{T\text{sat}} = \frac{\text{EIRP}_{\text{sat}}}{G_T}$$

$$\begin{aligned}
[P_{T\text{sat}}]_{\text{dBW}} &= [\text{EIRP}_{\text{sat}}]_{\text{dBW}} - [G_T]_{\text{dBi}} \\
&= 65\ \text{dBW} - 42\ \text{dBi} \\
&= 23\ \text{dBW} = 53\ \text{dBm}
\end{aligned}$$

From (19.7),

$$P_{R\text{sat}} = \frac{\text{SFD} \times G_R \times \lambda_R^2}{4\pi}$$

$$\begin{aligned}
[P_{R\text{sat}}]_{\text{dBW}} &= [\text{SFD}]_{\text{dBW/m}^2} + [G_R]_{\text{dBi}} + \left[\frac{\lambda_R^2}{4\pi}\right]_{\text{dBm}^2} \\
&= (-95 + 27 - 36.87)\ \text{dBW} = -104.87\ \text{dBW} \\
&= -74.87\ \text{dBm}
\end{aligned}$$

From (19.9),

$$\begin{aligned}
\left[G_{\text{Transponder}}\right]_{\text{dB}} &= [\text{EIRP}_{\text{sat}}]_{\text{dBW}} - [G_T]_{\text{dBi}} - [P_{R\text{sat}}]_{\text{dBW}} \\
&= 65\ \text{dBW} - 42\ \text{dBi} + 104.87\ \text{dBW} \\
&= 127.87\ \text{dB} \cong 128\ \text{dB}
\end{aligned}$$

Thus, the gain of the transponder is 128 dB.

Block diagram of a typical satellite communication transponder with power levels at different stages based on the previous examples is shown in Fig. 19.10. The transponder consists of a preselect filter (PSF) at the input of the transponder to select the desired frequency band. The PSF restricts the noise and interferences outside the desired bandwidth entering into the receiver. The signals are received by the receiver, amplified by a low-noise amplifier (LNA) and then frequency translated by mixer and further amplified by IF amplifiers. A bandpass filter (BPF) is used just after the mixer to suppress the spurious products generated by the mixer. The output signals from the receiver are divided into several channels by channelization filters and then amplified to high power level using different transmitter chains. The channelized high power signals then combined in frequency domain by using an output multiplexer. The multiplexed signal passes through a harmonic reject filter to provide sufficient rejection to the harmonics generated by the power amplifiers operated in their nonlinear region. Total gain of a transponder of about 130 dB is distributed among the receiver and transmitter. The gains of receive and transmit sections are about 50 and 84 dB, respectively, as shown in Fig. 19.10. The loss of the preselect filter and feeder cable in between antenna output and the receiver input needs to be taken into account to determine the overall system noise temperature. Similarly, loss of the output multiplexer, HRF and interconnecting cable up to antenna input are to be taken into account to determine final output power transmitted by the antenna to achieve required saturated EIRP.

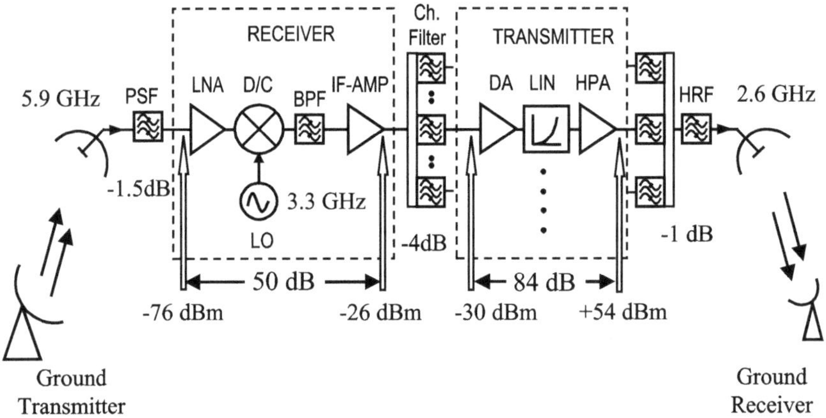

Fig. 19.10 Block diagram of a satellite communication system with power levels

Example 19.12 Calculate insertion loss and return losses of a 2-port network of S-parameter matrix:

$$[S] = \begin{bmatrix} 0.05\angle -80^\circ & 0.84\angle -25^\circ \\ 0.84\angle -25^\circ & 0.05\angle -80^\circ \end{bmatrix}$$

Calculate insertion loss and return losses when such two networks connected in cascade.

Solution

Insertion loss (IL) of the network is

$$\begin{aligned} \mathrm{IL_{dB}} &= -20 \times \log(|S_{21}|) \\ &= -20 \times \log(0.84) \\ &= 1.51 \text{ dB} \end{aligned}$$

Return loss (RL) of the network is

$$\begin{aligned} \mathrm{RL_{dB}} &= -20 \times \log(|S_{11}|) \\ &= -20 \times \log(0.05) \\ &= 26.02 \text{ dB} \end{aligned}$$

ABCD parameters of the network can be calculated using (7.124)

$$\begin{aligned} A &= \frac{(1+S_{11})(1-S_{22})+S_{12}S_{21}}{2S_{21}} \\ &= \frac{(1+0.05\angle -80^\circ)(1-0.05\angle -80^\circ)+0.84\angle -25^\circ \times 0.84\angle -25^\circ}{2 \times 0.84\angle -25^\circ} \end{aligned}$$

$$= \frac{(1 + 0.009 - 0.049i)(1 - 0.009 + 0.049i) + (0.761 - 0.355i) \times (0.761 - 0.355i)}{2 \times (0.761 - 0.355i)}$$
$$= 0.924\angle 4.662^\circ$$

$$B = Z_o \frac{(1 + S_{11})(1 + S_{22}) - S_{12}S_{21}}{2S_{21}}$$
$$= 50 \times \frac{(1 + 0.05\angle -80^\circ)(1 + 0.05\angle -80^\circ) - 0.84\angle -25^\circ \times 0.84\angle -25^\circ}{2 \times 0.84\angle -25^\circ}$$
$$= 50 \times \frac{(1 + 0.009 - 0.049i)(1 + 0.009 - 0.049i) + (0.761 - 0.355i) \times (0.761 - 0.355i)}{2 \times (0.761 - 0.355i)}$$
$$= 21.252\angle 63.159^\circ$$

$$C = Y_o \frac{(1 + S_{11})(1 - S_{22}) - S_{12}S_{21}}{2S_{21}}$$
$$= \frac{1}{50} \times \frac{(1 + 0.05\angle -80^\circ)(1 - 0.05\angle -80^\circ) - 0.84\angle -25^\circ \times 0.84\angle -25^\circ}{2 \times 0.84\angle -25^\circ}$$
$$= \frac{1}{50} \times \frac{(1 + 0.009 - 0.049i)(1 - 0.009 + 0.049i) - (0.761 - 0.355i) \times (0.761 - 0.355i)}{2 \times (0.761 - 0.355i)}$$
$$= 0.010\angle 75.463^\circ$$

$$D = \frac{(1 - S_{11})(1 + S_{22}) + S_{12}S_{21}}{2S_{21}}$$
$$= \frac{(1 - 0.05\angle -80^\circ)(1 + 0.05\angle -80^\circ) + 0.84\angle -25^\circ \times 0.84\angle -25^\circ}{2 \times 0.84\angle -25^\circ}$$
$$= \frac{(1 - 0.009 + 0.049i)(1 + 0.009 - 0.049i) + (0.761 - 0.355i) \times (0.761 - 0.355i)}{2 \times (0.761 - 0.355i)}$$
$$= 0.924\angle 4.662^\circ$$

Thus, ABCD parameters in matrix form of the 2-port network are

$$[ABCD] = \begin{bmatrix} 0.924\angle 4.662^\circ & 21.252\angle 63.159^\circ \\ 0.010\angle 75.463^\circ & 0.924\angle 4.662^\circ \end{bmatrix}$$

$[ABCD]$ parameters of such two 2-port networks connected in cascade will be given by

$$[A_cB_cC_cD_c] = [ABCD] \times [ABCD]$$
$$= \begin{bmatrix} 0.924\angle 4.662^\circ & 21.252\angle 63.159^\circ \\ 0.010\angle 75.463^\circ & 0.924\angle 4.662^\circ \end{bmatrix} \times \begin{bmatrix} 0.924\angle 4.662^\circ & 21.252\angle 63.159^\circ \\ 0.010\angle 75.463^\circ & 0.924\angle 4.662^\circ \end{bmatrix}$$
$$= \begin{bmatrix} 0.740\angle 21.976^\circ & 39.283\angle 67.821^\circ \\ 0.018\angle 80.125^\circ & 0.740\angle 21.976^\circ \end{bmatrix}$$

$[S]$ parameters of the cascaded 2-port network can be derived using (7.125)

$$\text{Cascaded } S_{11} = \frac{A_c + B_cY_o - C_cZ_o - D_c}{A_c + B_cY_o + C_cZ_o + D_c}$$

$$= \frac{0.740\angle 21.976^\circ + 0.786\angle 67.821^\circ - 0.910\angle 80.125^\circ - 0.740\angle 21.976^\circ}{0.740\angle 21.976^\circ + 0.786\angle 67.821^\circ + 0.910\angle 80.125^\circ + 0.740\angle 21.976^\circ}$$
$$= \frac{0.220\angle -50.338^\circ}{2.841\angle 50.049^\circ}$$
$$= 0.077\angle -100.387^\circ$$

Thus, return loss (RL) of the cascaded network is

$$\text{Cascaded RL (in dB)} = -20 \times \log(0.077)$$
$$= 22.22 \text{ dB}$$

$$\text{Cascaded } S_{21} = \frac{2}{A_C + B_C Y_O + C_c Z_O + D_C}$$
$$= \frac{2}{0.740\angle 21.976^\circ + 0.786\angle 67.821^\circ + 0.910\angle 80.125^\circ + 0.740\angle 21.976^\circ}$$
$$= \frac{2}{2.841\angle 50.049^\circ}$$
$$= 0.704\angle -50.049^\circ$$

Thus, insertion loss (IL) of the cascaded network is

$$\text{Cascaded IL (in dB)} = -20 \times \log(0.704) dB$$
$$= 3.05 \text{ dB} \quad !! \; (= \text{about} \, (1.51 + 1.51) \, \text{dB})$$

This example shows that the insertion loss of the cascaded network (3.049 dB) is equal to sum of the insertion losses of the individual networks.

Example 19.13 Calculate insertion loss and return losses of a 2-port network of S-parameter matrix:

$$[S] = \begin{bmatrix} 0.30\angle -90^\circ & 0.84\angle -26^\circ \\ 0.84\angle -26^\circ & 0.30\angle -90^\circ \end{bmatrix}$$

Calculate insertion loss and return losses when such two networks connected in cascade.

Solution

Insertion loss (IL) of the network is

$$\text{IL (in dB)} = -20 \times \log(|S_{21}|)$$
$$= -20 \times \log(0.84)$$
$$= 1.51 \text{ dB}$$

Return loss (RL) of the network is

$$\begin{aligned}\text{RL (in dB)} &= -20 \times \log(|S_{11}|)\\ &= -20 \times \log(0.30)\\ &= 10.46\ \text{dB}\end{aligned}$$

ABCD parameters of the network can be calculated using (7.124)

$$\begin{aligned}A &= \frac{(1+S_{11})(1-S_{22})+S_{12}S_{21}}{2S_{21}}\\ &= \frac{(1+0.3\angle-90^\circ)(1-0.3\angle-90^\circ)+0.84\angle-26^\circ \times 0.84\angle-26^\circ}{2\times 0.84\angle-26^\circ}\\ &= \frac{(1+0-0.3i)(1-0+0.3i)+(0.755-0.368i)\times(0.755-0.368i)}{2\times(0.755-0.368i)}\\ &= 0.966\angle 5.961^\circ\end{aligned}$$

$$\begin{aligned}B &= Z_o\frac{(1+S_{11})(1+S_{22})-S_{12}S_{21}}{2S_{21}}\\ &= 50\times\frac{(1+0.3\angle-90^\circ)(1+0.3\angle-90^\circ)-0.84\angle-26^\circ \times 0.84\angle-26^\circ}{2\times 0.84\angle-26^\circ}\\ &= 50\times\frac{(1+0-0.3i)(1+0-0.3i)+(0.755-0.368i)\times(0.755-0.368i)}{2\times(0.755-0.368i)}\\ &= 14.215\angle 20.717^\circ\end{aligned}$$

$$\begin{aligned}C &= Y_o\frac{(1+S_{11})(1-S_{22})-S_{12}S_{21}}{2S_{21}}\\ &= \frac{1}{50}\times\frac{(1+0.3\angle-90^\circ)(1-0.3\angle-90^\circ)-0.84\angle-26^\circ \times 0.84\angle-26^\circ}{2\times 0.84\angle-26^\circ}\\ &= \frac{1}{50}\times\frac{(1+0-0.3i)(1-0+0.3i)-(0.755-0.368i)\times(0.755-0.368i)}{2\times(0.755-0.368i)}\\ &= 0.015\angle 93.638^\circ\end{aligned}$$

$$\begin{aligned}D &= \frac{(1-S_{11})(1+S_{22})+S_{12}S_{21}}{2S_{21}}\\ &= \frac{(1-0.3\angle-90^\circ)(1+0.3\angle-90^\circ)+0.84\angle-26^\circ \times 0.84\angle-26^\circ}{2\times 0.84\angle-26^\circ}\\ &= \frac{(1-0+0.3i)(1+0-0.3i)+(0.755-0.368i)\times(0.755-0.368i)}{2\times(0.755-0.368i)}\\ &= 0.966\angle 5.961^\circ\end{aligned}$$

Thus, ABCD parameters in matrix form of the 2-port network are

$$[ABCD] = \begin{bmatrix} 0.966\angle 5.961^\circ & 14.215\angle 20.717^\circ \\ 0.015\angle 93.638^\circ & 0.966\angle 5.961^\circ \end{bmatrix}$$

$[ABCD]$ parameters of such two 2-port networks connected in cascade will be given by

$$\begin{aligned}[A_cB_cC_cD_c] &= [ABCD] \times [ABCD] \\ &= \begin{bmatrix} 0.966\angle 5.961^\circ & 14.215\angle 20.717^\circ \\ 0.015\angle 93.638^\circ & 0.966\angle 5.961^\circ \end{bmatrix} \\ &\quad \times \begin{bmatrix} 0.966\angle 5.961^\circ & 14.215\angle 20.717^\circ \\ 0.015\angle 93.638^\circ & 0.966\angle 5.961^\circ \end{bmatrix} \\ &= \begin{bmatrix} 0.911\angle 25.027^\circ & 27.459\angle 26.677^\circ \\ 0.029\angle 99.598^\circ & 0.911\angle 25.027^\circ \end{bmatrix}\end{aligned}$$

$[S]$ parameters of the cascaded 2-port network can be derived using (7.125)

$$\begin{aligned}\text{Cascaded } S_{11} &= \frac{A_c + B_cY_o - C_cZ_o - D_c}{A_c + B_cY_o + C_cZ_o + D_c} \\ &= \frac{0.911\angle 25.027^\circ + 0.549\angle 26.677^\circ - 1.437\angle 99.598^\circ - 0.911\angle 25.027^\circ}{0.911\angle 25.027^\circ + 0.549\angle 26.677^\circ + 1.437\angle 99.598^\circ - 0.911\angle 25.027^\circ} \\ &= \frac{1.380\angle -58.039^\circ}{3.090\angle 52.000^\circ} \\ &= 0.447\angle -110.039^\circ\end{aligned}$$

Thus, return loss (RL) of the cascaded network is

$$\begin{aligned}\text{Cascaded RL}_{\text{dB}} &= -20 \times \log(0.447) \\ &= 7.00 \text{ dB}\end{aligned}$$

$$\begin{aligned}\text{Cascaded } S_{21} &= \frac{2}{A_c + B_cY_o + C_cZ_o + D_c} \\ &= \frac{2}{0.911\angle 25.027^\circ + 0.549\angle 26.677^\circ + 1.437\angle 99.598^\circ - 0.911\angle 25.027^\circ} \\ &= \frac{2}{3.090\angle 52.000^\circ} \\ &= 0.647\angle -52.000^\circ\end{aligned}$$

Thus, insertion loss (RL) of the cascaded network is

$$\begin{aligned}\text{Cascaded IL (in dB)} &= -20 \times \log(0.647) \\ &= 3.78 \text{ dB} \quad \text{!! (different from } (1.51 + 1.51) \text{ dB)}\end{aligned}$$

This example shows that the insertion loss of the cascaded network (3.78 dB) is more than the sum of the insertion losses of the individual networks. This is due to the poor port return losses (10.46 dB) of the individual networks.

Example 19.14 Calculate insertion loss and return losses of the two 2-port networks P and Q of S-parameter matrixes:

$$[S]_P = \begin{bmatrix} 0.6\angle -91^\circ & 0.8\angle -1^\circ \\ 0.8\angle -1^\circ & 0.6\angle -91^\circ \end{bmatrix}$$

and

$$[S]_Q = \begin{bmatrix} 0.5\angle 26^\circ & 0.9\angle -63^\circ \\ 0.9\angle -63^\circ & 0.5\angle 26^\circ \end{bmatrix}$$

Calculate insertion loss and return losses when such two networks connected in cascade.

Solution

Insertion loss (IL) of the network P is

$$\begin{aligned} \text{IL (in dB)} &= -20 \times \log(|S_{21}|) \\ &= -20 \times \log(0.8) \\ &= 1.938 \text{ dB} \end{aligned}$$

Return loss (RL) of the network P is

$$\begin{aligned} \text{RL (in dB)} &= -20 \times \log(|S_{11}|) \\ &= -20 \times \log(0.6) \\ &= 4.44 \text{ dB} \end{aligned}$$

Insertion loss (IL) of the network Q is

$$\begin{aligned} \text{IL (in dB)} &= -20 \times \log(|S_{21}|) \\ &= -20 \times \log(0.9) \\ &= 0.92 \text{ dB} \end{aligned}$$

Return loss (RL) of the network Q is

$$\begin{aligned} \text{RL (in dB)} &= -20 \times \log(|S_{11}|) \\ &= -20 \times \log(0.5) \\ &= 6.02 \text{ dB} \end{aligned}$$

ABCD parameters of the network P can be calculated using (7.124)

$$\begin{aligned} A &= \frac{(1+S_{11})(1-S_{22})+S_{12}S_{21}}{2S_{21}} \\ &= \frac{(1+0.6\angle-91^\circ)(1-0.6\angle-91^\circ)+0.8\angle-1^\circ\times 0.8\angle-1^\circ}{2\times 0.8\angle-1^\circ} \\ &= \frac{(1-0.01-0.6i)(1+0.01+0.6i)+(0.8-0.014i)\times(0.8-0.014i)}{2\times(0.8-0.014i)} \\ &= 1.250\angle 0^\circ \end{aligned}$$

$$\begin{aligned} B &= Z_o\frac{(1+S_{11})(1+S_{22})-S_{12}S_{21}}{2S_{21}} \\ &= 50\times\frac{(1+0.6\angle-91^\circ)(1+0.6\angle-91^\circ)-0.8\angle-1^\circ\times 0.8\angle-1^\circ}{2\times 0.8\angle-1^\circ} \\ &= 50\times\frac{(1-0.01-0.6i)(1-0.01-0.6i)-(0.8-0.014i)\times(0.8-0.014i)}{2\times(0.8-0.014i)} \\ &= 36.409\angle-90^\circ \end{aligned}$$

$$\begin{aligned} C &= Y_o\frac{(1+S_{11})(1-S_{22})-S_{12}S_{21}}{2S_{21}} \\ &= \frac{1}{50}\times\frac{(1+0.6\angle-91^\circ)(1-0.6\angle-91^\circ)-0.8\angle-1^\circ\times 0.8\angle-1^\circ}{2\times 0.8\angle-1^\circ} \\ &= \frac{1}{50}\times\frac{(1-0.01-0.6i)(1+0.01+0.6i)-(0.8-0.014i)\times(0.8-0.014i)}{2\times(0.8-0.014i)} \\ &= 0.015\angle 90^\circ \end{aligned}$$

$$\begin{aligned} D &= \frac{(1-S_{11})(1+S_{22})+S_{12}S_{21}}{2S_{21}} \\ &= \frac{(1-0.6\angle-91^\circ)(1+0.6\angle-91^\circ)+0.8\angle-1^\circ\times 0.8\angle-1^\circ}{2\times 0.8\angle-1^\circ} \\ &= \frac{(1+0.01+0.6i)(1-0.01-0.6i)+(0.8-0.014i)\times(0.8-0.014i)}{2\times(0.8-0.014i)} \\ &= 1.250\angle 0^\circ \end{aligned}$$

Thus, ABCD parameters in matrix form of the 2-port P network are

$$[ABCD]_P = \begin{bmatrix} 1.250\angle 0^\circ & 36.409\angle-90^\circ \\ 0.015\angle 90^\circ & 1.250\angle 0^\circ \end{bmatrix}$$

In similar way, ABCD parameters in matrix form of the 2-port Q network are

$$[ABCD]_Q = \begin{bmatrix} 0.516\angle-3.535^\circ & 78.865\angle 90.039^\circ \\ 0.009\angle 92.513^\circ & 0.516\angle-3.535^\circ \end{bmatrix}$$

Thereforc, $[ABCD]$ parameter matrix of the cascaded networks P and Q can be written as

$$\begin{aligned}[ABCD]_{PQ} &= [ABCD]_P \times [ABCD]_Q \\ &= \begin{bmatrix} 0.984\angle-1.448^\circ & 79.817\angle 90.880^\circ \\ 0.020\angle 90.060^\circ & 0.575\angle 175.949^\circ \end{bmatrix}\end{aligned}$$

$[S]$ parameters of the cascaded 2-port network can be derived using (7.125)

$$\begin{aligned}\text{Cascaded } S_{11} &= \frac{A+BY_o-CZ_o-D}{A+BY_o+CZ_o+D} \\ &= \frac{0.984\angle-1.448^\circ+1.596\angle 90.880^\circ-0.981\angle 90.060^\circ-0.575\angle 175.949^\circ}{0.984\angle-1.448^\circ+1.596\angle 90.880^\circ+0.981\angle 90.060^\circ+0.575\angle 175.949^\circ} \\ &= \frac{1.658\angle 22.379^\circ}{2.540\angle 81.295^\circ} \\ &= 0.653\angle-58.916^\circ\end{aligned}$$

Thus, return loss (RL) of the cascaded network is

$$\begin{aligned}\text{Cascaded RL (in dB)} &= -20\times\log(0.653) \\ &= 3.70\text{ dB}\end{aligned}$$

$$\begin{aligned}\text{Cascaded } S_{21} &= \frac{2}{A+BY_o+CZ_o+D} \\ &= \frac{2}{0.984\angle-1.448^\circ+1.596\angle 90.880^\circ+0.981\angle 90.060^\circ+0.575\angle 175.949^\circ} \\ &= \frac{2}{2.540\angle 81.295^\circ} \\ &= 0.787\angle-81.295^\circ\end{aligned}$$

Thus, insertion loss (RL) of the cascaded network is

$$\begin{aligned}\text{Cascaded IL (in dB)} &= -20\times\log(0.787) \\ &= 2.08\text{ dB} \quad !!\ (\text{different from } (1.938+0.92)\text{ dB})\end{aligned}$$

This example shows that the insertion loss (2.08 dB) of the cascaded network is less than the sum of the insertion losses of the individual networks. This is due to the poor port return losses of the individual networks.

Examples 19.12, 19.13 and 19.14 show that in case of good return losses of the individual networks, insertion loss and gain of the combined (cascaded) network are the algebraic sum of the individual networks' gain (loss), whereas, in case of poor return losses of the individual networks, combined (cascaded) gain (loss) may increase or decrease.

19.4 Receiver

Function of on-board receiver of a communication satellite is to amplify the received signal linearly with minimum possible addition of noise and translating the frequency band of the received signal to the required downlink frequency band using a local oscillator. Most important parameters of a communication receiver are operating frequency, bandwidth, gain, noise figure, linearity, spurious levels and frequency translation error [2–4]. Block diagram of a typical receiver with operating gain of about 50 dB and overall noise figure of 1.7 dB is shown in Fig. 19.11. Front end of the receiver consists of a low-noise amplifier (LNA) to achieve overall low-noise temperature of the system. The input matching network of an LNA corresponds to its optimum noise figure which is different from complex conjugate matching; thus, input VSWR of an LNA is poor. An isolator is used at input of the LNA to avoid mismatch of the receive antenna output to the receiver. A high-gain LNA (28 dB in this example) is used to minimize the noise contributions due to the losses of the following elements such as filters and mixer. Here, a three-stage low-noise amplifier using pHEMT device is used for total gain of 28 dB and noise figure of 1.7 dB. In general, double-balanced mixer is used to downconvert the receive frequency band to transmit IF frequency band using a local oscillator. The frequency of the local oscillator is the frequency difference in between receive and transmit frequencies. A bandpass filter (BPF) at the input of the mixer is used to pass only the required frequency band and rejecting the unwanted out-of-band frequencies including the rejection of image frequency band. The BPF at the output of the mixer is used to reject various mixing products generated by the mixer. An IF amplifier is used to provide rest of the gain required for the receiver. The IF amplifier is designed to extract maximum power gain from the device, which operates in its linear region with moderate noise figure. Thus, if stability criterion allows, the IF amplifiers are designed with complex conjugate matching at its input as well as output for achieving maximum power gain from the used device. The amplifiers may be realized in hybrid microwave integrated circuit (HMIC) or in monolithic microwave integrated circuit (MMIC). In case of HMIC implementation, discrete components including active devices are mounted on printed substrate to realize the complete circuit, whereas, in MMIC implementation, all passive and active components including matching elements are built in a single substrate. Due to the absence of packaging of individual components and interconnecting elements in MMIC realization, the frequency of operation and achievable bandwidth are more. Photograph of a 3-stage microwave amplifier, realized using discrete components, is shown in Fig. 19.12. Here, packaged pHEMT, chip resistors and chip capacitors are used as discrete components which are mounted in an alumina substrate on which matching transmission microstrip elements are printed. Photograph of a double-conversion receiver is shown in Fig. 19.13 which is realized in HMIC.

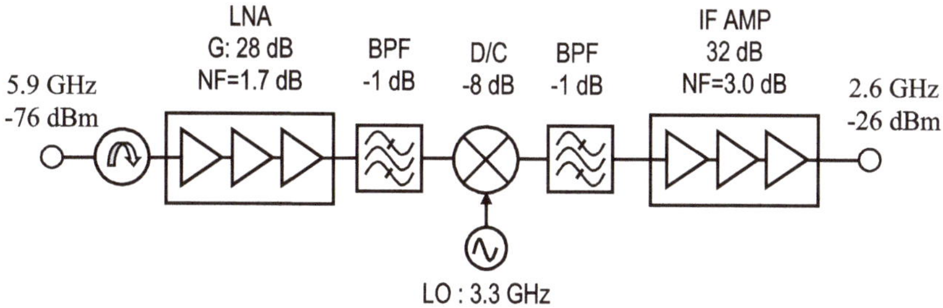

Fig. 19.11 Block diagram of a single conversion communication receiver

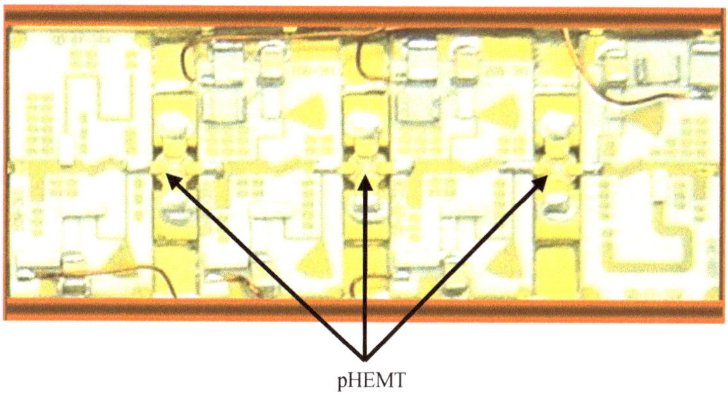

Fig. 19.12 Photograph of a microwave amplifier realized in HMIC

Example 19.3 Calculate overall noise figure (NF) and noise temperature of a receiver of block diagram as shown in Fig. 19.11. Consider loss of the isolator = 0.2 dB.

Solution

Isolator loss: 0.2 dB
LNA gain: 28 dB, LNA NF: 1.7 dB
Filters and mixer combined loss: (1 + 8 + 1) dB = 10 dB
IF amplifier gain: 32 dB, IF amplifier NF: 3 dB

Simplified block diagram of the receiver with gain and noise figures of individual modules is shown in Fig. 19.14. Using the following Friis formula for calculation of receivers overall noise figure (NF_{RX}),

$$\begin{aligned}\mathrm{NF}_{RX} &= \mathrm{NF}_1 + \frac{\mathrm{NF}_2 - 1}{G_1} + \frac{\mathrm{NF}_3 - 1}{G_1 \times G_2} + \frac{\mathrm{NF}_4 - 1}{G_1 \times G_2 \times G_3} \\ &= 1.047 + \frac{0.479}{0.955} + \frac{9.0}{0.955 \times 631} + \frac{0.995}{0.955 \times 631 \times 0.1} \\ &= 1.5803\end{aligned}$$

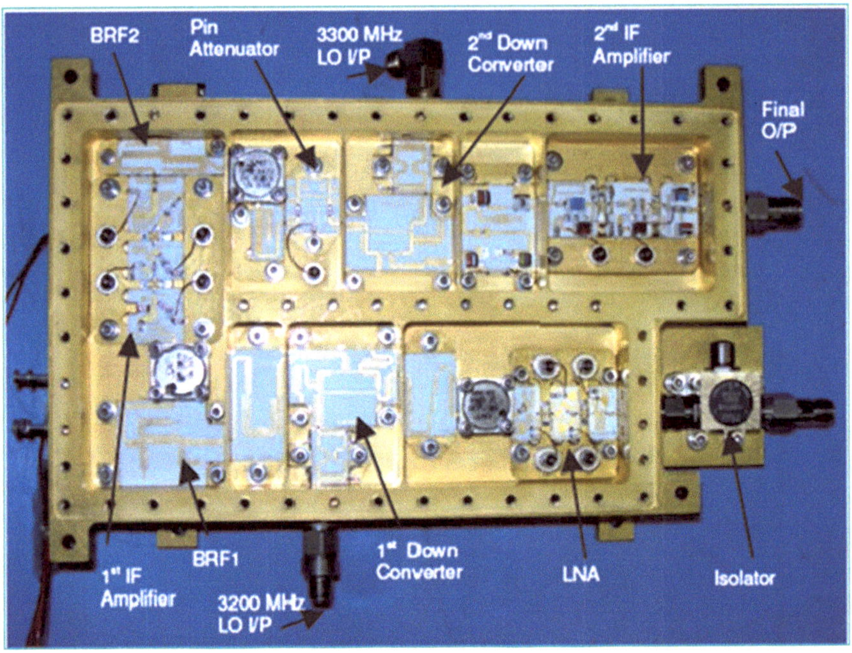

Fig. 19.13 Photograph of a communication receiver

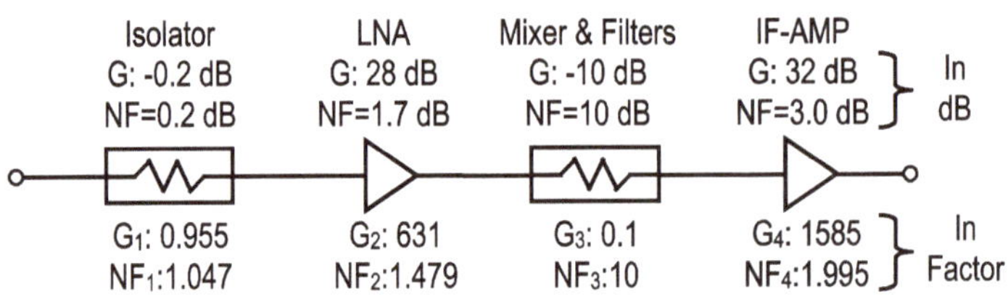

Fig. 19.14 Block diagram of the receiver for noise calculation

$$[\mathrm{NF}_{RX}]_{\mathrm{dB}} = 1.987 \text{ dB}$$

Noise temperature T_{Rx} of the receiver is

$$\begin{aligned} T_{RX} &= (\mathrm{NF}_{RX} - 1)T_o \\ &= (1.5803 - 1)290 \text{ K} = 168.3 \text{ K} \end{aligned}$$

Example 19.4 Calculate overall third-order intermodulation product ($IM3$) of a receiver (shown in the block diagram in Fig. 19.11) at input power level of −76 dBm. The $P_{o1\mathrm{dB}}$ (output power at 1-dB gain compression point) of LNA and IF amplifiers are 5 and +7 dBm, respectively. The P_{oIP3} (output third-order intercept point) of the mixer is 10 dBm.

Solution

$P_{o1\mathrm{dB}}$ of LNA: +5 dBm
$P_{o1\mathrm{dB}}$ of IF amplifier: +7 dBm
P_{oIP3} of mixer: +10 dBm

Considering the I–O characteristic of LNA, mixer and IF amplifier can be expressed by a power series up to third order, the third-order intercept output power level P_{oIP3} can be written in terms of $P_{o1\mathrm{dB}}$ as

$$P_{oIP3} = P_{o1\mathrm{dB}} + 10.63$$

Using this equation, P_{oIP3} of the LNA and IF amplifier are +15.63 and +17.63 dBm, respectively.

Output power levels of each element at the input carrier power level of −76 dBm are shown in Fig. 19.15, considering all the elements are operating with constant gain. The third-order intermodulation power level P_{oIM3} in dBc with respect to the output carrier level is given by (14.15b)

$$P_{oIM3} = P_{o2f_1-f_2} - P_{of_1} = 2\left(P_{of_1} - P_{oIP3}\right)$$

Here, $P_{of_1} = P_{\mathrm{out}}$. Therefore, the third-order intermodulation power level of LNA, mixer and IF amplifiers at the output of respective elements are

$$\begin{aligned} P_{oIM3}(\mathrm{LNA}) &= 2(-48.2 - 15.6)\,\mathrm{dBc} = -127.7\ \mathrm{dBc} \Rightarrow -175.9\ \mathrm{dBm} \\ P_{oIM3}(\mathrm{Mixer}) &= 2(-57.2 - 10)\,\mathrm{dBc} = -134.4\ \mathrm{dBc} \Rightarrow -191.6\ \mathrm{dBm} \\ P_{oIM3}(\mathrm{IF\ Amp}) &= 2(-26.2 - 17.6)\,\mathrm{dBc} = -87.7\ \mathrm{dBc} \Rightarrow -113.86\ \mathrm{dBm} \end{aligned}$$

Considering the linear gain of the IM_3 power levels by the following stages, the overall IM_3 level at the output of the receiver will be

$$\begin{aligned} P_{oIM3} &= (-175.9 + 22)\,\mathrm{dBm} + (-191.6 + 31)\,\mathrm{dBm} + (-113.86)\,\mathrm{dBm} \\ &= -113.86\ \mathrm{dBm} \end{aligned}$$

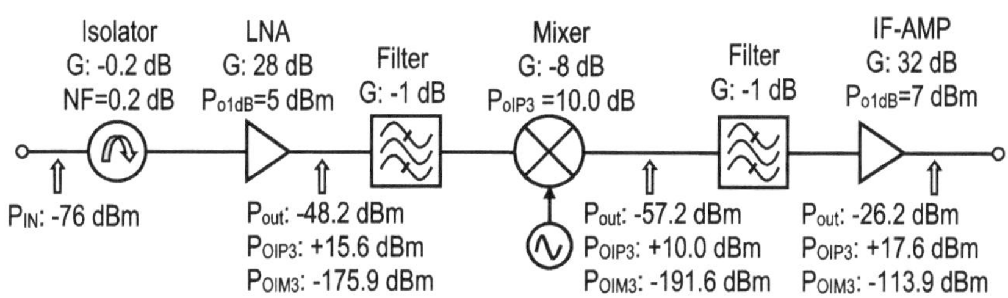

Fig. 19.15 Block diagram of the receiver for *IM*3 calculation

The IM_3 level in dBc with respect to the carrier level at the output is $(-113.86 + 26.2)$ dBc $= -87.7$ dBc.

This shows that for this receiver configuration, the contribution of LNA and mixer on the overall IM_3 is negligible. It is fully governed by the nonlinearity of the final stage, i.e. IF amplifier.

Therefore, if the receiver operates at higher input power levels say by 10 or 20 dB more than overall IM_3 level of the receiver will be more by 20 and 40 dB, respectively. Thus, for the receiver input power level of -66 and -56 dBm, the IM_3 level will be -67.7 and -47.7 dBc, respectively.

19.4.1 Local Oscillator

The role of a local oscillator in satellite transponder is to provide a stable reference RF frequency with sufficient output power level to drive mixer circuit in its nonlinear region for frequency translation. The frequency of the local oscillator is the difference between the centre frequency of the uplink band and the centre frequency of the downlink frequency band. Frequency stability, spectral purity in terms of phase noise, and spurious products and output power level are the most important parameters of a local oscillator. Generally, in a satellite application, local oscillators of the desired frequency are derived from a reference low-frequency source. The stability of local oscillator is determined by the stability of the reference frequency source.

Mostly, crystal-based oscillators are used as reference source for the local oscillators. Crystal cut in the form of a plate determines the fundamental frequency of oscillation. Frequency stability of a reference crystal oscillator is mainly influenced by the change of temperature and time. To minimize the influence of change of temperature, temperature-compensated crystal oscillator (TCXO) is used. Here, TCXO encases the oscillator circuit and temperature compensating networks in a closed container. Another option is the use of oven-controlled crystal oscillator (OCXO) as reference oscillator. Here, crystal oscillator and temperature-sensitive elements are kept in a thermally insulated container along with a heater. The heater maintained the inside temperature at oscillators' minimum sensitive region. The stability of a local oscillator is specified as long-term stability over the lifetime and short-term stability over the specified operating temperature range. In general, for satellite transponders, temperature-controlled crystal oscillator (TCXO) of short-term stability of ± 1 ppm (parts per million, i.e. error of 1 Hz in 1 MHz) and long-term stability of ± 10 ppm for 15 years is used. In some applications, where more stable frequency is required, oven-controlled crystal oscillator (OCXO) of one order that has better stability compared to TCXO is used. Two types of local oscillators are realized. One is based on frequency multiplier and another is using phase-locked loop (PLL).

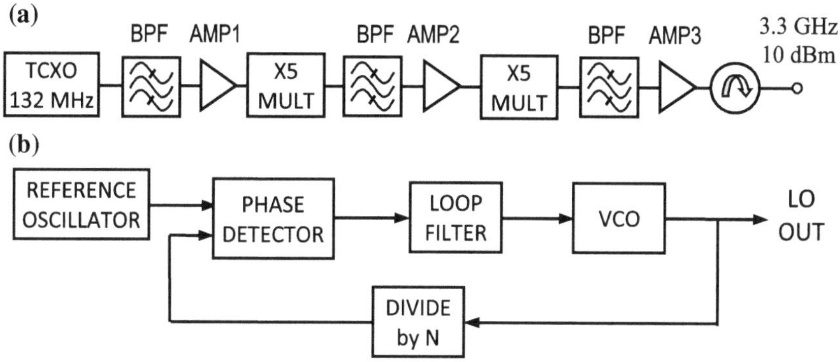

Fig. 19.16 **a** Block diagram of a multiplier-based local oscillator. **b** Simple block diagram of a PLL-based local oscillator

19.4.1.1 Multiplier-Based Local Oscillator

Block diagram of a multiplier-based LO is shown in Fig. 19.16a. Here, reference crystal oscillator frequency of 132 MHz and required LO output frequency of 3.3 GHz is considered. Two stages of X5 multiplication (overall X25) are done to generate 3.3 GHz frequency from the reference frequency of 130 MHz. Bandpass filters at different stages are used to suppress unwanted frequency components to a sufficient level (at least −60 dBc) with respect to the desired frequency component. Any unwanted frequency component of an LO acts as discrete frequency component and may severely affect the communication.

Amplifiers (AMP1 and AMP2) are used before each multiplier circuit to provide required power level to the multiplying device to operate it highly nonlinear condition. The final amplifier (AMP3) is used to increase the power level of the final frequency component to achieve the desired output level required for operating mixer in the receiver in its nonlinear region. In general, the final amplifier stage operates in saturation region to provide nearly constant output power level over the operating environmental (temperature, bias voltage variation, etc.) condition to ensure the operation of mixer in its fixed conversion gain/loss condition.

19.4.1.2 PLL-Based Local Oscillator

Simplified block diagram of a phase-locked loop (PLL)-based local oscillator is shown in Fig. 19.16b. It consists of a phase detector, loop filter, frequency divider and voltage-controlled oscillator (VCO). Generally, varactor diode is used to achieve voltage-dependent frequency of the VCO. Output frequency of the VCO is divided by a frequency divider and fed back to one input of the phase detector circuits. The phase detector compares the output frequency from the divider to the reference frequency of the reference oscillator. A phase error signal is generated by the phase detector and creates a signal whose magnitude is proportional to the phase error. This phase error signal is then low-pass filtered by the loop filter and fed to the control input of the VCO. The control signal controls the output frequency of the VCO. At the locked condition of the PLL, the two inputs to the phase detector are in-phase and the output frequency is equal to the reference oscillator frequency multiplied by the divider ratio, N.

19.5 Satellite Transmitter

The transmitter section amplified the channelized signal to the required RF power level before transmitting through a transmit antenna. As shown in Fig. 19.10, this section consists of a driver amplifier (DA) to boost the signal to proper drive level required for high power amplifier (HPA) which ultimately provides the required transmit power level. In a communication system, major contributions of nonlinearities are due to the final power amplifier that affects the overall communication performance severely [5–7]. A linearizer also used at input of the HPA to minimize the effect of nonlinearity of the HPA on transmitted signal. The driver amplifier (DA) is also called channel amplifier (CAMP) for its function to amplify the channelize frequency band. The CAMPs consist of several control systems for on-board controlling the transponder gain by ground command or automatically by sensing its RF power level.

The channel amplifier and linearizer are low-power systems, and thus these are realized very compactly using solid-state technology. However, depending on downlink frequency and required output power level, two types of high power amplifiers are used: Solid-state power amplifier (SSPA) and travelling wave tube amplifier (TWTA). Use of TWTAs is the only option in case of requirement of higher RF over high microwave frequency range. However, due to the advancement of solid-state device technology, it can provide required RF power level at least over lower microwave frequency range. Thus, SSPAs are preferable for its compact size and better linearity compared to TWTAs.

19.5.1 Driver Amplifier (DA)

Function of an on-board driver amplifier (DA) for a communication satellite is to amplify the channelized signal linearly, i.e. without much distortion, for providing sufficient drive level to the high power amplifier [8]. It has also provision for on-board gain setting of the transponder by ground command or automatically by sensing its RF power level. A driver amplifier for satellite transponder has two operating modes: fixed gain mode (FGM) and automatic level control (ALC) mode. Most important parameters of a driver amplifier are its operating frequency, bandwidth, gain, linearity, output power, gain setting range in FGM and dynamic range in ALC mode. In a

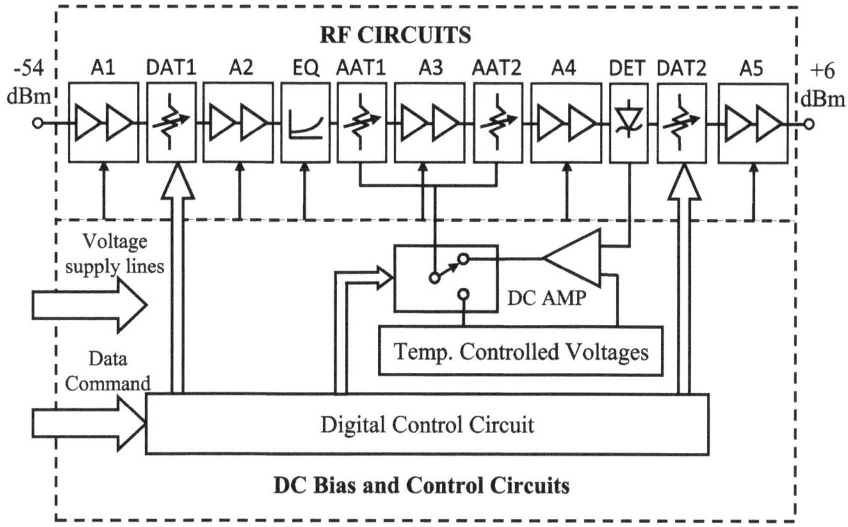

Fig. 19.17 Block diagram of a satellite channel amplifier

satellite transponder, a driver amplifier amplifies the signal after channelization, and thus it is also called channel amplifier (CAMP).

Block diagram of a channel amplifier is shown in Fig. 19.17. The CAMP consists of RF circuits and bias and control circuits. The RF lineup consists of several amplifier stages (A1–A5) to achieve required gain. Digital attenuators (DAT1 and DAT2) are used for commandable gain setting of the CAMP by issuing digital command. The analog attenuators (AAT1 and AAT2) are used to provide attenuation automatically by detecting the power level using RF power detector (DET). The detected voltage at the output of the detector amplified by a differential DC amplifier and applied to the control terminal of the analog attenuators through a commandable analog switch. In ALC mode of operation, the output of the DC amplifier will be connected to the analog attenuators. Another input of the differential DC amplifier is connected to a temperature-controlled voltage to keep the output RF power level constant in ALC mode over the operating temperature range. In FGM operation, the control terminals of the analog attenuators will be connected to temperature-controlled voltage to keep the constant gain over the operating temperature range. The digital control circuit processes the command data and generates appropriate control data to select the mode of operation (in between ALC and FGM) and to provide commandable gain setting in both the operation modes. An amplitude tilt active equalizer (EQ) is used to achieve broadband frequency response. The amplifiers and both types of attenuators are distributed in the lineup to achieve required noise figure and linearity of the CAMP over its entire gain (dynamic range) setting conditions for both the operation modes.

Photograph of a Ku-band CAMP of about 60 dB gain is shown in Fig. 19.18. The amplifier and attenuator modules are realized using MMIC technology. The full RF

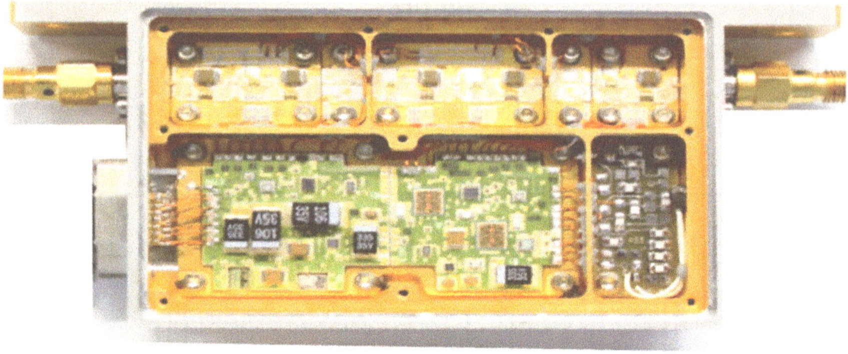

Fig. 19.18 Photograph of a Ku-band CAMP

Fig. 19.19 Frequency response of the CAMP with and without equalizer

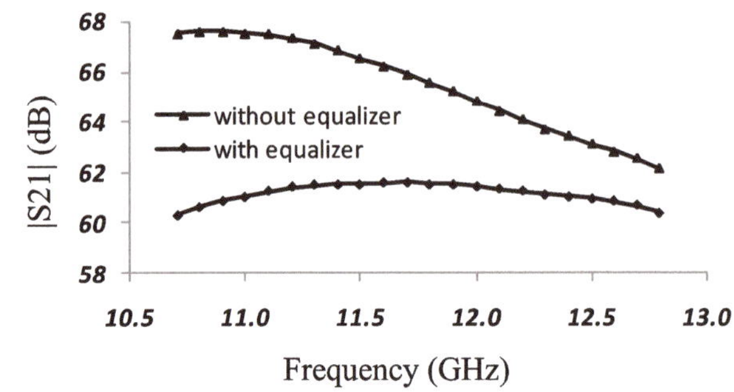

circuit is packaged in three compartments with narrow slits (0.7 mm × 2 mm) through which the circuits are interconnected using gold ribbons. The slit acts as waveguide of cut-off frequency below 10 GHz, and thus provides high isolations for the operating frequency in between two adjacent compartments, which prevents waveguide mode of propagation in operating frequency band, and thus ensures overall stability of the CAMP.

Frequency response of the CAMP with and without equalizer is shown in Fig. 19.19. It shows that the use of equalizer improves the gain flatness from 6 to 1.5 dB over the frequency range of 10.5–13 GHz. In fixed gain mode (FGM) operation, the digital attenuators DAT1 and DAT2 are used to control the gain of the CAMP to set the saturation flux density (SFD) of the transponder and also to operate the transponder at required power back-off condition. The typical gain setting is about 30-dB in steps of 1-dB. I–O characteristics of the channel amplifier in FGM operation for different gain setting conditions are shown in Fig. 19.20. The output of the CAMP saturated under higher input power level is due to the saturation of the final amplifier stage. Under nominal operating condition, the channel amplifier always operates in its linear I–O characteristic region.

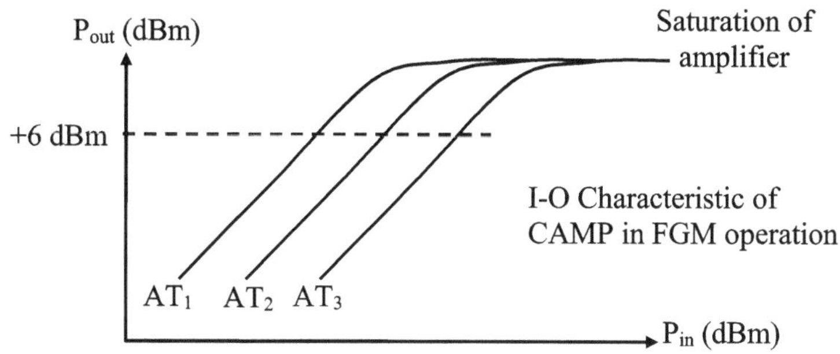

Fig. 19.20 I-O characteristic of CAMP in FGM operation for different gain setting condition

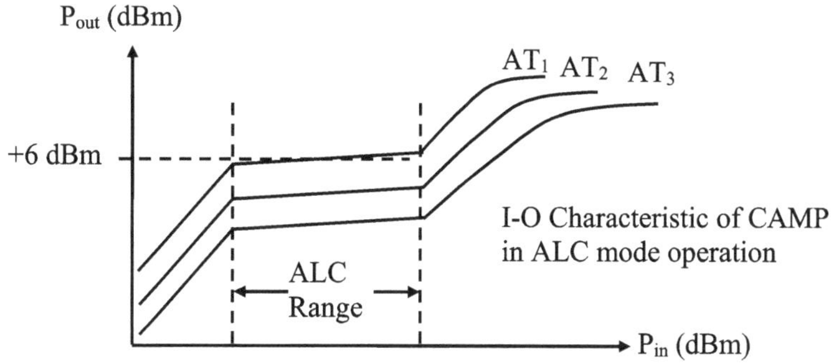

Fig. 19.21 I-O characteristic of CAMP in ALC mode operation for different attenuation in DAT2

In ALC mode of operation, the output power will remain constant irrespective of its input power level with the specified ALC dynamic range. Typical ALC dynamic range is about 30 dB. Here, the RF power level is detected before the digital attenuator DAT2 to control the output power level in ALC mode for operating the HPA back-off condition if required. Final amplifier stage is used after the detector and DAT2 to enable the use of low-power device for the final amplifier stage meeting the output power requirement. I–O characteristics of the CAMP in its ALC mode of operation are shown in Figs. 19.21 and 19.22 for setting of the digital attenuators DAT2 and DAT-1, respectively. In ALC mode of operation, the digital attenuator DAT2 is used to provide adjustable (variable) constant output power level as shown in Fig. 19.21. Typically, 15 dB attenuation range in steps of 0.5 dB is kept for this purpose. The digital attenuator DAT1 is used to slide the ALC range keeping the same ALC dynamic range as shown in Fig. 19.22. This provides the flexibility of SFD range setting of the transponder in ALC mode.

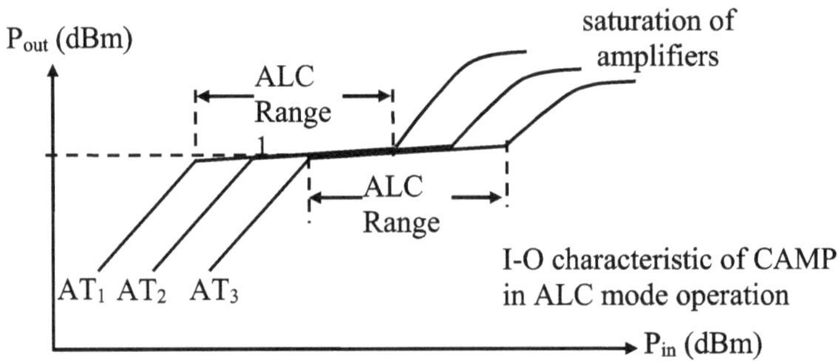

Fig. 19.22 I-O characteristic of CAMP in ALC mode operation for different attenuation in DAT1

19.5.2 Travelling Wave Tube Amplifier (TWTA)

Travelling wave tube amplifier (TWTA) is one of the most economically costliest subsystems which used most critical but highly matured technology for realization [9]. In most of the satellite transponders, TWTAs are used as high power amplifier (HPA) for its ability to provide higher output power with higher DC-to-RF efficiency at higher frequency of operation over broader frequency band compared to solid-state power amplifiers (SSPAs). A TWTA consists of travelling wave tube (TWT) and electronic power conditioner (EPC). The EPC provides required DC voltage and currents to the TWT and provides various controls and protection mechanisms of the TWT. In a satellite transponder, TWTAs amplify signal taking from the output of a channel amplifier and provides required output power. In most of the cases, a predistortion linearizer is used in between the channel amplifier and TWTA to minimize the effect of nonlinearity of a TWTA on communication system.

In travelling wave tube amplifier, amplification of microwave signal takes place due to the continued interaction between the wave and the high-energy electron beam travelling along the signal. Functional diagram of a TWT is shown in Fig. 19.23. Structurally, a TWT can be divided into three sections: electron gun, slow-wave structure and collector. The electron is generated by heating a cathode which travel towards anode due to high electric field generated by applying a very high potential difference in between anode and cathode. The electron beam after passing through the helix is collected at the collector. To fulfil the requirement of continued interaction of waves with the electron beam for long time, the microwave signal is passed through a slow-wave structure through which the electron beam flows and the electron beam is focused applying a longitudinal static magnetic field using permanent magnets as shown in Fig. 19.23. The helical slow-wave structure slowed down the microwave signal with its phase velocity of about $cp/2\pi r$, where p is the pitch of the helix of radius r. Due to the propagation of electromagnetic wave along the helix, a longitudinal electric field will be generated. This time-varying electric field results in velocity modulation in the electron beam passing through the helix. This velocity modulation will result in bunching of electrons in regular intervals of one wavelength

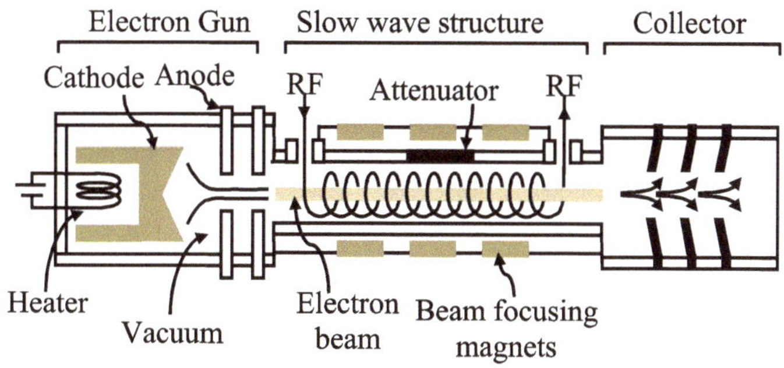

Fig. 19.23 Functional diagram of TWT

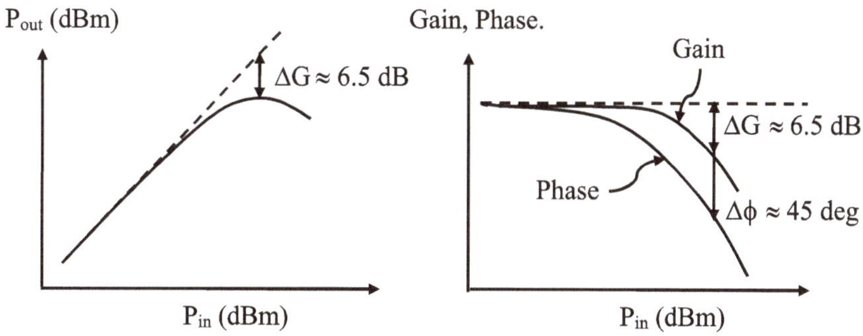

Fig. 19.24 Typical I–O, gain and phase characteristic of TWTA

of the applied signal. Thus, with the condition of equal phase velocity of microwave signal and electron beam, a continuous interaction takes place between the beams and the waves in the helix and bunches grow as the beam moves ahead. This continued interaction results in the amplification of microwave signal flowing through the helix by picking power from the electron beam. The amplified microwave signal is then coupled out of the helix at the output port. The collected electrons at the collector dissipate its rest of the energy at the collector.

Most important RF performance parameters of a TWTA are operating frequency, bandwidth, saturated output power, DC-to-RF efficiency and linearity. Capability to provide higher output power of about 250 W at *S*-band and 150 W at Ka-band with DC-to-RF efficiency about 65–70% for space grade TWTAs makes them very attractive for using as HPA in high-power satellite transponders. However, it has comparatively poor linearity, size and more mass compared to SSPAs. I–O characteristic, gain and phase dependency on RF power level are shown in Fig. 19.24. Typical values of gain compression and total phase shift are about 6.6 dB and 45°, respectively. These nonlinearities lead to distortion in amplified signal. One way to minimize the effect of these nonlinearities on amplified signal is to operate the TWTA at power back-off condition. However, efficiency decreases with the increase of back-off of TWTA. Practically, a linearizer is used with TWTA to minimize the effect of nonlinearity of TWTAs and thus minimizes the requirement of TWTA back-off.

19.5.3 Solid-State Power Amplifier (SSPA)

Solid-state power amplifiers (SSPAs) are used as high power amplifiers where output power requirement is less. SSPAs are advantageous for its better linearity, lower mass and smaller footprint area compared to TWTAs. Wherever SSPAs are capable to provide the required RF power level, it is preferable to use SSPAs. For example, at *L*-, *S*- and *C*-bands, SSPAs as well as TWTAs are used depending on the output power requirement. However, beyond *C*-band there is no option other than using TWTAs for space communication due to the non-availability of SSPAs with required power level. With the present technology, space qualified SSPA of output power up to 60 W is realized using GaAs-based MESFET/HFET power devices. With the advancement of GaN technology, presently devices are available to realize space qualified SSPAs of output power level about 250 W at *L*- and *S*-bands and 120 W at *C*-band.

Block diagram of a solid-state power amplifier of output RF power 53 dBm (200 W) is shown in Fig. 19.25. It consists of RF section, electronic power conditioner (EPC) and bias and temperature control circuits. The RF section consists of medium power amplifier and high power amplifier stages to provide output power level of 200 W with overall gain of 53 dB. Important parameters of an SSPA are operating frequency, bandwidth, output power level, DC-to-RF efficiency and linearity. To achieve higher efficiency, in general, the final two amplifier stages operate in Class-AB mode with harmonic tuning. However, the medium power amplifier stages operate in Class-A mode to achieve overall better linearity without sacrificing overall efficiency. The efficiency of a final stage amplifier at its maximum allowable operation power level is comparable to the efficiency of a TWTA which is about 70%. However, overall efficiency of an SSPA is significantly reduced due to the use of other amplifying stages to achieve gain comparable with a TWTA. For example, the efficiency of a *C*-band 40 W GaAs-based SSPA (including EPC) is about 45%. Another important aspect of design of a power amplifier is the protection of power devices under intentional or unintentional RF overdrive condition. Lineup of the power amplifier is so selected that full range of overdrive cannot pass to final amplifier stage. Another option is the use of open-loop limiter or closed-loop overdrive protection circuit to protect the devices from overdrive condition [10].

The bias and control circuits provide required bias voltages/currents to all the devices which are temperature-controlled to achieve temperature-compensated gain as well as output power level. Practically, it is required to increase the drain voltage of the final amplifying FET device with the increase of temperature to keep the output power constant over the operating temperature range. Photograph of an SSPA realized in hybrid microwave integrated circuit (HMIC) is shown in Fig. 19.26.

Near the saturation point, SSPA behaves differently from TWTA. Typical I–O, gain and phase characteristics of an SSPA are shown in Fig. 19.27. In case of TWTA, the output power decreases when it operated beyond its saturated point; however, for an SSPA, it remains nearly constant. SSPAs are not permitted to operate beyond its 2-dB gain compression point to ensure its reliability for space use. Excessive gate current may flow when a power FET operates beyond its 2-dB gain compression

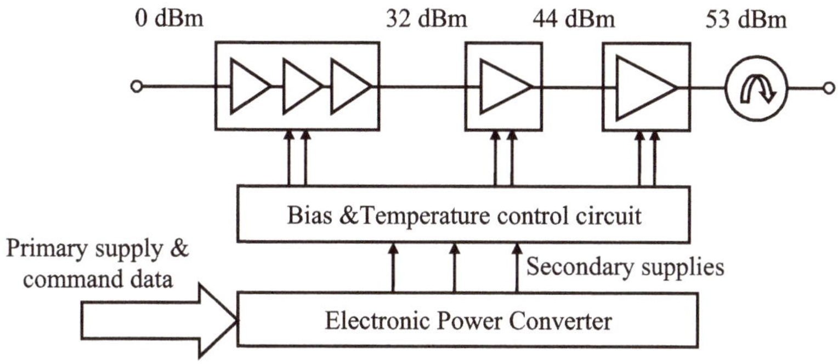

Fig. 19.25 Block diagram of a solid-state HPA

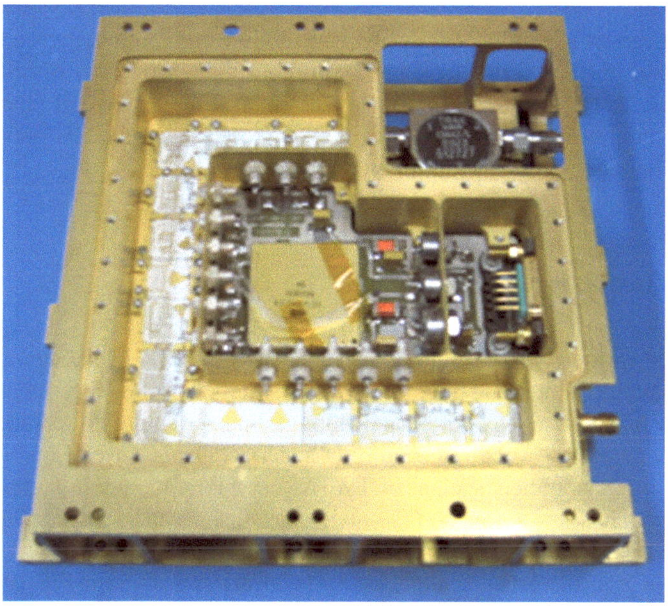

Fig. 19.26 Photograph of a solid-state power amplifier

point. Excessive flow of gate current for a power FET over a prolonged duration may lead to failure of the device. However, low-power FETs may be operated at more gain compression without compromising its life. Typical total phase shift for an SSPA is about 20° up to its 2-dB gain compression point. Another important point is to be noted that in case of SSPAs the phase of the output signal increases with the increase of power level as shown in Fig. 19.27, whereas it decreases in case of TWTAs as seen in Fig. 19.24.

Thus, it is clear that SSPAs are in general more linear than TWTAs when both are operated at their respective rated output power conditions. In practice, an SSPA

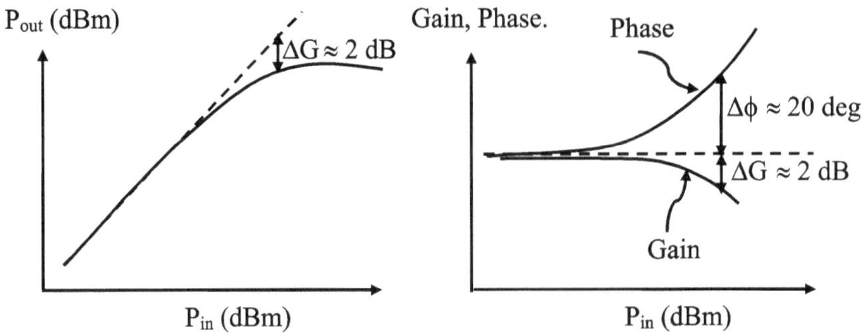

Fig. 19.27 Typical I–O characteristics of an SSPA

without linearizer has about equally good overall linearity as linearity of a linearized TWTA. Thus, there is very less scope for improvement of linearity of SSPAs using a linearizer. More precisely, using linearizer, there is large scope of phase as well amplitude nonlinearity improvement for TWTAs. However, there is only scope of phase nonlinearity improvement for an SSPA.

19.6 Linearizer

Among the various types of linearizers, predistortion (PD) linearizers are mostly used for satellite communication due to their simplicity, low power consumption, and ability to linearize over wide bandwidth of a power amplifier operating at near saturation condition [11–15]. Predistortion linearizer creates inverse nonlinearity of the transmitting amplifiers such as of TWTAs or SSPAs in order to compensate for the distortion. It is able to function as standalone unit and can be cascaded in between CAMP and HPA with proper input and output power level matching.

Block diagram of a broadband predistortion linearizer is shown in Fig. 19.28. Heart of the linearizer is the distortion generator module. In general, it is realized using Schottky and p-i-n diodes in a vector modulator configuration as discussed in Chap. 16 to generate required amplitude and phase nonlinearities. Two amplitude tilt active equalizers (EQ1 and EQ2) are used to make the linearizer broadband. An analog attenuator (AAT1) is used to compensate overall gain variation of the equalizer over its operating temperature range. The amplifier modules (A1, A2 and A3) are used to match the linearizer's input and output power levels with the output power level of CAMP and input power level of HPA in addition to compensate loss of the distortion generator, equalizers and attenuator. Photograph of a linearizer operating in Ku-band is shown in Fig. 19.29. The linearizer is realized using MMIC amplifier modules and other circuits realized on alumina substrate using HMIC technology. The distortion generator unit is realized as vector modulator with one arm nonlinear circuit using Schottky barrier diodes and the other arm with a linear circuit using p-i-n diodes. The

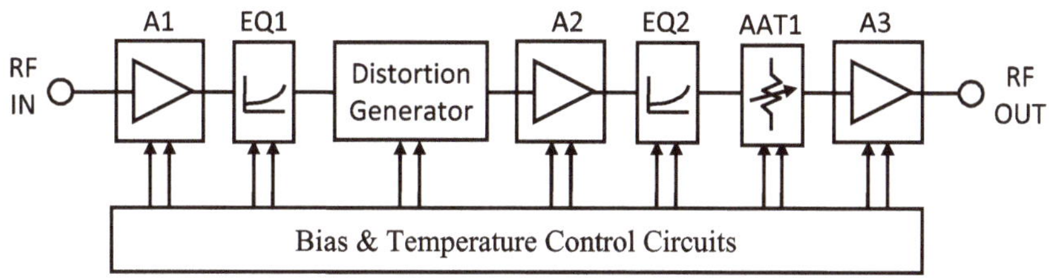

Fig. 19.28 Block diagram of a broadband predistortion linearizer

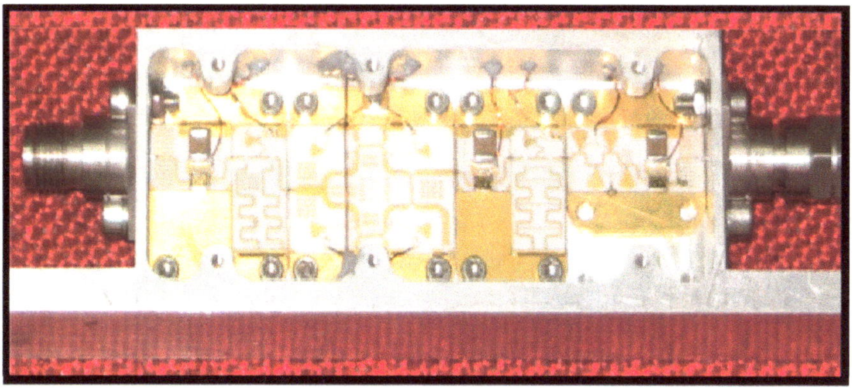

Fig. 19.29 Photograph of a Ku-band linearizer

amplitude tilt active equalizers and analog attenuator are realized using p-i-n diodes as voltage/current variable resistors. Temperature compensation of the unit is done using optimum load-line bias technique of Schottky barrier and p-i-n diodes as well as providing temperature-controlled voltage/current to the analog attenuator (AAT1).

Typical nonlinear amplitude and phase performances of a Ku-band linearizer cascaded with a TWTA are shown in Fig. 19.30. It shows that gain compression of a nonlinearized TWTA improves from 6.5 to ± 0.7 dB when it is cascaded and optimized with linearizer. Similarly, nonlinearized TWTA's total phase shift improves from 45° to $\pm 4^\circ$. Thus, amplitude and phase nonlinearity of linearized TWTA (LTWTA) improves significantly and comparable, even better than linearity of a nonlinearized SSPA. Two-tone third-order intermodulation levels (IM_3) of TWTA and LTWTA are shown in Fig. 19.31. It shows that there is no significant improvement of IM_3 level near saturation region of the TWTA. However, there is a significant IM_3 improvement at around 7 dB input back-off (IBO) of each carrier which corresponds to 4 dB IBO of the TWTA with respect to its saturation point.

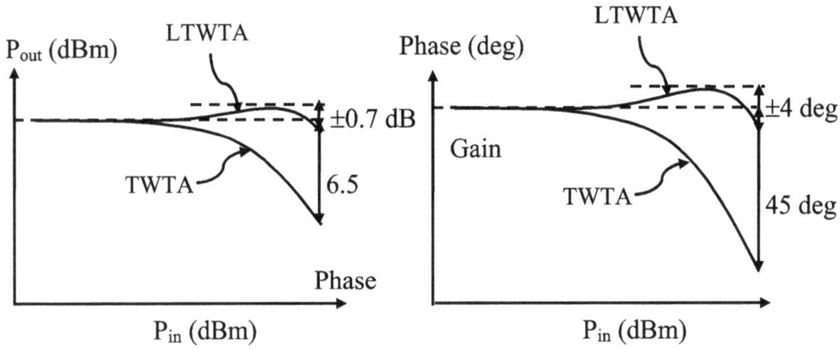

Fig. 19.30 Typical I–O, gain and phase characteristic of TWTA and Linearized TWTA

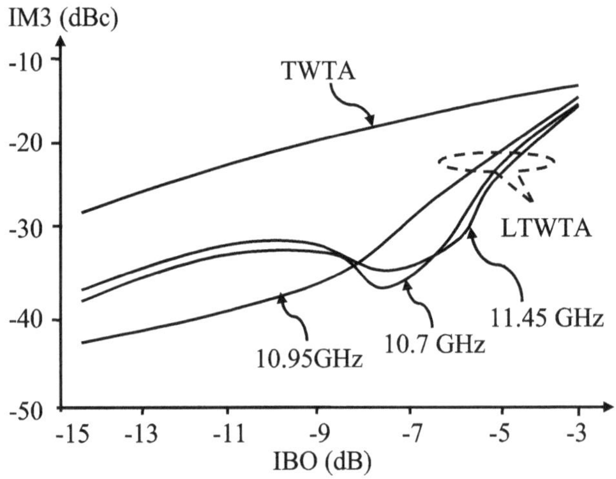

Fig. 19.31 Typical 3rd order IM levels of TWTA and LTWTA

19.7 Microwave Power Module (MPM)

Microwave power module (MPM) is the technology where both solid-state and vacuum tube technologies are combined to realize very compact with low mass transmitter module to provide highest possible output power with highest efficiency [16, 17]. The solid-state technology is advantageous for its miniaturized configuration using MMIC technology with excellent linearity. However, its output power giving capability and efficiency is less compared to TWTA. Presently, space qualified MMICs with medium output power up to 10 W are available. On the other hand, TWTAs are capable of providing highest output power level with high DC-to-RF efficiency. But its linearity is poor and also size is more in case of high-gain TWTAs. In MPM, short-length TWT is used as final power amplifier with lower gain which needs input power about 1–10 W which is achievable from solid-state MMIC configuration. Thus, rest

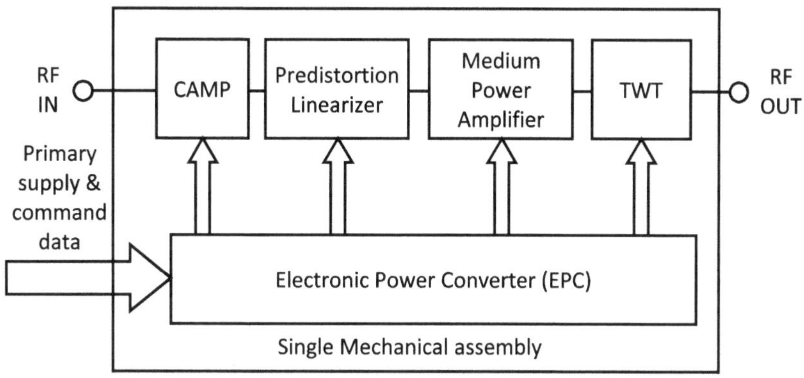

Fig. 19.32 Block diagram of microwave power module (MPM)

of the transmitter gain with medium-level output power is achieved using solid-state technology in MMICs. To improve the linearity of the total system, a predistortion linearizer is realized in solid-state technology cascaded at suitable position as shown in Fig. 19.32. To minimize the overall mass and footprint area, a single electronic power conditioner is used for all these microwave subsystem and a single mechanical assembly including all the RF units and the EPC. To realize a microwave power module, challenge is the thermal management due to the very compact assembly of the unit. EMI/EMC-related issues are also critical due to the proximity of the small signal and high power units in a single package.

19.8 Multiport Amplifier (MPA)

Multiport amplifiers (MPAs), also known as matrix amplifiers, have multiple inputs and multiple outputs with several high power amplifiers connected in parallel. These are used to provide flexibility in terms of power allocation, since combined output power from parallel-connected several amplifiers is shared between the output ports. Here, RF power can be allocated among the output ports as per requirements of a communication system. Therefore, the combined power of all the high power amplifiers is available for any output port provided that the other ports do not require any power at the same time. The multiport amplifier configurations are very useful in case of multibeam satellite communication systems, where beam-to-beam service requirements such as number of users/data rate changes dynamically over time. In case of multiport amplifier configuration, all the power amplifiers provide same RF output power irrespective of different powers required by the different output ports. Thus, operating conditions (input/output back-off) of all the amplifiers remain the same though RF powers taken from different output ports are different.

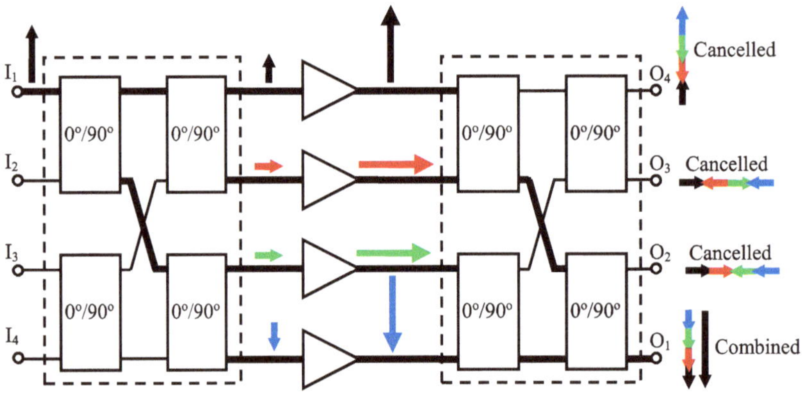

Fig. 19.33 Block diagram of a 4 × 4 MPA with signal flow paths (bold lines) when power fed to only Port-I_1

A multiport amplifier (MPA) consists of an array of power amplifiers (PAs) in parallel and a pair of complementary Butler matrix networks that consist of 90° hybrid networks [18]. Schematic diagram of a 4 × 4 MPA is shown in Fig. 19.33. The 4 × 4 MPA consists of four high power amplifiers connected in parallel using 4 × 4 input and 4 × 4 output networks. The signal at each input in the MPA is divided into four signals (in general n signals) with particular phase relationships. These signals are amplified separately in each power amplifier and are recombined in the output network. Thus, signal at each input is amplified by all the power amplifiers, however, assembled at the corresponding output ports. Figure 19.33 shows the flow of signal when it is applied to the input port-1. Thus, the signal is applied in port-1 amplified by all the amplifiers and then recombined at the output port-1. In a similar way, applied signal at input ports 2, 3 and 4 is amplified by all the amplifiers and recombined at the respective output ports 2, 3 and 4. In ideal condition, signal applied to the port-1 becomes available at the output port-1 only. Thus, no part of the signal applied at the input port-1 will be available at the output ports 2, 3 and 4. To achieve this ideal performance from an MPA, it is important to equalize the amplitude and to synchronize the transmitted phase among the signals. Practically, infinite isolation is not possible due to various nonideal electrical characteristics of each signal path. The finite isolation leads to signal loss as well as interference among signals coming from different input ports. Another drawback is the unwanted multicarrier operation of the power amplifiers. This multicarrier operation reached as all the input signals are amplified at each power amplifier even when a single carrier is fed at each input. The multiport amplifier is adapted to applications that require flexibility in terms of power allocation at its different output ports and is really advantageous if the input signals are already multicarrier.

Design of compact input and output Butler matrix networks with minimum insertion loss and proper amplitude and phase matching is crucial. The input network operates at lower RF power level, and thus loss of the input network can be easily

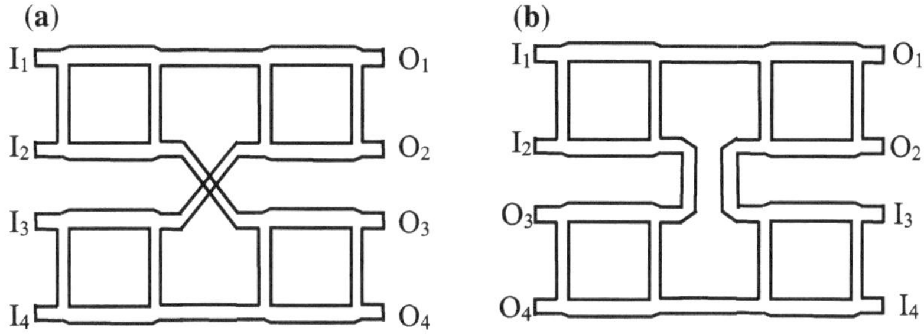

Fig. 19.34 4 × 4 Butler matrix using 3-dB 90° hybrid couplers **a** with crossover, **b** planner structure without crossover

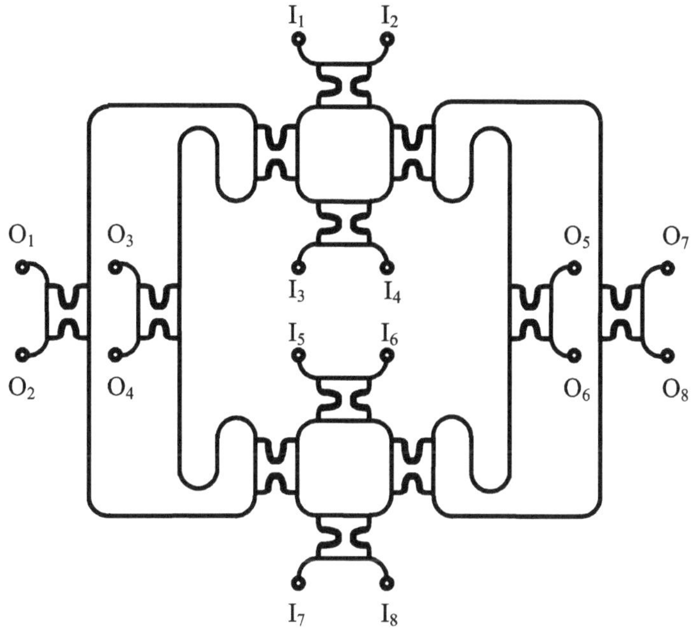

Fig. 19.35 8 × 8 Butler matrix using 3-dB 90° hybrid couplers

compensated by providing more gain at the driver amplifier stages. It is always preferable to realize input network in planar transmission line configuration very compactly with the compromise of insertion loss. Figure 19.34 shows planner configuration of a 4 × 4 Butler matrix with and without crossover connection and Fig. 19.35 shows planner configuration of an 8 × 8 Butler matrix suitable for realization in microstrip line configuration.

Any loss of the output butler network decreases total available output power. Generally, waveguide-based output network is designed to achieve low insertion

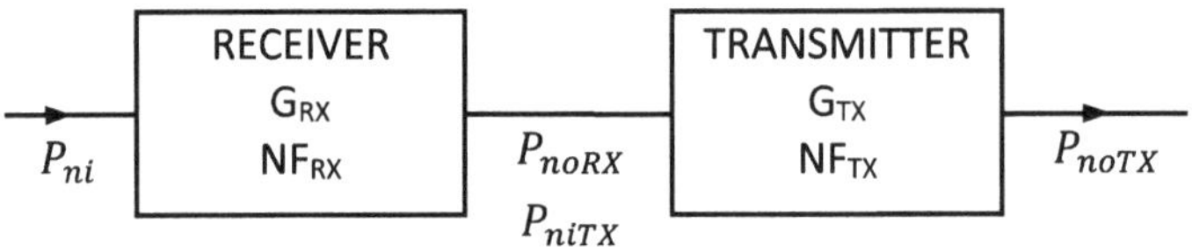

Fig. 19.36 Block diagram of the system of Example 19.5

loss, though it is bulky. Followings are the various examples related to microwave communication systems and subsystems.

Example 19.5 Derive the overall noise figure of a communication system which consists of a receiver and transmitter as shown in Fig. 19.36. Calculate the overall system noise figure for gain and noise figure of the receiver: 50 and 2 dB, respectively, and gain and noise figure of the transmitter: 80 and 20 dB, respectively.

Solution

Gain of the receiver $= G_{RX} = 50$ dB $= 10^5$
Noise figure of the receiver $= \mathrm{NF}_{RX} = 2$ dB $= 1.585$
Gain of the transmitter $= G_{TX} = 80$ dB $= 10^8$
Noise figure of the transmitter $= \mathrm{NF}_{TX} = 20$ dB $= 100$

Suppose the available noise power at the input of the receiver is P_{ni}; this is the thermal noise over the noise bandwidth B of the receiver and is given by

$$P_{ni} = kT_oB$$

Thus, the noise power output, P_{noRX}, of the receiver can be written as

$$\begin{aligned} P_{noRX} &= P_{niTX} = (P_{ni}) \times (G_{RX}\mathrm{NF}_{RX}) \\ &= P_{ni} \times G_{RX} + (\mathrm{NF}_{RX} - 1)P_{ni} \times G_{RX} \end{aligned}$$

The first term is the output noise due to the amplification of the input available thermal noise, and the second part is the noise added by the receiver (Fig. 19.36).

In the similar way, the total noise power at the output of the transmitter can be written as

$$P_{noTX} = P_{niTX} \times G_{TX} + (\mathrm{NF}_{TX} - 1)P_{ni} \times G_{TX}$$

$$P_{noTX} = P_{ni} \times G_{RX} \times \mathrm{NF}_{RX} \times G_{TX} + (\mathrm{NF}_{TX} - 1)P_{ni} \times G_{TX} \qquad (19.10a)$$

Suppose NF_{RXTX} is the overall noise figure of the system. The overall gain of the system is $G_{RX} \times G_{TX}$. Thus, the noise power at the output of the transmitter can be written as

$$P_{noTX} = (P_{ni}) \times (G_{RX} \times G_{TX} \times \mathrm{NF}_{RXTX}) \qquad (19.10b)$$

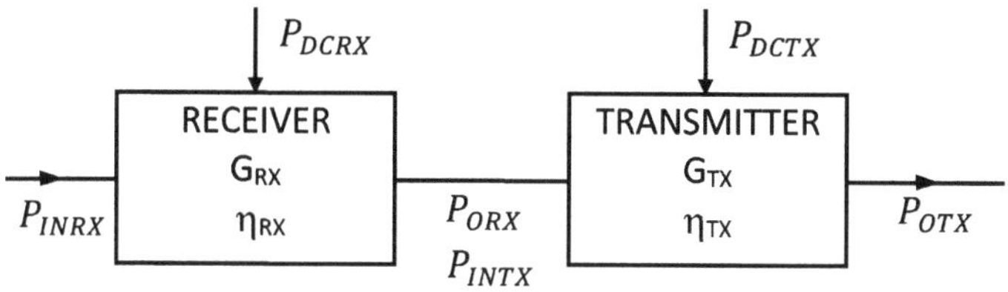

Fig. 19.37 Block diagram of the system of Example 19.6

Comparing (19.10a) and (19.10b), the overall system noise figure can be written as

$$\mathrm{NF}_{RXTX} = \mathrm{NF}_{RX} + \frac{(\mathrm{NF}_{TX} - 1)}{G_{RX}} \tag{19.11}$$

This is known the Friis's formula, already derived in Chap. 14. This formula shows that in case of sufficiently high receiver gain, the overall noise factor is dominated by the noise factor of the receiver.

Putting the values, the overall noise factor of the system

$$\begin{aligned}\mathrm{NF}_{RXTX} &= 1.585 + \frac{(100 - 1)}{100{,}000} = 1.586 \\ &= 2.003\ \mathrm{dB}\end{aligned}$$

This shows that though the noise figure of the transmitter is poor, its effect on the overall system is negligible due to the high gain of the receiver.

Example 19.6 Derive overall power-added efficiency of a communication system which consists of a receiver and transmitter as shown in Fig. 19.37. Calculate the overall power-added efficiency of the system for gain- and power-added efficiency of the receiver: 50 dB and 1%, respectively, and gain- and power-added efficiency of the transmitter: 80 dB and 50%, respectively.

Solution

Gain of the receiver = $G_{RX} = 50\ \mathrm{dB} = 10^5$
Power-added efficiency of the receiver = $\eta_{RX} = 1\% = 0.01$
Gain of the transmitter = $T_{RX} = 80\ \mathrm{dB} = 10^8$
Power-added efficiency of the transmitter = $\eta_{TX} = 50\% = 0.5$

Suppose

DC power to the receiver = P_{DCRX}
DC power to the transmitter = P_{DCTX}
Input RF power of the receiver = P_{INRX}
Input RF power of the transmitter = $P_{INTX} = P_{ORX}$
Output RF power of the transmitter = P_{OTX}

From the definition of power-added efficiency,

$$\eta_{RX} = \frac{P_{ORX} - P_{INRX}}{P_{DCRX}} \quad (19.12a)$$

$$\eta_{TX} = \frac{P_{OTX} - P_{INTX}}{P_{DCTX}} \quad (19.12b)$$

The overall power-added efficiency of the system η_{RXTX} can be written as

$$\eta_{RXTX} = \frac{P_{OTX} - P_{INRX}}{P_{DCRX} + P_{DCTX}} \quad (19.12c)$$

Putting $G_{RX} = P_{ORX}/P_{INRX}$, $G_{TX} = P_{OTX}/P_{INTX}$ and using (19.12a), (19.12b), the overall efficiency of the system can be written as

$$\frac{1}{\eta_{RXTX}} = \frac{G_{RX} - 1}{G_{RX}G_{TX} - 1} \times \frac{1}{\eta_{RX}} + \frac{(G_{TX} - 1)G_{RX}}{G_{RX}G_{TX} - 1} \times \frac{1}{\eta_{TX}} \quad (19.13)$$

Putting $G_{RX} = 10^5$, $\eta_{RX} = 0.01$, $T_{RX} = 10^8$ and $\eta_{TX} = 0.5$

$$\frac{1}{\eta_{RXTX}} = \frac{10^5 - 1}{10^{13} - 1} \times \frac{1}{0.01} + \frac{\left(10^8 - 1\right)10^5}{10^{13} - 1} \times \frac{1}{0.5}$$

or

$$\eta_{RXTX} = 0.5 = 50\%$$

This shows that though the power-added efficiency of the receiver is poor, its effect on the overall system is negligible due to the high gain of the transmitter and receiver.

Considering $G_{RX} \gg 1$ and $G_{TX} \gg 1$, the overall power-added efficiency of the system (19.13) can be written as

$$\frac{1}{\eta_{RXTX}} = \frac{G_{RX}}{G_{RX}G_{TX}} \times \frac{1}{\eta_{RX}} + \frac{G_{TX}G_{RX}}{G_{RX}G_{TX}} \times \frac{1}{\eta_{TX}} \quad (19.14a)$$

$$\frac{1}{\eta_{RXTX}} = \frac{1}{G_{TX}} \times \frac{1}{\eta_{RX}} + \frac{1}{\eta_{TX}} \quad (19.14b)$$

or

$$\eta_{RXTX} = \eta_{TX} \quad (19.14c)$$

Example 19.7 Derive worst-case overall two-tone third-order intermodulation level of a communication system which consists of a receiver and transmitter as shown in Fig. 19.38. Calculate the overall two-tone third-order intermodulation level of the system for IM_3 of the receiver and transmitter, which are (a) 20 and 10 dBc (b) 10 and 10 dBc, respectively.

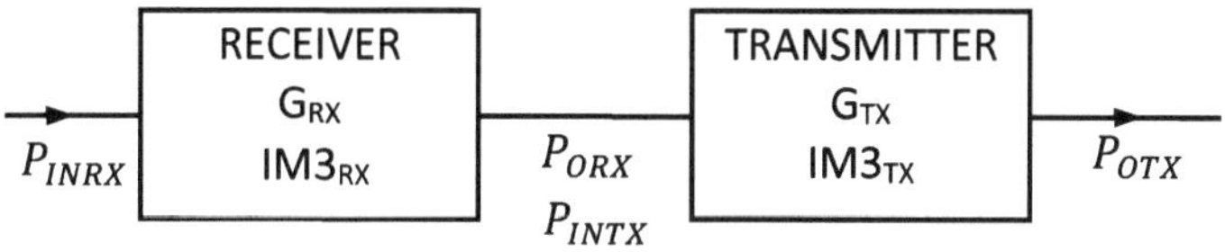

Fig. 19.38 Block diagram of the system of Example 19.7

Solution
Suppose

Output RF power of the receiver $= P_{ORX} = P_{INTX}$
Output RF power of the transmitter $= P_{OTX}$

Therefore, level of IM_{3RX} at the output of the receiver is

$$P_{ORX} \times 10^{-(IM_{3RX}/10)} \tag{19.15a}$$

Here, consider that the third-order intermodulation signal generated by the receiver will be linearly amplified by the transmitter. Thus, at the output of the transmitter, the third-order intermodulation signal level will be

$$10^{-(IM_{3RX}/10)} \times P_{ORX} \times G_{TX} + P_{OTX} \times 10^{-(IM_{3TX}/10)} \tag{19.15b}$$

$$= P_{OTX} \times \left(10^{-(IM_{3RX}/10)} + 10^{-(IM_{3TX}/10)}\right) \tag{19.15c}$$

Here, the first term is the contribution by the receiver and the second term is the contribution by the transmitter.

Thus, the overall IM_3 level

$$IM_{3RXTX} = \frac{P_{OTX} \times \left(10^{-(IM_{3RX}/10)} + 10^{-(IM_{3TX}/10)}\right)}{P_{OTX}}$$

$$= 10^{-(IM_{3RX}/10)} + 10^{-(IM_{3TX}/10)}$$

$$[IM_{3RXTX}]_{\text{dB}} = 10\log\left(10^{-(IM_{3RX}/10)} + 10^{-(IM_{3TX}/10)}\right) \tag{19.15d}$$

(a) Putting the value of $IM_{3RX} = 20$ dBc and $IM_{3TX} = 10$ dBc

$$[IM_{3RXTX}]_{\text{dB}} = 10\log\left(10^{-(20/10)} + 10^{-(10/10)}\right)$$
$$= 10\log(0.01 + 0.1)$$
$$= 10\log(0.01 + 0.1) = 9.59 \text{ dB}$$

(b) Putting the value of $IM_{3RX} = 10$ dBc and $IM_{3TX} = 10$ dBc

$$[IM_{3RXTX}]_{\text{dB}} = 10\log\left(10^{-(10/10)} + 10^{-(10/10)}\right)$$

$$= 10\log(0.1 + 0.1)$$
$$= -6.99 \text{ dB}$$

This example shows that effect of nonlinearity of receivers and transmitters equally affects the overall linearity of the system.

Example 19.8 Derive expressions for output powers of a transmitter of I-O characteristic governed by $v_o = a_1 v_i + a_3 v_i^3$ for a two-tone carrier input, $v_i = A\cos\omega_1 t + B\cos\omega_2 t$. Considering the transmitter is matched at its input and output ports with $R_o = 50\ \Omega$, and $a_1 = 10, a_3 = -0.04$, calculate the output power levels corresponding to the fundamental and third-order harmonics for the following cases:

(a) For total input power level of 0 dBW of equal power levels of two carriers.
(b) For total input power level of 0 dBW of unequal power levels of two carriers by 6 dB.
(c) For total input power level of 0 dBW of unequal power levels of two carriers by 10 dB.

Also, plot gains of both the carriers over the total input power level of −20 to 0 dBW for the case-b and case-c.

Solution
I–O characteristic of the transmitter is given by

$$v_o = a_1 v_i + a_3 v_i^3 \tag{19.16}$$

The two carriers' input excitation is

$$v_i = A\cos\omega_1 t + B\cos\omega_2 t \tag{19.17}$$

Therefore, the output is given by

$$v_o = a_1(A\cos\omega_1 t + B\cos\omega_2 t) + a_3(A\cos\omega_1 t + B\cos\omega_2 t)^3 \tag{19.18}$$

$$\begin{aligned} v_o = {} & \frac{3}{2}a_3A^2B + \frac{3}{2}a_3AB^2 \\ & + \left(a_1A + \frac{3}{4}a_3A^3 + \frac{3}{2}a_3AB^2\right)\cos\omega_1 t \\ & + \left(a_1B + \frac{3}{4}a_3B^3 + \frac{3}{2}a_3A^2B\right)\cos\omega_2 t \\ & + \left(\frac{1}{4}a_3A^3\right)\cos 3\omega_1 t + \left(\frac{1}{4}a_3B^3\right)\cos 3\omega_2 t \\ & + \left(\frac{3}{4}a_3A^2B\right)[\cos(2\omega_1 - \omega_2)t + \cos(2\omega_1 + \omega_2)t] \\ & + \left(\frac{3}{4}a_3AB^2\right)[\cos(2\omega_2 - \omega_1)t + \cos(2\omega_2 + \omega_1)t] \end{aligned} \tag{19.19}$$

Amplitude of the output voltage corresponding (v_{o1}) to carrier ω_1 is given by

$$v_{o\omega_1} = a_1 A + \frac{3}{4} a_3 A^3 + \frac{3}{2} a_3 A B^2$$

Thus, the output power corresponding to carrier ω_1 is given by

$$P_{o\omega_1} = \frac{1}{R_o}\left(\frac{v_{o\omega_1}}{\sqrt{2}}\right)^2 = \left(a_1 A + \frac{3}{4} a_3 A^3 + \frac{3}{2} a_3 A B^2\right)^2 / 2R_o \tag{19.20a}$$

Similarly, the output power corresponding to carrier ω_2 is given by

$$P_{o\omega_2} = \frac{1}{R_o}\left(\frac{v_{o\omega_2}}{\sqrt{2}}\right)^2 = \left(a_1 B + \frac{3}{4} a_3 B^3 + \frac{3}{2} a_3 A^2 B\right)^2 / 2R_o \tag{19.20b}$$

The output power corresponding to each third-order intermodulation product ($2\omega_1 \pm \omega_2$) is given by

$$P_{o(2\omega_1 \pm \omega_2)} = \left(\frac{3}{4} a_3 A^2 B\right)^2 / 2R_o \tag{19.21a}$$

Similarly, the output power corresponding to each third-order intermodulation product ($2\omega_1 \pm \omega_2$) is given by

$$P_{o(2\omega_2 \pm \omega_1)} = \left(\frac{3}{4} a_3 A B^2\right)^2 / 2R_o \tag{19.21b}$$

The input power corresponding to carrier ω_1 is given by

$$P_{i\omega_1} = \left(A/\sqrt{2}\right)^2 / R_o = A^2 / 2R_o \tag{19.22a}$$

The input power corresponding to carrier ω_2 is given by

$$P_{i\omega_2} = \left(B/\sqrt{2}\right)^2 / R_o = B^2 / 2R_o \tag{19.22b}$$

Therefore, total input power is given by

$$P_{i\mathrm{Total}} = \left(A^2 + B^2\right) / 2R_o \tag{19.23}$$

(a) For equal power levels of both the carriers with $P_{i\mathrm{Total}} = 0$ dBW $= 1$ W. Thus, from (19.23),

$$A = B = 7.071\ \mathrm{V}$$

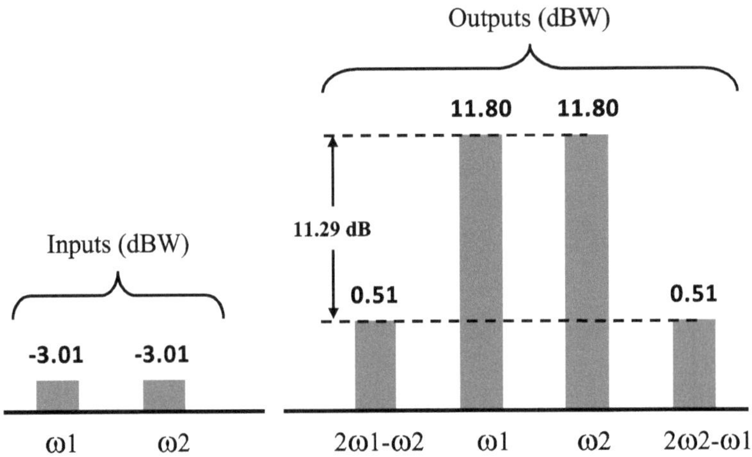

Fig. 19.39 Power levels under two-tone excitation of equal input power levels of Example 19.8

Here, $A = B = 7.071$ V and $a_1 = 10$, $a_3 = -0.04$ and $R_o = 50\ \Omega$.
From (19.22a) and (19.22b),

$$P_{i\omega_1} = P_{i\omega_2} = A^2/2R_o = 7.071^2/100 = 0.5\ \text{W} = -3.01\ \text{dBW}$$

From (19.20a) and (19.20b),

$$\begin{aligned} P_{o\omega_1} = P_{o\omega_2} &= \left(a_1A + \frac{3}{4}a_3A^3 + \frac{3}{2}a_3AB^2\right)^2 /2R_o \\ &= \left(10 \times 7.071 - \frac{3}{4} \times 0.04 \times 7.071^3 - \frac{3}{2} \times 0.04 \times 7.071 \times 7.071^2\right)^2 /100 \\ &= 15.13\ \text{W} = 11.80\ \text{dBW} \end{aligned}$$

From (19.21a), (19.21b),

$$\begin{aligned} P_{o(2\omega_1 \pm \omega_2)} = P_{o(2\omega_2 \pm \omega_1)} &= \left(\frac{3}{4}a_3A^2B\right)^2 /2R_o \\ &= \left(\frac{3}{4} \times 0.04 \times 7.071^3\right)^2 /100 \\ &= 1.12\ \text{W} = 051\ \text{dBW} \end{aligned}$$

The input and output power levels of the two equal power carriers are shown in Fig. 19.39. Under the two-tone carrier excitation of equal power levels, the output power levels for both the fundamental frequency components are same and also the power levels of both the third-order IMD components are same. The levels of the third-order IMD components are

$$
\begin{aligned}
P_{o(2\omega_1 \pm \omega_2)} - P_{o\omega_1} &= P_{o(2\omega_2 \pm \omega_1)} - P_{o\omega_2} \\
&= (0.51)\,\text{dBW} - 11.80\ \text{dBW} = -11.29\ \text{dBc}
\end{aligned}
$$

(b) For unequal power levels by 6 dB with $P_{i\text{Total}} = 0\ \text{dBW} = 1\ \text{W}$.

From (19.22a) and (19.22b),

$$
P_{i\omega_1} - P_{i\omega_2} = 10\log\left(A^2/2R_o\right) - 10\log\left(B^2/2R_o\right) = 6\ \text{dB} \tag{19.24a}
$$

And from (19.23),

$$
P_{i\text{Total}} = 10\log\left(A^2 + B^2\right)/2R_o = 0\ \text{dBW} \tag{19.24b}
$$

From (19.24a) and (19.24b),

$$
A = 8.94\ \text{V}, \quad B = 4.48\ \text{V}
$$

The input powers corresponding to carriers ω_1 and ω_2 are given by,

$$
\begin{aligned}
P_{i\omega_1} &= A^2/2R_o = 8.94^2/100 = 0.799\ \text{W} = -0.973\ \text{dBW} \\
P_{i\omega_2} &= B^2/2R_o = 4.48^2/100 = 0.201\ \text{W} = -6.973\ \text{dBW}
\end{aligned}
$$

From (19.20a),

$$
\begin{aligned}
P_{o\omega_1} &= \left(a_1 A + \frac{3}{4}a_3 A^3 + \frac{3}{2}a_3 A B^2\right)^2 /2R_o \\
&= \left(10 \times 8.94 - \frac{3}{4} \times 0.048.94^3 - \frac{3}{2} \times 0.04 \times 8.94 \times 4.48^2\right)^2 /100 \\
&= 32.71\ \text{W} = 15.15\ \text{dBW}
\end{aligned}
$$

From (19.20b),

$$
\begin{aligned}
P_{o\omega_2} &= \left(a_1 B + \frac{3}{4}a_3 B^3 + \frac{3}{2}a_3 A^2 B\right)^2 /2R_o \\
&= \left(10 \times 4.48 - \frac{3}{4} \times 0.04 \times 4.48^3 - \frac{3}{2} \times 0.04 \times 8.94^2 \times 4.48\right)^2 /100 \\
&= 4.25\ \text{W} = 6.29\ \text{dBW}
\end{aligned}
$$

From (19.21a),

$$
P_{o(2\omega_1 \pm \omega_2)} = \left(\frac{3}{4}a_3 A^2 B\right)^2 /2R_o
$$

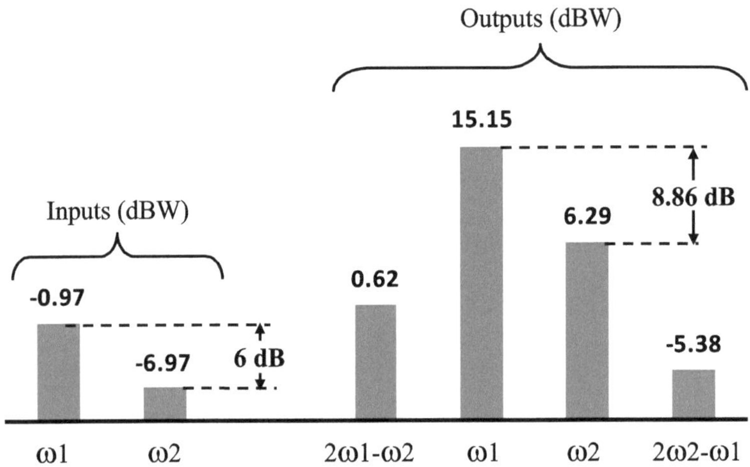

Fig. 19.40 Power levels under two-tone excitation of unequal input levels by 6 dB of Example 19.8

$$= \left(\frac{3}{4} \times 0.04 \times 8.94^2 \times 4.48\right)^2 /100$$
$$= 1.15\ \text{W} = 0.62\ \text{dBW}$$

From (19.21b),

$$P_{o(2\omega_2 \pm \omega_1)} = \left(\frac{3}{4} a_3 A B^2\right)^2 /2R_o$$
$$= \left(\frac{3}{4} \times 0.04 \times 8.94 \times 4.48^2\right)^2 /100$$
$$= 0.29\ \text{W} = -5.38\ \text{dBW}$$

The input and output power levels of the two unequal power level by 6 dB are shown in Fig. 19.40. The difference of output power levels among the fundamental carriers increases to 8.86 dB from the difference of 6 dB at input. This is due to the power transfer to the various harmonics and intermodulation components. This phenomenon is known as ‘power robbing’ when multiple carriers amplified by an amplifier operating in its nonlinear (gain compression) region. Though both the carriers pass through the same amplifier, they experience different levels of amplification.

Gain responses of both the carriers are shown in Fig. 19.41 over the input power level of −20 to 0 dBW. At lower power levels where the amplifier operates in linear region, the gains of both the carriers are same. But over the nonlinear region of the

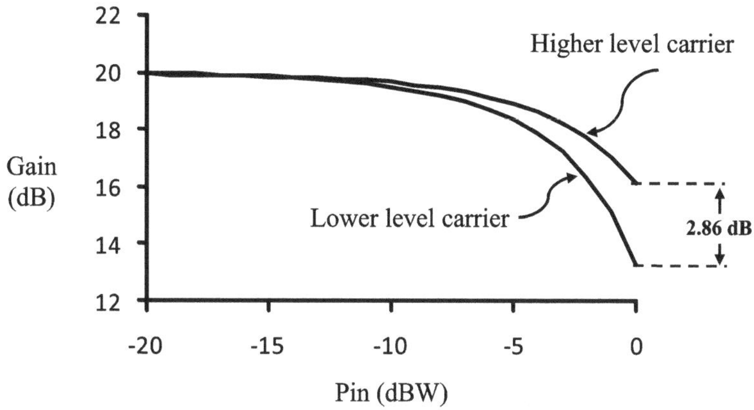

Fig. 19.41 Gain of the carriers with unequal input power levels by 6 dB of Example 19.8

amplifier, the weaker signal amplifies less compared to the stronger signal as shown in Fig. 19.41.

(c) For unequal power levels by 10 dB with $P_{i\text{Total}} = 0$ dBW = 1 W.

From (19.22a) and (19.22b),

$$P_{i\omega_1} - P_{i\omega_2} = 10\log\left(A^2/2R_o\right) - 10\log\left(B^2/2R_o\right) = 10\text{ dB} \tag{19.24a}$$

And from (19.23),

$$P_{i\text{Total}} = 10\log\left(A^2 + B^2\right)/2R_o = 0\text{ dBW} \tag{19.24b}$$

From (19.24a) and (19.24b),

$$A = 9.535\text{ V},\quad B = 3.015\text{ V}$$

The input powers corresponding to carriers ω_1 and ω_2 are given by

$$P_{i\omega_1} = A^2/2R_o = 9.535^2/100 = 0.909\text{ W} = -0.414\text{ dBW}$$
$$P_{i\omega_2} = B^2/2R_o = 3.015^2/100 = 0.091\text{ W} = -10.414\text{ dBW}$$

From (19.20a),

$$\begin{aligned} P_{o\omega_1} &= \left(a_1A + \frac{3}{4}a_3A^3 + \frac{3}{2}a_3AB^2\right)^2/2R_o \\ &= \left(10 \times 9.535 - \frac{3}{4} \times 0.04 \times 9.535^3 - \frac{3}{2} \times 0.04 \times 9.535 \times 3.015^2\right)^2/100 \\ &= 41.142\text{ W} = 16.14\text{ dBW} \end{aligned}$$

From (19.20b),

$$P_{o\omega_2} = \left(a_1 B + \frac{3}{4}a_3 B^3 + \frac{3}{2}a_3 A^2 B\right)^2 /2R_o$$
$$= \left(10 \times 3.015 - \frac{3}{4} \times 0.04 \times 3.015^3 - \frac{3}{2} \times 0.04 \times 9.535^2 \times 3.015\right)^2 /100$$
$$= 1.660\ \text{W} = 2.20\ \text{dBW}$$

From (19.21a),

$$P_{o(2\omega_1 \pm \omega_2)} = \left(\frac{3}{4}a_3 A^2 B\right)^2 /2R_o$$
$$= \left(\frac{3}{4} \times 0.04 \times 9.535^2 \times 3.015\right)^2 /100$$
$$= 0.68\ \text{W} = -1.70\ \text{dBW}$$

From (19.21b),

$$P_{o(2\omega_2 \pm \omega_1)} = \left(\frac{3}{4}a_3 A B^2\right)^2 /2R_o$$
$$= \left(\frac{3}{4} \times 0.04 \times 9.535 \times 3.015^2\right)^2 /100$$
$$= 0.07\ \text{W} = -11.70\ \text{dBW}$$

The input and output power levels of the two carriers with unequal power levels by 10 dB are shown in Fig. 19.42. Here, it can be noted that due to the power robbing phenomenon, the difference of output power levels of the fundamental carrier increases to 13.94 dB compared to the difference of 10 dB at input. Gain responses of both the carriers are shown in Fig. 19.43 over the input power level of −20 to 0 dBW. It can be noted that the gain of the weaker carrier becomes 3.94 dB lower compared to the gain of the stronger carrier at 0 dBW total input power level. Thus, weak signal becomes further weak when it passes through a nonlinear (gain compression region) amplifier in the presence of stronger signal.

Example 19.9 Write the expressions for output powers of a transmitter of IO characteristic governed by $v_o = a_1 v_i + a_3 v_i^3$ for a two-tone carrier input, $v_i = A\cos\omega_1 t + B\cos\omega_2 t$. Considering the transmitter is matched at its input and output ports with $R_o = 50\ \Omega$, and $a_1 = 4, a_3 = 0.02$, calculate the output power levels corresponding to fundamental components and third-order harmonics for the following cases:

(a) For total input power level of 0 dBW of equal power levels of two carriers.
(b) For total input power level of 0 dBW of unequal power levels of two carriers by 10 dB.

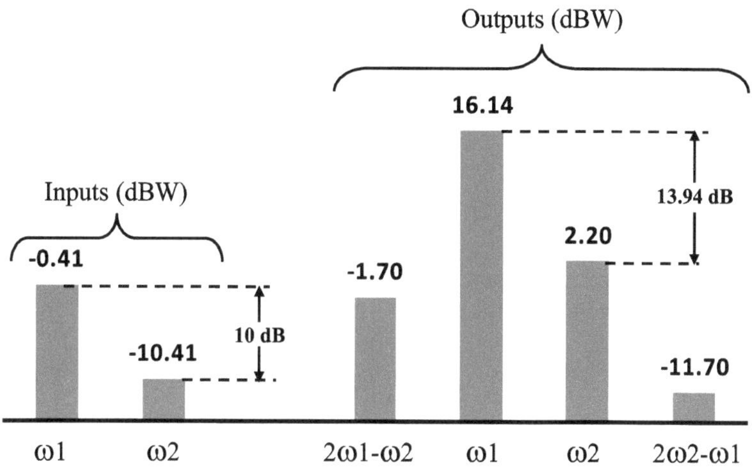

Fig. 19.42 Power levels under two-tone excitation of unequal input levels by 10 dB of Example 19.8

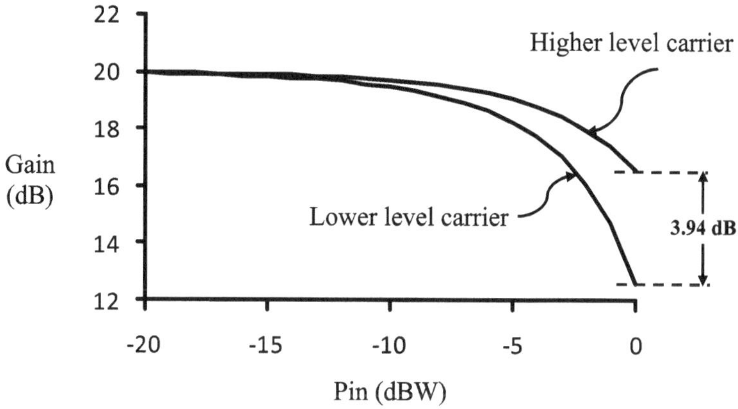

Fig. 19.43 Gain of the carriers with unequal input levels by 10 dB of Example 19.8

Also, plot gains of both the carriers over the total input power level of −20 to 0 dBW for the case-b.

Solution

I–O characteristic of the transmitter is given by

$$v_o = a_1 v_i + a_3 v_i^3$$

Under the two-carrier input excitation of $v_i = A\cos\omega_1 t + B\cos\omega_2 t$, the output power corresponding to carrier ω_1 is given by (19.20a)

$$P_{o\omega_1} = \frac{1}{R_o}\left(\frac{v_{o\omega_1}}{\sqrt{2}}\right)^2 = \left(a_1A + \frac{3}{4}a_3A^3 + \frac{3}{2}a_3AB^2\right)^2/2R_o$$

Similarly, the output power corresponding to carrier ω_2 is given by (19.20b)

$$P_{o\omega_2} = \frac{1}{R_o}\left(\frac{v_{o\omega_2}}{\sqrt{2}}\right)^2 = \left(a_1B + \frac{3}{4}a_3B^3 + \frac{3}{2}a_3A^2B\right)^2/2R_o$$

The output power corresponding to each third-order intermodulation product $(2\omega_1 \pm \omega_2)$ is given by (19.21a)

$$P_{o(2\omega_1 \pm \omega_2)} = \left(\frac{3}{4}a_3A^2B\right)^2/2R_o$$

The output power corresponding to each third-order intermodulation product $(2\omega_1 \pm \omega_2)$ is given by (19.21b)

$$P_{o(2\omega_2 \pm \omega_1)} = \left(\frac{3}{4}a_3AB^2\right)^2/2R_o$$

The input power corresponding to carrier ω_1 is given by (19.22a)

$$P_{i\omega_1} = \left(A/\sqrt{2}\right)^2/R_o = A^2/2R_o$$

The input power corresponding to carrier ω_2 is given by (19.22b)

$$P_{i\omega_2} = \left(B/\sqrt{2}\right)^2/R_o = B^2/2R_o$$

Therefore, total input power is given by (19.23)

$$P_{i\text{Total}} = \left(A^2 + B^2\right)/2R_o$$

(a) For equal power levels of both the carriers with $P_{i\text{Total}} = 0$ dBW $= 1$ W. From (19.23),

$$A = B = 7.071 \text{ V}$$

Here, $A = B = 7.071$ V and $a_1 = 4$, $a_3 = +0.02$ and $R_o = 50\ \Omega$.
From (19.22a) and (19.22b),

$$P_{i\omega_1} = P_{i\omega_2} = A^2/2R_o = 7.071^2/100 = 0.5 \text{ W} = -3.01 \text{ dBW}$$

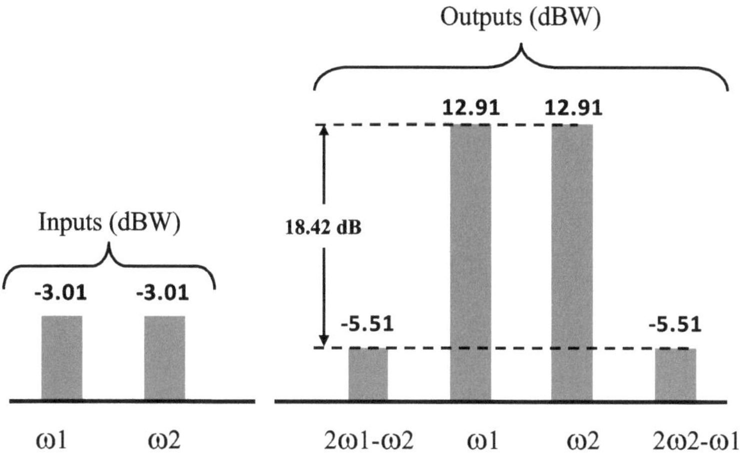

Fig. 19.44 Power levels under two-tone excitation of equal input levels of Example 19.9

From (19.20a) and (19.20b),

$$P_{o\omega_1} = P_{o\omega_2} = \left(a_1A + \frac{3}{4}a_3A^3 + \frac{3}{2}a_3AB^2\right)^2 / 2R_o$$

$$= \left(4 \times 7.071 + \frac{3}{4} \times 0.02 \times 7.071^3 + \frac{3}{2} \times 0.02 \times 7.071 \times 7.071^2\right)^2 / 100$$

$$= 19.53\ \text{W} = 12.91\ \text{dBW}$$

From (19.21a) and (19.21b),

$$P_{o(2\omega_1 \pm \omega_2)} = P_{o(2\omega_2 \pm \omega_1)} = \left(\frac{3}{4}a_3A^2B\right)^2 / 2R_o$$

$$= \left(\frac{3}{4} \times 0.02 \times 7.071^3\right)^2 / 100$$

$$= 0.28\ \text{W} = -5.51\ \text{dBW}$$

The input and output power levels of the two equal power carriers are shown in Fig. 19.44. Under the two-tone carrier excitation of equal power levels, the output power levels for both the fundamental frequency components are same and also the power levels of both the third-order IMD components are same. The levels of the third-order IMD components are

$$P_{o(2\omega_1 \pm \omega_2)} - P_{o\omega_1} = P_{o(2\omega_2 \pm \omega_1)} - P_{o\omega_2}$$

$$= -5.51\ \text{dBW} - 12.91\ \text{dBW} = -18.42\ \text{dBc}$$

(b) For unequal power levels by 10 dB with $P_{i\text{Total}} = 0$ dBW = 1 W.

Thus, from (19.22a) and (19.22b),

$$P_{i\omega_1} - P_{i\omega_2} = 10\log\left(A^2/2R_o\right) - 10\log\left(B^2/2R_o\right) = 10 \text{ dB}$$

And from (19.23),

$$P_{i\text{Total}} = 10\log\left(A^2 + B^2\right)/2R_o = 0 \text{ dBW}$$

From (19.24a) and (19.24b),

$$A = 9.535 \text{ V}, \quad B = 3.015 \text{ V}$$

The input powers corresponding to carriers ω_1 and ω_2 are given by

$$P_{i\omega_1} = A^2/2R_o = 9.535^2/100 = 0.909 \text{ W} = -0.414 \text{ dBW}$$
$$P_{i\omega_2} = B^2/2R_o = 3.015^2/100 = 0.091 \text{ W} = -10.414 \text{ dBW}$$

From (19.20a),

$$\begin{aligned} P_{o\omega_1} &= \left(a_1A + \frac{3}{4}a_3A^3 + \frac{3}{2}a_3AB^2\right)^2/2R_o \\ &= \left(4 \times 9.535 + \frac{3}{4} \times 0.02 \times 9.535^3 + \frac{3}{2} \times 0.02 \times 9.535 \times 3.015^2\right)^2/100 \\ &= 28.88 \text{ W} = 14.61 \text{ dBW} \end{aligned}$$

From (19.20b),

$$\begin{aligned} P_{o\omega_2} &= \left(a_1B + \frac{3}{4}a_3B^3 + \frac{3}{2}a_3A^2B\right)^2/2R_o \\ &= \left(4 \times 3.015 + \frac{3}{4} \times 0.02 \times 3.015^3 + \frac{3}{2} \times 0.02 \times 9.535^2 \times 3.015\right)^2/100 \\ &= 4.28 \text{ W} = 6.32 \text{ dBW} \end{aligned}$$

From (19.21a),

$$\begin{aligned} P_{o(2\omega_1 \pm \omega_2)} &= \left(\frac{3}{4}a_3A^2B\right)^2/2R_o \\ &= \left(\frac{3}{4} \times 2.02 \times 9.535^2 \times 3.015\right)^2/100 \\ &= 0.17 \text{ W} = -7.72 \text{ dBW} \end{aligned}$$

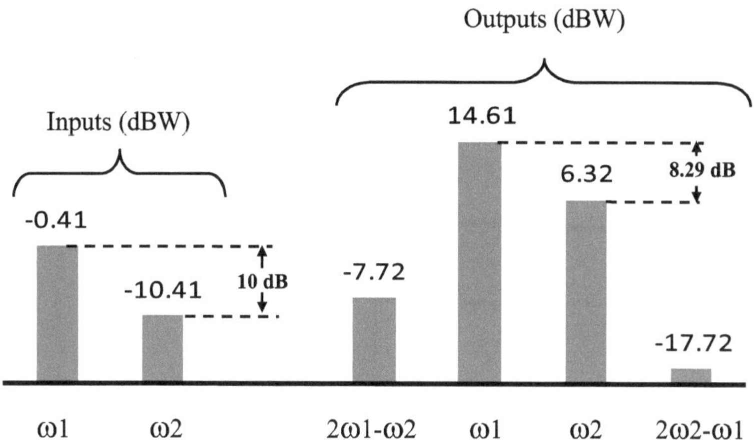

Fig. 19.45 Power levels under two-tone excitation of unequal input levels by 10 dB of Example 19.9

From (19.21b),

$$\begin{aligned} P_{o(2\omega_2 \pm \omega_1)} &= \left(\frac{3}{4} a_3 A B^2\right)^2 / 2R_o \\ &= \left(\frac{3}{4} \times 0.02 \times 9.535 \times 3.015^2\right)^2 / 100 \\ &= 0.02 \text{ W} = -17.72 \text{ dBW} \end{aligned}$$

The input and output power levels of the two unequal power levels by 10 dB are shown in Fig. 19.45. Here, the difference of output power levels of the fundamental carrier deceases to 8.29 dB from the difference of 10 dB at input. Gain responses of both the carriers are shown in Fig. 19.46 over the input power level of −20 to 0 dBW. It can be noted that the gain of the weaker carrier becomes 1.71 dB higher compared to the gain of the stronger carrier at 0 dBW total input power level.

Example 19.10 A transceiver consists of a receiver, ALC driver amplifier and high power amplifier as shown in Fig. 19.47. The ALC driver amplifier controls the gain of the transponder from 110 to 140 dB depending on its input power level to operate the transponder at saturation condition over its 30 dB dynamic range.

Calculate the noise and carrier powers at the transponder output when the transponder operates in saturated condition for (a) minimum and (b) maximum gain considering the following parameters:

System noise temperature (T_s)	650 K
Transponder noise bandwidth (B_N)	100 MHz
Saturated gain of the HPA	50 dB
Saturated output power of HPA	140 W

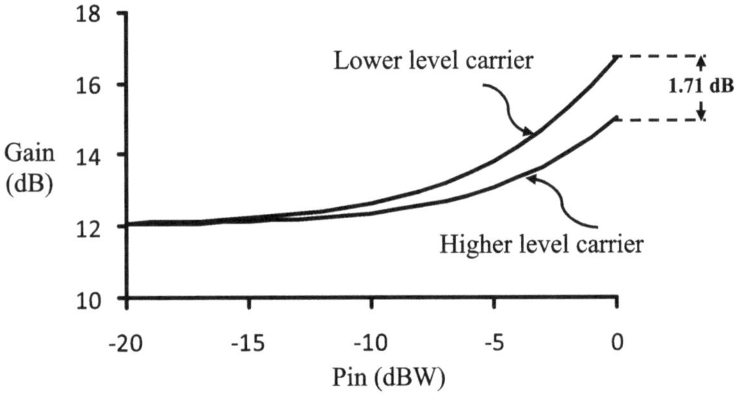

Fig. 19.46 Gain of the carriers with unequal input levels by 10 dB of Example 19.9

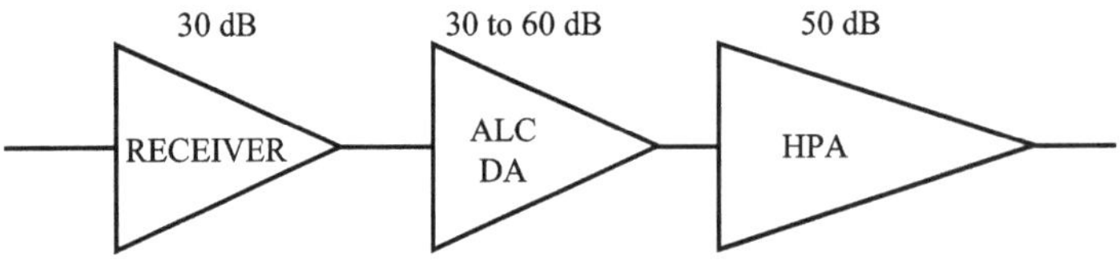

Fig. 19.47 Gain of the carriers with unequal input levels by 6 dB

Plot carrier and noise power at the output of the transponder over its gain of 110–140 dB considering the transponder are operating at saturation condition.

Solution

The system noise temperature (T_s): 650 K

Input noise power ($P_{N\text{in}}$) over the frequency band of 100 MHz:

$$\begin{aligned} P_{N\text{in}} &= kT_sB_N \\ &= 1.38 \times 10^{-23} \times 650 \times 100 \times 10^6\ \text{W} \\ &= -90.47\ \text{dBm} \end{aligned} \tag{19.25}$$

(a) At minimum gain of the transponder:
Total input noise power $P_{N\text{in}} = -90.47$ dBm
Gain of the transponder: 110 dB
Thus,

$$\begin{aligned} \text{Total output noise power}\, P_{N\text{out}} &= (-90.47 + 110)\,\text{dBm} \\ &= 19.53\ \text{dBm} \\ &= 0.09\ \text{W} \end{aligned}$$

Saturated output power of the transponder: = 140 W

Considering saturated output power of the HPA consists of amplified noise and carrier power only (neglecting intermodulation powers), the carrier power can be written as

$$
\begin{aligned}
&= \text{output saturated power} - \text{output noise power} \\
&= 140\ \text{W} - 0.09\ \text{W} \\
&= 139.91\ \text{W} \\
&= 51.46\ \text{dBm}
\end{aligned}
$$

(b) At maximum gain of the transponder:
Total input noise power $P_{N\text{in}} = -90.47$ dBm
Gain of the transponder: 140 dB
Thus,

$$
\begin{aligned}
\text{Total output noise power } P_{N\text{out}} &= (-90.47 + 140)\ \text{dBm} \\
&= 49.53\ \text{dBm} \\
&= 89.70\ \text{W}
\end{aligned}
$$

Saturated output power of the transponder: = 140 W

Considering saturated output power of the HPA consists of amplified noise and carrier power only (neglecting contribution of intermodulation powers) and assuming saturated power of HPA is same for single and multicarrier conditions, the carrier power can be written as

$$
\begin{aligned}
&= \text{output saturated power} - \text{output noise power} \\
&= 140\ \text{W} - 89.70\ \text{W} \\
&= 50.30\ \text{W} \\
&= 47.02\ \text{dBm}
\end{aligned}
$$

Plot of the output carrier and noise power of the transponder over its entire gain range of 110–140 dB is shown in Fig. 19.48.

Example 19.11 Show the use of 3-dB 90° hybrid coupler to separate RHCP and LHCP signals from an antenna.

Solution
Electric field of a right-hand circular polarized (RHCP) signal can be expressed as

$$E(\vec{r}, t) = E_a \sin(\omega_a t - k_a z)\hat{e}_x + E_a \cos(\omega_a t - k_a z)\hat{e}_y \tag{19.26}$$

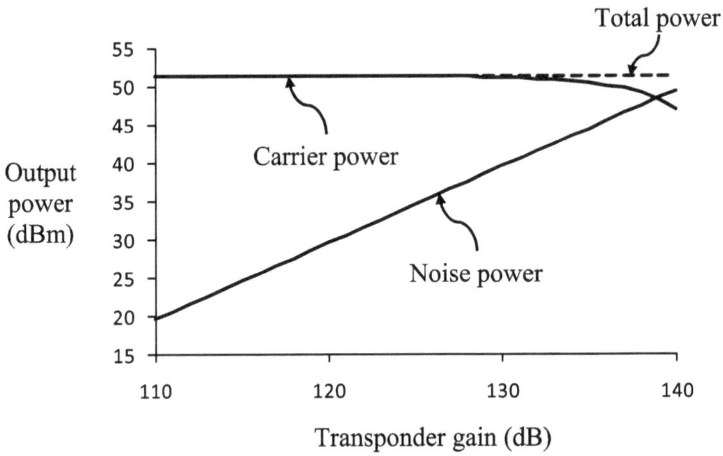

Fig. 19.48 Carrier and noise powers versus gain at output of the transponder

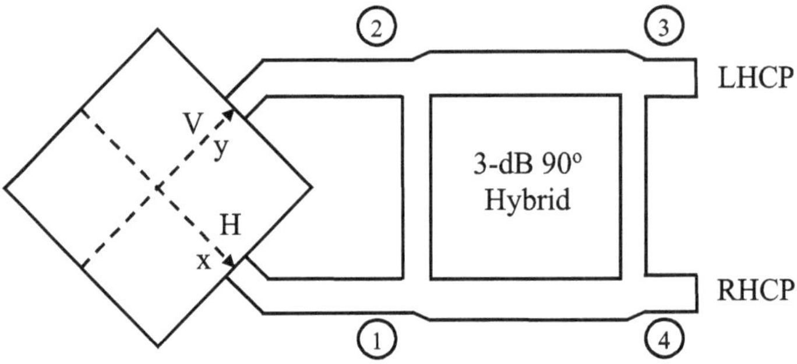

Fig. 19.49 Scheme for the separation of LHCP and RHCP signals using 3-dB 90° hybrid

And the electric field of another left-hand circular polarized (LHCP) signal can be expressed as

$$E(\vec{r}, t) = -E_b \sin(\omega_b t - k_b z)\hat{e}_x + E_b \cos(\omega_b t - k_b z)\hat{e}_y \tag{19.27}$$

Two input ports of a 3-dB 90° hybrid coupler are connected to the two ports of an antenna as shown in Fig. 19.49. Here, horizontal and vertical feeds are connected to the two inputs (1 and 2) of the hybrid. At the port-1 (say, $z = 0$), the signal voltage can be written as

$$v_1(t) = v_a \sin(\omega_a t) - v_b \sin(\omega_b t) \tag{19.28}$$

Similarly, at the port-2 (say, $z = 0$), the signal voltage can be written as

$$v_2(t) = v_a \cos(\omega_a t) + v_b \cos(\omega_b t) \tag{19.29}$$

Thus, the signal voltage at port-3 can be written as

$$\begin{aligned} v_3(t) &= \frac{v_a}{\sqrt{2}}\sin(\omega_a t - \pi) - \frac{v_b}{\sqrt{2}}\sin(\omega_b t - \pi) \\ &\quad + \frac{v_a}{\sqrt{2}}\cos\left(\omega_a t - \frac{\pi}{2}\right) + \frac{v_b}{\sqrt{2}}\cos\left(\omega_b t - \frac{\pi}{2}\right) \\ &= -\frac{v_a}{\sqrt{2}}\sin(\omega_a t) + \frac{v_b}{\sqrt{2}}\sin(\omega_b t) + \frac{v_a}{\sqrt{2}}\sin(\omega_a t) + \frac{v_b}{\sqrt{2}}\sin(\omega_b t) \\ &= \frac{v_b}{\sqrt{2}}\sin(\omega_b t) + \frac{v_b}{\sqrt{2}}\sin(\omega_b t) \\ &= \sqrt{2}v_b\sin(\omega_b t) \end{aligned} \tag{19.30}$$

This shows that the output port-3 of the hybrid contains only the frequency component corresponding to LHCP.

Thus, the signal voltage at the port-4 can be written as

$$\begin{aligned} v_3(t) &= \frac{v_a}{\sqrt{2}}\sin\left(\omega_a t - \frac{\pi}{2}\right) - \frac{v_b}{\sqrt{2}}\sin\left(\omega_b t - \frac{\pi}{2}\right) \\ &\quad + \frac{v_a}{\sqrt{2}}\cos(\omega_a t - \pi) + \frac{v_b}{\sqrt{2}}\cos(\omega_b t - \pi) \\ &= -\frac{v_a}{\sqrt{2}}\cos(\omega_a t) + \frac{v_b}{\sqrt{2}}\cos(\omega_b t) - \frac{v_a}{\sqrt{2}}\cos(\omega_a t) - \frac{v_b}{\sqrt{2}}\cos(\omega_b t) \\ &= -\frac{v_a}{\sqrt{2}}\cos(\omega_a t) - \frac{v_a}{\sqrt{2}}\cos(\omega_a t) \\ &= \sqrt{2}v_a\sin(\omega_a t) \end{aligned} \tag{19.31}$$

This shows that the output port-4 of the hybrid contains only the frequency component corresponding to RHCP.

In a similar way, it can be shown that two signals of different frequency bands can be transmitted one in LHCP and other in RHCP through a single antenna using a 3-dB 90° hybrid.

References

1. Wang C-X et al (2014) Cellular architecture and key technologies for 5G wireless communication networks. IEEE Commun Mag, Feb 2014
2. Pandian JD, Baker L, Cortes G, Goldsmith PF, Deshpande AA, Ganesan R, Hagen J, Locke L, Wadefalk N, Weinreb S (2006) Low-noise 6-8 GHz receiver. IEEE Microw Mag 7:74–84
3. Garg VK, Singh RV, Jain VK, Bera SC (1999) New amplifier design eliminates chip capacitors. IETE Tech Rev 16(6):197–201
4. Bera SC, Shah LB, Raval DU, Pandey S, Kumar V, Singh S, Das DK (2012) Design and development of V&W band amplifier modules. In: Proceedings of international conference on microwaves, antenna, propagation & remote sensing, 2012, Jodhpur, India

5. Buch SD, Bera SC (2013) Transponder nonlinearity characterization & mitigation techniques: present scenario & future trends. In: International workshop on sensor network and wireless communication, ADIT, Oct 2013
6. Yadav SP, Bera SC (2014) Nonlinearity effect of high power amplifiers in communication systems. In: International conference on advances in communication and computing technologies (ICACACT), Aug 2014
7. Yadav SP, Bera SC (2014) Nonlinearity effects of power amplifiers in wireless communication systems. In: Proceedings of IEEE international conference on electronics, communication and computational engineering (ICECCE 2014), Hosur, India, pp 1011–1016, Nov 2014
8. Bera SC, Singh RV, Garg VK (2006) Design and temperature compensation of a Ku-band channel amplifier with ALC for a satellite transponder. Microw J 49(4):68–82
9. Hu Y, Feng J (2016) The development and new trends of microwave vacuum electronic devices. In: IEEE international conference on emerging technologies (ICET), pp 1–5, Oct 2016
10. Bera SC, Singh RV (2004) A temperature-compensated closed loop overdrive level controller for microwave solid-state power amplifiers. Microw J 47(4):114–122
11. Yamauchi K, Mori K, Nakayama M, Mitsui Y, Takagi T (1997) A microwave miniaturized linearizer using a parallel diode with a bias feed resistance. IEEE Trans Microw Theory Tech 45(12):2431–2435
12. Bera SC, Bhardhwaj PS, Singh RV, Garg VK (2003) A diode linearizer for microwave power amplifiers. Microw J 46(11):102–113
13. Bera SC, Singh RV, Garg VK (2004) A compact Ku-band linearizer for space application. In: Proceedings of Asia Pacific microwave conference, 2004, pp 37–38
14. Bera SC, Singh RV, Garg VK (2008) Diode-based predistortion lineariser for power amplifiers. IEE Electron Lett 44(2):125–126
15. Bera SC, Kumar V, Singh S, Das DK (2013) Temperature behavior and compensation of diode-based predistortion linearizer. IEEE Microw Wirel Compon Lett 23(4):211–213
16. Abrams RH, Parker RK (1993) Introduction to the MPM: what it is and where it might fit. IEEE MTT-S Int Microw Symp Dig 1:107–110
17. Kowalczyk R, Zubyk A, Meadows C et al (2016) High efficiency E-band MPM for communications application. In: 17th IEEE international vacuum electronics conference, pp 513–514, 2016
18. Bera SC, Singh RV, Garg VK (2008) Modified Wilkinson power divider with harmonic suppression characteristic. Microw J, Nov 2008

Chapter 20
Multiple Choice Questions with Answers

20.1 Multiple Choice Questions

(1) Electron mobility of the following undoped elements in descending order is

(A) **GaAs, Ge, Si**
(B) GaAs, Si, Ge
(C) Si, Ge, GaAs
(D) Ge, Si, GaAs

(2) Common-base current gain of a p-n-p bipolar transistor is 0.99. The common-emitter current gain of the transistor is

(A) 101
(B) 0.01
(C) **99**
(D) 1.0

(3) Over a wide operating temperature range, around ambient temperature, equiresistance curves for a practical p-i-n diode can be considered as

(A) **Straight line**
(B) Parabolic
(C) Elliptic
(D) Circular

(4) Open-circuit voltage for temperature-invariant RF resistance of a silicon p-i-n diode is approximately

(A) 0.61 V
(B) **1.15 V**
(C) 2.25 V
(D) 3.8 V

S. C. Bera, *Microwave Active Devices and Circuits for Communication*, Lecture Notes in Electrical Engineering 533, https://doi.org/10.1007/978-981-13-3004-9_20

(5) Open-circuit voltage for temperature-invariant RF resistance of a GaAs p-i-n diode is approximately

(A) 0.61 V
(B) 1.12 V
(C) **1.42 V**
(D) 3.8 V

(6) An attenuator circuit consists of two silicon p-i-n diodes. The diodes are connected in series for DC bias. The open-circuit voltage for temperature-invariant attenuation of the attenuator is approximately

(A) 0.61 V
(B) 1.12 V
(C) 1.42 V
(D) **2.3 V**

(7) An attenuator circuit consists of two silicon p-i-n diodes. The diodes are connected in parallel for DC bias. The open-circuit voltage for temperature-invariant attenuation of the attenuator is approximately

(A) 0.61 V
(B) **1.15 V**
(C) 1.42 V
(D) 2.3 V

(8) 1 W power in dBm and dBW, respectively, are

(A) **30 and 0**
(B) 0 and 30
(C) 1 and 0
(D) 0 and 1

(9) 10 mW power in dBm is

(A) **10**
(B) 1
(C) 0.1
(D) 0

(10) 50 dBm power in Watt is

(A) 10
(B) **100**
(C) 1000
(D) 10,000

(11) 30 dBW power in Watt is

(A) 10
(B) 100
(C) **1000**
(D) 10,000

(12) 30 dBm power in Watt is

(A) 100
(B) 10
(C) **1**
(D) 0

(13) 0 dBW power in Watt is

(A) 10
(B) **1**
(C) 100
(D) 0.1

(14) Common-emitter DC current gain of a Si transistor is 100. The current through the 10 V Zener diode is

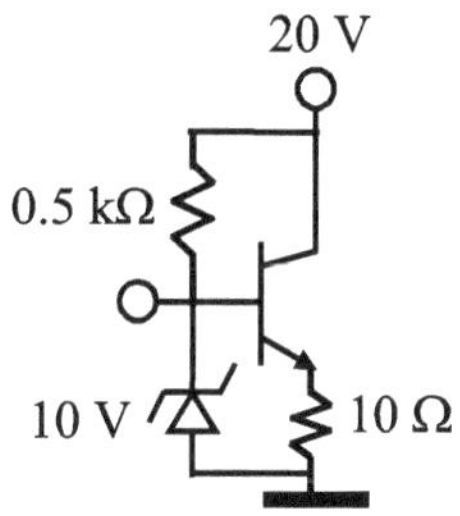

(A) **10.7 mA**
(B) 19.3 mA
(C) 20 mA
(D) 40 mA

(15) The circuit has an ideal operational amplifier with ±10 V supply. The output voltage is

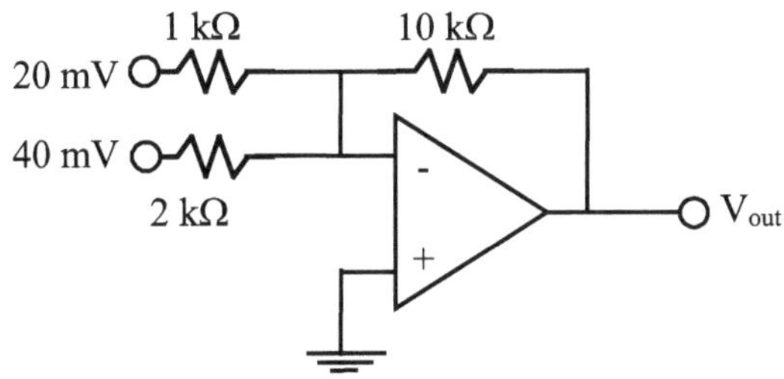

(A) −200 mV
(B) **−400 mV**
(C) −600 mV
(D) −300 mV

(16) The effective resistance faced by the voltage source is

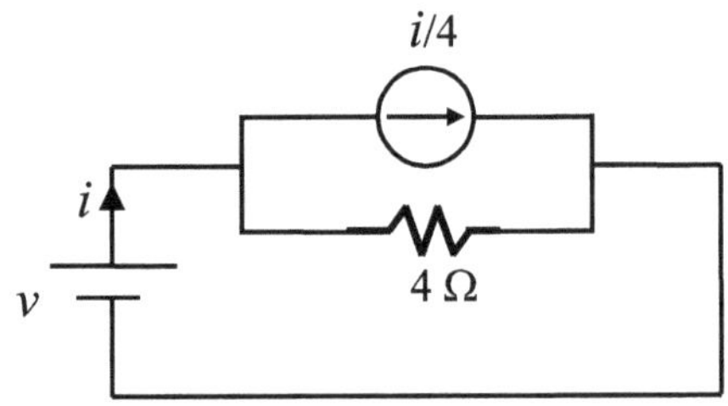

(A) 4 Ω
(B) 12 Ω
(C) **3 Ω**
(D) 16 Ω

(17) The voltage across the resistor *R* is

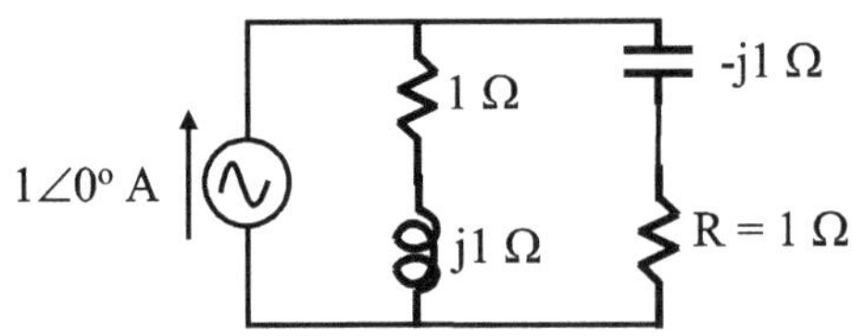

(A) $1/\sqrt{3}\ \angle 0°$ V
(B) $1/\sqrt{2}\ \angle 0°$ V
(C) $1/2\ \angle 45°$ V
(D) **$1/\sqrt{2}\ \angle 45°$ V**

(18) A step voltage of 1 V is applied to the circuit at $t = 0$. The currents in mA through the resistor R just after $t = 0$ and at steady state, respectively, are

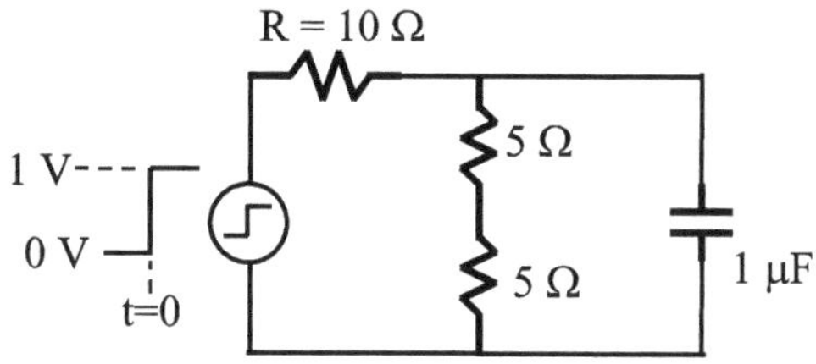

(A) **100, 50**
(B) 50, 50
(C) 100, 100
(D) 75, 75

(19) Under constant voltage bias condition, how does the bias current of a Schottky diode change with temperature?

(A) Current decreases with the increase in temperature
(B) **Current increases with the increase in temperature**
(C) Current remains constant with the change in temperature
(D) At low temperature, current increases and at higher temperature, it decreases with the increase in temperature

(20) Which of the following diodes is the most suitable to realize a voltage-variable microwave attenuator?

(A) Schottky barrier diode
(B) Varactor diode
(C) **p-i-n diode**
(D) P-N junction diode

(21) A BJT is having common-emitter current gain 100. Considering 10 V supply, $V_{BE} = 0.7$ V and $R_E = 100\ \Omega$, what is the value of R_B and R_C to set the quiescent point at $I_C = 10$ mA and $V_{CE} = 8$ V.

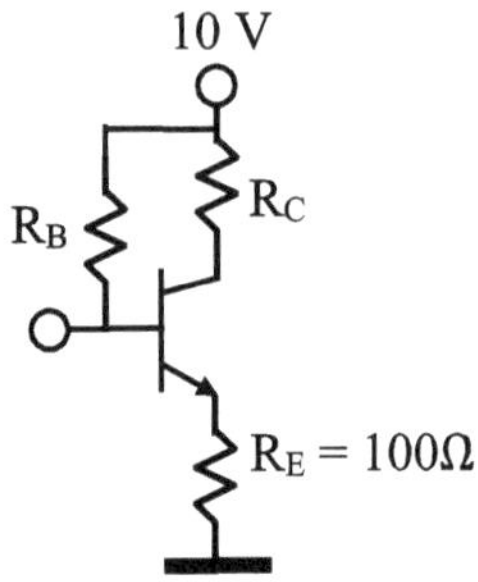

(A) $R_B = 83\ \text{k}\Omega, R_C = 200\ \Omega$
(B) $R_B = 93\ \text{k}\Omega, R_C = 200\ \Omega$
(C) $R_B = 93\ \text{k}\Omega, R_C = 100\ \Omega$
(D) $\boldsymbol{R_B = 83\ \text{k}\Omega, R_C = 100\ \Omega}$

(22) Amplifier-A has 6 dB gain and 3.8 dB NF. Amplifier-B has 20 dB gain and 4 dB NF. What should be the sequence of the cascaded two-stage amplifier for achieving the lowest NF? What will be the value of the cascaded lowest NF?

(A) **Input stage: amplifier-B, output stage: amplifier-A, 4.02 dB**
(B) Input stage: amplifier-A, output stage: amplifier-B, 3.85 dB
(C) Input stage: amplifier-A, output stage: amplifier-B, 4.44 dB
(D) Input stage: amplifier-B, output stage: amplifier-A, 3.85 dB

(23) Input impedance of a $\lambda/8$ lossless transmission line having characteristic impedance 13 Ω and terminated with a load impedance of $(5 + j12)\ \Omega$ is

(A) 13 Ω
(B) **65 Ω**
(C) $(18+j12)\ \Omega$
(D) $(18+j13)\ \Omega$

(24) In an ideal Class-F microwave power amplifier, time domain voltage and current waveforms of the device have

(A) Maximum overlap
(B) 50% overlap
(C) **No overlap**
(D) Less than 80% overlap

(25) The equivalent circuit of a FET is shown, where $r_g = 100\ \Omega, c_{gs} = 0.1$ pF, and $g_m = 0.1$ mho. For load resistance $r_L = 500\ \Omega$,, the voltage gain at 1 GHz is

(A) 30
(B) 20
(C) 100
(D) **50**

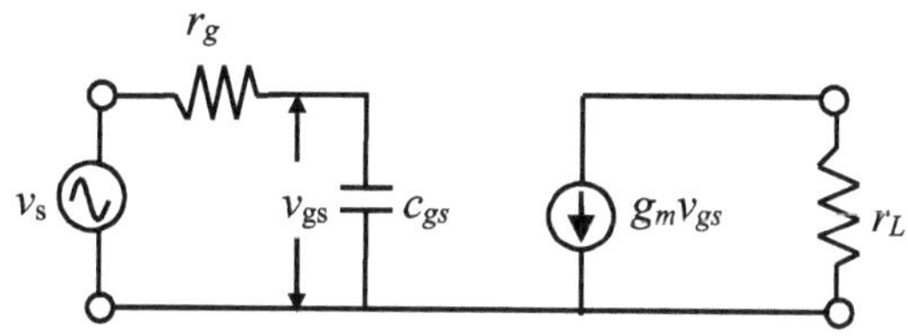

(26) A FET power amplifier operating in Class-A with 10 V drain supply and 2 A drain current bias. The RF load current is of amplitude 1 A. What is the DC to RF efficiency for the load resistance of 5 Ω?

(A) **12.5%**
(B) 25%
(C) 35%
(D) 50%

(27) A FET power amplifier operating in Class-B with 15 V drain supply. The amplifier provides 12 V peak signal to a 15 Ω load. DC to RF efficiency of the amplifier is

(A) 8.5%
(B) **62.8%**
(C) 50%
(D) 75%

(28) A high-gain MESFET packaged device of case-to-channel thermal resistance has 4.5 °C/W and provides 8 W RF output, taking 20 W DC power. What is the channel temperature of the device if the case temperature of the device maintained at 55 °C?

(A) 145 °C
(B) 59.5 °C
(C) **109 °C**
(D) 91 °C

(29) Main function of helix structure in a travelling wave tube (TWT) amplifier is

(A) To reduce the axial velocity of electron
(B) To properly focus the electron beam
(C) To reduce noise figure of TWT
(D) **To reduce the axial velocity of RF field**

(30) A load of the reflection coefficient Γ_L is transformed to Γ_{IN} by suitably connecting a circuit element as shown in the figure. The circuit element is

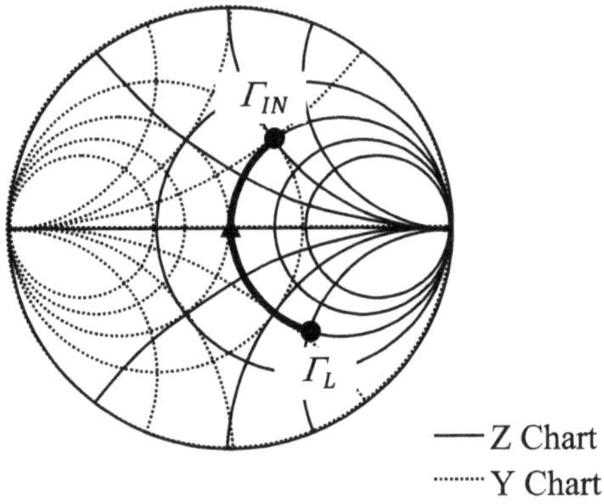

(A) **Series-connected inductor**
(B) Series-connected capacitor
(C) Series-connected the transmission line
(D) Shunt-connected inductor

(31) S-parameter matrix of a lossless transmission line of length $\lambda/8$ is

(A) $\begin{bmatrix} \mathbf{0} & \mathbf{1\angle{-45^\circ}} \\ \mathbf{1\angle{-45^\circ}} & \mathbf{0} \end{bmatrix}$

(B) $\begin{bmatrix} 0 & 1\angle{-90^\circ} \\ 1\angle{-90^\circ} & 0 \end{bmatrix}$

(C) $\begin{bmatrix} 0 & 1\angle{-22.5^\circ} \\ 1\angle{-22.5^\circ} & 0 \end{bmatrix}$

(D) $\begin{bmatrix} 0 & 1\angle{-180^\circ} \\ 1\angle{-180^\circ} & 0 \end{bmatrix}$

(32) A unilateral transistor has output impedance $Z_{out} = (32 + j30)\ \Omega$. Value of the series and shunt components of the matching network (as shown in the figure) for the complex conjugate match at the output of the device to a 50 Ω load is

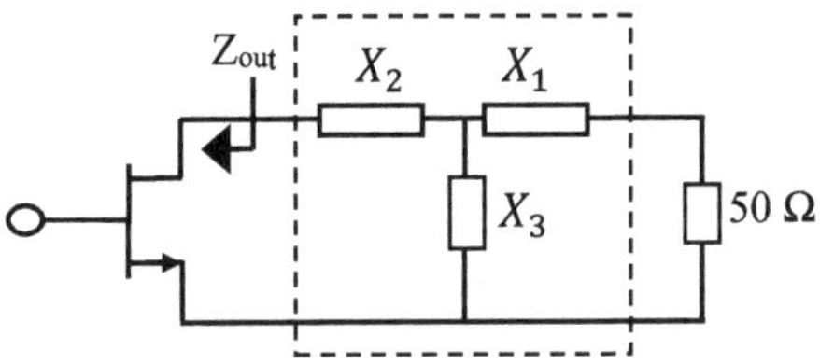

(A) $\boldsymbol{X_1 = j40, X_2 = j10, X_3 = -j40}$
(B) $X_1 = j40, X_2 = j10, X_3 = +j40$

(C) $X_1 = -j40, X_2 = j10, X_3 = -j40$
(D) $X_1 = j40, X_2 = -j40, X_3 = -j40$

(33) Transmission coefficient 'S_{21}' of the 2-port network shown in the figure, for a 50 Ω system is

(A) **2/3∠0°**
(B) 1/2∠0°
(C) 2/3∠180°

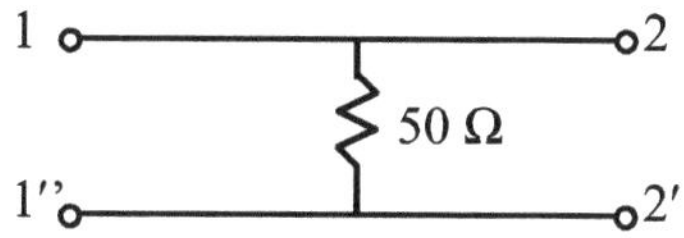

(D) 1/2∠180°

(34) Input impedance of a quarter-wavelength 50 Ω transmission line when terminated by 25 Ω is

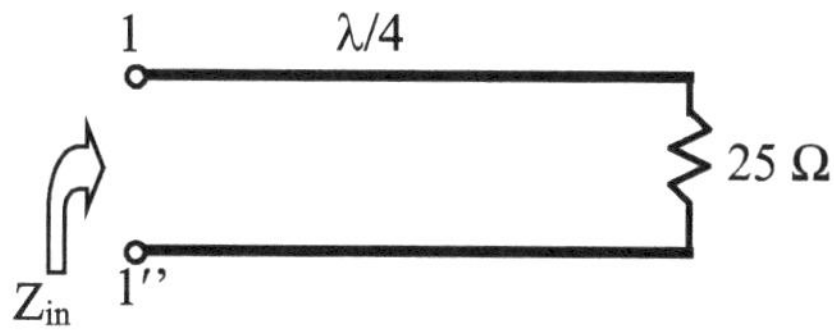

(A) **100 Ω**
(B) 200 Ω
(C) 50 Ω
(D) 25 Ω

(35) A unilateral transistor is having input and output reflection coefficients 0.498∠121° and 0.225∠−150°, respectively. Its optimum source and load reflection coefficients for minimum noise figure and maximum output power are 0.455∠−160° and 0.275∠+130°, respectively. To design a low-noise amplifier, source and load reflection coefficient should be, respectively,

(A) **0.455∠−160° and 0.225∠+150°**
(B) 0.498∠ − 121° and 0.225∠ + 150°
(C) 0.498∠ − 121° and 0.275∠ + 130°
(D) 0.455∠ + 160° and 0.225∠ + 150°

(36) The plot shows the operating load line and I-V characteristic of a Schottky diode at two different RF power levels +10 and −10 dBm. Under this bias condition, RF resistance of the diode

(A) **Increases with the increase in RF power level**
(B) Decreases with the increase in RF power level

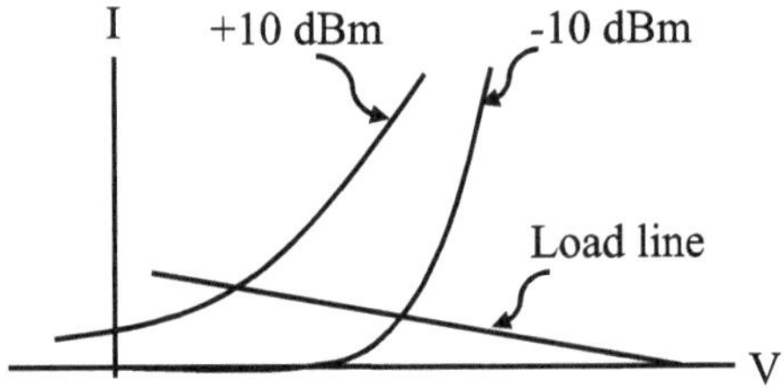

(C) Remains constant with the RF power level
(D) None of the above

(37) Which of the following devices is the most suitable for higher order RF frequency multiplier?

(A) **Step recovery diode**
(B) P-N junction diode
(C) Varactor diode
(D) P-I-N diode

(38) Which of the following diodes is the most suitable as linear resistance at microwave frequency?

(A) **P-I-N diode**
(B) Schottky diode
(C) Varactor diode
(D) P-N junction diode

(39) Input reflection coefficient 'S_{11}' of the 2-port network for a 50 Ω system is

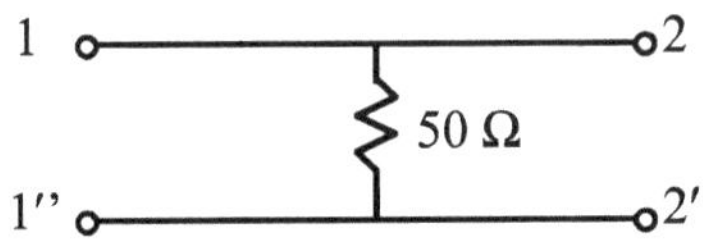

(A) **$1/3\angle 180°$**
(B) $1/2\angle 180°$
(C) $1/3\angle 0°$
(D) $1/2\angle 0°$

(40) Input reflection coefficient 'S_{11}' of the 2-port network for a 50 Ω system is

(A) **$1/5\angle 0°$**
(B) $1/2\angle 0°$

(C) $1/5\angle 0°$
(D) $1/2\angle 0°$

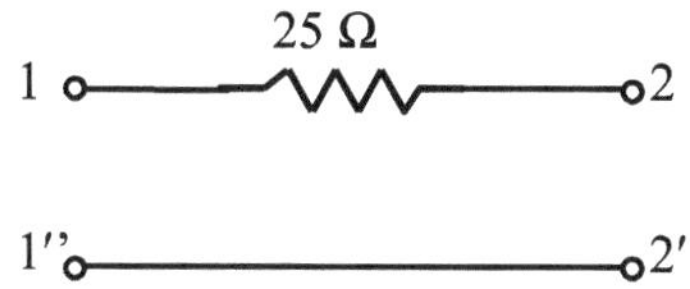

(41) Transmission coefficient 'S_{21}' of the 2-port network for a 50 Ω system is

(A) **$4/5\angle 0°$**
(B) $1/2\angle 0°$
(C) $4/5\angle 180°$
(D) $1/2\angle 180°$

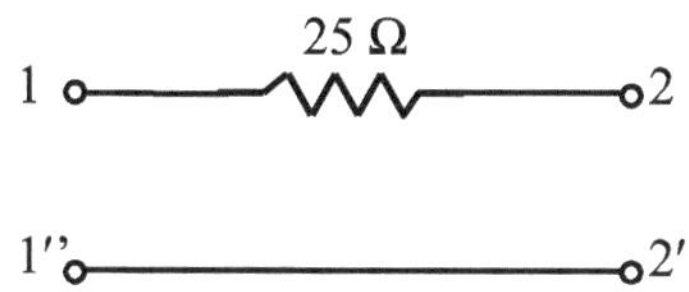

(42) Which of the following diodes is most suitable for detection of microwave signal?

(A) **Schottky barrier diode**
(B) p-i-n diode
(C) Varactor diode
(D) P-N junction diode

(43) Reason for superior high-frequency performance of metal–semiconductor diode compared to p-n junction diode is

(A) **No minority carrier storage effect in metal–semiconductor diode**
(B) Higher conductivity of metal compared to semiconductor
(C) Metal–semiconductor junction does not rectify RF signal
(D) Hole is the majority carrier in Schottky barrier diode operation

(44) Which statement is false about microstrip line over stripline?

(A) **Less radiative**
(B) Easier for component integration
(C) One-sided ground plane
(D) More interaction with neighbouring circuit element

(45) A 50 Ω transmission line is terminated by a load impedance of $(100+j100)$ Ω. Magnitude of reflection coefficient at the load is

(A) $\boldsymbol{\sqrt{5/13}}$
(B) $\sqrt{1/13}$
(C) $\sqrt{5/7}$
(D) $5/7$

(46) A unilateral transistor has output impedance $Z_{out} = (10 - j10)$ Ω. Value of the series and shunt components of the matching network for the complex conjugate match at the output of the device, for a 50 Ω load, are

(A) $\boldsymbol{X_1 = -j25, X_2 = +j30}$
(B) $X_1 = +j25,\ X_2 = +j30$
(C) $X_1 = -j25,\ X_2 = -j30$
(D) $X_1 = +j25, X_2 = -j30$

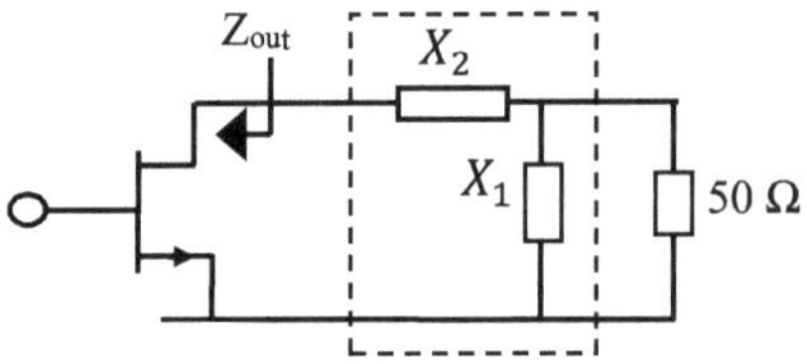

(47) A unilateral transistor has output impedance $Z_{out} = (10 - j10)$Ω. Value of the series and shunt components of the matching network for the complex conjugate match at the output of the device considering a 50 Ω load are

(A) $\boldsymbol{X_1 = +j25, X_2 = -j30}$
(B) $X_1 = -j25, X_2 = -j10$
(C) $X_1 = +j25, X_2 = +j10$
(D) $X_1 = -j25, X_2 = +j10$

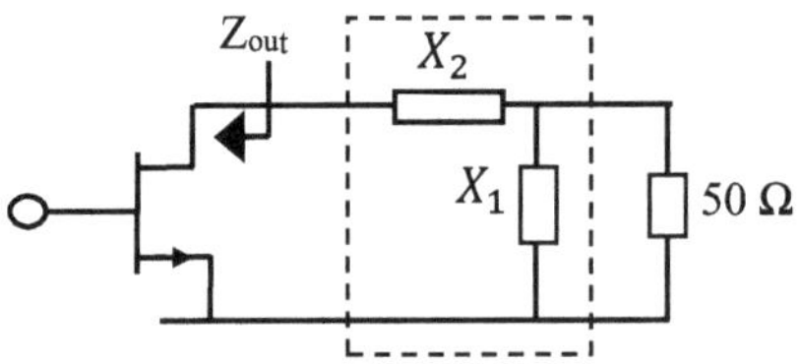

(48) A unilateral transistor has output impedance $Z_{out} = (8 - j8)$Ω. Values of the series and shunt components of the matching network for complex conjugate match at the output of the device considering a 50 Ω load are

(A) $\mathbf{X_1 = +j20, X_2 = +j28, X_3 = -j20}$

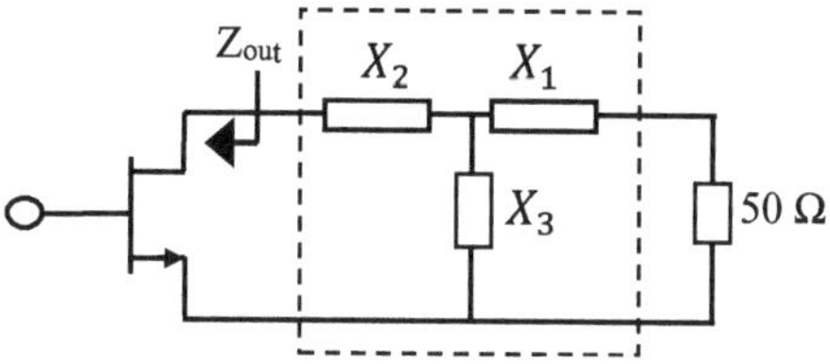

(B) $X_1 = +j20, X_2 = +j28, X_3 = +j40$
(C) $X_1 = -j20, X_2 = +j28, X_3 = -j40$
(D) $X_1 = +j20, X_2 = -j28, X_3 = -j40$

(49) DC to RF efficiency of an ideal Class-F amplifier is

(A) **100%**
(B) 78.4%
(C) 50%
(D) <80%

(50) *S*-parameter matrix of a lossless transmission line of length λ/4 is

(A) $\begin{bmatrix} \mathbf{0} & \mathbf{1\angle{-90^\circ}} \\ \mathbf{1\angle{-90^\circ}} & \mathbf{0} \end{bmatrix}$

(B) $\begin{bmatrix} 0 & 1\angle{-45^\circ} \\ 1\angle{-45^\circ} & 0 \end{bmatrix}$

(C) $\begin{bmatrix} 0 & 1\angle{-22.5^\circ} \\ 1\angle{-22.5^\circ} & 0 \end{bmatrix}$

(D) $\begin{bmatrix} 0 & 1\angle - 180^\circ \\ 1\angle - 180^\circ & 0 \end{bmatrix}$

(51) Magnitude of the input impedance of a λ/8 lossless 50 Ω transmission line terminated with 25 Ω is

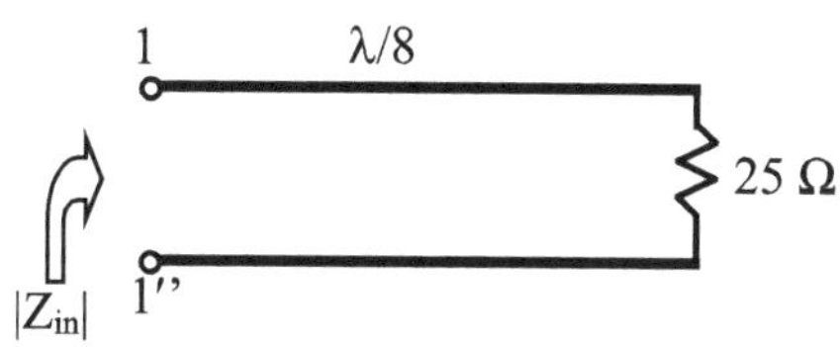

(A) **50 Ω**
(B) 25 Ω

(C) 100 Ω
(D) None of the above

(52) Gain and NF of a single-stage amplifier are 10 and 3 dB, respectively. When two such amplifiers are cascaded, then gain and NF of the cascaded amplifier will be

(A) **Gain: 20 dB, NF: 3.2 dB**
(B) Gain: 20 dB, NF: 4.2 dB
(C) Gain: 13 dB, NF: 3.6 dB
(D) Gain: 13 dB, NF: 3.7 dB

(53) A load of the reflection coefficient Γ_L is transformed to Γ_{IN} as shown in the figure by suitably connecting a circuit element. The circuit element is

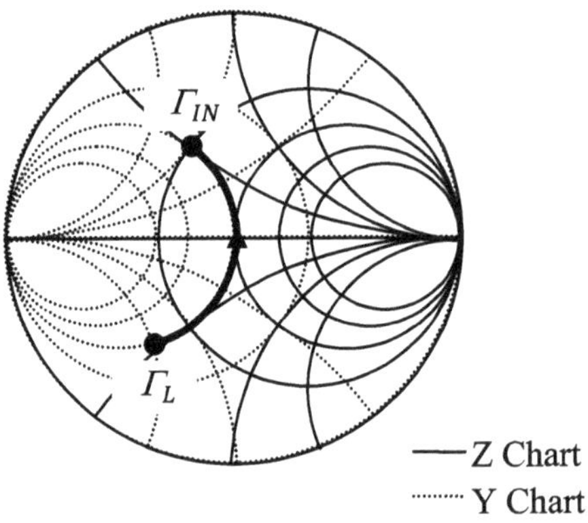

(A) **Short circuit transmission line stub or shunt connected inductor**
(B) Series connected capacitor
(C) Series connected transmission line
(D) Shunt connected capacitor

(54) A load of the reflection coefficient Γ_L is transformed to Γ_{IN} as shown in the figure by suitably connecting a circuit element. The circuit element is

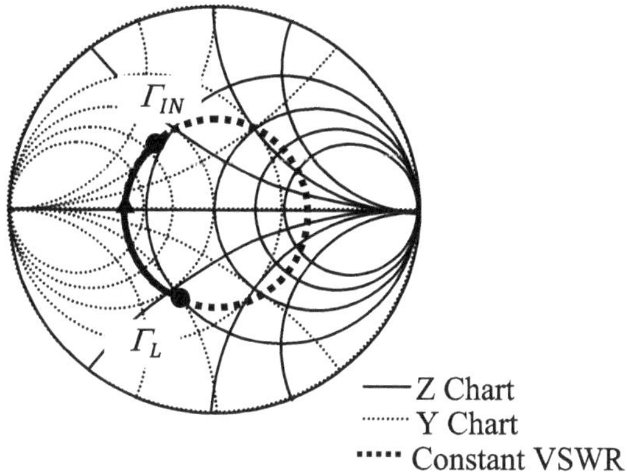

(A) **Series connected transmission line**
(B) Shunt connected inductor
(C) Series connected capacitor
(D) Shunt connected capacitor

(55) A load of the reflection coefficient Γ_L is transformed to Γ_{IN} as shown in the figure by suitably connecting a circuit element. The circuit element is

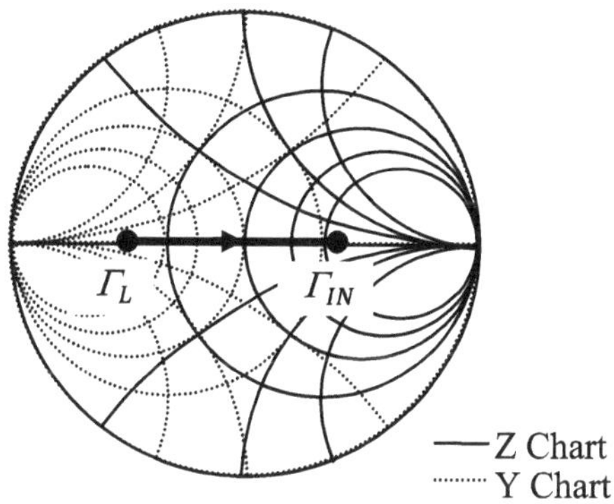

(A) **Series connected resistor**
(B) Shunt connected resistor
(C) Series connected transmission line
(D) Shunt connected capacitor

(56) A load of the reflection coefficient Γ_L is transformed to Γ_{IN} as shown in the figure by suitably connecting a circuit element. The circuit element is

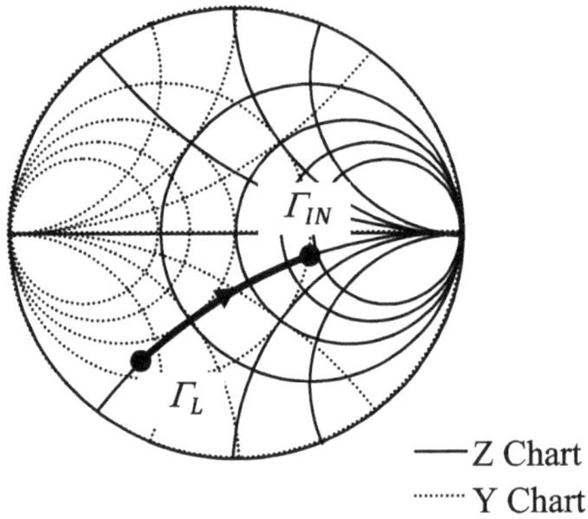

(A) **Series connected resistor**
(B) Shunt connected inductor
(C) Series connected transmission line
(D) Shunt connected capacitor

(57) A load of the reflection coefficient Γ_L is transformed to Γ_{IN} as shown in the figure by connecting a circuit element. The circuit element is

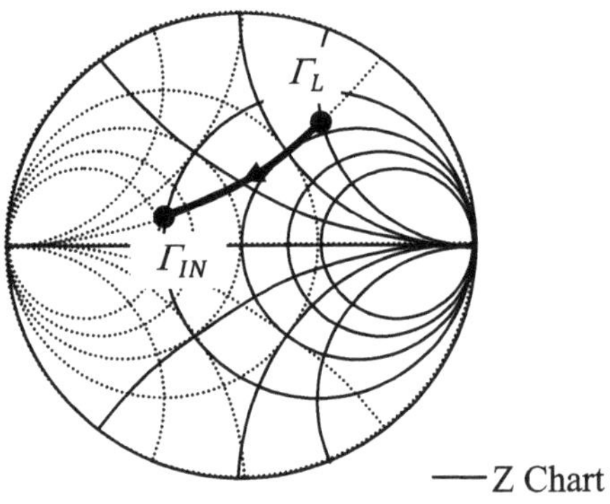

Fig. Q30

(A) **Shunt connected resistor**
(B) Shunt connected inductor
(C) Series connected transmission line
(D) Shunt connected capacitor

(58) A unilateral transistor is having input and output reflection coefficients $0.498\angle 121°$ and $0.225\angle -150°$, respectively. Its optimum source and load reflection coefficients for minimum noise figure and maximum output power are $0.455\angle -160°$ and $0.275\angle +130°$, respectively. To design a high-gain amplifier, source and load reflection coefficient should be, respectively,

(A) **$0.498\angle{-121^\circ}$ and $0.225\angle{+150^\circ}$**
(B) $0.455\angle{-160^\circ}$ and $0.225\angle{+150^\circ}$
(C) $0.498\angle{-121^\circ}$ and $0.275\angle{+130^\circ}$
(D) $0.455\angle{+121^\circ}$ and $0.225\angle{-150^\circ}$

(59) A unilateral transistor is having input and output reflection coefficients $0.498\angle 121^\circ$ and $0.225\angle{-150^\circ}$, respectively. Its optimum source and load reflection coefficients for minimum noise figure and maximum output power are $0.455\angle{-160^\circ}$ and $0.275\angle{+130^\circ}$, respectively. To design a power amplifier with highest possible output power and moderate gain, the source and load reflection coefficient should be, respectively,

(A) **$0.498\angle{-121^\circ}$ and $0.275\angle{+130^\circ}$**
(B) $0.498\angle{-121^\circ}$ and $0.225\angle{+150^\circ}$
(C) $0.455\angle{-160^\circ}$ and $0.225\angle{+150^\circ}$
(D) $0.455\angle{+121^\circ}$ and $0.225\angle{-130^\circ}$

(60) A BJT is having common-emitter current gain 100. Considering 10 V supply and $V_{BE} = 0.7$ V, the value of R_C and R_B to set the quiescent point at $I_C = 10$ mA and $V_{CE} = 8$ V is

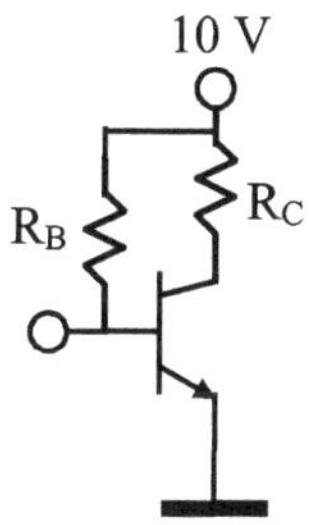

(A) **$R_C = 200\ \Omega$, $R_B = 93\ \text{k}\Omega$**
(B) $R_C = 2\ \text{k}\Omega$, $R_B = 100\ \text{k}\Omega$
(C) $R_C = 200\ \Omega$, $R_B = 100\ \text{k}\Omega$
(D) $R_C = 20\ \Omega$, $R_B = 93\ \text{k}\Omega$

(61) A power MESFET operates with 10 V drain supply and 2 A drain current. What is the maximum output power when operated in Class-A?

(A) **10 W**
(B) 20 W
(C) 15 W
(D) 5 W

(62) A lossless transmission line of characteristic impedance 50 Ω is terminated in a resistive load of impedance 100 Ω. The VSWR is

(A) 1:2
(B) **2:1**
(C) 4:1
(D) 1:4

(63) A linear amplifier of gain 30 dB is fed with 0.1 μW power, the output power of the amplifier

(A) 1 W
(B) 0 dBm
(C) 30 dBm
(D) **−10 dBm**

(64) 10 W RF power is transmitted with a circular polarized antenna having a gain of 10 dB. A omni-directional receiving antenna has vertical polarization. The path loss is 100 dB. The receiving signal is

(A) **−83 dBW**
(B) −80 dBW
(C) −88 dBW
(D) +80 dBW

(65) If the short-circuit and open-circuit impedance of a transmission line are 5 and 20 Ω, respectively. Characteristic impedance of the line is

(A) 100 Ω
(B) **10 Ω**
(C) 15 Ω
(D) 1000 Ω

(66) If each stage has a gain of 10 dB and NF of 10 dB, then the overall noise figure of two-stage cascaded amplifier will be

(A) 10 dB
(B) 10.9 dB
(C) **10.37 dB**
(D) 1.09 dB

(67) Transmission coefficient 'S_{21}' of the 2-port network for 50 Ω system is

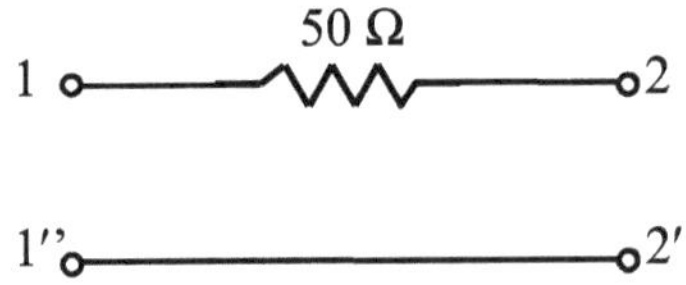

(A) 4/5 ∠0°

(B) **1/3 ∠0°**
(C) 4/5 ∠180°
(D) 1/2 ∠180°

(68) Transmission coefficient 'S_{21}' of the 2-port network for 50 Ω system is

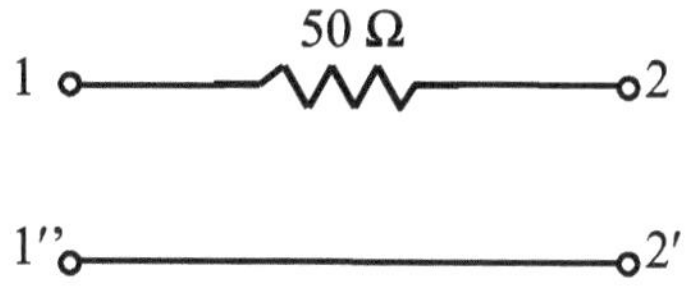

(A) 4/5 ∠0°
(B) 1/2 ∠0°
(C) 4/5 ∠180°
(D) **2/3 ∠0°**

(69) The value of shunt resistor of a T-type matched attenuator in a 50 Ω system is 100 Ω. The attenuation and the value of series resistors are, respectively,

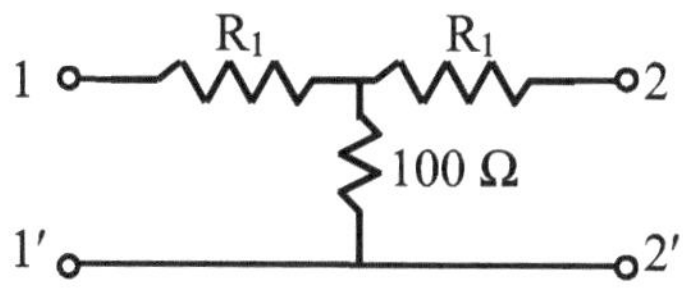

(A) 2.09 dB, 11.803 Ω
(B) 2.09 dB, 12 Ω
(C) **4.18 dB, 11.803 Ω**
(D) 4.18 dB, 6.02 Ω

(70) Attenuation of this T-type attenuator in a 50 Ω system is

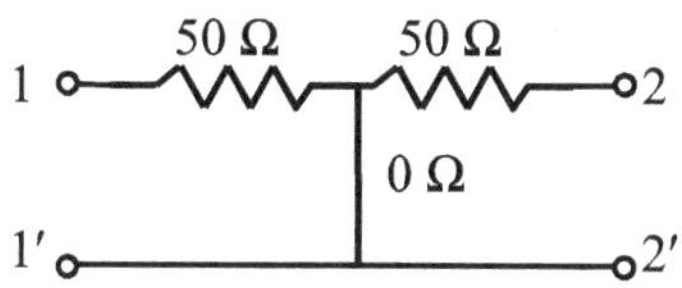

(A) 0 dB
(B) ∝ **dB**
(C) 10 dB
(D) 1 dB

(71) In a 50 Ω system, attenuation and return losses of this π attenuator in a 50 Ω system are

(A) ∝ dB, 12.04 dB
(B) **12.04 dB, 12.04 dB**
(C) 12.04 dB, ∝ dB
(D) 0 dB, 12.04 dB

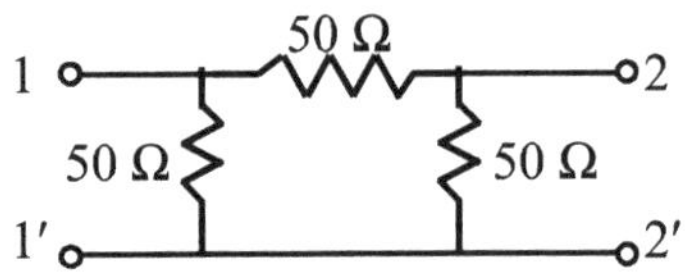

(72) Attenuation and return losses of the 2-port network in a 50 Ω system are, respectively,

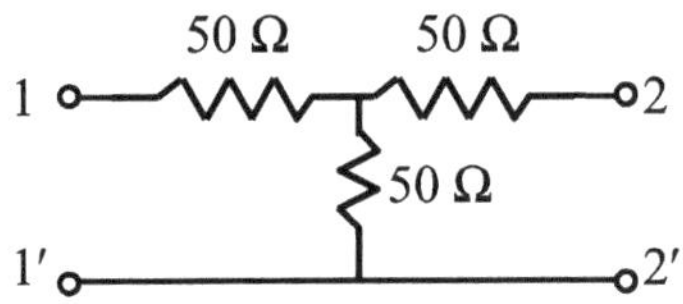

(A) ∝ dB, 12.04 dB
(B) 0 dB, 12.04 dB
(C) **12.04 dB, 12.04 dB**
(D) 0 dB, ∝ dB

(73) Attenuation and return losses of the 2-port network in a 50 Ω system are, respectively,

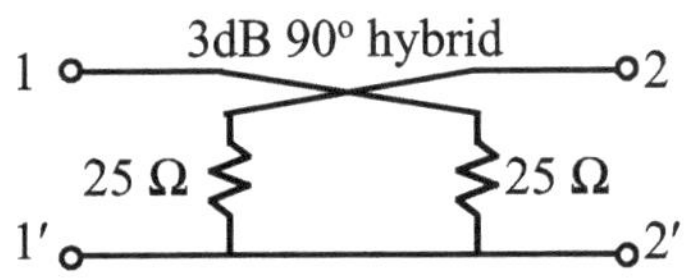

(A) ∝ dB, 9.54 dB
(B) 9.54 dB, ∝ dB
(C) 9.54 dB, 9.54 dB
(D) **9.54 dB, ∝ dB**

(74) Attenuation and return losses of the 2-port network in a 50 Ω system are, respectively,

(A) 20.98 dB, ∝ dB
(B) **9.39 dB, 20.98 dB**

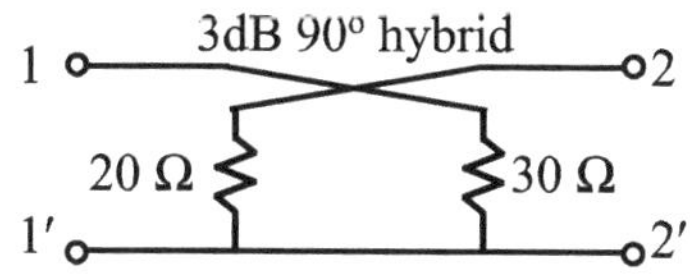

(C) 9.39 dB, 9.39 dB
(D) 20.98 dB, 20.98 dB

(75) Attenuation and return losses of the 2-port network in a 50 Ω system are, respectively,

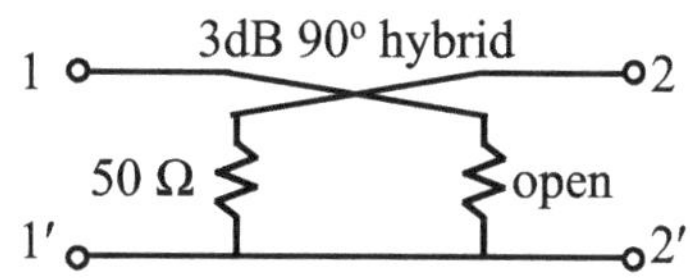

(A) ∝ dB, 0 dB
(B) 9.03 dB, 9.03 dB
(C) **6.02 dB, 6.02 dB**
(D) 0 dB, ∝ dB

(76) Attenuation and return losses of the 2-port network in a 50 Ω system are, respectively,

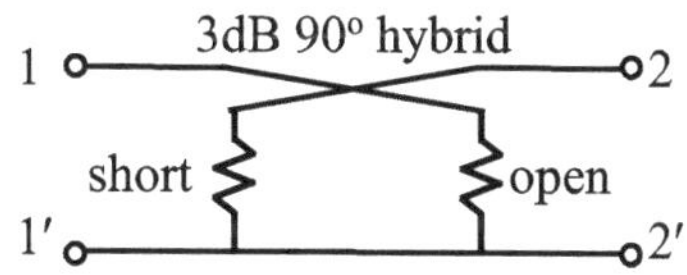

(A) **∝ dB, 0 dB**
(B) 9.03 dB, 9.03 dB
(C) 6.02 dB, 6.02 dB
(D) 0 dB, ∝ dB

(77) Optimum open-circuit voltage for temperature-invariant intrinsic layer resistance of a silicon (bandgap potential of Si is 1.12 V) p-i-n diode at the temperature of 25 °C is

(A) **1.17 V**

(B) 5.0 V
(C) 2.24 V
(D) 0.506 V

(78) Operating temperature dependency of optimum open-circuit voltage for temperature-invariant intrinsic layer resistance of p-i-n diodes:

(A) Highly dependent
(B) Not dependent at all
(C) **Weakly dependent**
(D) Exponentially decreases

(79) Output power at 1-dB gain compression point of an amplifier operating in Class-AB having third-order output intercept point 1-W is about

(A) 29.4 dBm
(B) **19.4 dBm**
(C) 0.5 W
(D) 0.75 W

(80) Two-tone IM_3 at 1-dB gain compression point of an amplifier operating in Class-AB is about

(A) **−21.3 dBc**
(B) −10.4 dBc
(C) −15.5 dBc
(D) −30.3 dBc

(81) Two-tone third-order intermodulation power at 1-dB gain compression point of an amplifier operating in Class-AB having third-order output intercept point 2-W is

(A) −21.3 dBc
(B) **+1.1 dBm**
(C) −10.63 dBm
(D) +30 dBm

(82) Two-tone IM_3 of an amplifier at its 1 mW output power with $P_{o1\text{dB}} = 10$ mW is

(A) **−41 dBc**
(B) −10 dBc
(C) −10 dBc
(D) −20 dBc

(83) Maximum power added efficiency of a practical Class-A amplifier using a MESFET is always less than 50%. Some of the reasons are

(A) Finite loss of output network/matching circuit of the amplifier

(B) Finite gain of the amplifier
(C) Non-zero knee voltage, V_K
(D) **All of the above**

(84) Practical limitations of the output power of a power amplifier come from

(A) Limitation of applied DC voltage across the device
(B) Limitation of current through the device
(C) Thermal dissipation inside the device
(D) **All of the above**

(85) Two-tone IM_3 of an amplifier with $P_{oIP3} = 100$ mW at its 1 mW output power is about

(A) −20 dBc
(B) **−40 dBc**
(C) −10 dBc
(D) −30 dBc

(86) Gain and noise figure of an amplifier are 40 and 10 dB, respectively. Total noise power at the output of the amplifier over the bandwidth of 40 MHz will be ($T_o = 290$ K)

(A) −174 dBm
(B) −77 dBm
(C) **−48 dBm**
(D) −228 dBm

(87) Gain and noise figure of an amplifier are 40 and 3 dB, respectively. Part of the total available noise power at the output of the amplifier which is added by the amplifier over the bandwidth of 40 MHz will be ($T_o = 290$ K)

(A) −174 dBm
(B) −77 dBm
(C) **−58 dBm**
(D) −48 dBm

(88) Gain and noise figure of an amplifier are 40 and 10 dB, respectively. Part of the total available noise power at the output of the amplifier which is added by the amplifier over the bandwidth of 40 MHz will be ($T_o = 290$ K)

(A) −174.4 dBm
(B) −77.4 dBm
(C) −58.4 dBm
(D) **−48.4 dBm**

(89) Gain and noise figure of two identical amplifiers are 10 and 3.0 dB respectively. Noise figure of the two amplifiers connected in cascade is

(A) **3.2 dB**
(B) 3.0 dB
(C) 6 dB
(D) 4 dB

(90) Gain and noise figure of AMP1 and AMP2 are 6 dB, 3.0 dB and 20 dB 3.2 dB, respectively. Noise figure of the two amplifiers connected (AMP1-AMP2) in cascade is

(A) 3.25 dB
(B) **3.56 dB**
(C) 3.05 dB
(D) 4.06 dB

(91) Gain and noise figure of AMP1 and AMP2 are 20 dB, 3.2 dB and 6 dB, 3.0 dB, respectively. Noise figure of the two amplifiers connected (AMP1-AMP2) in cascade is

(A) 3.00 dB
(B) 6.20 dB
(C) **3.22 dB**
(D) 4.06 dB

(92) Gain and output power $P_{o1\text{dB}}$ of two identical amplifiers are 10 dB and 40 dBm, respectively. The amplifiers are connected in balanced amplifier configuration using power divider and combiner. The gain and output power $P_{o1\text{dB}}$ of the balanced amplifier will be (neglect loss of the power divider and combiner)

(A) 10 dB, 40 dBm
(B) 20 dB, 80 dBm
(C) 20 dB, 40 dBm
(D) **10 dB, 43 dBm**

(93) Gain and output power $P_{o1\text{dB}}$ of two identical amplifiers are 10 dB and 40 dBm, respectively. The amplifiers are connected in balanced amplifier configuration using power divider (Insertion Loss 1.0 dB) and combiner (Insertion Loss 0.5 dB). The gain and output power $P_{o1\text{dB}}$ of the balanced amplifier will be

(A) 10 dB, 40 dBm
(B) 20 dB, 80 dBm
(C) **8.5 dB, 42.5 dBm**
(D) 10 dB, 43 dBm

(94) A broadband low-noise amplifier of gain and noise figure, 30 and 2.5 dB, respectively, is connected in cascade with a mixer of negligible insertion loss. In case of single-sideband signal, the effective noise figure of the cascaded system in the absence of any filter in between the LNA and the mixer will be

(A) 2.5 dB
(B) 5 dB
(C) **5.5 dB**
(D) 6 dB

(95) A broadband low-noise amplifier of gain and noise figure, 30 and 2.5 dB, respectively, is connected in cascade with a mixer of negligible insertion loss with an image band reject filter in between. The effective noise figure of the cascaded system will be

(A) **2.5 dB**
(B) 5.0 dB
(C) 5.5 dB
(D) 6.0 dB

(96) Noise figure of a system is 5 dB. The equivalent noise temperature of the system is ($T_o = 290$ K)

(A) 290 K
(B) **627 K**
(C) 570 K
(D) 300 K

(97) A system consists of a low-noise amplifier and a passive filter of insertion loss 0.5 dB at the input of the LNA. The gain and noise figure of the LNA are 30 and 2.2 dB, respectively. The noise temperature of the system is ($T_o = 290$ K)

(A) 290 K
(B) 450 K
(C) **250 K**
(D) 300 K

(98) A system consists of a low-noise amplifier and a passive filter of insertion loss 0.5 dB at the output of the LNA. The gain and noise figure of the LNA are 30 and 2.2 dB, respectively. The noise temperature of the system is ($T_o = 290$ K)

(A) 290.3 K
(B) 450.8 K
(C) 250.5 K
(D) **191.3 K**

(99) A system consists of two amplifiers and a passive filter in between them. The gain and noise figures of input and output amplifiers are 30, 2.7 and 40, 3 dB, respectively. The insertion loss of the filter is 10 dB. The noise temperature of the system is ($T_o = 290$ K)

(A) 470.8 K
(B) 450.8 K

(C) 290.5 K
(D) **255.5 K**

(100) Attenuation of an attenuator is 30 dB. For an input power of 10 W, power at the output of the attenuator will be

(A) **10 mW**
(B) 10 dBm
(C) −20 dBW
(D) All of the above

(101) A linear amplifier has gain of 30 dB. For an input power of 1 mW, power at the output of the amplifier will be

(A) 1 W
(B) 30 dBm
(C) 0 dBW
(D) **All of the above**

(102) Linear gain of a travelling wave tube amplifier (TWTA) is 50 dB. It provides 100 W output power when the input is 6 dBm. Operating gain compression of the amplifier is

(A) 0 dB
(B) 4 dB
(C) **6 dB**
(D) 2 dB

(103) Predistortion linearizer cascaded with TWTA improves

(A) Output power of the system
(B) **Linearity of the system**
(C) DC to RF efficiency when it operates at saturation
(D) All of the above

(104) A low-noise amplifier has gain and noise figure 40 and 3 dB, respectively. The total noise power at the output of the LNA, over the bandwidth of 40 MHz, is ($T_o = 290$ K):

(A) −58 dBm
(B) −42 dBm
(C) **−55 dBm**
(D) −55 dBm

(105) A system consists of a low-noise amplifier (LNA) and an attenuator of attenuation 5 dB at the input of the LNA. The gain and noise figure of the LNA are 40 and 3 dB, respectively. The total noise power at the output of the system over the bandwidth of 40 MHz is ($T_o = 290$ K)

(A) −58 dBm
(B) −42 dBm
(C) **−55 dBm**
(D) −55 dBm

(106) A system consists of a low-noise amplifier (LNA) and an attenuator of attenuation 10 dB at the input of the LNA. The gain and noise figure of the LNA are 40 and 3 dB, respectively. The total noise power at the output of the system over the bandwidth of 40 MHz is ($T_o = 290$ K)

(A) −58 dBm
(B) −42 dBm
(C) **−55 dBm**
(D) −55 dBm

(107) A lossless transmission line of characteristic impedance 50 Ω is terminated in a load of 100 Ω resistance. The VSWR is

(A) 1.2
(B) **2:1**
(C) 4:1
(D) 1:4

(108) Short-circuit and open-circuit impedance of a transmission line are 25 and 100 Ω, respectively. The characteristic impedance of the line is

(A) 100 Ω
(B) 10 Ω
(C) **50 Ω**
(D) 200 Ω

(109) The input impedance of a short-circuited line of length l, where $\lambda/4 > l > \lambda/2$

(A) **Capacitive**
(B) Resistive
(C) Inductive
(D) None of these

(110) Noise figure of an amplifier depends on

(A) Bandwidth
(B) Input power
(C) Output power
(D) **None of these**

(111) VSWR of a transmission line is always

(A) **Greater than unity**
(B) Less than unity

(C) Zero
(D) Infinity

(112) Waveguide acts like a

(A) Bandpass filter
(B) **High-pass filter**
(C) All pass filter
(D) Low-pass filter

(113) The wavelength of a 1 GHz electromagnetic wave propagating through a perfect non-magnetic dielectric of relative permittivity $\varepsilon_r = 9$ is

(A) 3 cm
(B) 3 cm
(C) 100 cm
(D) **10 cm**

(114) TEM wave exists in

(A) A circular Waveguide
(B) A rectangular waveguide
(C) **A coaxial cable**
(D) None of the transmission lines

(115) When VSWR is 3, then the magnitude of the reflection coefficient is

(A) **1/2**
(B) 1
(C) 0
(D) 1/4

(116) Which of the following transmission lines is ideal for handling high RF power?

(A) Coaxial
(B) Microstrip
(C) Stripline
(D) **Rectangular waveguide**

(117) When a transmission line is terminated with a load equal to its characteristic impedance, the reflection coefficient is

(A) **Zero**
(B) +1
(C) −1
(D) Infinity

(118) What is the importance of the noise resistance of a transistor when it is used for LNA design?

(A) It gives the effective impedance offered by the input with respect to noise current
(B) It offers the criterion for conjugate matching of input impedance
(C) It indicates the resistance that would generate the samc amount of noise at room temperature
(D) **It indicated how the noise figure increases as we deviate from the optimum source impedance for the lowest noise figure**

(119) If a line is terminated to an open-circuit element, the VSWR of the line is

(A) 0
(B) 1
(C) −1
(D) **Infinity**

(120) A transmission line has characteristic impedance 50 Ω. The line is terminated at its one end by $+j50\ \Omega$. VSWR produced in the transmission line:

(A) +1
(B) 0
(C) −1
(D) **Infinity**

(121) Major disadvantage of microstrip line compared to stripline is

(A) **More likely to radiate**
(B) Bulkier
(C) More expensive
(D) More complicated to manufacture

(122) The disadvantage of single-stub matching as compared to double-stub matching is that

(A) **The stub position has to be adjusted**
(B) Only shunt stub can be used
(C) Only resistive load can be matched
(D) Useful only in two-wire transmission line

(123) A lossy open-circuit stub having a length of $1/32$ of wavelength, is approximately equivalent to

(A) A small capacitor in shunt with a small resistor
(B) **A small capacitor in shunt with a large resistor**
(C) A small inductor in shunt with a small resistor
(D) A small inductor in shunt with a large resistor

(124) A p-n junction diode in series with a 50 Ω resistor is forward biased so that a current of 100 mA flows. If the voltage across the combination is instantaneously reversed to 10 V at $t = 0$, the reverse current that flows through the junction at $t = 0_+$ is approximately given by

(A) 0 mA
(B) 200 mA
(C) 50 mA
(D) **200 mA**

(125) For a transistor, the manufacturer specifies maximum power dissipation $P_D\text{max} = 1$ W at 25 °C at and the maximum junction temperature, $T_J\text{max} =$ 125 °C. Its thermal resistance is

(A) 10 °C/W
(B) **100 °C/W**
(C) 1000 °C/W
(D) 10,000 °C/W

(126) In a Class-AB amplifier, the current flows through the active device for

(A) Less than half of the duration of input cycle
(B) Half duration of input cycle
(C) **More than half but less than full duration of the input cycle**
(D) Full duration of the input cycle

(127) An earth station transmitter having power amplifier of 40 dBW saturated output power operates at 3-dB output back-off. The total branching and feeder loss is 3 dB, and transmit gain of antenna is 40 dB. The operating EIRP of the transmitter is

(A) 40 dBW
(B) **74 dBW**
(C) 34 dBW
(D) 80 dBW

(128) A transistor amplifier with 85% efficiency is likely to be

(A) Class-A
(B) Class-B
(C) **Class-F**
(D) Class-AB

(129) A satellite transponder has a receive antenna facing towards earth with a gain of 30 dB facing towards earth. The receiver has gain of 50 dB, and equivalent noise temperature is 26 dBK. G/T of the satellite is

(A) **1.6 dB/K**

(B) −4 dB/K
(C) 26 dB/K
(D) −26 dB/K

(130) When VSWR is 3, the magnitude of the reflection coefficient will be

(A) 1/4
(B) **1/2**
(C) 1/3
(D) 3

(131) A transmission line of characteristic impedance 50 Ω is terminated by a load impedance of $50 - j50$. The magnitude of reflection coefficient at the load is

(A) $\sqrt{5}/3$
(B) $\sqrt{3}/5$
(C) $\mathbf{1/\sqrt{5}}$
(D) None of the above

(132) If the RF transmission line is terminated in its characteristic impedance Z_o, which of the following statements is wrong

(A) The input impedance of transmission line becomes Z_o
(B) The transmission line acts as an infinite long transmission line
(C) The VSWR becomes one
(D) **The VSWR becomes infinite**

20.2 Answers of MCQs with Explanations

S. No.	Answer	Explanation
1	A	$\mu_{GaAs} = 8500\ cm^2/V\ s$, $\mu_{Ge} = 3900\ cm^2/V\ s$, $\mu_{Si} = 1500cm^2/V\ s$
2	C	Common-emitter current gain β $= \alpha/(1-\alpha) = 0.99/(1-0.99) = 99$
3	A	Equiresistance curves are straight line over wide range of operating temperatures
4	B	Optimum open-circuit voltage for temperature-invariant resistance of a p-i-n diode is slightly higher than the bandgap potential of the semiconductor from which it is made. E_g of Si is 1.12 V
5	C	Optimum open-circuit voltage for temperature-invariant resistance of a p-i-n diode is slightly higher than the bandgap potential of the semiconductor from which it is made. E_g of GaAs is 1.42 V
6	D	For series-connected p-i-n diodes, the optimum open-circuit voltage is slightly higher than E_g multiplied by the number of diodes connected in series
7	B	For parallel connected p-i-n diodes, the optimum open-circuit voltage is same as single diode circuit, i.e. slightly higher than E_g
8	A	$P_{dBm} = 10\log(P_{mW}) = 10\log(1000) = 30$ $P_{dBW} = 10\log(P_W) = 10\log(1) = 0$
9	A	$P_{dBm} = 10\log(P_{mW}) = 10\log(10) = 10$
10	B	$P_{mW} = 10^{(P_{dBm}/10)} = 10^{(50/10)} = 100{,}000\ mW$
11	C	$P_W = 10^{(P_{dBW}/10)} = 10^{(30/10)} = 1000\ W$
12	C	$P_{mW} = 10^{(P_{dBm}/10)} = 10^{(30/10)} = 1000\ mW$
13	B	$P_W = 10^{(P_{dBW}/10)} = 10^{(0/10)} = 1\ W$
14	A	$I_{0.5k\Omega} = (20-10)/0.5\ mA = 20\ mA$ $I_{10\Omega} = (10-0.7)/10\ A = 930\ mA$ $I_B = 930/100\ mA = 9.3\ mA$ $I_Z = I_{0.5\ k\Omega} - I_B = (20-9.3)mA = 10.7\ mA$
15	B	Current through the 10 kΩ: $I_{10\ k\Omega} = (20/1 - 40/2)\ \mu A = 40\ \mu A$ Thus, output voltage: $V_{out} = -40\ \mu A \times 10\ k\Omega = -400\ mV$
16	C	Current through the 4 Ω resistor $(3i/4)$ The resistance of the upper branch $= 4\Omega \times ((3i/4)/(i/4)) = 12\ \Omega$ Thus resistance faced by the voltage source $= 4\ \Omega \| 12\ \Omega = 3\ \Omega$
17	D	Current through the resistor R $I_R = (1+j)/((1+j)+(1-j)) = (1+j)/2$ Amp $V_R = I_R \times R = (1+j)/2\ V = 1/\sqrt{2}\angle 45^\circ\ V$

(continued)

(continued)

S. No.	Answer	Explanation
18	A	At $t = 0_+$, capacitor can be considered as short. Thus, $I_R = 1/10$ A = 100 mA At steady state, capacitor can be considered as open. Thus, $I_R = 1/(10+5+5)A = 50$ mA
19	B	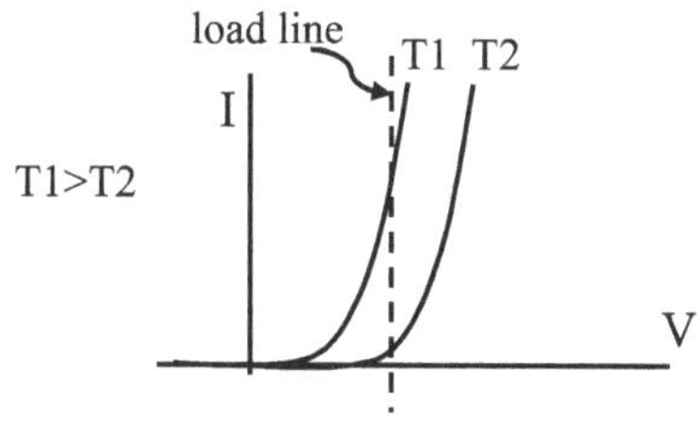 These are typical I-V characteristic of Schottky diode at two different temperatures. At constant voltage bias condition, load line is vertical. Thus, current increases with the increase in temperature
20	C	p-i-n diode has an intrinsic layer in between p and n layer. The stored charge in the I-layer is controllable by controlling the forward current, i.e. resistance is controllable Schottky diode rectifies microwave. Varactor and p-n junction diodes have high junction capacitance and cannot work at microwave frequency
21	D	$I_C = 10$ mA and $V_{CE} = 8$ V Voltage across R_E: 0.1 $k\Omega \times 10$ mA = 1 V Voltage across R_B: (10 − 0.7 − 1) V = 8.3 V $I_B = I_C/\beta_{dc} = 10/100 = 0.1$ mA Therefore, $R_B = (8.3/0.1)\ \text{k}\Omega = 83\ \text{k}\Omega$ $R_C = ((10-8-1)/10)\ \text{k}\Omega = 100\ \Omega$
22	A	AMP-A: Gain = G_A = 6 dB = 4 Noise figure = NF_A = 3.8 dB = 2.4 AMP-B: Gain = C = 20 dB = 100 Noise figure = NF_B = 4 dB = 2.5 In case of input stage: AMP-B $\text{NF}_{BA} = \text{NF}_B + (\text{NF}_A - 1)/G_B$ $= 2.5 + (2.4-1)/100 = 2.524 = 4.024$ dB In case of input stage: AMP-A $NF_{AB} = NF_A + (NF_B - 1)/G_A = 2.4 + (2.5-1)/4 = 2.775 = 4.43$ dB
23	B	$Z_{in(\lambda/8)} = Z_o \frac{Z_L + jZ_o}{Z_o - jZ_L}$ Here, $Z_o = 13$, $Z_L = 5 + j12$ $Z_{\text{in}(\lambda/8)} = 13\frac{5+j25}{1+j5} = 65\ \Omega$
24	C	In an ideal Class-F amplifier, there is no overlap between device voltage and its current in time domain. Thus, efficiency is 100%

(continued)

(continued)

S. No.	Answer	Explanation
25	D	At 1 GHz, capacitive reactance $= 1/2\pi f C_{gs} = 1592\ \Omega$ Thus, $r_g \ll 1/2\pi f C_{gs}$ Therefore, $v_{gs} = v_s$ Voltage gain $= g_m \times v_{gs} \times r_L/v_s = g_m \times r_L = 0.1 \times 500 = 50$
26	A	$P_{DC} = 10\ \text{V} \times 2\ \text{A} = 20\ \text{W}$ RF power $= \left(1/\sqrt{2}\right)^2 \times 5\ \text{W} = 2.5\ \text{W}$ Thus, Efficiency $= 2.5/20 = 12.5\%$
27	B	Efficiency in Class-B operation: $= (\pi/4) \times \left(V_{drf}/V_{dd}\right) = 62.8\%$
28	C	$R_{\text{th}} = 4.5\ °\text{C/W}$ $P_{\text{dissipation}} = (20 - 8)\ \text{W} = 12\ \text{W}$ Temperature increase $= 12 \times 4.5\ °\text{C} = 54\ °\text{C}$ Thus, channel temperature $= (55 + 54)\ °\text{C} = 109\ °\text{C}$
29	D	Helix structure in TWT is used to slow down the RF field along the axial direction to maximize interaction of RF field with the electron beam
30	A	Transformed through constant resistance circle: circuit element is series connected. Moving upwards: the component is inductor
31	A	Transmission line: lossless, thus $\|S_{21}\| = 1$ length $\lambda/8$, thus phase $= \angle - 45°$
32	A	For the complex conjugate match $Z_{\text{network}} = Z^*_{\text{out}} = 32 - j30$ In case of (A) $Z_{\text{network}} = X_2 + X_2\|\|(X_2 + 50)$ $= j10 - j40\|\|(j40 + 50)$ $= j10 + (32 - j40)$ $= 32 - j30$ $= Z^*_{\text{out}}$
33	A	S_{21} = Transmission coefficient $= \dfrac{2Z_a}{Z_o + 2Z_a}$ Here, $Z_a = 50\ \Omega$ and $Z_o = 50\ \Omega$ $S_{21} = 2 \times 50/(50 + 2 \times 50) = 100/150 = 2/3\angle 0°$
34	A	$Z_o = 50, Z_L = 25$ For quarter-wavelength transmission line $Z_{\text{in}} = Z_o^2/Z_L = 50^2/25\ \Omega = 100\ \Omega$
35	A	For low-noise amplifier: Input to be matched: with source reflection coefficient for minimum noise figure. Thus $\Gamma_S = 0.455\angle - 160°$ Output to be matched: complex conjugate for achieving maximum gain. Thus $\Gamma_L = \Gamma^*_{\text{out}} = 0.225\angle + 150°$
36	A	RF resistance is determined by $(\mathrm{d}V/\mathrm{d}I)$ of the I-V characteristic at the intercept with load line. Here, $\mathrm{d}V/\mathrm{d}I$ is more in case of +10 dBm compared to the case of −10 dBm RF power
37	A	SRD generates comblike spectrum, i.e. higher order harmonics contain more RF power level compared to other mentioned devices. Thus, it is most suitable for higher order frequency multiplier

(continued)

(continued)

S. No.	Answer	Explanation
38	A	p-i-n diode has intrinsic layer in between p and n layer. The stored charge in the I-layer is controllable by controlling the forward current, i.e. resistance is controllable Schottky diode rectifies the microwave. Varactor and p-n junction diodes have high junction capacitance, and thus cannot work at microwave frequency
39	A	S_{11} = input reflection coefficient $= (Z_L - Z_o)/(Z_L + Z_o)$ Here, $Z_L = 50\|\|50 = 25\ \Omega$ $S_{11} = (25 - 50)/(25 + 50) = -1/3 = 1/3\angle 180°$
40	A	S_{11} = input reflection coefficient $= (Z_L - Z_o)/(Z_L + Z_o)$ Here, $Z_L = 25 + 50 = 75\ \Omega$ $S_{11} = (75 - 50)/(75 + 50) = 25/125 = 1/5\angle 0°$
41	A	S_{21} = Transmission coefficient $= \frac{2Z_o}{2Z_o + Z_a}$ Here, $Z_a = 25\ \Omega$ and $Z_o = 50\ \Omega$ $S_{21} = 2 \times 50/(2 \times 50 + 25) = 100/125 = 4/5\angle 0°$
42	A	Schottky diode has no minority carrier storage effect thus it can rectify microwave signals efficiently. Other junction diodes have minority carrier storage effect and thus cannot rectify microwave signals efficiently
43	A	In case of metal–semiconductor diode, only majority carrier takes part for current conduction. Thus, there is no minority carrier storage effect
44	A	Top face of a microstrip line is open, i.e. only one side has ground plane. Thus, it radiates more RF/microwave power. Whereas, stripline has ground plane on both the sides, thus it radiates less
45	A	$\|S_{11}\| = (\|Z_L - Z_o\|)/(\|Z_L + Z_o\|)$ Here, $Z_L = (100 + j100)$, $Z_o = 50$ $\|S_{11}\| = \sqrt{50^2 + 100^2}/\sqrt{150^2 + 100^2}$ $= \sqrt{5/13}$
46	A	For the complex conjugate match $Z_{\text{network}} = Z^*_{\text{out}} = 10 + j10$ In case of (A) $Z_{\text{network}} = X_2 + X_1\|\|50 = j30 + (-j25\|\|50)$ $= j30 + (10 - j20) = 10 + j10 = Z^*_{\text{out}}$
47	A	For the complex conjugate match $Z_{\text{network}} = Z^*_{\text{out}} = 10 - j10$ In case of (A) $Z_{\text{network}} = X_2 + X_1\|\|50 = -j30 + (j25\|\|50)$ $= -j30 + (10 + j20) = 10 - j10 = Z^*_{\text{out}}$
48	A	For the complex conjugate match $Z_{\text{network}} = Z^*_{\text{out}} = 8 + j8$ In case of (A) $Z_{\text{network}} = X_2 + X_3\|\|(X_1 + 50) = j28 - j20\|\|(j20 + 50)$ $= j28 + (8 - j20) = 8 + j8 = Z^*_{\text{out}}$
49	A	In case of ideal Class-F amplifier, there is no overlap in between device voltage and with its current in time domain. Thus, efficiency is 100%

(continued)

(continued)

S. No.	Answer	Explanation
50	A	Transmission line: lossless, thus $\|S_{21}\| = 1$ length $\lambda/4$, thus phase $= \angle - 90°$
51	A	$Z_{\text{in}} = Z_o \frac{Z_L + jZ_o \tan \beta l}{Z_o + jZ_L \tan \beta l}$ Thus, $Z_{\text{in}(\lambda/8)} = Z_o \frac{Z_L + jZ_o}{Z_o + jZ_L}$ Thus, $\|Z_{\text{in}(\lambda/8)}\| = Z_o = 50\ \Omega$
52	A	NF = 3 dB = 2 (in factor) Gain = 10 dB = 10 (in factor) Cascaded NF = 2 + (2 − 1)/10 = 2.1 = 3.2 dB Cascaded Gain = (10 + 10) dB = 20 dB
53	A	Transformed through constant conductance circle: circuit element is shunt-connected Moving upward: the component is inductive, thus it may be inductor or short-circuited stub
54	A	Transformed through constant VSWR circle: circuit element is series-connected transmission line
55	A	Transformed through constant reactance and resistance increases: circuit element is series-connected resistor
56	A	Transformed through constant reactance and resistance increases: circuit element is series-connected resistor
57	A	Transformed through constant susceptance and conductance increases: circuit element is shunt-connected resistor
58	A	For high-gain amplifier: Input and output matching: complex conjugate Thus, $\Gamma_S = \Gamma_{\text{in}}^* = 0.498\angle{-121°}$ And $\Gamma_L = \Gamma_{\text{out}}^* = 0.225\angle{+150°}$
59	A	To design power amplifier: Input matching: complex conjugate Thus, $\Gamma_S = \Gamma_{in}^* = 0.498\angle{-121°}$ Output matching: corresponds to maximum output power And $\Gamma_L = 0.275\angle{+130°}$
60	A	$R_C = (10 - 8)/10\ \text{k}\Omega = 200\ \Omega$ $I_B = I_C/\beta_{dc} = 10/100 = 0.1\ \text{mA}$ $R_B = (10 - 0.7)/0.1\ \text{k}\Omega = 93\ \text{k}\Omega$
61	A	$P_{\text{DC}} = 10\ \text{V} \times 2\ \text{A} = 20\ \text{W}$ Class-A amplifier has maximum efficiency = 50% Thus, maximum output power = 20 × 50% W = 10 W
62	B	Reflection coefficient: $\Gamma = (Z_L - Z_o)/(Z_L + Z_o) = (100 - 50)/(100 + 50) = 1/3$ VSWR $= (1 + \|\Gamma\|)/(1 - \|\Gamma\|) = 2$
63	D	Output Power $= P_{\text{in}}(\text{dBm}) + G(\text{dB})$ $= -40 + 30$ $= -10$ dBm

(continued)

(continued)

S. No.	Answer	Explanation
64	A	Received Signal = P(Transmit) + G(Tx antenna) − Path loss − Polarization loss $= 40\text{ dBm} + 10\text{ dB} \quad 100\text{ dB} - 3\text{ dB} = -53\text{ dBm} = -83\text{ dBW}$
65	B	$Z_L = \sqrt{Z_{\text{open}} \times Z_{\text{short}}}$ $= \sqrt{20 \times 5} = 10\ \Omega$
66	C	$\text{NF}_{12} = \text{NF}_1 + (\text{NF}_2 - 1)/G_1$ $= 10 + (10 - 1)/10 = 10.9 = 10.37\text{ dB}$
67	B	S_{11} = input reflection coefficient $= (Z_L - Z_o)/(Z_L + Z_o)$ Here, $Z_L = (50 + 50)\Omega = 100\Omega$ $S_{11} = (100 - 50)/(100 + 50) = 50/150$ $= 1/3\angle 0°$
68	D	S_{21} = Transmission coefficient $= \frac{2Z_o}{2Z_o + Z_a}$ Here, $Z_a = 50\ \Omega$ and $Z_o = 50\Omega$ $S_{21} = 2 \times 50/(2 \times 50 + 50)$ $= 100/150 = 2/3\angle 0°$
69	C	For matched attenuator in a 50 Ω system, $50 = R_1 + 100\|\|(R_1 + 50)$ Thus, $R_1 = 11.803$ Attenuation is given by $\text{Attn} = 20 \times \log((50 + 11.803 + 100)/100) = 4.18\text{ dB}$
70	B	The attenuator terminates the full power. Thus, attenuation is infinite
71	B	$V_{\text{out}} = 0.125 \times V_S$ $V_{\text{in}} = 0.5 \times V_S$ $S_{21} = 20 \times \log(V_{\text{out}}/V_{\text{in}}) = -12.04\text{ dB}$ $Z_{\text{in}} = 30\Omega$ $S_{11} = 20 \times \log(0.25) = -12.04\text{ dB}$
72	C	$V_{\text{out}} = 0.125 \times V_S$ $V_{\text{in}} = 0.5 \times V_S$ $S_{21} = 20 \times \log(V_{\text{out}}/V_{\text{in}}) = -12.04\text{ dB}$ $Z_{\text{in}} = 83.33\Omega$ $S_{11} = 20 \times \log(0.25) = -12.04\text{ dB}$
73	D	Reflection Coefficient (Γ) = −0.3333 $\text{IL} = 20 \times \log(\|1/\Gamma\|) = 9.54\text{ dB}$
74	B	Reflection Coefficients: $\Gamma_1 = -0.4286$ $\Gamma_2 = -0.25$ $\text{IL} = 20 \times \log(2\lceil \Gamma/_1 + \Gamma_2 \rceil) = 9.39\text{ dB}$ $\text{RL} = 20 \times \log(2\lceil \Gamma/_1 - \Gamma_2 \rceil) = 20.98\text{ dB}$

(continued)

(continued)

S. No.	Answer	Explanation
75	C	Reflection Coefficients: $\Gamma_1 = 0$ $\Gamma_2 = 1$ $\text{IL} = 20 \times \log(2\lceil \Gamma_1 + \Gamma_2 \rceil) = 6.02\ \text{dB}$ $\text{RL} = 20 \times \log(2\lceil \Gamma_1 - \Gamma_2 \rceil) = 6.02\ \text{dB}$
76	A	Reflection Coefficients: $\Gamma_1 = -1$ $\Gamma_2 = 1$ $\text{IL} = 20 \times \log(2\lceil \Gamma_1 + \Gamma_2 \rceil) = \propto\ \text{dB}$ $\text{RL} = 20 \times \log(2\lceil \Gamma_1 - \Gamma_2 \rceil) = 0\ \text{dB}$
77	A	Optimum open-circuit voltage is given by $V_{\text{OPT}} = E_b + \frac{2kT_o}{q}$ $= 1.12 + \frac{2 \times 1.38 \times 10^{-23}}{1.6 \times 10^{-19}}(273 + 25)\ \text{V}$ $= 1.17\ \text{V}$
78	C	Optimum open-circuit voltage is given by $V_{\text{OPT}} = E_b + \frac{2kT_o}{q}$
79	B	$P_{o1\text{dB}} = (30 - 10.63) = 19.37\ \text{dBm}$
80	A	$P_{o1\text{dB}} = P_{o\text{IP3}} - 10.63$ Thus, $\text{IMD}_3(\text{at } P_{o1\text{dB}}) = 2 \times (-10.63) = -21.26\,\text{dBc}$
81	B	$P_{o\text{IP3}} = 2\ \text{W} = 33\ \text{dBm}$ $P_{o1\text{dB}} = P_{o\text{IP3}} - 10.63 = 33 - 10.63 = 22.37\ \text{dBm}$ $\text{IMD}_3(\text{at } P_{o1dB}) = 2 \times (-10.63) = -21.26\ \text{dBc}$ $\text{IMD}_3(\text{at } P_{o1dB})\text{in dBm} = 22.37 - 21.26 = 1.11\ \text{dBm}$
82	A	$P_{o1\text{dB}} = 10\,\text{dBm}$ $P_{o\text{IP3}} = 10 + 10.63 = 20.63\,\text{dBm}$ $P_{o(\text{operating})} = 0$ dBm, it is 20.63 dBc below $P_{o\text{IP3}}$ Thus, expected $\text{IM}_3(\text{operating}) = -2 \times (20.63) = -41.26\,\text{dBc}$
83	D	Power added efficiency is always less than 50% is due to the finite loss of output network, finite gain of the amplifier and non-zero value of knee voltage
84	D	Limitations of the output power of a power amplifier come from the limitation of applied DC voltage across the device, limitation of current through the device, thermal dissipation inside the device, etc
85	B	$P_{o\text{IP3}} = 20\,\text{dBm}$ $P_{o(\text{operating})} = 0$ dBm, it is 20 dBc below Thus, expected $\text{IM}_3(\text{operating}) = -2 \times (20) = -40\ \text{dBc}$
86	C	$kT = -174\ \text{dBm/Hz}$ $G_A = 40\ \text{dB}$, $\text{NF} = 10\,\text{dB}$, $B = 40\ \text{MHz} = 76\ \text{dBHz}$ $P_{no} = (-174 + 40 + 10 + 76)\ \text{dBm} = -48\ \text{dBm}$

(continued)

(continued)

S. No.	Answer	Explanation
87	C	$kT = -174$ dBm/Hz $G_A = 40$ dB, NF = 3dB, $B = 40$ MHz = 76 dBHz Output noise due to only input thermal noise: $= kTBG_A = (-174 + 76 + 40)$ dBm = −58 dBm Total output noise: $= kTBG_AF = (-174 + 76 + 40 + 3)$ dBm = −55 dBm Thus, part of the total noise which is added by the amplifier: = −55 dBm − 58 dBm = −58 dBm
88	D	$kT = -174$ dBm/Hz $G_A = 40$ dB, NF = 10 dB, $B = 40$ MHz = 76 dBHz Output noise due to only input thermal noise: $= kTBG_A = (-174 + 76 + 40)$ dBm = −58 dBm Total output noise: $= kTBG_AF = (-174 + 76 + 40 + 10)$ dBm = −48 dBm Thus, part of the total noise which is added by the amplifier: = −48 dBm − 58 dBm = −48.4 dBm
89	A	AMP1: $G_1 = 10$ dB = 10, $NF_1 = 3$ dB = 2 AMP2: $G_2 = 10$ dB = 10, $NF_2 = 3$ dB = 2 AMP12: $NF_{12} = NF_1 + (NF_2 - 1)/G_1 = 2 + (2 - 1)/10$ = 2.1 = 3.2 dB
90	B	AMP1: $G_1 = 6$ dB = 3.98, $NF_1 = 3$ dB = 2 AMP2: $G_2 = 20$ dB = 100, $NF_2 = 3.2$ dB = 2.1 AMP12: $NF_{12} = NF_1 + (NF_2 - 1)/G_1 = 2 + (2.1 - 1)/3.98$ = 2.27 = 3.56 dB
91	C	AMP1: $G_1 = 20$ dB = 100, $NF_1 = 3.2$ dB = 2.1 AMP2: $G_2 = 6$ dB = 3.98, $NF_2 = 3$ dB = 2 AMP12: $NF_{12} = NF_1 + (NF_2 - 1)/G_1 = 2.1 + (2 - 1)/100$ = 2.10 = 3.22 dB
92	D	$G_1 = 10$ dB, $P_{O1B} = 40$ dBm For balanced amplifier: Gain = 10 dB, $P_{O1B} = (40 + 3)$ dBm = 43 dBm
93	C	$G = 10$ dB, $P_{O1B} = 40$ dBm $IL_{divider} = 1.0$ dB, $IL_{combiner} = 0.5$ dB For balanced amplifier: Gain = $(G - IL_{divider} - IL_{combiner}) = (10 - 1.0 - 0.5) = 8.5$ dB $P_{o1B} = (40 + 3 - IL_{combiner})$ dBm = 42.5 dBm
94	C	In the absence of image band reject filter before mixer, noise corresponds to the image frequency band folded and added with the in band noise at the output of the mixer. Thus, total noise power will be doubled, i.e. 3-dB more at the output
95	A	In the presence of image band reject filter before mixer, noise corresponds to the image frequency band will be sufficiently rejected by the filter thus NF will be same as noise figure of the LNA

(continued)

(continued)

S. No.	Answer	Explanation
96	B	$\mathrm{NF}_r = 5\ \mathrm{dB} = 3.16$ $T_e = (\mathrm{NF} - 1)T_o = 2.16 \times 290\ \mathrm{K} = 627\ \mathrm{K}$
97	C	Noise figure of the system: $\mathrm{NF}_s = \mathrm{IL}_{\mathrm{filter}} + \mathrm{NF}_{\mathrm{LNA}} = (0.5 + 2.2)\ \mathrm{dB} = 1.862$ $T_e = (\mathrm{NF} - 1)T_o = 0.862 \times 290\ \mathrm{K} = 250\ \mathrm{K}$
98	D	Noise figure of the system: $\mathrm{NF}_s = \mathrm{NF}_{\mathrm{LNA}} + (\mathrm{IL}_{\mathrm{filter}} - 1)G_{\mathrm{LNA}} = 2.2\ \mathrm{dB} = 1.66$ $T_e = (\mathrm{NF} - 1)T_o = 0.66 \times 290\ \mathrm{K} = 191.3\ \mathrm{K}$
99	D	Input Amplifier: $G_1 = 30\ \mathrm{dB} = 1000$, $\mathrm{NF}_1 = 2.7\ \mathrm{dB} = 1.86$ Output Amplifier: $G_2 = 40\ \mathrm{dB} = 10{,}000$, $\mathrm{NF}_2 = 3\ \mathrm{dB} = 2.00$ Filter: $G_f = -10\ \mathrm{dB} = 10.1$, $\mathrm{NF}_f = 10\ \mathrm{dB} = 10$ Noise figure of the system: $\mathrm{NF}_s = \mathrm{NF}_1 + \left(\mathrm{NF}_f - 1\right)/G_1 + (\mathrm{NF}_2 - 1)/G_1 G_2 = 1.88$ $T_e = (\mathrm{NF} - 1)T_o = 0.88 \times 290\ \mathrm{K} = 255.50\ \mathrm{K}$
100	A	Attenuation = 30 dB Input power = 10 W = 40 dBm Output power = (40 − 30) dBm = 10 dBm
101	D	Linear Gain = 30 dB Input power = 1 mW = 0 dBm Output power = (0 + 30) dBm = 30 dBm = 0 dBW = 1 W
102	C	Linear Gain = 50 dB Input power = 6 dBm = 4 mW Output power = 100 W = 50 dBm Operating Gain = (50 − 6) dB = 44 dB Gain Compression = (linear gain − operating gain) = 6 dB
103	B	Predistortion linearizer is used to improve linearity of the system
104	C	Gain of the LNA, $G = 40$ dB Noise figure of the LNA, $F = 3$ dB Bandwidth, $B = 40$ MHz = 76 dBHz Output noise, $= kT_oBGF = (-174 + 76 + 40 + 3)\ \mathrm{dBm} = -55\ \mathrm{dBm}$
105	C	Gain of the system, $G = (40 - 5)$ dB = 35 dB Noise figure of the system, $F = (3 + 5)$ dB = 8 dB Bandwidth, $B = 40$ MHz = 76 dBHz Output noise, $= kT_oBGF = (-174 + 76 + 35 + 8)\ \mathrm{dBm} = -55\ \mathrm{dBm}$
106	C	Gain of the system, $G = (40 - 10)$ dB = 30 dB Noise figure of the system, $F = (3 + 10)$ dB = 13 dB Bandwidth, $B = 40$ MHz = 76 dBHz Output noise, $= kT_oBGF = (-174 + 76 + 30 + 13)\ \mathrm{dBm} = -55\ \mathrm{dBm}$
107	B	$\Gamma = \frac{Z_L - Z_o}{Z_L + Z_o} = \frac{100-50}{100+50} = \frac{1}{3}$
108	C	$Z_{in}(\mathrm{short}) = jZ_o \tan(\beta l)$, $Z_{in}(\mathrm{open}) = jZ_o \cot(\beta l)$ Thus, $Z_o = \sqrt{Z_{\mathrm{in}}(\mathrm{short}) \times Z_{\mathrm{in}}(\mathrm{open})} = \sqrt{25 \times 100}\ \Omega = 50\ \Omega$

(continued)

(continued)

S. No.	Answer	Explanation
109	A	Z_{in}(short) $= jZ_o \tan(\beta l)$ For, $\lambda/4 > l > \lambda/2, \pi/2 > \beta l > \pi$, $\tan(\beta l)$ is negative Thus, Z_{in}(short) is capacitive
110	D	Noise figure of an amplifier does not depend on its bandwidth, input power or output power
111	A	VSWR is defined as the ratio of maximum voltage to minimum voltage of the standing wave. Thus, it is always greater than unity
112	B	Waveguide does not allow low frequencies to pass. Thus, it acts as a high-pass filter
113	D	$\lambda = c/\left(f\sqrt{\varepsilon_r}\right) = 3 \times 10^{10}/\left(10^{10} \times \sqrt{9}\right)$ cm $= 10$ cm
114	D	TEM wave can exist in a coaxial cable and not in circular or rectangular waveguides
115	A	$\lvert\Gamma\rvert = (\rho - 1)/(\rho + 1) = 1/2$
116	D	Rectangular waveguide can handle more RF power than coaxial, microstrip or stripline
117	A	$\Gamma = (Z_L - Z_o)/(Z_L + Z_o) = 0$
118	D	Noise resistance of a transistor indicates how the noise figure increases as the source impedance move away from the optimum source impedance corresponds to lowest noise figure
119	D	Here, $= (Z_L - Z_o)/(Z_L + Z_o) = 1$
120	D	$\Gamma = (Z_L - Z_o)/(Z_L + Z_o) = (j50 - 50)/(j50 + 50) = (j - 1)/(j + 1)$ Thus, $\lvert\Gamma\rvert = 1$ Therefore, VSWR $= (1 + \lvert\Gamma\rvert)/(1 - \lvert\Gamma\rvert) = \propto$
121	A	The signal bearing conductor is open for microstrip line. This it is more likely to radiate. Otherwise, it is less expensive, easier to fabricate and less bulky compared to stripline
122	A	In case of single-stub matching, the stub position from a load end cannot be kept at any distance. It has to be adjusted as per the requirement to match a load to particular source impedance
123	B	An open-circuit stub having less than $\lambda/4$ line length can be approximated to a small capacitor and lossy characteristic can be represented by a large resistor in shunt with the capacitor
124	D	Due to the charge storage effect, just after altering the forward-biased diode to a reverse bias, it will remain in forward biased and the voltage across the diode will be negligible. Thus, the current through the diode will be 10 V/50 = 200 mA
125	B	$P_{D\max} = 1$ W at 25 °C and $T_{J\max} = 125$ °C Increase of temperature = (125 – 25) °C = 100 °C
126	C	In a Class-AB amplifier, device current flows for more than half but less than full duration of input cycle

(continued)

(continued)

S. No.	Answer	Explanation
127	B	EIRP = (40 dBW − 3 dB − 3 dB + 40 dB) = 74 dBW
128	C	Maximum efficiency of Class-A: 50% Maximum efficiency of Class-B: 78.5% Maximum efficiency of Class-AB: <78.5% Maximum efficiency of Class-F: 100% Thus, among Class-A, B, AB and F, only Class-F amplifier can provide 85% efficiency
129	A	Antenna gain: $G = 30$ dB Equivalent noise temperature of receiver: $T_e = 26$ dBK = 398.1 K Total noise temperature: $T = 398.1$ K + 290 K = 688.1 K = 28.4 dBK Thus, $G/T = (30 - 28.6)$ dB/K = 1.6 dB/K
130	B	$\|\Gamma\| = (\rho - 1)/(\rho + 1) = (3 - 1)/(3 + 1) = 2/4 = 1/2$
131	C	$\Gamma = (Z_L - Z_o)/(Z_L + Z_o) = (50 - j50 - 50)/(50 - j50 + 50)$ $= (-j50)/(100 - j50) = (-j)/(2 - j)$ Thus, $\|\Gamma\| = 1/\sqrt{5}$
132	D	A RF transmission line terminated by its characteristic impedance does not generate standing wave. Thus, VSWR is unity

Index

S. C. Bera, *Microwave Active Devices and Circuits for Communication*, Lecture Notes in Electrical Engineering 533, https://doi.org/10.1007/978-981-13-3004-9

MIX
Papier aus verantwortungsvollen Quellen
Paper from responsible sources
FSC® C105338

If you have any concerns about our products,
you can contact us on
ProductSafety@springernature.com

In case Publisher is established outside the EU,
the EU authorized representative is:
Springer Nature Customer Service Center GmbH
Europaplatz 3, 69115 Heidelberg, Germany

Printed by Libri Plureos GmbH
in Hamburg, Germany